Non-Equilibrium Air Plasmas at Atmospheric Pressure

Series in Plasma Physics

Series Editors: **Steve Cowley**, Imperial College, UK
Peter Stott, CEA Cadarache, France
Hans Wilhelmsson,
Chalmers University of Technology, Sweden

Series in Plasma Physics

Non-Equilibrium Air Plasmas at Atmospheric Pressure

K H Becker
Stevens Institute of Technology, Hoboken, NJ, USA

U Kogelschatz
ABB Corporate Research, Baden, Switzerland (retired)

K H Schoenbach
Old Dominion University, Norfolk, VA, USA

and

R J Barker
US Air Force Office of Scientific Research, Arlington, VA, USA

CRC Press
Taylor & Francis Group
Boca Raton London New York

CRC Press is an imprint of the
Taylor & Francis Group, an **informa** business

CRC Press
Taylor & Francis Group
6000 Broken Sound Parkway NW, Suite 300
Boca Raton, FL 33487-2742

First issued in paperback 2019

© 2005 by Taylor & Francis Group, LLC
CRC Press is an imprint of Taylor & Francis Group, an Informa business

No claim to original U.S. Government works

ISBN-13: 978-0-7503-0962-2 (hbk)
ISBN-13: 978-0-367-86417-0 (pbk)

British Library Cataloguing-in-Publication Data

A catalogue record for this book is available from the British Library.

ISBN 0 7503 0962 8

Library of Congress Cataloging-in-Publication Data are available

**Visit the Taylor & Francis Web site at
http://www.taylorandfrancis.com**

**and the CRC Press Web site at
http://www.crcpress.com**

Contents

Note:
A summary of references to Air Plasmas compiled by R Vidmar is available on the Web at:

http://bookmark.iop.org/bookpge.htm?&isbn=0750309628

Foreword

Air plasmas (lightning and aurora) and flames were probably the first plasmas to be studied. Until reliable vacuum pumps were developed, these complicated plasmas were the subject of mostly empirical studies. Up to the 1940s, studies were often made with what was a relatively poor vacuum. In the 1920s and 1930s the favorite discharge was the mercury vapor discharge because of the ubiquitous mercury diffusion pump, McLeod gauge and the interest in developing large rectifiers and the fluorescent lamp. Langmuir greatly advanced the understanding of many plasma phenomena using simple mercury vapor discharges. When vacuum techniques improved, most of the attention was on the rare gases or, at most, binary mixtures of these gases. After 1946, there was an initial interest in the real gas effects in air flows over blunt bodies moving at hypersonic speeds. At Mach numbers greater than about 12, modest dissociation and ionization effects already occur and air can no longer be considered as a mixture of just nitrogen, oxygen, and argon. At Mach numbers around 20, the gas temperature behind a normal shock for a blunt body reaches values higher than 6500 K and the effects of dissociation, ionization, radiation and recombination on heat transfer and radio wave communication become dramatic. The quality of the work performed at that time was very impressive and includes two of the now classical reports from F. R. Gilmore of the Rand Corporation, 'Equilibrium Composition and Thermodynamic Properties of Air to 24 000 K' and his often cited potential energy diagrams in 'Potential Energy Curves for N_2, NO, O_2 and Corresponding Ions' published in 1955 and 1964, respectively. There were excellent reports from several laboratories treating the problems of re-entry mostly using local thermodynamic equilibrium approaches. After the initial surge of interest, the aeronomy studies continued apace. However, it took some years for the non-equilibrium plasma tools to mature.

Plasmas generated and maintained at atmospheric pressure enjoyed a renaissance in the 1980s, mostly driven by applications such as high power lasers, opening switches, novel plasma processing applications and sputtering, EM absorbers and reflectors, remediation of gaseous pollutants, medical sterilization and biological decontamination and excimer lamps and other non-coherent vacuum–ultraviolet (VUV) light sources. Atmospheric-pressure plasmas in air are of particular importance as they do not require a vacuum enclosure and/or additional feed gases. This edited volume brings to the community the state-of-the-art in atmospheric-pressure air plasma research and its technological applications. Advances in atmospheric-pressure plasma source development, air plasma diagnostics and characterization, air plasma chemistry at atmospheric pressure, modeling and computational techniques as applied to atmospheric-pressure air plasmas, and an assessment of the status and prospects of atmospheric-pressure air plasma applications are addressed by a diverse group of experts in the field from all over the world.

While the book emphasizes atmospheric-pressure plasmas in air, many results presented will also be applicable, perhaps with modifications, to atmospheric-pressure plasmas in other gases and gas mixtures. This book is primarily directed to researchers and engineers in the field of plasmas and gas discharges, but it is also suitable as a pedagogical review of the areas for graduate and professional certificate courses. The extensive section on applications (in various states of technological maturity) makes this book also attractive for practitioners in many fields of application where technologies based on atmospheric-pressure air plasmas are emerging.

Alan Garscadden
February 2004

[Dr Alan Garscadden is the Chief Scientist of the Propulsion Directorate at the Air Force Research Laboratory, Wright-Patterson Air Force Base, Dayton, Ohio, USA. He has worked extensively in the areas of plasmas, optical and mass spectroscopy, laser kinetics and diagnostics, and propulsion and power technologies. He has authored or co-authored 160 publications in professional journals and he has given numerous invited talks at international conferences on topics relating to gas discharge and plasma physics and their applications. Among Dr Garscadden's many credentials are the Will Allis Prize of the American Physical Society (2001) and the Presidential Meritorious Award. He is a Fellow of the APS, IEEE, AIAA, and the Institute of Physics (UK).]

Chapter 1

Introduction and Overview

R J Barker

Interest continues to grow worldwide in practical applications of weakly ionized, low-temperature, sea-level air plasmas. This book is written for scientists, engineers, practitioners, and graduate students who seek a detailed understanding of 'cold' (non-equilibrium) atmospheric-pressure air plasmas; and their generation, sustainment, characterization, modeling, and practical application. Non-thermal, ambient temperature and pressure volumes of natural air plasmas avoid the restrictions imposed by costly, cumbersome vacuum chambers and by destructively high temperatures. At the same time, however, they vastly complicate the plasma physics and chemistry involved. This edited volume provides the technically savvy reader with the fundamental knowledge necessary to understand the science and the application of these non-equilibrium air plasmas at atmospheric pressure.

This first chapter sets the stage for all that follows and should be read carefully in order to maximize one's appreciation for the following chapters. It begins by explaining *why* this topic is important to researchers in the fields of defense, medicine, electronics, materials science, environmental health, and aviation. Equally important, it explains why this book is an excellent information source for this topic. After that, the second section carefully describes what portion of air plasma parameter space is treated in this book. This is crucial for determining the range of applicability of the information provided herein. Section 1.3 then digresses briefly to provide the reader with a natural reference frame from which to better view the subsequent discussions of man-made air plasmas; namely it describes where and how *nature* generates large volumes of plasma in air.

Section 1.4 presents the wealth of sources, both in publications as well as in conferences, from which a reader may gain further details of and updates to the air plasma information contained in this volume.

This first chapter ends with a complete chapter-by-chapter overview of this entire edited volume. The logic underlying the flow of the book is discussed and brief synopses of the material covered in each of the remaining chapters are presented. A reader can use section 1.5 to identify which chapters contain the most important information relating to his/her specific area of air plasma interest.

1.1 Motivation

One of the most important yet often underutilized questions facing any technical author is, 'Who should read this book and why?' This proper delineation of a book's target audience is crucial toward determining the ultimate 'usefulness' of the book. The two major characteristics of concern regarding the audience are (1) its educational level and (2) its technical interests.

At the earliest stage in the preparation of this book, the editors agreed that all material will be written under the assumption that it will be read by a scientist and/or engineer/practitioner who has completed at least a Master of Science or Engineering degree. The reader should have a familiarity with basic electromagnetics as well as concepts governing chemical rate equations. Completion of at least a basic course in plasma physics and/or plasma chemistry would be beneficial but not mandatory. This volume may be appropriate for classroom adoption as a graduate level text for a special-topics seminar course in high-pressure plasmas or for supplemental reading in a graduate level course on Gas Discharge Physics or Plasma Processing or for a continuing education or short course text. Nevertheless, it was *not* intentionally designed to be used as a textbook. (For example, it lacks end-of-chapter homework problems.) At the same time, its intended usefulness is specifically *not* limited to university air plasma researchers but rather broadly targeted to also include industrial and military applications and design engineers. For these reasons, not only are underlying theories discussed but also practical laboratory techniques are explained, with care being taken at the end to show how all can have important real-world applications.

This book was written to serve as a comprehensive source of detailed information for readers with a wide variety of technical interests. To begin with, this would make valuable reading for *anyone* in the fields of plasma physics and/or plasma chemistry. It covers parameter ranges of growing importance to the industrial community but which are normally omitted from traditional university plasma courses. However, the value of this volume is by no means limited to the plasma communities. On the contrary, pains were taken throughout to ensure its understanding by all scientific and engineering communities that have interests in atmospheric pressure 'cold' air plasmas for a growing list of applications. The technical fields involved include but are not limited to the following:

1. *Microwave propagation*. Volumes of lightly ionized air can act as extremely efficient and broadband absorbers of microwave radiation. The free electrons present act to collisionally convert the electromagnetic energy into thermal energy in the ambient gas (Vidmar 1990).
2. *Sterilization/decontamination*. Weakly ionized air is an extremely efficient killer of micro-organisms, including bacteria and even spores (Laroussi *et al* 2002, Birmingham and Hammerstrom 2000, Roth *et al* 2001, Montie *et al* 2000). This seems to be driven by the plasma chemistry of the ions and excited neutral species rather than any short-lived free electron population.
3. *Pollution control*. Air ionization systems are used to deposit electrical charge on particulate pollutants and then efficiently extract such particles from the airflow via oppositely-charged electrodes (White 1963, Parker 1997). More recent work has shown promise for using air plasma chemistry to neutralize chemical pollutants as well (Nishida *et al* 2001).
4. *Surface materials processing*. A brief exposure of certain types of materials to a volume of ionized air can significantly modify the surface properties of the material. For example, the water-repellent surfaces of certain plastics have been made wettable (Tsai *et al* 1997).
5. *Aerodynamics*. There is evidence that thin, weakly-ionized volumes of air flowing along airfoils can be electronically steered, thereby offering the possibility of achieving some level of flight control without hydraulic mechanical actuator servers (Roth 2003, Van Dyken *et al* 2004). There have also been claims of plasma-based supersonic shock-front mitigation although this remains controversial (Kuo *et al* 2000).
6. *High-speed combustion*. The 'flame-out' of jet engines in high-speed flight can be a disconcerting event even for experienced pilots. Furthermore, as military aircraft designers push toward hypersonic speeds, possibly driven by ramjet technology, they must be concerned even more about uniform combustion ignition and sustained 'flame holding'. Plasma-based combustors are being successfully tested and employed for such an application (Kuo and Bivolaru 2004, Liu *et al* 2004).
7. *Lightning discharge control*. Violent lightning strikes cause millions of dollars worth of damage every year to commercial power distribution systems. The sometimes extended power outages that can result cause even more millions of dollars worth of loss to industrial and private customers. It would be useful to create methodologies for the pre-planned establishment of air plasma channels through the atmosphere to harmlessly drain thunderstorm charge accumulations in a safe manner before lightning-strike conditions can even be achieved in sensitive locales.

Those applications will be discussed in chapter 9. Additional possible future air plasma applications will also be addressed there.

While there are other books available for scientists and engineers interested in the examination and application of air plasmas, this is the only book that *combines* the following three elements in its focus.

1. *Natural air* is treated herein, not only simple laboratory mixtures of oxygen and nitrogen.
2. Results center on *one-atmosphere-pressure* air.
3. The emphasis is on *non-equilibrium, 'cold'* air plasmas rather than their thermally equilibrated counterparts.

The combination of the above three characteristics make this book a unique technical resource and a valuable reference work to newcomers and experienced air plasma researchers alike. Of course, subject matter alone cannot ensure the value of this or of any book. The other crucial factor that makes this book an important work is the stature and recognized expertise of its international team of contributing authors. The authors are leaders in their respective fields, intimately familiar with the state-of-the-art as well as with likely future trends.

1.2   Parameter Space of Interest

Conducting plasma experiments on gases sealed in a chamber gives one the powerful advantage of controlling, or at least the ability to control, the precise pressure and chemical composition of those gases. For that reason, most of the empirical studies discussed in this book will deal with such chambered gases. A scientist seeks to understand complex phenomena by collecting data points for systems with as many knowns and as few variables as possible. In that way, solid data can form the solid foundation for complex predictions.

Such considerations highlight the ambitious goal of this book to focus on non-equilibrium atmospheric pressure air plasmas. What is sought here is an understanding of non-thermal plasma formation in 'open' air. One is here interested in creating a population of free electrons in whatever ambient air happens to be present in one's laboratory (or work-site). Since this laboratory may be situated in the humid, sea-level environment of Hamburg, Germany, as likely as in the high (1.52 km above sea level), dry environment of Albuquerque, New Mexico, USA, it is important to specify the known range of chemical constituents and pressures that may be encountered. Being mindful of such differences can prepare one for observed variations in air plasma results from place to place on the globe and even from season to season.

Although the deviations of ground level from sea level may seem large, nevertheless, every point on the surface of earth lies well within the lowest (and thinnest) layer of the atmosphere, namely the troposphere (see figure

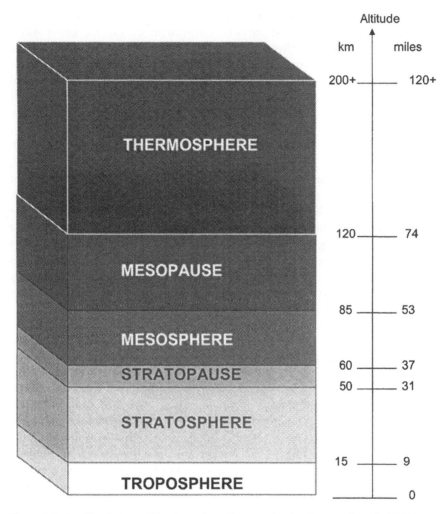

Figure 1.1. Profile of the earth's atmosphere from sea level to low earth orbit (LEO).

1.1). At any point on the earth's surface, the ambient dry air is composed of the following independent gases at approximately the respective volume percentages: nitrogen (N_2, 78.09%), oxygen (O_2, 20.95%), argon (Ar, 0.93%), carbon dioxide (CO_2, 0.03%), neon (Ne, 0.0018%), helium (He, 0.00053%), and krypton (Kr, 0.0001%). There are slight variations to those numbers from location to location, and of course experimental errors can creep into any such measurements. In addition to the gases listed above, relatively minute amounts of hydrogen and xenon are permanent constituents of air. Finally, trace amounts of radioactive isotopes, nitrogen oxides, and ozone may also be found in a given sample of dry surface air. By far the most variable constituent of surface air is water vapor. When one departs from the use of dry air, then one is subject to the ambient humidity of a given locale. Aside from obvious humidity variations due to the proximity of large bodies of water, there are measurable annual averages based on latitude that show a clear dependence on average air temperature. As an illustration, it is instructive to compare such annual averages as follows that show the relative volume percentages of $N_2/O_2/Ar/H_2O/CO_2$ for the equator, 50°N, and 70°N respectively: 75.99/20.44/0.92/2.63/0.02, 77.32/20.80/0.94/0.92/0.02, and 77.87/20.94/0.94/0.22/0.03. In a common misconception, it is often assumed that the concentration of the heavier molecules decreases with increasing altitude due to gravity. It would seem reasonable that lighter molecules would preferentially migrate upward. In reality, however, the powerful dynamics of solar heating cause such extensive mixing that relative molecular concentrations remain virtually unchanged from ground level up to about 20 km. The only large deviations occur in the relative concentration of water vapor since it depends critically on the local ambient temperature and that average temperature decreases with increasing altitude (Humphreys 1964).

Thus, the chemical composition of the air treated in this book is left to nature and, luckily, behaves quite well except for the few percent variations due to ambient water vapor. The question of gas temperature is one more closely controlled by the individual experimentalist and here there was indeed some divergence among this book's contributing authors. Fundamentally, there was unanimous agreement on the focus of *non-equilibrium* plasmas. The goal remained to discuss techniques for generating a much larger population of free electrons in air than could result from the simple brute-force heating of the background air. The rationale for that goal is twofold; first, thermal ionization implies minimum efficiency of plasma generation due to the 'wasted' heating of the background gas, and, second, the thousands of degrees of temperature necessary to achieve even a modest 10^{12} free electrons per cm^3 in a thermal plasma would be clearly destructive to many of the proposed beneficiaries of the previously listed air plasma treatments. At the same time, there is no ionization technique that can completely avoid any heating of the background gas. Thus, a truly 'cold' plasma in which the background air remains fixed at room temperature is not realistic for the practical applications that motivate this book. Therefore, it can best be stated that this book deals with 'warm' plasmas in which background air temperatures of several hundred Kelvin above 'room temperature' are considered quite acceptable.

The paragraph above touches on a subject that cannot be passed over so lightly, namely that of power consumption necessary for the generation of ambient air plasmas. This point is crucial for anyone seeking to apply air plasmas to real-world applications since this is the issue that drives the cost of the application. Over the past decades, several attractive technologies have been sidelined simply because they required the sustainment of electron densities on the order of 10^{13} per cm^3 and that required hundreds of mega-watts of electrical power per cubic meter. To some, this simply excluded the consideration of ambient air plasmas for a range of applications. To others, however, this signaled a challenge to explore hybrid ionization techniques that avoided the brute-force re-ionization of molecules on electron recombination timescales. The pioneering efforts of those forward-thinking researchers is captured herein. Luckily, a vast majority of air plasma applications require only very modest free electron populations to achieve. Those applications, and their required technologies for realization are likewise covered herein.

1.3   Naturally-occurring Air Plasmas

A more accurate title for this book would be '*Artificial* Non-equilibrium Air Plasmas at Atmospheric Pressure'. This books treats only non-thermal air plasmas that result from other-than-natural causes. From that perspective, it is worth a brief digression here to examine what types of plasmas (in the broadest sense) can be found in Nature. Sometimes a researcher can gain insights by first observing what Nature has wrought.

To begin with, sea-level air abhors free electrons. As will be discussed later in this book, at room temperatures three-body recombination of electrons with molecular oxygen limits electron lifetimes to only about 16 ns. The situation becomes friendlier for free electrons as one increases one's altitude in the atmosphere and, thereby encounters ever-decreasing air pressure. For example, at 30 000 and 60 000 ft the free electron lifetime increases to 119 ns and 1.83 µs respectively. Above about 60 km above sea level, one enters the ionosphere, where the copious flux of extreme ultraviolet (EUV) solar photons and, to a lesser extent, collisions with energetic particles (mostly electrons) that penetrate the atmosphere easily maintains free electron densities on the order of 10^2 to $10^7 cm^{-3}$ in the rarified background (Schunk and Nagy 2000). The dominant ion species balancing that electron charge consists primarily of H^+ and He^+ above 1000 km, O^+ from 300 to 500 km and molecular ions (NO^+, O_2^+, and N_2^+) below 200 km (NASA 2004). There exist some excellent reviews of the dominant ionospheric ionization processes (Hudson 1971, Stolarski and Johnson 1972) as well as complete lists of the major plasma chemistry reactions at work (Torr 1979). It should be noted, however, that there are numerous reactions that result in minor chemical constituents that are not well understood. Some of these involve metastable atomic states, negative ions, ionization by photoelectrons, energetic neutrals, and the vibrational states of molecules. Readers interested in the photo-ionization of air would be well advised to first familiarize themselves with such ionospheric chemistry.

When Nature seeks to generate high free-electron densities in the lower atmosphere, she resorts to thermal plasma generation via lightning discharges. The physics of natural lightning is fascinating and certainly worthy of its own text. Unfortunately, scientific details must generally be gleaned from sections of meteorology texts (Moran *et al* 1996). Nevertheless, readers interested in atmospheric 'arcs and sparks' would do well to examine such natural phenomena before embarking on a quest for laboratory imitations. Simply stated, a lightning discharge may be best described as 'a complex propagating gas breakdown process' (Jursa 1985). It is believed to be triggered when large amounts of space charge accumulate in small volumes in clouds and thus create locally intense electric fields of several hundred kV/m. The lightning channel progressively extends below the cloud base (in cloud-to-ground lightning) in what is termed a 'stepped leader'. In this process, each 'leader' breaks down the air in a sequence of (approximately) 50 m 'steps'. It is interesting to note that each step forms in only about 1 µs but there is an average of a 50 µs delay before the next step is formed. This ever-growing stepped leader continues extending toward the ground until the huge voltage (about 10^8 V) between its head and the earth's surface (or conducting projection from that surface) exceeds the air breakdown threshold. At that moment, there occurs a very rapid equalization of the charge in the channel at the amazing speed of about one-third the speed of light. It is this so-called 'return stroke' from the ground that is responsible for the most intense and rapid heating and expansion of a significant volume of air, thus producing the characteristic bright flash and loud thunder associated with a bolt of lightning. Typically, subsequent lightning strokes will follow the existing partially ionized channel. Overall, a given ground lightning 'event' lasts only 0.1–1.0 s with 0.5 s being a typical value. Most such events neutralize tens of coulombs of charge. Such individual events typically consist of three or four individual strokes, each lasting about 1 ms and separated by 40–100 ms.

Before ending this section, one may venture into murkier researches of science by considering the possible natural occurrence of 'ball lightning'. For newcomers to the air plasma arena, a caution must be voiced. While numerous claims of ball lightning sightings have been reported in the scientific and popular press, no reproducible laboratory experiments for the recreation of such phenomena (except for tiny manifestations) have been published. This fact unfortunately has relegated this to the status of 'borderline' science. It is instructive to note that the most comprehensive, recent text on this subject is largely anecdotal in nature (Stenhoff 1999). Still it is reasonable to deduce that there is some type of unexplained, plasma-related atmospheric phenomenon that underlies the 'ball lightning' sightings. One may hope that someday the proper scientific tools are brought to bear so that a true understanding may follow.

1.4 Sources of Additional Information

No book could hope to capture all of the technical details of so complex a subject as non-equilibrium atmospheric pressure air plasmas. This book rather serves as a comprehensive guide to the current state of knowledge regarding these phenomena. It surveys the rich history and details today's capabilities and opportunities regarding these plasmas and thus constitutes an ideal *starting point* for the non-equilibrium high pressure air plasma professional who has at his/her disposal a comprehensive library of reference works. In this section, suggestions are made regarding specific books and journals that would be most useful for such reference purposes. In addition, mention is made of particular professional meetings that may be most rewarding for pursuing specific topical areas. It is a certainty that not *every* relevant book, journal, and conference will be mentioned here. However, as one examines those that are referenced here, one can then branch out, as always, to explore the references that they reference. This is a natural process.

In order to understand many of the concepts covered in this book, a reader must have a firm foundation in electromagnetic theory and plasma physics. There are many excellent, comprehensive texts covering these subjects (Jackson 1998, Pollack and Stump 2002, Griffiths 1998, Chen 1984, Dendy 1995, Boyd and Sanderson 2003). The choice of 'favorite' texts will vary from scientist to scientist.

On the specific subject of non-equilibrium atmospheric pressure air plasmas, two other books stand out as excellent companion works to this book. The first is one co-edited by one of this book's editors (K.H.S.) and concentrates on non-equilibrium low temperature plasmas but not in air (Hippler *et al* 2001). That collection of papers deals with any and all species of lightly ionized, low temperature gases, although atmospheric applications are discussed in several of the papers. It has a strong bias toward industrial plasma processing and lighting applications. It spends little time on theory and modeling fundamentals but does give a good discussion of relevant diagnostic techniques that complements presentations in this book on that subject. It also gives good experimental details but mainly on industrial plasma reactor concerns.

A second excellent possible companion to this work is one that, instead of dealing directly with air, deals only with various mixtures of air's principal molecular constituents, namely oxygen, nitrogen, and major oxides of nitrogen (Capitelli *et al* 2000). That monograph focuses on theoretical (computational) analyses of basic kinetic theory and detailed investigation of kinetic processes of lightly ionized, low temperature, non-equilibrium plasmas in N_2, O_2, and their mixtures. It examines self-consistent solutions of the electron Boltzmann equation coupled to a system of vibrational and electronic state master equations, including dissociation and ionization reactions in conjunction with electrodynamics. The main target applications there are gas discharges and natural (e.g. ionospheric or spacecraft re-entry) plasmas, although sea-level applications are also discussed. It looks at ionization degrees ranging from 10^{-7} to 10^{-3} and mean electron energies from 0.1 to 10 eV. In effect, the book serves as an excellent 'how-to' book for a theoretician interested in understanding air plasma phenomena. Experimental data are cited, but only to benchmark theoretical treatments. In addition, there are several other books that concentrate on the fundamental physics and the applications of non-equilibrium gas discharge plasmas and mention in passing atmospheric-pressure plasmas (Raizer *et al* 1995, Roth 1995, Batenin *et al* 1994, Lieberman and Lichtenberg 1994, Chapman 1980, Mitchner and Kruger 1973).

Also worthy of note are several texts that explore specific subtopics covered herein. For those readers particularly interested in computer modeling and simulation of plasma phenomena, there are two primary reference texts, the first by Birdsall and Langdon (1991) and the second by Hockney and Eastwood (1988). For experimentalists most concerned with the difficult task of taking accurate data in complex plasma systems, an excellent reference may be found in Hutchinson's classic diagnostics text (Hutchinson 2002). Finally, readers focused on rapid plasma applications may benefit from referring to the second volume of Roth's industrial plasma text (Roth 2001).

In order to reap the many benefits of interacting with scientists and engineers with similar air plasma interests, there are a number of professional organizations a reader should consider joining. This is an excellent way for individuals who are new to the field to make necessary personal technical contacts with individuals already active in the field. An approximate ordering of these professional organizations in roughly decreasing order of air plasma involvement is as follows:

1. The Institute of Electrical and Electronics Engineers (IEEE) Nuclear and Plasma Sciences Society (NPSS).
2. The Institute of Physics (IOP), United Kingdom.
3. The American Vacuum Society (AVS) and its industrial affiliates.
4. American Institute of Aeronautics and Astronautics (AIAA).
5. The American Physical Society (APS) through the Division of Plasma Physics, the Division of Atomic, Molecular, and Optical Physics, and the Division of Chemical Physics.

6. The European Physical Society (EPS) through its Division of Atomics, Molecular, and Optical Physics and its Division of Plasma Physics.
7. Institute of Electrical Engineering (IEE), United Kingdom.
8. Corresponding societies in Japan and Korea.

For the same reasons given above, researchers and engineers who wish to be active in the field of air plasmas would be wise to participate in those technical meetings that at least have technical sessions devoted to this topical area. Again in approximately decreasing order of air plasma participants such meetings may be listed as follows:

1. The Gaseous Electronics Conference, GEC (annual).
2. The International Conference on Phenomena in Ionized Gases, ICPIG (bi-annual).
3. The IEEE International Conference on Plasma Science, ICOPS (annual).
4. The International Symposium on High Pressure Low Temperature Plasma Chemistry (also known as the 'Hakone Conference', named after the city of Hakone in Japan where the first meeting was held in 1987) is a bi-annual series of conferences devoted exclusively to high-pressure discharge plasmas and their applications.
5. The Eurosectional Conference on Atomic and Molecular Processes in Ionized Gases, ESCAMPIG, which is a bi-annual European conference on fundamental processes in ionized gases.
6. The AIAA Conference in Reno, Nevada, USA (every January) (only the 'Weakly Ionized Gas (WIG)' sessions are of interest there).
7. 'ElectroMed', International Symposium on Non-thermal Medical/ Biological Treatments Using Electromagnetic Fields and Ionized Gases (bi-annual).
8. The APS annual meetings of the Division of Plasma Physics and the Division of Atomic, Molecular, and Optical Physics.

Finally, researchers in the field of non-equilibrium, atmospheric pressure air plasmas should consider publications in any of the following professional journals:

1. *Plasma Sources, Science, and Technology* (IOP).
2. *IEEE Transactions on Plasma Science*.
3. *Plasma Chemistry and Plasma Processing* (Kluwer Academic/Plenum Publishers).
4. *Journal of Physics D: Applied Physics* (IOP).
5. *Plasma Processes and Polymers* (Wiley-VCH).
6. *Physics of Plasmas* (AIP).
7. *Physical Review Letters* and *Physical Review* (AIP).
8. *Applied Physics Letters/Journal of Applied Physics* (AIP).
9. *Review of Scientific Instruments* (AIP).
10. *Contributions to Plasma Physics* (Wiley).

1.5   Organization of this Book

This volume has been assembled using three cooperative levels of authorship consisting of Authors, Chapter Masters, and Editors. The Authors, as listed in the front of this book, are those who have written significant sections of one or more chapters. The Chapter Masters acted not only as Authors but were also responsible for the content of their specific chapters. In cooperation with the Editors, they established the detailed outlines of their respective chapters and determined which sections to write themselves and which sections to solicit from other expert authors. These Chapter Masters had the responsibility to modify contributed text in order to smooth the internal flow of the sections and to ensure consistency within their chapters. They worked with the Editors and with the other Chapter Masters to resolve issues of overlap and repetition. Finally, the Editors, in addition to their service as Authors and Chapter Masters for specific portions of this book, shared the responsibility of reviewing the entire volume. To ensure a coherent book with synergistic chapters, they iterated numerous changes with Authors and worked toward a common terminology throughout and a reduction of differences in writing styles between the various chapters.

There are three major groupings of chapters within this book. The first grouping consists of chapters 1–5 and is fundamentally introductory in nature. After the subject matter is delineated in this chapter, chapter 2 proceeds to present the rich history of this field. Chapters 3 and 4 then proceed to provide the reader with all necessary theoretical foundations in both plasma physics and plasma chemistry respectively. This first grouping ends with chapter 5 which shows how the theoretical formulations of the previous two chapters are integrated into computer simulations to better understand and eventually predict observed air plasma phenomena. The next grouping, this one consisting of three chapters, takes the reader into the plasma laboratory itself to examine actual air plasma experiments, including the demanding experimental diagnostics necessary to truly understand the ionized phenomena under study. The final chapter, chapter 9, is a group unto itself. It looks to the future, discussing first the remaining scientific challenges presented by these plasmas and then looking closely at the array of attractive practical applications for which they can be employed. In the remainder of this section, each chapter is examined one by one. The responsible Chapter Master as well as all the individual contributing Authors of each chapter are listed in their respective chapter's heading.

Chapter 2, 'History of Non-Equilibrium Air Discharges', presents the historical progression and development of cold-plasma generation techniques. First, the discovery and study of dielectric barrier discharges is covered, followed by corona discharges and pulsed air discharges. Electrical breakdown and spark formation, as well as much of the fundamentals of corona discharges and high pressure glow discharges, are all treated herein. The evolution of the concept of non-equilibrium plasma conditions is traced.

Chapter 3, 'Kinetic Description of Plasmas', not only captures the key points of the classic textbook by Mitchner and Kruger (1973), but also focuses on those elements crucial to the specific understanding of sea-level air plasmas. The characteristics of weakly ionized and weakly coupled plasmas are presented including the concepts of multi-body elastic and inelastic collisions, an explanation of total and differential collision cross sections and rate constants, surface interactions and other 'collision-like' processes, as well as characteristic lengths and time-scales. A complete kinetic description of electrons is presented, including the concepts of phase space and velocity distribution functions, the general form of kinetic equations, collision terms and their general properties, a comparison with the fluid-dynamic picture, and the impossibility of general analytic and numerical solutions.

Chapter 4, 'Air Plasma Chemistry', reviews relevant collision processes including electron, ion–molecule, three-body, and step-wise collisions. The key reactions and types of reactions governing air plasma chemistry are highlighted. Ion–molecule reactions at elevated temperatures are discussed, highlighting the inadequacy of using rate constants obtained over a limited temperature range at high temperatures where vibrational excitation is important. The chapter then turns to non-equilibrium ion chemistry with considerations of the vibrational energy dependence of ion–molecule reactions, collision-induced dissociation reactions, scaling approaches, and state-resolved experiments and results. The state-of-the-art in electron–ion recombination science is then explained, with emphasis on product distribution and energy dependencies as well as recent key measurements.

Chapter 5, 'Modeling', illustrates how the theoretical formulations of plasma physics and plasma chemistry that were presented in chapters 3 and 4 have been successfully incorporated into computational models. The chapter begins with a general discussion of the technical challenges one encounters when undertaking air plasma modeling. It then presents a successful effort dealing with non-equilibrium air discharges using a numerical technique based on finite-volume computational fluid dynamics. Then the modeling of the electrical properties of different plasma-based devices is discussed, beginning with dc glow discharges in atmospheric pressure air. This is followed in turn by models for a negative corona in pin-to-plane configurations, dielectric barrier discharges, and a surface-discharge-type plasma display panel. By examining the techniques employed for the range of successful models presented, a reader can gain valuable insight regarding solutions applicable to their particular area of interest.

Chapter 6, 'DC and Low Frequency Air Plasma Sources', begins with a discussion of plasma sources that are often termed 'self-sustained plasmas', but that term was not used here to avoid confusion on the part of those outside the plasma discharge community. Among the topics covered are filamentary breakdown in dielectric barrier discharges, homogeneous and regularly-patterned barrier discharges, overall discharge parameters of barrier discharges, hollow and micro-hollow cathode discharges, recently discovered cathode boundary layer discharges (CBDs), discharges with micro-structured electrodes (MSEs), capillary plasma electrode discharges (CPEDs), positive and negative corona discharges, pulsed streamer coronas, pulsed diffuse discharges, glow discharges, and ac torch discharges with pronounced non-equilibrium properties.

Chapter 7, 'High Frequency Air Plasmas', gives an overview of the various 'external' means used to generate an air plasma including lasers, flash-tubes, rf and microwave, pulsed power, and electron beams. A dominant theme in this chapter is the ability to ionize air 'at a distance' away from any driving electrodes, unlike the methodologies described in the previous chapter. The air plasma technologies presented in this chapter begin with those using the highest available frequencies, namely those using photons as the driving ionization source. Two classes of photo-ionization technique are presented, the first using lasers and the second using ultraviolet flashlamps. Both of those techniques require the addition of photo-ionization seedants. The next section turns to rf-sustained discharges, including a microwave torch, rf-sustainment of a laser-initiated plasma, and creation of a localized plasma defined by the intersection of two microwave beams. Repetitively pulsed discharges are then discussed in the fourth section, followed by a section detailing a successful electron-beam ionization experiment using laser excitation. The final section in this chapter summarizes specific research challenges and opportunities associated with various of these techniques.

Chapter 8, 'Plasma Diagnostics', discusses the scientific challenges associated with trying to apply proven low-pressure plasma measurement techniques to the far more complex realm of collisionally dominated atmospheric pressure plasmas. Some techniques can be carried over but others cannot, depending also upon the desired resolution. The treatment of individual techniques begins in the second section with elastic and inelastic laser scattering in air plasmas. The next two sections look at electron density measurements, the first using millimeter-wave interferometry and the second using infrared (IR) heterodyne interferometry. From there, the chapter turns to diagnostics employing plasma emission spectroscopy. The chapter concludes with a section detailing the powerful cavity ring-down spectroscopic diagnostic for measuring ion concentrations.

Chapter 9, 'Current Applications of Atmospheric Pressure Air Plasmas', presents a series of the most compelling established and emerging applications for air plasma technology. These include the subjects of electrostatic precipitation, ozone generation, microwave reflection and absorption, aerodynamic applications, plasma-aided combustion, surface treatment, chemical decontamination, biological decontamination, and medical applications. Common for most of these applications is the unique ability of non-equilibrium air plasma to generate high concentrations of reactive species, without the need for elevated gas temperatures.

Acknowledgments

This chapter represents a (hopefully) faithful summary of the contributed thoughts and motivations of all the editors and authors who have collaborated in the creation of this volume. Particular assistance was provided by the author's co-editors along with the generous patience of our Editor-in-Chief, Professor Kurt Becker.

References

Batenin V M, Klimovskii L I, Lysov G V and Troitskii V N 1994 *Superhigh Frequency Generators of Plasma* (Boca Raton: CRC Press)

Birdsall C K and Langdon A B 1991 *Plasma Physics via Computer Simulation* (Bristol: Institute of Physics Press)

Birmingham J and Hammerstrom D 2000 'Bactcrial decontamination using ambient pressure plasma discharges' *IEEE Trans. Plasma Science* **28**(1) 51–56

Boyd J M and Sanderson J J 2003 *The Physics of Plasmas* (Cambridge: Cambridge University Press)

Capitelli M, Ferreira C M, Gordiets B F and Osipov A I 2000 *Plasma Kinetics in Atmospheric Gases* (Berlin: Springer)

Chapman B 1980 *Glow Discharge Processes: Sputtering and Plasma Etching* (New York: John Wiley and Sons)

Chen F F 1984 *Introduction to Plasma Physics* (New York: Plenum Publishing Corp.)

Dendy R O 1995 *Plasma Physics: An Introductory Course* (Cambridge: Cambridge University Press)

Van Dyken R, McLaughlin T and Enloe C 2004 'Parametric investigations of a single dielectric barrier plasma actuator' *Proc. 42nd AIAA Aerospace Sciences Meeting and Exhibit* Reno, NV AIAA Paper 2004–846

Griffiths D J 1998 *Introduction to Electrodynamics* (New York: Prentice Hall)

Hippler R, Pfau S, Schmidt M and Schoenbach K H (eds) 2001 *Low Temperature Plasma Physics* (Berlin: Wiley-VCH)

Hockney R W and Eastwood J W 1988 *Computer Simulation Using Particles* (Bristol: Adam Hilger)

Hudson R D 1971 'Critical review of ultraviolet photoabsorption cross section for molecules of astrophysical and aeronomic interest' *Rev. Geophys. Space Phys.* **9** 305–406

Humphreys W J 1964 *Physics of the Air* (New York: Dover Publications) 67–81

Hutchinson I H 2002 *Principles of Plasma Diagnostics* (Cambridge: Cambridge University Press)

Jackson J D 1998 *Classical Electrodynamics* (New York: Wiley Text Books)

Jursa A S (ed) 1985 *Handbook of Geophysics and the Space Environment* (US Air Force Geophysics Laboratory, Hanscom Air Force Base, MA, USA) US Defense Technical Information Center (DTIC) Document Accession Number: ADA 167000

Kogelschatz U, Egli W and Gerteisen E A 1999 *ABB Rev.* 4/1999 33–42

Kuo S P and Bivolaru D 2004 'Plasma torch igniters for a scramjet combustor' *Proc. 42nd AIAA Aerospace Sciences Meeting and Exhibit* Reno, NV AIAA Paper 2004–839

Kuo S P, Kalkhoran I M, Bivolaru D and Orlick L 2000 'Observation of shock wave elimination by a plasma in a Mach-2.5 Flow' *Physics of Plasmas* **7**(5) 1345–1348

Laroussi M, Richardson J P and Dobbs F C 2002 'Effects of nonequilibrium atmospheric pressure plasmas on the heterotropic pathways of bacteria and on their cell morphology' *Appl. Phys. Lett.* **81**(4) 22

Lieberman M A and Lichtenberg A J 1994 *Principles of Plasma Discharges and Materials Processing* (New York: Wiley-Interscience)

Liu J, Wang F, Lee L, Theiss N, Romney P and Gundersen M 2004 'Effect of discharge energy and cavity geometry on flame ignition by transient plasma' *Proc. 42nd AIAA Aerospace Sciences Meeting and Exhibit*, Reno, NV, AIAA Paper 2004–1011

Mitchner M and Kruger C H 1973 *Partially Ionized Gases* (New York: John Wiley and Sons)

Montie T C, Kelly-Wintenberg K and Roth J R 2000 'Overview of research using a one atmosphere uniform glow discharge plasma (OAUGDP) for sterilization of surfaces and materials' *IEEE Trans. on Plasma Science* **28**(1) 41–50

Moran J M, Morgan M D, Pauley P M and Moran M D 1996 *Meteorology: The Atmosphere and Science of Weather* (New York: Prentice Hall)

NASA 2004 (http://nssdc.gsfc.nasa.gov/space/model/ionos/about_ionos.html)

Nishida M, Yukimura K, Kambara S and Maruyama T 2001 *J. Appl. Phys.* **90** 2672–2677

Parker K R (ed) 1997 *Applied Electrostatic Precipitation* (London: Blackie Academic & Professional)

Pollack G and Stump D 2002 *Electromagnetism* (New York: Prentice Hall)

Raizer Y P, Shneider M N and Yatsenko N A 1995 *Radio-Frequency Capacitive Discharges* (Boca Raton: CRC Press)

Roth J R 1995 *Industrial Plasma Engineering: Principles* (Bristol and Philadelphia: Institute of Physics Publishing)

Roth J R 2001 *Industrial Plasma Engineering: Applications to Non-Thermal Plasma Processing* (Bristol and Philadelphia: Institute of Physics Publishing)

Roth J R 2003 'Aerodynamic flow acceleration using paraelectric and peristaltic electrohydrodynamic (EHD) effects of a one atmosphere glow discharge plasma' *Physics of Plasmas* **10**(5) 2117–2126

Roth J R, Chen Z, Sherman D M, Karakaya F, Tsai P P-Y, Kelly-Wintenberg K and Montie T C 2001 'Increasing the surface energy and sterilization of nonwoven fabrics by exposure to a one atmosphere uniform glow discharge plasma (OAUGDP)' *International Nonwovens J.* **10**(3) 34–47

Schunk R W and Nagy A F 2000 *Ionospheres: Physics, Plasma Physics, and Chemistry* (Cambridge: Cambridge University Press)

Stenhoff M 1999 *Ball Lightning* (New York: Kluwer Academic/Plenum Publishers)

Stolarski R S and Johnson N P 1972 'Photoionization and photoabsorption cross sections for ionospheric calculations' *J. Atmos. Terr. Phys.* **34** 1691

Torr D G 1979 'Ionospheric chemistry' *Rev. Geophys. Space Phys.* **17** 510–521

Tsai P, Wadsworth L and Roth J R 1997 'Surface modification of fabrics using a one-atmosphere glow discharge plasma to improve fabric wettability' *Textile Research J.* **5**(65) 359–369

Vidmar R J 1990 'On the use of atmospheric pressure plasmas as electromagnetic reflectors and absorbers' *IEEE Trans. Plasma Science* **18**(4) 733–741

White H J 1963 *Industrial Electrostatic Precipitation* (Reading, MA: Addison Wesley)

Chapter 2

History of Non-Equilibrium Air Discharges

U Kogelschatz, Yu S Akishev and A P Napartovich

2.1 Introduction

Chapter 2 provides a short review of the historical development of non-equilibrium discharges with a tendency to focus on air plasmas at atmospheric pressure. The main physical mechanisms of breakdown and classifications of various discharges are discussed. The principal discharge configurations are presented and their main properties and applications are discussed. The fundamentals of corona discharges (Akishev, Napartovich) and dielectric-barrier discharges are presented. More detailed information and recent developments are treated in chapter 6.

2.2 Historical Roots of Electrical Gas Discharges

Until the beginning of the 18th century air like any other gas was believed to be an ideal electrical insulator. The fact that air can pass electrical charges was first established by Coulomb, who could show that two oppositely charged metal spheres gradually lost their charges (Coulomb 1785). In carefully designed experiments he could conclusively demonstrate that this loss of electrical charge was due to leakage through the surrounding air and not through imperfect insulation. In the middle of the 18th century Benjamin Franklin had shown experimentally that a laboratory spark and lightning were of common nature. Around 1800 V. V. Petrov in St. Petersburg and Humphry Davy in Britain started to investigate arc discharges in air. Davy suggested the name arc because the extremely bright discharge column is normally bent due to the buoyancy of hot air. Arcs can get very hot and were normally started by separating two carbon electrodes connected to a voltage supply. Powerful batteries were required to supply enough current to maintain the arc. In addition to these hot arc discharges, cold glow discharges were investigated. Major investigations on the passage of electricity through various gases and on fundamental properties of gas discharges were performed by Faraday (1839, 1844, 1855), Hittorf (1869), Crookes (1879), Stoletow (1890), Thompson (1903), and Townsend (1915), to name only the most important ones. Faraday was probably the first to realize that an ionized gas had unique properties and carefully documented his observations in three volumes of *Experimental Researches in Electricity* (1839, 1844, 1855).

Many experiments were carried out at reduced pressure. This had the advantage that only moderate voltages were required to start the discharge and that the whole discharge vessel could be filled with discharge plasma. The progress in gas discharge physics depended heavily on the development of vacuum pumps and the availability of adequate voltage sources. Of equal importance were the skills of a good glass blower. Faraday could already evacuate tubes to about 1 torr and apply voltages up to 1000 V. He introduced the concept of ions as carriers of electricity (in electrolytes) and distinguished between cathode and anode, even between cations moving to the cathode and anions passing to the anode. Crookes emphasized that a gas discharge actually constitutes a fourth state of the matter. The term plasma was coined much later, by Langmuir and Tonks, in 1928. Today the word plasma is mainly used to describe a quasi-neutral collection of free-moving electrons and ions.

More refined experiments with rarefied gases started at the beginning of the 20th century. For a long time the transport of electricity through gases had been treated like the flow of charges in electrolytes. Only about 1900, mainly due to the work of Wilson (1901) and Townsend (1904), it was established that conductivity in electrical gas discharges was due to ionization of gas atoms or molecules by collisions with electrons. In most gas discharges the current is mainly carried by electrons.

From the very beginning it was obvious that cold glow discharge plasmas had different properties than the hot arc discharges. For a long time it was believed that glow discharges which are characterized by hot electrons and essentially cold heavy particles (atoms, molecules, ions) could exist only at low pressure. It is one of the purposes of this book to describe recent developments showing that non-equilibrium plasma conditions with electron energies substantially higher than those of heavy particles, and properties resembling those of low pressure glow discharges, can exist also at much higher pressure, for example in atmospheric pressure air.

References

Crookes W 1879 *Phil Trans. Pt. 1* 135–164
Coulomb M 1785 *Mém. Acad. Royale des Sci. de Paris* 612–638

Faraday M 1839 *Experimental Researches in Electricity* vol. I (London: Taylor and Francis)

Faraday M 1844 *Experimental Researches in Electricity* vol. II (London: Taylor and Francis)

Faraday M 1855 *Experimental Researches in Electricity* vol. III (London: Taylor and Francis)

Hittorf W 1869 *Pogg. Ann.* **136** 1–31 and 197–235

Stoletow M A 1890 *J. de Phys.* **9** 468–472

Thompson J J 1903 *Conduction of Electricity through Gases* (Cambridge: Cambridge University Press)

Townsend J S 1915 *Electricity in Gases* (Oxford: Clarendon Press)

Townsend J S and Hurst H E 1904 *Phil. Mag.* **8** 738–753

Wilson C T R 1901 *Proc. Phys. Soc. London* **68** 151–161

2.3 Historical Progression of Generating Techniques for Hot and Cold Plasmas

From the early days of gas discharge physics it was apparent that, after ignition of the discharge, entirely different plasma states can be established in the same medium. One representative was the hot arc discharge, typically operated in air at atmospheric pressure, approaching conditions of local thermodynamic equilibrium (LTE). This thermodynamic state is characterized by the property that all particle concentrations are only a function of the temperature. In short, these plasmas are also referred to as thermal plasmas. Cold plasmas, on the other hand, are characterized by the property that the energy is selectively fed to the electrons leading to electron temperatures that can be considerably higher than the temperature of the heavy particles in the plasma. These non-equilibrium or non-LTE plasmas exhibit typical plasma properties such as electrical conductivity, light emission and chemical activity already at moderate gas temperatures, even at room temperature. Both hot and cold plasmas have found important and far-reaching technical applications. In the following sections the historical development of the discharge configurations used to produce hot or cold plasmas is briefly discussed with special emphasis on the properties of air plasmas.

2.3.1 Generation of hot plasmas

Typical examples of thermal plasmas are plasmas produced in high-intensity arcs, plasma torches or radio frequency (rf) discharges at or above atmospheric pressure. Figure 2.3.1 shows three simple configurations used to produce arcs or plasma jets in atmospheric pressure air.

The electrodes are either water-cooled metal parts or simply graphite rods. Typical currents range from 10 to 1000 A, typical temperatures from

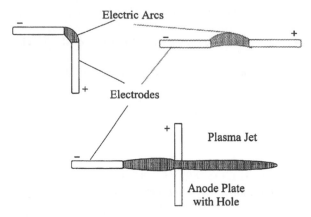

Figure 2.3.1. Principal arc configurations.

5000 to 50 000 K. In most arcs the degree of ionization lies between 1 and 100%. The high temperature of the arc column can be utilized for light emission as well as for melting materials and for initiating chemical reactions. The plasma plume extending several centimeters from the orifice in the anode plate in the lower part of figure 2.3.1 represents a neutral plasma with zero net current. When specially shaped nozzles are used, supersonic expansion into a low-pressure environment can produce pronounced non-equilibrium plasma conditions.

Around 1808 Humphry Davy invented the carbon-arc lamp, using an arc between two carbon electrodes, which later found applications in movie projection lamps, in searchlights and as a radiation standard for spectroscopy. Davy used arcs for melting (1815) and investigated the effects of magnetic fields on arcs (1821). But it wasn't until 1878 that Sir Charles William Siemens in Britain built and patented arc furnaces for steel making using direct-arc and indirect-arc principles. In France this technology was investigated by Moissan (1892, 1897) and by Héroult. Much of the early work on electric arcs is summarized in the monograph of Ayrton (1902). In 1901 Marconi used an electric arc for radio transmission across the Atlantic, and around 1910 already 120 arc furnaces of the Schönherr and Birkeland–Eyde design were installed in Southern Norway for nitrogen fixation. In this electric-arc process, proposed by Birkeland and Eyde in 1903, nitrogen and oxygen in air were combined to form nitrogen oxides, nitric acid, and finally artificial fertilizer (Norge salpeter, i.e. calcium nitrate). By 1917 the plant had been extended to use up to 250 MW of cheap hydro power. Arc welding was first demonstrated around 1910, and in its various forms is now responsible for the bulk of fusion welds.

Schönherr (1909) was the first to use a forced gas flow to stabilize long carbon arcs. Today various kinds of flow and vortex arc stabilization techniques are used in plasma torches. Many technological developments are

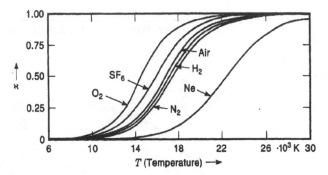

Figure 2.3.2. Degree of thermoionization in different atmospheric pressure gases (from Boeck and Pfeiffer (1999) p 130).

described in a book edited by Dresvin (1977), and in reviews by Pfender (1978) and Pfender *et al* (1987). The fundamentals and applications of thermal plasmas are discussed in Boulos *et al* (1994), in Heberlein and Voshall (1997) and in Pfender (1999). The most important applications include circuit breakers, lamps, plasma spraying, welding and cutting, metallurgical processing and waste disposal. Most arcs are approaching the state of local thermal equilibrium (LTE) and require high temperatures to maintain sufficient electrical conductivity by thermal ionization. From figure 2.3.2, showing the degree of ionization as a function of temperature for different gases including air, it is apparent that temperatures well in excess of 5000 K are required.

Figure 2.3.3 shows the temperature dependence of particle number concentrations of an LTE plasma in atmospheric pressure dry air. With rising temperature the molecules O_2 and N_2 are dissociated, new molecules like NO form, the atoms N and O prevail around 8000 K and, at higher temperatures, the charged the particle species e, N^+, and O^+ dominate.

2.3.2 Generation of cold plasmas

Besides thermal plasmas also cold non-equilibrium (non-LTE) plasmas are of increasing interest. In contrast to thermal plasmas, cold plasmas are characterized by a high electron temperature T_e and a rather low gas temperature T_g characterizing the heavy particles: atoms, molecules, and ions ($T_e \gg T_g$). The thermodynamic properties of the equilibrium and non-equilibrium states of plasmas were discussed by Drawin (1971). In extreme cases the electron temperature can reach well above 20 000 K while the gas temperature stays close to room temperature. Such non-equilibrium plasmas can be produced in various types of low-pressure glow and rf discharges (figure 2.3.4) as well as in corona, barrier, and hollow cathode discharges at atmospheric pressure (see sections 2.5 and 2.6 and chapter 6).

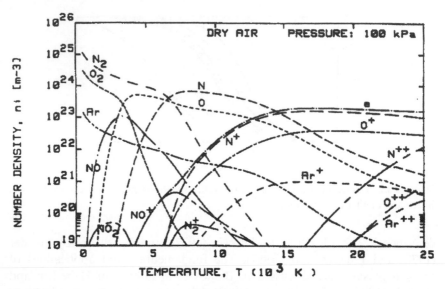

Figure 2.3.3. Composition of an atmospheric pressure dry air plasma versus temperature (from P. Fauchais, Summer School, ISPC–16 2003).

The glow discharge at reduced pressure, known since the days of Faraday, Hittorf and Crookes, has been thoroughly investigated experimentally as well as theoretically. Its main part, the positive column can provide large volumes of quasi-neutral non-LTE plasma. Glow discharges have found widespread applications in fluorescent lamps and as a processing medium for surface modification and plasma enhanced chemical vapor deposition (PECVD). The inductive rf plasma shown also in figure 2.3.4 was first observed by Hittorf (1884). It provides an elegant way of producing a plasma not in contact with metal electrodes. Thomson (1927) formulated a theory and Eckert (1974) published a detailed state of the art. The rf driven,

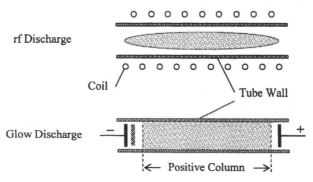

Figure 2.3.4. Principal configurations of rf discharges and dc glow discharges.

inductively-coupled plasma (ICP) has found a wide range of industrial uses, including spectroscopic diagnostic tools, plasma torches, and the heating of fusion plasmas. More recently, ICPs also found important applications in lamps and as processing tools in the semiconductor industry.

It should be mentioned that arcs can also be operated at reduced pressure and glow discharges at higher pressure. In addition to dc operation all types of discharges can be operated at various frequencies or in a pulsed mode. Special effects can be achieved if additional magnetic fields are used to influence electron motion: magnetron discharges and electron cyclotron resonance (ECR) sources.

Since collisions cause a continual exchange of energy between electrons of mass m_e and heavy particles of mass m_g with a tendency to equilibrate temperatures it is more difficult to maintain non-equilibrium conditions at elevated pressure with high collision rates and short mean free paths. For steady-state discharges the deviation from local thermodynamic equilibrium can be expressed by the following formula which was derived from an energy balance (Finkelnburg and Maecker 1956).

$$\frac{T_e - T_g}{T_e} = \frac{m_g}{4m_e} \frac{(\lambda_e e E)^2}{(\frac{3}{2}kT_e)^2}. \tag{2.3.1}$$

In this relation λ_e is the mean free path of electrons, the term $\lambda_e e E$ is the amount of directed energy an electron picks up along one free path in the direction of the electric field E and $\frac{3}{2}kT_e$ is the average thermal energy (e is the electronic charge, k is the Boltzmann constant). From relation (2.3.1) it is apparent that large mean free paths (low pressure or density), high electric fields and low electron energies favor deviations from LTE conditions. Figure 2.3.5 shows in a semi-schematic diagram how electron and gas temperatures separate in an electric arc with decreasing pressure (Pfender 1978).

Pronounced non-equilibrium conditions are obtained at reduced pressure, while in atmospheric pressure arcs columns the deviation from LTE is on the order of 1%. At high pressure, non-equilibrium conditions can be encountered when fast temporal changes occur (ignition and extinction of a discharge) and in regions of high field or concentration gradients. In many cases short high voltage pulses are used to preferentially heat electrons. In recent years also dc non-equilibrium air discharges at atmospheric pressure have been extensively investigated at reduced gas density (Kruger *et al* 2002, Yu *et al* 2002, Laroussi *et al* 2003). These experiments were performed at gas temperatures between 700 and 2000 K. Stable diffuse non-equilibrium air discharges were obtained with electron densities in excess of 10^{12} cm^{-3}. This value is roughly six orders of magnitude higher than the equilibrium value of $n_e = 3 \times 10^6$ cm^{-3} for an LTE air plasma at 2000 K (Yu *et al* 2002).

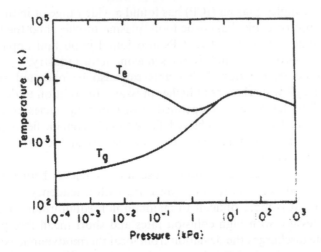

Figure 2.3.5. Electron temperature and gas temperature in an arc as a function of pressure (from Pfender (1978) p 302).

In the literature there exist a number of models treating non-LTE plasmas. Many of them are based on a fluid approach. In the simplest case a two-fluid model can be used with two different temperatures, T_e and T_g. The electron kinetics can be treated by determining the electron energy distribution function (EEDF) by means of the Boltzmann equation using, for example, a two-term approximation. The reaction rate coefficients can be obtained as functions of the average electron energy, which, in this local field approximation, is only a function of the reduced electric field E/N. Knowledge of all relevant electron impact cross sections is an important requirement.

2.3.3 Properties of non-equilibrium air plasmas

Air is a mixture of many constituents. The CRC *Handbook of Chemistry and Physics* (1997 edition) lists the following composition for the sea level dry air (in vol% at 15°C and 101 325 Pa):

Nitrogen	78.084%	Methane	0.0002%
Oxygen	20.9476%	Helium	0.000524%
Argon	0.934%	Krypton	0.000114%
Carbon dioxide	0.031%	Hydrogen	0.00005%
Neon	0.001818%	Xenon	0.0000087%

Electron collision cross sections have been measured and compiled for more than a century now. The cross sections for the three major air constituents

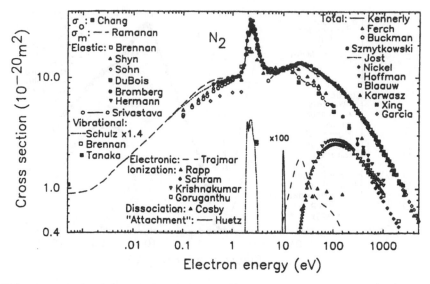

Figure 2.3.6. Integral cross sections for electron scattering of N_2 (from Zecca *et al* (1996) p 94). (Copyright Societa Italiana di Fisica.)

N_2, O_2 and Ar, taken from a critical review by Zecca *et al* (1996), are given in figures 2.3.6–2.3.8.

As a result of such Boltzmann computations figure 2.3.9 shows the monotonous relation between the mean electron energy and the reduced

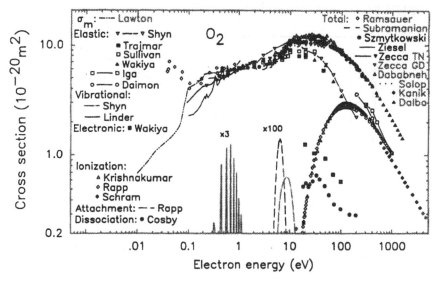

Figure 2.3.7. Integral cross sections for electron scattering of O_2 (from Zecca *et al* (1996) p 115). (Copyright Societa Italiana di Fisica.)

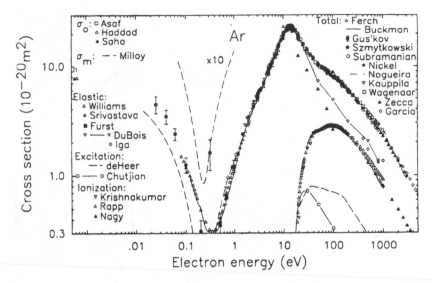

Figure 2.3.8. Integral cross sections for electron scattering of Ar (from Zecca *et al* (1996) p 31). (Copyright Societa Italiana di Fisica.)

electric field. Breakdown in a homogeneous electric field and wide gaps occur when a reduced field E/N of about 100 Td ($1\,\text{Td} = 10^{-21}\,\text{V m}^2$) is reached. According to figure 2.3.9 this will produce electrons of mean energy close to 3 eV, corresponding to an electron temperature of roughly 20 000 K. In narrow discharge gaps, pulsed discharges, and in front of the head of a propagating streamer these values can be higher.

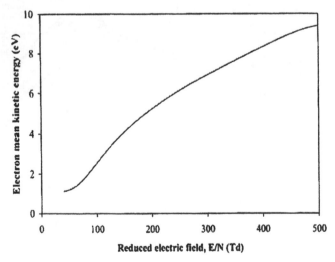

Figure 2.3.9. Mean electron energy in dry air as a function of the reduced field E/N (from Chen (2002) p 48).

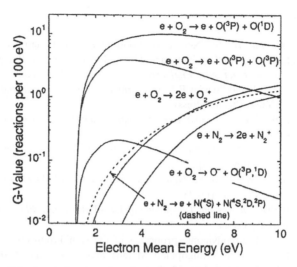

Figure 2.3.10. Calculated G-values (number of reactions per 100 eV of input energy) for dissociation and ionization reactions in dry air, shown as functions of the electron mean energy in a non-equilibrium discharge plasma (from Penetrante *et al* (1997) p 253).

Computations in non-equilibrium air plasmas have been carried out for applications in ozone generation and for pollution control. The efficiency of a particular electron impact reaction can be expressed in terms of the G-value, which gives the number of reactions per 100 eV of input power. Figure 2.3.10 shows computed values for the dissociation and ionization reactions in atmospheric pressure dry air.

In the electron energy range encountered in non-equilibrium gas discharges (typically 3–6 eV, in pulsed discharges up to 10 eV) oxygen dissociation is the most efficient reaction (highest G-value). This explains why non-equilibrium discharges in air invariably lead to the formation of ozone and nitrogen oxides.

Non-equilibrium plasmas are mainly used to generate chemically reactive species and for their electromagnetic properties. Their applications include the synthesis of thermally unstable compounds like ozone and the generation of intermediate free radicals for pollution control. Surface modification of polymer foils, thin film deposition and plasma etching in the electronic industry are further applications. Progress in the understanding and control of atmospheric pressure non-equilibrium discharges has led to increased activity in recent years which is manifested in several monographs and review papers devoted to this special subject (Capitelli and Bardsley 1990, Eliasson and Kogelschatz 1991, Lelevkin *et al* 1992, Penetrante and Schultheis 1993, Manheimer *et al* 1997, Capitelli *et al* 2000, Kunhardt 2000, Protasevich 2000, van Veldhuizen 2000, Hippler *et al* 2001, Kruger *et al* 2002).

References

Ayrton H 1902 *The Electric Arc* (New York, London: The Electrician Print. Publ. Co.)

Boeck W and Pfeiffer W 1999 'Conduction and breakdown in gases' in *Wiley Encyclopedia of Electrical and Electronics Engineering* (New York: Wiley) vol. 4 p 130

Boulos M I, Fauchais P and Pfender E 1994 *Thermal Plasmas: Fundamentals and Applications* (New York: Plenum Press)

Capitelli M and Bardsley J N (eds) 1990 *Nonequilibrium Processes in Partially Ionized Gases* (New York: Plenum)

Capitelli M, Ferreira C M, Gordiets B F and Osipov A-I 2000 *Plasma Kinetics in Atmospheric Gases* (Berlin: Springer)

Chen J 2002 *Direct current corona-enhanced chemical reactions*, PhD Thesis (Minneapolis: University of Minnesota) p 48

Dresvin S V (ed) 1977 *Physics and Technology of Low-Temperature Plasmas* (Ames: Iowa State University Press)

Drawin H W 1971 'Thermodynamic properties of the equilibrium and nonequilibrium states of plasmas' in Venugopalan M (ed) *Reactions under Plasma Conditions* (New York: Wiley), vol. 1 pp 53 –238

Eckert H U 1974 *High Temp. Sci.* **6** 99–134

Eliasson B and Kogelschatz U 1991 *IEEE Trans. Plasma Sci.* **19** 1063–1077

Finkelnburg W and Maecker H 1956 'Elektrische Bögen und thermisches Plasma' in Flügge S (ed) *Encyclopedia of Physics* (Berlin: Springer) vol. XXII p 307

Heberlein J V R and Voshall R E 1997 'Thermal plasma devices' in Trigg G L (ed) *Encyclopedia of Applied Physics* (New York: Wiley) vol. 21 pp 163–191

Hippler R, Pfau S, Schmidt M and Schoenbach K H (eds) 2001 *Low Temperature Plasma Physics* (Weinheim: Wiley-VCH)

Hittorf W 1884 *Wiedemann Ann. Phys. Chem.* **21** 90–139

Kruger C H, Laux C O, Yu L, Packan D L and Pierot L 2002 *Pure Appl. Chem.* **74** 337–347

Kunhardt E E 2000 *IEEE Trans. Plasma Sci.* **28** 189–200

Laroussi M, Lu X and Malott C M 2003 *Plasma Sources Sci. Technol.* **12** 53–56

Lelevkin V M, Otorbaev D K and Schram D C 1992 *Physics of Non-Equilibrium Plasmas* (Amsterdam: Elsevier)

Manheimer W, Sugiyama L E and Stix T H (eds) 1997 *Plasma Science and the Environment* (Woodbury: American Institute of Physics)

Moissan H 1892 *C. R. Acad. Sci. Paris* **115** 1031–1033

Moissan H 1897 *Le Four Électrique* (Paris: Steinheil)

Penetrante B M and Schultheis S E (eds) 1993 *Non-Thermal Plasma Techniques for Pollution Control* (Berlin: Springer) Part A and B

Penetrante B M, Hsiao M C, Bardsley J N, Merritt B T, Vogtlin G E, Kuthi A, Burkhart C P and Bayless J R 1997 *Plasma Sources Sci. Technol.* **6** 251–259

Pfender E 1978 'Electric arcs and arc gas heaters' in Hirsh M N and Oskam H J (eds) *Gaseous Electronics: Electrical Discharges* (New York: Academic) vol. 1 pp 291–398

Pfender E 1999 *Plasma Chem. Plasma Proc.* **19** 1–31

Pfender E, Boulos M and Fauchais P 1987 'Methods and principles of plasma generation' in Feinman J (ed) *Plasma Technology in Metallurgical Processing* (Warrendale: Iron and Steel Society) pp 27–47

Protasevich E T 2000 *Cold Non-Equilibrium Plasma* (Cambridge: Cambridge Int. Sci. Publ.)

Schönherr O 1909 *Elektrotechn. Zeitschr.* **30**(16) 365–369 and 397–402

Thomson J J 1927 *Phil. Mag.* Ser. 7, **4**(25) Suppl. Nov. 1927, 1128–1160
van Veldhuizen E M (ed) 2000 *Electrical Discharges for Environmental Purposes: Fundamentals and Applications* (Commack: Nova Science)
Yu L, Laux C O, Packan D M and Kruger C H 2002 *J. Appl. Phys.* **91** 2678–2686
Zecca A, Karwasz G P and Brusa R S 1996 *Rivista Nuovo Cim.* **19**(3) 1–146

2.4 Electrical Breakdown in Dense Gases

Electrical breakdown in dense gases like air at atmospheric pressure has been the object of many investigations. In high voltage engineering one of the major aspects is to avoid breakdown or flashover between adjacent conductors or between a conductor and ground. The subject of gaseous insulation has recently been reviewed by Niemeyer (1999). The physical phenomena occurring in the early phases of breakdown in atmospheric pressure air or in other compressed gases have many similarities with the ignition phase of a low pressure gas discharge. They all start with an initial electron growing into an electron avalanche under the influence of the electric field. In dense gases, however, the fate of an electron avalanche can be quite different, depending on the way the voltage is applied to the gas gap. A short overview of the physical processes involved in breakdown under different conditions and of the discharge types breakdown can lead to is given in the following sections.

2.4.1 Discharge classification and Townsend breakdown

Traditionally, many gas discharges have been operated at low or very low pressure compared to atmospheric conditions. In this context we consider, for the purpose of this book, atmospheric pressure as high pressure. Also at this pressure it is useful to characterize the type of discharge similar to the traditional classification at low pressure (figure 2.4.1). The diagram is a modified version of a graph from the famous paper by Druyvesteyn and Penning (1940). It originally related to a discharge in 1 torr Ne, an electrode area of $10\,cm^2$ and an electrode separation of 50 cm. Nevertheless many fundamental concepts also apply to a discharge in air at atmospheric pressure. Since there is always some natural radioactivity resulting in the production of 10–100 electrons per cm^3 per s we can always draw a minute base current if an electric field is applied. In air at atmospheric pressure the saturation value of the current density amounts to about $10^{-18}\,A\,cm^{-2}$ and is subjected to statistical fluctuations. It can be considerably increased if x-ray irradiation or ultraviolet illumination of the cathode is used to produce additional electrons (region A $\rightarrow$ A$'$). In this region the current

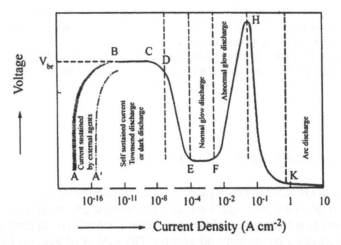

Figure 2.4.1. Discharge characterization (based on Druyvesteyn and Penning 1940).

drops to the base current if the external source of electrons is switched off (non-self-sustained region). Once the breakdown voltage V_{br} of the gas space is reached we get into the self-sustained discharge region, starting with a Townsend discharge. The range of the Townsend discharge is characterized by a negligible influence of space charge on the applied external field. This condition is normally fulfilled in the current density range $j = 10^{-15}$–10^{-6} A cm^{-2}.

According to an empirical relation found by Paschen in 1889 the value of the breakdown voltage for a given gas (and cathode material) is only a function of the product pressure p times electrode separation d, $V_{br} = f(pd)$, or, as we would formulate it today, $V_{br} = f(Nd)$, where N is the number density of the gas. The old relation is valid only for a given temperature, in most cases room temperature, while the second relation is more universal and does not depend on temperature. Some examples for Paschen breakdown curves in different gases are given in figure 2.4.2.

Since the isolation properties of atmospheric pressure air are of fundamental interest in high voltage engineering the Paschen curve of air is extremely well investigated and documented (figure 2.4.3).

It should be mentioned that humidity has an influence on the breakdown voltage of air. Small admixtures lower the breakdown voltage, which reaches a minimum at about 1% water vapor and then rises again (Protasevich 2000, p 69). There is also a pronounced frequency dependence of the breakdown voltage with a minimum value at about 1 MHz (Kunhardt 2000).

The Paschen curve can be obtained from the ionization coefficient α of the gas and the γ coefficient quantifying the number of secondary electrons produced at the cathode per ion of the primary avalanche. The first Townsend coefficient, the ionization coefficient α, defines the number of electrons

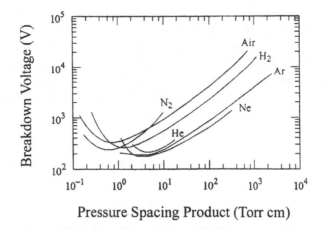

Pressure Spacing Product (Torr cm)

Figure 2.4.2. Paschen breakdown voltages for static breakdown in N_2, air, H_2, He, Ne, Ar (based on Vollrath and Thomer 1967 p 81).

produced in the path of a single electron traveling 1 cm in the direction of the field E. The second Townsend coefficient γ depends on the cathode material and the gas and includes contributions by positive ions, by photons, by fast atoms, and by metastable atoms and molecules. Theoretically also volume processes like photo-ionization of the background gas can produce

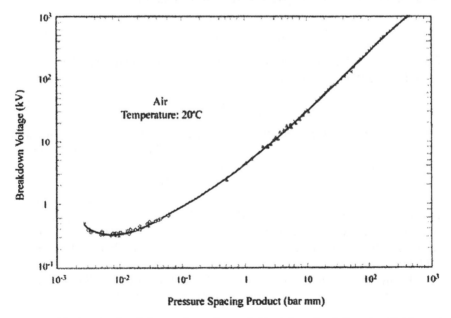

Pressure Spacing Product (bar mm)

Figure 2.4.3. Paschen breakdown voltages for static breakdown in air (based on Dakin *et al* 1974).

secondary electrons to meet the self-sustainment criterion. However, electrons released at the cathode travel the whole distance to the anode and produce more ionization than electrons created en route. For this reason the onset of breakdown is determined by γ-effects at the cathode.

Typical values of γ are in the range 10^{-4} to 10^{-1}. According to Townsend (1915) current amplification in the homogeneous field can be written as

$$I = I_0 \frac{e^{\alpha d}}{1 - \gamma(e^{\alpha d} - 1)} \tag{2.4.1}$$

and breakdown is reached when current amplification in a gap tends to infinity:

$$\gamma(e^{\alpha d} - 1) = 1. \tag{2.4.2}$$

This Townsend criterion for stationary self-sustainment of the current has been used ever since as a general criterion for stationary breakdown in homogeneous fields.

If the ionization coefficient α is approximated by a relation also suggested by Townsend

$$\frac{\alpha}{p} = A\,e^{-Bp/E} \tag{2.4.3}$$

where A and B are constants characterizing the gas under investigation. The breakdown voltage V_{br} is given by the simple relation

$$V_{br} = \frac{Bpd}{\ln(Apd) - \ln\ln[(1 + \gamma)/\gamma]}. \tag{2.4.4}$$

For rough calculations in dry air the ionization coefficient α can be approximated in modern writing as

$$\frac{\alpha}{N} = A\,e^{-BN/E} \tag{2.4.5}$$

where N is the number density of the molecules, $A = 1.4 \times 10^{-20}\,\mathrm{m}^2$, and $B = 660\,\mathrm{Td}$ (1 Td corresponds to $10^{-21}\,\mathrm{V\,m}^2$). This relation approximates experimental data by Wagner (1971) and Moruzzi and Price (1974) in the range $10\,\mathrm{Td} < E/N < 150\,\mathrm{Td}$ (Sigmond 1984). Experimental data for higher E/N ranges were provided by Raja Rao and Govinda Raju (1971) and by Maller and Naidu (1976). More sophisticated analytical approximations for ionization and attachment coefficients covering a wider E/N range in air can be found in Morrow and Lowke (1997) or Chen and Davidson (2003).

Using the characteristic values at the minimum of the Paschen curve ($V_{\min}$ and $\delta = pd/(pd)_{\min}$) equation (2.4.3) can be rewritten as

$$\frac{V_{br}}{V_{\min}} = \frac{\delta}{1 + \ln \delta}, \tag{2.4.6}$$

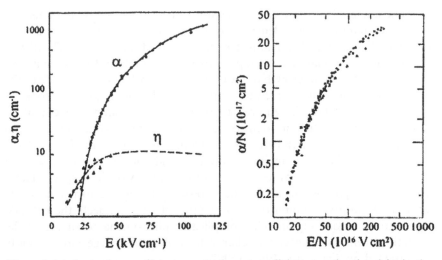

Figure 2.4.4. Ionization coefficient α, attachment coefficient η and reduced ionization coefficient α/N for dry air (left plots, Les Renardières Group 1972; right curve from Raja Rao and Govinda Raju 1971).

a simple formulation of the Paschen law which holds for an extended pd-range and can be used to get an estimate of the breakdown voltage in a homogeneous field. For air $V_{min} = 230\text{–}370\,\text{V}$, depending on the cathode material, $(pd)_{min} \approx 0.6\,\text{torr cm}$. As mentioned before, the original concept of gas breakdown by successive electron avalanches and a feed-back mechanism at the cathode was proposed by Townsend in 1915. Later, more detailed, descriptions can be found in Loeb (1939), Little (1956), Raether (1964), Hess (1976), Dutton (1978, 1983), Raizer (1986, 1991), and Boeck and Pfeiffer (1999). A detailed review on the relative contributions of different γ feedback mechanisms in argon was recently published by Phelps and Petrović (1999). An important extension of the simple Townsend breakdown criterion (2.4.1) for electronegative gases was formulated by Geballe and Reeves (1953). Introducing the attachment coefficient η the effective ionization coefficient becomes $\alpha_{\text{eff}} = \alpha - \eta$, and the self-sustainment condition (2.4.1) becomes

$$\frac{\gamma\alpha}{\alpha - \eta}[\exp(\alpha - \eta)d - 1] = 1. \qquad (2.4.7)$$

The ionization and attachment coefficients for room temperature dry air are plotted in figure 2.4.4. They cross at an E/p value about $25\,\text{kV cm}^{-1}\,\text{bar}^{-1}$ corresponding to an E/N value of about $100\,\text{Td}$. At this value the effective ionization coefficient of air equals zero because electron collisions leading to ionization are balanced by electron attachment reactions. At higher fields ionization dominates, at lower fields attachment.

The range of the Townsend discharge (dark discharge) is characterized by the fact that the current density and the charge density in the plasma is so low that it has practically no influence on the applied electric field. The degree of ionization is so small that no appreciable light is emitted. In this regime we observe an exponential growth of the electron density from the cathode to the anode, and practically the entire volume is filled with positive ions. A relatively high voltage is required to meet the self-sustainment condition (2.4.2). When the current density is increased beyond about 10^{-5} to $10^{-6}\,\mathrm{A\,cm^{-2}}$ the Townsend discharge changes to a glow discharge. Now space charge fields play an important role and the voltage necessary to sustain the discharge drops to a few hundred volts. A positive space charge region with high electric fields, the cathode fall region, forms near the cathode. A positive column of quasi-neutral plasma connects the cathode region to the anode region. The complicated phenomena occurring in the transition from a Townsend discharge to a glow discharge have recently been treated by Šijačič and Ebert (2002).

The theory of the normal glow discharge was formulated by von Engel and Steenbeck (1934) by applying the Townsend condition for self-sustainment to the cathode layer. For a wide pressure and current density range the parameters j/p^2, V_{cf} and pd_{cf} are constant, where j is the current density, V_{cf} is the voltage across the cathode fall region and d_{cf} is the thickness of the cathode fall region. The values of V_{cf} and pd_{cf} are of the same order of magnitude as those at the minimum of the Paschen curve. It turns out that the obtained combination of j/p^2 and V_{cf} corresponds to minimal power dissipation in the cathode layer (Steenbeck's minimum principle). Typical values for a glow discharge in air are $j/p^2 = 200\text{–}570\,\mu\mathrm{A}/(\mathrm{cm\,torr})^2$, $V_{cf} = 230\text{–}370\,\mathrm{V}$, and $pd_{cf} = 0.22\text{–}0.52\,\mathrm{torr\,cm}$, again depending heavily on the cathode material. From these relations it becomes apparent that glow discharges at atmospheric pressure can only operate at high current densities with extremely thin cathode layers.

A characteristic feature of the glow discharge is that the two cases of a normal cathode fall and that of an abnormal cathode fall must be distinguished. In the normal glow discharge the current covers only part of the cathode area, the surface area covered being proportional to the current. In this case the normal cathode fall voltage is practically independent of current and pressure. If the current is increased beyond the value required to cover the whole cathode surface, a region is entered in which the current density and the cathode fall voltage increase (abnormal glow discharge, section F → H in figure 2.4.1, sometimes also referred to as anomalous glow discharge). The abnormal glow discharge has attracted considerable attention for technical applications. Due to the positive current voltage characteristic many of such discharges can be operated in parallel without requiring individual ballast resistors.

When the current is increased beyond the stage of the abnormal glow discharge the required voltage drops considerably, to about $10\,\mathrm{V}$, and an arc discharge is established. At atmospheric pressure the plasma in most arc discharges is approaching local thermodynamic equilibrium (thermal plasma). Thermal plasmas are outside the scope of this book. It should be mentioned, however, that in the arc fringes, and especially in fast moving arcs (gliding arcs), non-equilibrium plasma conditions can also be found and can be utilized for technical applications (Fridman *et al* 1999, Mutaf-Yardimci *et al* 2000).

2.4.2 Streamer breakdown

As was pointed out by Rogowski (1928), breakdown in wide atmospheric-pressure air gaps subjected to pulsed voltages proceeds much faster than can be explained by the mechanism of successive electron avalanches supported by secondary cathode emission. An essential feature of this Townsend breakdown mechanism is that the space charge of a single electron avalanche does not distort the applied homogeneous electric field in the gap. This limits the number of electrons in the avalanche head to stay below a critical value N_{cr} (about 10^8):

$$e^{\alpha d} \leq N_{cr}. \tag{2.4.8}$$

When the amplification of the avalanche reaches this critical value before arriving at the anode, local space charge accumulation leads to a completely different breakdown mechanism. The concept of this 'Kanalaufbau' or 'streamer breakdown' was developed independently by Raether (1939, 1940), Loeb and Meek (1941) and Meek (1940). Streamer breakdown is a much faster process and results in a thin conductive plasma channel. Streamer breakdown can always be provoked by applying a certain over-voltage to the gap with fast pulsing techniques. The concept of streamer breakdown is based on the notion that a thin plasma channel can propagate through the gap by ionizing the gas in front of its charged head due to the strong electric field induced by the head itself. In air the conditions for Townsend breakdown or streamer breakdown are well established (figure 2.4.5).

Only close to the boundary line may both types of breakdown occur. From this curve it is apparent that at larger pd values a relatively modest overvoltage will result in streamer breakdown. It should also be pointed out that the often cited criterion originally derived by Raether (1940), that in air at pd values <1000 torr cm Townsend breakdown can be expected and above this value streamer breakdown is not always applicable in this generality. In dry air the Townsend mechanism must be invoked at low over-voltages at least to pd values up to $10\,000$ torr cm (Allen and Phillips 1963).

Following the early observations of the Loeb school in California and of Raether and his students in Hamburg many experimental investigations have been devoted to the observation of the streamer phase in different gases. The physical processes involved are discussed in review papers by Marshak

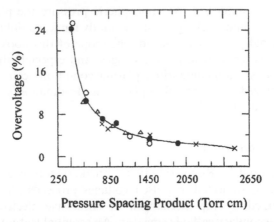

Figure 2.4.5. Curve separating conditions resulting in air breakdown by the Townsend mechanism (lower region) and by the streamer mechanism (upper region) (from Korolev and Mesyats 1998 p 65).

(1961), Lozanskii (1976), Kunhardt (1980), Kunhardt and Byszewski (1980), Dhali and Williams (1985, 1987) and in various handbook articles (Dutton 1978, 1983) and textbooks (Loeb and Meek 1940, Llewellyn-Jones 1957, 1967, Raether 1964, Meek and Craggs 1978, Kunhardt and Luessen 1983, Korolev and Mesyats 1998).

The numerical treatment of streamer propagation has become possible only later, starting with simplified one-dimensional models about 1970. Among the first computer simulations were those of Dawson and Winn (1965), Davies *et al* (1971), Kline and Siambis (1971, 1972), Gallimberti (1972), and Reininghaus (1973). An analytical approach to streamer propagation was proposed by Lozansky and Firsov (1973). They considered the streamer to be a conductive body having the shape of an oblong ellipsoid of revolution, placed in an external field E. For this configuration an analytical solution exists for the potential distribution around the body. In such models the streamer propagation velocity is determined by the drift of electrons in the enhanced field region at the streamer tip. Higher velocities can be obtained if processes are included that generate electrons in front of the streamer head or that assume a certain level of background ionization. There is still considerable debate about the major physical processes involved in streamer propagation and about the appropriate boundary conditions for numerical simulations. In air, or other oxygen nitrogen mixtures, photo-ionization in the gas volume in front of the streamer head is considered an important process that is included in many numerical simulations. Unfortunately there is only limited experimental evidence of this process (Penney and Hummert 1970, Zheleznyak *et al* 1982). Some authors claim that photo-ionization is a crucial feedback mechanism placing seed electrons

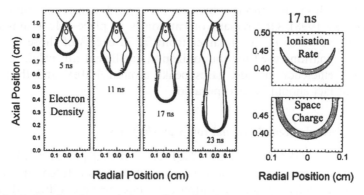

Figure 2.4.6. Results of numerical two-dimensional streamer simulations in atmospheric pressure dry air (Kulikovsky 1998).

ahead of the streamer front in order for the streamer to propagate (Morrow and Lowke 1995). As a matter of fact, in such models positive (cathode directed) streamers will not propagate if no photo-ionization and zero background ionization is assumed. It must be stated, however, that zero background charge density is not a realistic assumption in atmospheric air. Negative (anode directed) streamers, on the other hand, can propagate in numerical simulations without photo-ionization and without background electrons. Recent two-dimensional simulations of negative streamers starting from one initial electron obtain streamer propagation and even streamer branching without these additional assumptions (Arrayás *et al* 2002, Rocco *et al* 2002).

With the advent of faster computers and the availability of better numerical algorithms to cope with steep gradients and small time steps numerical two-dimensional simulations of streamer propagation were greatly improved. Recent developments were reviewed by Babaeva and Naidis (2000). Figure 2.4.6 shows some details of such simulations performed by Kulikovsky (1998) on the propagation of a positive streamer in a weak field in atmospheric pressure dry air. In the left part the electron density contours of the propagating streamer are plotted for 5, 11, 17, and 23 ns. The outer contour corresponds to 10^{11} cm^{-3}, the inner contour to 10^{13} cm^{-3}. The right part of figure 2.4.4 shows an enlargement of the region of high ionization rate and space charge density at 17 ns. In these two plots the region is defined by the contour line corresponding to 10% of the maximum value. These numerical simulations demonstrate that the ionization region at the streamer head is extremely thin, about 0.015 cm in thickness, and that the streamer body reaches appreciable electric conductivity with electron densities in excess of 10^{13} cm^{-3}. A comparison shows that the original Raether–Meek streamer criterion and analytical models based on the propagation of a highly charged sphere predict streamer properties close to those obtained in two-dimensional simulations (Kulikovsky 1998). One remark of caution is in place. Most simulations arrive at field value in the streamer front far in excess of those for which the approximations used for the ionization coefficient are valid. Since the ionization efficiency in most gases peaks around 100 eV and then drops again, this fact should be incorporated.

2.4.3   Pulsed air breakdown and runaway electrons

Rogowski *et al* (1927) and Buss (1932) reported that pulsed breakdown in atmospheric pressure air can also occur in two steps (Stufendurchschlag). Apparently, under certain conditions, an intermediate diffuse discharge phase can be established before complete breakdown of the gap. This phenomenon was later investigated in more detail (Chalmers 1971, Water and Stark 1975), and also in other atmospheric pressure gases. In addition to air, investigations were performed in nitrogen (Farish and Tedford 1966, Doran 1968, Chalmers 1971, Koppitz 1973), and in hydrogen (Edels and Gambling 1959, Doran and Meyer 1967, Meyer 1967, Cavenor and Meyer 1969). Spectroscopic investigations revealed that this transient diffuse discharge phase can be classified as a glow discharge with pronounced non-equilibrium plasma properties. The energy balance and the electrical characteristics of pulsed glow discharges have been investigated by Boeuf and Kunhardt (1986) and by Dhali (1989). At atmospheric pressure the duration of the pulsed glow discharge is normally restricted to less than 1 μs by instabilities, most likely originating in the cathode layer, that cause constriction or filamentation of the diffuse volume discharge. With the advent of transversely excited atmospheric (TEA) lasers this transient glow phase in high pressure gases gained immense practical importance (Rhodes 1979).

Following the early work of Felsenthal and Proud (1965) and Mesyats and Bychkov (1968) many experimental investigations have been devoted to nanosecond pulse breakdown in atmospheric pressure air. At certain over-voltages an intermediate diffuse non-equilibrium volume discharge phase with an electron density on the order of 10^{16} cm^{-3} can be obtained. The physical phenomena occurring in pulsed breakdown have been treated in several review articles (Kunhardt 1980, 1983, 1985) and in some monographs devoted to this special subject (Lozanzkii and Firsov 1975, Bazelian and Raizer 1998, Korolev and Mesyats 1998). Self-sustained volume discharges have recently been reviewed by Osipov (2000).

A special situation arises when extremely high electric fields are applied to a gas gap. Since the cross sections for all electron collisions (elastic, exciting, ionizing) have a maximum at a certain electron energy, high enough electric fields must lead to a situation where an electron picks up more energy between collisions than it loses by collisions with the background gas. This leads to a runaway situation in which electrons are continuously accelerated. For most gases the ionization cross section peaks around 100 eV. Wilson (1924) suggested this mechanism as a possible explanation for the lightning observed in thunderstorms. A rough estimate of the electric fields required to reach the transition from the streamer mechanism to continuous acceleration of electrons was formulated by Babich and Stankevich (1973). The requirement is to get to field values that correspond to about three times the value for stationary breakdown. Runaway electrons were also suggested as a conceivable mechanism for streamer propagation (Kunhardt and Byszewski 1980). A recent monograph on High-Energy Phenomena in Electric Discharges in Dense Gases (Babich 2003) treats the history of the concept of runaway electrons and the experimental evidence in detail. To establish runaway conditions in atmospheric pressure air in the laboratory, high voltage pulses of sub-nanosecond rise time and duration are required (Alekseev *et al* 2003, Tarasenko *et al* 2003).

References

Alekseev S B, Orlovskii V M and Tarasenko V F 2003 *Tech. Phys. Lett.* **29** 411–413
Allen K R and Philips K 1963 *Electr. Rev.* **173** 779–783
Arrayás M, Ebert U and Hundsdorfer W 2002 *Phys. Rev. Lett.* **88** 174502
Babaeva N Yu and Naidis G V 2000 'Modeling of streamer propagation' in van Veldhuizen E M (ed) *Electrical Discharges for Environmental Purposes* (Huntington: Nova Science) pp 21–48
Babich, L P 2003 *High-Energy Phenomena in Electric Discharges in Dense Gases: Theory, Experiment and Natural Phenomena* (Arlington: Futurepast)
Babich L P and Stankevich Yu 1973 *Sov. Phys.– Techn. Phys.* **12** 1333–1336
Bazelian E M and Raizer Yu P 1998 *Spark Discharge* (Boca Raton: CRC Press)
Boeck W and Pfeiffer W 1999 'Conduction and breakdown in gases' in Webster J G (ed) *Wiley Encyclopedia of Electrical and Electronics Engineering* (New York: Wiley) vol. 4 pp 123–172
Boeuf J P and Kunhardt E E 1986 *J. Appl. Phys.* **60** 915–923
Buss K 1932 *Arch. Elektrotech.* **26** 266–272
Cavenor M C and Meyer J 1969 *Austr. J. Phys.* **22** 155–167
Chalmers I D 1971 *J. Phys. D: Appl. Phys.* **4** 1147–1151
Chen J and Davidson J H 2003 *Plasma Chem. Plasma Process.* **23** 83–102
Dakin T W, Luxa G, Oppermann G, Vigreux J, Wind G and Winkelnkemper H 1974 *Electra* **32** 61–82
Davies A J, Davies C S and Evans C J 1971 *Proc. IEE* **118** 816–823
Dawson G A and Winn W P 1965 *Z. Phys.* **183** 159–171
Dhali S K 1989 *IEEE Trans. Plasma Sci.* **17** 603–611
Dhali S K and Williams P F 1985 *Phys. Rev. A* **31** 1219–1221
Dhali S K and Williams P F 1987 *J. Appl. Phys.* **62** 4696–4707
Doran A A 1968 *Z. Physik* **208** 427–440
Doran A A and Meyer J 1967 *Brit. J. Appl. Phys.* **18** 793–799
Druyvesteyn M J and Penning F M 1940 *Rev. Mod. Phys.* **12** 87–174

Dutton J 1978 'Spark breakdown in uniform fields' in Meek J M and Craggs J D (eds) *Electrical Breakdown of Gases* (Chichester: John Wiley) pp 209–318

Dutton J 1983 'Prebreakdown ionization in gases under steady-state and pulsed conditions in uniform fields' in Kunhardt E E and Luessen L E (eds) *Electrical Breakdown and Discharges in Gases* NATO ASI Series B: Physics, vol. 89a: (New York: Plenum Press) pp 207–240

Edels H and Gambling W A 1959 *Proc. Roy. Soc. A* **249** 225–236

von Engel A and Steenbeck M 1932, 1934 *Elektrische Gasentladungen* (Berlin: Springer) vol. 1 and 2

Farish O and Tedford D J 1966 *Brit. J. Appl. Phys.* **17** 965–966

Felsenthal P and Proud J M 1965 *Phys. Rev.* **139** 1796–1804

Fridman A, Nester S, Kennedy L A, Saveliev A and Mutaf-Yardimci O 1999 *Progr. Energy Combust. Sci.* **25** 211–231

Gallimberti I 1972 *J. Phys. D: Appl. Phys.* **5** 2179–2189

Geballe R and Reeves M L 1953 *Phys. Rev.* **92** 867–868

Hess H 1976 *Der elektrische Durchschlag in Gasen* (Braunschweig: Vieweg)

Kline L E and Siambis J G 1971 *Proc. IEEE* **59** 707–709

Kline L E and Siambis J G 1972 *Phys. Rev. A* **5** 794–805

Koppitz J 1973 *J. Phys. D: Appl. Phys.* **6** 1494–1502

Korolev Yu D and Mesyats G A 1998 *Physics of Pulsed Breakdown in Gases* (Yekatarinburg: URO-Press)

Kulikovsky A A 1998 *Phys. Rev. E* **57** 7066–7074

Kunhardt E E 1980 *IEEE Trans Plasma Sci.* **8** 130–138

Kunhardt E E 1983 'Nanosecond pulse breakdown of gas insulated gaps' in Kunhardt E E and Luessen L E (eds) *Electrical Breakdown and Discharges in Gases* NATO ASI Series B: Physics (New York: Plenum) vol. 89a pp 241–263

Kunhardt E E 1985 'Pulse breakdown in uniform electric fields' in *Proc. 17th Int. Conf. on Phenomena in Ionized Gases (ICPIG XVII)*, Budapest 1985, Invited Papers 345–360

Kunhardt E E 2000 *IEEE Trans. Plasma Sci.* **28** 189–200

Kunhardt E E and Byszewski WW 1980 *Phys. Rev.* **21** 2069–2077

Kunhardt E E and Luessen L E (eds) 1983 *Electrical Breakdown and Discharges in Gases*, NATO ASI Series B: Physics (New York: Plenum) vol. 89a and 89b

Les Renardières Group 1972 *Electra* **23** 53–157

Little P 1956 'Secondary effects' in Flügge S (ed) *Handbook of Physics* (Berlin: Springer) vol. 21 pp 574–563

Llewellyn-Jones F 1957 1966 *Ionization and Breakdown in Gases* (London: Methuen)

Llewellyn-Jones F 1967 *Ionization, Avalanches, and Breakdown* (London: Methuen)

Loeb L B 1939 *Fundamental Processes of Electrical Discharge in Gases* (New York: Wiley)

Loeb L B and Meek J M 1940 *J. Appl. Phys.* **11** 438–74

Loeb L B and Meek J M 1941 *The Mechanism of the Electric Spark* (Stanford: University Press)

Lozanskii E D 1976 *Sov. Phys. Usp.* **18** 893–908

Lozanzkii E D and Firsov O B 1975 *Theory of Spark* (Moscow: Atomizdat Publishers) (in Russian)

Lozanzky E D and Firsov O B 1973 *J. Phys. D: Appl. Phys.* **6** 976–981

Maller V N and Naidu M S 1976 *Indian J. Pure Appl. Phys.* **14** 733–737

Marshak I S 1961 *Sov. Phys. Usp.* **3** 624–651

Meek J M 1940 *Phys. Rev.* **57** 722–728

Meek J M and Craggs J D (eds) 1978 *Electrical Breakdown of Gases* (Chichester: Wiley)

Mesyats G A and Bychkov Y I 1968 *Sov. Phys. Tech. Phys.* **12** 1255–1260

Meyer J 1967 *Brit. J. Appl. Phys.* **18** 801–806

Morrow R and Lowke J J 1995 *Austr. J. Phys.* **48** 453–460

Morrow R and Lowke J J 1997 *J. Phys. D: Appl. Phys.* **30** 614–627

Moruzzi J L and Price D A 1974 *J. Phys. D: Appl. Phys.* **7** 1434–1440

Mutaf-Yardimci O, Savaliev A V, Fridman A A and Kennedy L A 2000 *J. Appl. Phys.* **87** 1632–1641

Niemeyer L 1999 'Gaseous insulation' in Webster J G (ed) *Wiley Encyclopedia of Electrical and Electronics Engineering* (New York: Wiley) vol. 8 pp 238–258

Osipov V V 2000 *Phys. Usp.* **43** 221–241

Paschen F 1889 *Wiedemann Ann. Phys. Chem.* **37** 69–96

Penney G W and Hummert G T 1970 *J. Appl. Phys.* **41** 572–577

Phelps A V and Petrovič Z L 1999 *Plasma Sources Sci. Technol.* **8** R21–R44

Protasevich E T 2000 *Cold Non-Equilibrium Plasma* (Cambridge: Cambridge International Science Publishing)

Raether H 1939 *Z. Phys.* **112** 464–489 (in German)

Raether H 1940 *Naturwissenschaften* **28** 749–750 (in German)

Raether H 1964 *Electron Avalanches and Breakdown in Gases* (London: Butterworths)

Raizer Yu P 1986 *High Temp.* **24** 744–754

Raizer Yu P 1991, 1997 *Gas Discharge Physics* (Berlin: Springer)

Raja Rao C and Govinda Raju G R 1971 *J. Phys. D: Appl. Phys.* **4** 494–503

Reininghaus W 1973 *J. Phys. D: Appl. Phys.* **6** 1486–1493

Rhodes Ch K (ed) 1979 1984 *Excimer Lasers* (New York: Springer)

Rocco A, Ebert U and Hundsdorfer W 2002 *Phys. Rev. E* **66** 035102-1 to 035102-4

Rogowski W 1928 *Arch. Elektrotech.* **20** 99–106 (in German)

Rogowski W, Flegler E and Tamm R 1927 *Arch. Elektrotech.* **18** 479–512 (in German)

Sigmond R S 1984 *J. Appl. Phys.* **56** 1355–1370

Šijačič D D and Ebert U 2002 *Phys. Rev. E* **66** 066410

TarasenkoV F, Yakovlenko S I, Orlovskii V M, Tkachev A N and Shumailov S A 2003 *JETP Lett.* **77** 611–615

Townsend J S 1915 *Electricity in Gases* (Oxford: Clarendon Press)

Vollrath K and Thomer G 1967 *Kurzzeitphysik* (Wien: Springer) p 81

Wagner K H 1971 *Z. Phys.* **241** 258–270

Water R T and Stark W B 1975 *J. Phys. D: Appl. Phys.* **8** 416–426

Wilson C T R 1924 *Proc. Cambridge Phil. Soc.* **22** 534–538

Zheleznyak M B, Mnatsakanyan A Kh and Sizykh S V 1982 *High Temp.* **20** 357–362

2.5 Corona Discharges

2.5.1 Phenomenology of corona discharges

Similar to lightning, corona discharges in ambient air can be observed under natural conditions, for instance, corposant or 'St. Elmo's Fire' in a thunderstorm. As a rule, a naturally occurring corona arises at points and wires having high electrical potential with respect to the environment and exhibits itself around sharp edges like a faint glow in the form of a crown. An appearance of a corona may produce useful or undesirable effects. For instance, a corona arising spontaneously around high-voltage wires of an electrical power transmission line results in a loss of electrical energy. On the other hand, coronas are widely used in many practical applications like dust collection with electrical precipitators, atmospheric pressure non-thermal plasma surface treatment of polymers, cleaning of exhausted gases, etc.

The corona discharge is a low-current discharge caused by partial (or local) breakdown of a gas gap with strongly inhomogeneous electric field. To form a non-uniform electric field distribution in the gap, at least one of the electrodes must be sharpened with a radius of curvature of far less than the length of the inter-electrode gap. The most typical configurations of electrode systems used in practice to generate corona discharges are pin-to-plane, multi-pin-to-plane, wire-to-pipe, wire-to-plane or wire between two planes, multi-wire-to-plane or multi-wire between two planes, coaxial wire-cylinder and so on (Goldman and Goldman 1978, Sigmond 1978, Goldman and Sigmond 1982).

Steady dc corona discharges exist in several forms depending on the polarity of the electric field, the electrode system and discharge current. A schematic view of different forms of dc coronas in static ambient air at atmospheric pressure in pin-to-plane gaps under positive and negative polarities of the high-voltage stressed pin electrode is shown in figure 2.5.1 (figure is taken from Chang *et al* 1991). The sequence of pictures in the left-to-right direction corresponds to increasing discharge current. A typical range of corona current, averaged in time, extends approximately from 1 to 200 μA per pin. Characteristic voltages applied to sustain dc coronas depend mainly on the geometrical parameters of the electrode system (such as the radius of the tip of a pin and the length of the inter-electrode gap) and ranges over several units or tens of kV. For the same electrode system, the onset voltage is roughly the same for positive and negative corona in air.

For a positive corona, the discharge, apparent already to the naked eye, starts with a burst corona. This regime exhibits seldom and non-regular current pulses accompanied with short and faint streamers originating away from the pin. The burst regime proceeds to the streamer corona, silent glow corona and finally the non-stationary spark as the applied voltage increases. However, in most cases, a glow regime of a positive corona precedes the streamer regime. The positive glow corona is known as the Hermstein glow (Hermstein 1960). It is similar to the low-pressure discharge in a Geiger tube. A steady current at a fixed voltage, quiet operation, and almost no sparking characterize this glow corona.

On the contrary, the streamer regime is non-steady, quite audio noisy and emits strong radio noise. This regime corresponds to the existence of numerous thin, short-living and repetitive current filaments (streamers)

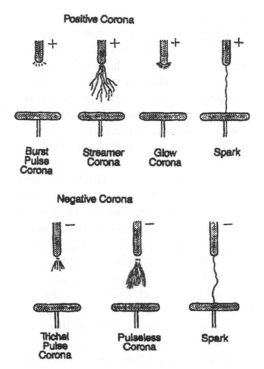

Figure 2.5.1. Schematic view of types of corona discharges (from Chang *et al* 1991).

originating from the pin. Due to branching of fast-moving streamers, their instant image in the gap shows up as a bush 'growing' from the tip of the pin. The streamer regime can be regarded as an uncompleted breakdown of the gas gap. Therefore it is the direct precursor to the spark: once the streamers bridge the gap, the spark occurs. However, the transition from corona to spark is not sharply defined. For a wire-to-pipe or wire-to-plate electrode configuration, the corona generated at the positively stressed electrode may appear as a tight glow sheath around and along the wire or as streamers moving away from different locations of the wire.

For the dc negative corona in a pin-to-plane geometry, the initial form of a discharge is a non-steady Trichel pulse corona (Trichel 1938), characterized by regular current pulses, glow luminosity around the tip of a pin and a dark inter-electrode gap. The repetition frequency of Trichel pulses increases linearly with corona current and ranges over 1–100 kHz. For static air at atmospheric pressure, the Trichel pulse regime is continued up to 120–140 μA and is followed by a pulse-less negative glow corona as the applied voltage increases. The negative glow usually requires clean, smooth electrodes to form. For parallel wire-to-cylinder, wire-to-plate or coaxial wire-cylinder electrodes, the corona generated at negative electrodes may take the form of rapidly moving glow spots or it may be concentrated into small active spots regularly placed along the wire, called 'tufts' or 'beads'. The glow corona often changes with time into the tuft form. The tuft corona is also noisy and has a sparking potential similar to that of the glow form (Lawless *et al* 1986). In static air, the steady pulse-less negative corona is followed directly by a spark. The sparking potential of the negative corona is much higher than that of the positive streamer corona. This is the reason why the negative corona is used in electrical precipitators (see section 9.2).

For a pin-to-plane geometry, Warburg (1899) found that the radial distribution of the current density j at the plane electrode follows an empirical relation $j(\theta) \cong j_0 \cos^n \theta \equiv j_0 (1 + \tan^2 \theta)^{-n/2}$, which was confirmed later by others (Jones *et al* 1990, Allibone *et al* 1993). Here j_0 is the current density at the plane at the axis of the corona, $\tan \theta = r/d$, r is the current radius, and $n \cong 5 \pm 0.5$ for different experiments. There are exceptions to this Warburg relation (Goldman *et al* 1988, 1992, Akishev *et al* 2003a).

A significant concentration of electric field exclusively around the sharpened electrode plays a key role in the formation of special properties of coronas in comparison with discharges in uniform electric fields. First, an inception voltage of the corona is far lower compared with Paschen's breakdown voltage, corresponding to uniform inter-electrode gap of the same length. Second, due to a minor contribution of ionization processes in the total balance of charged particles in the drift region of the corona, the inter-electrode gap is filled mainly with negative or positive space charge. This implies that the corona discharge is space-charge limited in magnitude, and that the volt–ampere characteristic (VAC) of the corona has a positive slope: an increase in current requires higher voltage to drive it. Third, intensive ionization processes at the point, accompanied by an intensive local energy deposition, provoke the development of ionization instabilities resulting in an appearance of streamers in the corona gap under conditions, in which the Meek's (or Reather's) criterion for streamer breakdown is not fulfilled.

Outstanding contributions to the development of the fundamentals of corona discharges were reported in the past century by researchers belonging to the scientific schools of Kaptsov (Kaptsov 1947, 1953) and of Loeb (Loeb 1965 and literature cited therein). Both of these schools used intensively a conception of the electron avalanches originally developed by Townsend (Townsend 1914). Indeed, from a physical point of view, the corona discharge belongs to the same class of self-sustained discharges as the extremely low-current (10^{-15}–10^{-7} A) dark Townsend discharge and the medium-current (3×10^{-4}–3×10^{-3} A) glow discharge. In these discharge types the emission of charged particles from electrode surfaces does not play an essential role in the transport of the electric current through the metal–gas boundary, but electron avalanches play a key role in sustaining the discharge.

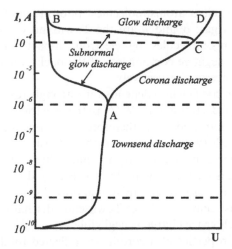

Figure 2.5.2. Schematic classification of self-sustained gas discharges.

It is well known for low-pressure discharges in a plane-to-plane geometry (von Engel and Steenbeck 1934, von Engel 1955, Brown 1959) that the dark Townsend discharge goes directly to the glow discharge (figure 2.5.2, path A–B).

In principle, for an electrode system with initial non-uniform distribution of the electric field there is also the chance to transit from the dark Townsend discharge to the glow discharge. For the positive corona, this transition is achieved only at lower pressures and it is accompanied by a non-monotonic behavior of the voltage drop across the inter-electrode gap (figure 2.5.2, path A–C–B): there is a reduction in the discharge voltage in the glow regime similar to that observed in discharges in a plane-to-plane geometry. For the negative corona, a transition to a glow discharge can be realized in air up to atmospheric pressure (Akishev *et al* 1993, 2000, 2001), and this transition is followed by a monotonic increase of the discharge voltage (figure 2.5.2, path A–C–D). For both cases, contrary to gas discharges in a plane-to-plane geometry, the corona discharge is an additional intermediate discharge stage between the dark Townsend discharge and the glow discharge.

For discharges sustained due to the development of electron avalanches, a self-sustained steady regime occurs if the replenishment criterion for electron avalanches in the gap is fulfilled:

$$\int_0^l (\alpha - \eta) \, \mathrm{d}x = \ln\left(1 + \frac{1}{\gamma}\right). \tag{2.5.1}$$

Here γ is the total coefficient of a positive feedback for electron avalanches due to surface and volume processes: emission of electrons from a cathode by positive ions, excited atoms/molecules, photons, and photo-ionization of the background gas; α and η are the Townsend coefficients for direct ionization of atoms/molecules by electron impact and attachment of electrons to electronegative components of the background gas due to two- and three-body processes, respectively.

The ionization coefficient α depends very strongly (exponentially) on the reduced electric field strength E/N (E is the electric field strength, N is density of the background gas), therefore discharge regions with a high electric field strength (where $\alpha \geq \eta$, and the intensity of ionization is very high) bring a major contribution to the total value of the integral written above. For this reason, the magnitude of l usually does not coincide with the length d of inter-electrode gap (commonly $l < d$ or $l \ll d$; the case $l = d$ corresponds only to dark Townsend discharge between plane electrodes, in which space charge is negligible.

The electric properties of a corona are reflected totally in the relation between the discharge current I and the applied voltage V. Therefore knowledge of the volt–ampere characteristic of a corona is desirable for many practical purposes. The initial inhomogeneous distribution of the electric field in the corona allows in some cases for substantial simplification of analytical and numerical calculations of the VAC. Indeed, the strong concentration of the electric field around the electrode with a small radius of curvature results in a division of the inter-electrode gap of coronas into two very different parts: a thin generation zone with intensive ionization located in the vicinity of the electrode with small curvature, and a drift zone with a space charge occupying the rest of the gap. As a rule, the voltage difference across drift region is higher than the voltage drop across the generation zone (about $0.5 \pm 0.2\,\text{kV}$). In this case, the VAC of the drift region can be attributed with good accuracy to the VAC of the corona in total.

Townsend (1914) was the first to use this idea and calculated the VAC for a steady corona in a coaxial wire-cylinder geometry:

$$I \cong \frac{8\pi\varepsilon_0\mu_\text{i}}{R^2 \ln R/r_0}\, U_0(U - U_0). \qquad (2.5.2)$$

Here μ_i is the mobility of carriers of the current in the drift region (for instance, negative or positive ions for negative and positive coronas in air, respectively); ε_0 is the permittivity of a vacuum; R and r_0 are the radii of the outer cylinder and the inner wire respectively; U_0 is so-called inception voltage of the corona, corresponding to an appearance of a very noticeable corona current (as a rule, $I > 0.1\,\mu\text{A}$) and luminosity around the wire or the sharpened electrode.

Equation (2.5.2) is obtained under the assumption that the space charge in a drift region is small enough. Therefore this formula describes only the initial current of a corona under the influence of an applied voltage

U not far from the inception voltage U_0. Loeb (1965) suggested that the time-averaged VAC of a corona discharge can be approximated by a universal parabolic dependence

$$I = kU(U - U_0) \tag{2.5.3}$$

which can describe the corona current in any geometry and at any voltage up to the spark transition. In this case, the proportionality factor k and the corona ignition voltage U_0 depend on the geometrical features of the electrode system (for instance, on the tip radius of the pin and the inter-electrode distance), the polarity of the applied voltage, the pressure and the mixture of the background gas) and has to be determined by experiment. This idea is very popular in the literature at present, and some results on fitting of the parabolic approximation with experiment can be found in Lama and Gallo (1974), Sigmond (1982), Vereshchagin (1985), and Akishev *et al* (2003).

2.5.2 Negative dc corona discharges

For definiteness, the emphasis in this section is on the physical properties of a negative corona for a pin-to-plane geometry mainly in air. The mechanism of Trichel pulses and the transition of the negative corona to the spark are discussed in detail.

Regularly pulsing corona

As mentioned above, while studying the negative point-to-plane corona in air, Trichel revealed the presence of regular relaxation pulses (Trichel 1938). The qualitative explanation given by him included some really important features like the shielding effect produced by a positive ion cloud in the vicinity of the cathode. In later work (Loeb *et al* 1941) it was stated that the Trichel pulses exist only in electronegative gases, and particular emphasis was put on the processes of electron avalanche triggering. It was also stressed that, usually, the time of the negative ion drift to the anode is much longer than the pulse period. More detailed measurements of the Trichel pulse shape demonstrated that the rise time of the pulse in air may be as short as 1.3 ns (Zentner 1970a), and a step on a leading edge of the pulse was observed (Zentner 1970b). Systematic studies of the electrical characteristics of Trichel pulses were undertaken (Fieux and Boutteau 1970, Lama and Gallo 1974), and relationships were found for the pulse repetition frequency, the charge per pulse and other properties.

Among attempts to give a theoretical explanation for the discussed phenomena the work of Morrow is most known (Morrow 1985a), in which the preceding theories were also reviewed. The continuity equations for electrons and for positive and negative ions in a one-dimensional form were numerically solved together with Poisson's equation. The negative corona in oxygen at a pressure of 50 torr was numerically simulated. Only the first pulse was computed, and extension of calculations for longer times showed only continuing decay of the current. In Morrow (1985a) the shape of the pulse was explained while practically ignoring the ion-secondary electron emission. In the following paper (Morrow 1985b) the step on the leading edge of the pulse was attributed to the inclusion of photon secondary emission, and the main peak was explained in terms of the ion-secondary emission. This explanation was criticized later by Černák and Hosokawa (1991), pointing at the importance of an ionization-wave-like evolution of the cathode layer at early stages.

A more detailed analysis of the mechanism of Trichel pulses based on numerical simulations was proposed by Napartovich *et al* (1997) with the use of a 1.5-dimensional numerical model. This numerical model, succeeding in reproducing the established periodical sequence of Trichel pulses in dry air in short-gap (<1 cm) coronas, was formulated for the first time. The three-component simplified kinetic model was used with only one type of negative ions, namely O_2^-, produced in an electron three-body attachment process. The electron–ion and ion–ion recombination may be neglected for the conditions of the corona discharge.

To describe the pulse mode of the negative point-to-plane corona it is sufficient to solve the continuity equations for electrons, positive and negative ions and Poisson's equation under the assumption that the current cross section discharge area $S(x)$ is a known function of coordinate x. The boundary conditions for positive and negative ions are self-evident: their number density is equal to zero at the anode and cathode, respectively. For electrons, in contrast to Morrow, only the ion secondary emission is included.

It was assumed that all physical quantities are constant in every cross section of the discharge current. The same approximation was used by Morrow, but he assumed unrealistically the form of discharge channel to be cylindrical. However, it is well known from numerous experiments that the discharge current is concentrated near to the point and occupies a comparatively large area on the anode surface. The ratio of the current spot radii on the anode and cathode is of the order of 10^4.

A sample of a calculated current pulses and the time dependence of the replenishment criterion integral $M = \int \alpha \, dx$ during pulsation are shown in figure 2.5.3 (α is the ionization coefficient).

To illustrate effects of non-linear evolution of the corona in the pulse regime, spatial distributions of physical quantities in the active zone at the moments listed in table 2.5.1 corresponding to the front of the pulse are presented in figure 2.5.4.

When the number density of positive ions becomes larger, it causes an increase of the electric field strength near the cathode. This increase is in turn followed by a rapid growth of the electron multiplication factor, and

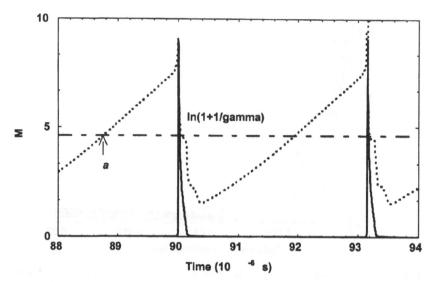

Figure 2.5.3. Calculated current pulses and replenishment criterion integral $M = \int \alpha \, dx$ as a function of time.

the ion number density. This feedback is strongly non-linear because of the exponential growth of the electron current with the ionization coefficient α, which is also a steep function of the electric field strength. As a result of the strong electron multiplication a plasma region is formed where the electric field strength diminishes due to high electron mobility (in other words, due to plasma shielding). This structure propagates at very high speed to the cathode (see the transition from curve 2 to 3). As a result of this wave propagation, the voltage drop across the active zone diminishes while the electric field strength at the cathode grows. This means that the dynamical differential resistance of the shrinking cathode layer is negative. At this phase the electric current at the cathode is predominantly the displacement current.

To illustrate in more detail the processes during the pulse decay, the spatial distributions of the physical quantities at the moments listed in table 2.5.2 are presented in figure 2.5.5.

Table 2.5.1.

	Moment						
	1	2	3	4	5	6	7
Time (μs)	89.99907	89.99952	90.00000	90.00050	90.00125	90.00325	90.00525
Current (μA)	319	548	2351	1728	1216	2249	2741

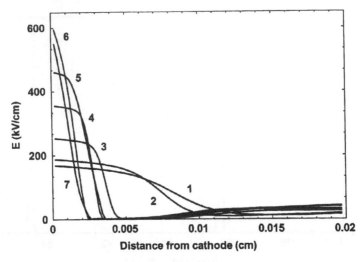

Figure 2.5.4. Time evolution of the electric field distribution in the active zone. Moments 1–7 correspond to table 1 (leading edge of current pulse).

In conclusion, the decay of the Trichel pulse is governed by the decay of a cathode layer formed in the course of the preceding evolution. This cathode layer is similar in many respects to the well-known cathode layer of the glow discharge. In particular, the cathode current density at the maximum is of the order of the so-called normal current density. However, certainly this layer does not coincide with the classical cathode layer. In particular, figure 2.5.5 demonstrates that the electric field distribution controlled initially by space charge (moments 1–4) evolves to the 'free-space' distribution (moment 5). Due to the strong increase of the 'free-space' cathode layer in thickness, the replenishment criterion integral $M = \int \alpha \, dx$ grows again, and the pulse process repeats.

More recent three-dimensional calculations of a negative corona with Trichel pulses (Napartovich *et al* 2002) revealed a new feature in the dynamics of the active zone (cathode layer) during the leading and trailing edges of a current pulse: the cathode layer shrinks in axial direction and

Table 2.5.2.

	Moment				
	1	2	3	4	5
Time (μs)	90.02200	90.06200	90.10005	90.20217	90.40041
Current (μA)	1579	678	367	12.1	0.67

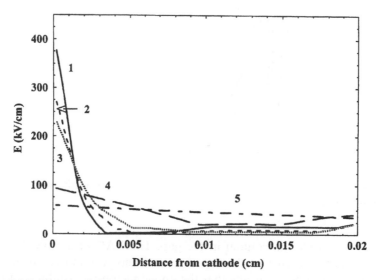

Figure 2.5.5. Time evolution of the electric field distribution in the active zone. Moments 1–5 correspond to table 2 (trailing edge of current pulse).

extends in radial direction when the current increases, and it shrinks in radial direction and extends in axial direction when current decreases.

The results presented above show that the negative ions do not play an essential role in the mechanism of Trichel pulses. This implies, in contraposition to popular opinion, that the pulsed regime can also be observed for a negative corona in electropositive gases like Ar, He, and N_2. Indeed, experiments performed by Akishev *et al* (2001b) proved this conclusion. Current oscillations caused by the existence of a negative differential resistance of the dynamic cathode layer at its formation were also observed in dielectric barrier discharge in He (Akishev *et al* 2001c).

Spark formation

There is scanty information on spark formation in negative coronas. For instance, for a pin-to-plane configuration Goldman *et al* (1965) stated that spark occurs due to development of ionization phenomena on both sides of the gap resulting in the propagation of a positive streamer originated at the plane anode if the critical electric field strength ($\sim$25 kV/cm) is reached at the anode. However, this general statement does not take into account in an explicit form the existence of glow discharge regime (see section 6.7), which follows the true negative corona and precedes the spark.

The corona-to-glow discharge transition is accompanied first by the appearance of an intensive light emission near the anode corresponding to the formation of an anode layer of the glow discharge, and second by the

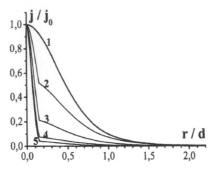

Figure 2.5.6. Evolution of radial distribution of anode current density with increase in current of a pin-plane discharge.

formation of a plasma column in the gap. The VAC of a glow discharge anode layer with a current density of several tens to hundreds of $\mu A/cm^2$ has a negative slope. It means that the anode region is unstable and tends to shrink into small current spot(s), which provoke glow discharge constriction and spark formation. Therefore, in order to understand adequately the mechanism of the corona-to-spark transition in a pin-plane geometry, it is necessary to take into account the physical properties of glow discharge, which is the intermediate stage of this transition. Experiments on the evolution of the current and light emission radial distribution under transient process true negative corona $\rightarrow$ glow discharge $\rightarrow$ spark were carried out by Akishev *et al* (2002, 2003a). Some data from these investigations are presented in figures 2.5.6–2.5.8. Experiments were performed in static air at 300 torr. The gap length was $d = 10$ mm, the radius of a pin tip was 0.06 mm.

One can see (figure 2.5.6), as the total current increases and the pin-plane discharge is switched from corona to glow discharge, that the electric current concentrates more and more around the pin-plane axis. The radial distribution of light emission near the anode exhibits a different behavior. At the initial currents of the glow discharge, light emission concentrates predominantly at the pin-plane axis. However, the effective radius of the glow region near the anode grows slowly with increasing total current. This tendency is seen in the glow discharge regime up to glow discharge-to-spark transition. Nevertheless, the effective radius of the current channel always exceeds the radius of glow column.

The corona-to-spark transition was induced by the superposition of a saw-tooth pulse on a steady corona at low current. The appropriate waveforms of current and voltage of the discharge in the course of its induced sparking are presented in figure 2.5.7. The data in figure 2.5.6 correspond to those in figure 2.5.7. The region of the oscillogram with low amplitude of discharge current corresponds to quasi-stationary true negative corona;

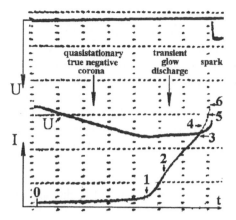

Figure 2.5.7. Time behavior of current (*I*) and voltage (*U*) under induced corona–spark transition. $[t] = 100\,\mu s/div$, $[I] = 2\,mA/div$, $[U] = 2\,kV/div$. Initial current $I = 100\,\mu A$.

the region with a rapidly growing current corresponds to the transient glow discharge, and an extremely short region with vigorously growing current corresponds to spark formation.

Some shots of a pin-plane discharge in the course of its induced sparking are presented in figure 2.5.8.

The five pictures in figure 2.5.8 present the development of spatial structure of the transient glow discharge from its forming up to the spark transition. The numbers of the pictures correspond to the moments indicated in this figure. No. 1 corresponds to the formation of an anode layer of the glow discharge; No. 2 corresponds to the formation of plasma column in the gap; No. 3 corresponds to constriction of anode layer into two

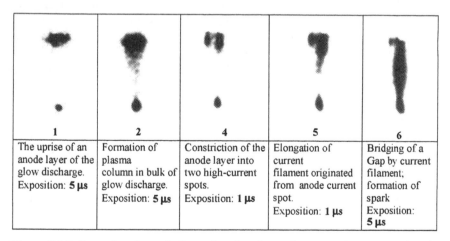

1	2	4	5	6
The uprise of an anode layer of the glow discharge. Exposition: **5 μs**	Formation of plasma column in bulk of glow discharge. Exposition: **5 μs**	Constriction of the anode layer into two high-current spots. Exposition: **1 μs**	Elongation of current filament originated from anode current spot. Exposition: **1 μs**	Bridging of a Gap by current filament; formation of spark Exposition: **5 μs**

Figure 2.5.8. Scenario of spark formation in pin-to-plane negative corona in air. $P = 300\,torr$.

high-current anode spots; No. 4 corresponds to the elongation of a current filament originated from one of the spots; No. 5 corresponds to bridging of the gap by the filament and formation of a spark.

Figure 2.5.8 shows that a sharpened cathode pin does not initiate sparking but that the plane anode does. The presented scenario of spark formation in a pin-to-plane negative corona is the same in principle as the constriction of a glow discharge observed in experiments with diffusive glow discharges in air flows at medium pressures (Velikhov *et al* 1982, Napartovich *et al* 1993, Akishev *et al* 1999a). The characteristic velocity of the current filament propagating towards the cathode pin through the plasma column of the glow discharge equals 10^4–10^5 cm/s. This is much slower than the velocity of 10^7–10^8 cm/s typical for classical positive streamers.

2.5.3   Positive dc corona discharges

Burst corona

The self-sustained Townsend regime of a positive corona ($I \leq 10^{-7}$ μA) is characterized by almost the same voltage compared with that of the negative corona. This regime exhibits so-called burst pulses, the frequency of which increases with current, and which disappears towards the end of Townsend regime to be followed by quiet glow corona. The burst corona is a difficult problem for quantitative description because of its statistical nature.

Glow corona

The generation zone of the glow corona consists of two regions: a very thin anode layer with negative space charge, and a positively charged glow or ionization zone. The anode layer has the VAC with negative slope. The glow zone is very similar to the cathode layer of a classical glow discharge. Once the corona current increases, the thickness of glow zone also grows. At lower pressure, the transition corona-to-glow discharge occurs when the glow zone of the corona occupies the whole inter-electrode gap. Subsequently, the glow zone breaks off from the wire or pin and attaches to the plane or cylindrical cathode in the form of a thin and uniformly extended glow cathode layer. This process is accompanied by oscillations of discharge current and reduction in discharge voltage. In static air (1 atm), the cathode layer and plasma column of the glow discharge at a current of several mA are very constricted ($\cong 1$ mm).

It is widely believed that self-sustaining of a positive corona is provided exclusively due to photo-ionization of the background gas. On the other hand, if the background gas is a pure mono-atomic or mono-molecular gas like pure He or N_2, it is hard to explain an emergence of the needed high-energy photons in such gases because information about electron–atom and electron–molecule collision processes, resulting in emission of quanta of energy greater than the ionization potential, is not known. However, there is no necessity to take into consideration the photo-ionization in the case of a steady or slowly changing corona. Indeed, the characteristic time of a positive feedback for the development of electron avalanches due to photoemission of secondary electrons from cathode equals the drift time for electrons, $\tau_e \approx 10^{-6}$–10^{-5} s across an inter-electrode gap filled with electropositive gas or the drift time of negative ions, $\tau_{in} \approx 10^{-4}$–$10^{-3}$ s for given electronegative components in the background gas mixture. In the latter case it is presumed that negative ions release electrons at the generation zone in the vicinity of a pin due to fast detachment processes in strong electric fields. For positive ion γ-emission of electrons from the plane electrode, the total time of the feedback is the sum $\tau_f = \tau_e + \tau_{ip} \cong \tau_{ip}$ and $\tau_f = \tau_{in} + \tau_{ip}$ in the case of electropositive and electronegative processing gases, respectively. So, for steady or slowly changing conditions (i.e. characteristic time in the changing of corona parameters exceeds τ_f) the positive corona can be sustained by a feedback mechanism identical to that in the negative corona. The VACs of positive coronas calculated with the use of this idea are in good agreement with the experimental ones (Akishev *et al* 1999b).

For a long time, it was believed that the electrical current of the positive corona in the glow mode is stable. It seems likely Colli *et al* (1954) were the first to report on oscillatory behavior of the glow corona current in a cylindrical geometry. In pioneering studies on non-linear oscillations (Fieux and Boutteau 1970, Beattie 1975, Boullound *et al* 1979, Sigmond 1997) it was revealed that the current and luminosity of the glow corona were in fact not constant, but oscillated regularly with a high frequency (10^5–10^6 Hz). It was also found that the waveform of the current self-oscillations had a relaxation type with a sharp increase of current at the leading edge of pulse and a slow decay at the pulse tail. The waveform of a light emission signal was more symmetrical. The maximum of the light emission signal was correlated with the maximum of the current pulse. According to Fieux and Boutteau (1970) and Beattie (1975), the period of self-oscillations fell with the decrease in radius of the corona electrode and practically did not depend on the average current of corona. The region of existence of free-running oscillations in plane of the *IP* parameters (current *I*, gas pressure *P*) for coaxial wire-cylinder glow coronas in N_2 is given in figure 2.5.9 (taken from Akishev *et al* 1999b).

For the description of the positive corona between a wire and a cylinder, the fluid model equations were solved by Akishev *et al* (1999b) on the assumption that the ionizing agent in the vicinity of a wire is the soft x-ray radiation produced in collisions of electrons accelerated in a strong electric field near wire with the wire surface. This is the so-called Bremsstrahlung radiation. The total electric current was a sum of displacement and conductivity currents. A numerical model developed in Akishev *et al* (1999b)

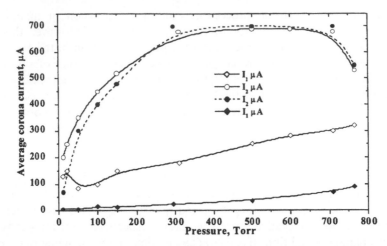

Figure 2.5.9. *IP*-region of existence of oscillations for a coaxial wire-cylinder corona in N_2. The oscillation region is bounded by curves I_1 and I_2. Radii of anode and cathode are 0.75 and 10 mm, respectively. Empty and filled markers correspond to a mesh and to a solid cathode.

provides a description of the VAC averaged in time and non-stationary effects in glow positive corona with a satisfactory accuracy.

Streamer corona

The quiet glow corona follows a noisy streamer regime. The threshold current depends on the degree of inhomogeneity of the electric field in the gap: in general, the greater the radius of curvature of the electrode, the lower the threshold current. As a rule, the streamer regime of the steady corona in fact is a regime with intermittent transitions between glow and streamers. The repetition frequency of the streamer appearance in the corona gap increases with total current.

First, it should be particularly emphasized that the mechanism of initiation of streamers in the steady glow corona is not the same as that in a non-pre-ionized gap stressed with a high-voltage pulse. In the latter case, a necessary condition for formation of a positive streamer is a high initial value of the replenishment integral $M = \int_0^l (\alpha - \eta)\,\mathrm{d}x \geq 18\text{--}20$ (Meek's or Raether's criterion). Recall that M is the resulting coefficient of ionization multiplication of an electron avalanche across inter-electrode gap. However, the value of M in a self-sustained glow corona always stays much lower ($M = \ln((1 + \gamma)/\gamma) \leq 3\text{--}6$) at any current. Therefore, it is not clear from the point of view of Meek's criterion how it is possible to induce streamers in glow corona if Meek's criterion is not met.

Figure 2.5.10. The sequence of eight frame pictures illustrating the chaotic dynamics of high-density current spots on the anode surface of a glow positive wire-cylinder corona. Air, $P = 30$ torr, radius of inner wire (anode) $r_a = 0.5$ mm, radius of cylinder $R_c = 10$ mm, reduced corona current per cm of its length $I = 80\,\mu A/cm$, $U = 1.6$ kV. Time exposition of each frame picture is $5\,\mu s$. The time interval between neighboring frames is $5\,\mu s$. A typical diameter of current spot is 0.5 mm.

Second, the streamers developing in the gap of a glow corona propagate through well pre-ionized gas with a marked concentration of charged particles (electrons and/or negative ions), which is higher or of the same order compared with the number density of the seed electrons obtained in the numerical calculations due to using photo-ionization in the model. This means that it is not necessary to engage a disputable photo-ionization process for the description of streamer development in a glow corona.

A search for the reasons responsible for initiation of streamers in a glow corona at low M was carried out in Akishev *et al* (2002b). The anode region of the glow corona appears to the naked eye as homogeneous, but in fact glow is not uniform. Akishev *et al* (2002b) revealed the formation of numerous and non-stationary small current spots on the glowing anode (figure 2.5.10).

To obtain controlled conditions in the experiment, they used a positive corona at lower pressure. The critical current for the appearance of spots decreases with pressure, and at atmospheric pressure it is close to the threshold current for the initiation of streamers (about $50-70\,\mu A$ per pin for a corona in air). The anode spots become more intensive and appear more frequently when the total corona current increases. This finding correlates with the same behavior of the streamers.

Akishev *et al* (2002b) suggest that the current spots arise due to development of an ionization instability in the anode region, and that these spots induce streamers in a glow corona. As a matter of fact, each current spot corresponds to a local breakdown of the glow generation zone. This breakdown releases a voltage drop of about $0.5-1$ kV, which results in an instantaneous and strong increase of the local reduced electric field that is sufficiently large to induce a streamer at the anode. The time it takes to develop an ionization instability depends on the mixture of the processing gas. The use of admixtures like Ar or CO_2 injected in the anode region results in an increase of intensity and frequency of streamers in a positive corona in air (Yan 2001 and literature cited therein). This is consistent with the idea mentioned above about provocation of streamers by anode current spots.

Streamers-to-spark transition

This phenomenon is presented here using the example of sparking of a positive steady corona in a pin-to-plane electrode configuration and based on experimental results obtained at different times by the teams of Loeb, Kaptsov, Goldman, Marode, Sigmond, Rutgers, Veldhuizen, Ono, Yamada, and many other groups.

An increase in the corona current precedes the elongation of the streamers and finally bridging of the gap by some of them. Each bridging results in a current pulse of several tens of mA (see figure 2.5.11 taken from Akishev *et al* 2002b), which is not yet a spark pulse. The amplitude of a streamer pulse is much higher compared with the average corona current. Such an amplitude is possible due to existence of stray capacitance in the external circuit.

For low current steady corona, a sequence of several bridging streamers is required for a spark to happen, with the time interval between two streamers not longer than about 100 μs. Such a short interval ensures that the local energy deposited by the foregoing streamer in a gas volume of tiny size (of the order of the streamer diameter) is not dispersed due to diffusion before the subsequent streamer occurs. In such a case, energy will accumulate in time within a small volume near the tip of the pin. The high

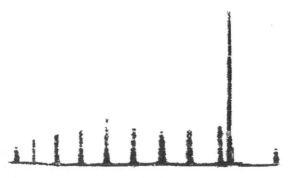

Figure 2.5.11. Waveform of a positive corona current under self-running streamers and regular streamers-to-spark transition. Horizontal and vertical scales are 50 μs and 10 mA in division. Air, 1 atm. Pin-to-plane gap, 17 mm.

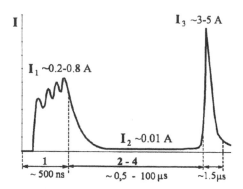

Figure 2.5.12. Generalized behavior in time of positive corona current under induced sparking. Each scale is an arbitrary one. Gap length 17 mm. Ambient air at atmospheric pressure. $U = 20.7\,\text{kV}$, $I = 55\,\mu\text{A}$, $\Delta U = 1.8\,\text{kV}$.

level of specific energy deposited in the gas will result in a dramatic intensification of ionization and detachment processes and in the creation at the pin of the embryo of a pre-spark current filament, which will elongate and propagate towards the cathode plane and eventually form a spark. Estimations of the specific energy locally deposited by streamers gives a minimal value of the order of $0.6\text{--}1\,\text{J/cm}^3$.

High-speed photography is used to investigate the spatio-temporal evolution of the discharge during sparking. Pioneering experiments were done with high over-voltage of a pin-plane gap with the use of streak cameras. It was revealed that spark formation takes two stages. The first is a fast propagation (with velocity about $10^8\,\text{cm/s}$) of the so-called primary streamer traveling from the pin towards the plane cathode. The second stage occurs with some delay, heavily depending on the magnitude of the over-voltage. At this stage, the so-called secondary streamer propagates slowly with a velocity of about $10^6\,\text{cm/s}$ along the same trajectory. Upon bridging of the gap by the secondary streamer, the discharge current increases abruptly, and spark formation is completed. The experiments with a steady corona under stepwise small change in applied voltage (low over-voltage) showed that several generations of primary streamers take place during the first stage (Akishev *et al* 2002b). The secondary streamer develops very slowly (with velocity about $10^5\text{--}10^4\,\text{cm/s}$) supported by a low magnitude of the discharge current (figures 2.5.12 and 2.5.13).

So, in contrast to a primary streamer developing due to intensive direct ionization in strong electric field around its head, the secondary streamer propagates due to an increase of the ionization processes associated mainly with a slow process of energy deposition into its body (gas heating, vibrational excitation, etc). In this respect, propagation of the secondary streamer is analogous to the non-homogeneous constriction of a pulsed glow discharge at atmospheric pressure and to steady glow discharge in gas

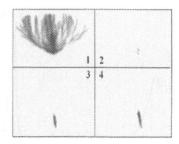

Figure 2.5.13. Typical temporal evolution of positive pin-plane corona morphology under induced sparking. Experimental conditions are the same as in figure 2.5.14. Time exposition for frames 1, 2, and frames 3, 4 is 0.2 and 0.5 μs respectively. Time interval between neighboring frames is 1 μs.

flows at sub-atmospheric pressure (Velikhov *et al* 1982, Napartovich *et al* 1993, Akishev *et al* 1999a). Finally, the mechanisms of propagation of both the secondary streamer (pre-spark filament) in the positive corona and pre-spark filament in the negative corona are based on the development of ionization instabilities in the discharge and therefore have much in common.

The completion of spark formation is the bridging of the gas gap and is accompanied by a dramatic growth of the discharge current (current amplitude of several amperes and slope of current rise $\partial I/\partial t \geq 10^7$ A/s). As a rule, the external circuit of a typical corona discharge includes a power supply delivering several units or tens of kV in output voltage and a ballast resistor of several units or tens of MΩ. It is clear that such huge current amplitudes of the spark can be sustained only by a displacement current in the external circuit. However, there is one problem. Calculations of the charge transferred by spark, require a capacitance much in excess of a static stray capacitance (about units or tens of pF) of an external circuit. A possible reason for this discrepancy is that the quasi-static approach commonly used for the analysis of the corona circuit does not work in the case of a spark with rapidly changing current generating a vorticity of the electric field.

2.5.4 AC corona discharges

Alternating voltages applied across a corona gap introduce new features in the physics of this discharge. First, due to low mobility of the charge carriers in air ($\mu_I \cong 2 \times 10^{-4}$ m^2 V s for positive and negative ions) and low concentration of ions in the bulk, the displacement current can be a marked or even dominant component of the total corona current at relatively low frequencies of the supply voltage. Indeed, from the condition $\varepsilon_0(\partial E/\partial t) \geq e\mu_i E n_i$ for $E(t) = E_0 \cos \omega t$, one can obtain an estimate for minimal circular ω and cyclic f frequencies satisfying this inequality: $\omega \geq 3 \times 10^{-7} n_i$ and $f \geq 5 \times 10^{-8} n_i$ (n_i is the local density of ions in cm^{-3}

in the bulk) of the gap, and which result in a displacement current being an essential component of the total current. For centimeter gaps of pin-plane coronas, the number density of the ion space charge may range over $n_i \approx (2 \times 10^9)-(2 \times 10^{10})\,\text{cm}^{-3}$ depending on the magnitude of the corona current. This means that the displacement current has to be taken into account in ac coronas at frequencies of applied voltages $f \geq 10^2-10^3$ Hz.

Second, the drift of ions across the inter-electrode gap takes a finite time of the order of $\tau_i \approx d/\mu_i E$ and governs the establishment of a unipolar positive or negative dc corona (for a negative corona in a electropositive gas, it is necessary to take the time of the electron drift). In the case $\tau_i > T/2$ ($T = 1/f$ is the period of the applied ac voltage), ions (say, positive ions) formed during the preceding half-period are trapped in the bulk of the gap by an electrical field of opposite direction in the succeeding half-period. The same situation will occur with negative ions. This means that the drift region of an ac corona is filled with ions of the opposite sign that tend to diminish the resultant space charge in the drift region and that are subjected to volume recombination. So, in some respect, an ac corona at frequencies $f \geq \mu_i E/2d$ is akin to the bipolar dc corona between two wires or sharpened pins. A quantitative estimate for the critical frequency of the supply voltage is

$$f \geq \frac{(2 \times 10^3)-(2 \times 10^4)}{d\,(\text{cm})} \quad \text{Hz.} \tag{2.5.4}$$

Finally, for high frequencies $f \geq 10^5$ Hz, the ac corona is called a torch corona, which has nothing in common with dc coronas. Detailed information about the properties of ac coronas can be found in Loeb (1965 ch 7D).

Interesting types of atmospheric pressure ac discharges for the generation of non-thermal plasma at/on dielectric surfaces were published recently by Akishev *et al* (2002c) and by Radu *et al* (2003). These discharges are sustained in the electrode configuration combining the electrode elements of both corona (metallic pin(s)) and dielectric barrier discharge (metallic plate covered with a thin dielectric layer) and called barrier corona or pin-to-plane barrier discharge. In Radu *et al* (2003), the authors investigated experimentally and theoretically the glow mode of a pin-to-plane barrier discharge in He at atmospheric pressure. In Akishev *et al* (2002c), the glow and streamer regimes of a barrier corona in ambient air, Ar, He, and N_2 are investigated. Some results of the latter investigation are presented below.

Properties of ac barrier corona (ACBC) in air

The properties of ACBC in air, widely used as a processing gas for the generation of non-thermal plasma at atmospheric pressure, are interesting in themselves, but the main goal here is a comparison of discharges in air and Ar, in order to show an important advantage of the latter. The presence

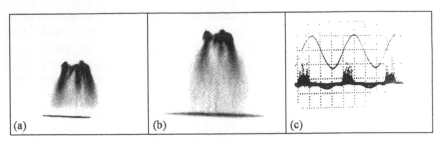

Figure 2.5.14. Side view of ac barrier corona in air with a sharpened electrode and barrier of *PE*-film, at a frequency of 50 Hz and different inter-electrode gaps, h: (a) $h = 1.5$ cm, $U = 25$ kV; (b) $h = 2.5$ cm, $U = 32$ kV. (c) Typical voltage (above) and current oscillograms of an ac barrier corona in air with a sharpened electrode and a barrier of PE-film. Frequency: 50 Hz. The time scale is 5 ms/div, the voltage amplitude is 32 kV.

of electron attachment processes in air results in great differences between discharge parameters and visual appearance observed for ACBC in air and argon. An ac discharge in air requires a substantially higher voltage (more than ten-fold) to sustain the discharge than that in Ar. Images of ACBC in ambient air are presented in figure 2.5.14(a) and (b), where the ac barrier corona appears almost homogeneous in the gas gap and above the surface of a dielectric film. In fact, the ACBC in air has two different current modes, depending on positive or negative polarity of the applied voltage. These modes clearly reveal themselves in the waveform of the ACBC current. Representative examples of current and voltage oscillograms are presented in figure 2.5.14(c).

During the positive half-period, the ACBC is non-uniform because it operates in the streamer regime. Streamers manifest themselves in the form of sharp spikes in the current oscillogram. The number and amplitude of spikes increases with rising voltage amplitude and inter-electrode gap length. As a rule, each current spike correlates with a separate group (or generation) of streamers. These streamers, which originate at the sharpened metal electrode, the anode during the positive half-period, are distributed randomly within a dome over the dielectric film. The diameter of this dome-shaped volume increases with the length of inter-electrode gap (figure 2.5.14(a), (b)). Each streamer strikes the surface and branches over it in the form of short sliding surface streamers. The streamer length in the bulk of the gap is much greater than those on the surface. Volume streamer characteristics are identical to those in the streamer regime of steady-state dc positive pin-plane corona in air with metallic electrodes, while the properties of the short surface streamers are close to those observed in classical ac barrier discharges (Eliasson *et al* 1987, Eliasson and Kogelschatz 1991).

The negative half-period of the ACBC corresponds to a homogeneous glow regime without any spikes in the current oscillogram. The discharge properties of ACBC in the gap (the magnitudes of average electric field and current density) during this half-period are practically the same as those in a steady-state dc negative pin-plane corona in air with metallic electrodes. The properties of the ACBC near the surface of the polymer film are similar to those of the anode region of both the classical barrier discharge in the low-current, uniform glow mode, and of the steady-state dc negative pin-plane corona with a resistive anode plate.

There are two reasons why streamers are absent during the ACBC negative half-period. First, pins do not provoke streamers in a negative corona, and second, the uniform anode region formed near the dielectric film is highly tolerant to streamer initiation as well.

In summary, low frequency ACBCs in air simultaneously exhibit properties that are inherent in both steady-state dc negative and positive pin-plane coronas with metallic electrodes, and in classical ac barrier discharges under uniform glow and streamer current regimes.

In comparison with air, Ar is an easily ionized gas. Therefore sliding surface streamers in Ar spread over a surface very readily. This is a distinctive property of ACBC in Ar, which is an extremely important property with respect to surface treatment. The cross-section of the surface occupied by the ac barrier corona in He was markedly smaller than that in Ar at the same frequency and voltage, but larger than the surface area in N_2.

2.5.5   Pulsed streamer corona discharges

Pulsed coronas are referred to as streamer discharges, which are used in practice to generate non-thermal plasma at atmospheric pressure. As a rule, plasma generators based on positive pulsed corona in air are used because of their higher efficiency in the generation of streamers compared with that of the negative pulsed corona. In the latter case less streamer branching is observed. Therefore, the main attention here is paid to experimental techniques and different properties of pulsed positive coronas.

Typical geometries of electrodes for the generation of positive pulsed coronas are coaxial wire-cylinder, multi-pins-to-plane and multi-wires-to-plane(s). For example, for a cylindrical geometry the outer electrode is a metallic tube about 2 m in length and 20–30 cm in diameter. The inner electrode is either a smooth wire or a rod with lots of small spikes designed to increase a number of streamers.

It is common that high-voltage pulses of 50–150 kV in amplitude and 100–1000 Hz repetition rate are used to generate streamer coronas. This amplitude ensures the fulfillment of Meek's criterion for streamer breakdown of the gap. The leading edge of the voltage pulse has to be short enough ($\leq 0.1\,\mu s$) with a current rise $dI/dt > 10^{10}$ A/s that guarantees a high amplitude of the current density per 1 cm along the wire (up to 10 A/cm) and correspondingly a high density of streamers (up to several streamers per

1 cm). The duration of the pulse trailing edge of the voltage pulse has to be kept short (<0.5 μs is common) to avoid a spark formation in the gap. It should be noted that the appearance of a spark in a pulsed corona device operating at huge peak currents of several hundreds amperes creates much more danger compared with a spark in a steady corona because of possible damage to the electrode system due to melting.

For a coaxial configuration, the excitation of the gas gap by a pulsed corona is non-uniform, because the density of streamers decreases with the distance from wire approximately as $1/r$. Because of this, an effective volume excited by streamers equals only 60–80% of the total volume of tube. The average deposited power is low ($\cong 1$ W/cm^3).

Simultaneous electrical and optical measurements of a pulsed positive corona in a cylindrical geometry were combined into one picture (see figure 2.5.15 taken from Blom 1997), which allows a direct comparison of the electrical and optical parameters of pulsed streamer discharge, and an observation of streamer and spark formation. A number of ICCD images (shutter time 5 ns) are recorded in the course of the development of the corona discharge. During a single pulse, only one image was recorded. Repetitive production of similar corona discharges, and variable delay between the images and the initial rise of the voltage pulse, allow an investigation of the temporal and spatial behavior of the pulsed corona. From each recorded image, an appropriate slice was taken, and figure 2.5.15 was constructed.

In figure 2.5.15 one can see that primary streamers arrive at the surface of cylindrical cathode after 120–140 μs. After this time, slow development of the secondary streamers begins at the wire (pre-spark embryos are apparent). So, the bridging of the gap by primary streamers is not a danger for the safety of the electrode system. To avoid in this experiment the undesirable development of any secondary streamer into a spark, the duration of the applied voltage pulse is restricted to 200 μs. Additional experimental information about streamer formation/propagation in pulsed coronas can be obtained from Marode (1975), Sigmond (1984), van Veldhuizen and Rutgers (2002), and Ono and Oda (2003).

Results of numerical modeling of positive streamers in air can be found in papers by Babaeva and Naidis (1996a,b, 2000), Kulikovsky (1997a,b, 1998), Morrow and Lowke (1997) and Naidis (1996). Numerical simulation of streamer formation and propagation is a rather complicated task. In general, the simulation of the negative streamer in N$_2$ at atmospheric pressure is a simpler task compared with that for positive streamers in air. A two-dimensional simulation model is used by Vitello *et al* (1993) for the description of the development of a negative streamer in short gap (0.5 cm) in N$_2$. This simplified model does not take into account the loss of charged particles in the body of streamer due to electron–ion recombination. This means that model does not describe a formation of a realistic state in plasma behind the head of a streamer.

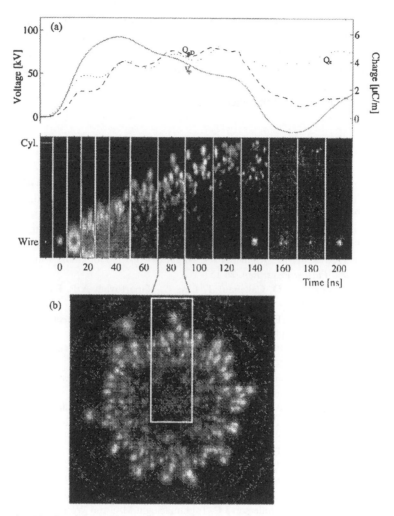

Figure 2.5.15. Combined presentation of the electrical and the optical measurements. (a) Electrical measurements, and image slices taken from a full ICCD image such as in (b). The electrical measurements are the voltage pulse V_p (solid curve, left axis), the external charge Q_e (dotted curve, axis), and the displacement charge Q_{gD} (dashed curve, axis). Discharge parameters: positive voltage pulse $V_p = 93\,kV$, air pressure $P = 360\,torr$, cylinder diameter 29 cm.

The modeling shows that head of a short negative streamer in N_2 tends to the deformation in spatial structure such as branch off. The same results were obtained recently by Arrayas *et al* (2002). Nevertheless, it is difficult to say unambiguously whether such simplified models describe a real branching of streamer because in fact the branching of negative streamers is as a rule observed in long gaps (as a rule, $d > 10\,cm$). In our opinion, the deformation in spatial structure of the developing electron avalanches in a short gap obtained by Vitello and Arrayas and their co-authors can be interpreted as the initial stage of the near-cathode process, which can result in the formation of several current cathode spots (consequently, of several streamers originating from the cathode) but not as branching of a single negative streamer in space.

References

Akishev Yu S and Leys C 1999a *J. Techn. Phys.* (Polish Acad. Sci., Warsaw) **40** 127–143

Akishev Yu S, Deryugin A A, Kochetov I V, Napartovich A P and Trushkin N I 1993 *J. Phys. D: Appl. Phys.* **26** 1630–1637

Akishev Yu S, Grushin M E, Deryugin A A, Napartovich A P and Trushkin N I 1999b *J. Phys. D: Appl. Phys.* **32** 2399–2409

Akishev Yu S, Grushin M E, Kochetov I V, Napartovich A P, Pan'kin M V and Trushkin N I 2000 *Plasma Phys. Rep.* **26** 157–163

Akishev Yu S, Goossens O, Callebaut T, Leys C, Napartovich A P, Pan'kin MV and Trushkin N I 2001a *J. Phys. D: Appl. Phys.* **34** 2875–2882

Akishev Yu S, Grushin M E, Karal'nik V B and Trushkin N I 2001b *Plasma Phys. Rep.* **27** 520–531 (part I) and 532–541 (part II)

Akishev Yu S, Dem'yanov A V, Karal'nik V B, Pan'kin M V and Trushkin N I 2001c *Plasma Phys. Rep.* **27** 164–171

Akishev Yu S, Napartovich A P and Trushkin N I 2002a *Bull. American Phys. Soc.* **47**(7) 55th Annual Gaseous Electronics Conference, 76

Akishev Yu S, Karal'nik V B and Trushkin N I 2002b *Proc. SPIE* **4460** 26–37

Akishev Yu S, Grushin M E, Napartovich A P and Trushkin N I 2002c *Plasmas and Polymers* **7** 261–289

Akishev Yu S, Grushin M E, Karal'nik V B, Monich A E and Trushkin N I 2003a, *Plasma Phys. Rep.* **29** 717–726

Akishev Yu S, Grushin M E, Karal'nik V B, Kochetov I V, Monich A E, Napartovich A P and Trushkin N I 2003b *Plasma Phys. Rep.* **29** 176–186.

Allibone T E, Jones J E, Saunderson J C, Taplamacioglu M C and Waters R T 1993 *Proc. R. Soc. Lond. A* **441** 125–146

Arrayas M, Ebert U and Hundsdorfer W 2002 *Phys. Rev. Lett.* **88** 1745

Babaeva N Yu and Naidis G V 1996a *J. Phys. D.: Appl. Phys.* **29** 2423 – 2431

Babaeva N Yu and Naidis G V 1996b *Phys. Lett. A* **215** 187–190

Babaeva N Yu and Naidis G V 2000 in van Veldhuizen E M (ed) *Electrical Discharges for Environmental Purposes: Fundamentals and Applications* (New York: Nova Science Publishers) pp 21–48

Beattie I 1975 PhD Thesis, University of Waterloo, Canada

Blom P P M 1997 High-Power Pulsed Corona, PhD Thesis, Eindhoven University of Technology

Boullound A, Charrier I and Le Ny R 1979 *J. Physique* **40**(C7) 241

Brown S C 1959 *Elementary Processes in Gas Discharge Plasma* (Cambridge, MA: MIT Press)

Brown S C 1966 *Basic Data of Plasma Physics* (Cambridge, MA: MIT Press)

Černák M and Hosokawa T 1991 *Phys. Rev. A* **43** 1107–1109

Chang J-S, Lawless P A and Yamamoto T 1991 *IEEE Trans. Plasma Sci.* **19** 1102–1166

Colli L, Facchii U, Gatti E and Persano A 1954 *J. Phys. D: Appl. Phys.* **25** 429–432

Eliasson B, Hirth M and Kogelschatz U 1987 *J. Phys. D: Appl. Phys.* **20** 1421–1437

Eliasson B and Kogelschatz U 1991 *IEEE Trans. Plasma Sci.* **19** 309–323

von Engel A V 1955 *Ionized Gases* (Oxford: Clarendon Press)

von Engel A V and Steenbeck M 1934 *Electrische Gasentladungen*, Berlin

Fieux R and Boutteau M 1970 *Bull. Dir. Etude Rech.* serie B, *Réseaux Electriques Matériels Electriques* **2** 55–88

Goldman A, Goldman M, Rautureau M and Tchoubar C 1965 *J. de Physique* **26** 486–489

Goldman A, Goldman M, Jones J E and Yumoto M 1988 *Proceedings of the 9th International Conference on Gas Discharges and their Applications*, Venice, Padova: Trip pp 197–200

Goldman A, Goldman M and Jones J E 1992 *Proceedings of the 10th International Conference on Gas Discharges and their Applications*, Swansea, pp 270–273

Goldman M and Goldman A 1978 in Hirsh M N and Oskam H J (eds) *Gaseous Electronics* vol. I (New York: Academic Press) pp 219–290

Goldman M and Sigmond R S 1982 *IEEE Trans. Electrical Insulation* **EI-17** 90–105

Hermstein W 1960 *Archiv für Electrotechnik* **45** 209–279

Jones J E, Davies M, Goldman A and Goldman M 1990 *J. Phys. D: Appl. Phys.* **23** 542–552

Kaptsov N A 1947 *Corona Discharge* (Moscow: Gostekhizdat), 1953 *Electronics* (Moscow: Gostekhizdat)

Kulikovsky A A 1997a *J. Phys. D: Appl. Phys.* **30** 441–450 and 1515–1522

Kulikovsky A A 1997b *IEEE Trans. Plasma Sci.* **25** 439–445

Kulikovsky A A 1998 *Phys. Rev. E* **57** 7066–7074

Lama W L and Gallo C F 1974 *J. Appl. Phys.* **45** 103–113

Lawless P A, McLean K J, Sparks L E and Ramsey G H 1986 *J. Electrostatics* **18** 199–217

Loeb L B 1965 *Electrical Coronas* (Berkeley–Los Angeles: Univ. of California Press)

Loeb L B, Kip A F, Hudson G G and Bennet W H 1941 *Phys. Rev.* **60** 714–722

Marode E 1975 *J. Appl. Phys.* **46** 2005–2015 (part I) and 2016–2020 (part II)

Marode E, Goldman A and Goldman M 1993 NATO ASI Series, vol. G 34 Part A, Penetrante B M and Schultheis S E (eds) (Berlin, Heidelberg: Springer) pp 167–190

Morrow R 1985a *Phys. Rev. A* **32** 1799–1809; 1985b, *Phys. Rev. A* **32** 3821–3824

Morrow R and Lowke J J 1997 *J. Phys. D: Appl. Phys.* **30** 3099–3144

Naidis G V 1996 *J. Phys. D: Appl. Phys.* **29** 779–783

Napartovich A P and Akishev Yu S 1993 *Proceedings XXI ICPIG*, vol. III, Ruhr-Universität Bochum pp 207–216

Napartovich A P, Akishev Yu S, Deryugin A A, Kochetov I V, Pan'kin M V and Trushkin N I 1997 *J. Phys. D: Appl. Phys.* **30** 2726–2736

Napartovich A P, Akishev Yu S, Kochetov I V and Loboyko A M 2002 *Plasma Physics Reports* **28** 1049–1059

Ono R and Oda T 2003 *J. Phys. D: Appl. Phys.* **36** 1952–1958

Radu I, Bartnikas R and Wertheimer M R 2003 *J. Phys. D: Appl. Phys.* **36** 1284–1291

Sigmond R S 1978 'Corona discharges' in Meek J M and Craggs J D (eds) *Electrical Breakdown of Gases* (New York: Wiley) pp 319–384

Sigmond R S 1982 *J. Appl. Phys.* **53** 891–898

Sigmond R S 1984 *J. Appl. Phys.* **56** 1355–1370

Sigmond R S 1997 in *Proceedings XXIII ICPIG* Toulouse **C4** 383–395

Townsend J S 1914 *Phil. Mag.* **28** 83–90

Trichel G W 1938 *Phys. Rev.* **54** 1078–1084

van Veldhuizen E M and Rutgers W R 2002 *J. Phys. D: Appl. Phys.* **35** 2169–2179

Velikhov E P, Golubev V S and Pashkin S V 1982 Glow discharge in gas flow, *Uspekhi Fizicheskikh Nauk*, Moscow, **137** 117–137

Vereshchagin I P 1985 *Corona Discharge in Electronic and Ionic Technologies* (Moscow: Énergoatomizdat)

Vitello P A, Penetrante B M and Bardsley J N 1993 *NATO ASI Series*, vol. G 34 Part A, Penetrante B M and Schultheis S E (eds) (Berlin, Heidelberg: Springer) pp 249–271

Warburg E 1899 *Wied. Ann.* **67** 68–93; 1927 *Handbuch der Physik* (Berlin: Springer) vol. 4 pp 154.

Yamada T, Kondo and Miyoshi Y 1980 *J. Phys. D: Appl. Phys.* **13** 411–417

Yan K 2001 PhD Thesis, Technische Universiteit Eindhoven

Zentner R 1970a *ETZ-A* **91**(5) 303–305

Zentner R 1970b *Z. Angew. Physik* **29** 294–301

2.6   Fundamentals of Dielectric-Barrier Discharges

2.6.1   Early investigations

In 1857 Siemens in Germany proposed an electrical discharge for 'ozonizing' air. The novel feature of this configuration was that no metallic electrodes were in contact with the discharge plasma. Atmospheric-pressure air or oxygen was passing in the axial direction through a narrow annular space in a double-walled cylindrical glass vessel (figure 2.6.1). Cylindrical electrodes inside the inner tube and wrapped around the outer tube were used to apply an alternating radial electric field, high enough to cause electrical breakdown of the gas inside the annular discharge gap.

Due to the action of the discharge, part of the oxygen in the gas flow was converted to ozone. If air was used as a feed gas traces of nitrogen oxides were also produced. The glass walls, acting as dielectric barriers, have a

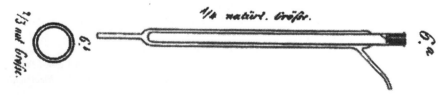

Figure 2.6.1. Siemens' historical ozone discharge tube of 1857 ('natürl. Grösse' means natural size).

strong influence on the discharge properties, which is therefore often referred to as the dielectric-barrier discharge (DBD) or simply barrier discharge (BD). Also the term 'silent discharge', introduced by Andrews and Tait (1860), is frequently used in different languages (stille Entladung, décharge silentieuse). It was soon realized that the Siemens tube was an ideal plasma chemical reactor in which many gases could be decomposed without using excessive heat (Thénard 1872, Berthelot 1876, Hautefeuille and Chappius 1881, 1882, Warburg 1903, 1904). Much of the older work was reviewed by Warburg (1909, 1927) in handbook articles on the silent discharge and in the books by Glockler and Lind (1939) and by Rummel (1951). Investigations on the mechanism of 'electrodeless' discharges, and especially on the influence of radiation on breakdown, were carried out by Harries and von Engel (1951, 1954) and by El-Bakkal and Loeb (1962).

An important observation about breakdown of atmospheric-pressure air in a narrow gap between two glass plates was made by the electrical engineer Buss (1932). He observed that breakdown occurred in many short-lived luminous current filaments, rather than homogeneously in the volume. He also obtained photographic Lichtenberg figures showing the footprints of individual current filaments and recorded oscilloscope traces of the applied high voltage pulse. Buss came up with fairly accurate information about the number of filaments per unit area, the typical duration of a filament and the transported charge in a filament. Further contributions to the nature of these current filaments were made by Klemenc *et al* (1937), Suzuki and Naito (1952), Gobrecht *et al* (1964) and Bagirov *et al* (1972). Today these current filaments are often referred to as microdischarges. They play an important role as partial discharges in voids of solid insulation under ac stress and in many DBD applications. The accomplishment of recent years was that microdischarge properties were tailored to suit desired applications and that the development of power electronics resulted in efficient, affordable and reliable power supplies for a wide frequency and voltage range. More recent investigations also showed that homogeneous or diffuse DBDs can be obtained under certain well-defined operating conditions (see chapter 6). Also regularly patterned DBDs can be obtained in different gases. The phenomenology and discharge physics of these different types of DBDs were reviewed by Kogelschatz (2002a).

Siemens referred to the process as an electrolysis of the gas phase. Today we call it a non-equilibrium discharge in which chemical changes are brought about by reactions of electrons, ions, and free radicals generated in the discharge. The main advantage of the dielectric barrier discharge is that controlled non-equilibrium plasmas can be generated in a simple and efficient way at atmospheric pressure. In addition to its original use for the generation of ozone (see section 9.3) many additional applications have evolved: pollution control, surface treatment, generation of ultraviolet radiation in excimer lamps and infrared radiation in CO_2 lasers, mercury-free fluorescent lamps

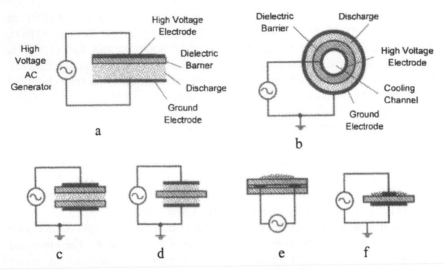

Figure 2.6.2. Different dielectric-barrier discharge configurations.

and flat plasma display panels (Kogelschatz *et al* 1997, 1999, Kogelschatz 2002b, 2003, Wagner *et al* 2003).

2.6.2 Electrode configurations and discharge properties

In addition to the original Siemens ozone discharge tube different electrode configurations have been proposed, all of which have in common that at least one dielectric barrier (insulator) is used to limit the discharge current between the metal electrode(s). Figure 2.6.2 shows a number of different dielectric-barrier discharge configurations covering volume discharges (a, b, c, d) as well as surface discharges (e, f). The presence of the dielectric barrier precludes dc operation because the insulating material cannot pass a dc current. AC or pulsed operation is possible, because any voltage variation dU/dt will result in a displacement current in the dielectric barrier(s).

DBDs are operated with electrode separations between 0.1 mm and several cm, frequency ranges from line frequency to microwave frequencies, and at voltages ranging from about 100 V to several kV. DBDs in different gases and gas mixtures have been studied at various pressure levels. In the context of this book we will concentrate on DBDs operating close to atmospheric pressure, mainly in air.

2.6.3 Overall discharge parameters

In the following sections some properties are described that are common to all DBDs. Although the current flow and power dissipation in most DBDs at about atmospheric pressure occurs in a large number of short-lived

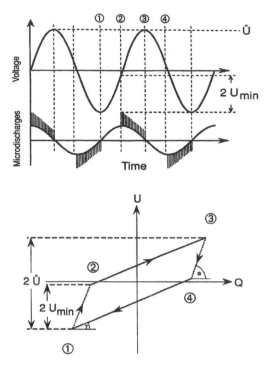

Figure 2.6.3. Applied sinusoidal voltage, schematic representation of microdischarge activity, and resulting voltage-charge Lissajous figure of a dielectric-barrier discharge.

microdischarges the overall discharge behavior, for many purposes, can be described by average quantities. If an ac voltage is applied to a DBD configuration we always have periods of discharge activity (when the voltage inside the gas gap is high enough to initiate breakdown and maintain a discharge) and pauses in between (when the gap voltage is below that value). According to the schematic diagram of figure 2.6.3 we observe alternating phases of discharge activity and discharge pauses. Only at high operating frequencies there may not be enough time for the charge carriers to recombine or be swept out of the gap between consecutive half-waves. In this case some electrical conductivity remains throughout the full voltage period.

The voltage-charge Lissajous figure given in the lower part of figure 2.6.3 is frequently used in ozone research and in investigations on partial discharges. In general it is a useful tool to study DBD properties.

For most DBDs the voltage charge diagram resembles a parallelogram (Manley 1943, Kogelschatz 1988, Falkenstein and Coogan 1997). This is true for large DBD installations used for ozone generation comprising hundreds of square meters of electrode area. It is also true for the tiny cells used in plasma displays (Kogelschatz 2003). It can easily be obtained by using a measuring condenser in the circuit to integrate the current and a high voltage probe to measure the voltage. Both signals are then displayed on a scope in x–y mode. As long as the peak to peak voltage is less than $2U_{min}$ we just see a straight line and have no discharge in the gap. The slope corresponds to the total capacitance of the electrode configuration: $C_{total} = 1/\tan\alpha$. After ignition we observe discharge pauses in the time intervals $1 \to 2$ and $3 \to 4$. During the time intervals $2 \to 3$ and $4 \to 1$ we have discharge activity in the gap and the slope corresponds to the capacity of the dielectric barriers: $C_D = 1/\tan\gamma$. This electrical behavior can be represented by a simple equivalent circuit in which the discharge is represented by two antiparallel Zener diodes which limit the discharge voltage at $\pm U_{Dis}$. The discharge voltage U_{Dis} represents the average gap voltage during discharge activity. It is a fictitious though useful quantity which can be obtained from the voltage charge diagram:

$$U_{Dis} = U_{min}/(1 + \beta) \qquad (2.6.1)$$

where $\beta = C_G/C_D$ is the ratio of the capacitances of the gap C_G and that of the dielectric(s) C_D. In the discharge pauses C_G and C_D act as a capacitive divider. An exact definition of the discharge voltage U_D can be derived from the power P:

$$P = \frac{1}{T}\int_0^T U(t)I(t)\,dt = \frac{1}{\Delta T}\,U_{Dis}\int_{\Delta T} I(t)\,dt \qquad (2.6.2)$$

where the first integral is extended over one period T of the voltage cycle and the second integral is extended only over the active phases during which the discharge is ignited.

All capacitances are linked by the relation

$$\frac{1}{C_{total}} = \frac{1}{C_G} + \frac{1}{C_D}. \qquad (2.6.3)$$

The well defined parallelogram in figure 2.6.4 with sharp corners is an indication that all microdischarges have similar properties. As long as the voltage in the gap is below U_{Dis} no microdischarges occur. Once we reach that value microdischarge activity starts and continues until the peak value

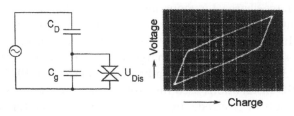

Figure 2.6.4. Equivalent circuit of a dielectric-barrier discharge and recorded voltage-charge Lissajous figure of an ozone discharge tube.

$\hat{U}$ of the external applied voltage is reached. At this point dU/dt is zero, which implies that the displacement current through the dielectric(s) stops. After voltage reversal, a certain swing of the external voltage is required before the value of U_{Dis} is reached in the gap again.

As was first derived by Manley in 1943, the enclosed area of the voltage charge Lissajous figure corresponds to the power dissipated during one discharge cycle. The average discharge power is obtained by multiplying with the frequency f:

$$P = 4fC_{\mathrm{D}}U_{\mathrm{Dis}}[\hat{U} - (1+\beta)U_{\mathrm{Dis}}]\begin{cases} \text{for } \hat{U} \geq (1+\beta)U_{\mathrm{Dis}}, \\ \text{otherwise } P = 0. \end{cases} \qquad (2.6.4)$$

This is the well-known power formula for ozonizers which has been used for the technical design of many DBD applications. Using the minimum external voltage U_{min} required to ignite the discharge, rather than the fictitious discharge voltage U_{Dis}, the power formula can be rewritten as

$$P = 4fC_{\mathrm{D}}(1+\beta)^{-1}U_{\mathrm{min}}[\hat{U} - U_{\mathrm{min}}]\begin{cases} \text{for } \hat{U} \geq U_{\mathrm{min}}, \\ \text{otherwise } P = 0. \end{cases} \qquad (2.6.5)$$

The somewhat surprising feature of this relation is that only the peak voltage $\hat{U}$ enters and not the form of the applied voltage. For a given peak voltage the power is proportional to the frequency. For a given discharge configuration (U_{min} fixed) and given frequency the discharge ignites at $U = U_{\mathrm{min}}$, and the power rises proportionally to the peak voltage with the slope $4fC_{\mathrm{D}}U_{\mathrm{min}}/(1+\beta)$. A special and simple operating case is arrived at when the voltage or current is adjusted until ignition occurs at zero external voltage, which is always possible. In this case two corners of the voltage charge diagram fall on the abscissa and $\hat{U} = 2U_{\mathrm{min}}$. For this special case fairly simple relations can be derived. For a sinusoidal feeding voltage,

$$P = \frac{C_{\mathrm{D}}}{1+\beta}\,\hat{U}^2 = \frac{2C_{\mathrm{D}}}{1+\beta}\,fU_{\mathrm{eff}}^2 \qquad (2.6.6)$$

$$I_{\mathrm{eff}} = \pi f\hat{U}^2 \frac{C_{\mathrm{D}}}{1+\beta}\sqrt{1+2\beta(1+\beta)}, \qquad U_{\mathrm{eff}} = \hat{U}/\sqrt{2} \qquad (2.6.7)$$

$$\text{Power factor:} \quad \overline{\cos\varphi} = \frac{\sqrt{2}}{\pi}\,\frac{1}{\sqrt{1+2\beta(1+\beta)}} \qquad (2.6.8)$$

$$U_{Dis} = \frac{\hat{U}}{2(1+\beta)} = \frac{U_{\mathrm{eff}}}{(1+\beta)\sqrt{2}}. \qquad (2.6.9)$$

Also for an impressed square-wave current simple relations can be derived

$$I_{\mathrm{eff}} = 2\hat{U}C_{\mathrm{D}}\frac{1+2\beta}{1+\beta} \qquad (2.6.10)$$

$$U_{\text{eff}} = \frac{\hat{U}}{\sqrt{3}} \tag{2.6.11}$$

$$\text{Power factor:} \quad \overline{\cos\varphi} = \frac{\sqrt{3}}{2(1 + 2\beta)}. \tag{2.6.12}$$

The time average power factor $\overline{\cos\varphi}$ is an important parameter the knowledge of which is required for matching the power supply to the DBD discharge. Contrary to the power itself the power factor does depend on the voltage form. Values for the power factors in the cases of sinusoidal feeding voltage and impressed square-wave currents are given by Kogelschatz (1988). As a consequence of the presence of the dielectric barrier(s), DBD configurations always present a capacitive load. The load acts as a pure capacitance when there is no discharge and still has a strong capacitive component at time intervals when the discharge is ignited. These phases alternate twice during each cycle of the driving voltage. While the discharge is ignited power is dissipated in the gas gap and the current is limited by the dielectric(s). The power factor is defined as an average quantity for a whole operating cycle of duration T:

$$\text{Power factor:} \quad \overline{\cos\varphi} = \frac{P}{U_{\text{eff}}I_{\text{eff}}} = \frac{1}{U_{\text{eff}}I_{\text{eff}}T} \int_0^T U(t)I(t)\,\mathrm{d}t. \tag{2.6.13}$$

In general it can be stated that square-wave current feeding results in higher power factors. For large DBD installations power factor compensation is mandatory. This can be achieved either by using matching boxes or by using an LC resonance where the apparent capacity of the DBD is compensated by an inductance L in the supply lines.

In this section the overall discharge behavior of DBDs was discussed and some important 'engineering formulae' describing the ignition, temporal behavior and power dissipation of the discharge were compiled. The physical processes inside the discharge gap of DBDs will be discussed in more detail in chapter 6 in sections 6.2 to 6.4.

References

Andrews T and Tait P G 1860 *Phil. Trans. Roy. Soc. London* **150** 113–131

Bagirov M A, Nuraliev N E and Kurbanov M A 1972 *Sov. Phys.–Tech. Phys.* **17** 495–498

Berthelot M 1876 *Compt. Rend.* **82** 1360–1366

Buss K 1932 *Arch. Elektrotech.* **26** 261–265

El-Bakkal J M and Loeb L B 1962 *J. Appl. Phys.* **33** 1567–1577

Falkenstein Z and Coogan J J 1997 *J. Phys. D: Appl. Phys.* **30** 817–825

Glockler G and Lind S C 1939 *The Electrochemistry of Gases and other Dielectrics* (New York: Wiley)

Gobrecht H, Meinhardt O and Hein F 1964 *Ber. Bunsenges. Phys. Chem.* **68** 55–63

Harries W L and von Engel A 1951 *Proc. Phys. Soc. (London)* B **64** 916–929

Harries W L and von Engel A 1954 *Proc. Royal Soc. (London)* A **222** 490–508

Hautefeuille P and Chappius J 1881 *Compt. Rend.* **92** 80–82

Hautefeuille P and Chappius J 1882 *Compt. Rend.* **94** 1111–1114

Klemenc A, Hinterberger H and Höfer H 1937 *Z. Elektrochem.* **43** 708–712

Kogelschatz U 1988 'Advanced ozone generation' in Stucki S (ed) *Process Technologies for Water Treatment* (New York: Plenum Press) pp 87–120

Kogelschatz U 2002a *IEEE Trans. Plasma Sci.* **30** 1400–1408

Kogelschatz U 2002b *Plasma Sources Sci. Technol.* **11**(3A) A1–A6

Kogelschatz U 2003 *Plasma Chem. Plasma Process.* **23** 1–46

Kogelschatz U, Eliasson B and Egli W 1997 *J. Phys. IV (France)* **7** C4–47 to C4–66

Kogelschatz U, Eliasson B and Egli W 1999 *Pure Appl. Chem.* **71** 1819–1828

Manley T C 1943 *Trans. Electrochem. Soc.* **84** 83–96

Rummel T 1951 *Hochspannungs-Entladungschemie und ihre industrielle Anwendung* (Munich: Verlag von R. Oldenbourg und Hanns Reich Verlag)

Siemens W 1857 *Poggendorffs Ann. Phys. Chem.* **102** 66–122

Suzuki M and Naito Y 1952 *Proc. Jpn. Acad.* **2** 469–476

Thénard A 1872 *Compt. Rend.* **74** 1280

Wagner H-E, Brandenburg R, Kozlov K V, Sonnenfeld A, Michel P and Behnke J F 2003 *Vacuum* **71** 417–436

Warburg E 1903 *Sitzungsber. der königl. Preuss. Akad. der Wissensch. (Math-Phys)* 1011–1015

Warburg E 1904 *Ann. der Phys. (4)* **13** 464–476

Warburg E 1909 'Über chemische Reaktionen, welche durch die stille Entladung in gasförmigen Körpern heibeigeführt werden' in Stark J (ed) *Jahrbuch der Radioaktivität und Elektronik* vol. 6 (Leipzig: Teubner) pp 181–229

Warburg E 1927 'Über die stille Entladung in Gasen' in Geiger H and Scheel K (eds) *Handbuch der Physik* vol. 14 (Berlin: Springer) pp 149–170

Chapter 3

Kinetic Description of Plasmas

Ralf Peter Brinkman

3.1 Particles and Distributions

Partially ionized plasmas of gas mixtures like air are complex systems. One may think of a plasma as a large collection of different particles that interact among each other and with external fields: ground-state and excited atoms and molecules, positive and negative ions, electrons, possibly dust. Also radiation—in the ray limit—has particle properties. (We will, however, refer by 'particle' only to matter. Photons are sufficiently different to justify separate treatment.)

- Heavy particles or baryons are species which have at least one nucleon (proton or neutron). They are either atomic (one nucleus) or molecular (several nuclei). Air, for example, consists of 78% N_2 (molecular nitrogen), 21% O_2 (molecular oxygen), 0.9% Ar (argon), and traces of CO_2 (carbon dioxide), H_2O (water), O_3 (ozone), He (helium), Kr (krypton), Xe (xenon) etc. Neutrals carry no charge, $q = 0$, positive ions (cations) with charge $q = Ze$ can be singly ($Z = 1$) or multiply ($Z > 1$) ionized. In electro-negative gases (for example oxygen and nitrogen), negative ions (anions) can also exist, mostly singly charged ($q = -e, Z = -1$). Species are denoted by the 'sum formula' (e.g. H_3O^+) which suffices for most purposes. (Isomer effects—sensitivities to structural differences of molecules having the same sum formula—are, for example, analyzed by Deutsch *et al* [3].) Later in this text it will be useful to view the sum formula as an integer vector $(R) = (R_Z, R_H, R_{He}, \ldots, R_U)$ of charge number and elementary content. H_3O^+, e.g., denotes $(H_3O^+) = (1, 3, 0, 0, 0, 0, 0, 0, 1, 0, 0, 0, \ldots, 0)$. Neglecting electron contributions, the mass of a heavy particle is $m_R = \sum_{n=H}^{U} R_n m_n \approx Aa$, where A is the total number of nucleons of the nuclei. Heavy particles are non-relativistic, i.e. at a given velocity $\vec{v}$ their momentum is $\vec{p} = m\vec{v}$ and their kinetic energy $E = \frac{1}{2}mv^2$. Except for fully

76

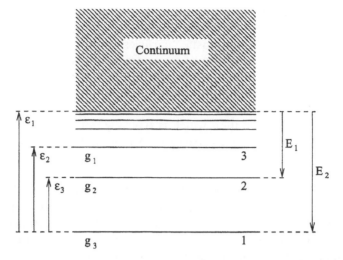

Figure 3.1. Schematic depiction of an energy level diagram of a heavy particle (taken from [7]). Levels of increasing energy are labeled by increasing integers. The lowest level is called the ground state and labeled 1. The energy of the first excited level is ε_2, the energy of the second level ε_3 etc. The level of mimimum energy above the ground level corresponding to a free electron is called the series limit and defines the ionization energy ε_i. Since all energies are possible for free particles, depending on their relative kinetic energies, the energy region above ε_i is called the continuum. The number of different quantum states corresponding to the same energy level ε is called the degeneracy or statistical weight of that level and denoted by g.

ionized cations, heavy particles have also internal structure and may therefore exist in different energy states ε_i. (For a schematic energy level diagram, see figure 3.1.) Atoms or atomic ions have only electronic excitations, with a typical scale of some eV. Molecules or molecular ions have also vibrational and rotational excitations, with energies of a few meV (rotation) or a few tens of meV (vibration). The energies of different species can be compared by accounting the standard enthalpy of formation ΔH_f° (e.g. [5]).

- A particular type of heavy particles is dust. Dust grains can have diameters up to the nanometer and micrometer scale and masses up to several 10^{12} amu. In a plasma environment, they are negatively charged and may represent a sizable fraction of the total charge density. The presence of dust considerably alters the dynamics of a plasma and gives rise to a whole set of new phenomena. Accordingly, the theory of such complex plasmas is very involved. In air plasmas, dust is mostly absent due to the oxidative nature of the medium.
- Electrons are particles with a mass m_e that is much smaller than the mass of the baryons. In this context, they are also non-relativistic. At a speed $\vec{v}$ their kinetic energy is $E = \frac{1}{2} m_e v^2$, and the momentum is $\vec{p} = m_e \vec{v}$. Electrons have

no internal structure, except for their spin which can be ignored in most plasma considerations. In the notation above, $(e) = (-1, 0, \ldots, 0)$.

- Photons are massless relativistic 'particles' which propagate with the speed of light c. For a photon of frequency ν and propagation direction $\vec{e}$, the energy is $E = h\nu$ and the momentum $\vec{p} = h\nu\vec{e}/c$, where h is Planck's constant. In plasma kinetics, the momentum carried by a photon is normally negligible. Photons also have no internal structure, except for their polarization which is typically not important in plasma dynamics (but may, of course, carry important information for diagnostic purposes).

Depending on the pressure, a plasma may contain from 10^{10} to 10^{22} particles per cubic centimeter. (At a temperature of $T = 300\,\text{K}$ and a pressure of $p = 10^5\,\text{Pa}$, it is $n = p/k_{\text{B}}T \approx 2 \times 10^{19}\,\text{cm}^{-3}$.) The task of plasma physics is to analyze and describe the dynamics of these particles under the influence of their mutual interaction and possibly external fields.

Quantum mechanics aside (for the moment), this could in principle be done by solving Newton's equation or their relativistic equivalents for all particles, plus Maxwell's equations for the fields. A short calculation, however, drastically shows that this 'in principle' actually means: 'not really'. The combined information storage available on all computers on earth would allow for a complete specification of roughly a picogram of air plasma in terms of the position $\vec{r}$, velocity $\vec{v}$ and inner state ε of all particles. (This does not even consider the problem of recording a temporal evolution, nor does it account for the computer power required to *solve* the equations of motion!)

There is of course a solution to this problem, well known under the heading 'statistical mechanics': instead of attempting a complete description, one considers the value of an incomplete description. Various decisions on which information is essential and which can be disposed of are possible. Kinetic theory denotes an approach which is particularly suited to describe collections of weakly interacting particles, such as the particles in a gas or plasma, or the 'quasi-particles' in a solid. The first example was developed 1877 by Boltzmann for a neutral gas; it is still the prototype (to the extent that 'Boltzmann equation' is a synonym for kinetic theory in general) [10].

Kinetic theory is based on the assumption that the essential information on the system is given by the one-particle distribution f, a real-valued, time-dependent function of the phase space μ, which is the set $\mu = V \times \mathbb{R}^3$ of all spatial and velocity positions $(\vec{r}, \vec{v})$ that a particle can assume. We assume that there are N different species present, counting as such also different internal states. They are distinguished by subscript indices, where we use the convention that indices s and r run over all species, α and β are charged species, a and b neutrals, e is the electron, i denotes ions. The distribution function states that, at a given time t, the expected number ΔN_s of particles

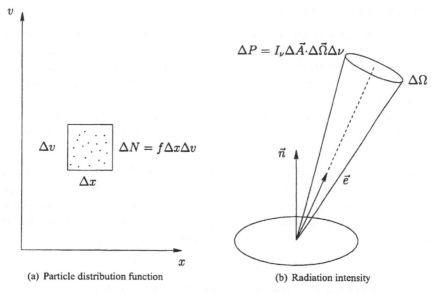

(a) Particle distribution function   (b) Radiation intensity

Figure 3.2. Visualization of the particle distribution function (left) and the radiation intensity (right). The distribution function gives the number of particles ΔN in the phase space volume $\Delta^3 r \Delta^3 v$ as $\Delta N = f(\vec{r}, \vec{v}, t) \Delta^3 r \Delta^3 v$, while the radiation intensity represents the energy flux ΔP per area $\Delta \vec{A}$ and frequency interval $\Delta \nu$ from the solid angle $\Delta \vec{\Omega}$ as $\Delta P = I_\nu(\vec{r}, \vec{e}, \nu, t) \, \Delta \nu \, \Delta \vec{A} \cdot \Delta \vec{\Omega}$.

of species s to be found in the volume $\Delta^3 r \Delta^3 v$ around the point $(\vec{r}, \vec{v})$ is given as

$$\Delta N_s = f_s(\vec{r}, \vec{v}, t) \Delta^3 r \Delta^3 v, \qquad [f_s] = s^3/m^6. \tag{3.1}$$

Alternatively, one may define the distribution function f as a suitably averaged ('coarse grained') form of the exact microscopic distribution of an ensemble of particles,

$$f_s(\vec{r}, \vec{v}, t) = \left\langle \sum_i \delta^{(3)}(\vec{r} - \vec{r}_i(t)) \, \delta^{(3)}(\vec{v} - \vec{v}_i(t)) \right\rangle. \tag{3.2}$$

Radiation can be described in similar terms. In the geometric limit, it is seen as a stream of massless photons propagating with the speed of light at a position $\vec{r}$ and time t in a given direction $\vec{e}$. The radiation intensity I describes the energy flux ΔP per frequency interval $\Delta \nu$ flowing out of a solid angle $\Delta \vec{\Omega}$ around $\vec{e}$ onto a surface element $\Delta \vec{A}$ as

$$\Delta P = I_\nu(\vec{r}, \vec{e}, \nu, t) \, \Delta \nu \, \Delta \vec{A} \cdot \Delta \vec{\Omega}, \qquad [I_\nu] = W/Hz\,m^2. \tag{3.3}$$

At first glance, the definitions (3.1) for the particles and (3.3) for the photons seem rather different. In actuality, the two concepts are quite similar, if the following is taken into account:

- The distribution function f makes reference to the particle number, while the radiation intensity I_ν does not count photons but refers to their energy.
- The distribution function is defined to account for the particle density *per volume* $\Delta^3 r$, while the radiation intensity represents the energy influx *per area* $\Delta \vec{A}$.
- The distribution function assumes non-relativistic behavior (particles can have any speed), while the radiation intensity sees the photons as 'ultra-relativistic' (their speed is c).

To compare the two concepts, a quantity is needed that is defined for both particles and photons. This can be found in the momentum $\vec{p}$, or—more convenient here—in the wave vector $\vec{k} = \vec{p}/\hbar$. The corresponding phase space distribution, a dimensionless quantity, shall be termed $\Phi(\vec{r}, \vec{k})$. It provides the number of particles in a volume $\Delta^3 r \Delta^3 k$ as

$$\Delta N = \Phi \Delta^3 r \Delta^3 k, \quad [\Delta N] = 1 \tag{3.4}$$

and the flux of particles $\Delta \Psi$ through a surface element $\Delta \vec{A}$ as

$$\Delta \Psi = \Phi \vec{v} \cdot \Delta \vec{A} \Delta^3 k, \quad [\Delta \Psi] = \text{s}^{-1}. \tag{3.5}$$

The corresponding energy flux (with a quantum E per particle) is

$$\Delta P = E \Phi \vec{v} \cdot \Delta \vec{A} \Delta^3 k, \quad [\Delta P] = \text{W}. \tag{3.6}$$

For a photon of frequency ν, the energy is $E = 2\pi\hbar\nu$, the speed is $\vec{v} = c\vec{e}$, and the wave number is $k = 2\pi\nu/c$. Using also the representation of the momentum element,

$$\Delta^3 k = k^2 \Delta k \Delta\Omega = 8\pi^3 c^{-3} \nu^2 \Delta\nu\Delta\Omega \tag{3.7}$$

one obtains for the energy flux

$$\Delta P = \frac{16\pi^4 \hbar\nu^3}{c^2} \Phi \vec{e} \cdot \Delta \vec{A} \Delta\nu\Delta\Omega. \tag{3.8}$$

The comparison with the definition above shows that I_ν can indeed, up to a factor, be identified with a distribution function. In particular, one has

$$I_\nu(\vec{r}, \vec{e}, \nu, t) = \frac{16\pi^4 \hbar\nu^3}{c^2} \Phi\left(\vec{r}, \frac{2\pi\nu}{c}\vec{e}, t\right). \tag{3.9}$$

Finally in this context, also the often used energy distribution function (EDF) will be discussed. When the distribution function $f(\vec{v})$ is isotropic (or the anisotropy cannot be resolved), it is convenient to introduce a distribution function F which depends only on the particle energy E, normalized so that the number of particles between E and $E + \Delta E$ is

$$\Delta N = f(v) 4\pi v^2 \Delta v \Delta^3 r = F(E)\Delta E \Delta^3 r. \tag{3.10}$$

Using the relation $E = \frac{1}{2}mv^2$, one arrives at

$$F(E) = 4\pi\sqrt{\frac{2E}{m^3}}f\left(\sqrt{\frac{2E}{m}}\right). \tag{3.11}$$

The distribution function allows calculation of a variety of other quantities, particularly the so-called moments, a systematic sequence of symmetric tensors depending on $\vec{r}$ and t,

$$M^n_{s\,\mu_1\mu_2\cdots\mu_n}(\vec{r},t) = \int v_{\mu_1}v_{\mu_2}\cdots v_{\mu_n}f_s(\vec{r},\vec{v},t)\,\mathrm{d}^3v. \tag{3.12}$$

Also of importance are the contracted moments, i.e. integrals of the moment type with two indices (or more generally, p index pairs) set equal and summed over. They have the structure

$$M^n_{s\,\mu_1\mu_2\cdots\mu_{n-2p}}(\vec{r},t) = \int v_{\mu_1}v_{\mu_2}\cdots v_{\mu_{n-2p}}v^{2p}f_s(\vec{r},\vec{v},t)\,\mathrm{d}^3v. \tag{3.13}$$

Connected to each moment is a moment of the next order, the corresponding flux

$$\vec{\Gamma}^n_{s\,\mu_1\mu_2\cdots\mu_n}(\vec{r},t) = \int v_{\mu_1}v_{\mu_2}\cdots v_{\mu_n}\vec{v}f_s(\vec{r},\vec{v},t)\,\mathrm{d}^3v \tag{3.14}$$

with a similar definition for the contracted moments,

$$\vec{\Gamma}^n_{s\,\mu_1\mu_2\cdots\mu_{n-2p}}(\vec{r},t) = \int v_{\mu_1}v_{\mu_2}\cdots v_{\mu_{n-2p}}v^{2p}\vec{v}f_s(\vec{r},\vec{v},t)\,\mathrm{d}^3v. \tag{3.15}$$

By summation over the species index s the moments are also defined for the plasma as a whole. The relative weights depend on the physical meaning of the quantities. They are unity, m_s or q_s, for quantities related to the particle number, mass, and charge, respectively.

Several of these moments have particular physical importance. The zeroth, the first and the contraction of the second moment directly relate to the conservation laws of mass, momentum, and energy. For each species, the zeroth moment defines the particle density

$$n_s(\vec{r},t) = \int f_s\,\mathrm{d}^3v. \tag{3.16}$$

A summation over the species yields the total densities of particle number, mass, and charge. (In accordance with the standard notation, the symbol ρ is used for both the mass density and the charge density. Whenever necessary, a superscript differentiates the two.)

$$n(\vec{r},t) = \sum_s \int f_s\,\mathrm{d}^3v = \sum_s n_s \tag{3.17}$$

$$\rho^M(\vec{r},t) = \sum_s \int m_s f_s\,\mathrm{d}^3v = \sum_s m_s n_s \tag{3.18}$$

$$\rho^C(\vec{r}, t) = \sum_s \int q_s f_s \, d^3 v = \sum_s q_s n_s. \tag{3.19}$$

The first moment defines the flux of particles

$$\vec{\Gamma}_s = \int \vec{v} f_s \, d^3 v. \tag{3.20}$$

An equivalent, but more frequently employed, definition is that of the average particle velocity $\vec{u}_s$, also referred to as the bulk speed

$$\vec{u}_s(\vec{r}, t) = \int \vec{v} f_s \, d^3 v / n_s = \vec{\Gamma}_s / n_s. \tag{3.21}$$

Summation over the species index s defines the fluxes of total particle number, charge, and mass. The latter two have an direct interpretation as current and momentum density

$$\vec{\Gamma}(\vec{r}, t) = \sum_s \int \vec{v} f_s \, d^3 v = \sum_s n_s \vec{u}_s \tag{3.22}$$

$$\vec{j}(\vec{r}, t) = \sum_s \int q_s \vec{v} f_s \, d^3 v = \sum_s q_s n_s \vec{u}_s \tag{3.23}$$

$$\vec{p}(\vec{r}, t) = \sum_s \int m_s \vec{v} f_s \, d^3 v = \sum_s m_s n_s \vec{u}_s. \tag{3.24}$$

The average velocity of the plasma is defined with reference to the center-of-mass motion,

$$\vec{u}(\vec{r}, t) = \sum_s \int m_s \vec{v} f_s \, d^3 v / \rho^M = \vec{p} / \rho^M \neq \vec{j} / \rho^C. \tag{3.25}$$

As a consequence, the momentum density can be written as

$$\vec{p} = \rho^M \vec{u}. \tag{3.26}$$

The difference of the species velocity $\vec{u}_s$ and the center-of-mass motion is the diffusion velocity

$$\vec{U}_s(\vec{r}, t) = \int (\vec{v} - \vec{u}) f_s \, d^3 v / n_s = \vec{u}_s - \vec{u}. \tag{3.27}$$

The higher moments are only important in mass-related form. The uncontracted moment of second order is the (full) pressure tensor and represents the flux of the momentum density

$$\mathbf{\Pi}_s = \int m_s \vec{v} \vec{v} f_s \, d^3 v \tag{3.28}$$

$$\mathbf{\Pi} = \sum_s \int m_s \vec{v} \vec{v} f_s \, d^3 v = \sum_s \mathbf{\Pi}_s. \tag{3.29}$$

The pressure tensor is also definable with respect to the center-of-mass velocity, then denoted **P**. Its isotropic part (a third of the trace) defines the pressure scalar p

$$\mathbf{P}_s(\vec{r}, t) = \int m_s(\vec{v} - \vec{u}_s)(\vec{v} - \vec{u}_s)f_s \, d^3v \tag{3.30}$$

$$\mathbf{P}(\vec{r}, t) = \sum_s \int m_s(\vec{v} - \vec{u})(\vec{v} - \vec{u})f_s \, d^3v = \sum_s (\mathbf{P}_s + m_s n_s \vec{U}_s \vec{U}_s) \tag{3.31}$$

$$p_s(\vec{r}, t) = \frac{1}{3} \int m_s(\vec{v} - \vec{u}_s)^2 f_s \, d^3v \tag{3.32}$$

$$p(\vec{r}, t) = \sum_s \frac{1}{3} \int m_s(\vec{v} - \vec{u})^2 f_s \, d^3v = \sum_s \left(p_s + \frac{1}{3} m_s n_s \vec{U}_s^2 \right). \tag{3.33}$$

The irreducible remainder is known as the stress tensor (**I** denotes the unit tensor)

$$\pi_s = \mathbf{P}_s - p_s \mathbf{I} \tag{3.34}$$

$$\pi = \mathbf{P} - p\mathbf{I}. \tag{3.35}$$

Using these definitions, the following identities arise:

$$\mathbf{\Pi}_s = \rho_s \vec{u}_s \vec{u}_s + \pi_s + p_s \mathbf{I} \tag{3.36}$$

$$\mathbf{\Pi} = \rho \vec{u} \vec{u} + \pi + p\mathbf{I}. \tag{3.37}$$

The contraction of the second moment gives the kinetic energy density (counting only translation, the rotational and vibrational degrees of freedom are part of the internal energies)

$$e_s(\vec{r}, t) = \int \frac{1}{2} m_s v^2 f_s \, d^3v \tag{3.38}$$

$$e(\vec{r}, t) = \sum_s \int \frac{1}{2} m_s v^2 f_s \, d^3v = \sum_s e_s. \tag{3.39}$$

The contracted second moment can also be used to define the so-called kinetic temperature T_s. In equilibrium, it coincides with the thermodynamic temperature (when measured in energy units). In situations far from equilibrium the notion still provides a convenient shorthand for 'two thirds of the average thermal energy'. (The kinetic temperature T of the whole plasma becomes a questionable concept when different species differ strongly in their thermal energy.)

$$T_s(\vec{r}, t) = \frac{1}{3n_s} \int m_s(\vec{v} - \vec{u}_s)^2 f_s \, d^3v = \frac{p_s}{n_s} \tag{3.40}$$

$$T(\vec{r}, t) = \frac{1}{3n} \sum_s \int m_v(\vec{v} - \vec{u})^2 f_s \, d^3v. \tag{3.41}$$

Each species of the plasma and the plasma as a whole obey the ideal gas equation

$$p_s = n_s T_s \tag{3.42}$$

$$p = nT \tag{3.43}$$

and the full kinetic energies can be expressed as

$$e_s(\vec{r}, t) = \tfrac{1}{2} m_s n_s \vec{u}_s^2 + \tfrac{3}{2} n_s T_s \tag{3.44}$$

$$e(\vec{r}, t) = \tfrac{1}{2} \rho \vec{u}^2 + \tfrac{3}{2} nT. \tag{3.45}$$

The flux of the energy is given by the contracted moment of the third order

$$\vec{\Gamma}_s^e(\vec{r}, t) = \int \frac{1}{2} m_s v^2 \vec{v} f_s \, \mathrm{d}^3 v \tag{3.46}$$

$$\vec{\Gamma}^e(\vec{r}, t) = \sum_s \int \frac{1}{2} m_s v^2 \vec{v} f_s \, \mathrm{d}^3 v. \tag{3.47}$$

The corresponding quantity in the co-moving system known as the heat flux

$$\vec{q}_s(\vec{r}, t) = \int \frac{1}{2} m_s (\vec{v} - \vec{u}_s)^2 (\vec{v} - \vec{u}_s) f_s \, \mathrm{d}^3 v \tag{3.48}$$

$$\vec{q}(\vec{r}, t) = \sum_s \int \frac{1}{2} m_s (\vec{v} - \vec{u})^2 (\vec{v} - \vec{u}) f_s \, \mathrm{d}^3 v = \sum_s \vec{q}_s. \tag{3.49}$$

This gives rise for the following identities for the energy flux

$$\vec{\Gamma}_s^e = (\tfrac{1}{2} \rho_s \vec{u}_s^2 + \tfrac{3}{2} n_s T_s) \vec{u}_s + \vec{q}_s + p_s \vec{u}_s + \pi_s \cdot \vec{u}_s \tag{3.50}$$

$$\vec{\Gamma}^e = (\tfrac{1}{2} \rho \vec{u}^2 + \tfrac{3}{2} nT) \vec{u} + \vec{q} + p\vec{u} + \pi \cdot \vec{u} = \sum_s \vec{\Gamma}_s^e. \tag{3.51}$$

Also the 'distribution of the photons', the radiation density I_ν, allows suitable moments to be defined. In the field of low temperature plasma physics, however, they are less frequently employed than their particle counterparts: non-equilibrium radiation has such a pronounced structure that spectral and other averages are not very meaningful. Also, in non-relativistic plasmas, the photon momentum is negligible; radiation pressure and related quantities are thus less important.

The radiation intensity I_ν gives the radiation from a solid angle element $\Delta \Omega$ around a direction $\vec{e}$, the corresponding spectral energy flux density is the integral of I_ν over all directions

$$\vec{F}_\nu(\vec{r}, \nu, t) = \int_\Omega \vec{e} I_\nu \, \mathrm{d}\Omega. \tag{3.52}$$

The energy density of a radiation field is more difficult to calculate. Either by geometric considerations (see figure 3.3), or by employing the representation

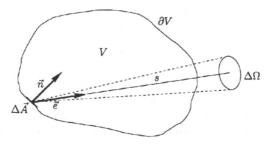

Figure 3.3. Geometric motivation of definition (53). The spectral energy flux ΔP_ν from the solid angle $\Delta\Omega$ around the direction $\vec{e}$ onto the surface element $\Delta\vec{A}$ equals $\Delta P_\nu = I_\nu(\vec{e}, \nu)\, \vec{e} \cdot \Delta\vec{A}\,\Delta\Omega$. The photons spend a travel time s/c in the volume V, which therefore has a total spectral energy content $\Delta U_\nu = \iint s I_\nu(\vec{e}, \nu)\, \vec{e} \cdot d\vec{A}\, d\Omega/c$. By vector analytic means, this expression can be transformed into the equivalent representation $\Delta U_\nu = V \int I_\nu(\vec{e}, \nu)\, d\Omega/c$.

(3.9) and equating the energy content in a volume element $\Delta U = u_\nu \Delta\nu \Delta^3 r$ with the expression $2\pi\hbar\nu \int_\Omega F\, d\Omega\, k^2 \Delta k \Delta^3 r$ one can motivate the definition of the spectral energy density,

$$u_\nu(\vec{r}, \nu, t) = \frac{1}{c}\int I_\nu\, d\Omega. \tag{3.53}$$

All spectrally resolved quantities also have integral counterparts. The integral radiation intensity I, radiation energy flux $\vec{F}$, and radiation energy density u, the total photon density n and the total photon flux $\vec{\Gamma}$ are given as

$$I(\vec{r}, \vec{e}, t) = \int I_\nu\, d\nu \tag{3.54}$$

$$\vec{F}(\vec{r}, t) = \int_0^\infty \vec{F}_\nu\, d\nu = \int_\Omega \vec{e} I\, d\Omega \tag{3.55}$$

$$u(\vec{r}, t) = \int_0^\infty u_\nu\, d\nu = \frac{1}{c}\int I\, d\Omega \tag{3.56}$$

$$n(\vec{r}, t) = \int_0^\infty \frac{1}{h\nu} u_\nu\, d\nu \tag{3.57}$$

$$\vec{\Gamma}(\vec{r}, t) = \int_0^\infty \frac{1}{h\nu} \vec{F}\, d\nu. \tag{3.58}$$

In general, distribution functions are very complex and cannot be given in simple analytical form. The following examples, however, represent certain model situations and are frequently useful. Their parameters correspond to the moments defined above; spatial homogeneity is assumed.

The first example is that of a mono-energetic beam, i.e. a collection of particles which have the same velocity and direction $\vec{u}$. Often this distribution is chosen to represent particles which enter the plasma from outside under carefully controlled experimental conditions,

$$f_B(\vec{v}) = n\,\delta^{(3)}(\vec{v} - \vec{u}). \tag{3.59}$$

The Maxwellian, on the other hand, arises when a plasma is allowed to relax into equilibrium. It can also be employed when no other information is available on the status of a plasma component other than the value of the first three moments; the justification for this is either information theory ('maximum entropy estimate') or pragmatism ('easy to handle'),

$$f_M(\vec{v}) = \frac{n}{(2\pi T/m)^{3/2}} \exp\left(-\frac{m(\vec{v} - \vec{u})^2}{2T}\right). \tag{3.60}$$

Finally, Druyvesteyn's distribution shall be mentioned which is met, for example, in certain simplified models of the electron component of a noble gas plasma. It has the form

$$f_D(\vec{v}) = C\exp\left(-\beta\vec{v}^4\right) \tag{3.61}$$

with the two parameters C and β related to the density and the kinetic temperature as

$$n = C\pi\Gamma_{3/4}\beta^{-3/4} \tag{3.62}$$

$$T = (\Gamma_{5/4}/\Gamma_{3/4})m\beta^{-1/2}. \tag{3.63}$$

Compared to a same temperature Maxwellian, it has a much steeper decrease at high energies. Very often, Maxwellian and Druyvesteyn calculations are compared to illustrate the sensitivity of certain results on the form of the distribution function. (See figure 3.4).

Also the spectral radiation intensities I_ν are generally complicated functions which do not follow a simple analytical form. But again, some explicit examples may be useful. Like the distribution functions f, they are given under the assumption of spatial homogeneity.

The first example is that of a monoenergetic radiation beam of photons with a frequency ν_B and a radiation intensity I_B, propagating into the direction $\vec{e}_B$. Its spectral radiation intensity is (with $\delta^{(2)}$ denoting the delta function with respect to the solid angle)

$$I_\nu(\vec{e}, \nu) = I_B\,\delta^{(2)}(\vec{e} - \vec{e}_B)\delta(\nu - \nu_B). \tag{3.64}$$

The spectral radiation flux and radiation energy density are

$$\vec{F}_\nu(\nu) = I_B\,\delta(\nu - \nu_B)\,\vec{e}_B \tag{3.65}$$

$$u_\nu(\nu) = \frac{I}{c}\,\delta(\nu - \nu_B). \tag{3.66}$$

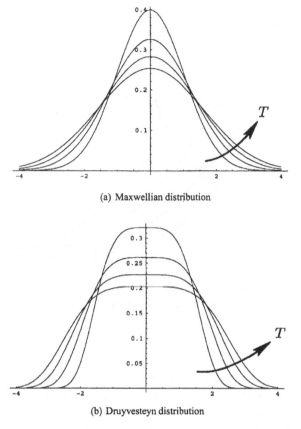

(a) Maxwellian distribution

(b) Druyvesteyn distribution

Figure 3.4. Normalized Maxwellian (top) and Druyvesteyn (bottom) distribution functions at the same density n, for different kinetic temperatures T. The Druyvesteyn distribution is flatter for small v, but has a much steeper decrease at high energies.

The second example is that of the well-known black body radiation, given by Planck's formula

$$I_\nu(\nu) = \frac{4\pi\hbar\nu^3}{c^2} \frac{1}{\exp\left(\dfrac{2\pi\hbar\nu}{T}\right) - 1}. \qquad (3.67)$$

As this radiation is isotropic, the radiation flux vanishes. The spectral energy density is

$$u_\nu(\nu) = \frac{16\pi^2\hbar\nu^3}{c^3} \frac{1}{\exp\left(\dfrac{2\pi\hbar\nu}{T}\right) - 1}. \qquad (3.68)$$

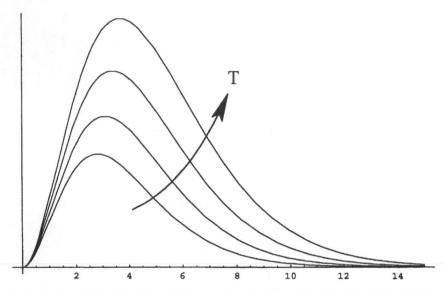

Figure 3.5. Radiation intensity $I_\nu(\nu)$ and energy density $u_\nu(\nu)$ of the Planck black body function for different normalized temperatures T. (In arbitrary units, they differ only by a factor $4/c$.)

The total radiation energy density of the black body radiation follows the well-known Stefan–Boltzmann T^4 law. (Note here that the temperature is given in energy units.)

$$u = \frac{\pi^2 T^4}{15 c^3 \hbar^3}. \tag{3.69}$$

Kinetic theory assumes the information in the distribution function as mathematically complete: if f is known at a time t_0, along with all external fields and the boundary conditions, then it can be calculated for all future times $t > t_0$. More explicitly, kinetic theory postulates the existence of a *closed* equation for f, called the kinetic equation (or Boltzmann equation, for its prototype). In this chapter, we will establish and discuss the kinetic description for the case of a complex, partially ionized plasma far from equilibrium, such as air.

As all mathematical models, kinetic theory has its limitations. First, it should be noted that we deal with a continuum theory that itself makes no reference to the atomistic nature of its system. The probabilistic relation (1) is a physical *interpretation*, not a strict mathematical *definition*. For it to make sense, the phase space volume $\Delta^3 r \Delta^3 v$ should be chosen small enough to resolve macroscopic structures but large enough so that statistical fluctuations $\sim \Delta N_s^{-1/2}$ are negligible. If the smallest macroscopic length is not much larger than the interparticle distance, the scales are not sufficiently separated and the kinetic model breaks down. (Definition (2) embodies similar problems because the invoked 'suitable average' also makes reference to an intermediate scale.) Second, kinetic theory assumes that higher order correlations are not dynamic, but can be calculated as functionals of the one-particle distribution f. This assumption generally holds when the interactions among the particles are sufficiently weak and/or rare.

Let us discuss the assumptions in more detail for the considered case of a weakly ionized plasma, presupposing some material of the next section. The neutral density is n_N, the electron density n_e, with the ionization degree $\alpha = n_e/n_N \ll 1$. The corresponding temperatures are T_N and T_e; typically T_e is much larger than T_N.

The neutral particles will interact when the relative distance becomes smaller than their diameter d; these interactions are rare when the particle distance $r_N = n_N^{-1/3}$ is large, $r_N \gg d$. Clearly, this condition is always met, it simply implies that the density of the neutral gas component is small compared to that of the condensed phase. For the Coulomb interaction, it is custom to introduce three characteristic distances, namely the average distance $r_e = n_e^{-1/3}$, the distance of closest approach for thermal particles, $r_c = e^2/(4\pi\varepsilon_0 T_e)$, and the Debye length $\lambda_D = (\varepsilon_0 T_e/e^2 n)^{1/2}$. The scales are not independent; using the plasma parameter $\Lambda = \lambda_D/r_c$ one has $\lambda_D/r_e = \Lambda/(4\pi)^{1/3}$. The condition that the Coulomb interaction is weak implies that the interparticle distance is large compared with the distance of closest approach, or equivalently, the total Coulomb interaction energy is small compared with the kinetic energy. Because of the relation between the scales, this is often stated as the condition that the number of particles in a Debye sphere be large, $\lambda_D^3 n_e \equiv \Lambda/3 \gg 1$. Plasmas that fulfill the condition are referred to as weakly coupled or 'ideal'.

The limitations of the kinetic description should not be overstated. For most practical applications, the approach is very satisfactory, as ideal plasmas cover the majority of cases under consideration. Important non-ideal plasmas are high pressure arcs; also dusty plasmas are non-ideal with respect to their dust component. From a pragmatic point of view, one may state that the difficulties in treating the kinetic model alone are so huge that one hardly ever is tempted to employ an even more general description. In other words, the real challenge is to reduce the kinetic description itself to a more tractable form. We will come to this later.

The rest of this chapter is organized as follows. In the next section, the various interactions of the particles in a plasma will be discussed and physically classified into 'forces' and 'collisions'. Then the mathematical form of the kinetic model will be established. The last section will briefly describe the possibilities of evaluating the kinetic representation. In particular, we will mention some simplifications that are based on the smallness of the electrons' mass and other approximations, and sketch the connection to the more elementary plasma descriptions.

3.2   Forces, Collisions, and Reactions

The particles of a plasma are subject to various types of interactions, among themselves, with the surrounding walls, with radiation, and with externally applied fields. All interactions are electromagnetic, except for a constant gravity which is sometimes included. In the final formulation of kinetic theory, however, they are represented by contributions of very different mathematical form. A discussion of the processes and their description is the subject of this section.

Kinetic theory regards the plasma constituents (except the photons, of course) as classical and here non-relativistic point particles. As such, they follow Newton equation of motion with the acceleration calculated from the Lorentz force (and possibly constant gravitation),

$$\frac{d\vec{r}}{dt} = \vec{v} \tag{3.70}$$

$$\frac{d\vec{v}}{dt} = \frac{q}{m}\left(\vec{E}(\vec{r}, t) + \vec{v} \times \vec{B}(\vec{r}, t)\right) + \vec{g}. \tag{3.71}$$

The electromagnetic fields may be externally generated, but typically include also contributions which arise from the charges and currents within the plasma itself. The 'self-consistent fields' can be calculated directly from Maxwell's equations, using the above expressions for ρ and $\vec{j}$:

$$\frac{1}{\mu_0}\nabla \times \vec{B} = \varepsilon_0 \frac{\partial \vec{E}}{\partial t} + \sum_s q_s \int \vec{v} f_s \, d^3 v \tag{3.72}$$

$$\nabla \times \vec{E} + \frac{\partial \vec{B}}{\partial t} = 0 \tag{3.73}$$

$$\varepsilon_0 \nabla \cdot \vec{E} = \sum_s q_s \int f_s \, d^3 v \tag{3.74}$$

$$\nabla \cdot \vec{B} = 0. \tag{3.75}$$

The self-consistent fields, however, do not account for all plasma interactions. As described above, the one-particle distribution function neglects information on the correlation of the particles, and processes related to the individual encounters of particles are therefore not included in (3.70)–(3.75). These 'collisions' (an obvious, but unfortunately misleading term) can be of very different type; they may be classified with the help of the following considerations.

Neutral particles, typically the majority in the plasma, interact when their electron shells overlap. The interaction vanishes rapidly when the particle separation becomes larger than a few Bohr radii. The Lenard-Jones model, e.g., assumes a form $\sim r^{-6}$ in the potential and $\sim r^{-7}$ in the force [4]. If one of the interaction partners is charged, it induces an electrical dipole moment in the partner, the corresponding interaction is attractive and behaves as r^{-5}. Only if both partners carry charges, a long range interaction arises which goes $\sim r^{-2}$.

The decrease of the forces with r must be compared to the increase in the number of interaction partners which scales $\sim r^2$ for large r. For neutral–neutral and neutral–charge interactions the accumulated interaction force is finite and, in fact, is dominated by the small distance contributions. These interactions are thus mainly few-body collisions, i.e. they can be understood as the interaction of two or three particles which asymptotically are before and after the collision free (for $t \to \pm\infty$). Charged particle interactions, on the other hand, have an accumulated field which formally diverges for large distances: charged particles are always under the simultaneous influence of (many) other charges and the 'collision' concept breaks down.

Let us first consider the few body collisions (figure 3.6). Practically speaking, 'few-body' means 'maximally three interaction partners', and two-body collisions are by far the most important. None the less, it is advantageous to start with a general discussion for an arbitrary number of collision partners. We consider a set of free particles and photons, the

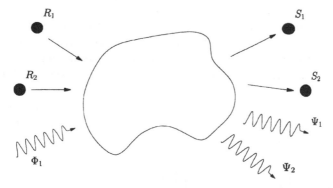

Figure 3.6. Schematic illustration of a few-body collision. Educt particles and photons enter the 'black box' reaction zone and are scattered into product particles and photons. The size of the reaction zone is small compared to the average interparticle distance, so that the particles can be considered as asymptotically free before and after the collision. Nothing specific is assumed about the interaction except the validity of the general laws of physics (conservation of nucleon identity, charge, momentum and total energy, principles of Galilei invariance and detailed balance).

educts $R_1, \ldots, R_M$ and $\Phi_1, \ldots, \Phi_{\bar{M}}$, referred to also by the indices $r_1, \ldots, r_M$ and $\phi_1, \ldots, \phi_{\bar{M}}$. They undergo an interaction for a finite time until they appear as particles or photons which are again free (the products $S_1, \ldots, S_N$ and $\Psi_1, \ldots, \Psi_{\bar{N}}$, referred to also by the indices $s_1, \ldots, r_N$ and $\psi_1, \ldots, \psi_{\bar{N}}$). Such a process reads in a chemical notation (a variety of other conventions exists):

$$R_1 + R_2 + \cdots + R_M + \Phi_1 + \cdots + \Phi_{\bar{M}}$$
$$\rightarrow S_1 + S_2 + \cdots + S_N + \Psi_1 + \cdots + \Psi_{\bar{N}}. \tag{3.76}$$

If the educts and products are the same particle set, one speaks of particle conserving collisions, otherwise of (chemical) reactions. If the sum of the kinetic energies before and after the collisions is the same, the collision is elastic, otherwise inelastic (subelastic for negative energy differences, super-elastic for positive ones). Chemical reactions are a particular kind of inelastic collisions.

The investigation of few-body collisions is the realm of scattering theory, which has been developed both within classical and quantum mechanics. In both descriptions, the scattering event is described by a certain probability p that is a function of the educt and product particle velocities. In classical mechanics, input and output states are considered as beams, and the stochastic character of the scattering is due to incomplete spatial information. In quantum theory, input and output are interpreted as eigenstates of the momentum operator, and the dynamic itself is genuinely stochastic. For the purpose of kinetic theory these differences do not matter: The scattering process is seen as a black box, subject only to the general laws of physics.

The stochastic view of the interaction implies that the scattering rate $\dot{n}$—the number of scattering events per volume and time, $[\dot{n}] = \mathrm{cm}^{-3}\,\mathrm{s}^{-1}$—is proportional to the density of the educt states. Assuming the absence of microscopic correlations ('molecular chaos'), this is the product of the phase space densities of the educt particles and the radiation intensities of the educt photons. (For each photon, a factor of $1/h\nu$ must be introduced to transform the radiation intensity into the corresponding photon flux.) Explicitly, the rate is calculated as

$$\dot{n} = \prod_{i=1}^{M} \int \mathrm{d}^3 v_{r_i} \prod_{i=1}^{\bar{M}} \int \mathrm{d}\Omega_{\phi_i} \int \mathrm{d}\nu_{\phi_i} \prod_{i=1}^{N} \int \mathrm{d}^3 v_{s_i} \prod_{i=1}^{\bar{N}} \int \mathrm{d}\Omega_{\psi_i} \int \mathrm{d}\nu_{\psi_i}$$

$$\times\, p(\vec{v}_{r_1}, \ldots, \nu_{\phi_{\bar{M}}}, \vec{e}_{\phi_{\bar{M}}}, \vec{v}_{s_1}, \ldots, \nu_{\psi_{\bar{N}}}, \vec{e}_{\psi_{\bar{N}}}) \prod_{i=1}^{M} f_{r_i}(\vec{v}_{r_i}) \prod_{i=1}^{\bar{M}} \frac{I_{\nu\phi_i}(\nu_{\phi_i}, \vec{e}_{\phi_i})}{h\nu}. \tag{3.77}$$

The physics of the scattering interaction is embodied in the factor p, which gives the probability of a certain educt state being scattered into a certain product state. Independent of the details of the interaction, one can state that it must equal zero for combinations of educt and product states that do not meet the laws of energy and momentum conservation. Utilizing that the momentum of the photons can be neglected, these laws read

$$\sum_{i=1}^{N}\left(\frac{1}{2}m_{s_i}\vec{v}_{s_i}^2 + \varepsilon_{s_i}\right) + \sum_{i=1}^{\bar{N}}h\nu_{\psi_i} = \sum_{i=1}^{M}\left(\frac{1}{2}m_{r_i}\vec{v}_{r_i}^2 + \varepsilon_{r_i}\right) + \sum_{i=1}^{\bar{M}}h\nu_{\phi_i} \quad (3.78)$$

$$\sum_{i=1}^{N}m_{s_i}\vec{v}_{s_i} = \sum_{i=1}^{M}m_{r_i}\vec{v}_{r_i}. \quad (3.79)$$

Consequently, the probability p must be the product of a kernel K and appropriate δ-functions,

$$p = K(\vec{v}_{r_1},\ldots,\vec{v}_{r_M},\nu_{\psi_1},\vec{e}_{\phi_1},\ldots,\nu_{\phi_{\bar{M}}},\vec{e}_{\phi_{\bar{M}}},\vec{v}_{s_1},\ldots,\vec{v}_{s_N},\vec{e}_{\psi_1},\ldots,\nu_{\psi_{\bar{N}}},\vec{e}_{\psi_{\bar{N}}})$$

$$\times \delta\left(\left(\sum_{i=1}^{M}\frac{1}{2}m_{r_i}\vec{v}_{r_i}^2 + \varepsilon_{r_i} + \sum_{i=1}^{\bar{M}}h\nu_{\phi_i}\right) - \left(\sum_{i=1}^{N}\frac{1}{2}m_{s_i}\vec{v}_{s_i}^2 + \varepsilon_{s_i} + \sum_{i=1}^{\bar{N}}h\nu_{\psi_i}\right)\right)$$

$$\times \delta^{(3)}\left(\sum_{i=1}^{M}m_{r_i}\vec{v}_{r_i} - \sum_{i=1}^{N}m_{s_i}\vec{v}_{s_i}\right). \quad (3.80)$$

The mathematical form of the kernel can be specified even further by noting that the scattering relation (3.77) must hold for every inertial system. In the non-relativistic formulation employed here, the kernel must be invariant against arbitrary Galilei transformations. These consist of rotations, which are given by an orthonormal matrix $\mathbf{T}$ and transform particles and photons as

$$\vec{v} \to \mathbf{T}\vec{v} \quad (3.81)$$

$$(\nu,\vec{e}) \to (\nu,\mathbf{T}\vec{e}). \quad (3.82)$$

and of translations by a velocity $\vec{V}$, which induce

$$\vec{v} \to \vec{v} + \vec{V} \quad (3.83)$$

$$(\nu,\vec{e}) \to \left(\nu(1 + \vec{e}\cdot\vec{V}/c),\vec{e}\right). \quad (3.84)$$

The form (3.80) and the invariances (3.81)–(3.84) are valid in the non-relativistic limit, i.e. for transformation with small speed and photon energies much below mc^2. In evaluating them, one typically encounters quadratic errors in v/c. This inconsistency may be healed, of course, by switching to a relativistic treatment of the kinematics and requiring invariance under Lorentz transformations. Traditionally, however, low temperature plasma physics employs Newtonian formulations.

Besides observing the energy and momentum conservation laws, the scattering probability must be symmetric against arbitrary permutations of the educt and the product variables among each other, and must obey the principle of detailed balance. (Quantum mechanically, the matrix elements of the reaction and the back reaction must be the same.) Furthermore, only those processes are possible where the total charge stays constant, and all atoms of the educt also appear in the products. These constraints do not appear as symmetries but are simply conditions for a non-vanishing K. Employing the integer vector view of the sum formula (see section 3.1), they can be formulated as

$$\sum_{i=1}^{N} S_{i,n} = \sum_{i=1}^{M} R_{i,n}, \qquad n = Z, H, \dots, U. \tag{3.85}$$

Very often, the kinematic state of the scattering educts is not important, only the event as such. It is then advantageous to introduce the absolute scattering probability P as the integral of the differential probability p,

$$P(\vec{v}_{r_1}, \dots, \nu_{\phi_{\bar{M}}}, \vec{e}_{\phi_{\bar{M}}}) = \prod_{i=1}^{N} \int d^3 v_{s_i} \prod_{i=1}^{\bar{N}} \int d\Omega_{\psi_i} \int d\nu_{\psi_i}$$

$$\times \, p(\vec{v}_{r_1}, \dots, \nu_{\phi_{\bar{M}}}, \vec{e}_{\phi_{\bar{M}}}, \vec{v}_{s_1}, \dots, \nu_{\psi_{\bar{N}}}, \vec{e}_{\psi_{\bar{N}}}). \tag{3.86}$$

In terms of this quantity, the scattering rate now reads

$$\dot{n} = \prod_{i=1}^{M} \int d^3 v_{r_i} \prod_{i=1}^{\bar{M}} \int d\Omega_{\phi_i} \int d\nu_{\phi_i} \, P(\vec{v}_{r_1}, \dots, \nu_{\phi_{\bar{M}}}, \vec{e}_{\phi_{\bar{M}}})$$

$$\times \prod_{i=1}^{M} f_{r_i}(\vec{v}_{r_i}) \prod_{i=1}^{\bar{M}} \frac{I_{\nu\phi_i}(\nu_{\phi_i}, \vec{e}_{\phi_i})}{h\nu}. \tag{3.87}$$

We will now leave the general discussion of the few-body collisions and proceed by describing the most important processes in some detail. The educts and products may be any combination of photons, electrons, neutrals, excited neutrals, and positive or negative ions. To refer to them, we employ the notation displayed in table 3.1. As the number of possible interactions increases drastically with the number of reactions partners, we will essentially restrict ourselves to one-body and two-body collisions. Collisions with three or even more interaction partners are relatively infrequent under normal conditions, and they will be addressed by only a few remarks.

The simplest case is the 'one-body collision', i.e. the spontaneous decay of an isolated particle. Such reactions are, of course, only possible for excited heavy particles; electrons and photons are stable. Typical examples are listed in table 3.2.

Table 3.1. Notation used for tables 3.2–3.5. Note the particular convention used for heavy particles. For example, AB refers to a molecule of constituents A and B, but A is not necessarily an atom but can be a molecule as well. (It may be also excited or ionized, for that matter.)

Symbol	Meaning	Remarks
e	Electron	
ϕ, ψ	Photon	
A, B, C	Heavy particle	Atomic or molecular, possibly excited or charged
AB	Molecule from constituents A, B	Possibly excited or charged
A^*	Electronically excited particle	Atomic or molecular, possibly additionally excited or charged
AB^v	Vibrationally excited molecule	Possibly additionally excited, possibly charged
A^+	Positive ion (cation)	Atomic or molecular, possibly excited
A^-	Negative ion (anion)	Atomic or molecular, possibly excited

Referring to the educt by the name R or the index r, the kinematic conservation rules of energy and momentum for a spontaneous decay are

$$\sum_{i=1}^{N} \left(\frac{1}{2} m_{s_i} \vec{v}_{s_i}^2 + \varepsilon_{s_i} \right) + \sum_{i=1}^{\bar{N}} h\nu_{\psi_i} = \frac{1}{2} m_r \vec{v}_r^2 + \varepsilon_r \qquad (3.88)$$

$$\sum_{i=1}^{N} m_{s_i} \vec{v}_{s_i} = m_r \vec{v}_r. \qquad (3.89)$$

The conservation laws of nucleon identity and charge read

$$\sum_{i=1}^{N} S_{i,n} = R_n, \qquad n = Z, H, \dots, U. \qquad (3.90)$$

Equation (3.86), specialized for the case of a spontaneous decay, states that the total scattering probability P can only depend on the particle velocity $\vec{v}$. There is, however, no possibility of constructing a Galilei invariant out of a

Table 3.2. Examples of 'one-body' or spontaneous decay processes.

Reaction	Description
$A^* \longrightarrow A + \phi$	Photonic de-excitation
$A^* \longrightarrow A^+ + e + \phi$	Auger effect (autoionization)
$AB^* \longrightarrow A + B$	Autodissociation
$AB^* \longrightarrow A + B + \phi$	Decay of excited dimers (e.g. in excimer lasers)
$AB^- \longrightarrow A + B + e$	Auto detachment

single velocity vector $\vec{v}$, and the dependence must actually vanish. This corresponds to the fact that the decay probability of an unstable particle is a constant, and a dimensional analysis shows that P must be identical to the inverse of the particle life time τ,

$$P = \frac{1}{\tau}. \tag{3.91}$$

The absolute decay rate can be calculated as

$$\dot{n} = \int \frac{1}{\tau} f_r(\vec{v}_r)\, \mathrm{d}^3 v_r = \frac{n_r}{\tau}. \tag{3.92}$$

To some extent, the differential scattering probability is determined from the constraints (3.88)–(3.90). When two educts result, their final energies are fixed (as are their momenta, up to an arbitrary rotation in the rest frame). Particularly in a photonic decay, the photon carries off (in an arbitrary direction) the full energy difference between the product and the educt state. (Doppler shift must be taken into account.) This is, of course, the basis of optical spectroscopy. If more than two educt particles are produced, their energies may have a statistical distribution.

Each spontaneous decay of an excited particle requires a preceding excitation. For some applications, it is reasonable to classify a process as spontaneous decay when the lifetime of the state is long enough so that the energy uncertainty $\Delta\varepsilon \approx h/\tau$ is negligible. In other situations, it may be advantageous to restrict the considerations to metastables. These are particles with a life-time long enough so that transport effects can occur; they exist for example in argon.

Next, we discuss the case of two-body interactions, where we distinguish between collisions of matter particles and interactions of a particle and a photon. We begin with the first, for which the conservation laws of energy and momentum read

$$\sum_{i=1}^{N} \left(\frac{1}{2} m_{s_i} \vec{v}_{s_i}^2 + \varepsilon_{s_i} \right) + \sum_{i=1}^{\bar{N}} h\nu_{\psi_i} = \frac{1}{2} m_{r_1} \vec{v}_{r_1}^2 + \varepsilon_{r_1} + \frac{1}{2} m_{r_2} \vec{v}_{r_2}^2 + \varepsilon_{r_2} \tag{3.93}$$

$$\sum_{i=1}^{N} m_{s_i} \vec{v}_{s_i} = m_{r_1} \vec{v}_{r_1} + m_{r_2} \vec{v}_{r_2} \tag{3.94}$$

and the conservation rules of charge and nucleon identity are

$$\sum_{i=1}^{N} S_{i,n} = R_{n,1} + R_{n,2}, \qquad n = Z, H, \ldots, U. \tag{3.95}$$

Equation (3.86) now states that the total scattering probability must be a function of $\vec{v}_{r_1}$ and $\vec{v}_{r_2}$. These velocities combine to only one possible Galilei invariant, namely the absolute value of their difference $g = |\vec{g}| = |\vec{v}_{r_1} - \vec{v}_{r_2}|$.

Dimensional considerations show that P must be a product of g and a factor σ which has the dimension of an area. This so-called total scattering cross section is in general a function of the difference velocity,

$$P(\vec{v}_{r_1}, \vec{v}_{r_2}) = |\vec{v}_{r_1} - \vec{v}_{r_2}| \sigma_t(|\vec{v}_{r_1} - \vec{v}_{r_2}|). \tag{3.96}$$

With the help of the total cross section, the reaction rate $\dot{n}$ can be calculated as

$$\dot{n} = \iint |\vec{v}_{r_1} - \vec{v}_{r_2}| \sigma_t f_{r_1}(\vec{v}_{r_1}) f_{r_2}(\vec{v}_{r_2}) \, \mathrm{d}^3 v_{r_1} \, \mathrm{d}^3 v_{r_2}. \tag{3.97}$$

The scattering relations become particularly transparent when the considered collisions are elastic. Switching to standard notation, two particles of mass m and M with initial velocities $\vec{v}$ and $\vec{V}$ are assumed to scatter into the final velocities $\vec{v}'$ and $\vec{V}'$. It is convenient to introduce as variables the center-of-mass velocity $\vec{w} = (m\vec{v} + M\vec{V})/(m + M)$ and difference velocity $\vec{g} = \vec{v} - \vec{V}$. Momentum is conserved when the center-of-mass velocities remain unchanged; energy conservation implies $|\vec{g}'| = |\vec{g}|$. The scattering probability p may thus be written as

$$p = \frac{\mathrm{d}\sigma}{\mathrm{d}\Omega}(g, \theta) \, \delta^{(3)}(\vec{w} - \vec{w}') \, \delta\left(\tfrac{1}{2}g^2 - \tfrac{1}{2}g'^2\right). \tag{3.98}$$

Galilei invariance demands that the differential cross section $\mathrm{d}\sigma/\mathrm{d}\Omega$ introduced by (3.98) may only depend on the absolute value g of the difference velocity and on the scattering angle $\theta = \angle(\vec{g}, \vec{g}')$. (See figure 3.7.) By inserting expression (3.98) into the two-body version of (3.86), and utilizing that the transformation from $(\vec{v}, \vec{V})$ to $(\vec{w}, \vec{g})$ has a Jacobian of unity, one arrives at

$$P = \int g \frac{\mathrm{d}\sigma}{\mathrm{d}\Omega} \, \mathrm{d}\Omega. \tag{3.99}$$

Comparison of this result with relation (3.96) shows that the total cross section of an elastic scattering process is the integral of the differential cross section over all scattering angles,

$$\sigma_t = \int \frac{\mathrm{d}\sigma}{\mathrm{d}\Omega} \, \mathrm{d}\Omega = 2\pi \int_0^\pi \frac{\mathrm{d}\sigma}{\mathrm{d}\Omega}(g, \theta) \sin\theta \, \mathrm{d}\theta. \tag{3.100}$$

The differential cross section represents the ratio of the scattering events (into a given solid angle element $\Delta\Omega$) to the incoming flux of collision partners. In general, $\mathrm{d}\sigma/\mathrm{d}\Omega$ is a complicated function of both arguments g and θ. For the limiting case of a 'hard sphere' potential (one that rises from zero to ∞ at a radius R), however, the cross section is constant and the scattering isotropic,

$$\frac{\mathrm{d}\sigma}{\mathrm{d}\Omega}(g, \theta) = \frac{\sigma_t}{4\pi} = \frac{\pi R^2}{4\pi}. \tag{3.101}$$

Isotropic scattering is a popular approximation for neutral–neutral interactions, where the potential is at least comparatively hard. The dependence

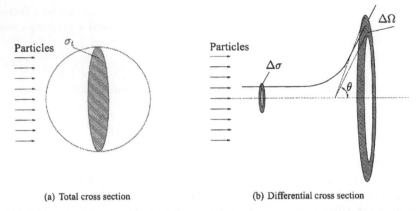

(a) Total cross section   (b) Differential cross section

Figure 3.7. Illustration of the total cross section σ_t (left) and the differential cross section $d\sigma/d\Omega$ (right). The total cross section is the ratio of the number of scattering events per particle, relative to the flux of incident interaction partners. The differential cross section measures the number of particles which are scattered into the solid angle element $\Delta\Omega = 2\pi \sin\theta\, \Delta\theta$. Note that $d\sigma/d\Omega$ is defined under more general conditions than σ_t but, if both exist, they are related by $\sigma_t = \int (d\sigma/d\Omega)\, d\Omega$.

on the velocity is often kept

$$\frac{d\sigma}{d\Omega}(g,\theta) = \frac{\sigma_t(g)}{4\pi}. \tag{3.102}$$

Softer potentials (which rise less drastically with decreasing distance) favor forward scattering. The extreme example is the very soft $\sim r^{-2}$ Coulomb potential. Using q and Q for the charges of the particles and m_R for their reduced mass, the corresponding Rutherford cross section reads

$$\frac{d\sigma}{d\Omega}(g,\theta) = \frac{q^2 Q^2}{(8\pi\varepsilon_0)^2 m_R^2 g^4 \sin^4(\theta/2)}. \tag{3.103}$$

The total cross sections calculated from this expression, however, are infinite, due to a divergence at small angles (large distances). This is the result again that charged particles are never really free. The proper treatment of Coulomb interactions will be discussed at the end of this section.

We now proceed to the inelastic two-body collisions, of which a large manifold of variants exist. The educt particles may be any combination of electrons, neutrals, excited neutrals, and positive or negative ions, only inelastic electron–electron collisions do not exist in the plasma energy range. The products may be an arbitrary number of particles plus possibly photons. Each of the 14 categories a–n in table 3.3 may be further divided into different reaction channels.

The following tables display a list of the most frequent types of inelastic two-body collisions, ordered with respect to their main source of energy.

Table 3.3. Overview on the possible inelastic two-body interactions. Except for inelastic electron–electron scattering which does not exist at non-relativistic energies, each combination is possible. Most of the categories actually represent several physically different reaction channels.

	Electron	Neutral	Excited	Cation	Anion
Electron	–	a	b	c	d
Neutral		e	f	g	h
Excited			i	j	k
Cation				l	m
Anion					n

Electron driven processes are contrasted with interactions that involve only heavy particles.

Processes driven by electron impact (a–d)

Electrons as the lightest, fastest and normally the most energetic particles are responsible for the bulk of the interactions in a plasma. The energy of the electrons is due to external fields (heating); sometimes also externally generated electrons (beams) play a role. Ionization is responsible for plasma generation, and, together with electronic excitation, dominates the energy balance. (Table 3.4.)

Table 3.4. Inelastic two-body interactions driven by electron impact. (For notation see table 3.1.)

Reaction	Description
$A + e \longrightarrow A^* + e$	Electron impact excitation
$AB + e \longrightarrow AB^v + e$	Electron impact vibrational excitation
$A^* + e \longrightarrow A + e$	Superelastic collision
$AB^v + e \longrightarrow AB + e$	Superelastic collision
$A + e \longrightarrow A^+ + e + e$	Electron impact ionization
$AB + e \longrightarrow A + B + e$	Electron impact dissociation
$AB + e \longrightarrow A^+ + B + e + e$	Electron impact dissociative ionization
$AB + e \longrightarrow A^+ + B^- + e$	Ion-pair production
$A + e \longrightarrow A^{-*} \longrightarrow A^- + \phi$	Dielectronic attachment
$AB + e \longrightarrow A^- + B$	Dissociative attachment
$A^+ + e \longrightarrow A + \phi$	Radiative recombination
$AB^+ + e \longrightarrow A + B$	Dissociative recombination
$A^- + e \longrightarrow A + e + e$	Electron detachment
$AB^- + e \longrightarrow A^- + B + e$	Electron impact dissociation of anions

Reactions among heavy particles (e–n)

Reactions among heavy particles can take many forms. A list—far from exhaustive—of inelastic and reactive heavy particle interactions is given in table 3.5. Some influence mainly the transport behavior and the energy content of the plasma, others alter the composition. Reactions that change the chemical identity of the particles are referred to as plasma chemistry.

Heavy particle reactions are typically driven by the internal energy of the reactants, sometimes (for example, during space craft reentry or in

Table 3.5. Inelastic and reactive two-body interactions between baryonic particles.

Reaction	Description
$A^+ + B \longrightarrow AB^+$	Polarization scattering (capture)
$A^- + B \longrightarrow AB^-$	Polarization scattering (capture)
$A^* + B \longrightarrow A + B + p$	Band resonance radiation, dipole radiation
$A + B \longrightarrow A + B^+ + e$	Ionization
$A + B \longrightarrow AB^+ + e$	Ionization
$A^* + B \longrightarrow A + B^+ + e$	Penning ionization
$A^* + B \longrightarrow AB^+ + e$	Penning ionization
$A + B \longrightarrow A^+ + B^-$	Electron capture
$A + B \longrightarrow A^* + B$	Excitation
$A^+ + B \longrightarrow A^+ + B^*$	Excitation
$AB + C \longrightarrow AB^v + C$	Vibrational excitation
$A^* + B \longrightarrow A + B^*$	Excitation exchange
$A^* + A \longrightarrow A + A^*$	Resonant excitation exchange
$AB^v + CD \longrightarrow AB + CD^v$	Vibrational excitation exchange
$A^+ + B \longrightarrow A + B^+$	Charge transfer
$A^+ + A \longrightarrow A + A^+$	Resonant charge transfer
$A^- + B \longrightarrow A + B^-$	Charge transfer
$A^- + A \longrightarrow A + A^-$	Resonant charge transfer
$A^+ + B^- \longrightarrow AB$	Positive-to-negative ion recombination
$A^+ + B^- \longrightarrow A + B$	Positive-to-negative ion recombination
$A^+ + B^- \longrightarrow A^* + B$	Positive-to-negative ion recombination with excitation
$AB^* + C \longrightarrow A + B + C$	Dissociation
$AB^+ + C \longrightarrow A + B + C^+$	Dissociation
$A^- + B \longrightarrow A + B + e$	Collisional detachment
$A^- + B \longrightarrow AB + e$	Associative detachment
$A^- + B^- \longrightarrow A^- + B + e$	Detachment
$A + BC \longrightarrow AB + C$	Chemical reaction
$A^+ + BC \longrightarrow AB^+ + C$	Chemical reaction
$AB + CD \longrightarrow AC + BD$	Chemical reaction
$AB + CD \longrightarrow ABC + D$	Chemical reaction
$A^* + B \longrightarrow AB$	Deexcitation, quenching, deactivation
$A^* + B \longrightarrow A + B$	Deexcitation, quenching, deactivation
$AB^v + C \longrightarrow AB + C$	Vibrational deactivation

other supersonic shocks) also by their kinetic energy. A particular case are the resonant reactions that occur between differently excited molecules of the same type (resonant excitation exchange), or between an ion and its parent molecule (resonant charge exchange). These processes do not require any reaction energy.

We now consider the two-body interactions of one particle, referred to as R, and one photon Ψ. The conservation laws of energy and momentum are

$$\sum_{i=1}^{N} \left(\frac{1}{2} m_{s_i} \vec{v}_{s_i}^2 + \varepsilon_{s_i} \right) + \sum_{i=1}^{\bar{N}} h\nu_{\psi_i} = \frac{1}{2} m_r \vec{v}_r^2 + \varepsilon_r + h\nu_\psi \qquad (3.104)$$

$$\sum_{i=1}^{N} m_{s_i} \vec{v}_{s_i} = m_r \vec{v}_r \qquad (3.105)$$

and the conservation rules of charge and nucleon identity read

$$\sum_{i=1}^{N} S_{i,n} = R_{l,n}, \qquad n = Z, II, \ldots, U. \qquad (3.106)$$

The total scattering probability depends on $\vec{v}_r$, ν_ψ, and $\vec{e}_\psi$. The only invariant combination of these quantities is $\nu(1 - \vec{e} \cdot \vec{v}/c)$, which is the frequency of the photon in the rest system of the particle. The scattering probability can thus be written in terms of a total cross section σ_t as

$$P(\vec{v}_r, \nu_\psi, \vec{e}_\psi) = \sigma_t \left(\left(1 - \frac{\vec{e}_\psi \cdot \vec{v}_r}{c} \right) \nu_\psi \right). \qquad (3.107)$$

Assuming that the particles are described by the distribution function $f_r(\vec{v}_r)$ and the photons by the radiation density $I_\nu(\nu_\psi, \vec{e}_\psi)$, the reaction rate $\dot{n}$ can be calculated. The result is easily understood: the reaction probability per particle is the flux of the incident photons times the cross section σ_t, integrated over all frequencies and directions. The reaction density is then obtained by integrating over the particle distribution. Note that the Doppler effect is correctly taken into account,

$$\dot{n} = \iiint \sigma_t \left(\left(1 - \frac{\vec{e}_\psi \cdot \vec{v}_r}{c} \right) \nu_\psi \right) f_r(\vec{v}_r) \frac{I_\nu(\nu_\phi, \vec{e}_\phi)}{h\nu} \, d^3 v_r \, d\Omega_\phi \, d\nu_\phi. \qquad (3.108)$$

Again, the kinematic relations become much more transparent for the case of elastic scattering. Consider a particle of mass m and velocity $\vec{v}$ scattering a photon of frequency ν and direction $\vec{e}$. The respective educt quantities are $\vec{v}'$, ν', and $\vec{e}'$. Evaluating (3.104) and (3.105) shows that the momentum of the particle and the energy of the photon are conserved; the only quantity that experiences a change is the direction of the photon. The differential scattering probability can thus be expressed in the following form, where the differential cross section $d\sigma/d\Omega$ may be a function of the reduced frequency and the scattering angle $\theta = \angle(\vec{e}, \vec{e}')$,

$$p = \frac{d\sigma}{d\Omega}\left(\left(1 - \frac{\vec{e}_\psi \cdot \vec{v}_r}{c}\right)\nu_\psi, \theta\right)\delta^{(3)}(\vec{v} - \vec{v}')\delta(\nu - \nu'). \tag{3.109}$$

The total cross section is again the angular integral of the differential cross section

$$\sigma_t = \int \frac{d\sigma}{d\Omega}\,d\Omega. \tag{3.110}$$

An example for elastic photon interaction is Thompson scattering at free electrons. With r_0 being the classical electron radius, the differential and the total cross section are

$$\frac{d\sigma}{d\Omega} = \frac{1}{2}r_0^2(1 + \cos^2\theta) \tag{3.111}$$

$$\sigma_t = \frac{8\pi}{3}r_0^2. \tag{3.112}$$

The momentum and the energy of massive particles remain, to a good approximation, uneffected by the elastic scattering of photons. Such unaffected processes thus have little dynamical influence in plasmas. They are, however, important for optical diagnostic methods. The scattering of photons by free electrons, e.g., underlies the method of Thompson scattering: the photons are provided by an external laser beam and the scattered light is measured with high angular and energy resolution. It is possible to determine the density and the distribution function of the free electrons in the plasma by evaluating the differential cross section (3.111) together with the second order in the photon energy shift, $\Delta\nu = \nu\vec{v}\cdot(\vec{e}' - \vec{e})/c$.

More important for the plasma dynamics itself are inelastic photon interactions, particularly the radiation driven reactions. Table 3.6 gives a selection of some important processes.

Table 3.6. Inelastic processes and reactions driven by radiation. (For notations see table 3.1.)

Reaction	Description
$A + \phi \longrightarrow A^*$	Photoexcitation, or bound-bound absorption
$AB + \phi \longrightarrow AB^v$	Vibrational photoexcitation
$A + \phi \longrightarrow A^+ + e$	Photoionization, or bound-free absorption
$AB + \phi \longrightarrow A + B$	Photodissociation
$AB + \phi \longrightarrow A^* + B + e$	Dissociative photoexcitation
$A + \phi \longrightarrow A + \phi'$	Luminescence, fluorescence, Raman scattering
$A^* + \phi \longrightarrow A + \phi + \phi$	Induced emission
$A^- + \phi \longrightarrow A + e$	Photo detachment
$AB^- + \phi \longrightarrow A + B + e$	Dissociative photo detachment

The processes listed in tables 3.2 to 3.6 are only a selection of the interactions possible in a plasma. When three-body (and higher) collisions are considered, the situation becomes even more complex. An exhaustive account which lists more than a hundred different types of many-body interactions is given in reference [7]. In plasmas that are maintained in gas mixtures such as air, the number of atomic and molecular species is typically large and the number of different scattering and reaction processes can easily be a few hundred.

The complete quantitative characterization of plasma dynamics is difficult. A first orientation may be provided by the following general rules. The principle of detailed balance states that the matrix elements of a reaction and its back reaction must coincide. If radiation is included, this extends to a relation between the coefficients of absorption, emission, and spontaneous emission. Typically, inelastic processes are less likely than elastic collisions (in a semi-classical picture, the motion of the nuclei is adiabatic). Radiative transitions are less likely than non-radiative ones. Three-body events are often negligible. (A counter-example is third-body assisted recombination; non-radiative two-body recombination is often suppressed by energy and momentum conservation.) Two-photon processes take place only at very high radiation densities.

For more specific information, one can either turn to theory or to experiment. True first principle calculations are difficult, and empirically found data are seldom complete. As a rule, one can state that angular resolved information on the products is difficult to obtain, so that the total reaction cross section σ_t becomes the preferred data format. Frequently, even that information is missing, and only empirical reaction rates are available, often expressed in terms of Arrhenius' formula. The lack of reaction data is a serious problem for all modeling efforts. The body of knowledge, however, is in rapid growth; many gases—particularly those of technical importance—are already well characterized, and new data are added on a regular basis. (See reference [11] for a start.)

We now turn to the Coulomb interactions which cannot be described as collisions in the strict sense. Instead, a charged particle is simultaneously influenced by many other charges. For a rough consideration these 'field charges' may be divided into three groups: (a) a small number of charges inside the strong interaction zone $r \approx r_c$ (on average less than one, the probability scales $\sim \Lambda^{-1}$), (b) a relatively large number that are in a Debye sphere $r_c < r < \lambda_D$ (this number scales like $\sim \Lambda^2$), and (c) the other charges beyond the Debye radius (in effect infinitely many).

Each group of field particles influences the test particle differently: the close encounters—set (a)—change its momentum vector drastically, similar to a hard sphere collision. The absolute frequency of these events is, however, not very high, and their effect is masked by the influence of group (b). Each of the (b) particles induces only a small velocity change, but their simultaneous action gives rise to a substantial stochastic acceleration, describable as a 'random walk in velocity space'. Particles (c) also have a measurable influence, but due to the large number their contribution loses its statistical nature. The resulting average is, in fact, a regular acceleration which is contained in the self-consistent fields calculated from (3.72) and (3.73).

As the final topic in the section, we discuss very briefly the interaction of plasma particles with material objects, such as electrodes, walls, or substrates. Their reaction rates are often substantial. Surfaces (solids or fluids) have a high density of available quantum states. In addition, surfaces are connected to a large sink of energy and momentum. This has the consequence that surface reactions are not subject to any selection rules. The detailed study of these processes is the subject of a separate science, plasma surface chemistry, which has established a huge body of knowledge (particularly within the past decade). See reference [2] for a start.

Electrons are always absorbed by the material surfaces. In metals, they enter the Fermi reservoir; in insulators, they occupy the surface states and accumulate. Typically, their flux is much higher than that of any other species, so that a negative 'floating potential' develops which is a few times the electron thermal voltage T_e/e. (It can be much higher when a dc or rf bias is applied.) The plasma, in turn, reacts to the wall potential with the formation of a plasma boundary sheath, a positive space charge zone with a strong wall-pointing electrical field. For details see [9].

Positive ions which enter the sheath are accelerated to the wall, and are very likely to reach it. Very often, the wall forms the most prominent sink. Close to the wall, the cations are neutralized by electrons tunneling into the unoccupied quantum states. (When unbound states are accessed, free electrons can be generated which escape into the plasma: this is secondary electron emission.) The former ion, now a fast neutral, continues its trajectory onto the surface.

Negative ions are repelled by the field of the sheath and reflected, as their energy per charge unit is typically much less than the floating potential. They tend to accumulate within the plasma, waiting to be neutralized either by recombination or by detachment. Only when the wall potential vanishes (for example, in the afterglow phase of pulsed plasmas), negative ions can recombine at the walls.

The fate of a neutral that reaches a surface depends strongly on the characteristics of both partners. One factor is the available energy. Excited species, radicals and particles with a high impact energy are relatively reactive; saturated or thermal ones are often simply reflected. A surface of high temperature is more reactive than a cold one. Other factors are of chemical nature.

Particles may also be emitted from a surface. Electrons can be liberated by ions, by radiation, by a strong electric field, or thermally from heated surfaces. Neutrals can be generated either by the impact of other particles (e.g. sputtering, desorption), they can appear as free products of a chemical reaction (e.g. etching), or the material may decompose due to thermal effects (evaporation). Ion production at the surfaces is usually not important.

3.3 The Kinetic Equation

Section 3.1 discussed the various particles that are present in a partially ionized plasma and introduced the one-particle distribution function f to describe their state (at a given time t). Section 3.2 gave a physical account of the forces that influence these particles, originating both from external fields and from their mutual interaction. This section now combines the two lines of thought and describes how the forces and interactions change the distribution function over time. The mathematical formulation of this is called kinetic equation. In its most compact form, it states that the convective or laminar term is equal to the collision term and reads as follows:

$$\frac{\mathrm{d}f_s}{\mathrm{d}t} = \langle f \rangle_s, \quad s = 1, \ldots, N. \tag{3.113}$$

This, of course, must be explained. The convective term on the left is given by a total derivative; it denotes the temporal change of f evaluated with respect to a moving frame of reference:

$$\frac{\mathrm{d}f_s}{\mathrm{d}t}(\vec{r}, \vec{v}, t) = \frac{\partial f_s}{\partial t} + \frac{\mathrm{d}\vec{r}}{\mathrm{d}t} \cdot \frac{\partial f_s}{\partial \vec{r}} + \frac{\mathrm{d}\vec{v}}{\mathrm{d}t} \cdot \frac{\partial f_s}{\partial \vec{v}}$$

$$= \frac{\partial f_s}{\partial t} + \vec{v} \cdot \frac{\partial f_s}{\partial \vec{r}} + \vec{a}_s \cdot \frac{\partial f_s}{\partial \vec{v}}. \tag{3.114}$$

The motion of the reference frame is defined by Newton's equations, evaluated with the external and the self-consistent fields, equations (3.70)–(3.75). For charged particles, it reads

$$\frac{\mathrm{d}f_\alpha}{\mathrm{d}t}(\vec{r}, \vec{v}, t) = \frac{\partial f_\alpha}{\partial t} + \vec{v} \cdot \frac{\partial f_\alpha}{\partial \vec{r}} + \left(\vec{g} + \frac{q_\alpha}{m_\alpha}(\vec{E} + \vec{v} \times \vec{B}) \right) \cdot \frac{\partial f_\alpha}{\partial \vec{v}}. \tag{3.115}$$

For neutrals it is simply

$$\frac{\mathrm{d}f_a}{\mathrm{d}t}(\vec{r}, \vec{v}, t) = \frac{\partial f_a}{\partial t} + \vec{v} \cdot \frac{\partial f_a}{\partial \vec{r}} + \vec{g} \cdot \frac{\partial f_a}{\partial \vec{v}}. \tag{3.116}$$

Equation (3.113) states that, up to the 'action of the collisions', the distribution function is temporally constant in the co-moving frame. This statement may be understood with the help of figure 3.8, which shows the temporal evolution of a phase space element ΔV according to the laminar term. Assuming that the equations of motion arise from a Hamiltonian—true for equations (3.70) and (3.71)—the volume of the phase space element is constant over time. (Its form, of course, will change!) Collisions absent, the particles also follow the equations (3.70) and (3.71), implying that all particles present in the element ΔV at t_0 will end up in $\Delta V'$ at t_1. Their total number ΔN is thus conserved. The phase space density, being the ratio of ΔN and ΔV, is then also a constant in time.

This fact is often stated by saying 'the phase space density behaves like an incompressible fluid'. For a more direct verification of this analogy, one may also note that the total derivative in (3.113) can be written as a partial derivative plus the divergence of a flux in the phase space,

$$\frac{df_s}{dt}(\vec{r}, \vec{v}, t) = \frac{\partial f_s}{\partial t} + \vec{v} \cdot \frac{\partial f_s}{\partial \vec{r}} + \vec{a}_s \cdot \frac{\partial f_s}{\partial \vec{v}}$$

$$= \frac{\partial f_s}{\partial t} + \frac{\partial}{\partial \vec{r}} \cdot (\vec{v} f_s) + \frac{\partial}{\partial \vec{r}} \cdot (\vec{a}_s f_s). \tag{3.117}$$

This is, if set equal to zero, similar to the fluid-dynamical equation of continuity. Its derivation uses the 'phase space analogy' of the incompressibility condition $\nabla \cdot \vec{v} = 0$, but this relation is, of course, not an equation of state but a consequence of the Hamiltonian nature of the dynamics,

$$\frac{\partial}{\partial \vec{r}} \cdot (\vec{v}) + \frac{\partial}{\partial \vec{v}} \cdot \left(\vec{g} + \frac{q_\alpha}{m_\alpha}(\vec{E} + \vec{v} \times \vec{B}) \right) = 0. \tag{3.118}$$

The term $\langle f \rangle_s$ on the right of equation (3.113) represents all forces and interactions that are not accounted for by the external and self-consistent forces. Summarily referred to as 'collisions', these interactions scatter particles in and out of the co-moving phase space volume. (See figure 3.8.) The scattering of a particle into the phase space element ΔV corresponds to a 'gain' process, a scattering out of the element counts as a 'loss'.

The collision term in (3.113) is just a symbol, in contrast to the explicitly displayed convective term. In reality, it is a quite complicated sum of several contributions, each of which corresponds to one of the interaction processes discussed in section 3.2. All contributions have in common that they are local in the spatial dependence, i.e. act only on the velocity part of f: only particles at the same position can collide, and they only experience a change in velocity, not a change in position. (That is, when they keep their identity. In chemical reactions they may locally appear or disappear.) The dependence on $\vec{r}$ and t will be suppressed in the further notation.

It is advantageous to divide the collision term contributions into three physically distinct groups. The first corresponds to the most frequent interactions, the elastic two-body collisions; the second represents all other few-body collisions (including the interaction with radiation); and the last represents the Coulomb interaction. We use the subscripts *el*, *in*, and *cb*, respectively,

$$\langle f \rangle_\alpha(\vec{v}) = \langle f \rangle_{el,\alpha} + \langle f \rangle_{cb,\alpha} + \langle f \rangle_{in,\alpha}. \tag{3.119}$$

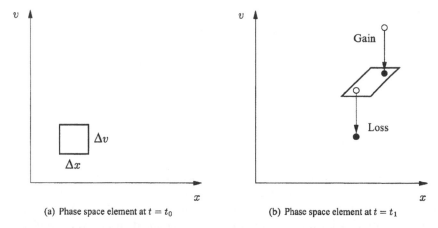

(a) Phase space element at $t = t_0$ (b) Phase space element at $t = t_1$

Figure 3.8. Schematic illustration of the kinetic equation (3.113). Shown is a phase space element ΔV which evolves according to the equations of motion, keeping its volume constant but not the shape. (This is a consequence of the Hamiltonian nature of the dynamics.) Under the action of the convection term, the particles move in the same fashion so that the phase space density is conserved. The collision term on the right of (3.113) scatters individual particles into or out of ΔV, giving rise to gains or losses, respectively. As shown in the figure, a particle-conserving collision is represented by a translation along the velocity axis; the spatial position remains unchanged. In addition, there are chemical reactions which create or destroy particles.

In the case of neutral particles there are, of course, no Coulomb interactions,

$$\langle f \rangle_a(\vec{v}) = \langle f \rangle_{\text{el},a} + \langle f \rangle_{\text{in},a}. \tag{3.120}$$

The radiation intensity I_ν was introduced above as the photon analog of the distribution function, with the particular situation of massless particles taken into account. The analogy can be carried further to the photon equivalent of the kinetic equation, termed the radiation transport equation. It is also a scalar partial differential equation of first order, with a somewhat different appearance. The differences are partially due to physics (photons propagate with a constant speed c, so that $\vec{v} = c\vec{e}$ and acceleration terms are missing) and partially due to convention (the radiation intensity refers to energy flux, not to photon number, and all terms are divided by c),

$$\frac{1}{c} \frac{\partial I_\nu}{\partial t} + \vec{e} \cdot \nabla I_\nu = \varepsilon_\nu - \kappa_\nu I_\nu. \tag{3.121}$$

Similar as in the kinetic equation, the terms on the left describe the propagation of the photons. The term $\vec{e} \cdot \nabla I_\nu$ is called the streaming term. The expressions on the right represent the interaction of the radiation with other plasma constituents. As stated, photon–photon interaction does not exist in the energy regime under discussion. The quantity ε_ν represents emission, κ_ν denotes absorption. These quantities are here defined with respect to the volume, $[\varepsilon_\nu] = \mathrm{W}/\mathrm{Hz}\,\mathrm{m}^3$, $[\kappa_\nu] = 1/\mathrm{m}$. A sometimes employed alternative definition introduces emission and absorption coefficients per mass element; practically, this corresponds to a substitution $\kappa_\nu \to \rho\kappa_\nu$ and $\varepsilon_\nu \to \rho\varepsilon_\nu$ in (3.121). Note that both the absorption and emission coefficient are functions which in general depend on the time t, the position $\vec{r}$, the propagation direction $\vec{e}$, and the frequency ν. Particularly the latter dependence often shows very narrow and complex structures.

As discussed in section 3.2, several different elementary processes contribute to the interaction of photons with matter. From the radiation transport point of view, one distinguishes between emission (a photon is generated), absorption (a photon is captured), and scattering (an absorption occurs but a secondary photon appears with negligible time delay). Scattering is further divided into elastic scattering and inelastic scattering. A particular type of scattering is induced emission, where the incident photon is replaced by two photons of the same direction and energy.

The physical meaning of the kinetic equation and the radiation transport equation can be illustrated with the help of the appropriate moment equations. For the particles, one multiplies the kinetic equation with the combination $v_{\mu_1}, v_{\mu_2}, \ldots, v_{\mu_n}$ and integrates over all velocities, to obtain

$$\frac{\partial}{\partial t} M^n_{s,\mu_1,\mu_2,\ldots,\mu_n} + \nabla \cdot \vec{\Gamma}^n_{s,\mu_1,\mu_2,\ldots,\mu_n}$$

$$= \sum_{i=1}^{n} \int v_{\mu_1}, \ldots, v_{\mu_{i-1}} a_{s\mu_i} v_{\mu_{i+1}}, \ldots, v_{\mu_n} f_s \, \mathrm{d}^3 v$$

$$+ \int v_{\mu_1}, v_{\mu_2}, \ldots, v_{\mu_n} \langle f \rangle_s \, \mathrm{d}^3 v. \tag{3.122}$$

The two terms on the left of (3.122) were already substituted using definitions (3.12) and (3.14). The first is the time derivative of the moment M^n_s of order n, the second is the divergence of the corresponding flux. By structure it is a derivative combination of moments of order $n + 1$. The first term on the right represents the action of the macroscopic field. It is a linear combination of moments of the order $n - 1$ and n. (The latter is only present when a magnetic field is included.) The second term on the right represents the change in the moment due to the action of the collisions. For this contribution we introduce the notation

$$\dot{M}_{s,\mu_1,\mu_2,\ldots,\mu_n} = \int v_{\mu_1}, v_{\mu_2}, \ldots, v_{\mu_n} \langle f \rangle_s \, \mathrm{d}^3 v. \tag{3.123}$$

Equation (3.122) shows that the time derivate of the moment M^n of order n is related to the divergence of the corresponding flux $\vec{\Gamma}^n$, i.e. to a moment of the order $n + 1$. The balances thus form an infinite chain of coupled equations which are together equivalent to the original equation itself. Only if the chain of equations is terminated after a certain stage (using additional assumptions), a simpler plasma model may be derived. We will return to this question in section 3.4.

Here we will employ the first three moment equations, corresponding to the balances of particle number, momentum, and energy. With the definitions of section 3.1 they read

$$\frac{\partial n_s}{\partial t} + \nabla \cdot \vec{\Gamma}_s^n = \dot{n}_s \tag{3.124}$$

$$\frac{\partial \vec{p}_s}{\partial t} + \nabla \cdot \mathbf{\Pi}_s = (m_s \vec{g} + q_s(\vec{E} + \vec{u}_s \times \vec{B}))n_s + \dot{\vec{p}}_s \tag{3.125}$$

$$\frac{\partial e_s}{\partial t} + \nabla \cdot \vec{\Gamma}_s^e = (m_s \vec{g} + q_s \vec{E}) \cdot \vec{u}_s n_s + \dot{e}_s. \tag{3.126}$$

As in the general form, the terms on the left are the derivative of the considered moment and the divergence of the corresponding flux. The field term vanishes for the particle balance; it represents the acceleration in the momentum balance and the related power density in the energy equation. The production densities of particle number, momentum, and energy are explicitly

$$\dot{n}_s(\vec{r}, t) = \int \langle f \rangle_s \, d^3 v \tag{3.127}$$

$$\dot{\vec{p}}_s(\vec{r}, t) = \int m_s \vec{v}_s \langle f \rangle_s \, d^3 v \tag{3.128}$$

$$\dot{e}_s(\vec{r}, t) = \int \frac{1}{2} m_s v^2 \langle f \rangle_s \, d^3 v. \tag{3.129}$$

In analogy to the balances of the particles we now derive the balance equations of the photons. By integrating the radiation transport equation over the total solid angle 4π and invoking the definitions of the spectral energy density u_ν and the energy flux $\vec{F}_\nu$—see (3.52) and (3.53)—we obtain the spectral energy balance

$$\frac{\partial u_\nu}{\partial t} + \nabla \cdot \vec{F}_\nu = \int \varepsilon_\nu \, d\Omega - \int \kappa_\nu I_\nu \, d\Omega. \tag{3.130}$$

Integrating this expression further over the full frequency range gives the total energy balance, with the emissions counting as gains and the absorption as losses,

$$\frac{\partial u}{\partial t} + \nabla \cdot \vec{F} = \dot{e}^G - \dot{e}^L \tag{3.131}$$

$$\dot{e}^G = \iint \varepsilon_\nu \, d\Omega \, d\nu \tag{3.132}$$

$$\dot{e}^L = \iint \kappa_\nu I_\nu \, d\Omega \, d\nu. \tag{3.133}$$

Integrating (3.130) with the weight $1/h\nu$ yields the photon number balance, with the corresponding gain and loss terms on the right,

$$\frac{\partial n}{\partial t} + \nabla \cdot \vec{\Gamma} = \dot{n}^G - \dot{n}^L \tag{3.134}$$

$$\dot{n}^G = \iint \frac{1}{h\nu} \varepsilon_\nu \, d\Omega \, d\nu \tag{3.135}$$

$$\dot{n}^L = -\iint \frac{1}{h\nu} \kappa_\nu I_\nu \, d\Omega \, d\nu. \tag{3.136}$$

Having established the framework of particle and radiation transport, we now proceed with an explicit discussion of the interaction terms. We first concentrate on the few-body collisions in general, of which elastic scattering is a special case. In section 3.2, the total scattering rate of M educt particles $R_1, \ldots, R_M$ and $\bar{M}$ educt photons $\Phi_1, \ldots, \Phi_{\bar{M}}$ (referenced by $r_1, \ldots, r_M, \phi_1, \ldots, \phi_{\bar{M}}$) into the product $S_1, \ldots, S_N, \Psi_1, \ldots, \Psi_{\bar{N}}$ (referenced by $s_1, \ldots, s_N, \psi_1, \ldots, \psi_{\bar{N}}$) was described by (3.77). It is repeated here for convenience:

$$\dot{n} = \prod_{i=1}^{M} \int d^3 v_{r_i} \prod_{i=1}^{\bar{M}} \int d\Omega_{\phi_i} \int d\nu_{\phi_i} \prod_{i=1}^{N} \int d^3 v_{s_i} \prod_{i=1}^{\bar{N}} \int d\Omega_{\psi_i} \int d\nu_{\psi_i}$$

$$\times \, p(\vec{v}_{r_1}, \ldots, \nu_{\phi_{\bar{M}}}, \vec{e}_{\phi_{\bar{M}}}, \vec{v}_{s_1}, \ldots, \nu_{\psi_{\bar{N}}}, \vec{e}_{\psi_{\bar{N}}}) \times \prod_{i=1}^{M} f_{r_i}(\vec{v}_{r_i}) \prod_{i=1}^{\bar{M}} \frac{I_{\nu\phi_i}(\nu_{\phi_i}, \vec{e}_{\phi_i})}{h\nu}.$$

The integrand of this expression can be interpreted as the rate of scattering per element of phase space $d^3 v$ (for the particles) and per frequency interval $d\nu$ and solid angle $d\Omega$ (for the photons). Each scattering event means a loss of educt particles and a gain of products, represented by a corresponding loss or gain term on the right side of the kinetic equation. For a particular educt R_k, the loss rate L in phase space due to a process is calculated by integrating the scattering rate over the velocity coordinates of *all other* educts and over *all* products,

$$L_{r_k}(\vec{v}_{r_k}) = \prod_{i=1, i \neq k}^{M} \int d^3 v_{r_i} \prod_{i=1}^{\bar{M}} \int d\Omega_{\phi_i} \int d\nu_{\phi_i} \prod_{i=1}^{N} \int d^3 v_{s_i} \prod_{i=1}^{\bar{N}} \int d\Omega_{\psi_i} \int d\nu_{\psi_i}$$

$$\times \, p(\vec{v}_{r_1}, \ldots, \nu_{\phi_{\bar{M}}}, \vec{e}_{\phi_{\bar{M}}}, \vec{v}_{s_1}, \ldots, \nu_{\psi_{\bar{N}}}, \vec{e}_{\psi_{\bar{N}}})$$

$$\times \prod_{i=1}^{M} f_{r_i}(\vec{v}_{r_i}) \prod_{i=1}^{\bar{M}} \frac{I_{\nu\phi_i}(\nu_{\phi_i}, \vec{e}_{\phi_i})}{h\nu}. \tag{3.137}$$

By dropping the factor $f_{r_k}(\vec{v}_{r_k})$ in this formula, one arrives at a notion which expresses the particle loss per educt particle R_k. This quantity is often termed the specific loss frequency

$$\nu^L_{r_k}(\vec{v}_{r_k}) = \prod_{i=1, i\neq k}^{M} \int d^3v_{r_i} \prod_{i=1}^{\bar{M}} \int d\Omega_{\phi_i} \int d\nu_{\phi_i} \prod_{i=1}^{N} \int d^3v_{s_i} \prod_{i=1}^{\bar{N}} \int d\Omega_{\psi_i} \int d\nu_{\psi_i}$$

$$\times \, p(\vec{v}_{r_1}, \ldots, \nu_{\phi_{\bar{M}}}, \vec{e}_{\phi_{\bar{M}}}, \vec{v}_{s_1}, \ldots, \nu_{\psi_{\bar{N}}}, \vec{e}_{\psi_{\bar{N}}})$$

$$\times \prod_{i=1, i\neq k}^{M} f_{r_i}(\vec{v}_{r_i}) \prod_{i=1}^{\bar{M}} \frac{I_{\nu\phi_i}(\nu_{\phi_i}, \vec{e}_{\phi_i})}{h\nu}. \tag{3.138}$$

The relation between the loss rate and the loss frequency is, of course,

$$L_{r_k}(\vec{v}_{r_k}) = \nu^L_{r_k}(\vec{v}_{r_k}) f_{r_k}(\vec{v}_{r_k}). \tag{3.139}$$

Both quantities can be utilized to calculate the total scattering rate:

$$\dot{n} = \int L_{r_k}(\vec{v}_{r_k}) \, d^3v_{r_k} = \int \nu^L_{r_k}(\vec{v}_{r_k}) f_{r_k}(\vec{v}_{r_k}) \, d^3v_{r_k}. \tag{3.140}$$

As the term L_{r_k} refers to the losses of particle species R_k, the respective contributions to the balances of particle number, momentum, and energy must be counted as *negative*,

$$\dot{n}^L_{r_k} = -\int L_{r_k} \, d^3v_{r_k} \equiv -\dot{n} \tag{3.141}$$

$$\dot{\vec{p}}^L_{r_k} = -\int m_{r_k} \vec{v}_{r_k} L_{r_k} \, d^3v_{r_k} \tag{3.142}$$

$$\dot{e}^L_{r_k} = -\int \frac{1}{2} m_{r_k} v^2_{r_k} L_{r_k} \, d^3v_{r_k}. \tag{3.143}$$

Similar considerations can be made for a product particle S_l. To calculate the total gain rate G, the integration must be performed over *all* educt variables and *all other* product variables. (Note that the definition of a specific gain frequency is not possible.)

$$G_{s_l}(\vec{v}_{s_l}) = \prod_{i=1}^{M} \int d^3v_{r_i} \prod_{i=1}^{\bar{M}} \int d\Omega_{\phi_i} \int d\nu_{\phi_i} \prod_{i=1, i\neq l}^{N} \int d^3v_{s_i} \prod_{i=1}^{\bar{N}} \int d\Omega_{\psi_i} \int d\nu_{\psi_i}$$

$$\times \, p(\vec{v}_{r_1}, \ldots, \nu_{\phi_{\bar{M}}}, \vec{e}_{\phi_{\bar{M}}}, \vec{v}_{s_1}, \ldots, \nu_{\psi_{\bar{N}}}, \vec{e}_{\psi_{\bar{N}}})$$

$$\times \prod_{i=1}^{M} f_{r_i}(\vec{v}_{r_i}) \prod_{i=1}^{\bar{M}} \frac{I_{\nu\phi_i}(\nu_{\phi_i}, \vec{e}_{\phi_i})}{h\nu}. \tag{3.144}$$

Performing the final integration gives again the scattering rate

$$\dot{n} = \int G_{r_l}(\vec{v}_{r_l}) \, d^3v_{r_l}. \tag{3.145}$$

As gains, the contributions to particle number, momentum, and energy are *positive*,

$$\dot{n}_{s_l}^G = \int G_{s_l} \, d^3 v_{s_l} \equiv \dot{n} \qquad (3.146)$$

$$\dot{\vec{p}}_{s_l}^G = \int m_{s_l} \vec{v}_{s_l} \, G_{s_l} \, d^3 v_{s_l} \qquad (3.147)$$

$$\dot{e}_{s_k}^G = \int \frac{1}{2} m_{s_l} v_{s_l}^2 \, G_{r_k} \, d^3 v_{s_l}. \qquad (3.148)$$

We now turn to the photons. By integrating the phase space resolved scattering rate over all product variables and over all educt variables but those of photon Φ_k and dividing by the factor $I_{\nu \phi_k}/h\nu_{\phi_k}$, we obtain the absorption coefficient of the considered process,

$$\kappa = \prod_{i=1}^{M} \int d^3 v_{r_i} \prod_{i=1, i \neq k}^{\bar{M}} \int d\Omega_{\phi_i} \int d\nu_{\phi_i} \prod_{i=1}^{N} \int d^3 v_{s_i} \prod_{i=1}^{\bar{N}} \int d\Omega_{\psi_i} \int d\nu_{\psi_i}$$

$$\times \, p(\vec{v}_{r_1}, \dots, \nu_{\phi_{\bar{M}}}, \vec{e}_{\phi_{\bar{M}}}, \vec{v}_{s_1}, \dots, \nu_{\psi_{\bar{N}}}, \vec{e}_{\psi_{\bar{N}}})$$

$$\times \prod_{i=1}^{M} f_{r_i}(\vec{v}_{r_i}) \prod_{i=1, i \neq k}^{\bar{M}} \frac{I_{\nu \phi_i}(\nu_{\phi_i}, \vec{e}_{\phi_i})}{h\nu}. \qquad (3.149)$$

Performing the missing integrations gives again the total reaction rate $\dot{n}$ according to equation (3.77). This justifies the interpretation of (3.149) as the coefficient κ.

$$\dot{n} = \iint \kappa \frac{I_{\nu \phi_i}(\nu_{\phi_i}, \vec{e}_{\phi_i})}{h\nu} \, d\Omega_{\phi_k} \, d\nu_{\phi_k}. \qquad (3.150)$$

In a similar way, by integrating the phase space resolved scattering rate over all educt variables and over all product variables but those of photon Ψ_k, we obtain the emission coefficient,

$$\varepsilon = \prod_{i=1}^{M} \int d^3 v_{r_i} \prod_{i=1}^{\bar{M}} \int d\Omega_{\phi_i} \int d\nu_{\phi_i} \prod_{i=1}^{N} \int d^3 v_{s_i} \prod_{i=1, i \neq k}^{\bar{N}} \int d\Omega_{\psi_i} \int d\nu_{\psi_i}$$

$$\times \, p(\vec{v}_{r_1}, \dots, \nu_{\phi_{\bar{M}}}, \vec{e}_{\phi_{\bar{M}}}, \vec{v}_{s_1}, \dots, \nu_{\psi_{\bar{N}}}, \vec{e}_{\psi_{\bar{N}}})$$

$$\times \prod_{i=1}^{M} f_{r_i}(\vec{v}_{r_i}) \prod_{i=1}^{\bar{M}} \frac{I_{\nu \phi_i}(\nu_{\phi_i}, \vec{e}_{\phi_i})}{h\nu}. \qquad (3.151)$$

The corresponding total reaction rate has the following form, also demonstrating that the interpretation of (3.151) as emission coefficient is correct:

$$\dot{n} = \int \varepsilon \, d\Omega_{\phi_i} \int d\nu_{\phi_i}. \qquad (3.152)$$

We now consider two special cases of the general few-body formalism, namely the elastic scattering of two particles and the elastic scattering of particle and a photon. The first situation is the one originally investigated by Boltzmann. Employing the probability formula (3.98) and combining the loss and gain term into one formula gives

$$
\langle f_r | f_s \rangle_{el}(\vec{v}) = \iiint g \frac{d\sigma}{d\Omega}\bigg|_{rs} f_r\left(\vec{v} - \frac{m_s}{m_r + m_s}\vec{g} - \frac{m_r}{m_r + m_s}\vec{g}'\right)
$$

$$
\times f_s\left(\vec{v} + \frac{m_r}{m_r + m_s}\vec{g} - \frac{m_r}{m_r + m_s}\vec{g}'\right) g^2 \, dg \, d\Omega \, d\Omega'
$$

$$
- \iiint g \frac{d\sigma}{d\Omega}\bigg|_{rs} f_r(\vec{v} - \vec{g}) f_s(\vec{v}) \, g^2 \, dg \, d\Omega \, d\Omega' \tag{3.153}
$$

In this expression, $\vec{g} = g\vec{e}$ and $\vec{g}' = g\vec{e}'$ are the difference velocities before and after the collision, θ is the scattering angle $\angle(\vec{e}, \vec{e}')$, and the cross section $(d\sigma/d\Omega)|_{rs}$ is a function of g and θ, symmetric with respect to the indices r and s,

$$
\frac{d\sigma}{d\Omega}\bigg|_{rs}(g, \theta) = \frac{d\sigma}{d\Omega}\bigg|_{sr}(g, \theta). \tag{3.154}
$$

The elastic collision term is subject to the conservation of particle number, momentum, and energy. Particle conservation holds for each species separately, as the elastic collisions do not affect the identity of each particle,

$$
\dot{n}_{rs} = \int \langle f_r | f_s \rangle_{el,rs} \, d^3v = 0. \tag{3.155}
$$

Energy and momentum, on the other hand, can be exchanged between the species, so that the conservation of these quantities is expressed as the anti-symmetry of the production terms,

$$
\dot{p}_{rs} = \int m_s \vec{v} \langle f_r | f_s \rangle_{el,rs} \, d^3v
$$

$$
= - \iint \frac{m_r m_s}{m_r + m_s} \vec{g} g \sigma_{m,rs} f_r\left(\vec{v} - \frac{m_s}{m_r + m_s}\vec{g}\right)
$$

$$
\times f_s\left(\vec{v} + \frac{m_r}{m_r + m_s}\vec{g}\right) d^3g \, d^3v = -\dot{p}_{sr} \tag{3.156}
$$

$$
\dot{e}_{rs} = \int \frac{1}{2} m_s v^2 \langle f_r | f_s \rangle_{el} \, d^3v
$$

$$
= - \iint \frac{m_r m_s}{m_r + m_s} \vec{v} \cdot \vec{g} g \sigma_{m,rs} f_r\left(\vec{v} - \frac{m_s}{m_r + m_s}\vec{g}\right)
$$

$$
\times f_s\left(\vec{v} + \frac{m_r}{m_r + m_s}\vec{g}\right) d^3g \, d^3v = -\dot{e}_{sr}. \tag{3.157}
$$

The $\sigma_{m,rs}$ is the cross section with respect to momentum transfer, calculated as defined above,

$$\sigma_{m,rs} = \int (1 - \cos\theta) \frac{d\sigma}{d\Omega}\bigg|_{rs} d\Omega. \tag{3.158}$$

The second special case, the elastic scattering of photons by particles, starts from expression (3.107). The particles are not affected: only the absorption and emission of photon must be represented. We assume that the cross section depends only weakly on the energy and neglect the Doppler shift. Inserting (3.107) into (3.149) and (3.151), carrying out all possible integrations, and streamlining the notation leads to the following expression for the combined absorption and emission processes

$$\varepsilon_\nu - \kappa_\nu I_\nu\big|_{\text{scattering}} = n\left(\int d\Omega_{\phi_i} \frac{d\sigma}{d\Omega}(\nu_\psi, \theta) \frac{I_{\nu\phi_i}(\nu_{\phi_i}, \vec{e}_{\phi_i})}{h\nu} - I\int d\Omega_{\psi_i} \frac{d\sigma}{d\Omega}(\nu_\psi, \theta)\right). \tag{3.159}$$

The net-effect of the scattering is that the photon only changes its direction. The photon number and the energy stay the same, so corresponding quantities vanish,

$$\dot{n}\big|_{\text{scattering}} = 0 \tag{3.160}$$

$$\dot{e}\big|_{\text{scattering}} = 0. \tag{3.161}$$

The remaining term to be discussed is the Coulomb term $\langle f \rangle_{cb}$, arising from the long range interactions of the charged electrons and ions. To a good approximation (see below), it is also a bi-linear term which couples all charged species,

$$\langle f \rangle_{cb,\alpha} = \sum_\beta \langle f_\beta | f_\alpha \rangle_{cb,\beta\alpha}. \tag{3.162}$$

Several different versions of the Coulomb interaction term are available; they differ in their special physical assumptions and/or in their mathematical complexity. Their general form, however, is the same, namely that of a differential operator of second order in velocity space. With two coefficients called the friction vector and the diffusion tensor, respectively, it reads

$$\langle f_\beta | f_\alpha \rangle_{cb,\beta\alpha} = -\frac{\partial}{\partial \vec{v}}(\vec{A}_{\alpha\beta} f_\alpha) + \frac{1}{2}\frac{\partial^2}{\partial \vec{v}\vec{v}}(\mathbf{B}_{\alpha\beta} f_\alpha). \tag{3.163}$$

This mathematical form can be understood from the remarks made above, namely that the action of the Coulomb collisions gives rise to a random walk motion in velocity space. The various theories for the Coulomb interaction differ in the exact expressions for the coefficients; they are, in general, complicated functionals of the distribution function.

We restrict ourselves to the simple case of a plasma which is not too inhomogeneous, not collision dominated, and not strongly magnetized. (The assumptions mean that the Debye length is smaller than the gradient length, the mean free path for collisions with neutrals, and the Larmor radius.) Using arguments that are essentially equivalent to the physical discussion above [1], one arrives at the so-called Landau collision term which expresses the dynamical coefficients as

$$\vec{A}_{\alpha\beta} = \frac{q_\alpha^2 q_\beta^2 (m_\alpha + m_\beta) \ln \Lambda}{4\pi\varepsilon_0^2 m_\alpha^2 m_\beta} \frac{\partial}{\partial \vec{v}} \int \frac{1}{|\vec{v} - \vec{v}'|} f_\beta(\vec{v}') \, d^3 v' \qquad (3.164)$$

$$\mathbf{B}_{\alpha\beta} = \frac{q_\alpha^2 q_\beta^2 \ln \Lambda}{4\pi\varepsilon_0^2 m_\alpha^2} \frac{\partial^2}{\partial \vec{v} \vec{v}} \int |\vec{v} - \vec{v}'| f_r(\vec{v}') \, d^3 v'. \qquad (3.165)$$

The parameter Λ in these equations is the Coulomb ratio defined above. Its appearance under the logarithm makes it insensitive to small alterations; it is customary to replace $\ln \Lambda$ in calculations by a value averaged over all species and spatial locations (or, even more drastically, to set it equal to 10 for low temperature plasmas and equal to 20 for fusion applications). The Coulomb interaction terms then become exactly bi-linear. Balescu [1] proposes the value

$$\ln \Lambda = \ln \frac{6\pi\varepsilon_0 (T_e + T_i)\lambda_D}{q_e q_i}. \qquad (3.166)$$

As the elastic collisions, Coulomb interactions conserve particle number, momentum, and energy. The first property holds for each species separately; the latter two follow again from the antisymmetry of the exchange of momentum and energy between the species,

$$\dot{n}_{\beta\alpha} = \int \langle f_\beta | f_\alpha \rangle_{cb,\beta\alpha} \, d^3 v = 0 \qquad (3.167)$$

$$\dot{p}_{\beta\alpha} = \int m_r \vec{v} \langle f_\beta | f_\alpha \rangle_{cb,\beta\alpha} \, d^3 v$$

$$= \frac{q_\alpha^2 q_\beta^2 \ln \Lambda}{4\pi\varepsilon_0^2} \frac{m_\alpha + m_\beta}{m_\alpha m_\beta} \iint \frac{\vec{v} - \vec{v}'}{|\vec{v} - \vec{v}'|^3} f_\alpha(\vec{v}) f_\beta(\vec{v}') \, d^3 v \, d^3 \vec{v}$$

$$= -\dot{p}_{\beta\alpha} \qquad (3.168)$$

$$\dot{e}_{\beta\alpha} = \int \frac{1}{2} m_r \vec{v}^2 \langle f_r | f_s \rangle_{cb,\beta\alpha} \, d^3 v$$

$$= \frac{q_\alpha^2 q_\beta^2 \ln \Lambda}{4\pi\varepsilon_0^2} \iint \frac{m_\beta \vec{v}'^2 - m_\alpha \vec{v}^2 + (m_\alpha - m_\beta)\vec{v}\vec{v}'}{m_\alpha m_\beta |\vec{v} - \vec{v}'|^3} f_\alpha(\vec{v}) f_\beta(\vec{v}') \, d^3 v \, d^3 \vec{v}$$

$$= -\dot{e}_{\alpha\beta}. \qquad (3.169)$$

Having discussed in some detail the propagation and interaction terms of particles and photons, we can assemble them to the final forms of the kinetic equation and the radiation transport equation. The terms L and G are the building blocks of the few-body collision terms of the kinetic equations. For a given species s, all loss terms (all instances where the particle appears as an educt) must be added with a negative, all gain terms (appearances of s as a product) with a positive sign. Summing over all processes (under restoration of the index P) , one gets

$$\langle f \rangle_{el,s} + \langle f \rangle_{in,s} = \sum_{\text{processes}} G_s^{(P)}(\vec{v}) - \sum_{\text{processes}} L_s^{(P)}(\vec{v}). \tag{3.170}$$

For neutral particles, the kinetic equation is thus

$$\frac{\partial f_a}{\partial t} + \vec{v} \cdot \frac{\partial f_a}{\partial \vec{r}} + \vec{g} \cdot \frac{\partial f_a}{\partial \vec{v}} = \sum_{\text{processes}} G_s^{(P)}(\vec{v}) - \sum_{\text{processes}} L_s^{(P)}(\vec{v}). \tag{3.171}$$

For charged particles, the action of the electromagnetic field and the Coulomb collisions have to be taken into account, so that their equation reads

$$\frac{\partial f_\alpha}{\partial t} + \vec{v} \cdot \frac{\partial f_\alpha}{\partial \vec{r}} + \left(\vec{g} + \frac{q_\alpha}{m_\alpha} (\vec{E} + \vec{v} \times \vec{B}) \right) \cdot \frac{\partial f_\alpha}{\partial \vec{v}}$$

$$= \sum_{\text{processes}} G_s^{(P)}(\vec{v}) - \sum_{\text{processes}} L_s^{(P)}(\vec{v}) + \sum_{\beta} \langle f_\beta | f_\alpha \rangle_{cb,\beta\alpha}. \tag{3.172}$$

Similarly, the emission and absorption terms are building blocks of the radiation transport equation. All appearances of the photon as an educt count as absorptions; all appearances as a product contribute to the emissions. They are added corresponding to the rule

$$\varepsilon - \kappa I_\nu = \sum_{\text{processes}} \varepsilon^{(P)} - \sum_{\text{processes}} \kappa^{(P)} I_\nu. \tag{3.173}$$

The particle production densities of all species are identical, up to the sign which is negative for educts (losses) and positive for products (gains). This reflects the conservation of chemical identity known as Dalton's law,

$$-\dot{n}_{r_k}^L = \dot{n}_{s_l}^G = \dot{n}. \tag{3.174}$$

From the arguments of the delta functions embodied in the scattering probability P in (3.87), one can deduce the balance laws of momentum and energy. Momentum is strictly conserved, energy only when the internal contributions are included:

$$\sum_{l=1}^{M} \dot{\vec{p}}_{r_l} - \sum_{k=1}^{N} \dot{\vec{p}}_{r_k} = 0 \tag{3.175}$$

$$\left(\sum_{l=1}^{M} \dot{e}_{r_l} + \sum_{l=1}^{\bar{M}} \dot{e}_{\phi_l} \right) - \left(\sum_{k=1}^{N} \dot{e}_{r_k} + \sum_{l=1}^{\bar{N}} \dot{e}_{\psi_l} \right) = \left(\sum_{l=1}^{M} \varepsilon_{r_l} - \sum_{k=1}^{N} \varepsilon_{r_k} \right) \dot{n}. \quad (3.176)$$

Adding all terms, we can finally state that the plasma as a whole obeys the conservation rules of particle number, momentum and energy.

3.4   Evaluation and Simplification of the Kinetic Equation

Reviewing the material of the preceding sections, the reader might get the impression that kinetic theory is a mathematical construction of overwhelming complexity. This impression is true: as coupled sets of nonlinear integro-differential equations in $6+1$ dimensions, coupled to another system of partial differential equations (Maxwell's), kinetic models are indeed difficult to solve, both analytically and numerically. For all but the most simple situations, exact solutions will remain elusive in the foreseeable future. (This statement also applies to particle-in-cell simulations, which are sometimes referred to as stochastic solutions of the kinetic equation: they only provide satisfactory results under very limited conditions.)

In this situation, why bother with kinetic theory at all?

To this (rhetorical) question, there are basically two answers. The first one was already given above: kinetic theory provides a general conceptual framework, i.e. a formalism in terms of which (nearly) all relevant plasma phenomena, in particular non-equilibrium features, can be understood. In the last analysis, the underlying reason for the wide applicability of kinetic theory lies in the fact that its sole assumption is met in (nearly) all plasmas of practical interest: the particles in low temperature plasmas are weakly bound, and their average potential energy is much smaller than their thermal energy. The one-particle distribution function thus captures the essence of the dynamics; higher order correlations are not of importance.

Other frameworks, like the one-particle picture, fluid dynamics, or the traditional drift-diffusion model, are much more limited than kinetic theory. Accordingly, kinetic argumentations have become very popular in recent years. It has even been stated that 'all plasma physics must be reformulated kinetically' [8].

The second possible answer to the rhetorical question will occupy us for the rest of this section: kinetic theory, even if it is 'unsolvable' itself, forms the foundation of simpler plasma models which are accessible to solution or simulation. These simplified models can be formally derived from kinetic theory, but, of course, only by invoking certain additional assumptions or neglections. The derived descriptions are thus less general and less accurate than the original kinetic model. Several such descriptions are available which differ in their level of accuracy and complexity; choosing the right one requires physical judgement and insight into the situation.

This is not the place to give a systematic overview of all the different derived plasma descriptions and their relation to the underlying kinetic theory. Some important examples, however, may serve as an illustration of the various possibilities and the typical arguments that are employed.

One important class of model simplifications arises when symmetry arguments can be invoked. Invariance with respect to time leads to steady state situations. Invariance with respect to a spatial direction may appear as Cartesian or cylindrical symmetry, reducing the distribution function in suitable coordinates to the form $f = f(x, y, v_v, v_y, v_z, t)$ or $f = f(r, z, v_r, v_\phi, v_z, t)$. (Often stated as 'the kinetic description is reduced to 2d3v1t dimensions'.) Two simultaneous spatial symmetries are also possible; they reduce the kinetic description to 1d2v1t dimensions. A frequent example is planar symmetry, where the distribution function turns $f = f(x, v_x, v_\perp, t)$. Spherical symmetry with $f = f(r, v_r, v_\perp, t)$ is rare. The assumption that three invariant directions exist is equivalent to assuming spatial homogeneity. In this case, the distribution function reduces to 0d2v1t dimensions, i.e. to the form $f = f(v_\parallel, v_\perp, t)$, where the notions $\parallel$ and $\perp$ refer to the direction of the electrical field. (Note that these dimensionality arguments have implicitly assumed that the magnetic field B is weak; magnetized plasmas require a more elaborate discussion.)

Another important class of simplifications deserves discussion. It arises when the components of the kinetic equation can be separated into groups of different magnitude (which in the following will be termed the 'dominating interaction' and 'a small perturbation'). Under certain conditions, the resulting dynamics assumes a characteristic two-phase structure, where the dominating interaction induces a 'violent relaxation' on a fast time scale, which is followed by a perturbation-induced 'secular evolution' on a slow time scale. Frequently only the latter phase is of physical interest, and it is generally possible to describe it by a reduced model which is both mathematically and conceptually simpler than the original kinetic equation.

The classic example, of course, concerns the dynamics of a neutral gas, for which it is often possible to replace the gas kinetic description by the simpler Navier–Stokes equations. (See, e.g., [10].) For low temperature plasmas, a similar reasoning is possible, when one excludes the electrons and restricts oneself to the heavy particles (ions and neutrals). In this subsystem, one finds that the frequency of the elastic (two-body) collisions is typically much larger than the frequency of all other events, such as chemical reactions, electron-induced ionization and excitation, or recombination. The collision terms of the heavy particle kinetics can be therefore split into two separate groups, the dominant elastic interaction and the inelastic perturbation, with the corresponding collision frequencies ν_{el} and ν_{in}, related by $\nu_{el} \gg \nu_{in}$. Also the laminar parts of the kinetics are accounted for under 'perturbation', implying that scale lengths are large against the elastic mean free path λ_{el}.

The violent relaxation phase in this situation takes place on the time scale $\sim \nu_{el}^{-1}$, where the elastic collisions are dominant and the perturbation interaction is negligible. Boltzmann's H-theorem states that under these conditions the particle system relaxes into local thermodynamic equilibrium, i.e. it maximizes its local entropy under the constraints of particle number, momentum and energy. Correspondingly, all heavy particle distributions become close to local Maxwellians,

$$f_s(\vec{r}, \vec{v}, t) \overset{\text{fast}}{\Longrightarrow} \frac{n_s}{(2\pi T/m_s)^{3/2}} \exp\left(-\frac{m_s(\vec{v} - \vec{u})^2}{2T}\right). \qquad (3.177)$$

The open parameters in these expressions, the fluid dynamic variables n_s (particle species density), $\vec{u}$ (common speed), and T (common temperature), are arbitrary but slowly varying functions of $\vec{r}$. (Abitrary means that they are not determined by the relaxation process but formally enter as its initial conditions; slowly varying refers to the implicit assumption that their gradients are small compared to the mean free path λ_{el}.) Physically, of course, the fluid variables are not arbitrary; they just evolve on the time scale $\sim \nu_{in}^{-1}$. The equations which determine this secular evolution are referred to as the fluid transport equations; after some algebra they assume the form of coupled partial differential equations for n_s, T, and $\vec{u}$. They are, in fact, formally similar to the moment equations discussed in section 3.3 (summed over the species index s if applicable):

$$\frac{\partial n_s}{\partial t} + \nabla \cdot \vec{\Gamma}_s^n = \dot{n}_s \qquad (3.178)$$

$$\frac{\partial \vec{p}}{\partial t} + \nabla \cdot \mathbf{\Pi} = \rho^M \vec{g} + \rho^C \vec{E} + \vec{j} \times \vec{B} \qquad (3.179)$$

$$\frac{\partial e}{\partial t} + \nabla \cdot \vec{\Gamma}^e = \vec{g} \cdot \vec{p} + \vec{E} \cdot \vec{j} + \dot{e}. \qquad (3.180)$$

The equations, however, are now closed. In particular, the fluxes $\vec{\Gamma}_s^n$, $\mathbf{\Pi}$, and $\vec{\Gamma}^e$ can be calculated from the gradients of n_s, T, and $\vec{u}$. (The terms $\dot{n}_s$ and $\dot{e}$ contain interactions with the electrons. Their evaluation requires, of course, knowledge about the distribution f_e.) Details of the related algebra can be found in reference [7]. Strictly speaking, the resulting fluid dynamic transport theory leaves the realm of kinetic modeling and thus lies beyond the scope of this section.

A second important application of the relaxation/evolution scenario concerns the plasma electrons. Unlike heavy particle transport theory, the resulting reduced model stays kinetic and shall be outlined here in more detail. The argument starts again from a physically motivated separation of the terms of the kinetic equation into two groups. Under typical conditions, the predominant interaction of the electrons is elastic scattering at neutrals. Because of the small mass ratio m_e/m_N and the small thermal speed of the neutrals, this process is much more likely to change an electron's direction $\vec{v}/v$ than its speed $v = |\vec{v}|$. Accordingly, the dominating interaction is that of a pure isotropization, mathematically described as

$$\langle f_e \rangle_{\text{el}} = \int \frac{d\nu_{\text{el}}}{d\Omega} f_e(|\vec{v}|\vec{e}) \, d\Omega - \int \frac{d\nu_{\text{el}}}{d\Omega} \, d\Omega \, f_e(\vec{v}). \tag{3.181}$$

The residuum of the approximation $m_e \ll m_N$ and all other collision terms are grouped into an inelastic collision term $\langle f_e \rangle_{\text{in}}$. This term is viewed as a perturbation; its absolute value is considered small compared to the elastic scattering frequency ν_{el},

$$\langle f_e \rangle_{\text{in}} \approx \nu_{\text{in}} f_e \ll \langle f_e \rangle_{\text{el}} \approx \nu_{\text{el}} f_e. \tag{3.182}$$

Also the laminar contributions to the kinetic equation must be small. This requires the gradients to be small compared to the inverse mean free path $\lambda = v_{th}/\nu_{\text{el}}$, and the electrical field compared to $T_e/e\lambda$:

$$\left| \vec{v} \cdot \frac{\partial f}{\partial \vec{r}} \right| \ll \nu_{el} f_e \tag{3.183}$$

$$\left| \frac{e}{m_e} \vec{E} \cdot \frac{\partial f_e}{\partial \vec{v}} \right| \ll \nu_{el} f_e. \tag{3.184}$$

The scaling conditions (3.182)–(3.184) determine the absolute magnitude of the perturbation terms (with respect to the elastic collisions), but not their relative magnitude (with respect to each other.) This still leaves some ambiguity, and, in fact, different 'regimes' are possible which arise from subtle differences in the relative scaling of the perturbations. A particularly simple regime—suited for many applications—results from the assumption that the inelastic collisions are comparable with laminar terms that are *quadratic* in the gradients or fields. These scaling assumptions can be conveniently expressed by ordering the kinetic equation as follows, with ε being a formal smallness parameter (of value unity) to indicate the size of the respective terms:

$$\frac{\partial f_e}{\partial t} + \varepsilon \vec{v} \cdot \frac{\partial f_e}{\partial \vec{r}} - \varepsilon \frac{e}{m_e} \vec{E} \cdot \frac{\partial f_e}{\partial \vec{v}} = \langle f_e \rangle_{\text{el}} + \varepsilon^2 \langle f_e \rangle_{\text{in}}. \tag{3.185}$$

Obviously, the dynamics separates indeed again into a fast relaxation and a slow evolution phase. The relaxation takes place on a time scale ν_{el}^{-1} and involves only the action of the elastic conditions. It leads to an angular isotropization of the initial distribution:

$$f_e(\vec{r}, \vec{v}, t) \stackrel{\text{fast}}{\Longrightarrow} f_0(\vec{r}, v, t) \equiv \frac{1}{4\pi} \int_\Omega f_e(\vec{r}, |\vec{v}|\vec{e}\,', t) \, d^2\Omega'. \tag{3.186}$$

To focus on the subsequent slow evolution phase (which acts on a scale $\sim \varepsilon^2$), we introduce the substitution $t \to \varepsilon^{-2} t$ and write the kinetic equation

$$\varepsilon^2 \frac{\partial f_e}{\partial t} + \varepsilon \vec{v} \cdot \frac{\partial f_e}{\partial \vec{r}} - \varepsilon \frac{e}{m_e} \vec{E} \cdot \frac{\partial f_e}{\partial \vec{v}} = \langle f_e \rangle_{\mathrm{el}} + \varepsilon^2 \langle f_e \rangle_{\mathrm{in}}. \tag{3.187}$$

This equation can conveniently be treated by means of a power expansion. We write all quantities as power series with respect to the formal smallness parameter ε,

$$f_e(\vec{r}, \vec{v}, t) = \sum_{n=0}^{\infty} \varepsilon^n f(\vec{r}, \vec{v}, t) \tag{3.188}$$

$$\vec{E}(\vec{r}, \vec{v}, t) = \sum_{n=0}^{\infty} \varepsilon^n \vec{E}^{(n)}(\vec{r}, \vec{v}, t) \tag{3.189}$$

and compare the coefficients of (3.187) in powers of ε. This procedure leads to an infinite hierarchy of equations, out of which we need only the first three:

$$0 = \langle f_0 \rangle_{\mathrm{el}} \tag{3.190}$$

$$\vec{v} \cdot \frac{\partial f_0}{\partial \vec{r}} - \frac{e}{m_e} \vec{E}_0 \cdot \frac{\partial f_0}{\partial \vec{v}} = \langle f_1 \rangle_{\mathrm{el}} \tag{3.191}$$

$$\frac{\partial f_0}{\partial t} + \vec{v} \cdot \frac{\partial f_1}{\partial \vec{r}} - \frac{e}{m_e} \vec{E}_0 \cdot \frac{\partial f_1}{\partial \vec{v}} - \frac{e}{m_e} \vec{E}_1 \cdot \frac{\partial f_0}{\partial \vec{v}} = \langle f_2 \rangle_{\mathrm{el}} + \langle f_0 \rangle_{\mathrm{in}}. \tag{3.192}$$

The first of these equations contains only the information that f_0 is isotropic; this was already expressed in (3.186). The second equation can be solved explicitly as

$$f_1 = -\frac{1}{\nu_m} \left(\vec{v} \cdot \frac{\partial f_0}{\partial \vec{r}} - \frac{e}{m_e} \vec{E}_0 \cdot \frac{\partial f_0}{\partial \vec{v}} \right) \tag{3.193}$$

where ν_m is the momentum transfer frequency defined as

$$\nu_m(v) = \int (1 - \cos \theta) \frac{\mathrm{d}\nu_{\mathrm{el}}}{\mathrm{d}\Omega}(v, \vartheta) \, \mathrm{d}\Omega. \tag{3.194}$$

From equation (3.192) only the angular average is used. Applying $\int \mathrm{d}\Omega$ on it and utilizing all previous information directly leads to the desired closed evolution equation for f_0,

$$\frac{\partial f_0}{\partial t} - \nabla \cdot \left(\frac{v^2}{3\nu_m} \nabla f_0 - \frac{v e \vec{E}}{3\nu_m m_e} \cdot \frac{\partial f_0}{\partial v} \right)$$

$$- \frac{1}{v^2} \frac{\partial}{\partial v} v^2 \left(-\frac{v e \vec{E}}{3\nu_m m_e} \cdot \nabla f_0 + \frac{e^2 \vec{E}^2}{3\nu_m m_e^2} \frac{\partial f_0}{\partial v} \right) = \langle f_0 \rangle_{\mathrm{in}}. \tag{3.195}$$

Under slightly different assumptions—particularly suited for the analysis of rf-driven plasmas—one can directly employ the so-called two-term-expansion

$f(\vec{v}) \approx f_0(v) + \vec{f}_1(v) \cdot \vec{v}/v$ to get

$$\frac{\partial f_0}{\partial t} + \frac{1}{3} v \nabla \cdot \vec{f}_1 - \frac{1}{3} \frac{e}{m_e} \frac{1}{v^2} \frac{\partial}{\partial v}(v^2 \vec{E} \cdot \underline{f}_1) = \langle f_0 \rangle_{\text{in}} \qquad (3.196)$$

$$\frac{\partial \vec{f}_1}{\partial t} + v \nabla f_0 - \frac{e}{m_e} \vec{E} \frac{\partial f_0}{\partial v} = -\nu_m \vec{f}_1. \qquad (3.197)$$

Apart from the two examples given above, other utilizations of the general ideas are also possible. In particular, one can systematically expand the distribution function into spherical harmonics,

$$f_e(\vec{r}, \vec{v}) \equiv f_e(\vec{r}, v, \theta, \phi) = \sum_{l=0}^{\infty} \sum_{m=-l}^{l} f_{lm}(\vec{r}, v) Y_{lm}(\theta, \phi). \qquad (3.198)$$

Formally, this procedure requires no assumptions on the gradients or the fields, but the series only converges quickly when the conditions (3.182)–(3.184) are met.

The various reduced kinetic theories have in common that they formulate equations (or systems of equations) for functions of $\vec{r}$, v, and t. In other words, they are generally of 3d1v1t dimensions. Compared to the original kinetic equation which was of type 3d3v1t, the numerical effort is hence reduced by two dimensions. Assuming for example that 100 grid points are necessary to resolve a velocity axis properly, one can estimate that the amount of storage is reduced by a factor of 10^4. (The numerical effort, which scales nonlinearly, is probably reduced even more.)

The efficiency gained by switching from the original to a reduced kinetic theory thus is dramatic. Particularly when combined with other methods of reducing the numerical effort, it can bring the kinetic models into the range of today's computers. Reviewing the current literature, it seems, for example, that mathematically three-dimensional problems have become sufficiently easy to handle. Reduced kinetic models are now studied for time dependent situations with planar or spherical geometry (1d1v1t), or for steady state situations with cylindrical or Cartesian symmetry (2d1v0t). Particularly when combined with appropriate transport models for the heavy species, such reduced kinetic descriptions can be used as powerful tools to analyze and simulate situations with high physical and technical complexity. A good overview over this exciting development and many references can be found in [6].

References

[1] R Balescu 1988 *Transport Processes in Plasmas* (Amsterdam: North-Holland)
[2] M E Barone and D B Graves 1966 *Plasma Sources Sci. Technol.* **5** 187

[3] H Deutsch, K Becker, R K Janev, M Probst and T D Märk 2000 *J. Phys. B Letters* **33** 865

[4] A Kersch and W J Morokoff 1995 *Transport Simulation in Microelectronics* (Basel: Birkhauser)

[5] M A Lieberman and A J Lichtenberg 1994 *Principles of Plasma Discharges and Material Processing* (New York: Wiley)

[6] D Loffhagen and R Winkler 2001 *J. Phys. D: Appl. Phys.* **34** 1355

[7] M Mitchner and Ch Kruger 1973 *Partially Ionized Gases* (New York: Wiley)

[8] L Tsendin 1999 private communication

[9] K-U Riemann 1991 *J. Phys. D: Appl. Phys.* **24** 491

[10] L Waldmann 1958 *Handbuch der Physik Bd XII, Transporterscheinungen in Gasen von mittlerem Druck* (Berlin, Göttingen, Heidelberg: Springer)

[11] NIST Online Data Base *Electron-Impact Cross Sections for Ionization and Excitation* http://physics.nist.gov/PhysRefData/Ionization/index.html.

Chapter 4

Air Plasma Chemistry

K Becker, M Schmidt, A A Viggiano, R Dressler and S Williams

4.1 Introduction

In a thermal plasma, all three major plasma constituents (electrons, ions, neutrals) have the same average energy or 'temperature' and for polyatomic species the rotational, vibrational and translational temperatures are in equilibrium. The temperature of thermal plasmas may range from a few thousand Kelvin (e.g. for plasma torches) to a few million Kelvin (in the interior of stars or in fusion plasmas). In contrast, non-thermal or cold plasmas are characterized by the fact that the energy is preferentially channeled into the electron component of the plasma and/or vibrational non-equilibrium of the polyatomic species. In non-thermal plasmas, the electrons may be much hotter (with temperatures in the range of tens of thousands up to a hundred thousand Kelvin) than the ions and neutrals, whose translational temperatures are essentially equal and typically range from room temperature to a few times the room temperature. Non-thermal plasmas thus represent environments where very energetic chemical processes can occur (via the plasma electrons) at low ambient temperatures (defined by the neutrals and ions in the plasma).

The processes that determine the properties of non-thermal plasmas are collisions involving the plasma electrons and other plasma constituents. Tables of relevant collision processes can be found in chapter 3 of this book. Electron collisions are particularly important because of the high mean energy of the plasma electrons. Ionizing collisions and, in molecular plasmas, dissociative electron collisions are of particular relevance. Ionizing collisions determine the charge carrier production by (i) direct ionization of ground state atoms and/or molecules in the plasma and by (ii) step-wise ionization of an atom/molecule through intermediate excited states. Ionization of ground state atoms/molecules, which have a high number density in the plasma, requires a minimum energy which is (for most species) above

124

10 eV. Thus, only the high-energy tail of the electron energy distribution function is capable of contributing to this process. Even though the density of metastable species in a plasma is typically much smaller than the ground-state density, the ionization cross section out of a metastable state is much larger than the ground-state ionization cross section and the energy required to ionize a metastable atom or molecule is much smaller than the ground-state ionization energy. As the number of low-energy electrons is typically much larger than the number of electrons with energies above 10 eV (see above), stepwise ionization processes can contribute significantly to the ionization balance in a non-thermal plasma.

The generation of chemically reactive free radicals by electron impact dissociation in molecular plasmas is an important precursor for plasma chemical reactions. As an example, fluorocarbons such as CF_4 and C_2F_6 are comparatively inert and will not react *per se* with Si or SiO_2. Etching of these materials in plasmas containing fluorocarbon compounds in the feed gas proceeds via F and CF_x radicals formed in the plasma by dissociation of the parent molecules by the plasma electrons.

As discussed in detail in the previous chapter, the probability for a particular electron collision process to occur is expressed in terms of the corresponding electron-impact cross section σ, which is a function of the energy of the electrons. All inelastic electron collision processes have minimum energies (thresholds) below which the process is energetically not possible. In plasmas, the electrons are not mono-energetic, but have an energy or velocity distribution, $f(E)$ or $f(v)$, where E and v refer to the energy and velocity of the colliding electron, respectively. In those cases, it is convenient to define a rate coefficient k for each two-body collision process

$$k(v) = \int \sigma(v) v f(v) \, dv \tag{4.1.1}$$

where $\sigma(v)$ denotes the corresponding velocity dependent cross section. In principle, the velocity v in equation (4.1.1) refers to the relative velocity between the two colliding particles. As the electron velocity is much larger than the velocity of the heavy particles (which are essentially at rest relative to the fast moving electrons), the quantity v in (4.1.1) is nearly identical to the electron velocity. Sometimes it is more convenient to express the rate coefficient as a function of electron energy E. As discussed in chapter 3, realistic electron velocity/energy distribution functions exhibit complicated shapes.

The concept of a rate coefficient is used in a similar fashion to describe reactive collisions between the randomly moving heavy particles, where the reaction probability is determined by the relative velocity between the colliding heavy particles. At equilibrium conditions, the velocity distribution is determined by the heavy-particle temperature, T, and the temperature dependence of the rate coefficient can be described by an Arrhenius law. However, equilibrium models of chemical kinetic systems depend on rate coefficients which are usually given by a modified Arrhenius dependence on temperature:

$$k(T) = AT^n \exp(E_a/k_B T) \tag{4.1.2}$$

where A is a scaling parameter, E_a is the chemical reaction activation energy, k_B is the Boltzmann constant, and n is a curvature parameter describing the growth of the rate coefficient with temperature.

The time scales of the processes in a reactive plasma span a wide range (Eliasson *et al* 1994). Electron-induced processes such as excitation and ionization occur in the range of picoseconds or less. The electron energy distribution function reaches equilibrium with the externally applied electric field also within picoseconds (Eliasson *et al* 1994). Electron-induced dissociative ionization and dissociation processes, in which the molecular target breaks up, take nanoseconds to micro-seconds. At atmospheric pressure, the time scale for chemical reactions involving ground-state species is in the range from milliseconds to seconds, while the free radical reactions occur in the range between micro-seconds and milliseconds.

The atmospheric-pressure air plasmas that are the subject of this book are weakly ionized. Their degree of ionization, α, defined as

$$\alpha = n_e/(n_e + n_o) \tag{4.1.3}$$

where n_e and n_o denote the density of respectively the plasma electrons and the plasma neutrals, is of the order of 10^{-5}, that is only one in every 100 000 plasma neutrals is ionized. The degree of dissociation is typically significantly higher. Despite the low degrees of ionization, both neutral–neutral and ion–neutral processes are important processes in the plasma chemistry of weakly ionized, non-thermal molecular plasmas. Equation (4.1.3) assumes that negative ions do not contribute significantly to the total number of negative charge carriers, which may not be true in air plasmas; in that case equation (4.1.3) must be modified to include negative ions.

In the following sections, we will summarize the state of our current knowledge of the most important plasma chemical reactions in atmospheric-pressure air plasmas for both reactions involving only neutral species ('neutral air plasma chemistry') and ionic species ('ionic air plasma chemistry'). In section 4.2, we discuss reactions of neutrals. As there is a larger number of such reactions, we will not discuss selected reactions in great detail, but rather give a survey summarizing the most important reactions between neutrals in terms of their known reaction rate coefficients and, to the extent available, the temperature dependence of the reaction rates. In the case of ion–molecule reactions in high-pressure air plasmas, the number of processes that have been studied extensively is much smaller and we will cover those reactions in more detail in section 4.3. Section 4.4 discusses the challenge of modeling non-equilibrium air plasma chemical systems where the relative velocity distributions of heavy-body collisions is not described by a temperature. Dissociative recombination, a principal electron loss mechanism, is discussed in section 4.5.

4.2   Air Plasma Chemistry Involving Neutral Species

4.2.1   Introduction

Chemical reactions in an air plasma are initiated by electron impact on the main air plasma constituents N_2 and O_2. Electron-driven processes with N_2 and O_2 include

$$e^- + X_2 \longrightarrow X_2^* + e^- \qquad (4.2.1.1a)$$

$$e^- + X_2 \longrightarrow X + X + e^- \qquad (4.2.1.1b)$$

$$e^- + X_2 \longrightarrow X^* + X + e^- \qquad (4.2.1.1c)$$

$$e^- + X_2 \longrightarrow X_2^+ + 2e^- \qquad (4.2.1.1d)$$

$$e^- + X_2 \longrightarrow X_2^{+*} + 2e^- \qquad (4.2.1.1e)$$

$$e^- + X_2 \longrightarrow X^+ + X + 2e^- \qquad (4.2.1.1f)$$

$$e^- + X_2 \longrightarrow X^+ + X^* + 2e^- \qquad (4.2.1.1g)$$

$$e^- + X_2 \longrightarrow X^- + X \qquad (4.2.1.1h)$$

$$e^- + X_2 + M \longrightarrow X_2 + M \qquad (4.2.1.1i)$$

(X: N_2, O_2; the asterisk denotes an excited state, which may be short-lived or metastable.)

We note that reactions (4.2.1.1h) and (4.2.1.1i) involve primarily O_2 as N_2 is not an electronegative gas. Furthermore, a third body 'M' is required in reaction (4.2.1.1i) in order to satisfy energy and momentum conservation simultaneously. The most recent compilation of measured electron impact cross sections for the molecules N_2 and O_2 as well as for the atoms N and O and for the most important molecular and atomic reaction products and impurities in air plasmas (H_2O, CO_2, CO, CH_4, NO, NO_2, N_2O, O_3, H, C, Ar, ...) can be found in the compilations of Zecca and co-workers (Zecca *et al* 1996, Karwasz *et al* 2001a,b). For subsequent chemical reactions, ground-state neutrals and ions are important, as are electronically excited species in low-lying states that are metastable. Short-lived excited species that can decay radiatively via optically allowed dipole transitions on a time scale of nanoseconds do not have a sufficiently long residence time in the plasma to contribute significantly to the plasma chemical processes (even though at atmospheric pressure their lifetime may become comparable to the inverse collision frequency, in which case their reactivity must also be considered). In the case of molecular species, rotational and vibrational excitation of the reactants can have a profound effect on the reaction pathways and reaction rates of these species, as will be discussed in more detail later. Several extensive compilations of gas phase processes relevant to air plasmas have been published since 1990 including those by Mätzing (1991), Kossyi *et al* (1992), Akishev *et al* (1994), Green *et al* (1995), Herron (1999), Chen and Davidson (2002), Herron and Green (2001), Herron (2001), Stefanovic *et al* (2001), and Dorai and Kushner (2003) (see also the NIST Chemical Kinetics Database, version 2Q98 (NIST Chemkin) and the online version (NIST index)).

4.2.2 Neutral chemistry in atmospheric-pressure air plasmas

This section deals with plasma chemical reactions in atmospheric-pressure air plasma that involve only neutral species. Processes involving ions will be discussed in subsequent chapters. Neutral chemistry and ion chemistry are connected through ion recombination processes in the gas phase or at surfaces as well as dissociative and associative ionization processes. A complete summary of all chemical reactions in an air plasma cannot be given here, because there are simply too many possible reactions. Thus, we will limit the discussion in this section to what we believe are the most important reactions. For a more detailed discussion of the various other chemical reactions we refer the reader to the above-mentioned original references including the NIST database. The examples presented here are limited to reactions involving oxygen and nitrogen atoms and molecules, ozone, and the NO_x reaction products. Table 4.1 lists the most important low-lying,

Table 4.1. Low-lying metastable states of N_2, O_2, N, and O (Radzig and Smirnov 1985).

Species	State	Energy (cm^{-1})	Energy (eV)
N_2	$A\,^3\Sigma_u^+$	50203.6	6.22
N_2	$B\,^3\Pi_g$	59618.7	7.39
N_2	$a'\,^1\Sigma_u^-$	69152.7	8.57
N_2	$C\,^3\Pi_u$	89136.9	11.05
O_2	$a\,^1\Delta_g$	7928.1	0.98
O_2	$b\,^1\Sigma_g^+$	13195	1.64
N	$^2D_{5/2}^0$	19224.5	2.384
N	$^2D_{3/2}^0$	19233.2	2.385
N	$^2P_{1/2}^0$	28839.9	3.576
O	$1D_2$	15867.9	1.967
O	$^1S_0$	33792.6	4.190

long-lived energy levels of the neutral species (N_2, O_2, N, and O) relevant to the neutral chemistry in air plasmas (Kossyi *et al* 1992) in terms of the energy required for their formation via electron collisions (Radzig and Smirnov 1985).

The electron impact dissociation of nitrogen and oxygen molecules into the reactive atomic radicals is an important step for the initiation of chemical processes. The electron impact neutral dissociation of N_2 requires a higher minimum energy as the dissociation of O_2 (Cosby 1993a,b, Stefanovic *et al* 2001). Furthermore, the O_2 dissociation cross section in the low energy range is significantly higher than that for N_2 (Cosby 1993a). For instance, at an electron energy of 18.5 eV, the neutral O_2 dissociation cross section has a value of 52.9×10^{-18} cm^2 (Cosby 1993b) compared to 17.4×10^{-18} cm^2 for N_2 (Cosby 1993a). However, both neutral dissociation processes are important in the initiation of the neutral air plasma chemistry. We note that the dissociative electron attachment to O_2 leading to the formation of $O^- + O$ has a threshold near 5 eV and a maximum cross section of about 1.5×10^{-18} cm^2 around 7 eV. Even though this cross section is comparatively low, the process is quite effective because of the higher electron density in this energy range compared to the energy required for neutral dissociation. Non-dissociative attachment to O_2 leading to the formation of O_2^- (in the presence of a third collision partner) occurs for electron energies near 0.1 eV (Christophorou *et al* 1984).

Figure 4.1 presents schematically the main plasma chemical reaction pathways in an air plasma starting with the electron-driven reactions

$$O_2 + e^- \longrightarrow O_2^* + e^- \tag{4.2.2.1a}$$

$$\longrightarrow O + O + e^- \tag{4.2.2.1b}$$

$$\longrightarrow O^* + O + e^- \tag{4.2.2.1c}$$

$$\longrightarrow O^- + O \tag{4.2.2.1d}$$

$$O_2 + e^- + M \longrightarrow O_2^- + M \tag{4.2.2.1e}$$

and at higher electron energies

$$N_2 + e^- \longrightarrow N_2^* + e^- \tag{4.2.2.2a}$$

$$\longrightarrow N + N + e^- \tag{4.2.2.2b}$$

$$\longrightarrow N^* + N + e^- \tag{4.2.2.2c}$$

(where the asterisk denotes one of the low-lying excited states listed in table 4.1), which are followed by the neutral heavy particle processes:

$$N^* + O_2 \longrightarrow NO + O \tag{4.2.2.3a}$$

$$O + O_2 + O_2 \longrightarrow O_3 + O_2 \tag{4.2.2.3b}$$

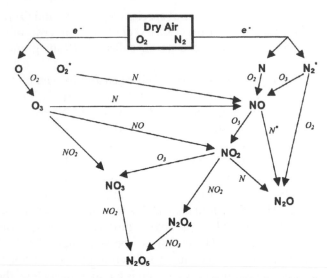

Figure 4.1. Schematic diagram of the primary chemical reactions in an air plasma (dry air) following electron impact on N_2 and O_2. Only the formation reactions up to the formation of N_2O_5 are shown.

It is interesting to note that reactions involving ground-state and excited species can have rate coefficients that differ by orders of magnitude. For instance, the rate coefficient of reaction (4.2.2.3a) involving an excited N atom has a value of $5 \times 10^{-12}\,\text{cm}^3/\text{s}$ (see table 4.5), whereas the rate coefficient for the corresponding ground state reaction is $7.7 \times 10^{-17}\,\text{cm}^3/\text{s}$ (see table 4.3). The required activation energy for the reaction involving the excited particle is lowered by the potential energy of the excited reaction partner (Eliasson and Kogelschatz 1991).

4.2.3 Summary of the important reactions for the neutral air plasma chemistry

The following tables summarize the most important neutral chemical reactions in an air plasma starting with two-body reactions involving O atoms (table 4.2) and N atoms (table 4.3) in the ground states. Table 4.4 presents three-body reactions involving ground-state species. Reactions with electronically excited species are presented in table 4.5 and in table 4.6 reactions are listed involving ozone molecules. To the extent known from the literature, we also list the temperature dependence of the rate constants. For the three-body reactions, the rate constants are given as the product of the temperature-dependent part and the gas density per cm^3 (of the 'third' body) at atmospheric pressure. This facilitates a meaningful comparison of these rate coefficients with rate coefficients for two-body reactions. All rate constants are given in units of cm^3/s except for the data for three-body

Table 4.2. Ground-state, two-body reactions involving O atoms.

Reaction	k_{298} ($cm^3\ mol^{-1}\ s^{-1}$)	Temperature dependence $k(T)$ ($cm^3\ mol^{-1}\ s^{-1}$)	Temperature range (K)	Reference
$O + O_3 \rightarrow O_2 + O_2$	8×10^{-15}	$2.0 \times 10^{-11} \exp(-2300/T)$ $8.0 \times 10^{-12} \exp(-2060/T)$ $1.9 \times 10^{-11} \exp(-2300/T)$	200–400	Kossyi *et al* (1992) Herron and Green (2001) Akishev *et al* (1994)
$O + NO_2 \rightarrow O_2 + NO$	9×10^{-12}	$6.5 \times 10^{-12} \exp(120/T)$ $5.6 \times 10^{-12} \exp(180/T)$ $1.13 \times 10^{-11} (T/1000)^{0.18}$ $5.21 \times 10^{-12} \exp(+202/T)$	250–350	Herron and Green (2001) Chen and Davidson (2002) Kossyi *et al* (1992) Mätzing (1991)
$O + NO_3 \rightarrow O_2 + NO_2$	1.7×10^{-11} 1.0×10^{-11}			Herron and Green (2001) Chen and Davidson (2002) Akishev *et al* (1994)
$O + N_2O_3 \rightarrow$ products	$\leq 3 \times 10^{-16}$			Herron and Green (2001)
$O + N_2O_5 \rightarrow 2NO_2 + O_2 \rightarrow$ products	1.0×10^{-16} $<3 \times 10^{-16}$			Chen and Davidson (2002) Kossyi *et al* (1992)

Table 4.3. Ground-state, two-body reactions involving N atoms.

Reaction	k_{298} ($cm^3 mol^{-1} s^{-1}$)	Temperature dependence $k(T)$ ($cm^3 mol^{-1} s^{-1}$)	Reference
$N + O_2 \rightarrow NO + O$	7.7×10^{-17}	$4.4 \times 10^{-12} \exp(-3220/T)$	Dorai and Kushner (2003)
$N + O_3 \rightarrow NO + O_2$	5.7×10^{-13}	$5 \times 10^{-12} \exp(-650/T)$	Stefanovic *et al* (2001)
	$\leq 2 \times 10^{-16}$		Herron (2001)
$N + NO \rightarrow N_2 + O$	3.2×10^{-11}	$3.4 \times 10^{-11} \exp(-24/T)$	Dorai and Kushner (2003)
$N + NO_2 \rightarrow N_2O + O$	1.2×10^{-11}	$5.8 \times 10^{-12} \exp(-220/T)$	Herron and Green (2001)
$N + NO_2 \rightarrow NO + NO$	2.3×10^{-12}		Kossyi *et al* (1992)
$N + NO_3 \rightarrow NO + NO_2$	3×10^{-12}		Herron and Green (2001)
$N + NO_2 \rightarrow N_2 + O + O$	9.1×10^{-13}		Kossyi *et al* (1992)

Table 4.4. Ground-state three-body reactions.

Reaction	k_{300}^* (cm^3 mol^{-1} s^{-1})	Temperature dependence $k(T)^*$	Temperature range (K)	Reference
$O + O + M \rightarrow O_2 + M$	9.8×10^{-14}	$4.5 \times 10^{-34} \exp(630/T)$ [N$_2$]	200–400	Herron and Green (2001)
$O + N + M \rightarrow NO + M$	2.7×10^{-13}	$6.3 \times 10^{-33} \exp(140/T)$ [N$_2$]	200–400	Herron and Green (2001)
$O + O_2 + M \rightarrow O_3 + M$	1.6×10^{-14}	$6.0 \times 10^{-34} (T/300)^{-2.8}$ [O$_2$]	100–300	Herron and Green (2001)
$O + O_2 + M \rightarrow O_3 + M$	1.5×10^{-14}	$5.6 \times 10^{-34} (T/300)^{-2.8}$ [N$_2$]	100–300	Herron and Green (2001)
$O + NO + M \rightarrow NO_2 + M$	2.7×10^{-12}	$1 \times 10^{-31} (T/300)^{-1.6}$ [N$_2$]	200–300	Herron and Green (2001)
$O + NO_2 + M \rightarrow NO_3 + M$	2.4×10^{-12}	$9.0 \times 10^{-32} (T/300)^{-2.0}$ [N$_2$]	200–400	Herron and Green (2001)
$N + N + M \rightarrow N_2 + M$	1.2×10^{-13}	$8.3 \times 10^{-34} \exp(500/T)$ [N$_2$]	100–600	Herron and Green (2001)
$NO + NO + O_2 \rightarrow NO_2 + NO_2$		$3.3 \times 10^{-39} \exp(526/T)$		Akishev *et al* (1992)
$NO + NO_2 + M \rightarrow N_2O_3 + M$	8.3×10^{-15}	$3.1 \times 10^{-34} (T/300)^{-7.7}$ [N$_2$]	200–300	Herron and Green (2001)
$NO_2 + NO_2 + M \rightarrow N_2O_4 + M$	3.8×10^{-14}	$1.4 \times 10^{-33} (T/300)^{-3.8}$ [N$_2$]	300–500	Herron and Green (2001)
$NO_2 + NO_3 + M \rightarrow N_2O_5 + M$	7.4×10^{-11}	$2.8 \times 10^{-30} (T/300)^{-3.5}$ [N$_2$]	200–400	Herron and Green (2001)

* The rate constants of Herron and Green (2001) are those in the low-pressure limit. The low-pressure third-order limit is characterized by a second-order rate constant $k_{300} = Af(T) \times 2.68 \times 10^{19}$ (cm^3 mol^{-1} s^{-1}) (Herron and Green 2001).

Table 4.5. Two-body reactions involving electronically excited species.

Reaction	k_{298} (cm^3 mol^{-1} s^{-1})	Temperature dependence $k(T)$ (cm^3 mol^{-1} s^{-1})	Reference
$O(^1D) + O_3 \rightarrow 2O + O_2$	1.2×10^{-10}		Herron and Green (2001)
$O(^1D) + O_3 \rightarrow 2O_2(3\Sigma_g^-)$	1.2×10^{-10}		Herron and Green (2001)
$O(^1D) + N_2O \rightarrow 2NO$	7.2×10^{-11}		Herron and Green (2001)
$O(^1D) + N_2O \rightarrow N_2 + O_2$	4.4×10^{-11}		Herron and Green (2001)
$O(^1D) + NO_2 \rightarrow NO + O_2$	1.4×10^{-10}		Herron and Green (2001)
$N(^2D) + O_2 \rightarrow O(^3P, {}^1D) + NO$	5×10^{-12}	$1.0 \times 10^{-11} \exp(-210/T)$	Herron and Green (2001)
$N(^2D) + O_3 \rightarrow NO + O_2$	1×10^{-10}		Herron and Green (2001)
$N(^2D) + NO \rightarrow N_2 + O(^3P, {}^1D, {}^1S)$	4.5×10^{-11}		Herron and Green (2001)
$N(^2D) + N_2O \rightarrow N_2 + NO$	2.2×10^{-12}	$1.5 \times 10^{-11} \exp(-570/T)$	Herron and Green (2001)
$N(^2P) + O_2 \rightarrow O(^3P, {}^1D, {}^1S) + NO$	2×10^{-12}	$2.5 \times 10^{-12} \exp(-60/T)$	Herron and Green (2001)
$O_2(^1\Delta_g) + N \rightarrow NO + O$	$\leq 9 \times 10^{-17}$		Herron and Green (2001)
$O_2(^1\Delta_g) + O_3 \rightarrow 2O_2 + O$	3.8×10^{-15}		Herron and Green (2001)
$O_2(^1\Sigma_g^-) + O_3 \rightarrow 2O_2 + O$	2.2×10^{-11}		Herron and Green (2001)
$N_2(A^3\Sigma_u^+) + O_2 \rightarrow N_2 + 2O$	2.5×10^{-12}	$5.0 \times 10^{-12} \exp(-210/T)$	Herron and Green (2001)
$N_2(A^3\Sigma_u^+) + O_2(^1\Delta_g) \rightarrow N_2 + 2O$	$< 2 \times 10^{-11}$		Herron and Green (2001)
$N_2(A^3\Sigma) + O_2 \rightarrow N_2O + O$	4.6×10^{-15}		Stefanovic et al (2001)
$N_2(A^3\Sigma_u^+) + O_3 \rightarrow N_2 + O_2 + O$	4.2×10^{-11}		Herron and Green (2001)
$N_2(A^3\Sigma_u^+) + NO_2 \rightarrow N_2 + NO + O$	1.3×10^{-11}		Herron and Green (2001)

Table 4.6. Reactions including O_3, mainly two-body reactions.

Reaction	k_{300} (cm^3 mol^{-1} s^{-1})	Temperature dependence $k(T)$	Reference
$O + O_2 + M \rightarrow O_3 + M$		$6.0 \times 10^{-34}(T/300)^{-2.8}$ [O_2]	Herron (2001b)
$O + O_2 + O_2 \rightarrow O_3 + O_2$		$8.6 \times 10^{-31}T^{-1.25}$	Stefanovic et al (2001)
$O + O_2 + M \rightarrow O_3 + M$		$5.6 \times 10^{-34}(T/300)^{-2.8}$ [N_2]	Herron and Green (2001)
$O + O_2 + N_2 \rightarrow O_3 + N_2$		$5.6 \times 10^{-29}T^{-2}$	Stefanovic et al (2001)
$N + O_3 \rightarrow NO + O_2$	$\leq 2 \times 10^{-16}$		Herron and Green (2001)
	1×10^{-16}		Chen and Davidson (2002)
	1×10^{-15}		Akishev et al (1994)
$O + O_3 \rightarrow O_2 + O_2$	8×10^{-15}	$8.0 \times 10^{-12}\exp(-2060/T)$	Herron and Green (2001)
		$1.9 \times 10^{-11}\exp(-2300/T)$	Akishev et al (1994)
$O + O_3 \rightarrow O_2(a^1\Delta) + O_2$	3×10^{-15}	$6.3 \times 10^{-12}\exp(-2300/T)$	Stefanovic et al (2001)
$O + O_3 \rightarrow O_2(b^1\Sigma) + O_2$	1.5×10^{-15}	$3.2 \times 10^{-12}\exp(-2300/T)$	Stefanovic et al (2001)
$O(^1D) + O_3 \rightarrow O_2 + 2O$	1.2×10^{-10}		Stefanovic et al (2001)
$O(^1D) + O_3 \rightarrow O_2 + O_2$	2.3×10^{-11}		Stefanovic et al (2001)
	1.2×10^{-10}		Akishev et al (1994)
$O(^1D) + O_3 \rightarrow 2O_2(^3\Sigma_g)$	1.2×10^{-10}		Herron and Green (2001)
$O(^1D) + O_3 \rightarrow O_2(a^1\Delta) + O_2$	1.5×10^{-11}		Stefanovic et al (2001)
$O(^1D) + O_3 \rightarrow O_2(b^1\Sigma) + O_2$	7.7×10^{-12}		Stefanovic et al (2001)
	3.6×10^{-11}		Akishev et al (1994)
$O(^1D) + O_3 \rightarrow O_2(4.5) + O_2{}^*$	7.4×10^{-11}	$5 \times 10^{-11}\exp(-2830/T)$	Stefanovic et al (2001)
$O_2(a^1\Delta) + O_3 \rightarrow O + O_2 + O_2$	4×10^{-15}		Stefanovic et al (2001)
$O_2(b^1\Sigma) + O_3 \rightarrow O_2 + O_2 + O$	1.5×10^{-11}		Stefanovic et al (2001)
$O_2(b^1\Sigma) + O_3 \rightarrow O_2(a^1\Delta) + O_2 + O$	7×10^{-12}		Stefanovic et al (2001)
$NO + O_3 \rightarrow NO_2 + O_2$	1.8×10^{-14}	$1.8 \times 10^{-12}\exp(-1370/T)$	Herron and Green (2001)
	1.6×10^{-14}	$9 \times 10^{-13}\exp(-1200/T)$	Stefanovic et al (2001)
$NO_2 + O_3 \rightarrow NO_3 + O_2$	3.5×10^{-17}	$1.4 \times 10^{-13}\exp(-2470/T)$	Herron and Green (2001)
	3.4×10^{-17}	$1.2 \times 10^{-13}\exp(-2450/T)$	Stefanovic et al (2001)

The rate constants of Herron and Green (2001) for the three-body reactions are the values in the low-pressure limit. See also table 4.4.

* O_2 (4.5): O_2 electronic levels near 4.5 eV, O_2 ($c^1\Sigma$, $C^3\Delta$, $A^3\Sigma$).

reactions, which are in units of cm^6/s. The results of modeling calculations and simulations involving such processes, their rate coefficients, and the temporal behavior of the concentrations of various chemically reactive species and reaction products can be found in the paper by Kossyi *et al* (1992) and to some extent also in other chapters in this book.

In addition to the gas-phase processes, heterogeneous processes such as surface reactions should also be taken into account. Deactivation reactions of excited particles as well as recombination processes of atomic species and chemical reactions are important in this context. The reaction probability for a given process depends on the surface material and the state of the surface in terms of its purity and temperature. In general, surface processes at atmospheric pressure are less important than at lower pressure. The modeling of a microwave atmospheric-pressure discharge in air (Baeva *et al* 2001) included the de-excitation of N$_2$, O$_2$, N, and O as well as the wall recombination of O atoms. A comprehensive discussion of the chemical reactions of the various air plasma components with a polypropylene surface is given by Dorai and Kushner (2003) (see also chapter 9 in this book). Other data for surface processes were given by Gordiets *et al* (1995).

4.3 Ion–Molecule Reactions in Air Plasmas at Elevated Temperatures

4.3.1 Introduction

Ion chemistry is a mature though continually evolving field. A wide variety of techniques have been exploited to measure ion reactivity over a large range of conditions (Farrar and Saunders 1988). In compilations of ion–molecule kinetics, there are over 10 000 separate entries (Ikezoe *et al* 1987) and the number of reactions studied continues to be impressive. This large body of work has led to many insights into reactivity and numerous generalities have emerged. In spite of the large number of studies, there are still several areas of ion kinetics that are largely unexplored, one of which is the study of ion–molecule reactions at elevated temperatures relevant to air plasma conditions.

The vast majority of the work on ion–molecule kinetics has been performed at room temperature (Ikezoe *et al* 1987). Temperature dependent studies have been mostly limited to the 77–600 K range. Outside of this temperature range, significant technical difficulties are encountered, e.g. the stability of materials and reactants at high temperature or condensation of the reactant species at low temperature. Most of the effort to extend the temperature range has focused on low temperatures (Smith 1994) due to the fact that many of the molecular species made in interstellar clouds are synthesized by ion–molecule reactions at extremely low temperatures (Smith and Spanel 1995). The techniques used to study low-temperature chemistry have been quite successful and have provided good tests of theory, especially with regard to ion–molecule collision rates (Adams *et al* 1985, Rebrion *et al* 1988, Troe 1992).

In contrast, the number of studies made at high temperature (>600 K) is very limited. Previous work on ion–molecule reactivity above 600 K was performed in the early 1970s and was limited to temperatures of 900 K and below (Chen *et al* 1978, Lindinger *et al* 1974). The impetus for those studies focused on reactions of the low density air plasma of the ionosphere that can reach temperatures as high as 2000 K range (Jursa 1985). A further limitation was that branching fractions could not be measured. Nevertheless, the technically challenging measurements provided useful and interesting data on how temperature affected rate constants. However, the conclusions were limited because only 10 reactions were studied in total.

The gap between the previous maximum laboratory operating temperature and relevant plasma temperatures was covered in other ways. In particular, the reactions were studied as a function of ion translational energy in drift tubes and beam apparatuses (Farrar and Saunders 1988). This allowed effective temperature dependencies to be calculated assuming translational energy, E_t, was equivalent to internal, rotational and vibrational, in controlling the reactivity (McFarland *et al* 1973a–c). As will be shown later, this approach can lead to large errors although it was the only reasonable way to extrapolate to higher temperature conditions at the time.

In high temperature air plasmas, most of the chemistry involves only monatomic and diatomic ions and neutrals, and, therefore, very little vibrational excitation is present at temperatures below 900 K due to the high vibrational frequencies of the respective diatomic molecules or molecular ions. Thus, the impact of both rotational and vibrational energy was not seriously considered. One notable exception, however, was the reaction

$$O^+ + N_2 \longrightarrow NO^+ + N. \qquad (4.3.1)$$

For this reaction, a separate study on the vibrational temperature dependence of the N_2 reactant was made (Chiu 1965, Schmeltekopf *et al* 1968). However, in that study both the ion center of mass (CM) translational energy and the rotational temperature were 300 K. While this was an obviously important step, no true temperature dependent study was made over 900 K. Note that true temperature here refers to the case where the translational, rotational, and vibrational degrees of freedom of the reactants are in equilibrium and can be represented by a single temperature.

The lack of measurements over an extended temperature range was one of several drivers leading to the development of a flowing afterglow apparatus capable of reaching temperatures of 1800 K. This apparatus will be hereafter referred to as the high temperature flowing afterglow (HTFA, Hierl *et al* 1996). While ionospheric plasma chemistry was an important driver for the development of the HTFA, there are other plasmas that require accurate ion–molecule kinetic measurements at high temperature. Examples include plasma sheathing around high speed vehicles during re-entry or hypersonic flight, spray coating and materials synthesis, microwave reflection/absorption, sterilization and chemical neutralization, shock-wave mitigation for sonic boom and wave-drag reductions in supersonic flights, and plasma igniters and pilots for subsonic to supersonic combustion engines.

In this section, high temperature air plasma reactions studied to date are discussed and compared to available results from different experiments. Most often the comparisons are between data taken in high temperature flow tubes and drift tubes, but in certain cases comparisons are also made to data taken in ion-beam experiments. The ensuing sections give a discussion of the derivation of internal energy dependencies which allow the results of different experiments to be compared. Then the results for relevant air plasma reactions are presented.

The fate of an ion in an air plasma depends critically on whether it is atomic or molecular. While atomic ions recombine slowly with electrons through three-body recombination reactions (see table 4.1), molecular ions undergo much more rapid dissociative recombination reactions. Consequently, reactions that convert atomic ions such as O^+ and N^+ to diatomic ions, speed up recombination, and are therefore important in controlling the ionization fraction of the plasma. Atomic ion reactions with N_2, O_2, and NO are discussed first. While nothing inherently prevents negative ion systems from being studied, relatively few reactions have been studied to date. Of these negative-ion reactions, the temperature dependence of O^- with NO and CO are discussed. As the number of atoms in a reaction increases, the detailed derivation of how temperature affects the reactivity becomes less clear, i.e. attributing the reactivity to a particular form(s) of energy. The larger reaction systems discussed include $N_2^+ + O_2$, $O_2^+ + NO$, $Ar^+ + CO_2$, and N_2^+ with CO_2.

4.3.2   Internal energy definitions

The average reactant rotational energy, $\langle E_{rot} \rangle$, is $\frac{1}{2}k_B T$ for each rotational degree of freedom, and the average reactant vibrational energy, $\langle E_{vib}^{neutral} \rangle$, is an ensemble average over a Boltzmann distribution of vibrational energy levels. The average translational energy, $\langle E_{trans} \rangle$, is $\frac{3}{2}k_B T$ in flow tube experiments and is the nominal CM collision energy in drift tube and ion beam experiments.

In the HTFA all degrees of freedom are thermally excited by heating the apparatus, i.e. the rotational, translational, and vibrational temperatures are in equilibrium. In a drift tube or beam apparatus, the translational energy of the ion is increased by the use of electric fields. Fortunately, the translational energy distribution in a drift tube operated with a He buffer gas can be approximated by a shifted Maxwellian distribution (Albritton *et al* 1977, Dressler *et al* 1987, Fahey *et al* 1981a,b). The average translational energy can be converted to an effective translational temperature by $E_t = \frac{3}{2}k_B T_{eff}$ and can be directly compared to the HTFA data since the translational energy distributions are similar. The internal energy dependence is derived by comparing data taken at the same translational temperature or average energy but with the neutrals at different temperatures. The internal energy dependence is most easily observed by plotting the data as a function of translational energy or temperature. In this type of plot, differences along the vertical, rate coefficient axis reflect the effect of internal energy on reactivity. Comparison to beam data is done in the same way but differences in translational energy distributions complicates the analysis.

The analysis of atomic ions reacting with diatomic neutrals is relatively straightforward. For most diatomics, little or no vibrational excitation occurs below *ca.* 1000 K. Therefore, at lower temperatures, any internal energy dependence is due solely to the rotational excitation of the reactant neutral. To elucidate the energy effects further, it is useful to plot the data as a function of average translational plus rotational energy, i.e. $\frac{5}{2}k_B T$. For drift tube data at 300 K, a constant value of $k_B T = 0.026\,eV$ is added to the translational energy, and the average translational energy in the HTFA is multiplied by $\frac{5}{3}$. As will be shown in the results section, plots of this type often have the drift tube and HTFA data overlapping below 1000 K or 0.2 eV. This agreement suggests that rotational and translational energy control the reactivity equally, at least in an average sense.

If rotational and translational energy are found to be equivalent at lower temperatures, it is assumed that they are equivalent at higher temperatures and that any differences between sets observed at higher temperatures are due to vibrational excitation. In this case, the HTFA rate constants can be written as

$$k = \sum_i \text{pop}(i) \times k_i \qquad (4.3.2)$$

where *i* represents the vibrational level, pop(*i*) is the fraction of the molecules in the *i*th state, and k_i is the rate constant (see equation (4.1.2)) for the *i*th state. The populations of the various states can be calculated assuming a Boltzmann distribution. Assuming all excited states react at the same rate, the $v \geq 1$ rate constant can be extracted with the aid of equation (4.3.2). In most cases, the derived $v \geq 1$ rate constant represents the $v = 1$ rate constant, because even at the temperatures achieved in the HTFA, most of the vibrational excitation is limited to $v = 1$. For some systems, either the HTFA or drift tube data are multiplied by a constant near unity to account for systematic errors between the systems.

In the case of diatomic ions reacting with diatomic molecules, the rotational energy of the reactant ion must also be included in the analysis. The rotational temperature of the ionic reactant in a drift tube is calculated from the CM energy with respect to the buffer (Anthony *et al* 1997, Duncan *et al* 1983). Vibrational excitation also occurs in both reactants and can only be separated if independent information exists regarding how vibrational excitation of one of the reactants affects the reactivity. In practice if such information is available, it is likely to be the vibrational dependence of the primary reactant ion.

For atomic and polyatomic ions reacting with polyatomic molecules, it is often useful to plot the data as a function of total energy, i.e. the sum of vibrational, rotational, and translational energy. This analysis does not allow for separation of the effects resulting from the various types of energy, but it does provide a test to determine if all types of energy control the reactivity similarly. Thus, there are three types of plots used to facilitate the discussion: reactivity versus (1) translational energy or temperature, (2) rotational plus translational energy, and (3) total energy. Each plot type yields useful information and examples of each type are given in the next section.

4.3.3   Ion–molecule reactions

4.3.3.1 $O^+ + N_2$

The reaction of O^+ with N_2 produces NO^+ and N as the primary reaction products as shown in reaction (4.3.1). This reaction has been thoroughly studied in the 1960s and 1970s (Albritton *et al* 1977, Chen *et al* 1978, Johnsen and Biondi 1973, Johnsen *et al* 1970, McFarland *et al* 1973b, Rowe *et al* 1980, Schmeltekopf *et al* 1968, Smith *et al* 1978). During that time period, the temperature dependence of this reaction has been measured up to 900 K (Chen *et al* 1978, Lindinger *et al* 1974). However, at 900 K only 2% of the N_2 molecules are vibrationally excited. To overcome this shortcoming both the translational energy dependence and the dependence on the N_2 vibrational temperature were measured independently (Schmeltekopf 1967, Schmeltekopf *et al* 1968). Figure 4.2 shows HTFA measurements (Hierl *et al* 1997) up to 1600 K along with the one of the previous temperature-dependent studies (Lindinger *et al* 1974) and a drift tube study of the energy dependence (Albritton *et al* 1977). The data from the drift tube study is converted to an effective temperature by assuming that the average translational energy equals $\frac{3}{2} k_B T_{eff}$. The two thermal experiments agree very well, and the other temperature-dependent study (Chen *et al* 1978) (not shown) is similar and shows the rate constants decreasing to 900 K. The

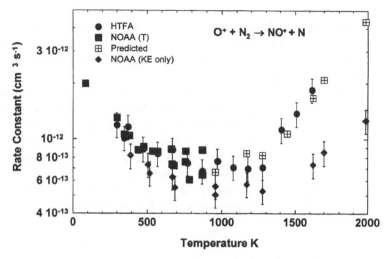

Figure 4.2. Plot of the rate constants for the reaction of O^+ with N_2 as a function of temperature. The HTFA (Hierl *et al* 1997), the NOAA (T) (Lindinger *et al* 1974), and NOAA (KE) (Albritton *et al* 1977) data are shown as circles, squares and diamonds, respectively. See the text for a description of the predicted values.

drift tube study also shows good agreement in this range, although the values are slightly below the thermal rate constants. This may be due in part to the difficulty of measuring such slow rate constants, which are approaching the lower limit that can be measured accurately in low-pressure flow tubes. The agreement between the drift tube data and the thermal data shows that rotational energy does not have a big effect on the reactivity. Above 1200 K, the HTFA and drift tube data start to increase with increasing temperature although the thermal data increase at a lower temperature and increase more rapidly. This shows that vibrational excitation increases the rate constants substantially.

There is a previous study on the effect of the vibrational temperature of N_2 on the rate constant (Schmeltekopf 1967, Schmeltekopf *et al* 1968). The combination of the translational energy dependence of the drift tube data with the vibrationally excited N_2 data provides an interesting comparison to the present data. The vibrational temperature data were reported relative to the 300 K rate constant. Scaling these data to the drift tube translational temperature ($T_{vib} = T_{trans}$), however, allows a thermal rate constant to be predicted with both vibrational and translational effects included, i.e. each drift tube translational energy data point is scaled according to the vibrational energy dependence at the corresponding effective temperature. This procedure ignores the effects of rotational excitation, which is small at temperatures below 900 K. This also assumes that the translational energy dependence of the vibrationally excited species is similar to that for $v = 0$.

The results of this prediction are shown in the figure 4.2. Very good agreement is found with the thermal rate constants. Unsatisfactory agreement is obtained (not shown) if the vibrational temperature data are plotted relative to the 300 K rate constant. The agreement between the data indicates that the above assumptions are good.

The large upturn in the rate constant above 1200 K is due to vibrational excitation. At first glance one would assume that it was due to N_2 ($v = 1$). However, the NOAA group has shown that $v = 1$ reacts at almost the same rate as $v = 0$ and that it is $v = 2$ and higher that react much faster, a factor of 40 faster than the lower energy states (Schmeltekopf 1967, Schmeltekopf *et al* 1968). Thus, the rather large difference between the HTFA and drift tube data is due to the less than 2% of the N_2 molecules that are excited to $v = 2$ or higher in the HTFA experiments.

4.3.3.2 $O^+ + O_2$

The rate constants for the reaction of O^+ with O_2 are shown in figure 4.3 as a function of temperature (Hierl *et al* 1997). This is one of only two reactions which was studied up to the full temperature range of 1800 K. The data decrease with temperature up to about 800 K, go through a minimum about 300 K wide and increase dramatically above that point. Two other datasets are shown for comparison (Ferguson 1974a, Lindinger *et al* 1974, McFarland *et al* 1973b). The previous temperature dependent data taken

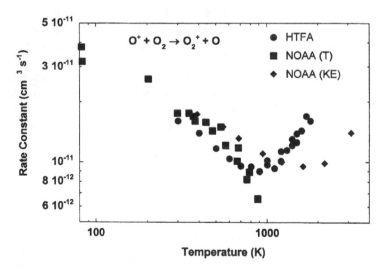

Figure 4.3. Plot of the rate constants for the reactions of O^+ with O_2 as a function of temperature. The HTFA (Hierl *et al* 1997), the NOAA (T) (Lindinger *et al* 1974), and NOAA (KE) (McFarland *et al* 1973b) data are shown as circles, squares and diamonds, respectively.

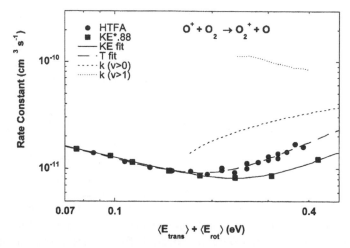

Figure 4.4. Plot of the rate constants for the reaction of O^+ with O_2 as a function of average translational plus rotational energy. The HTFA (Hierl *et al* 1997) and the NOAA (KE) (Ferguson 1974a, Lindinger *et al* 1974) data are shown as circles and squares, respectively. See the text for a description of the fits and predicted rate constants.

up to 900 K are in good agreement with the present data except for the 900 K point, which still agrees within the combined error limits. Only the NOAA drift tube data are shown and are slightly higher than the present values at low temperature with the difference increasing with higher temperatures. The drift tube study also has a much wider minimum and increases more slowly. Another drift tube study found values somewhat higher but with similar trends (Johnsen and Biondi 1973).

Figure 4.4 shows a plot for the HTFA and NOAA drift tube data versus rotational plus translational energy for the reaction of $O^+ + O_2$, the NOAA data have been scaled by 0.88 to better match the lowest energy HTFA points. This is a small correction, considerably less than the error limits, which accounts for a small systematic difference between the datasets. The data agree almost perfectly up to almost 0.2 eV. In this range very little of the O_2 is vibrationally excited. Since the two datasets have considerably different contributions from the two types of energy, the agreement indicates that rotational and translational energy affect reactivity similarly, at least in an average sense. At higher energies, the HTFA rate constant is significantly greater than the drift tube data. The separation between the two curves occurs at the temperature where an appreciable fraction of O_2 starts to be vibrationally excited.

For most of the high temperature range, only $v = 0$ and $v = 1$ of O_2 are significantly populated (Huber and Herzberg 1979). This allows for a determination of the rate constant for O_2 in the $v = 1$ state. To facilitate the derivation, the two datasets are fitted to a power law plus Arrhenius type exponential. The results of the fits are shown in figure 4.4 and are excellent representations of the data. The rate constants for vibrational excited O_2 can then be derived, by assuming that all excited vibrational states of O_2 react at the same rate. Since most of the excited population is in $v = 1$, this appears to be a reasonable assumption. The populations of $v = 0$ and $v > 0$ are calculated using the harmonic oscillator approximation, and the rate constant for $v = 0$ is taken as the drift tube rate constant. Equation (4.3.2) is then solved for k_1. The result is shown in figure 4.4 as the dashed line. The vibrationally excited rates are about 2–3 times higher than the ground state rate. Note this analysis is different from our original paper (Hierl *et al* 1997) where rotational energy was assumed not to influence the rate constant. The increase in rate constant may be attributed to changes in Franck–Condon factors. For near-resonant states the Franck–Condon factors are larger for the $v = 1$ state than the $v = 0$ state (Krupenie 1972, Lias *et al* 1988). As an alternative, rate constants were also derived for the assumption that $v = 1$ reacts similarly to $v = 0$. This is shown in figure 4.4 as $k (v > 1)$.

4.3.3.3 $O^+ + NO$

The last of the O^+ reactions to be discussed is the charge transfer reaction of O^+ with NO (Dotan and Viggiano 1999). Figure 4.5 shows the rate constants for this reaction plotted as a function of average rotational and translational

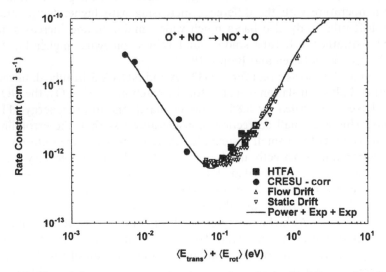

Figure 4.5. Plot of the rate constants for the reaction of O^+ with NO as a function of average translational plus rotational energy. The HTFA (Dotan and Viggiano 1999), CRESU (Le Garrec *et al* 1997), flow drift tube (Albritton *et al* 1977), and static drift tube data (Graham *et al* 1975) are shown as squares, circles, triangles and inverted triangles, respectively.

energy, as well as previous drift tube (Albritton *et al* 1977, Graham *et al* 1975) and ultra-low temperature data (Le Garrec *et al* 1997) corrected as described in our original paper. The combined datasets fit on one curve, showing the equivalence of rotational and translational energy in controlling the reactivity. The agreement between the highest temperature points and the drift tube data indicate that vibrational excitation to $v = 1$ does not substantially increase the rate constant.

Only by combining several datasets can the typical behavior for a slow ion–molecule reaction be observed, i.e. an initial decline in the rate constants followed by an increase at higher temperature/energy. The minimum does not show up clearly in any one dataset. The combined data look as though they could be fitted to a power law plus exponential, similar to what was done for the O_2 reaction. However, this does not fit the data well, but a power law plus two exponentials does. This fit is shown in figure 4.5. The slowness of the reaction has been attributed to a spin forbidden process (Ferguson 1974b). The lower activation energy (0.25 eV) appears well correlated with the $^3A_1$ and $^3B_1$ states of the NO_2^+ intermediate (Bundle *et al* 1970). Production of $NO^+(^3S)$ is endothermic by approximately 2 eV, correlating well with the 2.3 eV second activation energy.

4.3.3.4 $N^+ + O_2$

The above systems all have concise stories as to how different types of energy affect reactivity. In contrast, the reaction of N^+ with O_2 is more complicated. Three drift tube studies show flat translational energy dependencies with the rate constant approximately half the collision rate (Howorka *et al* 1980, Johnsen *et al* 1970, McFarland *et al* 1973b). In contrast, both the early HTFA data (Dotan *et al* 1997) and NOAA temperature dependence (Lindinger *et al* 1974) found the rate constant to increase with increasing temperature until the rate saturated at approximately the collision limit at 1000 K as shown in figure 4.6. Little vibrational excitation occurs at lower temperatures where the difference occurs. An upper limit for the $v > 0$ rate constant is shown (k_{max}) and cannot explain the difference. This rate constant is derived assuming that the v = 0 rate constant is given by the NOAA drift tube data and that all vibrationally excited O_2 reacts at the Langevin capture rate. Another possibility is that N^+ has three spin–orbit states. However, the equilibrium distributions of the three states in the two types of experiments are not different enough to completely explain the data, leaving rotational energy as the likely explanation. This conclusion would indicate that rotational energy is more efficient than translational energy in driving this reaction.

However, in writing a recent review on internal energy dependencies derived from comparisons of the HTFA data to kinetic energy data (Viggiano and Williams 2001), it became clear that this reaction was an

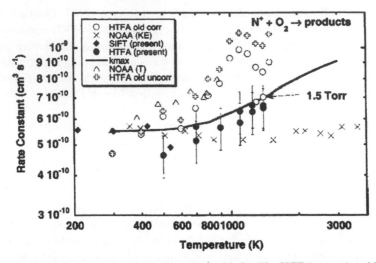

Figure 4.6. Rate constants for the reaction of N^+ with O_2. The SIFT (present) and HTFA (present) points are from the most recent study (Viggiano *et al* 2003). The NOAA kinetic energy (KE) data are from McFarland *et al* (1973b), the temperature data NOAA (*T*) are from Lindinger *et al* (1974). The HTFA old corr and HTFA old uncor refers to the published HTFA data (Dotan *et al* 1997) with and without the thermal transpiration correction. The error bars are ±15% on the present HTFA data. The old HTFA data taken at 1.5 torr are indicated by an arrow.

anomaly. Most of the difference between the temperature and kinetic energy data for this reaction had to be assigned to rotational energy. No other reaction of the dozens studied had a similar dependence on rotational energy. In all other cases involving species that do not have large rotational constants, rotational energy either behaved similarly to translational energy or had a negligible influence on reactivity. The unusual nature of the results prompted us to re-examine the kinetics in both the HTFA and the selected-ion-flow tube (SIFT) in our laboratory.

Figure 4.6 shows the rate constants as a function of temperature for different experiments, including the most recent HTFA and SIFT results (Viggiano and Williams 2001, Viggiano *et al* 2003), a previous drift tube measurement (McFarland *et al* 1973b) and the two previous studies at high temperature (Dotan *et al* 1997, Lindinger *et al* 1974). The previous HTFA study is plotted with and without a thermal transpiration correction for the capacitance monometer (Poulter *et al* 1983). The drift tube study shown in figure 4.6 is in good agreement with two other studies that are not shown for simplicity (Howorka *et al* 1980, Johnsen *et al* 1970). The drift tube studies show rate constants that are independent of kinetic energy. The SIFT data show no discernible temperature dependence from 200 to 550 K, in agreement with the drift tube results. The most recent

HTFA results (Viggiano and Williams 2001, Viggiano *et al* 2003) show a temperature dependence essentially equal to the relative error limits, i.e. very small. The two previous studies at high temperature found rate constants that increased with increasing temperature up to 1000 K. Above this temperature, the previous HTFA study found a leveling off at the collision rate. Thus, the new HTFA temperature studies are in disagreement with the previous ones.

Part of the discrepancy is due to thermal transpiration (Poulter *et al* 1983) as can be seen in figure 4.6. However, this is only a small part of the disagreement. Due to the disagreement between the two sets of HTFA data, a number of checks were performed on the most recent HTFA data. The SIFT data are in excellent agreement with the new HTFA measurements in the overlapping range and both new datasets lack a strong temperature dependence. In addition to remeasuring the rate constants, the original HTFA data have been re-examined. Data run at 1300 and 1400 K have both been taken at elevated pressure (1.5 torr versus 1 torr). The high pressure points are indicated with an arrow in figure 4.6 and agree with the present measurements. They are shown in the figure as the small circles on the solid line. The difference between the 1 and 1.5 torr rate constants results from incomplete source chemistry at the lower pressure. In other words, not enough N_2 was added to quench all the He^+ and He before the beginning of the reaction zone in the low pressure data. Because He^+ reacts with N_2 to produce both N^+ and N_2^+, insufficient N_2 will lead to a situation where He^+ is the dominant ion at the start of the reaction zone and N^+ and N_2^+ are dominant at the end of the reaction zone, i.e. at the mass spectrometer. Therefore, the disappearance of N^+ with the addition of O_2 was due to He^+ reacting with O_2 rather than N_2 as well as from the reaction of N^+ with O_2. The reaction of He^+ with O_2 is faster than for N^+ and proceeds with a rate constant equal to those in the plateau region of the previous measurements (Ikezoe *et al* 1987). It is not possible to speculate if this was also a problem in the NOAA temperature data as well. Due to the above problem, selected points for O^+ and N_2^+ reacting with O_2 were also measured. The rate constants were very slightly lower than the original values mainly due to the thermal transpiration correction. The small differences are not enough to change any of the original conclusions. No measurements of N^+ reactions with other neutrals have been made in the HTFA.

From a chemical dynamics viewpoint, the new data are easier to interpret. The old data required rotational energy to drive the reactivity much more efficiently than translational energy. No other system studied to date shows such a behavior (Viggiano and Williams 2001). Most systems studied show that rotational and translational energy have the same influence on reactivity. The drift tube data overlap within the error with the new HTFA data except at the highest temperatures. The good agreement between the SIFT and HTFA data with drift tube data implies that neither rotational nor translational energy have a large influence on the rate constants. At higher temperatures, the HTFA data are larger than the drift tube data although just slightly above the 15% relative error limits shown in figure 4.6. This indicates that vibrational excitation probably promotes reactivity. The line in figure 4.6 labeled k_{max} is calculated by taking the $v = 0$ rate constant as the drift tube data and assuming that the rate constants for vibrationally excited O_2 react at the collision rate. The line is in excellent agreement with the present data. This agreement suggests that O_2 ($v \geq 0$) reacts at close to the collision rate, but the small differences between the datasets makes definitive conclusions impossible.

4.3.3.5 $O^- + NO, CO$

The reactions of O^- with NO and CO are associative detachment reactions, forming an electron and NO_2 or CO_2. The data are shown in figure 4.7 (Miller *et al* 1994). While the trends in the data mimic previous work, the scatter is larger. Relative errors of 30% are probably more appropriate and comparisons of translational and rotational energy are inconclusive. Some of this scatter is a result of unwanted chemistry in the flow tube. O^- is normally made in flowing afterglows from electron attachment to N_2O. At low temperature, N_2O does not attach electrons. However, at high

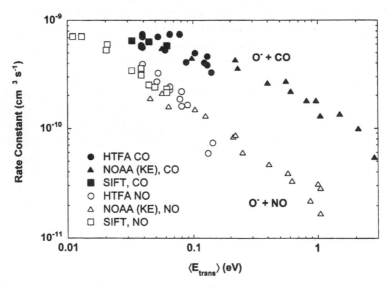

Figure 4.7. Rate constants for the reactions of O^- with CO and NO as a function of average translational energy. Closed and open circles refer to HTFA data for CO and NO (Miller *et al* 1994). Closed and open triangles refer to NOAA drift tube data for CO and NO (McFarland *et al* 1973c). Closed and open squares refer to SIFT data for CO and NO (Viggiano *et al* 1990b; Viggiano and Paulson 1983).

temperatures a distributed source of O^- was found, which was believed to be the result of the electrons from the detachment reactions re-attaching to N_2O in the flow tube. To circumvent this problem, CO_2 was used as the source of O^-, and SF_6 was used to scavenge electrons. In retrospect, the scatter in the data probably indicates that small problems remained. In addition, since the time these measurements were made, it was realized that NO reacts on hot ceramics and the possibility exists that NO may have also reacted on hot stainless steel. In particular, the highest temperature point is lower than the data trends which indicates that NO was destroyed on the surface. Taken at face value, these reaction-rate data seem to indicate that rotational energy does not change the rate constants.

4.3.3.6 $Ar^+ + O_2$, CO

Other interesting examples of vibrational enhancement are the reactions of Ar^+ with CO and O_2 which are very similar (Midey and Viggiano 1998). The rate constants for both reactions are in the $10^{-11}\,cm^3\,s^{-1}$ range and initially decrease with temperature, have minimums at about 1000 K, and increase at higher temperatures. Comparing rate constants from the HTFA to drift tube experiments (Dotan and Lindinger 1982a) at the same sum of translational and rotational energy shows good agreement before the minimum, indicating that the two forms of energy control the reactivity in a similar manner.

The higher temperature data for these two reactions not only indicate that vibrational excitation increases the rate constants but also that vibrational energy changes the rate constants faster than does other forms of energy. In deriving state specific rates from comparisons to translational energy data, it is usually assumed that all vibrationally excited states react at the same rate. However, a couple of observations lead one to believe that $v = 1$ reacts more like $v = 0$ and that $v = 2$ has the larger effect. Little or no enhancement of the rate constants occurs at temperatures where appreciable excitation of the $v = 1$ state occurs. Fits to a power law plus exponential yields activation energies (41.8 and 57.4 kJ/mol for O_2 and CO, respectively) in line with two quanta of vibrational excitation (37.8 and 51.84 kJ/mol for O_2 and CO, respectively) (Huber and Herzberg 1979). If one assumes that only states in $v \geq 2$ enhance the rate constants, one finds the values about a factor of 100 greater than the $v = 0$ rate constants and very close to the collisional limit and independent of temperature. When assuming that all states in $v \geq 1$ react at the same rate, one finds about a factor of 5 enhancement and rates that increase with increasing temperature. In either case the enhancement is much greater than can be explained by energy arguments. The production of $O_2^+(a)$ and $CO^+(A)$ states may lead to the observed behavior. The O_2 reaction will be compared to the similar reaction of N_2^+ below.

4.3.3.7 $N_2^+ + O_2$

The charge transfer reaction of N_2^+ with O_2 provides another example of the equivalency of translational and rotational energy in controlling the reactivity (Dotan *et al* 1997). This reaction is of lesser importance since it only converts one diatomic ion to another. Figure 4.8 shows a plot of the rate constants versus temperature. From room temperature to the minimum value at 1000 K, the rate constants decrease over a factor of 4, and increase by a factor of 2 from 1000 to 1800 K. Excellent agreement is found between the HTFA results and the previous study up to 900 K (Lindinger *et al* 1974). The drift tube study is distinctly different (McFarland *et al* 1973b). The rate constants decrease with increasing translational energy but quite a bit more slowly. The minimum is at a distinctly higher energy. At the minimum, the drift tube rate constants are a factor of 2 larger than those measured in the HTFA, a large difference. A power law plus exponential fits the data well, with all residuals less than 11% of the rate value. The activation energy is 0.29 eV.

The data are shown replotted as a function of rotational plus translational energy in figure 4.9. In this plot there is excellent agreement between the two datasets up to the minimum in the HTFA rate constants. This shows that rotational energy and translational energy are equivalent in

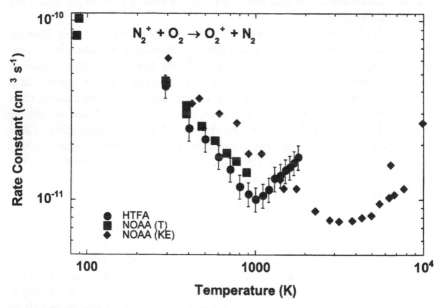

Figure 4.8. Plot of the rate constants for the reaction of N_2^+ with O_2 as a function of temperature. The HTFA (Dotan *et al* 1997), the NOAA (T) (Lindinger *et al* 1974), and NOAA (KE) (McFarland *et al* 1973b) data are shown as circles, squares and diamonds, respectively.

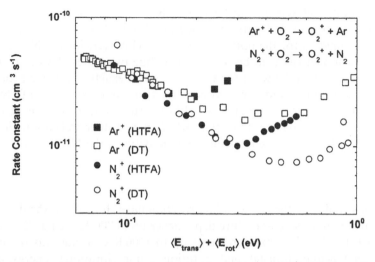

Figure 4.9. Plot of the rate constants for the reactions of Ar^+ and N_2^+ with O_2 as a function of average translational and rotational energy. The HTFA data for Ar^+ and N_2^+ are shown as solid squares (Midey and Viggiano 1998) and circles (Dotan *et al* 1997), respectively. Drift tube data for Ar^+ and N_2^+ are shown as open squares (Dotan and Lindinger 1982a) and circles (McFarland *et al* 1973b), respectively.

controlling the reactivity. The factor of two difference between the two datasets in figure 4.8 disappears. The fact that the rotational effect is so large is due in part to both reactants having rotational energy as opposed the reactions described above where only one reactant had rotational energy. This is one of the few cases for which conclusions about the rotational energy of the ion were able to be made. Above the minimum in the HTFA data, the two datasets diverge due to vibrational excitation in the HTFA experiment.

Several previous studies have shown that vibrational excitation of N_2^+ does not affect the reactivity (Alge and Lindinger 1981, Ferguson *et al* 1988, Kato *et al* 1994, Koyano *et al* 1987). This is probably a result of the fact that there is good Franck–Condon overlap between N_2^+ and N_2 in the same vibrational levels. These studies suggest that the differences above 0.3 eV are due exclusively to O_2 vibrations. If this assumption is correct, then the reaction of Ar^+ with O_2, which has similar energetics, should behave similarly. A power law plus exponential fit to the HTFA data yields an activation energy between the values for one and two quanta of O_2 vibrations. Therefore, rate constants for two cases were derived assuming (1) that the rate constant for $v = 1$ equals $v = 0$ and (2) all vibrationally excited states react at the same rate. The latter assumption yields rate constants a factor of 6 higher than those for $v = 0$ while the former assumption yields rate constants about a factor of 20 higher. In both the Ar^+ and N_2^+ reactions, the upturn has been attributed to the production of the $O_2^+(a\,\Pi_u)$

state (Schultz and Armentrout 1991), which is endothermic in both reactions. For the reaction of Ar^+ with O_2 it appeared that O_2 ($v \geq 2$) was the most likely explanation for the upturn in the data. However, for the N_2^+ reaction the activation energy is in between that for the two states. This also shows up in the minimum between the two datasets. If exactly the same processes are occurring the minimum between the two curves should shift by the recombination energy difference of 0.178 eV. However, the difference in the minimums is slightly less than this, which is a further indication that O_2 ($v = 1$) must already be enhancing the reactivity for the N_2^+ reaction.

4.3.3.8 $O_2^+ + NO$

Previous studies of the O_2^+ with NO reaction have shown that the drift tube dependence and the temperature dependence up to 900 K are flat (Lindinger *et al* 1974, 1975). The measurements up to 1400 K continue this trend and show that neither translational, rotational, nor vibrational energy has a large effect on the reactivity (Midey and Viggiano 1999).

4.3.3.9 $Ar^+, N_2^+ + CO_2$

Ar^+ and N_2^+ have similar recombination energies and for some reactions have similar reactivity, although one is atomic and the other diatomic. The similarities and differences in the reactions of these two ions with O_2 was described above. The reactions of these ions with CO_2 and SO_2 have also been studied in the HTFA (Dotan *et al* 1999, 2000). The reactions with CO_2 proceed exclusively by charge transfer and the SO_2 reaction is mainly charge transfer except at high temperature/energy, where SO^+ is produced by dissociative charge transfer which is endothermic at room temperature. Only CO_2 reactions are discussed here.

   Plots of rate constants versus temperature show clear differences between real temperature and kinetic temperature for the reactions of both Ar^+ and N_2^+ with CO_2 (Dotan and Lindinger 1982b, Dotan *et al* 2000), showing that internal energy has some effect on reactivity. The ability to separate rotational effects diminishes for molecules with three heavy atoms since vibrations are excited at low temperatures. Therefore, the data are replotted as a function of average rotational, translational, and vibrational energy instead of just rotational and translational energy. Such a plot for both reactants with CO_2 is shown in figure 4.10. The Ar^+ data fall on the same line up to energies of 0.4 eV, after which the temperature data are lower than the drift tube data. To test the high temperature behavior, data were taken in both the ceramic and quartz flow tubes, and similar results were found. In contrast, the N_2^+ temperature data are lower than the drift tube dependencies at all energies. Therefore, in both of these reactions internal excitation hinders the reactivity more than translational excitation.

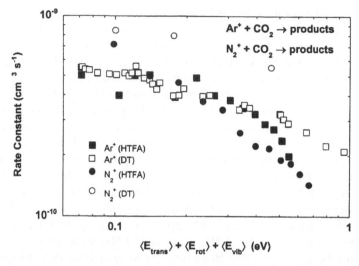

Figure 4.10. Plot of the rate constants for the reactions of Ar^+ and N_2^+ with CO_2 as a function of average translational and rotational energy. The Ar^+ HTFA (Dotan *et al* 1999), N_2^+ HTFA (Dotan *et al* 2000), Ar^+ drift tube (Dotan and Lindinger 1982b), and N_2^+ drift tube (Dotan *et al* 2000) data are shown as solid squares, solid circles, open squares and open circles, respectively.

As shown above, rotational energy only occasionally has a different effect than translational energy and the differences are probably due to CO_2 vibrations since N_2^+ vibrations are mostly unexcited (5% at 1400 K). The data show that the rate constant difference is bigger for the N_2^+ reaction.

4.3.4   Summary

One goal of high temperature experiments is to measure reactions at conditions relevant to air plasma environments. The data so far have demonstrated the importance of making 'true' high temperature measurements. However, it is not always possible to measure every reaction due to experimental and time constraints. Thus, it is useful to look for trends in the data so that better extrapolations of lower temperature data can be made for modeling applications. Trends are also important from a fundamental point of view. The study of internal energy effects has been summarized previously and several trends were noted (Viggiano and Morris 1996, Viggiano and Williams 2001). Some of the relevant conclusions of that work are outlined below.

In most ion–molecule reactions, rotational and translational energy are equivalent in controlling reactivity, at least in the low energy range where most of the data have been taken. This is true for both the ion and neutral rotational energy, although the conclusion has been tested for only a few systems for ion rotations. In the higher energy range, the data are too sparse to make a conclusion. There has been much more work on the effect of vibrational excitation on ion reactivity and much of the work up to 1992 has been summarized in two books (Baer and Ng 1992, Ng and Baer 1992). For diatomic and a few triatomic molecules, it has been possible to detect the product ion vibrational state by chemical means, the so-called monitor ion method (Durup-Ferguson *et al* 1983, 1984, Ferguson *et al* 1988, Lindinger 1987). The most detailed work on internal energy effects is often done using resonance enhanced multiphoton ionization (REMPI) to prepare ions in specific vibrational states. This technique was used extensively by Zare and coworkers (Conaway *et al* 1987, Everest *et al* 1998, 1999, Guttler *et al* 1994, Poutsma *et al* 1999, 2000, Zare 1998) and Anderson and coworkers (Anderson 1991, 1992a, 1997, Chiu *et al* 1992, 1994, 1995a,b, 1996, Fu *et al* 1998, Kim *et al* 2000a,b, Metayer-Zeitoun *et al* 1995, Orlando *et al* 1989, 1990, Qian *et al* 1997, 1998, Tang *et al* 1991, Yang *et al* 1991a,b) in guided-ion beams. Leone and Bierbaum (Frost *et al* 1994 1998, Gouw *et al* 1995, Kato *et al* 1993, 1994, 1996a,b, 1998, Krishnamurthy *et al* 1997) have used LIF to monitor vibrational excited N_2^+ ions in a selected ion flow tube to study collisional deactivation and vibrational enhancement of the charge transfer rate constant of $N_2^+(v = 0\text{–}4)$.

The ability to predict the behavior of complex reaction systems is particularly important for modeling applications, which often require extrapolation of a limited amount of existing data to conditions of practical interest. While the effect of rotational energy seems to be generally predictable, there are enough exceptions to warrant caution in making extrapolations. Furthermore, vibrational energy often displays state-specific effects both in overall reactivity and formation of new products. Therefore, it is still very difficult to predict reactivity at high temperature by extrapolating translational energy dependencies obtained at low temperature. In light of this fact, the next section outlines recent experimental and theoretical efforts aimed at developing a detailed understanding of the vibrational energy dependence of chemical reactivity.

4.4   Non-Equilibrium Air Plasma Chemistry

4.4.1   Introduction

In the present section, we consider the plasma chemical dynamics of a domain that is not in chemical equilibrium within a certain timescale and volume. This can be the case in high E/N conditions, where ion velocity distributions can be highly skewed with respect to a Maxwellian. Plasma kinetic models for non-equilibrium chemical systems are significantly more challenging because kinetics based on equilibrium rate coefficients, $k(T)$, described by some Arrhenius form as discussed in section 4.1, are no longer applicable. Instead, the models depend on knowledge of the non-Maxwellian heavy-body velocity distributions, the relative velocity dependence of chemical reaction cross sections, as well as the molecular vibrational distributions and the related vibrational state-to-state cross sections. In the following it is assumed that rotational energy is equivalent to translational energy at the total collision energies encountered in air plasmas. This assumption has been shown to be valid in several reactions presented in section 4.3.

As we have learned in the preceding sections, when molecular ions are formed through electron-impact ionization, photo-ionization or chemical processes such as atomic ion reactions with molecules (e.g. $O^+ + N_2 \longrightarrow NO^+ + N$, $O^+ + H_2O \longrightarrow O + H_2O^+$) and three-body association, they are formed in translational, rotational and vibrational energy distributions that differ greatly from Boltzmann distributions. This is particularly the case for three-body recombination processes:

$$A^+ + B + M \longrightarrow AB^+ + M \qquad (4.4.1)$$

which are important contributors to molecular ion formation in high pressure plasmas. In process (4.4.1), the nascent vibrational distributions of AB^+ are highly skewed towards vibrational levels near the AB^+ dissociation limit. If the system does not equilibrate, an understanding of the plasma dynamics requires knowledge of the chemical fate of these highly excited molecular ions. The vibrational energy dependence of competing dissociative, chemically reactive and relaxation collisions dynamics then becomes a critical component of a plasma kinetic model. The vibrational effects are particularly strong for endothermic processes such as collision-induced dissociation (CID). The latter is the reverse process of reaction (4.4.1), and microscopic reversibility arguments suggest that if reaction (4.4.1) favors product molecular ions in high vibrational states, the reaction probability of the reverse reaction should also be enhanced by vibrational excitation of the reactant molecular ion. Note that for endothermic processes, vibrational enhancement, or *vibrational favoring*, signifies a greater increase in reactivity due to vibrational energy than an equivalent amount of translational energy. Vibrational effects of chemical processes tend to decrease as the number of atoms of the participating molecules increases because the propensity to randomize the vibrational energy in a collision increases with the number of vibrational modes. Vibrational effects, however, cannot be neglected in air plasmas, given the preponderance of diatomic molecular species.

The determination of the vibrational energy dependence of chemical reactivity has been a particular challenge to experimentalists and theorists. State-selected chemical dynamics studies have to a large degree been limited to low vibrational levels where the vibrational energy represents only a small fraction of the molecular dissociation energy. Meanwhile, accurate, fully three-dimensional quantum dynamics calculations at the current state-of-the-art are rarely applied at total energies above 2 eV, even for simple triatomic systems, due to the rapidly increasing number of accessible product quantum channels with energy (Clary 2003). The demand for knowledge of kinetics at high levels of vibrational excitation has been particularly high in the rarefied gas dynamics community, which is the source of a considerable body of work dedicated to finding vibrational scaling laws for chemical reactivity and energy transfer that cover vibrational energy ranges comparable with bond dissociation energies. In section 4.4.2, concepts applied to model the translational and vibrational energy dependence of chemical processes will be presented. It is impossible to provide a satisfactory synopsis of the field which encompasses the vast research area of chemical reaction dynamics. The purpose of this section is to familiarize the reader with the generally accepted theories of the reaction dynamics community and to align them with the needs of the community that model non-equilibrium environments on a molecular level, such as non-equilibrium air plasmas. Arguments will be presented to adopt a universally applicable model with minimal adjustable parameters based on the work by Levine and coworkers (Levine and Bernstein 1972, 1987, Rebick and Levine 1973). In section 4.4.3, recent advances will be presented on theoretical and experimental efforts to study chemical dynamics at high levels of vibrational excitation.

4.4.2 Translational and vibrational energy dependence of the rates of chemical processes

Equilibrium models of chemical kinetic systems as discussed in the previous section depend on rate coefficients which are usually given by a modified Arrhenius dependence on temperature defined in equation (4.1.2). In non-equilibrium conditions, the temperature, T, no longer describes the energy distributions of the system, and it becomes more practical in describing the chemical kinetics in terms of cross sections as a function of the relative velocity and reactant vibrational and rotational quantum states, $\sigma_{v,J}(v)$, which are related to the equilibrium rate coefficient through an extension of equation (4.1.1):

$$k(T) = \sum_{v,J} f_T(v)f_T(J) \int_{v_a}^{\infty} f_T(v)\sigma_{v,J}(v)v\,dv \qquad (4.4.2)$$

where v and J refer to vibrational and rotational quantum numbers of the reactants (note that each reactant, if polyatomic, has multiple vibrational quantum numbers for each vibrational mode), the functions f_T refer to the normalized velocity and quantum state Boltzmann equilibrium distributions at a temperature T, and v_a is the threshold relative velocity,

$$v_a = \sqrt{2(E_a - E_v - E_J)/\mu} \qquad (4.4.3)$$

where μ is the reduced mass of the reactants and E_v and E_J represent the vibrational and rotational energy, respectively, for the specific set of quantum states. The complete, accurate non-equilibrium model must also account for the reaction product state distributions, and a rigorous model thus requires state-to-state cross sections, $\sigma_{v' \to v'', J' \to J''}(v)$, where $'$ and $''$ refer to the reactant and open product channel quantum states, respectively. It is easily seen that the master equations of a non-equilibrium plasma model can require thousands of state-to-state cross sections. The problem is somewhat reduced by assuming that rotational energy has the same effect as translational energy on cross sections.

Regrettably, there is not a one-glove-fits-all approach to modeling the translational and vibrational energy dependence of chemical reaction cross sections and associated product state distributions. Each bimolecular collision system is governed by its own unique set of $(3N - 6)$-dimensional potential energy surfaces, where N is the number of atoms of a particular chemical system, as well as by the respective atomic masses and associated kinematics. Meanwhile, there are no air plasma chemical processes that have been comprehensively studied over the pertinent energy range using either exact quantum scattering methods or state-resolved experiments. Efforts to model non-equilibrium environments thus rely on approximate approaches that recover some of the physical properties of chemical processes as retrieved from existing physical chemical research.

Historically, the energy dependence of chemical reaction and inelastic collision cross sections, and the determination of product energy distributions, has been treated using statistical approaches. This approach assumes that molecular collisions form an intermediate complex that redistributes the translational, rotational, vibrational, and in some instances electronic energy equally among all quantum levels of the complex (Levine and Bernstein 1987). Vibrational or electronic effects, as discussed earlier, are then regarded as a deviation from this so-called prior or statistical case. The development of statistical chemical reaction models followed two separate schools of thought: the rarefied gas dynamics community has used the semi-empirical analytical Total Collision Energy (TCE) (Bird 1994) cross section and Borgnakke–Larsen energy disposal models (Borgnakke and Larsen 1975), while in the chemical physics community statistical models were spearheaded through the information theoretical approaches by Levine (Levine 1995, Levine and Manz 1975), phase space theory (Chesnavich and Bowers 1977a,b, Light 1967, Pechukas *et al* 1966), and transition-state theories such as the RRKM theory (Marcus 1952, Marcus and Rice 1951). While the Borgnakke–Larsen approach targets computational efficiency and uses parameterization based on viscosity and transport properties determined for the gases, the physical chemical statistical models use known spectroscopic molecular constants. The methods of the rarefied gas community, as applied to direct simulation Monte Carlo (DSMC) methods, have been described (Bird 1994) and more recently reviewed by Boyd (2001) The cross section models derived by Levine (Levine and Bernstein 1972, 1987, Rebick and Levine 1973) based on statistical arguments and calculations have been used in both communities, and have found great utility in the interpretation of countless experiments of chemical reaction dynamics.

In a statistical approach, barrier free, exothermic reactions involving reactants in their ground electronic and rovibrational states occur with a probability of 1 if an encounter occurs. At low translational energies, E_t, an encounter can be defined by a capture collision associated with spiraling trajectories induced by an attractive interaction potential, $V(R)$ (Levine and Bernstein 1972):

$$V(R) = -C_s R^{-s}.   (4.4.4)$$

where R is the distance between reaction partners. The capture cross section is then given by

$$\sigma = A E_t^{-2/s}   (4.4.5)$$

where A, as in equation (4.4.2), is a scaling parameter. In the case of an ion-neutral encounter, the long-range attractive potential is given by a polarization potential with $s = 4$, thus yielding the well-known Langevin–Gioumousis–Stevenson (Gioumousis and Stevenson 1958) cross section energy dependence with $A = \pi q (2\alpha)^{0.5}$, where α is the polarizability of the neutral and q is the ion charge.

Assuming microscopic reversibility, the translational energy dependence of the cross section for the reverse, endothermic process at translational energies above the activation energy E_a is given by (Levine and Bernstein 1972):

$$\sigma(E_t) = A' \frac{(E_t - E_a)^{1-2/s}}{E_t}   (4.4.6)$$

where A' is again a scaling factor. Unfortunately, microscopic reversibility cannot be applied to integral cross sections blindly since the preferred mechanism (e.g. direct or indirect) can vary significantly between the forward and reverse reactions. Thus, for ion–molecule CID processes, equation (4.4.6) is only adhered to when this process proceeds via a complex (indirect) mechanism. This, however, is normally only the case at very low activation energies. Equation (4.4.6) has been applied more frequently in its more general form:

$$\sigma(E_t) = A' \frac{(E_t - E_a)^n}{E_t}   (4.4.7)$$

where A' and n are adjustable parameters. It is worth noting that $n = 1$ corresponds to the line-of-centers (LOC) hard-sphere model that assumes straight-line trajectories and is readily derived from

$$\sigma = 2\pi \int_0^{R_1 + R_2} P(b)b \, \mathrm{d}b \qquad (4.4.8)$$

where b is the collision impact parameter and R_1 and R_2 are the reactant radii, and the reaction probability $P(b)$ is 1 for all impact parameters where the translational energy associated with the relative velocity component along the line-of-centers when the hard spheres collide exceeds the activation energy, and 0 for larger impact parameters (Levine and Bernstein 1987). Equation (4.4.7) is usually referred to as the modified LOC model, where $n < 1$ is typical for highly indirect, complex forming processes, while $n > 2$ usually signifies a direct, impulsive mechanism. n has also been related to the character of the reaction transition state (Armentrout 2000, Chesnavich and Bowers 1979). It has been shown (Levine and Bernstein 1971) that under the assumption that CID follows a reverse three-body recombination (process (4.4.1)) mechanism, $n = 2.5$ can be expected.

The workings of the modified LOC model are nicely demonstrated in figure 4.11 that compares collision-induced dissociation cross sections as a function of translational energy of the $Ar_2^+ + Ar$ and $Ar_2^+ + Ne$ systems (Miller *et al* 2004). Ar_2^+ has an accurately known dissociation energy of 1.314 eV (Signorell and Merkt 1998, Signorell *et al* 1997). The solid lines are nonlinear least-squares fits of equation (4.4.7) convoluted with the experimental broadening mechanisms (ion energy distribution, target gas motion) to the experimental data. The figure also provides the derived parameters. In case of the $Ar_2^+ + Ar$ system, a threshold or activation energy in good agreement with the spectroscopic dissociation energy (Signorell and Merkt

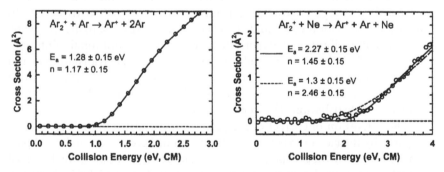

Figure 4.11. Guided-ion beam measurements (Miller *et al* 2003) of the translational energy dependence of collision-induced dissociation cross sections of the $Ar_2^+ + Ar$ and $Ar_2^+ + Ne$ collisions systems. Solid lines are modified line-of-centers (MLOC, equation (4.4.7)) fits to the experimental data. The fits take experimental broadening due to ion energy distributions and target gas motion into account. Activation energies, E_a, and curvature parameters, n, derived from the fits are also provided.

1998, Signorell *et al* 1997) is obtained and the small curvature parameter is close to the hard-sphere case. This relatively indirect behavior is not surprising considering that this collision system is highly symmetric, involving both resonant charge-exchange interactions as well as strongly coupled vibrational modes within the complex. This is also consistent with the low vibrational effects observed (Chiu *et al* 2000). In the $Ar_2^+ + Ne$ case, the onset is considerably more gradual. A free fit of the modified LOC model results in $E_a = 2.46 \pm 0.15\,eV$, considerably higher than the dissociation energy; however, the fit does not recover the weak signal just above the dissociation energy. A second, dashed curve in figure 4.11 is an alternative fit in which the threshold energy was frozen at the spectroscopic value of 1.3 eV. This fit, although not optimal, provides a curvature parameter of 2.46 ± 0.15 which is characteristic of a highly direct dissociation mechanism. The difference in dynamics in comparison with the $Ar_2^+ + Ar$ system can be attributed to the significantly weaker $ArNe^+$ interaction and the lighter mass of Ne.

Using statistical theory, Rebick and Levine (Rebick and Levine 1973) extended equation (4.4.7) to include the effect of vibrational excitation of the reactants:

$$\sigma(E_t, E_{vib}) = A' \frac{(E_t + E_{vib} - E_0)^n}{E_t} \exp(-\lambda F) \qquad (4.4.9)$$

where E_0 is the activation energy not including zeropoint vibrational energy of the reactants, F is the fraction of the total energy in vibration:

$$F = \frac{E_{vib}}{E_t + E_{vib}} \qquad (4.4.10)$$

and λ is the so-called surprisal parameter and determines the degree of vibrational enhancement of the respective reaction. $\lambda = 0$ corresponds to equivalence of vibrational and translational energy (statistical), while $\lambda < 0$ signifies a vibrational enhancement and $\lambda > 0$ a vibrational inhibition. Equation (4.4.9) thus can provide a description of the translational and vibrational energy dependence of reaction cross sections based on three adjustable parameters. Similarly, surprisal analyses can be applied to product state distributions (Levine and Bernstein 1987).

The derivation of correct non-equilibrium chemistry models is severely hampered by the lack of experimentally determined cross section data. Apart from some shock-tube experiments that suffer from poor knowledge of molecular vibrational energy distributions (Appleton *et al* 1968, Johnston and Birks 1972), there have been no experiments to validate the applied scaling laws. Recently, Wysong *et al* (2002) have made a first attempt to compare the various vibrational scaling laws applied in DSMC models for dissociation collisions to experiments on the $Ar_2^+ + Ar$ system (Chiu *et al* 2000). This system was studied with diatomic internal energies generated in the non-equilibrium conditions of a supersonic jet and was observed to exhibit essentially no vibrational effects. The expression by Rebick and Levine (equation (4.4.9), $\lambda \approx 0$) as well as the simple TCE model (which cannot account for deviations from the statistical result) provided the best agreement with the observations while other models, such as the classical threshold-line model with no adjustable parameters by Macharet and Rich (Macharet and Rich 1993) and the maximum entropy model (Gallis and Harvey 1996, 1998, Marriott and Harvey 1994) fared very badly. Other attempts to validate vibrational scaling models of chemical reactions have involved comparison with quasiclassical trajectory (QCT) calculations (Esposito and Capitelli 1999, Esposito *et al* 2000, Wadsworth and Wysong 1997). As will be further iterated in the following section, this is an incomplete description since QCT calculations based on a single potential energy surface do not capture the fact that molecules like N_2, NO, O_2, and their respective ions all have electronically excited states with equilibrium positions well below the dissociation limits. These states can be expected to interfere in the dynamics of the reaction at elevated excitation energies.

4.4.3 Advances in elucidating chemical reactivity at very high vibrational excitation

Most of the work on the dynamics of highly vibrationally excited molecules has focused on vibrational energy transfer. There is a considerable body of experimental work where molecules are prepared in high vibrational states using laser techniques such as stimulated emission pumping (SEP) (Dai and Field 1995, Silva *et al* 2001), and their decay is probed while the molecules undergo collisions in a buffer gas or in a crossed-beam configuration. Note that most SEP experiments do not probe the fate of the highly-excited molecules, merely the removal from the respective quantum state. The theory of vibrational energy transfer of highly vibrationally excited molecules is also extensive, ranging from three-dimensional quantum scattering studies, to semi-classical methods (Billing 1986), as well as analytical models such as the Schwartz–Slawsky–Herzfeld (SSH) theory (Schwartz *et al* 1952) and more recently the nonperturbative model of Adamovich and Rich (1998). One of the most intensively studied systems is the $O_2(\upsilon) + O_2$ systems, where stimulated emission pumping experiments in the group of Wodtke (Jongma and Wodtke 1999, Mack *et al* 1996, Price *et al* 1993, Rogaski *et al* 1993, 1995) discovered relaxation rates in excellent agreement with quantum dynamics calculations (Hernandez *et al* 1995) up to $\upsilon = 25$, above which the relaxation rates increase rapidly with υ and dramatically diverge from the theoretical values. The discrepancy has been interpreted to be due to an electronic interaction associated with the $O_2\,(b^1\Sigma_g^+)$ state, for which the respective potential energy surface was not included in the calculations. A probe of the final state distribution for these high υ states found a large fraction of multiquantum vibrational relaxation (Jongma and Wodtke 1999), consistent with an electronic mechanism. This example demonstrates nicely that surprises can be expected at vibrational energies in the proximity of excited electronic states which can participate in the dynamics.

The derived V–V and V–T rate coefficients determined for pure CO and $CO + N_2$ and O_2 collisions have been successfully used to derive highly non-equilibrium vibrational distributions of CO optically pumped by a CO laser at near atmospheric pressures (Lee 2000). The literature on chemical reaction dynamics at high levels of vibrational excitation is considerably sparser than that of vibrational energy transfer. The main experimental problem is producing sufficient quantities of state-selected reactants in order to be able to probe reaction products. As mentioned in section 4.3, the field of ion–molecule reaction dynamics has provided the most extensive studies of state-to-state reaction dynamics at controlled translational energies where absolute cross sections have been produced (Ng 2002, Ng and Baer 1992b). The significant body of work comes from the straightforward means of controlling the translational energy of reactants, the high sensitivity of mass spectrometric means to detect reactively scattered ionic products, and the ability to prepare molecular ions in selected vibrational levels using Resonance Enhanced Multiphoton Ionization (REMPI) (Anderson 1992b, Boesl *et al* 1978, Zandee and Bernstein 1979) or direct VUV (Koyano and Tanaka 1992, Ng 1992) techniques. Ion–molecule reaction studies using photo-ionization ion sources have provided an extensive understanding of state-to-state chemical reaction dynamics; however, the reactant vibrational levels have been limited to low excitation energies representing a small fraction of the dissociation energy of the respective molecular ions.

Very recently, Ng and coworkers have succeeded in preparing H_2^+ beams in all but the two highest vibrational states of the ground state (Qian *et al* 2003a,b, Zhang *et al* 2003). Their approach is based on recent advances in high-resolution photoelectron spectroscopy using a synchrotron light source (Jarvis *et al* 1999). A schematic of their apparatus is shown in figure 4.12, which is situated at the Lawrence Berkeley Advanced Light Source (ALS) synchrotron facility. Monochromatic ($\sim$10 cm^{-1} FWHM) VUV of the Chemical Dynamics Endstation 2 is used to promote hydrogen molecules to high-*n* Rydberg states just below the ionization limit of a targeted excited rovibrational state of the ion. In the multibunch mode of the ALS storage ring, there is a 104 ns dark-gap at the end of every 656 ns ring period. Approximately 10 ns after the onset of this dark gap, a pulsed electric field of approximately 10 V/cm and 200 ns duration is applied to the electrodes spanning the photo-ionization region. This pulsed field causes field-ionization of the resonantly populated high-*n* Rydberg molecules. This form of ionization is called pulsed-field ionization (PFI). The PFI photo-ion (PFI-PI) is accelerated towards an ion beam apparatus, while the associated, zero-kinetic energy photoelectron, or PFI-PE, is

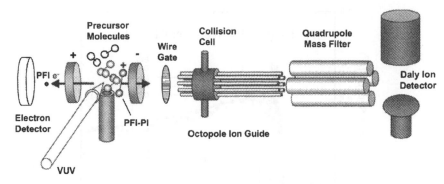

Figure 4.12. Schematic representation of the Pulsed-Field Ionization Photoelectron Secondary Ion Coincidence (PFI-PESICO) apparatus constructed at Endstation 2 of the Chemical Dynamics Beamline at the Lawrence Berkeley Advanced Light Source (Qian *et al* 2003a).

accelerated towards an electron detector. As the Rydberg states are excited, however, a significant number of ions in lower ionic states are also produced with associated electrons that have excess energies, $E_{hv} - E_{vJ}^+$, where E_{hv} and E_{vJ}^+ arc the photon and ionic internal energy, respectively. In order to get state selection, the electron detector is gated to accept PFI-PEs within a narrow time-window at a fixed delay with respect to the pulsed electric field. If a PFI-PE is detected, a fast, interleaved comb wire gate (Bradbury and Nielsen 1936, Vlasak *et al* 1996) is opened at a specific delay with respect to the PFI-PE pulse for $\sim$100–200 ns to allow the associated PFI-PI to pass. This approach suppresses signal due to false coincidences by orders of magnitude.

Ions transmitted through the wire gate enter a guided-ion beam (GIB) apparatus (Gerlich 1992, Teloy and Gerlich 1974) that has the virtue of examining ion–molecule collisions within the guiding fields of an rf octopole, thereby ensuring 100% collection of all scattered ions. Qian and co-workers (Qian *et al* 2003a,b, Zhang *et al* 2003) used a tandem octopole set-up, where the first octopole guides the ions through a collision cell containing the target gas. The second octopole transports reactant and product ions to a quadrupole mass filter for mass analysis prior to detection using a Daly ion detector (Daly 1960). Cross sections are determined from the primary and secondary ion true coincidence signals and the measured target gas density.

Zhang *et al* (2003) used this new coincidence approach in a systematic study of the vibrational energy dependence of the $H_2^+ + Ne$ proton-transfer reaction ($NeH^+ + H$ products) which is endothermic by 0.54 eV. Figure 4.13 shows the translational energy dependence of the cross section for the ground vibrational state of H_2^+. The dashed line is a fit to the data points of equation (4.4.7) including a convolution of the experimental

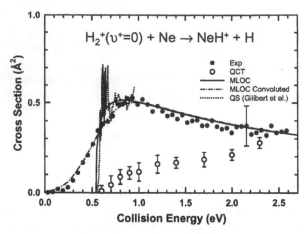

Figure 4.13. Translational energy dependence of the $H_2^+ + Ne$ proton transfer reaction for reactant ions in the ground vibrational state. A modified line-of-centers (MLOC) fit including convolution of experimental broadening mechanisms is applied to the data (dash-dot line). The deconvoluted fit (solid line) is also shown. The experimental data are compared with QCT and quantum scattering (QS) calculations by Gilibert *et al* (1999).

broadening mechanisms, primarily governed by the ion energy distribution with full-width at half maximum (FWHM) of ~0.3 eV. The solid line is the deconvoluted best-fit function with parameters $A' = 0.66\,\text{Å}^2\,\text{eV}^{1-n}$, $n = 0.353$. The very low curvature parameter signifies an almost vertical onset at the threshold of 0.54 eV, which is characteristic of long-lived intermediates. Fully three-dimensional quantum-theoretical studies (Gilibert *et al* 1999, Huarte-Larrañaga *et al* 1998, 2000) have discovered the existence of a dense spectrum of resonances for this system that greatly enhances the reactivity near threshold. The calculations of Gilibert *et al* are also shown in figure 4.13, exhibiting excellent agreement with the measurements of Zhang *et al*. Also shown are quasiclassical trajectory calculations by Zhang *et al* (2003), demonstrating that classical methods do not capture the mechanism near threshold.

Figure 4.14 shows $H_2^+ + Ne$ proton-transfer cross sections using the PFI-PESICO approach measured for a large number of reactant vibrational levels at three translational energies, 0.7, 1.7 and 4.5 eV. The proton-transfer reaction becomes exothermic for $\upsilon^+ = 2$. The measurements are compared with QCT calculations, which also include the dissociation channel. The latter could not be measured with the current experimental set-up of Zhang *et al* (2003). The cross sections are shown on a vibrational energy scale. At a translational energy of 0.7 eV, Zhang *et al* succeeded in measuring cross sections for all vibrational levels from $\upsilon^+ = 0$–17. All states were produced in the $N^+ = 1$ rotational level. The $\upsilon^+ = 17$, $N^+ = 1$ level is a

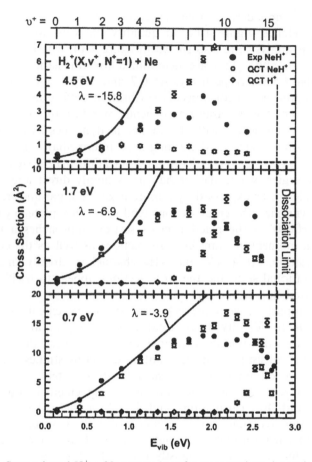

Figure 4.14. State-selected H_2^+ + Ne proton transfer cross sections determined using the PFI-PESICO approach. The measurements at three translational energies are shown on a vibrational energy scale and are compared with QCT calculations that also include cross sections for the dissociation channel. The respective vibrational quantum states are shown at the top of the figure. Also shown are the results of a surprisal analysis based on equation (4.4.9) (solid lines).

mere 0.03 eV below the dissociation limit, also indicated in the figure. Previous attempts to measure state-selected dynamics of H_2^+ using ion beams (Ng and Baer 1992) were limited to $v^+ = 0$–4. The PFI-PESICO measurements by Zhang *et al* provide the first glimpse of chemical reactivity of molecules excited to levels near the dissociation limit.

At low vibrational levels, a significant enhancement of the reaction cross section is observed at all translational energies. A surprisal analysis was conducted at low vibrational energies based on equation (4.4.9) and the parameters A' and n derived from the ground vibrational state translational energy dependence (figure 4.14) to quantify the vibrational effects. The results of the analysis are also shown in figure 4.14, where parameters, λ, of -3.9, -6.9, and -15.8 are determined for translational energies of 0.7, 1.7, and 4.5 eV, respectively. At 0.7 and 1.7 eV it is seen that this approach allows good predictions of the vibrational effects at low vibrational levels; however, the λ parameter depends significantly on translational energy. This is consistent with the change in dynamics as one goes from low translational energies, where long-lived intermediates associated with resonances that cause some energy randomization play a significant role, to higher energies, where the mechanism is highly direct and vibrational effects are higher, as expressed by a more negative λ parameter. At higher vibrational energies, the cross sections tend to reach a plateau due to both saturation effects as the reaction cross section approaches a total cross section (e.g. momentum transfer cross section) as well as the competition with dissociation. At 4.5 eV, the CID channel is already open for the ground vibrational state and the cross sections appear to oscillate outside of the reported statistical errors.

The comparison with quasiclassical trajectory calculations allowed Zhang *et al* to identify three total energy ranges: at low energies, $E_{tot} < 1$ eV, the state-selected experimental values exceed the QCT predictions, which is consistent with the quantum scattering studies that identified the importance of quantum resonances for this system; at intermediate energies, 1 eV $< E_{tot} < 3$ eV, very satisfactory agreement is found between experiment and QCT calculations, and the vibrational enhancement of the proton-transfer reaction can be quantified with a surprisal formalism according to equation (4.4.9), at high energies, $E_{tot} > 3$ eV, the measured proton-transfer cross sections mostly exceed QCT cross sections. This is particularly marked at 0.7 eV, where the measurements exhibit significant reactivity for states nearest the dissociation limit, while the QCT calculations predict more suppression of reaction due to competition with the dissociation channel. It is possible that QCT significantly overpredicts the dissociation cross section for high vibrational levels. At 1.7 eV, the high v^+ state cross sections vary dramatically from one vibrational quantum state to the other. The authors attribute the failure of the QCT calculations in capturing the dynamics at the highest energies to inadequacies of the applied H_2Ne^+ potential energy surface (Pendergast *et al* 1993) near the dissociation limit, and/or the increased importance of nonadiabatic effects and excited-state potential energy surfaces. So far, quantum studies of this benchmark system have not been conducted at total energies exceeding 1.1 eV.

The experimental results for the $H_2^+(v^+) + Ne$ system demonstrate again that, even for such a simple system, QCT can provide some answers, but substantial deviations can occur at energies where quantum effects are important and at energies where additional electronic states become accessible and the dynamics, therefore, is rendered more complicated by dynamics involving multiple potential energy surfaces. This is usually the case for dissociation channels because multiple states usually converge to a dissociation limit. Multi-surface QCT calculations involving surface hopping have in fact provided good agreement with state-selected experiments for the $H_2^+ +$ He CID system (Govers and Guyon 1987, Sizun and Gislason 1989). Both experimental and theoretical results provided evidence for the importance of a non-adiabatic mechanism involving electronic excitation to the surface associated with the repulsive $H_2^+ (^2\Sigma_u^+)$ state. The situation is far more complicated for dissociation systems involving air plasma neutrals O_2, N_2, and NO or ions O_2^+ and NO^+, since all of these molecules have excited electronic states with equilibrium energies substantially below the first dissociation limit. From these arguments, it must be considered doubtful that QCT calculations based only on the ground-state potential energy surface (and thus excluding surface-hopping mechanisms) can provide realistic dissociation and reaction cross sections for such systems. However, Capitelli and coworkers (Esposito and Capitelli 1999, Esposito *et al* 2000) have conducted extensive QCT calculations on the $N_2(\upsilon) + N$ dissociation system using a semi-empirical potential energy surface (Lagana *et al* 1987) and the resulting state-specific dissociation rate coefficients, when converted to global dissociation rates, were in good agreement with shock-tube measurements of the temperature dependence of the dissociation rate as provided by Appleton *et al* (1968). Esposito *et al* (2000) suggested that dissociation rates from high vibrational levels of the ground state would be similar to those of near-resonant low vibrational levels of electronic states. While this may be the case for the $N_2(\upsilon) + N$ system, the work by Wodtke and co-workers (Mack *et al* 1996, Price *et al* 1993, Rogaski *et al* 1993, 1995, Silva *et al* 2001) on $O_2(\upsilon) + O_2$ discussed earlier provided evidence of marked interference by excited electronic states. The day has yet to come when exact quantum approaches can address such complicated systems at high levels of excitation.

Finally, we conclude that equations (4.4.7) and (4.4.9) provide a good start to describe endothermic chemical processes, at least in the cross section growth phase of energy. Cross section parameters can be obtained from fits to measurements or calculations of the translational energy dependence of cross sections for ground state reactants, or from the temperature dependence of rate coefficients and an appropriate transformation. The latter approach, however, can only be reliably applied at low temperatures, where vibrational excitation of the reactants is insignificant. Vibrational effects, however, as quantified through the λ parameter, need a more careful consideration of the dynamics. The recent PFI-PESICO measurements (Qian *et al* 2003a,b, Zhang *et al* 2003) provide hope that similar studies will soon be applied to larger diatomic systems of relevance to air plasmas, such as O_2^+ and NO^+.

4.5 Recombination in Atmospheric-Pressure Air Plasmas

An important loss process for total charge density in atmospheric plasmas is the recombination of electrons with positive ions. In situations where negative ions are present, ion–ion recombination will also occur. However, the focus of this section is on electron–ion recombination. Atomic ions recombine exceptionally slowly with electrons since the large amount of energy gained during a recombination event must be emitted as a photon or removed via an interaction with a third body (McGowan and Mitchell 1984). The exothermicity is equal to the ionization potential of the atom. These processes, introduced in section 4.1, are called radiative or dielectric recombination and three-body recombination, respectively. In molecular ion recombination, energy can also be released as kinetic and internal energy, and the rate constants associated with this mechanism are usually extremely fast. This mechanism is called dissociative recombination and for a diatomic species is represented as

$$AB^+ + e^- \longrightarrow A + B + \text{kinetic energy.} \tag{4.5.1}$$

For polyatomic species, formation of three neutral particles is common (Larsson and Thomas 2001). Process (4.5.1) is the major electron loss process unless all positive ions are atomic or negative ions are present in concentrations of a factor of ten or greater than electrons. For air plasmas at temperatures of a few thousand Kelvin, the dissociative recombination loss process is dominant and involves mainly O_2^+, NO^+, N_2^+, and H_3O^+ and its hydrates (Jursa 1985, Viggiano and Arnold 1995). These systems are the only ones discussed here. Note, however, that in low temperature air plasmas, electron attachment to O_2 to produce negative ions is a very important electron loss mechanism.

Rate constants for dissociative recombination have been measured for decades under thermal conditions and as a function of electron energy for a variety of stable species (Adams and Smith 1988, McGowan and Mitchell 1984, Mitchell and McGowan 1983). In contrast, little was known about the product distributions of such reactions until the recent advent of storage ion rings (Larsson *et al* 2000, Larsson and Thomas 2001). Now, not only can product speciation for polyatomic species be measured, but also the product states for small systems, especially diatomic molecules. Very recently, measurements of both cross sections and product distributions for vibrationally and electronically excited species have been made (Hellberg *et al* 2003, Petrignani *et al* 2004). This is extremely important since theoretical calculations of dissociative recombination kinetics are very difficult and often fail to match experiment, although the agreement is improving for small systems. In this section, recent work done in storage rings is emphasized since those experiments yield the most detailed information.

4.5.1 Theory

Guberman (2003a) has recently reviewed the important mechanisms for dissociative recombination. Historically, two mechanisms are usually described. They have been termed direct and indirect (McGowan and Mitchell 1984). Direct recombination was originally proposed by Bates and Massey (1947) to explain the almost complete disappearance of the ionosphere at night. Indirect processes were first attributed to Bardsley (1968). Dissociation is efficient when there is a repulsive state of the neutral molecule in the vicinity of the ionic state, although mechanisms presently exist for which there is no curve crossing. Figure 4.15 illustrates the direct and indirect processes for a particular channel of O_2^+ recombination (Guberman and Giusti-Suzor 1991). Here the $^1\Sigma_u^+$ state of O_2 intersects the $X\,^2\Pi_g$ state of O_2^+. In the direct mechanism shown in figure 4.15, an electron with energy ε is captured from $O_2^+(v=1)$ into the $^1\Sigma_u^+$ dissociative state of the neutral and the dissociation occurs directly on the repulsive potential. This type of process is rapid if the neutral state crosses near a turning point of a vibrational level of the ion so that the Franck–Condon factor between the states is large. The nuclei separate rapidly on the repulsive curve if the auto-ionization lifetimes are smaller than those for dissociation. Direct recombination leads to cross sections that vary as E^{-1} (McGowan and Mitchell 1984).

The indirect mechanism involves the electron being captured into a vibrationally excited Rydberg state. In figure 4.15, an electron of energy ε' is captured into the $v=5$ level of the $^1\Sigma_u^+$ Rydberg state. Either vibronic

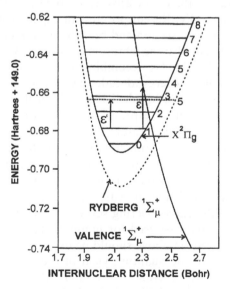

Figure 4.15. Potential energy curves involved in O_2^+ dissociative recombination. Terms are defined in the text (Guberman and Giusti-Suzor 1991).

or electronic coupling leads to predissociation on the repulsive curve. Since the Rydberg levels are discrete, indirect recombination results in resonances. For ions with many atoms, the resonances are usually not detectable except that the cross section changes with energy differently than E^{-1}.

4.5.2 $O_2^+ + e^-$

Dissociative recombination of O_2^+ can proceed to produce two O atoms in a variety of states. They are listed below in order of decreasing exothermicity,

$$O_2^+(X\,^2\Pi_g) + e^- \longrightarrow O(^3P) + O(^3P) + 6.54\,\text{eV} \qquad (4.5.2a)$$

$$\longrightarrow O(^3P) + O(^1D) + 4.99\,\text{eV} \qquad (4.5.2b)$$

$$\longrightarrow O(^1D) + O(^1D) + 3.02\,\text{eV} \qquad (4.5.2c)$$

$$\longrightarrow O(^3P) + O(^1S) + 2.77\,\text{eV} \qquad (4.5.2d)$$

$$\longrightarrow O(^1S) + O(^1D) + 0.8\,\text{eV}. \qquad (4.5.2e)$$

Both excited states of O are known to fluoresce in the atmosphere, the $O(^1D) \longleftarrow O(^1S)$ transition leads to what is referred to as the green line (at 5577 Å) (Guberman 1977, Kella *et al* 1997, Peverall *et al* 2000), a prominent component of atmospheric and auroral airglows. Red emissions (6300 and 6364 Å) are obtained from the $O(^3P_J) \longleftarrow O(^1D)$ transitions (Guberman 1988). Due to the importance of these atmospheric emissions, much effort has gone into studying the dissociative recombination of O_2^+, both experimentally and theoretically. Recent progress in experimental techniques has allowed not only for cross section and branching ratio data to be measured for the ground state but also for vibrationally excited states.

Rate constants for this O_2^+ recombination have been measured versus temperature and kinetic energy decades ago. The early work has been summarized (McGowan and Mitchell 1984, Mitchell and McGowan 1983) and the rate constant can be expressed as $1.9 \times 10^{-17}(300/T_e)^{0.5}\,\text{cm}^3\,\text{s}^{-1}$, where T_e is the electron temperature. More recent work has resulted in very detailed cross sections as a function of energy (Kella *et al* 1997, Peverall *et al* 2001). In the Peverall *et al* (2001) experiment only ground state O_2^+ was present. Figure 4.16 shows cross sections versus collision energy from that work. Resonances were found at 0.01, 0.2, 0.25, 1.4, and 1.8 eV, but do not show well on this graph covering several orders of magnitude in cross section. Such data should be used for non-equilibrium plasmas, otherwise the thermal rate expression above should be used.

A measurement of the quantum yield of the reaction versus collision energy was reported by Peverall *et al* (2001). At most energies, $O(^1D)$ is the most abundant product followed closely by $O(^3P)$. This indicates that channel b is dominant, followed by c and a. While the $O(^1S)$ yield is small,

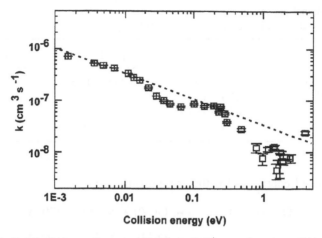

Figure 4.16. Rates constants for recombination of O_2^+ as a function of kinetic energy (Peverall *et al* 2001).

it is important since it is the source for the green airglow line (Guberman 1977, Guberman and Giusti-Suzor 1991, Peverall *et al* 2000). Its quantum yield decreases with energy at low energy and increases at high energy. The production of the $O(^1S)$ and $O(^1D)$ states has been discussed theoretically (Guberman 1977, 1987, 1988, Guberman and Giusti-Suzor 1991, Peverall *et al* 2000).

The most recent work on this reaction reports the vibrational level dependence for the cross sections and branching ratios at near 0 eV collision energy (Petrignani *et al* 2004). Vibrational excitation of the ion has been postulated to explain the abundance of the green airglow (Peverall *et al* 2000). The relative cross sections for $v = 0$, 1, and 2 are 14.9, 3.7, and 12.4 at *ca.* 0 eV (2 meV FWHM). It is interesting that the cross section for $v = 1$ is much smaller than for $v = 0$ or 2. Some of the resonances are enhanced with vibrational excitation, but the cross section versus energy data have not been derived as yet from the raw data. The branching data versus vibrational state are listed in table 4.7. The production of $O(^1S)$ increases with vibrational level, which indicates that the vibrational distribution of O_2^+ will be critical in determining airglow as has been predicted.

4.5.3 $NO^+ + e^-$

NO^+ is another important ion in air plasmas and excellent new studies have yielded detailed information on numerous aspects of the dissociative recombination reaction. Recombination of the ground state $(X^1\Sigma^+)$ can lead to three channels and seven more channels are possible for $NO^+(a^3\Sigma^+)$, a long-lived metastable species, or for high energy collisions. The channels

Table 4.7. Branching percentage for various channels as a function of vibrational state for O_2^+ dissociative recombination (from Petrignani *et al* 2004)

Channel	$v = 0$	$v = 1$	$v = 2$
$O(^1D) + O(^1S)$	4.7 ± 2.5	19.9 ± 10.5	10.7 ± 5.9
$O(^1D) + O(^1D)$	23.9 ± 12.0	28.8 ± 24.1	8.3 ± 7.6
$O(^3P) + O(^1D)$	47.9 ± 23.7	28.1 ± 36.7	63.4 ± 38.6
$O(^3P) + O(^3P)$	23.4 ± 11.7	23.3 ± 30.4	17.6 ± 20.5

and associated energetics for the ground state are

$$NO^+(X\,^1\Sigma^+) + e^- \longrightarrow O(^3P) + N(^4S) + 2.70\,eV \qquad (4.5.3a)$$

$$\longrightarrow O(^1D) + N(^4S) + 0.80\,eV \qquad (4.5.3b)$$

$$\longrightarrow O(^3P) + N(^2D) + 0.38\,eV \qquad (4.5.3c)$$

$$\longrightarrow O(^3P) + N(^2P) - 0.81\,eV \qquad (4.5.3d)$$

$$\longrightarrow O(^1S) + N(^4S) - 1.42\,eV \qquad (4.5.3e)$$

$$\longrightarrow O(^1D) + N(^2D) - 1.59\,eV \qquad (4.5.3f)$$

$$\longrightarrow O(^1D) + N(^2P) - 1.59\,eV \qquad (4.5.3g)$$

$$\longrightarrow O(^1S) + N(^2D) - 3.81\,eV \qquad (4.5.3h)$$

$$\longrightarrow O(^1S) + N(^2P) - 5.00\,eV \qquad (4.5.3i)$$

$$\longrightarrow O(^5S) + N(^4S) - 6.38\,eV. \qquad (4.5.3j)$$

Production of $O(^1D)$ from this reaction is another source for the red airglow and the $N(^4S) \longleftarrow N(^2D)$ radiation is responsible for the 5200 Å airglow line (Jursa 1985). As for O_2^+, rate constants for the sum of all channels have been known for years. The recommended rate from swarm experiments is $4.3 \times 10^{-7}(300/T_e)^{0.37}\,cm^3\,s^{-1}$, where T_e is the electron temperature (McGowan and Mitchell 1984, Mitchell and McGowan 1983). More recent work has yielded product state distributions and detailed cross section measurements for both the ground $(X\,^1\Sigma^+)$ state at several energies and for the $a\,^3\Sigma^+$ state at low energy (Hellberg *et al* 2003).

Table 4.8 gives the branching fractions for reaction (4.5.3) for several energies for the ground state and also for the metastable. At low energy, channel c accounts for nearly 100% of the reactivity and remains dominant at 1.25 eV collision energy, although kinetic energy is seen to drive channel d. At 5.6 eV collision energy, many other channels also become important with channel f being the most abundant. Finally, results for the $NO^+(a\,^3\Sigma^+)$ state

Table 4.8. Branching percentage for various channels as a function of energy and state for NO^+ dissociative recombination. Both experimental and statistical theoretical results are shown. Blanks indicate that the state is not accessible and a dash ('–') indicates that channel was not able to be derived experimentally (Hellberg *et al* 2003).

Channel	$NO^+(X\,^1\Sigma^+)$, 0 eV Exp't	Theory	$NO^+(X\,^1\Sigma^+)$, 1.25 eV Exp't	Theory	$NO^+(X\,^1\Sigma^+)$, 5.6 eV Exp't	Theory	$NO^+(a\,^3\Sigma^+)$, 0 eV Exp't	Theory
(4.5.3a)	5	17	10	11	3	3	1	6
(4.5.3b)	0	0	10	0	0	0	12	7
(4.5.3c)	95	83	70	57	15	20	23	32
(4.5.3d)			10	32	11	11	18	19
(4.5.3e)					0	0	4	1
(4.5.3f)					31	32	11	17
(4.5.3g)					21	20	7	10
(4.5.3h)					9	12	10	3
(4.5.3i)					10	3	13	2
(4.5.3j)							–	2

at low collision energy is included. Numerous channels are open and most are observed with channel c again being the most abundant, although only slightly.

Also included in table 4.8 are the results for a simple statistical model calculation. The model includes two effects. (1) The number of states connected to a dissociation limit determines its probability. For instance for channel a, one takes the product of spin and angular momentum multiplicities, i.e. $(3 \times 3 \times 4 \times 1) = 36$. Doing this for each open channel and then normalizing yields the probability for that channel. (2) The results are corrected so that spin-forbidden channels are not allowed, e.g. channel b. The agreement is quite good, especially considering the difficulty of doing detailed calculations.

4.5.4 $N_2^+ + e^-$

The final diatomic ion to be discussed is N_2^+. Again rate constants have long been measured and can be represented as $1.8 \times 10^{-7}(300/T_e)^{0.39}$ cm^3 s^{-1}, where T_e is the electron temperature. Thus, all the atmospherically important ions recombine at approximately the same rate and have about the same dependence on electron temperature. Details of this reaction have been studied recently in storage rings (Kella *et al* 1996, Peterson *et al* 1998) and theoretically (Guberman 2003b). The reaction can proceed by several channels:

$$N_2^+(v = 0^+) + e^- \longrightarrow N(^4S) + N(^4S) + 5.82\,eV \qquad (4.5.3a)$$

$$\longrightarrow N(^4S) + N(^2D) + 3.44\,eV \qquad (4.5.3b)$$

$$\longrightarrow N(^4S) + N(^2P) + 2.25\,eV \qquad (4.5.3c)$$

$$\longrightarrow N(^2D) + N(^2D) + 1.06\,eV \qquad (4.5.3d)$$

$$\longrightarrow N(^2D) + N(^2P) - 0.13\,eV. \qquad (4.5.3e)$$

The last channel is endothermic but is accessible for $v = 1$ and higher. The recombination of N_2^+ produces airglow at 5200, 3466, and 10 400 Å, the latter two from $N(^2P)$. The large exothermicity combined with the mass difference causes isotopic fractionation in the Mars atmosphere (Fox 1993). The lighter mass neutral, ^{14}N, can escape at the maximum energy allowed, but not ^{15}N.

The storage ring experiments found that rate constants are in good agreement with the early measurements and that vibrational excitation decreased the rate slightly, although they could not quantify the reduction for individual states (Peterson *et al* 1998). At an electron temperature of 300 K, Guberman calculated rate constants for $v = 0$, 1, and 2 to be 2.1×10^{-7}, 2.9×10^{-7}, and $1.1 \times 10^{-7}\,cm^3\,s^{-1}$, respectively. Since the $v = 1$ rate is calculated to increase and the $v = 2$ to decrease, the relatively insensitive nature of the experimental value to vibrational excitation could be a cancellation of the two effects.

Channel a, the lowest energy pathway, was found not to occur. The branching for the other channels for $v = 0$ are (b) $37 \pm 8\%$, (c) $11 \pm 6\%$, and (d) $52 \pm 4\%$ when the coldest source was used. For a higher temperature source, more of channel (c) was observed at the expense of the other two channels. The probability of producing the endothermic channel was found to increase with rotational quantum number.

4.5.5 $H_3O^+(H_2O)_n$

In wet atmospheres, production of H_3O^+ and its hydrates are likely. From 80 km and below in the atmosphere, these are either the dominant ions or an important intermediary in the production of other clusters (Viggiano and Arnold 1995). Since these molecules are polyatomic the detailed state information on the dissociative recombination process is not available, but complete product distributions are known.

The rate constants for dissociative recombination for these ions are extremely rapid. H_3O^+ rate constants can be expressed at $6.3 \times 10^{-7}(300/T_e)^{0.5}\,cm^3\,s^{-1}$ for $T_e < 1000\,K$ and $7.53 \times 10^{-7}(800/T_e)^{0.5}\,cm^3\,s^{-1}$ for $T_e > 1000\,K$ (McGowan and Mitchell 1984). The rate constants for the clusters are even faster. Johnsen gives the rate constants as $(0.5 + 2n)(300/T)^{0.5} \times 10^{-6} \, \text{cm}^3 \, \text{s}^{-1}$ for $n = 1$–4 (Johnsen 1993). Thus, in wet atmospheres, it is very hard to maintain a plasma unless negative ions are formed.

H_3O^+ can dissociate four ways. The pathways and the percentage of each channel are listed below for both H and D (Neau *et al* 2000),

$$H_3O^+ + e \; \longrightarrow \; H_2O + H + 6.4 \, \text{eV} \quad (\text{H: } 18 \pm 5\%)(\text{D: } 17 \pm 5\%) \qquad (4.5.4a)$$

$$\longrightarrow \; HO + H_2 + 5.7 \, \text{eV} \quad (\text{H: } 11 \pm 5\%)(\text{D: } 13 \pm 3\%) \qquad (4.5.4b)$$

$$\longrightarrow \; OH + 2H + 1.3 \, \text{eV} \quad (\text{H: } 67 \pm 6\%)(\text{D: } 70 \pm 6\%) \qquad (4.5.4c)$$

$$\longrightarrow \; O + H_2 + H + 1.4 \, \text{eV} \quad (\text{H: } 4 \pm 6\%)(\text{D: } 0 \pm 4\%). \qquad (4.5.4d)$$

No statistical difference was observed between the two isotopes. At the time of the measurements the preponderance of the channel producing three neutrals was surprising. This can obviously be an important source of radicals.

$H_3O^+(H_2O)$ can dissociate into a variety of pathways. The channel producing $2H_2O + H$ is by far the dominant channel ($94 \pm 4\%$) (Nagard *et al* 2002). The experiment was performed with deuterium for better separation of the channels. The only other channel definitely produced within error is the channel producing H_2O, $OH + H_2$ ($4 \pm 2\%$).

4.5.6 High pressure recombination

The above discussion refers to dissociative recombination in the low-pressure limit. At high pressures, heavy-body collisions occur while an electron is within the orbiting capture radius. This obviously can change the energy of the collision and lead to different kinetics. It has been shown that larger rate constants are found with increasing pressure. At pressures greater than an atmosphere, rate constants of $10^{-4} \, \text{cm}^3 \, \text{s}^{-1}$ have been measured (Armstrong *et al* 1982, Cao and Johnsen 1991, Morgan 1984, Warman *et al* 1979). Theoretical examinations have been made to explain the increase (Bates 1980, 1981, Morgan and Bardsley 1983). However, the number of such processes that has been studied is limited and at present no information is known about how pressure would effect product distributions. It may be expected that high pressure would result in less fragmentation, especially if the three-body fragmentation is sequential.

Acknowledgments

KB and MS would like to acknowledge invaluable discussion with and help from U. Kogelschatz. SW, AAV, and RD thank numerous colleagues who have contributed to sections 4.3, 4.4, and 4.5 of this chapter: John Paulson, Robert Morris, Thomas Miller, Jeff Friedman, Peter Hierl, Itzhak Dotan, Melani Menendez-Barreto, John Seeley, John Williamson, Fred Dale, Paul Mundis, Susan Arnold, Tony Midey, Jane Van Doren, Svetoza Popovic, Yu-Hui Chiu, Dale Levandier and Michael Berman. The authors thank Dick Zare, Scott Anderson, and Steve Leone for helpful discussions.

References

Adamovich I V and Rich J W 1998 *J. Chem. Phys.* **109** 7711

Adams N G and Smith D 1988 in Millar T J and Williams D A (eds) *Rate Coefficients in Astrochemistry* (The Netherlands: Kluwer Academic) p 173

Adams N G, Smith D and Clary D C 1985 *Astrophys. J.* **296** L31

Akishev Yu S, Deryugin A A, Karalnik V B, Kochetov I V, Narpatovitch A P and Trushkin N L 1994 *Plasma Physics Reports* **20** 511

Albritton D L, Dotan I, Lindinger W, McFarland M, Tellinghuisen J and Fehsenfeld F C 1977 *J. Chem. Phys.* **66** 410

Alge E and Lindinger W 1981 *J. Geophys. Res.* **86** 871

Anderson S L 1991 in Jennings K R (ed) *Gas Phase Ion Chemistry* (Dordrecht: Kluwer Academic) p 183

Anderson S L 1992a in Ng C and Baer M (eds) *State-Selected and State-to-State Ion-molecule Reaction Dynamics: Part 1. Experiment* (New York: John Wiley)

Anderson S L 1992b *Adv. Chem. Phys.* **82** 177 (Part I)

Anderson S L 1997 *Accts. of Chem. Res.* **30** 28

Anthony E B, Schade W, Bastian M J, Bierbaum V M and Leone S R 1997 *J. Chem. Phys.* **106** 5413

Appleton J P, Steinberg M and Liquornick D S 1968 *J. Chem. Phys.* **48** 599

Armentrout P B 2000 *Int. J. Mass Spectrom. Ion Phys.* **200** 219

Armstrong D A, Sennhauser E S, Warman J M and Sowada U 1982 *Chem. Phys. Lett.* **86** 281

Baer M and Ng C-Y (eds) 1992 *State-Selected and State-to-State Ion–Molecule Reaction Dynamics, Part 2* (New York: John Wiley)

Baeva M, Gier H, Pott A, Uhlenbusch J, Höschele J and Steinwandel J 2001 *Plasma Chem. Plasma Proc.* **21** 225–247

Bardsley J N 1968 *J. Phys. B* **1** 365

Bates D R 1980 *J. Phys. B* **13** 2587

Bates D 1981 *J. Phys. B* **14** 3525

Bates D R and Massey H S W 1947 *Proc. Roy. Soc. A* **192** 1

Billing G D 1986 in *Non-Equilibrium Vibrational Kinetics* (Berlin: Springer) p 85

Bird G A 1994 *Molecular Gas Dynamics and the Direct Simulation of Gas Glows* (Oxford: Clarendon Press)

Boesl U, Neusser H J and Schlag E W 1978 *Z. Naturforsch.* **33A** 1546

Borgnakke C and Larsen P S 1975 *J. Comp. Phys.* 18 405

Boyd I D 2001 in Dressler R A (ed) *Chemical Dynamics in Extreme Environments* (Singapore: World Scientific) p 81

Bradbury N E and Nielsen R A 1936 *Phys. Rev.* **49** 388

Bundle C R, Neumann D, Price W C, Evans D, Potts A W and Streets D G 1970 *J. Chem. Phys.* **53** 705

Cao Y S and Johnsen R 1991 *J. Chem. Phys.* **95** 5443

Chen J and Davidson J H 2002 *Plasma Chem. Plasma Process.* **22** 495

Chen A, Johnsen R and Biondi M A 1978 *J. Chem. Phys.* **69** 2688

Chesnavich W J and Bowers M T 1977a *J. Amer. Chem. Soc.* **99** 1705

Chesnavich W J and Bowers M T 1977b *J. Chem. Phys.* **66** 2306

Chesnavich W J and Bowers M T 1979 *J. Phys. Chem.* **83** 900

Chiu Y 1965 *J. Chem Phys.* **42** 2671

Chiu Y, Fu H, Huang J-T and Anderson S L 1994 *J. Chem. Phys.* **101**

Chiu Y, Fu H, Huang J and Anderson S L 1995a *J. Chem. Phys.* **102** 1199

Chiu Y, Fu H, Huang J-T and Anderson S L 1996 *J. Chem. Phys.* **105** 3089

Chiu Y, Pullins S, Levandier D J and Dressler R A 2000 *J. Chem. Phys.* **112** 10880

Chiu Y, Yang B, Fu H and Anderson S L 1995b *J. Chem. Phys.* **102** 1188

Chiu Y, Yang B, Fu H, Anderson S L, Schweizer M and Gerlich D 1992 *J. Chem. Phys.* **96** 5781

Christophorou L G, McCorkle D L and Christodoulides A A 1984 'Electron Attachment Processes' in L G Christophorou (ed) *Electron Molecule Interactions and their Applications* vol I pp 477–617 (Orlando: Academic Press)

Clary D C 2003 *Ann. Rev. Phys. Chem.* **54** 493

Conaway W E, Ebata T and Zare R N 1987 *J. Chem. Phys.* **87** 3447

Cosby P C 1993a *J. Chem. Phys.* **98** 9544

Cosby P C 1993b *J. Chem. Phys.* **98** 9560

Dai H-L and Field R W 1995 *Molecular Dynamics and Spectroscopy by Stimulated Emission Pumping* vol. 4 (Singapore: World Scientific)

Daly N R 1960 *Rev. Sci. Instrum.* **31** 264

Dorai R and Kushner M J 2003 *J. Phys. D: Appl. Phys.* **36** 666

Dotan I and Lindinger W 1982a *J. Chem. Phys.* **76** 4972

Dotan I and Lindinger W 1982b *J. Chem. Phys.* **76** 4972

Dotan I and Viggiano A A 1999 *J. Chem. Phys.* **110** 4730

Dotan I, Hierl P M, Morris R A and Viggiano A A 1997 *Int. J. Mass Spectrom. Ion Phys.* **167/168** 223

Dotan I, Midey A J and Viggiano A A 1999 *J. Am. Soc. Mass Spectrom.* **10** 815

Dotan I, Midey A J and Viggiano A A 2000 *J. Chem. Phys.* **113** 1732

Dressler R A, Meyer H, Langford A O, Bierbaum V M and Leone S R 1987 *J. Chem. Phys.* **87** 5578

Duncan M A, Bierbaum V M, Ellison G B and Leone S R 1983 *J. Chem. Phys.* **79** 5448

Durup-Ferguson M, Bohringer H, Fahey D W and Ferguson E E 1983 *J. Chem. Phys.* **79** 265

Durup-Ferguson M, Bohringer H, Fahey D W, Fehsenfeld F C and Ferguson E E 1984 *J. Chem. Phys.* **81** 2657

Eliasson B, Egli W and Kogelschatz U 1994 *Pure and Appl. Chem.* **66** 1279

Eliasson B and Kogelschatz U 1991 *IEEE Trans. Plasma Sci.* **19** 1063

Esposito F and Capitelli M 1999 *Chem. Phys. Lett.* **302** 49

Esposito F, Capitelli M and Gorse C 2000 *Chem. Phys.* **257** 193

Everest M A, Poutsma J C, Flad J E and Zare R N 1999 *J. Chem. Phys.* **111** 2507

Everest M A, Poutsma J C and Zare R N 1998 *J. Phys. Chem.* **102** 9593

Fahey D W, Dotan I, Fehsenfeld F C, Albritton D L and Viehland L A 1981a *J. Chem. Phys.* **74** 3320

Fahey D W, Fehsenfeld F C, Ferguson E E and Viehland L A 1981b *J. Chem. Phys.* **75** 669

Farrar J M and Saunders Jr. W H (eds) 1988 *Technique for the Study of Ion–Molecule Reactions* (New York: John Wiley)

Ferguson E E 1974a *Rev. Geophys. Space Phys.* **12** 703

Ferguson E E 1974b in Ausloos P (ed) *Interactions Between Ions and Molecules* (New York: Plenum) p 320

Ferguson E E, Fehsenfeld F C and Schmeltekopf A L 1969 in Bates D R (ed) *Advances in Atomic and Molecular Physics* (New York: Academic) p 1

Ferguson E E, Richter R and Lindinger W 1988 *J. Chem. Phys.* **89** 1445

Fox J. L. 1993 in Rowe B R (ed) *Dissociative Recombination Theory, Experiment and Application* (New York: Plenum Press)

Frost M J, Kato S, Bierbaum V M and Leone S R 1994 *J. Chem. Phys.* **100** 6359

Frost M J, Kato S, Bierbaum V M and Leone S R 1998 *Chem. Phys.* **231** 145

Fu H, Qian J, Green R J and Anderson S L 1998 *J. Chem. Phys.* **108** 2395

Gallis M A and Harvey J K 1996 *J. Fluid Mech.* **312** 149

Gallis M A and Harvey J K 1998 *Phys. Fluids* **10** 1344

Gerlich D 1992 *Adv. Chem. Phys.* **82** 1

Gilibert M, Giménez X, Huarte-Larrañaga F, González M, Aguilar A, Last I and Baer M 1999 *J. Chem. Phys.* **110** 6278

Gioumousis G and Stevenson D P 1958 *J. Chem. Phys.* **29** 294

Gordiets B F, Ferreira C M, Guerra V L, Loureiro A H, Nahorny J, Pagnon D, Touzeau M and Vialle M 1995 *IEEE Trans. Plasma Sci.* **23** 750

Gouw J A D, Ding L N, Frost M J, Kato S, Bierbaum V M and Leone S R 1995 *Chem. Phys. Lett.* **240** 362

Govers T R and Guyon P-M 1987 *Chem. Phys.* **113** 425

Graham E, Johnsen R and Biondi M A 1975 *J. Geophys. Res.* **80** 2338

Green D S, Sieck L W, Herron J T 1995 'Characterization of chemical processes in non-thermal plasmas for the destruction of volatile organic compounds' 12th International Symposium on Plasma Chemistry, Minneapolis 1995, *Proceedings* vol. 2 pp 965–969

Guberman S 1977 *Science* **278** 1276

Guberman S and Giusti-Suzor A 1991 *J. Chem. Phys.* **95** 2602

Guberman S L 1987 *Nature* **327** 408

Guberman S L 1988 *Planet. Space Sci.* **36** 47

Guberman S L 2003a in Guberman S L (ed) *Dissociative Recombination of Molecular Ions with Electrons* (New York: Plenum Press) p 1

Guberman S L 2003b in Guberman S L (ed) *Dissociative Recombination of Molecular Ions with Electrons* (New York: Kluwer/Plenum Academic Press) p 187

Guttler R D, Jones Jr., Posey L A and Zare R N 1994 *Science* **266**

Hellberg F, Rosen S, Thomas R, Neau A, Larsson M, Petrignani A and van der Zande W 2003 *J. Chem. Phys.* **118** 6250

Hernandez R, Toumi R and Clary D C 1995 *J. Chem. Phys.* **102** 9544

Herron J T 1999 *J. Chem. Ref. Data* **28** 1453

Herron J T 2001 *Plasma Chem. Plasma Process.* **21** 581

Herron J T and Green D S 2001 *Plasma Chem. Plasma Process.* **21** 459

Hierl P M, Dotan I. Seeley J V, Van Doren J M, Morris R A and Viggiano A A 1997 *J. Chem. Phys.* **106** 3540

Hierl P M *et al* 1996 *Rev. Sci. Inst.* **67** 2142

Howorka F, Dotan I, Fehsenfeld F C and Albritton D L 1980 *J. Chem. Phys.* **73** 758

Huarte-Larrañaga F, Giménez X, Lucas J M, Aguilar A and Launay J M 1998 *Phys. Chem. Chem. Phys.* **1** 1125

Huarte-Larrañaga F, Giménez X, Lucas J M, Aguilar A and Launay J M 2000 *J. Phys. Chem.* **104** 10227

Huber K P and Herzberg G 1979 *Molecular Spectra and Molecular Structure. IV. Constants of Diatomic Molecules* (New York: Van Nostrand Reinhold)

Ikezoe Y, Matsuoka S, Takebe M and Viggiano A A 1987 *Gas Phase Ion–Molecule Reaction Rate Constants Through 1986* (Tokyo: Maruzen Company Ltd.)

Jarvis G K, Song Y and Ng C Y 1999 *Rev. Sci. Instrum.* **70** 2615

Johnsen R 1993 *J. Chem. Phys.* **98** 5390

Johnsen R and Biondi M A 1973 *J. Chem. Phys.* **59** 3504

Johnsen R, Brown H L and Biondi M A 1970 *J. Chem. Phys.* **52** 5080

Johnston H and Birks J 1972 *Acc. Chem. Res.* **5** 327

Jongma R T and Wodtke A M 1999 *J. Chem. Phys.* **111** 10957

Jursa A S (ed) 1985 *Handbook of Geophysics and the Space Environment* (Springfield VA: National Technical Information Service)

Karwasz G P, Brusa R S, Zecca A 2001a *Rivista del Nuovo Cimento* **24**(1) 1

Karwasz G P, Brusa R S, Zecca A 2001b *Rivista del Nuovo Cimento* **24**(2) 1

Kato S, Bierbaum V M and Leone S R 1998 *J. Phys. Chem. A* **102** 6659

Kato S, Frost M J, Bierbaum V M and Leone S R 1993 *Rev. Sci. Instrum.* **64** 2808

Kato S, Frost M J, Bierbaum V M and Leone S R 1994 *Can. J. Chem.* **72** 625

Kato S, Gouw J A D, Lin C-D, Bierbaum V M and Leone S R 1996a *Chem. Phys. Lett.* **256** 305

Kato S, Lin G-D, Bierbaum V M and Leone S R 1996b *J. Chem. Phys.* **105** 5455

Kella D, Johnson P J, Pederson H B, Velby-Christensen L and Andersen L H 1996 *Phys. Rev. Lett.* **77** 2432

Kella D, Vejby-Christensen L, Johnson P J, Pedersen H B and Anderson L H 1997 *Science* **276** 1530

Kim H-T, Green R J and Anderson S L 2000a *J. Chem. Phys.* **112** 10831

Kim H-T, Green R J, Qian J and Anderson S L 2000b *J. Chem. Phys.* **112** 5717

Kossyi I A, Kostinki A Yu, Matveyev A A, Silakov V P 1992 *Plasma Sources Sci. Technol.* **1** 207

Koyano I and Tanaka K 1992 *Adv. Chem. Phys.* **82** 263

Koyano I, Tanaka K, Kato T and Suzuki S 1987 *Faraday Discuss. Chem. Soc.* **84** 265

Krishnamurthy M, Bierbaum V M and Leone S R 1997 *Chem. Phys. Lett.* **281** 49

Krupenie P H 1972 *J. Phys. Chem. Ref. Data* **1** 423

Lagana A, Garcia E and Ciccarelli L 1987 *J. Phys. Chem.* **91** 312

Larsson M and Thomas R 2001 *Phys. Chem. Chem. Phys.* **3** 4471

Larsson M, Mitchell J B A and Schneider I F (eds) 2000 *Dissociative Recombination: Theory, Experiment, and Applications IV* (Singapore: World Scientific)

Le Garrec J L, Lepage V, Rowe B R and Ferguson E E 1997 *Chem. Phys. Lett.* **270** 66

Lee W, Adamovic I V and Lempert W R 2001 *J. Chem. Phys.* **114** 1178.

Levine R D 1995 in Yurtsever E (ed) *Frontiers in Chemical Dynamics* (Kluwer Academic Publishers) 195

Levine R D and Bernstein R B 1971 *Chem. Phys. Lett.* **11** 552

Levine R D and Bernstein R B 1972 *J. Chem. Phys.* **56** 2281

Levine R D and Bernstein R B 1987 *Molecular Reaction Dynamics and Chemical Reactivity* (New York: Oxford University Press)

Levine R D and Manz J 1975 *J. Chem. Phys.* **63** 4280

Lias S G, Bartmess J E, Liebman J F, Holmes J L, Levine R D and Mallard W G 1988 *J. Phys. Chem. Ref. Data* **17** Supplement 1 p 1

Light J C 1967 *Disc. Faraday Soc.* **44** 14

Lindinger W 1987 *Int. J. Mass Spectrom. Ion Proc.* **80** 115

Lindinger W, Albritton D L, Fehsenfeld F C and Ferguson E E 1975 *J. Geophys. Res.* **80** 3725

Lindinger W, Fehsenfeld F C, Schmeltekopf A L and Ferguson E E 1974 *J. Geophys. Res.* **79** 4753

Macharet S O and Rich J W 1993 *Chem. Phys.* **174** 25

Mack J A, Mikulecky K and Wodtke A M 1996 *J. Chem. Phys.* **105** 4105

Marcus R A 1952 *J. Chem. Phys.* **20** 359

Marcus R A and Rice O K 1951 *J. Phys. Colloid Chem.* **55** 894

Marriott P M and Harvey J K 1994 in Weaver D P (ed) *Rarefied Gas Dynamics: Experimental Techniques and Physical Systems* (Washington DC: AIAA) p 197

Mätzing H 1991 *Adv. Chem. Phys.* **80** 315

McFarland M, Albritton D L, Fehsenfeld F C, Ferguson E E and Schmeltekopf A L 1973a *J. Chem. Phys.* **59** 6610

McFarland M, Albritton D L, Fehsenfeld F C, Ferguson E E and Schmeltekopf A L 1973b *J. Chem. Phys.* **59** 6620

McFarland M, Albritton D L, Fehsenfeld F C, Ferguson E E and Schmeltekopf A L 1973c *J. Chem. Phys.* **59** 6629

McGowan J W and Mitchell J B A 1984 in Christophorou L G (ed) *Electron–Molecule Interactions and their Applications* (New York: Academic Press) p 65

Metayer-Zeitoun C, Alcarez C, Anderson S L, Palm H and Dutuit O 1995 *J. Phys. Chem.* **99** 15523

Midey A J and Viggiano A A 1998 *J. Chem. Phys.* **109** 5257

Midey A J and Viggiano A A 1999 *J. Chem. Phys.* **110** 10746.

Miller J S, Chiu Y, Levandier D J and Dressler R A 2004 in preparation.

Miller T M, Friedman J F, Menendez-Barreto M, Viggiano A A, Morris R A, Miller A E S and Paulson J F 1994 *Phys. Scripta* **T53** 84

Mitchell J B A and McGowan J W 1983 in Brouillard F and McGowan J W (eds) *Physics of Ion–Ion and Electron–Ion Collisions* (New York: Plenum Press) p 279

Morgan W L 1984 *J. Chem. Phys.* **80** 4565

Morgan W L and Bardsley J N 1983 *Chem. Phys. Lett.* **96** 93

Nagard M B *et al* 2002 *J. Chem. Phys.* **117** 5264

Neau A *et al* 2000 *J. Chem Phys.* **113** 1762

NIST chemkin NIST Chemical Kinetics Database: Version 2Q98 (*http://www.nist.gov/srd/chemkin.htm*)

NIST index NIST Chemical Kinetics Database on the Web. A compilation of kinetics data on gas-phase reactions (http://kinetics.nist.gov/index.php)

Ng C Y 1992 *Adv. Chem. Phys.* **82** 401

Ng C Y 2002 *J. Phys. Chem.* **106** 5953

Ng C-Y and Baer M E (eds) 1992 *State-Selected and State-to-State Ion–Molecule Reaction Dynamics* Part 1 (New York: John Wiley)

Orlando T M, Yang B and Anderson S L 1989 *J. Chem. Phys.* **90** 1577

Orlando T M, Yang B, Chiu Y-H and Anderson S L 1990 *J. Chem. Phys.* **92** 7356

Pechukas P, Light J C and Rankin C J 1966 *J. Chem. Phys.* **44** 794

Pendergast P, Heck J M, Hayes E F and Jaquet R 1993 *J. Chem. Phys.* **98** 4543

Peterson J R *et al.* 1998 *J. Chem. Phys.* **108** 1978

Petrignani A, van der Zande W J, Cosby P, Hellberg F, Thomas R and Larsson M. private communication (2004) and in preparation

Peverall R *et al.* 2000 *Geophys. Res. Lett.* **27** 481

Peverall R *et al.* 2001 *J. Chem. Phys.* **114** 6679

Poulter K F, Rodgers M-J, Nash P J, Thompson T J and Perkin M P 1983 *Vacuum* **33** 311

Poutsma J C, Everest M A, Flad J E, Jones Jr. G C and Zare R N 1999 *Chem. Phys. Lett.* **305** 343

Poutsma J C, Everest M A, Flad J E and Zare R N 2000 *Appl. Phys. B* **71** 623

Price J M, Mack J A, Rogaski C A and Wodtke A M 1993 *Chem. Phys.* **175** 83

Qian J, Fu H and Anderson S L 1997 *J. Chem. Phys.* **101** 6504

Qian J, Green R J and Anderson S L 1998 *J. Chem. Phys.* **108** 7173

Qian X *et al.* 2003a *Rev. Sci. Instr.* **74** 4881

Qian X, Zhang T, Chiu Y, Levandier D J, Miller J S, Dressler R A and Ng C Y 2003b *J. Chem. Phys.* **118** 2455

Radzig A A and Smirnov B B 1985 *Reference Data on Atoms, Molecules and Ions* (Berlin: Springer)

Rebick C and Levine R D 1973 *J. Chem. Phys.* **58** 3942

Rebrion C, Marquette J B, Rowe B R and Clary D C 1988 *Chem. Phys. Lett.* **143** 130

Rogaski C A, Mack J A and Wodtke A M 1995 *Faraday Disc. Chem. Soc.* **100** 287

Rogaski C A, Price J M, Mack J A and Wodtke A M 1993 *Geophys. Res. Lett.* **20** 2885

Rowe B R, Fahey D W, Fehsenfeld F C and Albritton D L 1980 *J. Chem. Phys.* **73** 194

Schmeltekopf A L 1967 *Planet. Space Sci.* **15** 401

Schmeltekopf A L, Ferguson E E and Fehsenfeld F C 1968 *J. Chem. Phys.* **48** 2966

Schultz R H and Armentrout P B 1991 *J. Phys. Chem.* **95** 121

Schwartz R, Slawsky Z and Herzfeld K 1952 *J. Chem. Phys.* **20** 1591

Sieck L W, Herron J T, Green D S 2000 *Plasma Chem. Plasma Process.* **20** 235

Signorell R and Merkt F 1998 *J. Chem. Phys.* **109** 9762

Signorell R, Wüest A and Merket F 1997 *J. Chem. Phys.* **107** 10819

Silva M, Jongma R, Field R W and Wodtke A M 2001 *Ann. Rev. Phys. Chem.* **52** 811

Sizun M and Gislason E A 1989 *J. Chem. Phys.* **91** 4603

Smith D, Adams N G and Miller T M 1978 *J. Chem. Phys.* **69** 308

Smith D and Spanel P 1995 *Mass Spectrom. Rev.* **14** 255

Smith M A 1994 in Ng C-Y, Baer T and Powis L (eds) *Unimolecular and Bimolecular Ion–Molecule Reaction Dynamics* (New York: John Wiley) p 183

Stefanovic I, Bibinov N K, Deryugin A A, Vinogradov I P, Narpatovich A P and Wiesemann K 2001 *Plasma Sources Sci. Technol.* **10** 406

Tang B, Chui Y, Fu H and Anderson S L 1991 *J. Chem. Phys.* **95** 3275

Teloy E and Gerlich D 1974 *Chem. Phys.* **4** 417

Troe J 1992 in Baer M and Ng C-Y (eds) *State-Selected and State-to-State Ion–Molecule Reaction Dynamics: Theory* (New York: John Wiley) p 485

Viggiano A A and Arnold F 1995 in Volland H (ed) *Atmospheric Electrodynamics* (Boca Raton: CRC Press) p 1

Viggiano A A and Morris R A 1996 *J. Phys. Chem.* **100** 19227

Viggiano A A and Paulson J F 1983 *J. Chem. Phys.* **79** 2241

Viggiano A A and Williams S 2001 in Adams N G and Babcock L M (eds) *Advances in Gas Phase Ion Chemistry* (New York: Academic Press) p 85

Viggiano A A, Knighton W B, Williams S, Arnold S T, Midey A J and Dotan I 2003 *Int. J. Mass Spectrom.* **223–224** 397

Viggiano A A, Morris R A, Dale F, Paulson J F, Giles K, Smith D and Su T 1990a *J. Chem. Phys.* **93** 1149

Viggiano A A, Morris R A, Deakyne C A, Dale F and Paulson J F 1990b *J. Phys. Chem.* **94** 8193

Vlasak P R, Beussman D J, Davenport M R and Enke C G 1996 *Rev. Sci. Instrum.* **67** 68

Wadsworth D C and Wysong I J 1997 *Phys. Fluids* **9** 3873

Warman J M, Sennhauser E S and Armstrong D A 1979 *J. Chem. Phys.* **70** 995

Wysong I J, Dressler R A, Chiu Y and Boyd I D 2002 *J. Thermophys. Heat Transfer* **16** 83

Yang B, Chiu Y-H and Anderson S L 1991 *J. Chem. Phys.* **94** 6459

Yang B, Chiu Y-H, Fu H and Anderson S L 1991 *J. Chem. Phys.* **95** 3275

Zandee L and Bernstein R B 1979 *J. Chem. Phys.* **71** 1359

Zare R N 1998 *Science* **279** 1875

Zecca A, Karwasz G P and Brusa R S 1996 *Rivista del Nuovo Cimento* **19** 1

Zhang T, Qian X, Chiu Y, Levandier D J, Miller J S, Dressler R A and Ng C Y 2003 *J. Chem. Phys.* **119** 10175

Chapter 5

Modeling

Osamu Ishihara, Graham Candler, Christophe O Laux,
A P Napartovich, L C Pitchford, J P Boeuf
and John Verboncoeur

5.1 Introduction

This chapter deals with the state-of-the-art in computer modeling of the
theoretical formulations that were presented in the previous two chapters.
Air plasmas are inherently complex, a situation made worse by the presence
of molecular ions and electro-negative species. The air plasma consists of a
high-temperature mixture of nitrogen and oxygen. With higher gas temperature, dissociation and recombination of N_2 and O_2 will produce more
neutrals like N, O and NO. Further increase of the temperature prompts
the ionization process to take place, producing electron population in the
air. The resulting ionic species include N_2^+, N^+, O_2^+, O^+ and NO^+, while
electro-negative species are negligibly small in the amount relative to the
concentration of electrons for sufficiently high temperature. Detailed
mechanisms of ionization and recombination in atmospheric pressure air
plasmas are yet to be fully understood. The thermal state of the air plasma
may not be straightforward to describe because of the variety of populations
of atoms, molecules, and diatomic molecules involved. The energy of the
particles is characterized by their modes of motion, i.e. translation, vibration
and rotation. The thermal state may be well described by the electron
temperature and separate independent temperatures for heavy particles,
since free electrons are heated rapidly by external means while heavy particles
are much slower in changing their energy. A combination of computer
modeling in conjunction with experiments is expected to play an essential
role in filling in the gaps of our understanding and thereby lay the ground-
work for our eventual mastery over air plasmas.

The specific topics included in this chapter span the gamut of numerical
techniques used for the modeling of everything from glow discharges, to

diffuse discharges, multi-dimensional flow, Trichel pulses, dielectric barrier discharges, and the initial air breakdown.

It is worth mentioning that the determination of the electron energy distribution function is important since the ionization source term and transport coefficients are derived from this function. A model with the assumption of a Maxwell–Boltzmann distribution for electrons provides an accurate description of collisional air plasmas where it is possible to parameterize the electron energy distribution as a function of the local reduced field strength or the electron average energy. The models without the assumption of a Maxwell-Boltzmann distribution for electrons, although applied only to the electron–ion non-thermal plasma, are described in sections 5.4 and 5.5. Full kinetic models, while harder to apply to the complexity of the air plasma, offer the advantage of providing the electron energy distribution function as a function of space and time. A description of a full kinetic model and a novel application to gas breakdown (although limited in species) in certain geometries is given in section 5.6. The self-consistent calculation of the space charge electric field in the modeling is a challenging task in the air plasma. The electrical properties of the discharges depend on the cathode region where the charge neutrality fails to fulfill. The models described in sections 5.2 and 5.3 are focused on the air plasma without boundary effect and neglecting the coupling of the non-equilibrium plasma chemistry to the flowing air stream, while latter sections, although limited in ion species, concern the effect of boundaries.

Section 5.2, by G. V. Candler, deals with non-equilibrium air discharges and discusses approximations and numerical solutions to the governing equations. As a basis for modeling the atmospheric-pressure plasma, the governing equations are described in detail. Those are conservations of mass, momentum, and energy, supplemented by equations of vibration-electron energy and electron translational energy. The model involves 11 species air plasma, including five neutral species (N_2, N, O_2, O, NO), five ionic species (N_2^+, N^+, O_2^+, O^+, NO^+), and the electrons, with finite-rate chemical reactions and coupling between the energy modes and transport processes. A numerical technique based on finite-volume computational fluid dynamics is introduced.

Section 5.3, prepared by C. Laux, describes the modeling of dc glow discharges in atmospheric pressure air. The air plasma is modeled by two temperatures: electron temperature T_e and gas temperature T_g. The numerical solution of the two-temperature chemical kinetic model with 40 reactions of the 11 species, where the electron temperature is elevated with respect to the gas temperature, is studied. This section includes a brief description of experiments conducted to validate the modeled mechanism of ionization in two-temperature atmospheric pressure air plasmas.

Section 5.4, written by A. P. Napartovich, addresses the challenging problem of modeling Trichel pulses characterized by regular current pulses in a negative corona for pin-to-plane configurations. The proposed multi-dimensional model is found to be essential in demonstrating the regular current oscillations that are observed in Trichel pulses.

Section 5.5, contributed by L. C. Pitchford and J. P. Boeuf, provides an overview of electrical models of plasmas created in gas discharges such as dielectric barrier discharges (DBDs) and microdischarges associated with the study of non-thermal, atmospheric pressure plasmas. The state-of-the-art in modeling DBDs is advanced, but relatively few of the previous works have dealt with DBDs in air. However, the formulation of a suitable model and the understanding of the evolution of the plasma in DBDs is independent of gas mixture, and conclusions derived from model results are reviewed in this section. Models have helped us understand the different modes observed in DBDs and have clarified the underlying physical nature of atmospheric pressure glow discharges. Modeling of discharges in small geometries is now under way, and further work in this area should soon lead to a better understanding of scaling issues.

The final section, authored by J. Verboncoeur, then discusses a model for the initiation of breakdown in a surface-discharge-type PDP (plasma display panel) cell in which a gas mixture is ionized. The modified particle-in-cell (PIC) Monte Carlo (known collectively by the acronym, 'PIC-MC') collision model is described and a technique to measure Paschen-like curves is proposed.

5.2 Computational Methods for Multi-dimensional Nonequilibrium Air Plasmas

5.2.1 Introduction

There has been considerable interest in recent years in finding methods for reducing the power budget required to generate large volumes of atmospheric pressure air plasmas at temperatures below 2000 K with electron number densities of the order of 10^{13} cm^{-3}. These reactive air plasmas potentially have numerous applications. In order to increase the electron density without significantly heating the gas, the energy must be added in a targeted fashion. One method is to add energy to the free electrons with a dc discharge. This approach was successfully demonstrated at Stanford University in a series of experiments in atmospheric pressure air at temperatures between 1800 and 3000 K. The experiments showed that it is possible to obtain stable diffuse glow discharges with electron number densities of up to 2×10^{12} cm^{-3} from a 250 mA power supply, which is up to six orders of magnitude higher than in the absence of the discharge. In principle, the electron number density could be increased to higher values with a power supply capable of delivering more current. No significant degree of gas heating was observed, as the measured gas temperature remained within a few hundred Kelvin of its value without the discharge applied.

In this section, we present a computational approach for simulation of this type of non-equilibrium air plasma. First, we present the multi-dimensional governing equations that describe the atmospheric-pressure plasma generated in the Stanford experiments. We then describe how to solve the equations with a finite-volume computational fluid dynamics approach. The model presented assumes a three-temperature, 11-species air plasma. Finite-rate chemical reactions and coupling between the energy modes and all of the relevant transport processes are included. Such an approach may be extended to model many multi-dimensional air discharges.

5.2.2 Basic assumptions

The thermal state of the gas is assumed to be described by separate and independent temperatures. The energy in the translational mode of all the heavy particles is assumed to be characterized by a single translational temperature. The rotational state of the diatomic molecules is taken to be equilibrated with the translational temperature.

The vibration-electronic state of the gas is described by a separate vibration-electronic temperature. This approach is taken by Gnoffo *et al* (1989), and is based on the rapid equilibration of the vibrational mode of molecular nitrogen and the electronic states of heavy particles. The translational energy of the free electrons is characterized by a separate electron temperature, T_e. This implies that the translational energies of free electrons can be characterized by a Maxwell–Boltzmann distribution at that temperature. Additional specific assumptions are made and these will be discussed in conjunction with the derivation of the governing equations.

5.2.3 The conservation equations

The flow within the plasma experiment test-section is described by the Navier–Stokes equations that have been extended to include the effects of non-equilibrium thermo-chemistry. In this section the individual species' mass, momentum and the energy conservation equations are discussed.

The mass conservation equation for chemical species s is given by

$$\frac{\partial \rho_s}{\partial t} + \nabla \cdot (\rho_s \vec{u}_s) = w_s$$

where ρ_s is the species mass density, $\vec{u}_s$ is the species velocity vector and w_s represents the generation rate of species s. We define the mass-averaged velocity, $\vec{u}$, as

$$\vec{u} = \sum_{s=1}^{n} \frac{\rho}{\rho_s} \vec{u}_s$$

where the sum is over the n species present in the plasma. The total mass density, ρ, is

$$\rho = \sum_{s=1}^{n} \rho_s.$$

Then we define the diffusion velocity, $\vec{v}_s$, to be the difference between the species velocity and the mass-averaged velocity, $\vec{v}_s = \vec{u}_s - \vec{u}$. The species mass conservation equation becomes

$$\frac{\partial \rho_s}{\partial t} + \nabla \cdot (\rho_s \vec{u}) = -\nabla \cdot (\rho_s \vec{v}_s) + w_s$$

where the first term on the right hand side is the flux due to diffusion.

The electron conservation equation is more commonly written as

$$\frac{\partial n_e}{\partial t} + \nabla \cdot \vec{j}_e = \omega_e$$

where ω_e is the rate of formation of electrons by ionization reactions. The electron number flux, $\vec{j}_e$, is obtained from the electron momentum equation by neglecting inertia. This gives

$$n_e \vec{v}_e = \vec{j}_e = -n_e \mu_e \vec{E} - \frac{D_e}{T_e} \nabla(n_e T_e)$$

where D_e is the electron diffusion coefficient and μ_e is the electron mobility. These are given by

$$D_e = \frac{\mu_e k T_e}{e}, \qquad \mu_e = \frac{e}{m_e \nu_M}.$$

Now, for numerical reasons (Hammond *et al* 2002), it is more convenient to write the electron velocity in terms of the logarithmic derivative of the electron number density:

$$\vec{v}_e = -\mu_e \vec{E} - \frac{D_e}{T_e} \nabla T_e - D_e \nabla(\ln n_e).$$

This form results in significantly less numerical error in regions where the electron number density is changing rapidly.

The mass-averaged momentum equation is

$$\frac{\partial}{\partial t} (\rho \vec{u}) + \nabla \cdot (\rho_s \vec{u} \vec{u}) + \nabla p = -\nabla \cdot \tau + \sum_{s=1}^{n} N_s e Z_s \vec{E}$$

where p is the pressure, and τ is the shear stress tensor.

The total energy conservation equation is the total energy equation for the mixture,

$$\frac{\partial E}{\partial t} + \nabla \cdot ((E + p)\vec{u})$$

$$= -\nabla \cdot (\vec{q} + \vec{q}_{\text{v-el}} + q_e) - \nabla \cdot (u \cdot \tau) - \nabla \cdot \sum_{s=1}^{n} N_s e Z_s \vec{E}(\vec{u} + \vec{v}_s).$$

The heat conduction vector, $\vec{q} + \vec{q}_{\text{v-el}} + \vec{q}_e$, has been expressed in component form, where each term is due to gradients of the different temperatures.

In addition to the total energy equation, we require an equation for each independent energy mode. The vibration-electronic energy of a given species is defined to be the difference between that species' internal energy computed from the Gordon–McBride (1994) data and the sum of its translational–rotational energy and heat of formation. For example, for a diatomic molecule the specific vibration-electronic energy at the vibration-electronic temperature $T_{\text{v-el}}$ is given by

$$e_{\text{v-el},s}(T_{\text{v-el}}) = e_s(T_{\text{v-el}}) - \tfrac{5}{2} R_s T_{\text{v-el}} - h_s^\circ$$

where e_s is the species specific internal energy, $R_s = R/M_s$ is the specific heat, and h_s° is the heat of formation. For atoms the translational energy is removed from the enthalpy.

The vibration-electronic energy equation is similar to the total energy equation, and may be written as

$$\frac{\partial E_{\text{v-el}}}{\partial t} + \nabla \cdot (E_{\text{v-el}}\vec{u})$$

$$= -\nabla \cdot \sum_{s=1}^{n} \vec{v}_s E_{\text{v-el},s} - \nabla \cdot \vec{q}_{\text{v-el}} + Q_{\text{T-v-el}} + Q_{\text{e-v-el}} + \sum_{s=1}^{n} w_s e_{\text{v-el}}.$$

The various energy transfer mechanisms to the vibrational energy modes have been represented here. $Q_{\text{T-v-el}}$ and $Q_{\text{e-v-el}}$ are the rates of translation–vibration–electronic and electron–vibration–electronic energy exchange, respectively.

The conservation of the electron translational energy, $E_e = \tfrac{3}{2} n_e k T_e$, can be written as

$$\frac{\partial}{\partial t} (\tfrac{3}{2} n_e k T_e) + \nabla \cdot (\tfrac{3}{2} n_e k T_e \vec{v}_e) = -n_e e \vec{E} \cdot \vec{v}_e - Q_{\text{T-e}} - w_e I - \nabla \cdot \vec{q}_e$$

where $Q_{\text{T-e}}$ is the translation-electron energy exchange rate. The term $w_e I$ is due to the loss of electron energy due to ionization, where the ionization energy is I.

These differential equations describe the flow of a time-dependent, multi-component, multi-temperature gas. The solution of these equations yields the dynamics of the conserved quantities of mass, momentum, and energy. A detailed description of the applied electric field and the conservation of the current is given below.

5.2.4 Equations of state

Equations of state are required to derive the required non-conserved quantities of pressure and the temperatures. The total energy, E, is made up of the separate components of energy, namely the kinetic energy and the internal modes of energy constituting the thermal energy. It is written as

$$E = \sum_{s \neq e}^{n} \rho_s c_{vs} T + \frac{1}{2} \sum_{s \neq e}^{n} \rho_s \vec{u} \cdot \vec{u} + E_{v\text{-el}} + E_e + \sum_{s \neq e}^{n} \rho_s h_s^{\circ}.$$

This expression may be inverted to yield the energy in the translational–rotational modes, and consequently T. The constants of specific heat at constant volume, c_{vs}, are the sum of the specific heat of translation and the specific heat of rotation. Thus, for diatomic molecules $c_{vs} = 5R/2M_s$ and for atoms $c_{vs} = 3R/2M_s$. The vibration-electronic temperature is computed using a Newton method to find the root of the expression given above for the vibrational-electronic energy. The electron temperature is determined by simply inverting the relation between the electron energy, E_e, and the energy contained in the electron thermal energy

$$E_e = \rho_e c_{ve} T_e = \tfrac{3}{2} n_e k T_e.$$

The total pressure is the sum of the partial pressures

$$p = \sum_{s \neq e}^{n} p_s + p_e = \sum_{s \neq e}^{n} \rho_s \frac{R}{M_s} T + \rho_e \frac{R}{M_e} T_e.$$

5.2.5 Electrodynamic equations

The electric field can be computed from the Poisson equation for the electric potential:

$$\vec{E} = -\nabla \phi, \qquad \nabla^2 \phi = -4\pi e(n_+ - n_e).$$

However, we choose to take advantage of the experimental geometry, and assume that the field only varies in the direction along the axis of the flow. In this case, there is no forced diffusion in the radial direction, which simplifies the implementation of the numerical method outlined above. In addition, we can determine the local electric field from the known total current of the discharge. Fundamentally, we know

$$i = -\int_A e n_e v_{ex} \, dA = \int_A e n_e \left(\mu_e E_x + \frac{D_e}{T_e} \frac{\partial T_e}{\partial x} + D_e \frac{\partial \ln n_e}{\partial x} \right) dA$$

where A is the cross-sectional area of the discharge and v_{ex} is the axial electron velocity. Now, since the total current is a parameter set by the experimental conditions, we can compute the electric field at each axial location from the above equation. This ensures that the discharge carries the correct current at every location in the discharge. This concept is supported by previous work cited in Raizer (1997).

5.2.6 Transport properties

5.2.6.1 *Shear stresses and heat fluxes*

The shear stresses are assumed to be proportional to the first derivative of the mass-averaged velocities and the Stokes assumption for the bulk viscosity is made. This results in the conventional expression for the shear stress tensor. The heat conduction vectors are given by the Fourier heat law

$$\vec{q} = -\kappa_t \nabla T, \qquad \vec{q}_v = -\kappa_v \nabla T_{v\text{-el}}, \qquad \vec{q}_e = -\kappa_e \nabla T_e$$

where κ_t, κ_v, and κ_e are the translational–rotational, vibration–electronic and the electron translational conductivities.

5.2.6.2 *Viscosity and thermal conductivity*

The plasma flow is far from chemical equilibrium and properties based on local thermodynamic equilibrium cannot be used. Thus a general multi-component approach for transport properties is necessary. The collision cross section method of Gupta *et al* (1990) accounts for the transfer of momentum and energy by collision by means of a non-dimensional factor, which is a function of the molecular weights of the species pairs, as well as the collision cross sections.

5.2.6.3 *Collision cross section method*

The collision cross section method was developed for high temperature non-equilibrium conditions. It permits efficient computation in the numerical flow field and provides accurate non-equilibrium properties. The average collision cross sections $\bar{\Omega}_{rs}^{1,1}$ and $\bar{\Omega}_{rs}^{2,2}$ are evaluated per species from the Chapman–Enskog first approximation formulas and curve fits as a function of temperatures. Here it must be pointed out that if one of the colliding partners is an electron, the electron temperature T_e must be used in the curve fits.

Viscosity for the gas mixture is given below, where $\Delta_{rs}^{(2)}$ is a function of the collision cross sections evaluated at the appropriate temperatures

$$\mu = \sum_{s=1}^{n} \frac{(M_s/N_A)X_s}{\sum_r X_r \Delta_{rs}^{(2)}}.$$

The thermal conductivity components, translational thermal conductivity κ_{tr}, rotational thermal conductivity κ_{rot}, and vibrational thermal conductivity are defined as follows:

$$\kappa_{tr} = \frac{15}{4} k \sum_{s \neq e}^{n} \frac{X_s}{\sum_{r \neq e} \alpha_{rs} X_r \Delta_{rs}^{(2)}}$$

$$\kappa_{rot} = k \sum_{s = mol}^{n} \frac{X_s}{\sum_{r \neq e} \alpha_{rs} X_r \Delta_{rs}^{(1)}}$$

$$\kappa_{v\text{-el}} = k \sum_{s \neq e}^{n} \frac{X_s}{\sum_{r \neq e} \alpha_{rs} X_r \Delta_{rs}^{(1)}}.$$

Here α_{rs} are functions of the collision cross sections,

$$\alpha_{rs} = 1 + \frac{\left(1 - \frac{M_r}{M_s}\right)\left(0.45 - 2.54 \frac{M_r}{M_s}\right)}{\left(1 + \frac{M_r}{M_s}\right)^2}$$

and

$$\Delta_{rs}^{(1)} = \frac{8}{3} \sqrt{\frac{2\mu_{rs}}{\pi k T}} \bar{\Omega}_{rs}^{1,1}, \qquad \Delta_{rs}^{(2)} = \frac{16}{5} \sqrt{\frac{2\mu_{rs}}{\pi k T}} \bar{\Omega}_{rs}^{2,2}.$$

The electron thermal conductivity κ_e is given by

$$\kappa_e = \frac{15}{4} k \frac{X_e}{\sum_r 1.45 X_r \Delta_{er}^{(2)}}.$$

5.2.6.4 *Electrical conductivity*

The electrical conductivity is defined using the electron mobility, μ_e. In the discharge region, the charged particles are acted upon by the electric field. The electrons and ions move in opposite directions under the influence of the electric field. The force acting on the electrons due to collisions with other particles can be given as

$$\vec{F}_e = m_e \bar{\nu}_{eH} \vec{v}_e$$

where $\vec{v}_e$ is the diffusion velocity of the electrons. Here it has been assumed that the average collision frequency of electrons with ions is negligible compared with that of electrons with all heavy particles, $\bar{\nu}_{eH}$. The electron diffusion velocity can now be given by

$$\vec{v}_e = -\frac{e}{m_e \bar{\nu}_{eH}} \vec{E}.$$

The electron current density $\vec{j}_e$, defined as the average flux density of electron charge, is

$$\vec{j}_e = -en_e\vec{v}_e = \frac{n_e e^2}{m_e \bar{\nu}_{eH}} \vec{E}$$

where

$$\sigma_e = \frac{n_e e^2}{m_e \bar{\nu}_{eH}}$$

is the electron electrical conductivity.

5.2.6.5   Ordinary diffusion

Ramshaw's (1990) method is the basis of the multi-temperature multi-component ordinary mass diffusion modeling in this work. Recent comparisons (see Desilets and Proulx 1995) between an exact method, with effective binary, linear and Ramshaw's approximations show that only Ramshaw's method is adequate to model diffusion fluxes in the context of plasma flows with temperature gradients. Since the energy transfer between components is much slower than momentum transfer, a multi-temperature diffusion formulation is needed.

Correct treatment of ordinary diffusion in multi-component gas mixtures requires the solution of a linear system of equations for the diffusive mass fluxes relative to the mass-averaged velocity of the mixture. However, their solution presents unwelcome and costly complications in many situations, particularly in the present multi-dimensional numerical simulation where the diffusional fluxes are required at each mesh point and at every time step in the calculation. For this reason effective binary diffusion approximations are often used to avoid solving these equations. However, most formulations suffer from lack of mass conservation. Ramshaw (1990) correctly identified the origin of this inconsistency and developed a rational procedure for self-consistently removing it. Thus, Ramshaw's self-consistent effective binary diffusion approximation is used to model the ordinary diffusion fluxes. The reader is referred to the work of Ramshaw (1990) and Ramshaw and Chang (1991, 1993) for further details.

5.2.6.6   Energy exchange mechanisms

The energy exchange mechanisms that appear on the right hand side of the internal energy equations must be modeled. The models that have been proposed are simplifications of the complicated energy exchange processes that occur on a molecular level. The models used in this work are outlined below.

Translation–vibration electronic energy exchange. The rate of energy exchange between the vibration-electronic and translational modes is well described by the Landau–Teller formulation where it is assumed that the vibration-electronic level of a molecule can change by only one quantum level at a time. In this work we use the relaxation rates of Millikan and White (1963).

Translation and vibration–electron energy exchanges. The energy transfer rate between the heavy-particle and electron translational modes, $Q_{T\text{-e}}$, was originally derived by Appleton and Bray (1964).

$$Q_{T\text{-e}} = n_e \sum_h 3k(T_e - T) \frac{m_e}{m_h} \bar{\nu}_{eH}.$$

Appleton and Bray modeled the energy exchange for elastic collisions between electrons and atoms and between electrons and ions. However, the heating of electrons by interactions with the vibrational energy modes is important under the present conditions. This exchange is modeled using the inelastic energy factor δ_{eh} :

$$Q_{\text{e-v-el}} = n_e \sum_h 3k(T_e - T_{\text{e-v-el}}) \frac{m_e}{m_h} (\delta_{eh} - 1) \bar{\nu}_{eH}.$$

Expressions for $\bar{\nu}_{eH}$ and δ_{eh} are taken from the work of Laux *et al* (1999).

5.2.7 Chemical kinetics

As the plasma exits the torch and flows through the nozzle and discharge regions, chemical reactions occur and mass transfer between species takes place. As the characteristic times for the chemical reactions and fluid motion are far apart, equilibrium predictions cannot be used to determine the individual species concentrations. As a consequence, finite rate chemistry is introduced to determine individual species concentrations.

The plasma consists of a high-temperature mixture of nitrogen and oxygen. The species considered in the flow are the neutral species (N_2, O_2, NO, N, O), the ionic species (N_2^+, O_2^+, NO^+, N^+, O^+), and the electrons, e^-. A 38 reaction finite-rate chemical kinetics model (Laux *et al* 1999) is employed to describe the chemistry in the flow. Backward reaction rates in the law of mass action are computed from the equilibrium constants obtained from the Gordon–McBride (1994) data.

5.2.8 Numerical method

The electron number density varies by many orders of magnitude in the flow field, and therefore the numerical method must be designed to be stable and accurate under these conditions. Hammond *et al* (2002) developed a numerical method for glow discharges that reduces numerical error for this type of flow. In one dimension, the numerical representation of the electron conservation equation is written as

$$n_{e,i}^{n+1} = n_{e,i}^n - \frac{\Delta t}{\Delta x}(n_{e,i+1/2}^n v_{e,i+1/2}^n - n_{e,i-1/2}^n v_{e,i-1/2}^n) + \Delta t \omega_{e,i}^n$$

where $n_{e,i+1/2}$ is the average electron number density, and $v_{e,i+1/2}$ is computed using the electron temperature and number density at grid points i and $i+1$. This approach is easily extended to multiple dimensions. We use this approach for the electron conservation equation and a similar approach for the electron energy conservation equation.

The most difficult part of simulating the discharge flows is the huge range of time scales that govern the flow. The discharge energy relaxation has a time scale of a nanosecond or less, while the total flow time through the discharge region is of the order of 100 µs. Therefore, the time integration method must be designed to increase the stable time step size to the maximum extent possible.

Under the conditions of the present dc discharge experiments, the energy relaxation processes are very fast relative to the fluid motion time scales and the chemical kinetic processes. To handle this large disparity in characteristic time scales, we would usually use an implicit time integration method. However, for this problem a complete linearization of the problem is itself very expensive. (We solve 17 conservation equations, and the cost of evaluating the Jacobians and inverting the system scales with the square of the number of equations.) Therefore we linearize only those terms that are relatively fast, which results in a simple and inexpensive semi-implicit method that very substantially reduces the cost of the calculations.

The relatively fast terms are the internal energy relaxation and the Joule heating terms in the source terms for the three energy equations. Therefore, we split the source vector, W, into these terms, W_{fast}, and all of the other terms, W_{slow}. The conservation equations are then written as

$$\frac{\partial U}{\partial t} + \frac{\partial F}{\partial x} + \frac{1}{r}\frac{\partial rG}{\partial r} = W_{fast} + W_{slow}$$

where U is the vector of conserved variables, F is the axial direction flux vector and G is the radial direction flux vector. We then linearize W_{fast} in time

$$W_{fast}^{n+1} = W_{fast}^n + C_{fast}^n \delta U^n + O(\Delta t^2)$$

where C_{fast} is the Jacobian of W_{fast} with respect to U, and $\delta U^n = U^{n+1} - U^n$. Because of the form of W_{fast}, C_{fast} is a simple matrix that can be inverted analytically. Then the solution is integrated in time using

$$\delta U^{n+1} = (I - \Delta t C_{fast}^n)^{-1}\left(\Delta t(W_{fast} + W_{slow}) - \Delta t\left(\frac{\partial F}{\partial x} + \frac{1}{r}\frac{\partial rG}{\partial r}\right)\right).$$

This approach increases the stable time step by a factor of 50 compared to an explicit Euler method. This results in a very large reduction in the computer time required to obtain a steady-state solution.

A two-block grid is used to facilitate the implementation of the boundary conditions. The first grid block represents the nozzle section, and the second grid block represents the discharge region as well as a portion of the open air which acts as a large constant-pressure exhaust reservoir at one atmosphere.

The inflow boundary conditions are set by choosing the inflow static pressure to give the experimental mass flow rate of 4.9 g/s. The inflow is assumed to be in LTE at the measured temperature profile. This results in a consistent representation of the inflow conditions. The boundary conditions along the test-section surface are straightforward. The velocity is zero at the surface, the temperature is specified, and the normal-direction pressure gradient is zero. We assume that the metallic surface is highly catalytic to ion recombination. Otherwise, the surface is assumed to be non-catalytic to recombination for neutrals.

The computation is initialized as follows: first, the inflow conditions are specified as above. Then the test-section and reservoir are all initialized at atmospheric pressure, and at each axial location the temperature profiles and chemical concentration profiles are set identical to the inflow boundary profiles. Once a converged solution is obtained for the flow in LTE, the discharge is ignited by injecting a flux of electrons at the cathode and applying the Joule heating source term to the energy equations. Then a steady-state solution for the dc discharge is obtained.

5.2.9 Simulation results

In this section we present numerical simulations of the dc discharge experiment. Figure 5.2.1a shows the $\log_{10}$ of the electron number density contours in the computational domain. The dc discharge region can be observed in this figure. This is the bright region where the electron number density is several orders of magnitude higher than in the region upstream of the cathode where there is no discharge. It can be observed that the electron number density is slightly higher than $10^{12}\,\mathrm{cm}^{-3}$ in most of the discharge region. This is in good agreement with the experimental measurements for the electron number density. The electron number density falls off gradually downstream of the anode region. The shape of the discharge is similar to that observed in the experiments, which also shows that the discharge is constricted at the cathode and diffuses radially outward, away from the cathode. The simulations capture this behavior.

Figure 5.2.1b plots the electron temperature contours in the computational domain. It shows that the electron temperature is about 12 000 K in

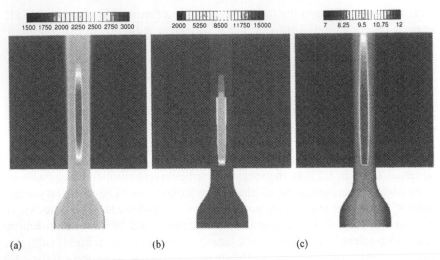

| 1500 1750 2000 2250 2500 2750 3000 | 2000 5250 8500 11750 15000 | 7   8.25   9.5   10.75   12 |

(a)   (b)   (c)

Figure 5.2.1. $\log_{10}$ of (a) the electron number density, (b) the electron temperature and (c) the translational temperature contours in the discharge region.

the discharge region. The computed electron temperatures are consistent with the experimental predictions. Figure 5.2.1b also shows that the electron temperature drops off sharply just downstream of the anode because the electrons rapidly equilibrate with the heavy particles due to their strong coupling with the heavy species.

Figure 5.2.1c shows contours of the translational temperature in the domain. It shows that the temperature in the discharge is about 3000 K in the discharge region. The computed temperatures are generally higher than the experimental measurements.

Figure 5.2.2 plots the axial variation of the centerline electron number density and the temperatures along with the experimental values. This figure quantitatively shows the variation of the electron concentration and the three temperatures along the centerline of the discharge. From the figure it can be seen that the electron number density remains slightly above $10^{12}\,\text{cm}^{-3}$ in the discharge region. It falls off gradually downstream of the anode. The computed electron temperature is very high in the cathode region and falls to about 12 000 K in most of the discharge region, which is close to the two-temperature kinetic model prediction. As observed in the contour plot for the electron temperature, the electron temperature falls off abruptly in the region downstream of the anode. The translational temperature increases from about 2200 K at the cathode to about 3000 K in the discharge region. This is higher than the measured translational temperature. However, the computed vibrational temperature is slightly lower than the experimentally measured value.

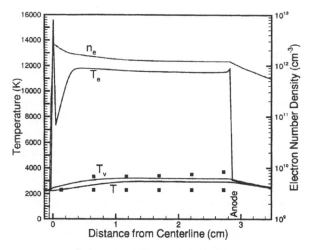

Figure 5.2.2. Computed electron number density and temperatures along the dc discharge centerline. Symbols denote experimentally measured values.

Figure 5.2.3 plots the radial profiles of the electron number density at two locations in the discharge. Near the cathode it can be seen that the diameter of the discharge is small and the electron number density is elevated in a region which is nearly equal to that of the cathode area. Near the center of the discharge the electron density is more diffuse and the diameter of the discharge is about 4 mm, which compares well with the experimentally observed diameter.

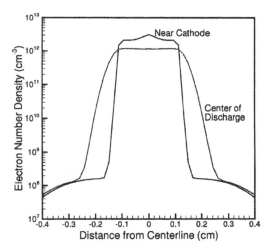

Figure 5.2.3. Computed radial profiles of electron number density.

5.2.10   Conclusions

The present work and that presented in section 5.3 demonstrates that stable, diffuse discharges with electron number densities approaching 10^{13} cm^{-3} at gas temperatures below 2000 K can be produced in atmospheric pressure air. This result stands in sharp contrast to the widespread belief that these diffuse discharges cannot exist without arcing instabilities or high levels of gas heating. A computational fluid dynamics code for the simulation of flowing non-equilibrium air plasmas including the presence of a dc discharge was developed and compared to the dc experiments conducted at Stanford University. The code uses a detailed two-temperature chemical kinetic mechanism, along with appropriate internal energy relaxation mechanisms. The discharge region was modeled by generalizing the channel model of Steenbeck, and a new semi-implicit time integration method was developed to reduce the computational cost. The computational results show good agreement with the experimental data; however, the heat loss is more rapid in the experiment than predicted by the computations.

Acknowledgments

This work was funded by the Director of Defense Research and Engineering (DDR&E) within the Air Plasma Ramparts MURI program managed by the Air Force Office of Scientific Research (AFOSR). Computer time was provided by the Minnesota Supercomputing Institute.

References

Appleton J P and Bray K N C 1964 'The conservation equations for a nonequilibrium plasma' *J. Fluid Mech.* **20** 659–672

Gnoffo P A, Gupta R N and Shinn, J L 1989 'Conservation equations and physical models for hypersonic air flows in thermal and chemical non-equilibrium' NASA TP-2867

Gordon S and McBride B J 1994 'Computer program for calculation of complex chemical equilibrium compositions and applications' NASA RP-1311

Gupta R N, Yos J M, Thompson R A and Lee K 1990 'A review of reaction rates and ther-modynamic and transport properties for an 11-species air model for chemical and thermal non-equilibrium calculations to 30 000 K' NASA RP-2953

Hammond E P, Mahesh K and Moin P 2002 'A numerical method to simulate radio-frequency plasma discharges' *J. Computational Phys.* **176** 402

Laux C, Pierrot L, Gessman R and Kruger C H 1999 'Ionization mechanisms of two-temperature plasmas' AIAA Paper No. 99-3476

Laux C O, Yu L, Packan D M, Gessman R J, Pierrot L and Kruger C H 1999 'Ionization mechanisms in two-temperature air plasmas' AIAA Paper 99-3476

Millikan R C and White D R 1963 'Systematics of vibrational relaxation' *J. Chem. Phys.* **39** 3209

Raizer Y P 1997 *Gas Discharge Physics* (Berlin: Springer) pp 275–287

Ramshaw J D 1990 'Self-consistent effective binary diffusion in multicomponent gas mixtures' *J. Non-Equilibrium Thermodynamics* **15** 295

Ramshaw J D 1993 'Hydrodynamic theory of multicomponent diffusion and thermal diffusion in multitemperature gas mixtures' *J. Non-Equilibrium Thermodynamics* **18** 121

Ramshaw J D and Chang C H 1991 'Ambipolar diffusion in multicomponent plasmas' *Plasma Chem. Plasma Proc.* **11**(3) 395

Ramshaw J D and Chang C H 1993 'Ambipolar diffusion in two-temperature multi-component plasmas' *Plasma Chem. Plasma Proc.* **13**(3) 489

Ramshaw J D and Chang C H 1996 'Friction-weighted self-consistent effective binary diffusion approximation' *J. Non-Equilibrium Thermodynamics* **21**

5.3   DC Glow Discharges in Atmospheric Pressure Air

5.3.1   Introduction

We present experimental and numerical investigations to determine whether and to what extent the electron number density can be increased in air plasmas by means of dc discharges. The strategy is to elevate the electron temperature, T_e, relative to the gas temperature, T_g, with an applied dc electric field.

Section 5.3.2 describes numerical investigations of two-temperature air plasma chemical kinetics. We present first a two-temperature kinetic mechanism to predict electron number density in air at a given gas temperature, as a function of the electron temperature. Close attention has been paid to the influence of the electron temperature on the rate coefficients, because collisions with energetic electrons can affect the vibrational population distribution of molecules, thereby the rates of ionization and dissociation.

Section 5.3.3 discusses the implications of this analysis for the generation of nonequilibrium air plasmas by means of electrical discharges. We determine in section 5.3.3.1 the relation between electron number density and current density, and between electron temperature and electric field. This is accomplished with Ohm's law and the electron energy equation, as discussed in section 5.3.3.2. A key quantity in the electron energy equation is the rate of electron energy lost by inelastic collisions. To predict inelastic losses in air plasmas, we have developed a detailed collisional-radiative model. This model is presented in section 5.3.3.3.

Section 5.3.4 describes experiments with dc glow discharges in air. We demonstrate that stable diffuse glow discharges with electron densities of up to $\sim 2 \times 10^{12} \mathrm{cm}^{-3}$ can be sustained in flowing preheated atmospheric pressure air. The electrical characteristics and thermodynamic parameters of the glow discharges are measured.

Section 5.3.5 compares the measured electrical characteristics of dc glow discharges in air with those obtained with the two-temperature and collisional-radiative model. This comparison validates the two-temperature model theoretical predictions. In addition, it enables us to establish the power requirements of dc discharges in air plasmas. This fundamental understanding forms the basis for the power budget reduction strategy using repetitively pulsed discharges presented in chapter 7 section 7.4.

5.3.2 Two-temperature kinetic simulations

This section presents results of numerical investigations to determine whether and to what extent electron number densities can be increased in air plasmas by elevating the electron temperature, T_e, relative to the gas temperature, T_g. The two-temperature kinetic mechanism and rates used for this work are presented in section 5.3.2.1. In section 5.3.2.2, the kinetic model is used to predict the temporal evolution and steady-state species concentrations in an atmospheric pressure air plasma with constant gas temperature of 2000 K and with electron temperatures varied from 4000 to 18 000 K. In section 5.3.2.3, the key reactions controlling ionization and recombination processes are identified. An analytical model based on the set of controlling reactions is then used to predict steady-state species concentrations in two-temperature air. As will be seen, the analytical model not only reproduces the CHEMKIN solution but also predicts an additional range of steady-state electron number densities.

5.3.2.1   *Two-temperature kinetic model*

The rate coefficients required for the two-temperature kinetic model depend on the relative velocities of collision partners (related to T_g for reactions between heavy particles and to T_e for electron-impact reactions) and on the population distributions over internal energy levels of atoms and molecules. Thus, these rate coefficients correspond to the weighted average of elementary rates over internal energy states of atoms and molecules. This forms the basis of the Weighted Rate Coefficient (WRC) method described in references [1–4]. The method assumes that the internal energy levels of atoms and molecules are populated according to Boltzmann distributions at the electronic temperature T_{el}, the vibrational temperature T_v, and the rotational temperature T_r. Elementary rate coefficients are calculated from cross-section data assuming Maxwellian velocity distribution functions for electrons and heavy particles at T_e and T_g, respectively. It is further assumed that $T_{el} = T_e$ and $T_r = T_g$. The remaining parameter, T_v, can only be determined in the general case by solution of the master equation for all vibrational levels by means of a collisional-radiative (CR) model that incorporates vibrationally specific state-to-state kinetics. We have recently developed such a model for nitrogen plasmas [1, 4] that provides insight into the relation between T_v and T_g and T_e in atmospheric pressure plasmas. The nitrogen CR model accounts for electron and heavy-particle impact ionization (atoms and molecules) and dissociation (molecules), electron-impact vibrational excitation, V–T and V–V transfer, radiation, and predissociation. Through comparisons between the results of the CR model and of a two-temperature kinetic model of nitrogen that assumed either $T_v = T_g$ or $T_v = T_e$, we have shown [4, 5] for the case of a nitrogen plasma at $T_g = 2000\,K$ that the steady-state species concentrations determined with the two-temperature kinetic model are in close agreement with the CR model predictions if one assumes (1) that $T_v = T_g$ for electron temperatures $T_e \leq 9500\,K$ and electron number densities $n_e \leq {\sim}10^{11}\,cm^{-3}$, and (2) that $T_v = T_e$ or $T_v = T_g$ for $T_e > 9500\,K$ and $n_e \geq {\sim}10^{15}\,cm^{-3}$ (in the latter range, best agreement is obtained with $T_v = T_e$ but assuming $T_v = T_g$ leads to electron number densities that are underestimated by at worse a factor of 5). It should be noted that the often-used assumption $T_v = T_e$ produces steady-state electron number densities that are several orders of magnitude greater than those obtained with the CR model for electron temperatures $T_e \leq 9500\,K$ and electron number densities $n_e \leq {\sim}10^{11}\,cm^{-3}$. We extend these results to atmospheric pressure air by calculating all WRC rate coefficients with the assumption $T_v = T_g$.

The full 11-species (O_2, N_2, NO, O, N, O_2^+, N_2^+, NO^+, N^+, O^+, and electrons), 40-reaction mechanism and rate coefficients for the case $T_g = 2000\,K$ are summarized in table 5.3.1. Electron attachment reactions can be neglected in atmospheric pressure air at temperature $>1500\,K$ because the equilibrium concentrations of O_2^- or O^- are negligibly small relative to the concentration of electrons above ${\sim}1500\,K$ (figure 5.3.1). For reactions between nitrogen species, the rate coefficients are taken from Yu [5]. This set is supplemented by two-temperature rate coefficients determined using the WRC method for electron-impact dissociation and ionization of O_2 and NO. For electron-impact ionization of O, we adopt the two-temperature rate of Lieberman and Lichtenberg [6]. Rate coefficients for $O^+ + N_2 \Leftrightarrow NO^+ + N$ and $O^+ + O_2 \Leftrightarrow O_2^+ + O$ are taken from Hierl *et al* [7], and the rate coefficient of the charge transfer reaction between O^+ and NO is calculated using the experimental cross-section reported by Dotan and Viggiano [8]. The remaining reactions involve collisions between heavy particles and thus mostly depend on the gas kinetic temperature (as we assume $T_r = T_v = T_g$). For these reactions, the rate coefficients of Park [9, 10] are employed.

The two-temperature kinetic calculations presented in the rest of this section were made with the CHEMKIN solver [11] modified [12] so as to allow a different temperature (T_e) to be specified for the rates of particular

Table 5.3.1. Two-temperature kinetic model of air plasmas. The temperature entering the Arrhenius-type expressions is either the gas (T_g) or the electron (T_e) temperature, as indicated in columns k_f (forward rate) and k_r (reverse rate). The present mechanism is for gas temperatures greater than 1500 K.

Reaction	Temperature dependence		Rate coefficient, $k = AT^b \exp(-E/RT)$			Ref. (see footnotes)
	k_f	k_r	A (mole cm s)	b	E/R (K)	
O_2 Dissociation/recombination						
1. $O_2 + O_2 = 2O + O_2$	T_g	T_g	2.00×10^{21}	-1.5	59 500	a
2. $O_2 + NO = O + O + NO$	T_g	T_g	2.00×10^{21}	-1.5	59 500	a
3. $O_2 + N_2 = O + O + N_2$	T_g	T_g	2.00×10^{21}	-1.5	59 500	a
4. $O_2 + O = O + O + O$	T_g	T_g	1.00×10^{22}	-1.5	59 500	a
5. $O_2 + N = O + O + N$	T_g	T_g	1.00×10^{22}	-1.5	59 500	a
6f. $O_2 + e \Rightarrow O + O + e$	T_e		2.85×10^{17}	-0.6	59 500	b
6b. $O + O + e \Rightarrow O_2 + e$	T_e		4.03×10^{18}	-0.4	0	b
NO dissociation/recombination						
7. $NO + O_2 = N + O + O_2$	T_g	T_g	5.00×10^{15}	0.0	75 500	a
8. $NO + NO = N + O + NO$	T_g	T_g	1.10×10^{17}	0.0	75 500	a
9. $NO + N_2 = N + O + N_2$	T_g	T_g	5.00×10^{15}	0.0	75 500	a
10. $NO + O = N + O + O$	T_g	T_g	1.10×10^{17}	0.0	75 500	a
11. $NO + N = N + O + N$	T_g	T_g	1.10×10^{17}	0.0	75 500	a
12f. $NO + e \Rightarrow N + O + e$	T_e		3.54×10^{16}	-0.2	75 500	b
12b. $N + O + e \Rightarrow NO + e$	T_e		8.42×10^{21}	-1.1	0	b
N_2 Dissociation/recombination						
13. $N_2 + O_2 = N + N + O_2$	T_g	T_g	7.00×10^{21}	-1.6	113 200	a
14. $N_2 + NO = N + N + NO$	T_g	T_g	7.00×10^{21}	-1.6	113 200	a
15. $N_2 + N_2 = N + N + N_2$	T_g	T_g	7.00×10^{21}	-1.6	113 200	a
16. $N_2 + O = N + N + O$	T_g	T_g	3.00×10^{22}	-1.6	113 200	a
17. $N_2 + N = N + N + N$	T_g	T_g	3.00×10^{22}	-1.6	113 200	a
18f. $N_2 + e \Rightarrow N + N + e$	T_e		1.18×10^{18}	-0.7	113 200	b
18b. $N + N + e \Rightarrow N_2 + e$	T_e		1.36×10^{23}	-1.3	0	b
Zeldovich reactions						
19. $N_2 + O = NO + N$	T_g	T_g	6.40×10^{17}	-1.0	38 400	a
20. $NO + O = O_2 + N$	T_g	T_g	8.40×10^{12}	0.0	19 400	a
Associative ionization/dissociative recombination						
21f. $N + O \Rightarrow NO^+ + e$	T_g		8.80×10^{08}	1.0	31 900	a
21b. $NO^+ + e \Rightarrow N + O$	T_e		9.00×10^{18}	-0.7	0	c
22f. $N + N \Rightarrow N_2^+ + e$	T_g		6.00×10^{07}	1.5	67 500	b
22b. $N_2^+ + e \Rightarrow N + N$	T_e		1.53×10^{18}	-0.5	0	b
23f. $O + O \Rightarrow O_2^+ + e$	T_g		7.10×10^{02}	2.7	80 600	a
23b. $O_2^+ + e \Rightarrow O + O$	T_e		1.50×10^{18}	-0.5	0	c

Table 5.3.1. *(Continued)*

Reaction	Temperature dependence k_f	k_r	Rate coefficient, $k = AT^b \exp(-E/RT)$ A (mole cm s)	b	E/R (K)	Ref. (see footnotes)
Electron impact ionization/three-body recombination						
24f. $O + e \Rightarrow O^+ + e + e$	T_e		7.74×10^{12}	0.7	157760	d
24b. $O^+ + e + e \Rightarrow O + e$	T_e		2.19×10^{21}	-0.8	0	e
25f. $N + e \Rightarrow N^+ + e + e$	T_e		5.06×10^{19}	0.0	168200	b
25b. $N^+ + e + e \Rightarrow N + e$	T_e		5.75×10^{26}	-1.3	0	b
26f. $O_2 + e \Rightarrow O_2^+ + e + e$	T_e		5.03×10^{12}	0.5	146160	b
26b. $O_2^+ + e + e \Rightarrow O_2 + e$	T_e		8.49×10^{23}	-1.9	0	b
27f. $N_2 + e \Rightarrow N_2^+ + e + e$	T_e		2.70×10^{17}	-0.3	181000	b
27b. $N_2^+ + e + e \Rightarrow N_2 + e$	T_e		2.05×10^{21}	-0.8	0	b
28f. $NO + e \Rightarrow NO^+ + e + e$	T_e		2.20×10^{16}	-0.3	107400	b
28b. $NO^+ + e + e \Rightarrow NO + e$	T_e		2.06×10^{25}	-2.0	0	b
Charge exchange/charge transfer						
29f. $N^+ + N_2 \Rightarrow N_2^+ + N$	T_g		4.60×10^{11}	0.5	12200	b
29b. $N_2^+ + N \Rightarrow N_2 + N'$	$T_g = 2000\,K$		1.93×10^{13}	0.0	0	b
30. $NO^+ + O = N^+ + O2$	T_g	T_g	1.00×10^{12}	0.5	77200	a
31. $NO + O^+ = N^+ + O2$	T_g	T_g	1.40×10^{05}	1.9	15300	a
32. $O^+ + N_2 = NO^+ + N$	T_g	T_g	4.40×10^{13}	0.0	5664	f
33. $O^+ + N_2 = N_2^+ + O$	T_g	T_g	9.00×10^{11}	0.4	22800	a
34. $NO^+ + N = N_2^+ + O$	T_g	T_g	7.20×10^{13}	0.0	35500	a
35. $O_2^+ + N = N^+ + O_2$	T_g	T_g	8.70×10^{13}	0.1	28600	c
36. $O_2^+ + N_2 = N_2^+ + O_2$	T_g	T_g	9.90×10^{12}	0.0	40700	c
37. $NO^+ + O_2 = O_2^+ + NO$	T_g	T_g	2.40×10^{13}	0.4	32600	c
38. $NO^+ + O = O_2^+ + N$	T_g	T_g	7.20×10^{12}	0.3	48600	c
39. $O^+ + O_2 = O_2^+ + O$	T_g	T_g	3.26×10^{13}	0.0	2064	f
40. $O^+ + NO = NO^+ + O$	T_g	T_g	2.42×10^{13}	0.0	902	g

a. Park [10].
b. WRC [3, 4]. These rates were calculated at $T_g = T_v = 2000$ K. The present fitting formulas are valid for $6000\,K \leq T_e \leq 20\,000\,K$
c. Park [9].
d. Lieberman [6].
e. Detailed balance
f. Hierl *et al* [7].
g. Dotan and Viggiano [8].

reactions. The extended code functions in a similar manner to CHEMKIN. For thermal reactions, reverse rate coefficients are computed from equilibrium thermodynamic functions (detailed balance). Reverse rate coefficients with a dependence on T_e were determined with the WRC model by detailed balance.

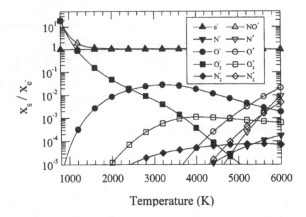

Figure 5.3.1. Charged species concentrations relative to the electron concentration in equilibrium air ($P = 1$ atm).

5.3.2.2   Results

We consider first the case of an air plasma taken to be in equilibrium ($T_g^0 = T_e^0 = 2000$ K, $P = 1$ atm) at time zero when an elevated electron temperature is instantaneously prescribed, in an idealized way modeling an electrical glow discharge in a reactor section. In the example shown in figure 5.3.2 the gas temperature is held constant at 2000 K and the electron

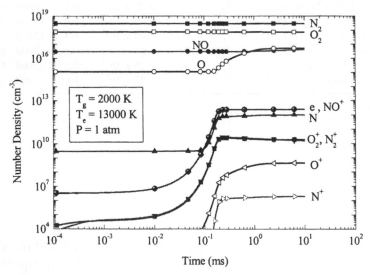

Figure 5.3.2. Temporal evolution of species concentrations in atmospheric pressure air at constant gas temperature (2000 K) and constant electron temperature (13 000 K).

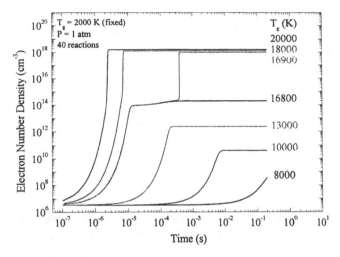

Figure 5.3.3. Temporal evolution of the electron number density in a two-temperature air plasma. Initial conditions are equilibrium air at 2000 K ($n_e^{(t=0)} = 3.3 \times 10^6 \, \text{cm}^{-3}$).

temperature is increased to 13 000 K at time zero. The time evolution of species concentrations computed with the two-temperature CHEMKIN solver is shown in figure 5.3.2. The electron number density rises from its initially low value of $3.3 \times 10^6 \, \text{cm}^{-3}$ to a steady-state value of $\sim 4 \times 10^{12} \, \text{cm}^{-3}$ in about 0.1 ms. The dissociation fraction of oxygen atoms increases from $\sim 0.03\%$ at time zero to $\sim 1\%$ at steady-state. NO$^+$ is the dominant ion at all times.

Additional calculations were made for various electron temperatures while keeping the gas temperature constant at 2000 K. The predicted temporal evolutions of the electron number density are shown in figure 5.3.3. Practically no increase in the electron number density is observed for electron temperatures below a threshold value of $T_e \cong 6000 \, \text{K}$, which corresponds to the temperature where electron-impact ionization reactions begin to dominate over heavy particle impact dissociation. As the electron temperature is further increased, the steady-state electron concentration increases significantly, with a very abrupt change at $T_e \cong 16\,800 \, \text{K}$.

Figure 5.3.4 shows the steady-state electron number densities predicted with CHEMKIN as a function of the electron temperature. At $T_e = 16\,800 \, \text{K}$ where the predicted steady-state electron number density suddenly increases from $\sim 10^{14}$ to $\sim 10^{18} \, \text{cm}^{-3}$ over a few Kelvin. It is interesting to examine the reverse case where steady-state electron concentrations are calculated for an initial composition given by the steady-state solution corresponding to $T_g = 2000 \, \text{K}$ and $T_e = 20\,000 \, \text{K}$ (corresponding to $n_e^{(t=0)} = 1.7 \times 10^{18} \, \text{cm}^{-3}$). As can be seen from figure 5.3.5 in this case the predicted steady-state electron number densities start by decreasing

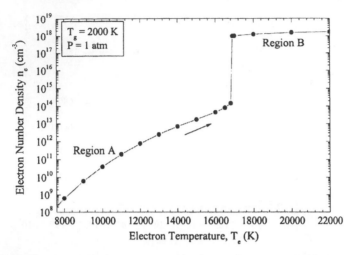

Figure 5.3.4. Steady-state electron number density predicted by CHEMKIN for air at $T_g = 2000\,\text{K}$, as a function of T_e. For each steady-state calculation, initial conditions are equilibrium air at 2000 K.

along the same curve as in figure 5.3.4 but, instead of the abrupt decrease at 16 800 K, continue their slow decrease until the electron temperature reaches $\sim$14 300 K. When the electron temperature is further decreased, the steady-state electron number density abruptly decreases to the level of the curve

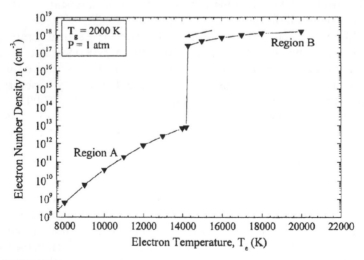

Figure 5.3.5. Steady-state electron number density predicted by CHEMKIN for air at $T_g = 2000\,\text{K}$, as a function of T_e. For each steady-state calculation, the initial condition corresponds to the steady-state composition predicted by CHEMKIN at $T_g = 2000\,\text{K}$ and $T_e = 20\,000\,\text{K}$ ($n_e^{(t=0)} = 1.7 \times 10^{18}\,\text{cm}^{-3}$).

of figure 5.3.4. Thus a hysteresis occurs as the electron temperature is increased and decreased in a cyclical fashion.

5.3.2.3 Analysis of the ionization mechanisms

Through detailed examinations of the reactions and rates, we found that the behavior in each of the two regions A and B of the curve in figure 5.3.4 can be explained in terms of the following simplified reaction mechanisms:

(A) Ionization mechanism in region A

In region A, the initial rapid electron concentration rise (see figure 5.3.2 for the case $T_e = 13\,000$,K) is the result of electron-impact ionization of N_2 and O_2 via three-body reactions:

$$O_2 + e \Rightarrow O_2^+ + e + e$$

$$N_2 + e \Rightarrow N_2^+ + e + e$$

and via electron-impact dissociation of O_2 followed by electron-impact ionization of O:

$$O_2 + e \Rightarrow O + O + e$$

$$O + e \Rightarrow O^+ + e + e$$

The charged species produced by these processes undergo rapid charge transfer to NO^+, via

$$O^+ + N_2 \Rightarrow NO^+ + N$$

$$N_2^+ + O_2 \Rightarrow N_2 + O_2^+$$

$$O_2^+ + NO \Rightarrow NO^+ + O_2$$

The main path for electron recombination is the two-body dissociative recombination reaction:

$$NO^+ + e \Rightarrow N + O$$

When the concentration of NO^+ becomes sufficiently large, the rate of dissociative recombination balances the rate of electron production and the plasma reaches steady-state. Thus, in region A, the termination step for the ionization process is the two-body recombination of a molecular ion.

(B) Ionization mechanism in region B

An example of the temporal evolution of species concentrations in region B is shown in figure 5.3.6 where the electron temperature is fixed at $T_e = 18\,000$ K. As in region A, the electron number density initially increases

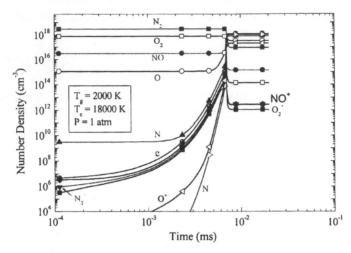

Figure 5.3.6. Temporal evolution of species concentrations in atmospheric pressure air at constant gas temperature (2000 K) and constant electron temperature (18 000 K).

by electron-impact ionization of N_2 and O_2 via

$$O_2 + e \Rightarrow O_2^+ + e + e$$

$$N_2 + e \Rightarrow N_2^+ + e + e$$

and via electron-impact dissociation of O_2 followed by electron-impact ionization of O:

$$O_2 + e \Rightarrow O + O + e$$

$$O + e \Rightarrow O^+ + e + e$$

The difference with region A is that the charge transfer reactions are not fast enough to produce NO^+ at a high enough rate. This is because these reactions are controlled by the gas temperature, whereas electron impact ionization reactions are controlled by T_e. The critical electron temperature that defines the limit between regions A and B corresponds approximately to the electron temperature for which the rate of the transfer reaction $O^+ + N_2 \Rightarrow NO^+ + N$ is comparable with the rate of avalanche ionization by electron impact. Above this critical electron temperature, the avalanche ionization process continues until all molecular species are dissociated. Eventually the rates of three-body electron recombination reactions balance the rate of ionization, and steady-state is reached.

It is noted that, in region A, electron impact dissociation of N_2 (or NO) is negligible because the dissociation energy of N_2 (9.76 eV) is much larger than that of O_2 (5.11 eV), and the concentration of NO is small relative to the concentration of O_2. It is only above the critical temperature that electron impact dissociation of N_2 starts having a noticeable effect.

(C)   Analytical solution

The kinetics in regions A and B can be described with a simplified subset of reactions that takes into account the dominant channels discussed in the foregoing section. With this simplified mechanism, the steady-state concentrations of major species are obtained by solving the following system of equations:

- Steady-state for e^-:

$$O_2 + e \Rightarrow O_2^+ + e + e$$

$$N_2 + e \Rightarrow N_2^+ + e + e$$

$$O + e \Leftrightarrow O^+ + e + e$$

$$NO^+ + e \Rightarrow N + O$$

- Steady-state for O_2:

$$O_2 + e \Rightarrow O + O + e$$

$$O + O + M \Rightarrow O_2 + M, \quad \text{with } M = O_2, N_2, O$$

- Steady-state for NO^+:

$$NO^+ + e \Rightarrow N + O$$

$$O^+ + N_2 \Rightarrow NO^+ + N$$

$$O_2^+ + NO \Rightarrow NO^+ + O_2$$

- Steady-state for O^+:

$$O + e \Leftrightarrow O^+ + e + e$$

$$O^+ + N_2 \Rightarrow NO^+ + N$$

$$O^+ + O_2 \Rightarrow O_2^+ + O$$

- Steady-state for O_2^+:

$$O_2 + e \Rightarrow O_2^+ + e + e$$

$$N_2^+ + O_2 \Rightarrow O_2^+ + N_2$$

$$O_2^+ + NO \Rightarrow NO^+ + O_2$$

- Steady-state for N_2^+:

$$N_2 + e \Rightarrow N_2^+ + e + e$$

$$N_2^+ + O_2 \Rightarrow O_2^+ + N_2$$

In writing the corresponding steady-state relations, we make the approximation that the change of plasma volume due to the increase of the total number

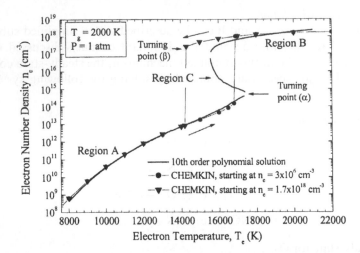

Figure 5.3.7. Steady-state electron number densities predicted by CHEMKIN and by the tenth-order polynomial analytical solution.

of moles at constant T and P is negligible, and that the concentration of N_2 remains constant and equal to its initial value. Furthermore, we consider that the dominant neutral species in the plasma are O_2, O, and N_2.

By elimination of n_O, n_{O_2}, n_{O^+}, $n_{O_2^+}$, $n_{N_2^+}$ and n_{NO^+}, we obtain a tenth-degree polynomial in n_e with coefficients that only depend on T_g, T_e, $n_{O_2}^{(0)}$, $n_{N_2}^{(0)}$ and the rate coefficients. The roots of this polynomial were extracted with the mathematical package MATLAB. The roots that give negative values for n_O are omitted from the solution. The remaining roots of the polynomial are plotted in figure 5.3.7 along with the CHEMKIN predictions corresponding to the full mechanism of table 5.3.1. As can be seen from figure 5.3.7, the approximate solution obtained with the simplified mechanism is in very good agreement with the CHEMKIN predictions in regions A and B. The small discrepancy in region B is due to the neglecting in our simplified model of nitrogen dissociation. Furthermore, the tenth-degree polynomial exhibits an extra solution that could not be attained with CHEMKIN (region C). The limits of region C are the turning points labeled (α) and (β) in figure 5.3.7. If we initialize CHEMKIN with the plasma composition corresponding to a point in region C of figure 5.3.7, CHEMKIN produces a new steady-state electron number density located on either the lower (region A) or upper (region B) limb of the steady-state curves. Thus region C of figure 5.3.7 cannot be obtained by fixing the electron temperature.

Figure 5.3.8 shows the concentrations of dominant species as predicted by the analytical model. We see that the concentration of NO^+ increases up to turning point (α), and then stays approximately constant as charge transfer becomes slower than oxygen ionization. The concentration of oxygen atoms steadily increases throughout regions A and C. Beyond

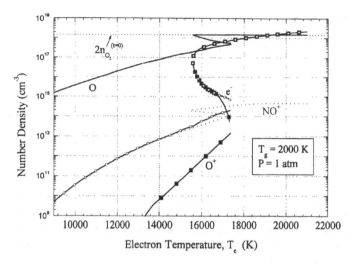

Figure 5.3.8. Steady-state species concentrations at $T_g = 2000\,\mathrm{K}$, as predicted by the analytical solution.

turning point β, molecular oxygen is nearly fully dissociated. The electron concentration is approximately equal to the concentration of NO^+ in region A and to the concentration of O^+ in regions B and C.

5.3.3 Predicted electric discharge characteristics

The electron number density and electron temperature can be related to the current density and electric field, respectively, by means of Ohm's law and the electron energy equation. The current density and electric field values provide guidance for the design of non-equilibrium dc discharges, as well as an estimate of the power requirements of such discharges.

5.3.3.1 Ohm's law

Ohm's law relates the current density j to the electric field E:

$$j = \sigma E = \frac{n_e e^2}{m_e \sum_h \bar{\nu}_{eh}} E \tag{1}$$

where σ is the electrical conductivity of the plasma, m_e the electron mass, and $\bar{\nu}_{eh}$ the average frequency of collisions between electrons and heavy particles. The total collision frequency is the sum of the collision frequencies of electrons with neutrals, n, and ions, i:

$$\bar{\nu}_{eh} = \bar{\nu}_{en} + \bar{\nu}_{ei}. \tag{2}$$

The electron–neutral collision frequency $\bar{\nu}_{en}$ can be expressed in terms of the number densities of neutral species, n_n, the electron velocity, $g_e = \sqrt{8kT_e/\pi m_e}$, and the energy-averaged momentum transfer cross-sections $\overline{Q_{en}}$ as

$$\bar{\nu}_{en} \cong n_n\, g_e\, \overline{Q_{en}}.$$

The energy-averaged electron–neutral momentum transfer collision cross-sections $\bar{Q}_{en}$ are calculated by:

$$\bar{Q}_{en} = \frac{2}{3} \frac{1}{T_e^3} \int_0^\infty \varepsilon^2 \exp\left(-\frac{\varepsilon}{T_e}\right) Q_{en}(\varepsilon)\, d\varepsilon \tag{3}$$

where T_e is the electron temperature, ε is the electron impact energy, and $Q_{en}(\varepsilon)$ is the momentum collision cross-section. For N_2, O_2 and O, these cross-sections have been taken from Brown [13], Shkarofsky *et al* [14], and Tsang *et al* [15]. The resulting average energy cross-sections $\bar{Q}_{en}$ are presented in figure 5.3.9.

In region A, where the dominant neutral species are N_2 and O_2, and where ions have negligible concentrations, the total collision frequency is well approximated by the electron–neutral collision frequency.

In region B, where the plasma is almost fully ionized, the total collision cross-section is approximately equal to the electron–ion collision frequency,

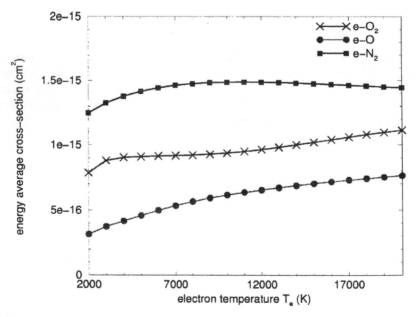

Figure 5.3.9. Energy-averaged momentum transfer cross-sections for collisions between electrons and N_2, O_2, O.

which can be expressed as [16]:

$$\bar{\nu}_{ei} = 3.64 \times 10^{-6} n_i \frac{\ln \Lambda}{T_e^{3/2}} \quad \sec^{-1} \tag{4}$$

where Λ represents the ratio of the Debye length to the impact parameter for 90° scattering and is approximately equal to 2.5 for electron number densities of 10^{18} cm^{-3} (Mitchner and Kruger [16]).

5.3.3.2 *Electron energy equation*

(A) *Introduction*
For a stationary plasma in a dc electric field, the electron energy equation can be written as

$$\sigma E^2 = n_e \sum_h \frac{3}{2} k (T_e - T_h) \frac{2m_e}{m_h} \bar{\nu}_{eh} + \sum_j \varepsilon_j \dot{n}_j + \dot{R}. \tag{5}$$

In equation (5), k is the Boltzmann constant and T_h the kinetic temperatures of heavy species (assumed to be the same for all heavy species and thus equal to T_g). The term on the left hand side of equation (5) represents the volumetric power for Joule heating of electrons by the electric field. The first and second terms on the right-hand side represent the volumetric power lost by free electrons through elastic and inelastic collisions, respectively, and the last term on the right-hand side stands for volumetric radiative losses.

In region A of the S-shaped curve, inelastic energy losses dominate the elastic and radiation losses by at least two orders of magnitude. In region B, however, inelastic losses are negligible relative to elastic and radiation losses. In the rest of this analysis, we limit ourselves to the lower limb of the S-shaped curve (region A), and therefore we neglect radiation losses.

The basis for calculations of inelastic losses in atmospheric pressure air plasmas is summarized below. Electrons lose energy through the following inelastic processes: vibrational excitation of molecular species (VE transfer), ionization, electronic excitation, dissociation of molecules, electronic excitation, and ionization of atoms. At electron temperatures below about 17 000 K, the main channel for inelastic electron energy loss in air is via electron impact vibrational excitation of nitrogen. This process is particularly important for the ground state of molecular nitrogen because vibrational excitation by electron impact of this state occurs via resonant transitions to the ground state of the unstable negative ion N_2^-:

$$N_2(X, v'') + e \rightarrow N_2^-(X, v) \rightarrow N_2(X, v') + e.$$

The net rate of energy lost by this process is:

$$\dot{\varepsilon}_{v''v'} = (K_{v''v'}[N_2(X, v'')][e] - K_{v'v''}[N_2(X, v')][e]) \Delta E_{v''v'}$$

where $K_{v''v'}$ and $K_{v'v''}$ are the rate coefficients of electron-impact vibrational excitation and de-excitation, and $\Delta E_{v''v'}$ stands for the difference of energy between the two vibrational levels. The net rate of inelastic losses is a function of the gas and electron temperatures, the electron number density, and the vibrational population distribution of the ground state of N_2. In the limiting case where the vibrational level populations follow a Boltzmann distribution at the electron temperature, the net rate of energy loss by VE transfer is equal to zero because the energy lost by VE excitation reactions is exactly balanced by the energy gained from de-excitation. In the other limiting case where the vibrational levels follow a Boltzmann distribution at the gas temperature, the rate of excitation is much larger than the rate of de-excitation. In the general case, the vibrational population distribution is intermediate between the two previous cases. The vibrational population distribution is then governed by the relative importance of the rates of vibrational excitation by electron impact, and the rates of de-excitation which are mostly determined by collisions with heavy species. The dominant de-excitation processes are vibrational–vibrational transfer between two N_2 molecules (V–V transfer), vibrational–vibrational transfer between one N_2 molecule and other molecular species such as O_2 and NO (V–V' transfer), and vibrational–translation (V–T) relaxation by collisions of N_2 with other heavy species (N_2, O_2, NO, N and O). Here the main relaxation processes are V–T relaxation by O and N_2. To calculate inelastic energy losses in air, we must therefore predict the vibrational population distribution of the nitrogen ground state by taking into account the aforementioned processes. This is a complex calculation that requires the use of a vibrationally specific collisional-radiative (CR) model. We have developed such a model for pure nitrogen plasmas [1, 2, 4, 17] and have recently extended it to air plasmas [18].

(B)   Rate coefficients controlling the vibrational distribution of N_2 levels
The rate coefficients for VE transfer are calculated [4] using the cross section calculation method of Kazansky and Yelets [19]. This method reproduces available experimental cross-sections relative to the low-lying vibrational levels within about 10%.

For electron temperatures up to $17\,000\,\text{K}$ (turning point α in the S-shaped curve), the dominant N_2 V–T, V–V and V–V' relaxation processes are:

$$N_2(Y, v + 1) + M \leftrightarrow N_2(Y, v) + M \tag{6}$$

$$N_2(X, v_1 + 1) + AB(X, v_2) \leftrightarrow N_2(X, v_1) + AB(X, v_2 + 1). \tag{7}$$

In equation (6), M represents the heavy particle collision partner $M = N_2$, O_2, NO or O and in equation (7), AB is the diatomic molecule N_2, O_2 or NO.

To our knowledge, no experimental data have been reported for the V–T relaxation of N_2 by collisions with N. However, Kozlov *et al* [20] experimentally determined an upper limit value of the V–T relaxation rate for

$v = 1 \rightarrow v = 0$. They showed that for temperatures between 2500 and 4500 K, this rate is about one order of magnitude lower than the rate of V–T relaxation of N_2 by collisions with O atoms. Since for our conditions the concentration of N atoms is at least four orders of magnitude lower than the concentration of O atoms, we neglect the V–T relaxation of N_2 by collisions with N atoms.

Most measured vibrational relaxation rate coefficients are for transitions between vibrational levels $v = 0$ and $v = 1$. When available, experimental rates have been preferred over theoretical ones. The existing experimental rates have been compared and critically selected. Rates for transitions between higher levels have been calculated using scaling functions derived from SSH theory, which is a reasonable approximation when the gas temperature is below 3000 K. The reverse rates have been determined by application of the detailed balance method.

For each V–T transfer process, the rates $k_{1,0}$ corresponding to transition $v = 1 \rightarrow v = 0$ have been calculated with the following analytical expression:

$$k_{1,0} = A T_g^n \exp\left(-\frac{B}{T_g^{1/3}} + \frac{C}{T_g^m}\right)\left[1 - D\exp\left(-\frac{E_{10}}{T_g}\right)\right]^{-1} \qquad (8)$$

where $k_{1,0}$ is expressed in $\mathrm{cm^3\,s^{-1}}$, and T_g and E_{10} (energy of the N_2 transition $v = 1 \rightarrow v = 0$) are expressed in Kelvin. The parameters A, B, C, D, m and n were determined in reference [18] and are listed in table 5.3.2.

The rate coefficients $k_{v+1,v}$ for transitions $v + 1 \rightarrow v$ between upper vibrational levels have been calculated using appropriate scaling laws from the measured k_{10} rates:

$$k_{v+1,v} = k_{1,0} G(v + 1). \qquad (9)$$

Using SSH theory [21] and some approximations for the Morse oscillator model [22], $G(v + 1)$ can be expressed as

$$G(v + 1) \simeq \frac{(v + 1)(1 - x_e)}{1 - x_e(v + 1)} \frac{F(y_{v+1,v})}{F(y_{1,0})} \qquad (10)$$

where x_e is the anharmonicity of the N_2 molecule, and $y_{v+1,v}$ is given by

$$y_{v+1,v} = 0.32 E_{v+1,v} L \sqrt{\frac{\mu}{T_g}} \qquad (11)$$

Table 5.3.2 Parameters for the V–T rates $k_{1,0}$.

M =	A	B	C	D	m	n
O	1.07×10^{-10}	69.9	0	0	0	0
N_2, O_2, NO	7.8×10^{-12}	218	690	1	1	1

where $E_{v+1,v}$ is the energy of the $v+1 \to v$ transition in Kelvin, L is the characteristic parameter of the short range repulsive potential in Å, μ is the reduced mass of the two colliding particles in atomic units, and T_g is the gas temperature in Kelvin. We have taken $L = 0.25$ Å for all V–T processes.

The function F in equation (10) is given by [21]

$$
\begin{cases}
F(y) = \dfrac{1}{2}\left[3 - \exp\left(-\dfrac{2y}{3}\right)\right] \exp\left(-\dfrac{2y}{3}\right) & \text{for } 0 \le y \le 20 \\[4mm]
F(y) = 8\left(\dfrac{\pi}{3}\right)^{1/2} y^{7/3} \exp(-3y^{2/3}) & \text{for } y > 20.
\end{cases}
\tag{12}
$$

The calculated forward and reverse rates are plotted in figure 5.3.10. The rates for V–T relaxation of $N_2(v)$ by collision with O atoms are between two and three orders of magnitude higher than the rates of V–T relaxation of $N_2(v)$ by collision with N_2.

The rates $k_{1,0}^{0,1}$ corresponding to the V–V and V–V' processes (7) with $v_1 = 0$ and $v_2 = 0$, are calculated using the following expression:

$$
k_{1,0}^{0,1} = A T_g^n \exp\left(-\frac{B}{T_g^{1/3}}\right)
\tag{13}
$$

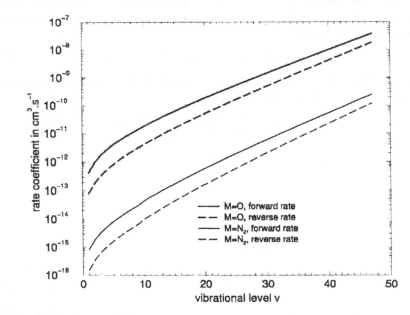

Figure 5.3.10. Rate coefficient for V–T relaxation: $N_2(X,v) + M \to N_2(X, v-1) + M$, with M = O, N_2 at $T_g = 2000$ K.

Table 5.3.3. Parameters for the V–V and V–V′ rates $k_{1,0}^{0,1}$.

V–V or V–V′ process	A	B	n
N_2–N_2	1.27×10^{-17}	0	1.483
N_2–O_2	1.23×10^{-14}	104	1
N_2–NO	4.22×10^{-10}	86.35	0

where the gas temperature T_g is expressed in Kelvin and the parameters A, B, and n are listed in table 5.3.3.

Note that the rate of V–V transfer for N_2–N_2 collisions was recently measured by Ahn *et al* [23]. The measured rates are about one order of magnitude lower than the values adopted here. However, this rate has practically no influence on the results presented here.

The rate coefficients $k_{v_1+1,v_1}^{v_2,v_2+1}$ for exothermic transitions between upper vibrational levels have been calculated using the relation

$$k_{v_1+1,v_1}^{v_2,v_2+1} = k_{1,0}^{0,1} G(v_1+1, v_2+1) \tag{14}$$

where $G(v_1+1, v_2+1)$ is an appropriate function which can be expressed using SSH theory [21] and some approximations for the Morse oscillator model as

$$G(v_1+1, v_2+1) \simeq \frac{(v_1+1)(1-x_{e1})}{1-x_{e1}(v_1+1)} \frac{(v_2+1)(1-x_{e2})}{1-x_{e2}(v_2+1)} \frac{F(y_{v_1+1,v_1}^{v_2,v_2+1})}{F(y_{1,0}^{0,1})} \tag{15}$$

where x_{e1} and x_{e2} are the anharmonicities of the two molecules involved. $F(y)$ is given by equation (12) with $y_{v_1+1,v_1}^{v_2,v_2+1}$ defined as

$$y_{v_1+1,v_1}^{v_2,v_2+1} = 0.32[E_{v_1+1} - E_{v_1} + E_{v_2} - E_{v_2+1}]L \sqrt{\frac{\mu}{T_g}} \tag{16}$$

where E_{v_i} are the energies of the initial and final levels in Kelvin. We have taken $L = 0.25$ Å for all V–V and V–V′ processes. Note that $y_{v_1+1,v_1}^{v_2,v_2+1}$ must always be positive since we are considering the reaction in the exothermic direction.

The calculated forward and reverse rates are plotted in figure 5.3.11 as a function of the vibrational number v_1 for $v_2 = 0$. For the N_2–O_2 process, the rates increase up to the resonance point at $v_1 = 27$, and decrease after this value. We observe the same behavior for the N_2–NO process but the resonance appears at a lower value of v_1 ($v_1 = 16$) because the spacing between NO levels is larger than between O_2 levels. For the N_2–N_2 process, the rates increase until $v_1 = 5$ and then decrease because of the increasing vibrational energy gap between the two N_2 molecules.

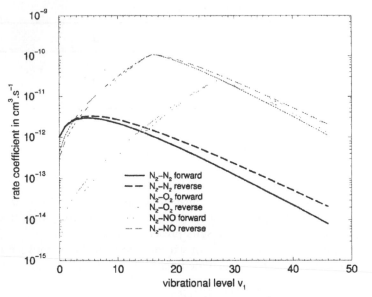

Figure 5.3.11. Rate coefficients for V–V and V–V' exchange: $N_2(X, v_1 + 1) + AB(X, 0) \rightarrow N_2(X, v_1) + AB(X, 1)$, with $AB = N_2$, O_2 and NO, at $T_g = 2000$ K.

Three sections with results pertaining to section 5.3.3 were inadvertently omitted from the manuscript. They have been added as an Appendix to the book at the proof stage. The Editors.

5.3.4 Experimental dc glow discharges in atmospheric pressure air plasmas

5.3.4.1 Introduction

Experiments have been conducted to validate the mechanisms of ionization in two-temperature atmospheric pressure air plasmas in which the electron temperature is elevated with respect to the gas temperature. To test the predicted S-shaped dependence of steady-state electron number density on the electron temperature and its macroscopic interpretation in terms of current density versus electric field, dc glow discharges have been produced in flowing low temperature, atmospheric pressure air plasmas. The flow velocity is around 400 m/s, and the gas temperature is varied between 1800 and 2900 K. These experiments show that it is feasible to create stable diffuse glow discharges with electron number densities in excess of 10^{12} cm^{-3} in atmospheric pressure air plasmas. Electrical characteristics were measured and the thermodynamic parameters of the discharge were obtained by spectroscopic measurements. The measured gas temperature is not noticeably affected by whether or not the dc discharge is applied. The discharge area was determined from spatially resolved optical measurements of plasma emission during discharge excitation. The measured discharge characteristics are compared in section 5.3.5 with the predicted electrical characteristics.

5.3.4.2 DC discharge experimental set-up

The ionization process in the discharge region is accompanied by energy transfer to the gas through collisions between electrons and heavy particles. Electrons lose more than 99.9% of the energy gained from the electric field to molecular N_2 through vibrational excitation, and the vibrationally excited N_2 transfers energy to translational modes through vibrational relaxation. Thus the degree of gas heating (ΔT_g) is a function of the volumetric power, jE, deposited into the plasma by the discharge and the competition of the vibrational relaxation time and the residence time τ of the plasma in the discharge region. To limit gas heating to acceptable levels for given volumetric power, it is desirable to flow the plasma at high velocity through the discharge region.

The experimental set-up is shown schematically in figure 5.3.12. Atmospheric pressure air is heated with a 50 kW rf inductively coupled plasma torch operating at a frequency of 4 MHz. A 2 cm exit diameter nozzle is mounted at the exit of the torch head. The flow rate injected in the torch was approximately 96 standard liters per minute (slpm) (64 slpm radial and 32 slpm swirl) and the plate power settings were 8.9 kV × 4.1 A, with approximately 14 kW of power deposited into the plasma. Under these conditions, the temperature of the plasma at the exit of the 2 cm diameter nozzle is about 5000 K and its velocity is ~100 m/s.

The plasma then enters a quartz test-section where it is cooled to the desired temperature by mixing with an adjustable amount of cold air injected into the plasma stream through a radial mixing ring. The quartz test-section length of 18 cm ensures that the flow residence time (approximately 1.6 ms here) is greater than the characteristic time for chemical and thermal equilibration of the plasma (<1 ms). Thus at the exit of the quartz test-section the air flow is close to local thermodynamic equilibrium (LTE) conditions. Finally, a 1 cm exit diameter converging water-cooled copper nozzle is mounted at the exit of the mixing test-section. This nozzle is used to control the velocity, hence the residence time, of the flow within the discharge region. Two-dimensional computational fluid dynamics (CFD) calculations performed at the University of Minnesota (see section 5.2 by Candler) show that the axial velocity at the entrance of the discharge region is approximately 445 m/s [24]. The discharge itself is produced between two platinum pin electrodes of 0.5 mm diameter held along the axis of the air stream by two water-cooled $\frac{1}{16}$ inch (1.6 mm) stainless-steel tubes placed crosswise to the plasma flow. The bottom electrode is mounted on the copper nozzle and the upper electrode is affixed to a Lucite ring, itself mounted on a vertical translation stage in order to provide adjustable distance between electrodes. The interelectrode distance was set to 3.5 cm.

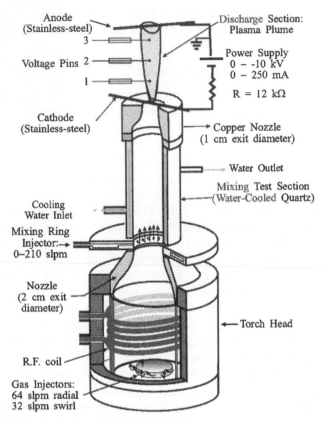

Anode
(Stainless-steel)

3

Voltage Pins   2

1

Cathode
(Stainless-steel)

Cooling
Water Inlet

Mixing Ring
Injector:→
0–210 slpm

Nozzle
(2 cm exit
diameter)

R.F. coil

Gas Injectors:
64 slpm radial
32 slpm swirl

Discharge Section:
Plasma Plume

Power Supply
0 – -10 kV
0 – 250 mA

R = 12 kΩ

Copper Nozzle
(1 cm exit diameter)

Water Outlet

Mixing Test Section
(Water-Cooled Quartz)

Torch Head

Figure 5.3.12. Set-up (not to scale) for discharge experiments showing the torch head, the injection ring, the 2 cm diameter, 18 cm long water-cooled quartz mixing test-section, the 2 → 1 cm converging nozzle, electrodes, voltage pins, and electrical circuit.

The discharge was driven by a Del Electronics Model RHSV10-2500R power supply with reversible polarities, capable of operation in control current or control voltage mode, with current and voltage outputs in the ranges 0–250 mA and 0–10 kV, respectively. For the present experiments, the cathode (bottom electrode) was biased to negative potentials with respect to ground.

The electric field within the discharge region is measured from the potential on a high purity platinum wire (0.02 inch (0.5 mm) diameter) that extends to the center of the discharge region. The platinum wire is held by a small ceramic tube installed on a two-way (horizontal and vertical) translation stage. Horizontal translation moves the pin into the discharge region for electric field measurements, and out of the discharge during spectral emission measurements. Vertical translation moves the pin along the discharge axis to determine the electric field from potential measurements. Although pure platinum melts at ~2045 K, radiation cooling prevents melting of the platinum wires for plasma temperatures up to at least 3000 K. The voltage measurements reported here were made with a Tektronix Model P6015A high voltage (20 kV dc, 40 kV peak pulse) probe and a Hewlett-Packard Model 54510A digitizing oscilloscope. The current was measured from the voltage drop across the 12 kΩ ballast resistor of the dc circuit.

The set-up for optical emission spectroscopy diagnostics includes a SPEX model 750M, 0.75 m monochromator fitted with a 1200 lines/mm grating blazed at 200 nm and a backthinned Spectrum One thermoelectrically cooled charge-coupled device (CCD) camera. The 30 × 12 mm CCD chip contains 2000 × 800 pixels of dimension 15 × 15 µm. The dispersion of the optical system is ~1.1 nm/mm. The monochromator entrance slit width was set at 200 µm, and 26 columns of 800 pixels were binned to produce an equivalent exit slit width of 390 µm. The spatial resolution was ~0.5 mm as determined by the monochromator entrance slit width and the magnification of the optical train (2.5 for two lenses of focal length 50 and 20 cm). Absolute spectral intensity calibrations were obtained with an Optronics model OL550 tungsten filament lamp and a 1 kW argon arcjet, with radiance calibrations traceable to National Institute of Standards and Technology (NIST) standards.

5.3.4.3   Spectroscopic measurements

(A)   Measurements without dc discharge applied
The gas temperature (rotational temperature) without dc discharge applied was measured by emission spectroscopy of the OH $(A \rightarrow X)$ transition. The OH $(A \rightarrow X)$ transition is one of most intense emission systems in low temperature $(T \leq 4000 \text{ K})$ air plasmas containing even a small amount (~1%) of H_2 or H_2O. In the present experiments, the water content of the air injected into the torch was sufficient to produce intense OH radiation. Rotational temperatures were obtained as described in chapter 8 section 8.5.

Line-of-sight OH emission spectra were recorded with the discharge off. The amount of cold air mixing was adjusted to vary the temperature of the preheated air. The measured OH spectra were later fitted with SPECAIR. As shown in figure 5.3.13 the gas temperature can be varied from 1800 to 2900 K by adjusting the amount of cold air mixing.

It is expected that the plasma conditions are close to LTE at the entrance of the discharge section for the 3000 K case. However, for the lowest temperature cases (T close to 2000 K), the electron density may be elevated with respect to LTE because electron recombination is small at these low temperatures. Nevertheless, the electron density entering the discharge section is expected in all cases to be less than 10^{10} cm^{-3}.

(B)   Measurements with dc discharge applied
Emission spectra were also measured with the discharge applied (discharge current of 150 mA). A typical spectrum is shown in figure 5.3.14. As can

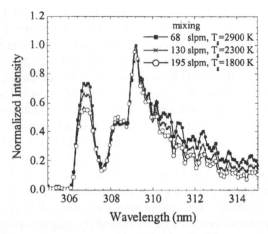

Figure 5.3.13. Measured OH $A \rightarrow X$ emission spectra without discharge applied as a function of the amount of cold air mixing.

be seen in the figure, a factor of $\sim 10^4$ enhancement of the emission due to NO gamma ($A \rightarrow X$) and a factor of $\sim 10^5$ enhancement of the emission due to N_2 ($C \rightarrow B$) bands were observed. Figure 5.3.14 also shows that the N_2 second positive system ($C \rightarrow B$) bands overlap the OH ($A \rightarrow X$) feature around 308 nm. This overlap precludes accurate measurements of the rotational temperature from OH ($A \rightarrow X$) transition. Therefore the gas temperature (rotational temperature) was measured by means of emission spectroscopy of the (0,0) band of the N_2 second positive system—the N_2 ($C \rightarrow B$) transition. Line-of-sight N_2 emission spectra were recorded along lateral chords of the plasma. The spectra were fitted with SPECAIR to obtain the rotational temperature T_r and the vibrational temperature T_v of the C state of N_2. The

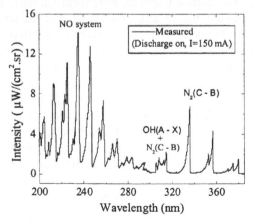

Figure 5.3.14. Line-of-sight emission spectra measured at a discharge current $I = 150\,\text{mA}$.

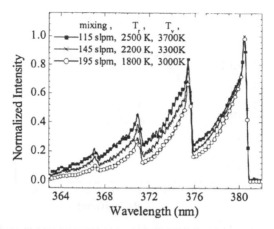

Figure 5.3.15. Measured N_2 second positive ($C \rightarrow B$) bands with discharge as a function of the amount of cold air mixing ($I = 150\,\text{mA}$).

analysis procedure is described in chapter 8 section 8.5. Finally, the absolute intensity of the spectrum was used to determine the population of the N_2 C electronic state.

Additional discharge experiments were conducted with different gas temperatures. Figure 5.3.15 shows the measured N_2 second positive system spectra and the rotational and vibrational temperatures at a discharge current of 150 mA, as a function of the amount of mixing air. It can be seen in the figure that both the rotational and vibrational temperatures are lower with a higher amount of mixing cold air. Figure 5.3.16 shows the measured spectrum as a function of the discharge current for 145 slpm of mixing air. As can be seen from the figure, the rotational temperature

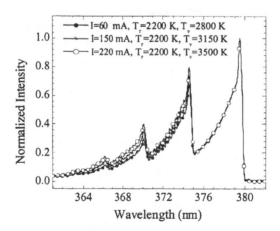

Figure 5.3.16. Measured N_2 second positive ($C \rightarrow B$) bands with discharge as a function of discharge current for the case of 145 slpm cold air mixing.

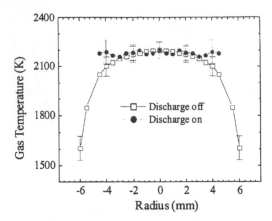

Figure 5.3.17. Rotational temperature profiles with and without the applied dc discharge at 1.5 cm downstream of the bottom electrode. The temperature profile without the discharge was measured from rotational lines of the OH $(A \to X)$ transition. With the discharge applied, the rotational temperature was measured from lines of the N_2 $(C \to B)$ transition in the ultraviolet.

remains the same at all currents, but the vibrational temperature increases with increasing discharge current.

Radial rotational temperature profiles with and without the discharge applied were measured along chords of the plasma from Abel-inverted N_2 second positive system emission spectra and OH emission spectrum, respectively. Figure 5.3.17 shows the measured radial temperature profiles at a distance of 1.5 cm downstream of the cathode (i.e. midway between the two electrodes). As can be seen from the figure, the applied discharge does not noticeably increase the rotational temperature of the plasma at this location. Figure 5.3.18 shows the radial N_2 C state electronic and vibrational temperature profiles. On the axis of the discharge, the electronic temperature of the N_2 C state reaches about 5000 K, and the vibrational temperature is about 3000 K.

Figure 5.3.19a shows a photograph of the air plasma plume at a temperature of approximately 2200 K in the region between the two electrodes without the discharge applied. Figure 5.3.19b shows the same region when a dc discharge of 5.2 kV and 200 mA is applied between the two electrodes. In these experiments, the distance between electrodes is 3.5 cm. The bright region in figure 5.3.19b corresponds to the discharge-excited plasma. Thus the plasma plume without discharge applied appears to be homogeneous over a larger diameter than the plasma plume with the discharge applied. However, figure 5.3.17 showed that the gas temperature profile is practically the same as in the discharge applied case. The increased brightness in figure 5.3.19b is due to the emission of excited electronic states of molecular NO and N_2 (see figure 5.3.18). Thus the applied discharge

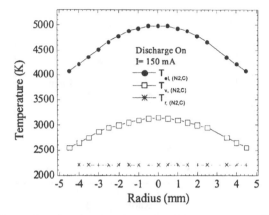

Figure 5.3.18. Electronic, vibrational and rotational temperature profiles of N_2 C state with an applied discharge current of $I = 150\,mA$.

increases excited state populations without significantly increasing the gas temperature.

5.3.4.4 Current density measurements

The current density at the center of the plasma was determined by dividing the measured current by the effective discharge area A^*, i.e. $j(r = 0) = I/A^*$. The effective discharge area is obtained from the following relation:

$$A^* = \int_0^R 2\pi r j(r)\, dr \Big/ j(r = 0) \qquad (17)$$

(a)   (b)

Figure 5.3.19. (a) Air plasma at 2000 K without electrical discharge. (b) Air plasma at 2000 K with applied discharge (1.4 kV/cm 200 mA). Interelectrode distance, 3.5 cm. The measured electron number density in the bright discharge region is around $10^{12}\,cm^{-3}$.

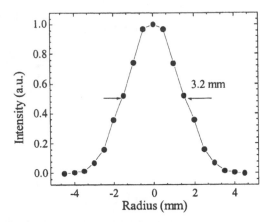

Figure 5.3.20. Spatial extent of the plasma produced by the discharge.

where $j(r)$ is the local current density. As shown in reference [12], $j(r)$ is approximately proportional to $n_e(r)$. Thus, A^* can be calculated as

$$A^* = \left(\int_0^R n_e(r) 2\pi r \, dr \right) \Big/ n_e(r = 0). \qquad (18)$$

In separate discharge experiments conducted with a nitrogen plasma [5], the electron number density profile n_e was measured using various techniques (from H_β Stark broadening and N_2 first-positive emission spectra) to calculate A^* using equation (18). The discharge area A^* was also estimated from the full width at half maximum (FWHM) of the N_2 $C \rightarrow B$ (0,0) band-head intensity profile. In these nitrogen discharge experiments, the effective area A^* obtained with equation (18) and the measured n_e profile was found to be close to the effective area obtained from the N_2 $C \rightarrow B$ (0,0) emission intensity measurement. Thus for the present air plasma discharge experiments, we estimate the effective discharge area from the spatially resolved optical measurements of N_2 C state emission. Spectroscopic measurements of N_2 $C \rightarrow B$ (0,0) emission with the applied discharge are shown in figure 5.3.20. It can be seen from the figure that the diameter (FWHM) of the discharge is approximately 3.2 mm. This diameter was monitored and found to be constant for all discharge currents ranging from 5 to 250 mA. The discharge diameter was therefore taken to be 3.2 mm and assumed constant along the axis of the discharge.

5.3.4.5 Electric field measurements

Electrode and pin potentials were measured as a function of the applied discharge current which was varied from 0 to 250 mA. The cathode current was measured from the voltage drop across the 12 kΩ ballast resistor

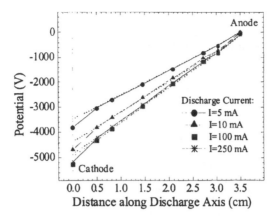

Figure 5.3.21. Measured potentials as a function of applied current in the discharge section.

placed in series with the discharge (see figure 5.3.12). There is a small difference of 7 mA between the anode and the cathode currents that was found to be due to a current leak through the water cooling circuit of the anode. All results reported below are based on the measured cathode current, which is not affected by current losses to the cooling circuit.

Figure 5.3.21 shows the measured pin voltage as a function of the applied current along the axis of the discharge. The potential varies approximately linearly along the axis of the discharge, indicating that the electric field is approximately uniform in the discharge region. The electric field measurements reported here were determined from the slope of a linear fit of the pin potentials. In the vicinity of the cathodes, voltage falls of up to several hundred volts were observed. These values are typical of the cathode fall voltage in glow discharges [25].

The total voltage across the discharge was also measured as a function of electrode separation, by translating the top electrode (anode) vertically. The voltage-length characteristic for a discharge current of 150 mA is shown in figure 5.3.22. The lowest voltage reading as the electrodes are brought within less than 0.2 mm from one another provides an approximation to the discharge voltage at zero gap length [26, 27]. The value of this voltage is found to be 285 V and is independent of the current in the current range investigated (10–250 mA). This value agrees with the cathode fall voltage reported in the literature [6, 28] for glow discharges in air with a platinum cathode. The voltage gradient in the positive column, given by the slope of the voltage-length characteristic, is constant as the discharge length is increased. For a discharge current of 150 mA, the gradient is about 1400 V/cm (see figure 5.3.22) and is consistent with the electric field value determined from the pin measurements.

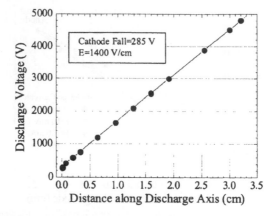

Figure 5.3.22. Voltage–length characteristic in the discharge region.

5.3.5   Electrical characteristics and power requirements of dc discharges in air

The experimental discharge characteristics presented in section 5.3.4 for plasma temperatures ranging from 1800 to 2900 K are shown in figure 5.3.23. They are also compared with the predicted characteristics at the corresponding gas temperatures. The method employed to predict the discharge characteristics was discussed in section 5.3.3. As can be seen from figure 5.2.23, good agreement is obtained between the measured and

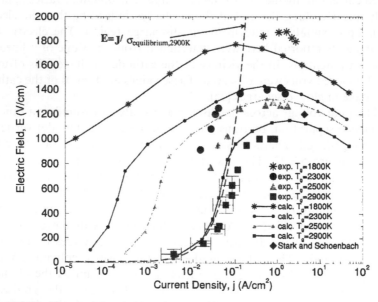

Figure 5.3.23. Measured (symbols) and predicted (solid lines) electrical discharge characteristics in atmospheric pressure air plasmas generated by dc electric discharges.

predicted discharge characteristics over a range of experiments spanning over three orders of magnitude in current density.

Figure 5.3.23 also shows (dashed curve) the resistive characteristic of equilibrium air at 2900 K, given by the relation

$$E = \frac{j}{\sigma_{\text{equilibrium, 2900 K}}} \qquad (19)$$

where $\sigma_{\text{equilibrium, 2900 K}}$, the electrical conductivity of equilibrium air at 2900 K, is calculated as

$$\sigma_{\text{equilibrium, 2900 K}} = \frac{n_e^{\text{equilibrium, 2900 K}} e^2}{m_e \bar{\nu}_{\text{e-air}}} \qquad (20)$$

where $n_e^{\text{equilibrium, 2900 K}} = 4 \times 10^{10} \, \text{cm}^{-3}$ and the collision frequency $\bar{\nu}_{\text{e-air}}$ is well approximated by

$$\bar{\nu}_{\text{e-air}} = \left(\frac{p}{kT_g} \right) g_e (1.5 \times 10^{-15} \, \text{cm}^2) \qquad (21)$$

where p is the pressure (1 atm), $T_g = 2900 \, \text{K}$ is the gas temperature, and $g_e = \sqrt{8kT_e/\pi m_e}$ is the electron thermal velocity. For $T_g = 2900 \, \text{K}$, as can be seen from figure 5.3.23 the predicted E versus j characteristic is close to the resistive equilibrium characteristic for current densities below 0.2 A/cm^2. In this current density range, the predicted electron temperature remains below approximately 8000 K and electron-impact reactions are inefficient in ionizing the plasma. Thus the electron number density increases only by a few percent. As the electron temperature increases, the frequency of collisions increases with $\sqrt{T_e}$, resulting in a decrease of the electrical conductivity of the plasma. This explains why the E versus j characteristic is higher than the resistive equilibrium characteristic for j below $\sim$0.2 A/cm^2. At higher values of the current density, where the discharge produces a significant increase in the electron density, the conductivity increases dramatically and the slope of the E versus j characteristic decreases. Thus the region to the right of the resistive equilibrium characteristic is where the discharge increases the electron number density. The experimental data at T_g between 1800 and 2900 K all show the turning trends of the non-resistive discharge characteristics. We note, however, that the predicted resistive part appears to be shifted to lower current densities relative to the experimental curves. This difference may be due to the fact that the electron number density was slightly above the equilibrium value in the incoming air stream. We recall that the 'equilibrium' air was produced by cooling of an air stream initially heated to high temperatures. Slow electron recombination could therefore explain the differences at low current densities.

Figure 5.3.23 also shows experimental data obtained by Stark and Schoenbach [29] in an atmospheric pressure glow discharge in air. The

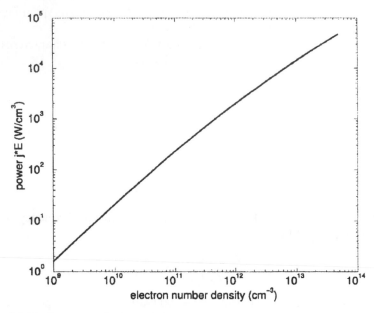

Figure 5.3.24. Power required to produce an elevated electron density in atmospheric pressure air at 2000 K by means of dc discharges.

discharge was produced between a microhollow cathode and a positively biased electrode, as described in reference [29]. The gas temperature was measured to be around 2000 K, and the center electron number density is reported to be above 10^{12} cm^{-3} [29, 30]. This measurement adds further support to the kinetic mechanism predictions.

We conclude this section with the power required to produce a given electron density in air at 2000 K by means of dc glow discharges. The results are shown in figure 5.3.24. We predict that the production of 10^{13} electron/cm^3 in air at 2000 K requires about 14 kW/cm^3. The corresponding electric field is ~1.35 kV/cm, and the current density is ~10.4 A/cm^2.

This level of Joule heating may not lead to significant overall gas heating in small scale stationary dc discharges where conduction to ambient air and to the electrodes is high. This was the case for instance in Gambling and Edels's experiments [27] where the positive column was a few millimeters in length and approximately 0.2 mm^2 in area. In larger volume dc discharges, however, it is necessary to control the effect of Joule heating of the gas, for instance by flowing the gas through the discharge at high velocities. For air at 2000 K flowing through a 1 cm diameter region of length 3.5 cm at a velocity of 450 m/s, the residence time is 78 µs. The vibrational relaxation times τ reported by Park [9] indicate that the fastest vibrational relaxation rate of molecular N_2 is through collisions with atomic oxygen. The rate constant is given by $p_0\tau = 10^{-6}$ atm s^{-1}), where p_0 is the partial pressure of atomic oxygen. In the present discharge experiments, the atomic oxygen mole fraction is less than 1%, according to the two-temperature kinetic model predictions. Thus, the vibrational relaxation time τ ($>100\,\mu$s) is larger than the flow time (78 μs). This is consistent with the observation that little gas heating was observed in the experiments. To limit gas heating to acceptable levels for a given volumetric power, it is desirable to flow the plasma at high velocity through the discharge region.

5.3.6 Conclusions

Investigations have been made of the mechanisms of ionization in two-temperature air plasmas with electron temperatures elevated with respect to the gas temperature. Numerical simulations of these mechanisms yield the notable result that the electron number density exhibits an S-shape dependence on the electron temperature at fixed gas temperature. This S-shaped behavior is caused by competing ionization and charge transfer reactions. The characteristic of electric field versus current density also exhibits a non-monotonic dependence.

Discharge experiments were conducted in air at atmospheric pressure and temperatures ranging from 1800 to 3000 K. In these experiments, a dc electric field was applied to flowing air plasmas with electron concentrations initially close to equilibrium. These experiments have shown that it is possible to obtain stable diffuse glow discharges in atmospheric pressure air with electron number densities of up to $2.5 \times 10^{12}\,cm^{-3}$, which is up to six orders of magnitude higher than in the absence of the discharge. The value of $2.5 \times 10^{12}\,cm^{-3}$ corresponds to the maximum current that can be drawn from the 250 mA power supply used in these experiments. The diffuse discharges are approximately 3.5 cm in length and 3.2 mm in diameter. No significant degree of gas heating was noticed as the measured gas temperature remained within a few hundred Kelvin of its value without the discharge applied. Results from these experiments are in excellent agreement with the predicted E versus j characteristics. Additional comparisons were made with results from glow discharge experiments in atmospheric pressure ambient air by Gambling and Edels [27] and Stark and Schoenbach [29]. The measurements of these authors are also consistent with the predicted E versus j characteristics. As these measurements were made in the reactive region of the E versus j curve, they support our proposed mechanism of ionization for two-temperature air.

As the power budget for dc electron heating is higher than desired for the practical use of air plasmas in many applications, methods to reduce the power budget are currently being explored in our laboratory. Based on the predictions of our chemical kinetics and electrical discharge models, we have found that a repetitively pulsed electron heating strategy can provide power budget reductions of several orders of magnitude with respect to dc electron heating. Repetitively pulsed discharges are presented in chapter 7 section 7.4.

Acknowledgment

The authors would like to acknowledge the contributions of Lan Yu, Denis Packan, Laurent Pierrot, Sophie Chauveau, J Daniel Kelley and Charles Kruger.

References

[1] Pierrot L, Laux C O and Kruger C H 1998 'Vibrationally-specific collisional-radiative model for non-equilibrium nitrogen plasmas' *Proc. 29th AIAA Plasmadynamics and Lasers Conference, AIAA 98-2664*, Albuquerque, NM

[2] Pierrot L, Laux C O and Kruger C H 1998 'Consistent calculation of electron-impact electronic and vibrational rate coefficients in nitrogen plasmas' *Proc. 5th International Thermal Plasma Processing Conference* (Begell House, New York), pp 153–160, St. Petersburg, Russia

[3] Yu L, Pierrot L, Laux C O and Kruger C H 1999 'Effects of vibrational non-equilibrium on the chemistry of two-temperature nitrogen plasmas' *Proc. 14th International Symposium on Plasma Chemistry*, Prague, Czech Republic

[4] Pierrot L, Yu L, Gessman R J, Laux C O and Kruger C H 1999 'Collisional-Radiative Modeling of Nonequilibrium Effects in Nitrogen Plasmas' *Proc. 30th AIAA Plasmadynamics and Lasers Conference, AIAA 99-3478*, Norfolk, VA

[5] Yu L 2001 'Nonequilibrium effects in two-temperature atmospheric pressure air and nitrogen plasmas' PhD Thesis, Stanford University

[6] Lieberman M A and Lichtenberg A J 1994 *Principles of Plasma Discharges and Materials Processing* (New York: John Wiley)

[7] Hierl P M, Dotan I, Seeley J V, Van Doren J M, Morris R A and Viggiano A A 1997 'Rate Constants for the reaction of O^+ with N_2 and O_2 as a function of temperature (300–1800 K)' *J. Chem. Phys.* **106** 3540–3544

[8] Dotan I and Viggiano A A 1999 'Rate constants for the reaction of O^+ with NO as a function of temperature (300–1400 K)' *J. Chem. Phys.* **110** 4730–4733

[9] Park C 1989 *Nonequilibrium Hypersonic Aerothermodynamics* (New York: Wiley)

[10] Park C 1993 'Review of Chemical-Kinetic Problems of Future NASA Missions, I: Earth Entries' *J. Thermophysics and Heat Transfer* **7** 385–398

[11] Kee R J, Rupley F M and Miller J A 1989 'Chemkin-II: A Fortran chemical kinetics package for the analysis of gas phase chemical kinetics' Sandia National Laboratories, Report No. SAND89-8009

[12] Laux C O, Yu L, Packan D M, Gessman R J, Pierrot L, Kruger C H and Zare R N 1999 'Ionization Mechanisms in Two-Temperature Air Plasmas' *Proc. 30th AIAA Plasmadynamics and Lasers Conference, AIAA 99-3476*, Norfolk, VA

[13] Brown S C 1966 *Basic Data of Plasma Physics* (MIT Press)

[14] Shkarofsky I P, Johnston T W and Bachynski M P 1966 *The Particle Kinetics of Plasmas* (Addison-Wesley)

[15] Tsang W and Herron J T 1991 'Chemical kinetic database for propellant combustion. I. Reactions involving NO, NO_2, HNO, HNO_2, HCN and N_2O' *J. Phys. Chem. Ref. Data* **20** 609–663

[16] Mitchner M and Kruger C H 1973 *Partially Ionized Gases* (New York: John Wiley)

[17] Pierrot L 1999 'Chemical kinetics and vibrationally-specific collisional-radiative models for non-equilibrium nitrogen plasmas' Stanford University, Thermosciences Division

[18] Chauveau S M, Laux C O, Kelley J D and Kruger C H 2002 'Vibrationally specific collisional-radiative model for non-equilibrium air plasmas' *Proc. 33rd AIAA Plasmadynamics and Lasers Conference, AIAA 2002-2229*, Maui, Hawaii

[19] Kazansky Y K and Yelets I S 1984 'The semiclassical approximation in the local theory of resonance inelastic interaction of slow electrons with molecules' *J. Phys. B* **17** 4767–4783

[20] Kozlov P V, Makarov V N, Pavlov V A, Uvarov A V and Shatalov O P 1996 'Use of CARS spectroscopy to study excitation and deactivation of nitrogen molecular vibrations in a supersonic gas stream' *Tech. Phys.* **41** 882–889

[21] Bray K N C 1968 'Vibrational relaxation of anharmonic oscillator molecules: relaxation under isothermal conditions' *J. Phys. B* **1** 705–717

[22] Keck J and Carrier G 1965 'Diffusion theory of non-equilibrium dissociation and recombination' *J. Chem. Phys.* **43** 2284–2298

[23] Ahn T, Adamovich I V and Lempert W R 2003 'Stimulated Raman Scattering Measurements of Nitrogen V-V Transfer' *Proc. 41st Aerospace Sciences Meeting and Exhibit, AIAA 2003-132*, Reno, NV

[24] Nagulapally M, Candler G V, Laux C O, Yu L, Packan D M, Kruger C H, Stark R and Schoenbach K H 2000 'Experiments and simulations of dc and pulsed discharges in air plasmas' *Proc. 31st AIAA Plasmadynamics and Lasers Conference, AIAA 2000-2417*, Denver, CO

[25] Raizer, Y P 1991 *Gas Discharge Physics* (Berlin: Springer)

[26] Thoma H and Heer L 1932 *Z. Tech. Phys. (Leipzig)* **13** 464

[27] Gambling W A and Edels H 1953 'The high-pressure glow discharge in air' *Br. J. Appl. Phys.* **5** 36–39

[28] Von Engel A 1965 *Ionized Gases* (Oxford: Oxford University Press)

[29] Stark R H and Schoenbach K H 1999 'Direct current high-pressure glow discharges' *J. Appl. Phys.* **85** 2075–2080

[30] Leipold F, Stark R H, El-Habachi A and Schoenbach K H 2000 'Electron density measurements in an atmospheric pressure air plasma by means of infrared heterodyne interferometry' *J. Phys. D* **33** 2268–2273

5.4 Multidimensional Modeling of Trichel Pulses in Negative Pin-to-Plane Corona in Air

5.4.1 Introduction

Negative corona—low current discharge between a cathode (a wire or a point) and a plane anode—is a quite common object widely used in industry.

While studying the negative point-to-plane corona in air, Trichel (1938) revealed the presence of regular relaxation pulses. Qualitative explanation given by him included some really important features like shielding effect produced by a positive ion cloud in the vicinity of the cathode. The role of negative ions was practically ignored. In the following work (Loeb *et al* 1941) it was stated that the Trichel pulses exist only in electronegative gases, and a particular emphasis was put on the processes of electron avalanche triggering. It was stressed also that, usually, the time of the negative ion drift to the anode is much longer than the pulse period. More detailed measurements of the Trichel pulse shape demonstrated that the rise time of the pulse in air may be as short as 1.3 ns (Zentner 1970a) and a step on a leading edge of the pulse was observed (Zentner 1970b). Later, the systematic study of the electrical characteristics of the Trichel pulses was undertaken (Lama and Gallo 1974), and some empirical relationships were found for the pulse repetition frequency, a charge per pulse and so on.

Among attempts to give theoretical explanation for discussed phenomena the work of Morrow (1985) is most known, where the preceding theories are reviewed also. Continuity equations for electrons and positive and negative ions in a one-dimensional form were numerically solved together with the Poisson equation computed by the method of disks. It was supposed that the electrical charges occupy the cylinder of a given radius. One of electrodes, cathode, was spherical. The negative corona in oxygen at a pressure 50 torr was numerically simulated. Only the first pulse was computed, and extension of calculations on longer times showed only continuing decay of the current. In Morrow (1985a) the shape of the pulse was explained while practically ignoring the ion-secondary electron emission. In the following paper (Morrow 1985b) the step on the leading edge of the pulse was attributed to the input of the photon secondary emission, and the main peak was explained in terms of the ion–secondary emission.

In Napartovich *et al* (1997a) a so-called 1.5-dimensional model of the pin-to-plane negative corona in air was formulated, theoretically reproducing, for the first time, periodical Trichel pulses. Predictions of parameter dependences within 1.5-dimensional model were in good agreement with experiments and allow for achieving some insight into the origin of the pulse mode. A two-peak shape of the regular pulse was predicted and associated with formation of a cathode-directed ionization wave in the vicinity of the point. However, to derive equations of this 1.5-dimensional model it was necessary to make some assumptions, the validity of which cannot be proved within the formulated theory. Moreover, most probably these assumptions (preservation of the current channel shape in time; slow variation of the current cross section area in space) are strictly false, and one could only rely on anticipated secondary role of these effects in the formation of Trichel pulses. Evidently, a more accurate description of Trichel pulses requires that a three-dimensional model be developed. Taking into account the circular symmetry of the corona geometry, it is sufficient to make a model in two spatial variables: a distance along the discharge axis, x, and a radius, r. Such a model was developed by Napartovich *et al* (1997b). Later, results of numerical studies on Trichel pulses dynamics in ambient air for pin-to-plane configuration with usage of the three-dimensional model were reported in Akishev *et al* (2002a) and published in Akishev *et al* (2002b).

5.4.2 Numerical model

In literature much attention is paid to multi-dimensional numerical simulations of streamer propagation, e.g. Dhali and Williams (1987), Vitello *et al* (1993), Egli and Eliasson (1989), Pietsch *et al* (1993), Babaeva and Naidis (2000), and Kulikovsky (1997a,b). In contrast to streamers formation and propagation, Trichel pulses are induced by a strongly non-uniform electric field in the vicinity of the pin tip. It means that the location where the most important processes take place is known in advance. Moreover, sizes of this area are small for the fine point. Thus, it seems natural in calculations to use a non-uniform mesh with small cells only around the point, increasing the size of the cell when moving away from the point. Pietsch *et al* (1993) exploit a similar technique in modeling a single micro-discharge in a dielectric barrier discharge. The specific feature of this problem is an overall small dimension, which makes the problem of high spatial resolution easily solvable. In the case of negative corona discharge, it is necessary to describe the evolution of the discharge in a region of $1\,\mathrm{cm} \times 1\,\mathrm{cm} \times 1\,\mathrm{cm}$ sizes. However, an even greater difference in micro-discharge (streamer) computing and Trichel pulses computing is in range of physical time, where essential processes happen. The typical duration of micro-discharge or streamer propagation is on the order of tens of nanoseconds. A single Trichel pulse has a similar duration. However, to understand the mechanism of regular repetition of Trichel pulses it is necessary to simulate at least several pulses until the negative ions fill up the discharge gap. For short-gap coronas this time is on the order of tens of microseconds. The required enormous number of time steps is available only for a code possessing a very high calculation rate. The discussed differences in requirements to the mathematical algorithms for description of seemingly similar phenomena (streamers and Trichel pulses) dictate the necessity to develop new algorithms for multi-dimensional simulations of Trichel pulses.

5.4.2.1 *Basic equations and electrode configuration*

To describe the pulse mode of the negative point-to-plane corona it is sufficient to solve the known continuity equations for electrons:

$$\partial n_e/\partial t + \operatorname{div} n_e \mathbf{w}_e = (\nu_i - \nu_a)n_e + \nu_d n_n - \beta_{ei}n_e n_i \qquad (5.4.1)$$

positive ions

$$\partial n_p / \partial t + \text{div } n_p \mathbf{w}_p = \nu_i n_e - \beta_{ei} n_e n_i \tag{5.4.2}$$

negative ions

$$\partial n_n / \partial t + \text{div } n_n \mathbf{w}_n = \nu_a n_e - \nu_d n_n \tag{5.4.3}$$

and Poisson's equation

$$\text{div } \mathbf{E} = e(n_p - n_e - n_n)/\varepsilon_0 \tag{5.4.4}$$

where the indexes e, p and n refer to electrons, positive and negative ions, respectively, n_p, n_e and n_n are the positive ion, electron and negative ion number densities, $\mathbf{w}_p$, $\mathbf{w}_e$ and $\mathbf{w}_n$ their drift velocities, ν_i, ν_a, and ν_d are the ionization, attachment, and detachment frequencies, e is the electronic charge, β_{ei} is the electron–ion dissociative recombination coefficient, ε_0 is the permittivity of free space. The electron drift velocity generally can be determined from solving the electron Boltzmann equation. However, in the following it was taken to be proportional to the electric field; the ion drift velocities were calculated using the known ion mobilities. The current in the external circuit, I, is determined from the equation

$$V = U_0 - RI \tag{5.4.5}$$

where V and U_0 are the discharge and power supply voltages, and R is the ballast resistor.

Equations (1)–(4) should be accomplished by boundary conditions. The boundary conditions for positive and negative ions are self-evident: their number density is equal to zero at anode and cathode, respectively. For electrons the boundary condition was formulated in terms of the ion secondary emission coefficient, γ

$$\mathbf{j}_e(r_s, x_s, t) = \gamma \mathbf{j}_p(r_s, x_s, t) \tag{5.4.6}$$

where $\mathbf{j}_e = n_e \mathbf{w}_e$, $\mathbf{j}_p = n_p \mathbf{w}_p$, and r_s and x_s are space variables at the cathode surface. In calculations the fixed value of $\gamma = 0.01$ was used. An electrode configuration was taken as in the experiments of Napartovich *et al* (1997a): the cathode pin in a form of cylinder with radius 0.06 mm ended with a semi-sphere of the same radius, and cathode–anode spacing of 7 mm. Kinetic coefficients were taken correspondent to dry air.

5.4.2.2 *Numerical algorithm*

To combine the requirements of accurate discrete approximations with a high calculation rate a good choice is to do these calculations on a non-uniform grid, which should be well adjusted to the electrode configuration. Because the shape of the cathode is rather complicated, it is desirable to apply some generator of a grid automatically fitted to boundary conditions. The generated grid is to be nearly orthogonal, with some pre-described accuracy. Generation of

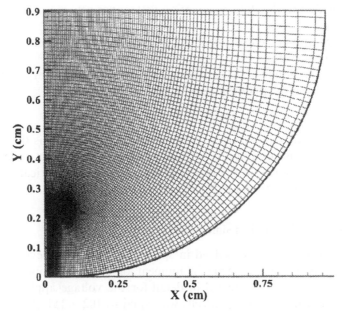

Figure 5.4.1. General view of the computational region and numerical grid. The minimum cell size in the cathode vicinity is 6×10^{-5} cm.

boundary-fitted meshes for curvilinear coordinate systems is a separate problem, and the details of its solving are omitted here. A differential mesh generation was employed, which locates the mesh points by solving an elliptic partial differential equation (Thompson *et al* 1985). The computation domain was bounded by a pin of length 2 mm, a flat anode 7 mm from the pin tip, and a dielectric sphere with a radius of 9.06 mm. The calculated mesh for the electrode configuration is shown in figure 5.4.1. An average deviation angle from orthogonality for this mesh is 0.48°, which may be considered as satisfactorily small.

An important point of controlling accuracy in numerical domain is the method of discretization of differential equations (1)–(4). In particular, certain geometric identities have to be satisfied accurately in the discrete form as well as in the continuous domain. A finite-volume approach yields more accurate conservative discrete approximations than the method based on the finite-differences approach. Therefore, a finite-volume discretization method (FVM) has been used with a consistent approximation of the geometric quantities in a curvilinear coordinate system. The global algorithm of calculations has the following steps:

1. The sources in continuity equations for charged particles are computed in cells, and drift fluxes are computed at the cell faces.
2. By virtue of the continuity equation solving the 'new' charged particle number densities are computed and then the total plasma conductivity is defined.

3. The solution to the Poisson equation determines new magnitudes for the potential.
4. The new total current is calculated by integration of its density over the respective surfaces.
5. The new magnitude of the cathode voltage is calculated from equation (5).
6. The condition for iteration convergence is checked:

$$|I^{s+1} - I^s| \leq \varepsilon_1 I^s \varepsilon_2$$

where ε_1 is the relative error, ε_2 the absolute error, and s is the iteration number. If this condition is still not satisfied, the iteration procedure is repeated starting from the first step. Details of the numerical algorithm developed can be found in Napartovich *et al* (1997b).

5.4.3   Results of numerical simulations

The equations above were solved in space and time giving evolution of a negative corona structure from a moment of step-wise applied voltage. This evolution will be analyzed in detail for the voltage applied (4.2 kV). The total number of numerical cells was equal to 102×151. The time step was variable and automatically selected to provide a good accuracy of calculations. Computing one period takes about 12 h of continuous operation on a Pentium 4 computer.

Figure 5.4.2 demonstrates evolution of discharge pulses after the initial voltage step 4.2 kV, and after the second step with amplitude 8.2 kV at the moment 40 μs. The height of the first peak is more than ten times higher than that of the following pulses. The regime with regular pulses at 4.2 kV

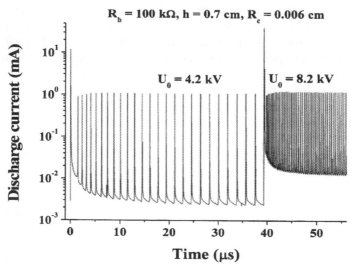

Figure 5.4.2. Discharge current evolution induced by two sequential voltage steps.

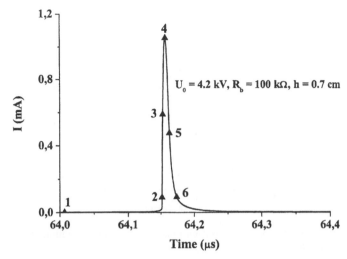

Figure 5.4.3. Fine structure of a regular pulse: 1, minimum current; 2, $0.1I_{max}$ leading edge; 3, about $0.5I_{max}$ leading edge; 4, peak of current pulse; 5, about $0.5I_{max}$ trailing edge; 6, $0.1I_{max}$ pulse trailing edge.

step is completely established to about the 25th pulse. The peak height stopped changing after four pulses, the minimum current between pulses is stabilized to about the 12th pulse, and the repetition period stabilizes about the 25th pulse. In the regime of regular pulses the ratio of peak to minimum current is equal to 442. With voltage increase the evolution proceeds faster. The height of the regular pulse is insensitive to the voltage applied, while the minimum current increases strongly. Such behavior agrees with experiment.

Figure 5.4.3 shows one of regular Trichel pulses on a nanosecond scale for the voltage applied (4.2 kV). The duration of a peak is about 12 ns, and the pulse has a smooth single-peaked shape with a trailing edge of about 20 ns length. In contrast to the prediction of the 1.5-dimensional model (Napartovich *et al* 1997a), there is no peculiarity in the pulse leading edge. To give an idea about pulse development and the dynamics of electrical current spatial distribution, a number of figures illustrate the behavior of some physical quantities for the moments marked in figure 5.4.3 by numerals.

Most strong variations of spatial distributions of charged particles and electric field happen in the immediate vicinity of the pin tip. To show deformation of electric field distribution induced by spatial charge and plasma produced just near the tip, the viewing region was limited in size by about 0.27 mm in axial and radial directions. Figures 5.4.4 and 5.4.5 demonstrate contour plots for the electric field strength at the moments corresponding to the minimum (figure 5.4.4) and maximum current (figure 5.4.5). The influence of the spatial charge remaining from the preceding pulse on the electric field is seen even at the minimum current. In the maximum, formation of a layer with high fields is clearly seen. This region resembles a classical cathode

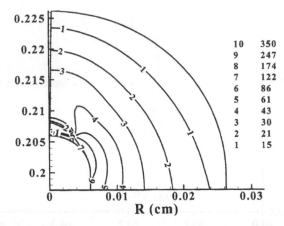

Figure 5.4.4. Electric field strength contour plot near the pin at the minimum current. The electric field strength in the legend is in kV/cm.

layer with the maximum electric field strength as high as 300 kV/cm. The thickness of this high-field region is about 7 μm. Near this high-field zone, a region appears with rather low fields (on the order of a few hundreds of V/cm). Potential curve leveling off indicates this zone. The transformation of an axial profile of electrical potential shown in figure 5.4.6 within an interval 40 μm from the pin tip demonstrates that already at 0.1 of the peak current (curve 2) something like a cathode layer is formed with a potential drop of about 180 V. Then this potential drop diminishes, approaching minimum at the current peak. It is seen that this layer broadens in the trailing edge of the pulse rather quickly (curves 5 and 6). Electron number density between pulses is lower than $10^8 \, \text{cm}^{-3}$ and approaches $4 \times 10^{15} \, \text{cm}^{-3}$ at the pulse peak.

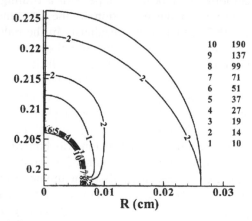

Figure 5.4.5. Electric field strength contour plot near the pin at the current peak. The electric field strength in the legend is in kV/cm.

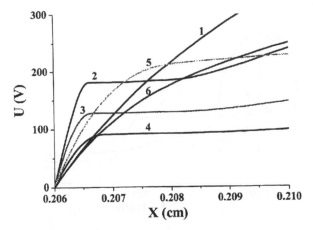

Figure 5.4.6. Electric potential distribution along the discharge axis in the vicinity of the pin tip at moments indicated in figure 5.4.3.

Negative ion distribution varies only close to the pin tip, and on the whole suffers only small changes. The contour plot for the negative ion concentration in the whole area is shown in figure 5.4.7 at the minimum current. The presented contour plots gave a rough idea about space–time evolution of discharge structure in regular pulses.

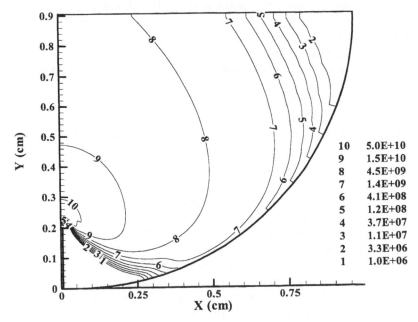

Figure 5.4.7. Negative ion number density contour plot at the minimum current for the computation region. Negative ion density in the legend is in cm^{-3}.

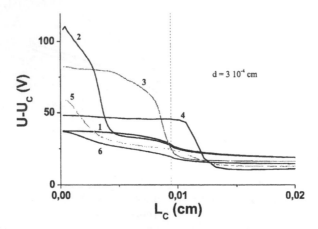

Figure 5.4.8. Distribution over the cathode surface of the voltage drop across the sheath adjacent to the cathode surface with thickness of 3 μm

Actually, the generation zone is the place where self-oscillations of the corona current are initiated. Therefore, it is of particular interest to look at the evolution of electric current at the cathode surface. Dynamics of the current distribution over the cathode is rather complicated. Generally, evolution of the total current profile can be described as an expansion over the cathode surface until the pulse peak with following fast contraction around the discharge axis. This feature of discharge evolution near the cathode is clearly seen in figure 5.4.8 drawn for the voltage drop across the sheath adjacent to the cathode surface with thickness of 3 μm. In the front of the pulse, the profile of this voltage drop looks like a shoulder, whose length grows and height goes down. In the trailing edge of the pulse, evolution proceeds in the reverse direction.

A time-average current radial profile on the anode is well known (Warburg 1899, 1927). Results of numerical simulations are compared with the Warburg profile in figure 5.4.9. The calculated radial current profile is narrowed against Warburg profile. It should be noted that, according to the Warburg distribution, the current density at the computation region boundary is about 0.1 of the maximum. This indicates that the dielectric spherical boundary imposed in calculations to restrict the computational region may influence the current distribution over the anode, and on the whole pulse dynamics. Indeed, experiments (Akishev *et al* 1996) demonstrated that restriction of the space occupied by the corona notably influences the amplitude of Trichel pulses and their repetition frequency (see section 6.7 in this book).

Numerical simulations for the same corona geometry performed for various voltages applied showed that the predicted charge per pulse is about three times smaller than experimental values for similar conditions. Theoretically predicted dependence of the pulse repetition period on the

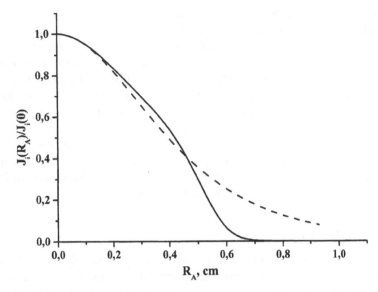

Figure 5.4.9. Average current density distribution over the the anode surface. Solid line, our simulations; dashed line, classical Warburg profile.

voltage applied in comparison with measurements (Akishev *et al* 2002a) agrees well for voltages higher than 6 kV. At 4.2 kV the predicted period is 2.5 times shorter than the measured one.

It is instructive to compare predictions made by the present multi-dimensional modeled Trichel pulses with the 1.5-dimensional model developed earlier (Napartovich *et al* 1997a). In the 1.5-dimensional model the current channel shape was assumed to be independent of time. It was taken corresponding on the whole to known experimental data, and depends on some parameters which were fitted to achieve better agreement between calculated and predicted characteristics of regular pulses. A specific feature of the current channel shape was a narrow (0.06 mm radius) cylinder adjacent to the cathode pin with length 0.2 mm. The present model free of fitting parameters predicts that the region with large gradients of particle densities and voltage is essentially shorter than assumed in the 1.5-dimensional model (tens of μm instead of hundreds of μm). Besides, the multidimensional model predicts strong variations of radial distributions. Nevertheless, the differences between the time histories of the integral quantities turned out to be not so strong. There are some details different in the two models. The 1.5-dimensional model predicts very fast propagation of a highly ionized region to the cathode at the front of the pulse. Besides, it predicts the formation of a very sharp subsidiary peak just prior to the main current peak. The present model predicts formation of a cathode layer (not coinciding with the normal cathode layer of glow discharge) first at the axis with following expansion over the cathode surface. Since the present model is free of arbitrary assumptions inherent to the 1.5-dimensional model, the scenario of pulse evolution predicted by it should be more realistic. However, we have to recognize that the problem of correct description of cathode layer formation still remains. Specifically, effects of non-locality of the electron energy distribution function were ignored, which may result in increase of ionization rate and lengthening of a region with significant ionization. The high ionization degree predicted by numerical simulations (up to 10^{-4} or greater) will influence the electron energy spectrum, too. Very high local power density in the pulse may lead to numerous processes becoming important in enhancing the ionization rate in low-field regions. All the listed effects can hardly be adequately accounted for at the present state of the theory.

5.4.4   Conclusions

The three-dimensional model with axial symmetry effectively reduced to the two-dimensional one is formulated and applied to numerical simulations of pulse evolution in a negative corona with a cathode in the form of a cylinder with a semi-spherical cap in dry air at atmospheric pressure. Calculations demonstrated that current oscillations became perfect regular after about 25 pulses. Space–time evolution of electric field and charged species densities within one cycle of regular pulses is described in detail. The model predicts fast formation of a cathode layer at the discharge axis followed by its quick expansion over the cathode surface at the leading edge of the current pulse.

For a higher power supply voltage, the peak current rises a little, while the current between pulses grows substantially. The predicted charge per pulse is about three times smaller than experimental values for similar conditions. The pulse repetition period is close to that observed at higher voltages, while it is shorter at a low voltage. In contrast to the simplified 1.5-dimensional model predicting a two-peak shape of a Trichel pulse, the exact three-dimensional model predicts single-peaked pulses when ion-induced secondary emission processes are included, and photo-emission is neglected. On the anode surface, radial profiles of electric current averaged over one cycle was calculated and compared with the experiments. Revealed discrepancies between experimental data on typical charge per pulse and current distribution over the anode clearly indicate the necessity to improve the model. A weak point in the model presented above is the oversimplified description of plasma kinetics formed near the cathode pin.

References

Akishev Yu S, Deryugin A A, Kochetov I V, Napartovich A P, Pan'kin M V and Trushkin N I 1996 *Hakone V Contr Papers* (Czech Rep.: Milovy) p 122

Akishev Yu S, Kochetov I V, Loboiko A I and Napartovich A P 2002a *Bulletin of the APS* **47** 76

Akishev Yu S, Kochetov I V, Loboiko A I and Napartovich A P 2002b *Plasma Phys. Rep.* **28** 1049

Babaeva N Yu and Naidis G V 2000 in van Veldhuizen E M (ed) *Electrical Discharges for Environmental Purposes: Fundamentals and Applications* (New York: Nova Science Publishers) pp 21–48

Dhali S K and Williams P F 1987 *J. Appl. Phys.* **62** 4696

Egli W and Eliasson B 1989 *Helv. Phys. Acta* **62** 302

Kulikovsky A A 1997a *J. Phys. D: Appl. Phys.* **30** 441

Kulikovsky A A 1997b *J. Phys. D: Appl. Phys.* **30** 1515

Lama W L and Gallo C F 1974 *J. Appl. Phys.* **45** 103

Loeb L B, Kip A F, Hudson G G and Bennet W H 1941 *Phys. Rev.* **60** 714

Morrow R 1985a *Phys. Rev. A* **32** 1799

Morrow R 1985b *Phys. Rev. A* **32** 3821

Napartovich A P, Akishev Yu S, Deryugin A A, Kochetov I V, Pan'kin M V and Trushkin N I 1997a *J. Phys. D: Appl. Phys.* **30** 2726

Napartovich A P, Akishev Yu S, Deryugin A A and Kochetov I V 1997b *Final report to the Contract between ABB Management Ltd. Corp. research, Baden, Switzerland and TRINITI*

Pietsch G J, Braun D and Gibalov V I 1993 in B M Penetrante and S E Schultheis (eds) *Non-thermal plasma techniques for pollution control*, Part A, NATO ASI Series pp 273–286

Thompson J F, Warsi Z U A and Mastin W C 1985 *Numerical Grid Generation* (New York: Elsevier)

Trichel G W 1938 *Phys. Rev.* **54** 1078

Vitello P A, Penetrante B M and Bardsley J N 1993 in Penetrante B M and Schultheis S E (eds) *Non-thermal plasma techniques for pollution control*, Part A, NATO ASI Series pp 249–271

Warburg E 1899 *Wied. Ann.* **67** 69

Warburg E 1927 'Charakteristik des Spitzenstormes' in *Handbuch der Physik 4* (Berlin: Springer) pp 154–155

Zentner R 1970a *ETZ-A* **91**(5) 303

Zentner R 1970b *Z. Angew. Phys.* **29**(5) 294

5.5 Electrical Models of DBDs and Glow Discharges in Small Geometries

5.5.1 Introduction

The purposes of our discussion here are to provide an overview of electrical models of plasmas created in gas discharges, to show how they have been used to improve our understanding of dielectric barrier discharges (DBDs), and to suggest where they could be used to help develop a better understanding of discharges in very small geometries (microdischarges). As discussed in greater detail in sections 2.6, 6.2, and 6.4 of this book, DBDs and micro-discharges are two approaches being investigated as means for producing non-thermal, atmospheric pressure plasmas.

In section 5.5.2 we describe briefly a physical model of the initiation and evolution of non-thermal plasmas in electrical discharges where the cathode region has a determining influence on the properties of the system. We then present a numerical model suitable for describing the electrical properties of such glow discharges. The same type of model has been used for essentially all studies on DBDs to date and for the few modeling studies of micro-discharges that have been published. We then summarize how modeling has contributed to our current understanding of DBDs and microdischarges (sections 5.5.3 and 5.5.4, respectively), using previously published results in oxygen and rare gas mixtures to illustrate the phenomena occurring during the transient evolution glow discharges in DBDs in general. The few previous modeling results on DBDs in air are discussed by Kogelschatz in section 6.2.3, and the conclusions from the studies in air are the same as those discussed below. A few concluding remarks are presented in the final section.

It is worth noting that the physical situation described in this section is different from those presented in sections 5.2 and 5.3. That is, for DBDs and discharges in small geometries, quasi-neutrality cannot be assumed; the space charge electric field must be calculated self-consistently with the charged particle transport and generation rate. The strong coupling between the space charge field distribution and the charged particle transport and genera-tion is a major issue here.

5.5.2   Model of plasma initiation and evolution

The physical situation we aim to describe is plasma initiation and evolution in an electrical discharge. The discharge geometry is arbitrary, although cylindrical or rectangular symmetry is often assumed in order to reduce the problem to two dimensions. A dc, pulsed or rf voltage is applied between two or more electrodes which may or may not be covered by dielectrics. The electrodes are separated by a gap filled with a gas at a pressure p and we are mostly interested in conditions appropriate to the generation of non-thermal plasmas at high pressure.

5.5.2.1   Physical model

For a sufficiently high applied voltage and gas pressure, free electrons in the gas gap gain enough energy from the electric field to produce ionization through collisions with neutral gas atoms or molecules. The ionization cascade due to one initial electron and its progeny is called an 'avalanche'. The electrons in each avalanche move rapidly to the anode and leave behind the slower ions that were also produced in ionization or attachment events. Gas breakdown [1] proceeds either via Townsend breakdown or via streamer breakdown. Townsend breakdown occurs when, on the average, each electron, before arriving at the anode, has produced enough ionization/excitation in the volume to replace itself through secondary emission processes at the cathode (e.g. via ion-induced secondary electron emission, photoemission, etc.). In contrast, 'streamer' breakdown occurs when the space charge in an avalanche produced by a single electron grows large enough to be self-propagating so that no secondary emission is needed. As shown below, the streamer breakdown mechanism is favored for large values of *pd* (the product of gas pressure *p* and gap spacing *d*) and for high overvoltage; therefore, for high electron multiplication conditions. This mechanism leads to thin, highly conducting channels.

Following Townsend breakdown, a 'glow' or 'transient glow' discharge results if the accumulated positive space-charge, resulting from successive generations of avalanches created by cathode-emitted electrons, becomes large enough in a given volume to trap the electrons there, thus forming a plasma. This plasma expands very quickly toward the cathode, not because of the transport of existing particles, since that would be too slow a process, but rather because the ionization produced by the cathode-emitted electrons is enhanced in the relatively higher electrical field on the cathode side of the expanding plasma. For dc discharges at steady-state, almost all the potential drop is squeezed into the cathode fall between the plasma and the cathode. In DBDs, the axial expansion of the plasma is limited because of the charging of the dielectric surfaces. The plasma then expands radially along the electrode surfaces until the local electric field is no longer sufficient to maintain the electron temperature needed for ionization. At that point, the discharge filament extinguishes.

Glow discharges resulting from Townsend breakdown can be uniform radially or filamentary, depending on the conditions. The discharge can be filamentary even in the absence of thermal effects or stepwise ionization which are usually associated with the onset of instabilities. As a general rule, when the radial dimension R of the electrodes is much larger than the radial extent, δr, of one electron avalanche in the gas gap, the discharge will tend to be filamentary. For typical discharge applications, $R/\delta r$ is much larger at higher pressure. This is the reason why the filamentary mode of glow discharges is often observed at high pressure even when the current can be limited, as in a dielectric barrier discharge.

Discharges resulting from streamer breakdown are filamentary in nature and thus, for applications requiring a uniform plasma, streamer breakdown must be avoided. Streamers tend to evolve into arcs due to the formation of hot spots on the electrodes and resultant thermal plasma channel. This evolution of arcs can be inhibited if the current density is limited by, for example, a dielectric coating on an electrode. Note that a high level of preionization can provide enough initial electrons for the streamers to overlap [2, 3]. This can result in a uniform plasma, at least for a time less than the time needed for the onset instabilities due to power loading of the gas.

5.5.2.2 Numerical model

The fundamental variables in a numerical model of plasma initiation and evolution are the electron and ion densities and the electric field, or potential. The equations for these variables, complemented by suitable boundary conditions, are solved self-consistently to yield charged particle densities and electric field distribution as functions of time and space. From these results, we can calculate most other quantities of interest.

The following equations provide a mathematical description charged particle and electric field evolution.

- Electron and *ion continuity equations*:

$$\frac{\partial n_e}{\partial t} + \nabla \cdot [n_e \overline{v_e}] = S_e \tag{1}$$

$$\frac{\partial n_i}{\partial t} + \nabla \cdot [n_i \overline{v_i}] = S_i \tag{2}$$

where $\overline{v_e}$ and $\overline{v_i}$ represent the mean velocity for electrons and ions respectively and $S_e(\mathbf{r}, t)$ and $S_i(\mathbf{r}, t)$ are the production rates for electrons and ions respectively. Each ion species is described with an equation in the form of equation (2).

- Equations for *conservation of momentum* for electrons and ions of sign, s, in the drift-diffusion approximation:

$$n_e \overline{v_e} = -n_e \mu_e \mathbf{E} - D_e \nabla n_e \tag{3}$$

$$n_i \overline{v_i} = s n_i \mu_i \mathbf{E} - D_i \nabla n_i \tag{4}$$

where $\mu_{e\,(i)}$ is the electron (ion) mobility and $D_{e\,(i)}$ is the electron (ion) free diffusion coefficient.

- The continuity and momentum transfer equations are coupled to *Poisson's equation*:

$$\nabla \cdot [\varepsilon \nabla V] = -e[n_+ - n_-] \tag{5}$$

where ε is the permittivity (in general a function of $\mathbf{x}$ to include the dielectric volumes), e is the unit charge, n^+ is the total positive charge density and n^- is the total negative charge density (volume and surface charge density). At the interface between the gas and any dielectric surface the charge density is calculated by integrating the charged particle current to the surface, during the evolution of each discharge pulse. Thus the spreading of the surface charge along a dielectric surface, due to radial field induced by the previous surface charge, can be taken into account.

The electric field, E, is calculated from the potential as

$$\bar{E} = -\nabla V. \tag{6}$$

With the assumption of rectangular or cylindrical symmetry, the problem becomes two-dimensional.

The system of equations (1)–(5) must be closed by some assumptions about the transport coefficients and source terms. In many models of high pressure discharges, the mobility, diffusion coefficients and ionization coefficient are assumed to be functions of the local reduced electric field. This is logically referred to as the 'local field approximation'. Often, the diffusion coefficients are assumed to be constant. This limits the occurrence of numerical instabilities. This local field approximation allows a simple and often realistic description of the discharge. However, a description of the electrons involving the first three moments of the Boltzmann equation (the electron energy equation in addition to the continuity and momentum conservation equations) is more satisfactory not only for a better quantitative description of the discharge but also, in some cases, for a better qualitative representation of the physical phenomena. When an energy equation is used, the electron mobility, diffusion coefficient, and ionization frequency are assumed to depend on the local mean electron energy. A good example of a high pressure dielectric barrier discharge model for plasma display panels (PDPs) can be found in Hagelaar *et al* [4].

Finally, the electron current leaving the cathode is related to the incident ion current and through the secondary electron emission coefficient, γ_k, as follows:

$$\varphi_e(\text{cathode}) = \sum_k \gamma_k \varphi_k(\text{cathode}) \tag{7}$$

where the sum is over all ion species, γ_k is the secondary electron emission coefficient due to the kth type of ion incident on the cathode, and φ_k is the flux of the kth type of ion to the cathode.

Note that photons and metastable atom bombardment of the cathode can also lead to secondary electron emission [5], and desorption of electrons from the dielectric layer has been proposed to account for some observations [6]. We return to this point below; however, it is important to emphasize now that the identification and quantification of the electron emission processes from the cathode are unresolved modeling issues.

To the extent that the degree of excitation is too low to influence the net rate of generation of charged particles, it is possible to neglect plasma chemistry in the electrical model. As the power deposited in the gas increases, two-step ionization (electron impact ionization of excited states) and associative or Penning ionization can start to play a role, in which cases a model of the plasma chemistry must be solved self-consistently with the electrical model. Gas heating is another consideration because the local value of E/N is high, and thus the ionization rate is high, where the gas temperature is high.

5.5.2.3 *Numerical methods*

Starting from the known or assumed initial conditions, equations (1)–(5) are integrated in time to yield the charged particle densities and the electric field as functions of space and time. Numerical methods for solutions of these equations are discussed, for example, by Kurata [7]. Nevertheless, there remain the following two particular numerical difficulties encountered in the modeling of high pressure plasmas.

1. *For dc or transient glow discharges (radially uniform or filamentary).* The simplest integration scheme for these equations is an explicit scheme in which the charged particle transport and Poisson's equations are solved sequentially. That is, Poisson's equation is solved at time t^k, and then the charged particles are transported for a time Δt in the field calculated at time t^k. Such an integration scheme is subject to the constraint that the time step Δt must be smaller than the dielectric relaxation (Maxwell) time, Δt_M, which is inversely proportional to the plasma density:

$$\Delta t_M = \frac{\varepsilon_0}{e(n_e \mu_e + n_i \mu_i)}. \tag{8}$$

Thus, for a plasma density of $10^{14} \, \mathrm{cm}^{-3}$, the integration time step in an explicit integration scheme is very approximately 10^{-12} s at 100 torr, and this simple integration scheme leads to impractically long computational times. Either semi-implicit [8] or fully implicit [7] schemes must be used.

2. *For streamers.* The modeling of streamer-type microdischarges is difficult numerically because streamers have two very different spatial scales that must be considered simultaneously, namely the streamer front with steep gradients and the streamer body with a nearly uniform plasma. Compounding this difficulty is that fact the streamer front propagates. There have been a large number of publications presenting results of modeling streamer formation and propagation (see, for example, Dhali and Williams [9]). In the context of DBDs in oxygen, Li and Dhali [10] have presented a method for solving these equations using an adaptive grid where the resolution is highest in the region of large density gradients.

In spite of the numerical complications, models have been developed that are very efficient. As an example, models of DBDs in typical PDP conditions [11] take about several seconds, several minutes and several hours, respectively, per pulse for one-, two-, or three-dimensional calculations using a $40 \times 40 \times 40$ grid running on a 2 GHz personal computer.

5.5.3 Dielectric barrier discharges

Orders of magnitude estimates for some of the DBD discharge properties are listed in table 5.5.1 for different operating modes at approximately atmospheric pressure and for the conditions indicated. We will briefly summarize results obtained from modeling these modes in the sections below, without attempting to be exhaustive in the list of references.

5.5.3.1 Random filament mode

The common discharge mode in atmospheric pressure DBDs is the random filament mode [14, 15] where as many as $10^6/cm^2/s$ transient glow discharge filaments occur at seemingly random locations, each being extinguished after bridging the gap. The filaments are random in the sense that we cannot predict where or when they will be initiated. We use this term to make explicit the difference between this mode and the self-organization (pattern formation, see below) sometimes observed in DBDs as the voltage is decreased.

Most all of the modeling for this type of discharge mode has concentrated on simulating the evolution of a single, isolated current filament. The early work of Eliasson *et al* [16] was developed to study the efficiency of ozone production in DBDs. This was later coupled to a two-dimensional electrical model consisting of plane parallel electrodes covered by dielectrics in which many aspects of DBD behavior [17] were quantified. These aspects include the spreading of the discharge along the dielectric surface due to accumulated surface charges, the dependence of the current pulse width on pressure and the total charge transferred per micro-discharge versus

Table 5.5.1

	Random filament mode [12]	PDP cells [11]	Atmospheric pressure glow discharge (APGD) [13]
Conditions	1 atm air/O_2 1 mm	560 torr, Xe/Ne 150 μm	1 atm He 0.5 cm
Current pulse duration	1–10 ns/filament	50–100 ns	>1 μs
Filament radius	~100 μm	100 μm	Uniform
Peak current density	~100–1000 A/cm^2	10 A/cm^2	~1 mA/cm^2
Total charge transferred	0.1–1 nC/filament	30 pC/pulse	13 nC/cm^2
Peak electron density	~10^{14}–10^{15} cm^{-3}	5×10^{13} cm^{-3}	<10^{11} cm^{-3}
Electron energy	1–10 eV	1–10 eV	1–10 eV

pressure. These authors found good agreement between their numerical results and experimental measurements of the latter two quantities. The former quantity was not measured. Results from the two-dimensional model were used to help define a simple model of the plasma chemistry in oxygen DBDs and in rare-gas mixtures such as those used for the generation of excimer radiation. Heating of the ions was identified by these authors as a mechanism limiting the efficiency for ozone generation in DBDs.

Kogelschatz (see sections 2.6 and 6.2) argues that the filaments observed in DBDs in this application are essentially transient high pressure glow discharges and that the current density and electron density are what would be expected based on j/p^2 (ratio of current density to the pressure squared) scaling of glow discharges to atmospheric pressure [18]. Note that pt, the product of the gas pressure and time, is also a scaling parameter. We assert that these scaling parameters cannot be applied to filaments resulting from streamer breakdown because of higher current density and narrower conducting channels following streamer breakdown, different distribution of energy in excitation, ionization, and dissociation channels, etc. Nevertheless, more work needs to be done to clearly identify the effects of the breakdown mode (Townsend or streamer) on the plasma properties of the filaments in DBDs.

Gibalov and Pietsch and their colleagues [19–21] have developed two-dimensional models of DBDs in different configurations (volume, coplanar and surface discharges) and for air/oxygen at atmospheric pressure. Braun *et al.* [19] describe the evolution of an isolated filament in DBDs with plane parallel electrodes, one of which is covered by a dielectric, in air for $pd = 76\,\text{torr cm}^{-1}$ and for a voltage near the breakdown voltage. These are not streamer conditions in the sense defined above, and secondary electron emission due to ion impact and as well as photo-ionization are considered in the model. According to this work, 'the microdischarges behave like transient high pressure glow discharges'; a plasma forms first near the anode and then expands towards the cathode. One could quibble with their persistent use of the term 'cathode-directed streamer', and we suggest that 'transient glow discharge filament' is a more appropriate term. Gibalov and Pietsch [20] studied the efficiency of ozone generation in DBDs in planar and surface discharge geometries. They modeled, for example, one dielectric-covered electrode and either a bare parallel electrode a short distance away or a bare perpendicular electrode touching the dielectric surface, respectively. They found nearly the same efficiency for both. Gibalov *et al* [21] have also studied DBDs in coplanar geometries and found reasonable agreement with experiment, noting that 75% of the energy losses are due to heating of the ions for the conditions of 100 μm coplanar electrode spacing, 1 mm gas gap, and 2 bar oxygen.

Results from the models of Gibalov and Pietsch and their colleagues in planar DBD configurations are generally consistent with Eliasson *et al* [16].

The minimum thickness of the cathode sheath was found to be large compared to a glow discharge. This was expected because the charging of the dielectric layer prevents a fully developed glow discharge from forming. The calculated discharge radius is about 200 µm in the volume, but the area covered by the surface charge is much larger than the channel diameter. This arises because the electron surface charge spreads more than the ion surface charge. At the peak value of the current, about 50% of the power is deposited in the ions and the remaining energy is distributed almost homogeneously in the electrons in the column. The increase in the mean gas temperature can be high near the cathode, depending on the dielectric capacitance, but is only a few Kelvin in the plasma column.

Other recent modeling work on single filaments in DBDs include that of Steinle *et al* [22] who published results from a two-dimensional simulation of filament evolution in DBDs in air at atmospheric pressure in a small gap with high dc applied voltage. Conditions for streamer formation were satisfied for the parameters chosen in this work, and the authors presented results showing that the efficiency for ultraviolet generation, but not the efficiency for generation of radicals, depends rather strongly on the applied voltage. In this work, secondary electron emission produced via photoemission from the cathode as well as photo-ionization in the volume are accounted for in a rather detailed way, but the role played by secondary electrons emitted from the cathode is not clear. Carman and Mildren [23] studied pulse-excited DBDs in xenon and found that the efficiency for generation of excimer radiation can be quite high in these conditions, namely greater than 60%, consistent with the experiments of Vollkommer and Hitzschke [24] who have developed a commercial lamp based on the concept of pulse-excited DBDs and for which streamer conditions were avoided. Xu and Kushner [25, 26] have reported one-dimensional radial calculations of interacting filaments in DBDs in N_2 and in N_2/O_2 mixtures at atmospheric pressure. They find that filaments affect their neighbors mainly through the charging of the dielectric and that the plasma chemistry in a given filament is otherwise very little affected by the presence of its near neighbors.

Golubovskii *et al* [27], and more recently Boeuf [28], have been investigating reasons for the formation of transient glow discharge filaments in DBDs following Townsend breakdown. This is an interesting question because Townsend breakdown can lead to either uniform or filamentary discharges, and more work needs to be done to clearly identify reasons for the occurrence of each.

5.5.3.2 *Plasma display panels (PDPs)*

Much modeling work has been done in dielectric barrier discharges for plasma display panels (PDPs) where the discharge dimensions are on the order of 100 µm and the gas pressure is about 500 torr for mixtures of rare gases containing xenon, and the applied frequency is a square wave with a frequency of 100 kHz or more. Most PDPs today use a 'coplanar' geometry where the main discharge occurs between parallel electrodes on the same substrate, at a position selected by applying a suitable low voltage to the third electrode (perpendicular) on the opposite substrate. 'Matrix' geometries, in which the electrodes are perpendicular stripes on opposite substrates, have also been studied for PDP applications.

The discharges in PDPs are at low values of pd (typically $5 \, torr \, cm^{-1}$) that are typical of glow discharges but lower than for other DBD applications. The ability to control each discharge separately and the reproducibility of the discharges are paramount in this application. In the sustaining mode, the applied voltage is less than the breakdown voltage, and it is the surface charge remaining from the discharge pulse on the last half cycle that makes operation at this low voltage possible. The operation at an applied voltage below breakdown is essential in the PDP application because it allows bi-stability, namely, the coexistence of discharge cells in the ON state and in the OFF state with the same sustaining voltage. To turn a discharge cell from the OFF state to the ON state, one must first apply an address-voltage pulse to the cell in order to deposit memory charges on the cell walls. These memory charges create a voltage drop across the dielectric layer that will add to the voltage across the electrodes when the sustaining voltage is applied. During the sustaining period, the voltage rise time in PDPs must be short enough that breakdown occurs during the plateau of the square wave voltage. Using a sinusoidal voltage, as found in many other DBD applications, does not allow adequate control of the voltage at which breakdown actually occurs. In the driving scheme of a PDP there is a period, called the 'priming period,' during which a very slowly rising voltage is applied between the electrodes in each discharge cell. This generates a low current discharge that will provide seed electrons in order to minimize the statistical time lag during addressing. The slowly rising voltage allows one to operate in a low current Townsend (or 'dark') discharge regime where the light emission is weak and does not significantly reduce the contrast. This shows that the rise-time of the voltage is a very important parameter and can help to control the discharge regime and the voltage at which breakdown occurs in a DBD. Using a sinusoidal voltage does not allow a simple control of the voltage rise-time because the only way to change the rise-time is to change either or both the amplitude or/and the frequency of the voltage waveform. There remains the need to study more systematically, both in experiments and simulations, the effects of the voltage wave-form on the properties of DBDs in general.

The first one-dimensional model published in 1978 [29] contained most of the elements of DBD operation. Since then one-dimensional, two-dimensional and recently three-dimensional fluid models have been used to study PDPs in considerably more detail. These simulations are described in the review of Boeuf [11]. State-of-the-art PIC-MC models have also been used to check assumptions and study such purely kinetic effects [30, 31] as the appearance of striations in the light intensity in the plasma spreading along the dielectric surface. These models have been used to quantify the characteristics of the plasmas produced in PDPs and, more recently as the models have been improved, to help guide the experimental optimization of these devices. Models were used to understand the reasons for the low luminous efficacy which is due to the energy wasted in accelerating ions in the sheath [32] and to suggest ways for improving efficiency such as modifying the electrode geometry and/or increasing the length of the transient positive column region [33–35]. For example, it has been seen that xenon is efficiently excited in the low field region accompanying the spreading of the discharge along the anode surface, and that enhancing this spreading increases the efficiency [11]. Note, however, that the radial field at the anode in ozonizers is not high enough to affect the desired chemistry and thus leads to a decrease in the efficiency in that application [21]. The addressing of individual cells in coplanar PDPs is accomplished by suitable application of voltage pulses between the electrodes and on the third 'addressing' electrode. Models have been used for parametric studies of different addressing schemes. Excellent agreement has been obtained between models and experiments of the electrical characteristics and with the space- and time-dependence of the emission intensity. There is also generally good agreement with available data for the efficiency for excitation of xenon and in the space and time evolution of the emission features.

5.5.3.3 *Atmospheric pressure glow discharge*

DBDs at high pressure are normally filamentary [14], but, in a limited range of conditions, it is possible to generate an atmospheric pressure glow discharge (APGD) [13, 36]. As mentioned above, the conditions leading to either a uniform plasma or filamentation in atmospheric pressure air are not yet completely understood [27]. However, it seems that the properties of any single filament which may arise are not very different from those of the APGD plasmas.

Modeling has been an important tool in gaining an understanding of the plasmas produced in APGDs. The models of Ségur and Massines [37], Tochikubo *et al* [38] and of Golubovskii and colleagues [6, 39] have been used to calculate the charged particle density and electric field distributions as functions of space (one-dimensional) and time in APGDs in helium and nitrogen. From these results, the time variations of gap voltage, memory voltage and current density have been obtained and compared with experiment. Model predictions agreed quite well with measured current waveforms and patterns of light emission intensity, although each found that the quantitative agreement required some additional ionization, which could be due to impurities or to the effects of metastables. Ségur and Massines and Golubovskii *et al* conclude that the uniform glow in nitrogen is in the Townsend regime in that little or no space charge distortion of the geometrical field and low charged particle densities are observed. However, they found higher plasma densities in helium APGDs.

One can conclude from comparisons of models and experiment [13] that the generation of an APGD depends on a slow growth of the avalanche, a high enough electron density at the beginning of each half cycle, and a high electron emission from the cathode. Essentially these same conclusions have been derived by Golubovskii and colleagues who find that the rate of voltage rise affects the discharge mode (see the discussion above for PDPs in section 5.5.3.2). Specifically, for a slow rise time, breakdown occurs near the Paschen minimum indicating a slow growth of the avalanche, and favors a uniform glow mode [39]. Golubovskii and colleagues have proposed a mechanism of desorption of electrons at the cathode to provide secondary electrons between current pulses in order to reproduce experimental results [6]. It is interesting to note that such additional electron emission was also needed to describe PDP operation at low frequency where the plasma has time to decay on each half cycle of the applied voltage [11]. Golubovskii *et al* [40] have also looked at the question of photoemission as a mechanism for discharge uniformity (photons strike the cathode at radial positions far from the axis of their parent filament) and other mechanisms that could lead to radial non-uniformities [27].

As emphasized by Aldea *et al* [41] and Tochikubo *et al* [38], the discharge cannot be uniform if the breakdown process itself is filamentary as indicated by streamer breakdown. Avoiding streamer formation is more difficult at high values of *pd* and depends on the gas composition as shown in figure 5.5.1. The minimum breakdown voltage, V_b, is plotted as a function of *pd*, which, as stated previously, is the product of the gas pressure *p* and gap spacing *d* in a parallel plate electrode geometry. The value of V_b can be determined through the self-sustaining condition [42]

$$M = \exp[\alpha(V_b, pd) \times d] = 1 + (1/\gamma) \tag{9}$$

where the electron multiplication, M, is related to the net ionization rate coefficient, α, which itself depends on V_b and pd, and γ is the secondary electron emission coefficient. A rough estimate of the voltage required for streamer formation, V_s, can be derived by supposing that streamers [43, 44] form when the electron multiplication in the gap exceeds 10^8. Using ionization and attachment rate coefficients from the SIGLO database [45], we calculated the ratio V_s/V_b shown in figure 5.5.1 by assuming a secondary electron emission coefficient of 0.3 in helium and 0.01 in air. Korolev and Mesyats [46] point out that the boundary between Townsend and streamer breakdown is not sharp, and thus the curves in figure 5.5.1 are only qualitative. Nevertheless, the conclusion is clear: avoiding streamer

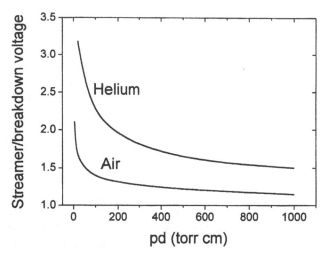

Figure 5.5.1. Ratio of the voltage required for streamer formation to the minimum breakdown voltage versus *pd* for air and helium, calculated assuming streamers are formed when the electron multiplication exceeds 10^8.

breakdown in air at high values of *pd* between parallel plate electrodes is difficult and requires the careful control of operating conditions. It is relatively easier to avoid streamer breakdown in helium. This simple comparison of minimum breakdown and streamer formation voltages suggests an explanation for the fact that APGDs in helium are so much easier to obtain than those in air. Note also that preionization [2, 3] can impede streamer formation and enhance discharge uniformity, and this may be provided in DBDs by charges remaining from the previous half cycle.

Finally, it should be mentioned there is no guarantee that plasma uniformity after breakdown can be maintained. Indeed there are numerical examples where an initially uniform plasma eventually reaches a steady-state where regular patterns appear (see section 5.5.3.4). Additional research is also needed in this area.

5.5.3.4 Pattern formation

Observations of the formation of patterns of regularly spaced, quasi-stationary filaments in DBDs have been summarized recently by Kogelschatz [18]. In general, there is a transition from the random filament mode in DBDs to a patterned discharge structure when the discharge voltage is decreased [47]. Reasons for this behavior are not completely clear, but some indications were obtained from a two-dimensional model [48] for a DBD in helium at 100 torr with a 0.5 mm gap spacing. It is interesting to note that thermal effects, stepwise ionization or other well known causes of instabilities were not included in these model calculations because they were not likely to be important under the simulated conditions. The model used periodic boundary conditions in the transverse direction and assumed uniform initial densities of the charged particles. A simple mathematical solution of the problem was therefore a series of radially uniform transient glow discharges at each half cycle of the applied voltage. However, the results showed that the uniform solution was not stable and degenerated within several cycles of the applied voltage into a non-uniform, filamentary solution very similar to the observed patterned discharge structure. A conclusion of this work was that if a local non-uniformity appears in the volume of the surface charge density, breakdown occurs faster at the radial location where the density is maximum. The charging of the surface occurs faster at this radial location and spreading the charges induces a decrease in the gap voltage around this location, resulting in the choking of the neighboring plasma. This explanation of filament formation suggests that the slope of the ionization coefficient as a function of the electric field is an important parameter in this process. One can expect that the tendency to form a filament will be smaller for conditions where the slope of the ionization coefficient as a function of the electric field is smaller.

5.5.4   Micro-discharges: discharges in small geometries

A build-up of the internal excitation or kinetic energy of the gas corresponds to an increase in the temperature of the gas and this can lead to instabilities [42, 49]. By 'instability' we mean that small perturbations or non-uniformities in the plasma conductivity tend to grow catastrophically and, if left unchecked, lead to a thermal plasma arc. Diffusion is a stabilizing mechanism, damping small fluctuations in the plasma density at low pressure. Since the diffusion rate decreases with gas pressure while rates for mechanisms leading to instabilities tend to increase, maintaining a stable non-thermal plasma is more difficult at high gas pressure. Concepts for the generation of non-thermal atmospheric pressure plasmas have been proposed recently which are based on the use of very small size geometries such that the value of pd is about that of typical glow discharges (e.g. less than about $10 \, torr \, cm^{-1}$) [50–53]. Provided streamer conditions are avoided at breakdown, a non-thermal plasma can be maintained in these 'micro-discharge' configurations apparently because diffusion effectively dissipates small fluctuations in the plasma density which could otherwise lead to constrictions of the current carrying channel.

An open question at this time is the extent to which phenomena in high-pressure microdischarges are the same as those in low pressure discharges with the same value of pd. For example, it has been suggested that micro-hollow cathode discharges are similar to those at low pressure [50, 52, 54] with the same pd. That is, the structure in the measured $V–I$ characteristic is attributed to the classical hollow cathode effect, namely, the penetration of the plasma into the hollow cathode cavity when the current density exceeds a certain value. This interpretation of the structure in the *V–I* characteristic is consistent with the calculations of Fiala *et al* [55] in similar geometries but with a pressure of 1 torr. While this interpretation may indeed be correct, more detailed analyses [56] including gas flow, thermal effects and power loading in the gas are needed to develop a better understanding of the behavior of this and other [53] micro-hollow cathode discharges.

Modeling work on discharges in very small geometries is under way. Recent examples are the work of Wilson *et al* [57] who compare experiments and model predictions in micro-hollow cathode discharges in nitrogen at about 10 torr; Kushner [58] who discusses issues of scaling in very small hollow cathode devices at 400–1000 torr; and Kothnur *et al* [59] examine the structure of dc discharges in very small gaps using a fluid model. The power density in the microdischarges can be quite high, and questions of thermal balance and the glow-to-arc transition are important for understanding the behavior of single microdischarges or arrays of microdischarges. More modeling and plasma diagnostics are needed to identify phenomena specific to microdischarges, and this will undoubtedly be a developing area in the coming years.

5.5.5 Conclusions

The purpose of our discussion here has been to provide an overview of electrical models of plasma created in gas discharges and to illustrate their application to DBDs and microdischarges. Over the past 20 years, modeling has proven to be a very powerful and useful tool for helping to understand the basic physics and for guiding the experimental optimization of different devices based on non-thermal plasmas at low pressure. Modeling has also contributed greatly to our current understanding of plasmas created in high pressure DBDs and will certainly be used more in the future to help understand the generation of non-thermal plasmas in micro-discharge configurations.

In all cases, models are most useful when used in combination with experiments, and they are dependent on experiments for validation and for determination of input data. Sophisticated diagnostic techniques are being used to identify plasma parameters in DBDs and other micro-discharge configurations. Examples of recent innovative applications of diagnostic tools include the detailed measurements of the argon excited state densities, plasma density and gas temperature in microdischarges [60] and detailed imaging of single DBD filaments [61]. These and many other recent experimental results give models valuable points for comparison with model predictions. There is also a continuing need for more systematic results of relatively simple electrical measurements and emission intensity measurements for results over a wide range of conditions in DBDs and microdischarges.

In conclusion, the following important issues can and should be addressed through modeling.

- In the context of DBDs, modeling can help define conditions for the formation of transient glow discharge filaments or for radially homogeneous glow discharges. Modeling could also be used to help optimize the excitation pulses for a given application. Volkommer and Hitzschke [24] have shown that very high efficiencies for the generation of excimer radiation can be obtained in DBDs in high pressure xenon with suitably tailored voltage pulses and with values of *pd* such that streamer conditions are avoided. Carman and Mildren [23] have addressed this problem through modeling. Similar studies in DBDs in air have not been performed to our knowledge. Through modeling it would also be possible to explore the question of how the energy deposition in transient glow discharge filaments in DBDs scales with operating conditions and how this scaling depends on the breakdown mechanism.
- In the context of microdischarges, models could be used to help clarify the physical mechanisms occurring in microdischarges operating at pressures up to one atmosphere and to evaluate the role of physical processes which are specific to high pressure/high power density conditions (e.g. gas heating, stepwise ionization, etc.). This, in turn, could be used to assess the validity or the range of validity of the usual similarity laws. Finally, a better understanding of conditions leading to the glow-to-arc transition due to hot spots on the electrodes or gas phase instabilities in micro-discharge configurations is needed in order to evaluate the upper limits on current density and plasma density possible in these devices.

References

[1] Meek J M and Craggs J D (eds) 1953 *Electrical Breakdown of Gases* (Oxford: Clarendon)
[2] Palmer J A 1974 *Appl. Phys. Lett.* **25** 138
[3] Levatter J I and Lin S C 1980 *J. Appl. Phys.* **51** 210
[4] Hagelaar G J M, Klein M H, Snijkers R J M M and Kroesen G M W 2001 *J. Appl. Phys.* **89** 2033
[5] Phelps A V and Petrivic Z M 1999 *Plasma Sci. Sources and Tech.* **8** R21
[6] Golubovskii Y B, Maiorov V A, Behnke J and Beknke J F 2002 *J. Phys. D: Appl. Phys.* **35** 751
[7] Kurata M 1982 *Numerical Analysis for Semiconductor Devices* (Lexington, MA: Heath)
[8] Ventzek P L G, Sommerer T J, Hoekstra R J and Kushner M J 1993 *Appl. Phys. Lett.* **63** 605
[9] Dhali S K and Williams P F 1987 *J. Appl. Phys.* **62** 4696
[10] Li J and Dhali S K 1997 *J. Appl. Phys.* **82** 4205
[11] Boeuf J P 2003 *J. Phys. D: Appl. Phys.* **36** R53
[12] Kogelschatz U, Eliasson B and Egli W 1997 *J. Phys. IV* **7** C4

[13] Massines F, Rabehi A, Decomps P, Gadri R B, Segur P and Mayoux C 1998 *J. Appl. Phys.* **83** 2950

[14] Kogelschatz U 2003 *Plasma Chem. and Plasma Proc.* **23** 1

[15] Kogelschatz U, see section 6.2 of this book

[16] Eliasson B, Hirth M and Kogelschatz U 1987 *J. Phys. D: Appl. Phys.* **20** 1421

[17] Eliasson B and Kogelschatz U 1991 *IEEE Trans. Plasma Sci.* **19** 309

[18] Kogelschatz U 2002 *IEEE Trans. Plasma Sci.* **30** 1400

[19] Braun D, Gibalov V and Pietsch G 1992 *Plasma Sources Sci. Technol.* **1** 166

[20] Gibalov V I and Pietsch G J 2000 *J. Phys D: Appl. Phys.* **33** 2618

[21] Gibalov V I, Murata T and Pietsch G J 2000 in MacGregor S J (ed) *XIII International Conference on Gas Discharges and their Applications*, Glasgow,

[22] Steinle G, Neundorf D, Hiller W and Pïetralla M. 1999 *J. Phys. D: Appl. Phys.* **32** 1350

[23] Carman R J and Mildren R P 2003 *J. Phys. D: Appl. Phys.* **36** 19

[24] Vollkommer F and Hitzschke L 1998 in Babucke G (ed) *8th International Symposium on the Sciences and Techniques of Light Sources*, Greifswald, p 51

[25] Xu X and Kushner M J 1998 *J. Appl. Phys.* **84** 4153

[26] Xu X P and Kushner M J 1998 *J. Appl. Phys.* **83** 7522

[27] Golubovskii Y B, Maiorov V A, Behnke J and Behnke J F 2003 *J. Phys. D: Appl. Phys.* **36** 975

[28] J P Boeuf 2003 in *Proceedings of the Plasma Technology Training School*, Glasgow, August

[29] Sahni O, Lanza C and Howard W E 1978 *J. Appl. Phys.* **49** 2365

[30] Ikeda Y, Suzuki K, Fukumoto H, Verboncoeur J P, Christenson P J, Birdsall C K, Shibata M and Ishigaki M 2000 *J. Appl. Phys.* **88** 6216

[31] Lee J K, Dastgeer S, Shon C H, Hur M S, Kim H C and Cho S 2001 *Jap. J. Appl. Phys.* **40** L528

[32] Meunier J, Belenguer P and Boeuf J P 1995 *J. Appl. Phys.* **78** 731

[33] Ganter R, Callegari T, Pitchford L C and Boeuf J P. 2002 *Appl. Surf. Sci.* **192** 299

[34] Ouyang J, Callegari T, Callier B and Boeuf J P 2003 *IEEE Trans Plasma Sci.* **31** 422

[35] Ouyang J T, Callegari T, Caillier B and Boeuf J P 2003 *J. Phys. D: Appl. Phys.* **36** 1959

[36] Okazaki S, Kogoma M, Uehara M and Kimura Y 1993 *J. Phys. D: Appl. Phys.* **26** 889

[37] Ségur P and Massines F 2000 in MacGregor S J (ed) *XIII International Conference on Gas Discharges and their Applications*, Glasgow, vol 1, p 15

[38] Tochikubo F, Chiba T and Watanabe T 1999 *Jpn. J. Appl. Phys.* **9A** 5244

[39] Golubovskii Y B, Maiorov V A, Behnke J and Behnke J F 2003 *J. Phys. D: Appl. Phys.* **36** 39

[40] Golubovskii Y B, Maiorov V A, Behnke J and Behnke J F 2003 in *Proceedings of the International Conference on Phenomena in Ionized Gases*, Greifswald

[41] Aldea E, Schrauwen C P G and van de Sanden M C M 2003 in *Proceedings of the International Symposium on Plasma Chemistry*, Taormina, Italy

[42] Raizer Y P 1991 *Gas Discharge Physics* (Heidelberg: Springer)

[43] Raether H 1939 *Z. Phys.* **112** 464

[44] Loeb L B and Meek J M 1940 *J. Appl. Phys.* **11** 438

[45] http://www.siglo-kinema.com

[46] Korolev Y D and Mesyats G A 1998 *Physics of Pulsed Breakdown in Gases* (Yekaterinburg: URO Press)

[47] Guikema J, Miller N, Niehof J, Klein M and Walhout M 2000 *Phys. Rev. Lett.* **85** 3817

[48] Brauer I, Punset C, Purwins H-G and Boeuf J P 1999 *J. Appl. Phys.* **85** 7569

[49] Kunhardt E E 2000 *IEEE Trans. Plasma Sci.* **28**

[50] Schoenbach K H, Verhappen R, Tessnow T, Peterkin F E and Byszewski W W 1996 *Appl. Phys. Lett.* **68** 13

[51] Shi W, Stark R H and Schoenbach K H 1999 *IEEE Trans. Plasma Sci.* **27** 16

[52] Frame J W, Wheeler D J, DeTemple T A and Eden J G 1997 *Appl. Phys. Lett.* **71** 1165

[53] Yu Z, Hoshimiya K, Williams J D, Polvinen S F and Collins G J 2003 *Appl. Phys. Lett.* **83** 854

[54] Sankaran R M and Giapis K P 2002 *J. Appl. Phys.* **92** 2406

[55] Fiala A, Pitchford L C and Boeuf J P 1995 in Becker K H, Carr W E and Kunhardt E E (eds) *International Conference on Phenomena in Ionized Gases* (Hoboken, NJ: Stevens Institute of Technology) vol 4, p 191

[56] Hsu D D and Graves D B 2003 *J. Phys. D: Appl. Phys.* **36** 2898

[57] Wilson C G, Gianchandani Y B, Arslanbekov R R, Kolobov V and Wendt A E 2003 *J. Appl. Phys.* **94** 2845

[58] Kushner M J 2004 *J. Appl. Phys.* **85** 846

[59] Kothnur P S, Yuan X and Raha L L 2003 *Appl. Phys. Lett.* **82** 629

[60] Penache C, Miclea M, Brauning-Demian A, *et al.* 2002 *Plasma Sources Sci. Technol.* **11** 476

[61] Kozlov K V, Wagner H-E, Brandenburg-Demian R, Hohn O, Schossler S, Jahnke T, Niemax K and Schmidt-Bocking H 2001 *J. Phys. D: Appl. Phys.* **34** 3164

5.6   A Computational Model of Initial Breakdown in Geometrically Complicated Ssystems

5.6.1   Introduction

In this section, a computational model for predicting the onset of breakdown for electrodes in arbitrary geometric configurations is described. Although the model described here is applied to a coplanar plasma display panel configuration using a mixture of noble gases, the model can be extended to many other configurations and gas chemistries in a straightforward manner.

Flat panel display technologies continue to increase in importance in the consumer market as well as in the computer market. The active matrix liquid crystal display (AMLCD) technology currently comprises the majority of flat panel displays at moderate sizes (6–19 inch (15–48 cm) diagonal measurement). Despite recent increases in size and resolution, sizes required for large screen applications such as high resolution computer monitors and high definition television (HDTV) remain challenging for AMLCD technology. Viewing angle and update speed remain problematic for AMLCD, although less severe than in the past. In addition, the brightness levels for a transmissive screen such as an AMLCD panel are presently inadequate for many applications and lighting conditions.

The emerging ac plasma display panel (PDP) technology provides a number of advantages. In contrast to the stringent sub-micron feature sizes of AMLCD panels, the PDP has super-micron features and can be manufactured with relatively simple process technology. An important limit on AMLCD size is the processing uniformity in manufacture, which drives the price up rapidly at larger sizes due to lower yields. The PDP scales well to large sizes (50 inches (127 cm) and up), and it is possible to build PDPs larger than cathode ray tubes (CRT) can be practically manufactured. PDPs are luminous devices, leading to higher brightness and contrast. The viewing angle in the PDP is also similar to that of a conventional CRT, and update speed is also comparable. Although the PDP is currently 3–4 times less efficient than the AMLCD, this is less important for HDTV and computer applications. The efficiency and cost of PDPs is expected to continue to decrease as production increases.

A typical three-electrode ac PDP is shown in figure 5.6.1. The panel consists of a rear glass substrate, with trenches etched or barriers deposited to separate neighboring pixels. The barriers are on the order of 10 μm in thickness, and the distance between barriers is on the order of 100 μm. Address electrodes are deposited in the bottom of the trenches, and covered with a dielectric material of about 10 μm in thickness. The dielectric and trench walls are then coated with phosphor, alternating red, green and

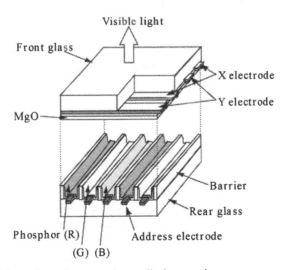

Figure 5.6.1. Schematic coplanar ac plasma display panel.

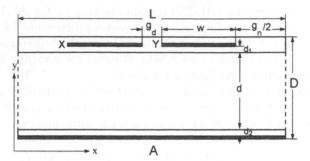

Figure 5.6.2. Schematic cross section of a coplanar PDP cell.

blue across rows. The top layer consists of a transparent glass substrate, with pairs of electrodes deposited perpendicular to the direction of the address electrodes. The upper electrodes, labeled X and Y in the figure, are the sustain electrodes. The surface of the substrate, as well as the X and Y electrodes, are then covered with a layer of dielectric on the order of 10 μm thick with dielectric constant typically about 10–15. The dielectric is coated with a copious secondary emitter, such as magnesium oxide. The region in the trenches is filled with a gas at a pressure of 300–700 torr, often a mixture of neon, xenon, and possibly other inert gases including helium and argon.

Cells are formed by the intersection of X–Y electrode pairs with address electrodes. The cross section of a cell is shown in figure 5.6.2. In the figure, the X and Y electrodes are perpendicular to the plane of the paper, while the address electrode extends to the left and right. Discharges are generated by various combinations of voltage applied to the X, Y, and A electrodes, as well as voltage due to charge accumulated at the surface of the dielectric from previous discharges.

The anatomy of a single discharge event is similar to that of a dc discharge, except that the walls charge up and eventually extinguish the discharge when the applied voltage is completely shielded by the wall charge. Ions are attracted to the positively charged cathode, accelerating through a non-neutral cathode fall region. Upon impact with the MgO-coated dielectric material enclosing the cathode, the ions generate secondary electrons. The secondary electrons accelerate through the cathode fall, undergoing many collision events with the background gas. Elastic scattering, electron impact ionization, and electron–neutral excitation, as well as many other collisional events play an important role in shaping the discharge. As the discharge current increases, charge builds up on the dielectric surfaces, decreasing the gap voltage in the cell. When the cathode fall no longer imparts sufficient energy to secondary electrons to generate ionization events, the plasma behaves like a decaying glow discharge and slowly extinguishes.

A priming pulse is applied to all cells in the PDP to initialize all cells with a specified charge on the dielectric surface. The priming pulse typically consists of a few hundred volts applied to the X and Y electrodes, and often a supplemental voltage on the order of 100 V to the address electrode. After all cells are primed, a refresh pulse continuously sweeps all cells. The refresh pulse is applied to an entire row of cells which share the same X and Y electrodes. The refresh pulse consists of a voltage difference applied between the X and Y electrodes which is insufficient for breakdown. In cells which discharged in the previous cycle, charge deposited on the walls augments the applied voltage such that it is sufficient for breakdown. This behavior, referred to as the 'memory effect', is a principal advantage of the PDP since it obviates the need to address every cell during each refresh cycle, leading to lower cost driving circuitry. In the PDP, only cells which must undergo a change in state are addressed. The state of a cell is changed by augmenting the X–Y voltage of the refresh pulse with a voltage on the address electrode. For cells which were previously off, a write pulse is applied which results in initiation of breakdown. For cells which were previously on, an erase pulse drains the excess charge and returns the cell to it post-priming pulse state.

In this work, the initiation of breakdown in a surface discharge type PDP cell is examined. The breakdown may correspond to the priming pulse, a refresh pulse in an activated cell, or a write pulse, depending on the applied voltages and the wall charge configuration. Specifically, we seek a spatial map of discharge current amplification, which indicates the strength of the local breakdown process. This is analogous to constructing a spatial map of the Paschen curve.

In section 5.6.2, the model for initial breakdown is described, including the algorithm for the analysis of the initial breakdown. In section 5.6.3, breakdown for specific configurations is described. In section 5.6.3.1, the case of equally spaced electrodes and neighbor cells is discussed. In section 5.6.3.2, a case with large separation from neighbor cells is discussed. Finally, conclusions are presented in section 5.6.4.

5.6.2 The numerical model

Consider the two-dimensional model shown in figure 5.6.2. Assume the gap, d, is filled with a gas comprising neon and xenon at a pressure p. The spacing between the sustain electrodes is g_d. The dielectric thickness between the sustain electrode and gap is given by d_1, and the thickness of the dielectric between the address electrode and the plasma gap is given by d_2. The width of the sustain electrodes is w, and the distance from the Y electrode to the cell edge is $g_n/2$. The cell is periodic in the length, L.

The PDP cell configuration is modeled using a modified version of the XOOPIC particle-in-cell (PIC) code [1]. XOOPIC is a two-dimensional

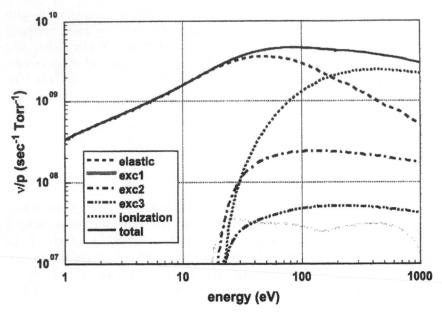

Figure 5.6.3. Normalized collision frequency for electron–neon collisions.

PIC code which includes both electrostatic and electromagnetic models in both axisymmetric and Cartesian coordinates. XOOPIC also includes a Monte Carlo collision model which can handle non-interacting gas mixtures, including elastic, excitation, ionization, and charge exchange collisions. For the work here, XOOPIC is operated in electrostatic mode in Cartesian coordinates.

The code includes a Monte Carlo collision (MCC) model including electron–neutral elastic scattering, electron–neutral excitation, and electron–neutral impact ionization [2]. The electron-neon momentum transfer cross section at low energies is from [3], and at high energies from [4]. The electron–xenon momentum transfer cross section at low energies was taken from [5], and at high energies from [6]. The electron–neon and electron–xenon excitation cross sections are taken from [7], except the grouped neon metastable level ($^3P_2$ and $^3P_0$) is taken from [8]. The electron–neon ionization cross sections are from [9] at low energy and [10] at high energy. The electron–xenon ionization cross sections are from [11]. Only direct ionization of the ground state is modeled here. The normalized electron–neutral collision frequencies in neon are shown in figure 5.6.3, and those for xenon are shown in figure 5.6.4.

The usual PIC-MCC scheme was modified to perform the calculation here, as illustrated in figure 5.6.5. First, Laplace's equation was solved for the vacuum configuration with specified electrode potentials to obtain

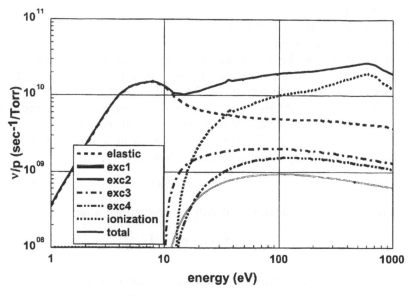

Figure 5.6.4. Normalized collision frequency for electron–xenon collisions.

$\Phi(x, y)$. The resulting electric field, $E(x, y) = -\nabla\Phi(x, y)$, was held fixed. Next, secondary electrons were released from a single point x_0 along the dielectric surface below the positively biased y electrode. The initial release point was scanned across the surface bounded by the midpoint between positively and negatively biased electrodes, $x_1 \leq x_0 \leq x_2$, as shown in

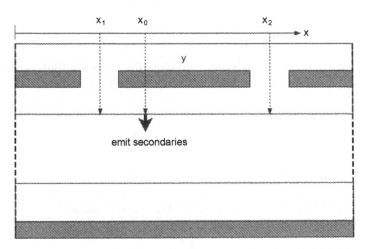

Figure 5.6.5. Schematic of a single coplanar PDP cell used for the initial breakdown calculation.

figure 5.6.5. The orbits were integrated for the released secondary electrons, also applying the MCC model. However, the space charge of the electron population was neglected during the calculation, since the density is low during the onset of the discharge. The integration of the equations of motion and MCC operation are performed until all the resulting particles have been collected at the surface to obtain the transfer function $f_i(x_0, x)$. No further secondary electrons are generated, although electrons and ions generated in ionization events are included in the calculation.

The ion distribution of species, f_i, collected at x due to the initial generation of secondary electron emission from x_0 is

$$\beta_{i,0}(x_0, x) = f_i(x_0, x).\tag{1}$$

We can write an approximate condition for breakdown when

$$\sum_i \gamma_i \beta_{i,0}(x_0, x_0) > 1\tag{2}$$

where γ_i is the secondary emission coefficient for impact of ion species, i, with the wall.

When equation (2) is satisfied, each secondary electron emitted at x_0 generates sufficient return ion flux at x_0 to emit more than one secondary in the next generation, leading to net growth of the discharge current at the point x_0. While satisfying equation (2) is sufficient to initiate breakdown, it is not necessary; a more complete breakdown condition should include not just the next generation, but all future generations in the secondary electron–ionization-ion wall flux cycle.

For a secondary coefficient, γ_i, the flux of the next generation of secondaries at x due to an initial emission at x_0 is

$$\Gamma_1(x_0, x) = \sum_i \gamma_i \beta_{i,0}(x_0, x).\tag{3}$$

These electrons then accelerate through the cathode fall, generating additional ionization events. The ions return to the dielectric surface, coating the cathode, with a distribution corresponding to the point of emission. This leads to the collection of the next generation of ions at the dielectric due to emission from the initial point x_0 returning back to the point x:

$$\beta_{i,1}(x_0, x) = \int_{x_1}^{x_2} \left(\sum_i \gamma_i \beta_{i,1}(x_0, x) \right) f_i(x', x)\, dx'.\tag{4}$$

Similarly, the flux of the second generation of secondaries at x due to the initial emission from x_0 is given by

$$\Gamma_2(x_0, x) = \sum_i \gamma_i \beta_{i,1}(x_0, x).\tag{5}$$

We can now generalize the nth generation of secondary electrons emitted at x due to the initial emission from x_0:

$$\Gamma_n(x_0, x) = \sum_i \gamma_i \beta_{i,n-1}(x_0, x). \tag{6}$$

Similarly, the nth generation of ions collected per secondary electron emitted from x_0 can be written

$$\beta_{i,n}(x_0, x) = \int_{x_1}^{x_2} \left(\sum_i \gamma_i \beta_{i,n-1}(x_0, x') \right) f_i(x', x) \, dx'. \tag{7}$$

Breakdown occurs due to emission at x_0 when successive generations of secondary flux at x_0 are increasing:

$$\frac{\Gamma_{n+1}(x_0, x_0)}{\Gamma_n(x_0, x_0)} > 1. \tag{8}$$

5.6.3 Simulation results

The initial breakdown model was first applied to coplanar ac plasma display panel cells [12, 13]. Here we consider the initial breakdown in coplanar ac plasma display panel cells with a narrow neighbor gap and a wide neighbor gap. The geometric configuration of interest is the three-electrode cell, shown schematically in figure 5.6.2. The addressing electrode is labeled A, while the other electrodes are labeled x and y, respectively. The dimensions of the cell are length $L = 440 \, \mu m$ and height $d = 110 \, \mu m$. The dielectric coating on the address electrode was taken to be $d_2 = 25 \, \mu m$, with $\varepsilon_r = 7.9$. The x and y electrodes are embedded a distance $d_1 = 25 \, \mu m$ into a dielectric with $\varepsilon_r = 11$. The x and y electrodes are separated by a distance $g_d = 80 \, \mu m$.

A Neumann boundary condition is used at the top edge of the cell, so at the plane, $y = D$, the normal component of the electric field, $E_y = 0$. The left and right edges of the cell, $x = 0$ and $x = L$, are periodic. Between the top boundary and the x and y electrodes is $25 \, \mu m$ of dielectric. The secondary emission coefficients were taken to be $\gamma_{Ne} = 0.5$ and $\gamma_{Xe} = 0.05$.

The boundary condition at the bottom of the cell, $y = 0 \, \mu m$, is fixed by the address electrode voltage. The neighbor gap, g_n, was varied along with an opposite variation in the electrode width w such that the cell size, L, remains a constant. For the symmetric case, $w/g_d = 4.4$ and $g_n/g_d = 1$, which leads to equal spacing among all cells as shown in figure 5.6.6. For the asymmetric

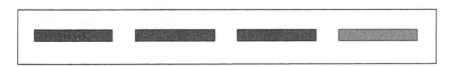

Figure 5.6.6. Schematic of symmetric spacing of X and Y electrodes.

Figure 5.6.7. Schematic of asymmetric spacing of X and Y electrodes.

case, $w/g_d = 2.9$ and $g_n/g_d = 4$, which leads to the spacing shown schematically in figure 5.6.7. Arbitrary electrode widths and neighbor gap separations can be studied using this technique. In both cases, the electrode voltages were $V_X = 160\,\mathrm{V}$, $V_Y = -160\,\mathrm{V}$, and $V_A = -80\,\mathrm{V}$.

First, the fields are solved for the initial (vacuum) condition to obtain $\Phi(x, y)$; in this case the fields are fixed throughout the run. This assumption is valid during the initial stages of breakdown, when the space charge is small. The Monte Carlo simulation is run, with the initial condition of 10^4 secondary electrons emitted from the location x_0 at cathode. These electrons are advanced in the fixed (vacuum) fields, undergoing collisions using the Monte Carlo algorithm. The electrons and ions created in ionizing collisions are also followed. When ions are absorbed at the cathode, they do not emit secondary electrons. Instead, the spatial distribution of the ion fluxes, $f_i(x_0, x)$, are collected along the dielectric surface beneath the cathode.

This process is repeated for initial emission points $x_1 \le x_0 \le x_2$. Hence, a map of the ion flux at the wall due to secondary electron emission from each point along the surface is generated.

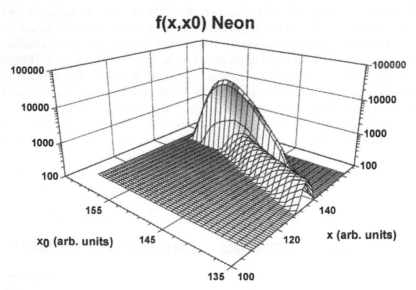

Figure 5.6.8. Neon ion flux distribution on the surface for the symmetric case.

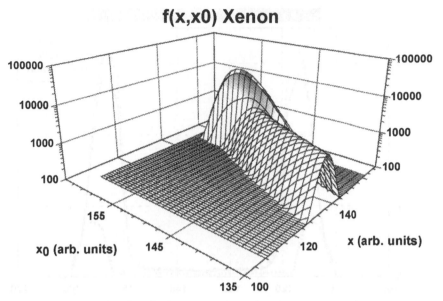

Figure 5.6.9. Xenon ion flux distribution on the surface for the symmetric case.

5.6.3.1 The case of symmetric gaps

The results of the Monte Carlo calculation for $f_i(x_0, x)$ for the symmetric case are shown in figures 5.6.8 and 5.6.9 for neon and xenon respectively. The plots can be understood by considering slices for a constant x_0, which indicate the returning ion distribution for emission from x_0. The ratio of β_1/β_0 is shown for the symmetric case in figure 5.6.10. Note that, for the specified conditions, the breakdown is initiated symmetrically at the edges between the neighboring electrodes.

5.6.3.2 The case of asymmetric gaps

The results of the Monte Carlo calculation for $f_i(x_0, x)$ for the asymmetric case are shown in figures 5.6.11 and 5.6.12 for neon and xenon respectively. As before, the plots can be understood by considering slices for a constant x_0, which indicate the returning ion distribution for emission from x_0.

The ratio of β_1/β_0 is shown for the asymmetric case in figure 5.6.13. Note that, for the specified conditions, the breakdown is initiated between the X and Y electrodes only, since the gaps between neighboring cells effectively eliminate inter-cell breakdown.

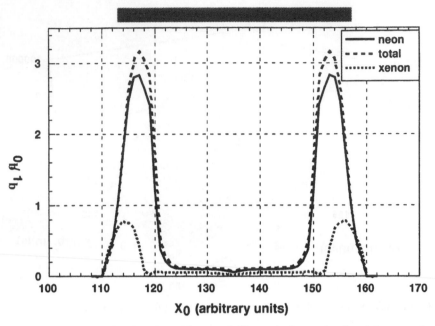

Figure 5.6.10. β ratio for the symmetric case. $\beta_1/\beta_0 > 1$ indicates that breakdown can be initiated from the position x_0. The electrode is shown schematically to scale above the figure.

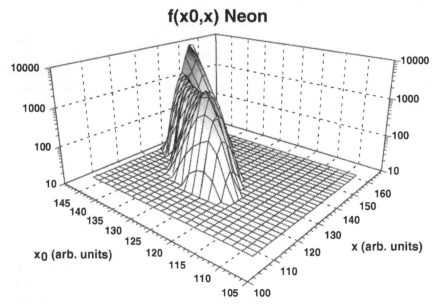

Figure 5.6.11. Neon ion flux distribution on the surface for the asymmetric case.

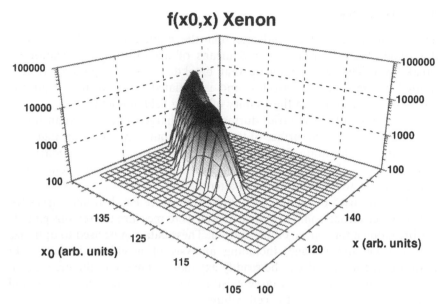

Figure 5.6.12. Xenon ion flux distribution on the surface for the asymmetric case.

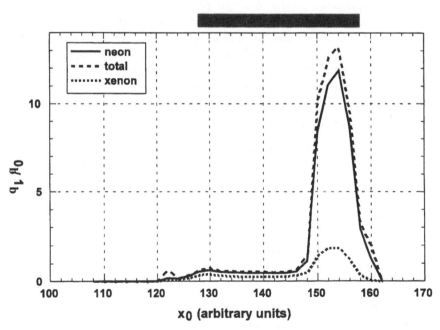

Figure 5.6.13. β ratio for the asymmetric case. $\beta_1/\beta_0 > 1$ indicates that breakdown can be initiated from the position x_0. The electrode is shown schematically to scale above the figure.

5.6.4   Discussion

The results of this study indicate that the numerical modeling method described above provides a rapid means of determining the location of breakdown. The results indicate that breakdown is only possible over a limited region of the electrodes, and is initiated most strongly near the edges of the electrodes in the vicinity of strong field gradients.

Charging of the dielectrics during the discharge will cause expansion of the discharge along the surface of the dielectric, but only within the region in which the amplification factor exceeds the inverse of the secondary coefficient. Note that this result may be modified when sufficient space charge and/or wall charge exists to alter $\Phi(x, y)$.

It is proposed to use this technique to measure Paschen-like curves for particular electrode configurations, as well as to measure the regions eligible for breakdown for a given configuration. These data can be used to optimize gap spacing and voltage, including analysis of neighbor discharge. In addition, the technique can be readily expanded to measure the breakdown conditions for a cell with charge existing on the dielectric surface, as well as fixed charge density in the cell volume.

The initial breakdown method described here can be extended to arbitrary geometric constructions as well as arbitrary gas chemistries. Extending the initial breakdown model to an air plasma, for example, would require adding a model for the air-plasma reactions which contribute to significant electron and ion energy loss as well as ionization paths. Inclusion of the full set of reactions is in principle possible, although the computation can become significant compared to the present calculation which can be done in less than an hour on a commodity computer.

5.6.5   Acknowledgments

This work supported in part by Hitachi Ltd. The author gratefully acknowledges the advice and support of C K Birdsall, Y Ikeda, and P J Christenson.

References

[1] Verboncoeur J P, Langdon A B and Gladd N T 1995 'An object-oriented electromagnetic PIC code' *Computer Phys. Commun.* **87** 199
[2] Vahedi V and Surendra M 1995 'Monte Carlo collision model for particle-in-cell method: Application to argon and oxygen discharges', *Computer Phys. Commun.* **87** 179
[3] Robertson A G 1972 *J. Phys.* B **5** 648
[4] Shimamura I 1989 *Scientific Papers Inst. Phys. Chem. Res.* **82**
[5] Hunter S R, Carter J G and Christophorou L G 1988 *Phys. Rev. A* **38** 5539
[6] Hayashi M 1983 *J. Phys. D* **16** 581

[7] Peuch V and Mizzi S 1991 *J. Phys. D* **24** 1974

[8] Mason N J and Newell W R 1987 *J. Phys. B* **20** 1357

[9] Wetzel R C, Baiocchi F A, Hayes T R and Freund R S 1987 *Phys. Rev. A* **35** 559

[10] de Heer F J, Jansen R H J and van der Kaay W 1979 *J. Phys. B* **12** 979

[11] Rapp D and Englander-Golden P 1965 *J. Chem. Phys.* **43** 1464

[12] Verboncoeur J P, Christenson P J and Cartwright K L 1997 'Breakdown in a 3-electrode ac plasma display panel'. *Proc. 50th Annual Gaseous Electronics Conf.* **42** 1739

[13] Verboncoeur J P 1998 'Initiation of breakdown in a 3-electrode plasma display panel cell', *25th IEEE ICOPS*, Raleigh, NC

Chapter 6

DC and Low Frequency Air Plasma Sources

U Kogelschatz, Yu S Akishev, K H Becker, E E Kunhardt,
M Kogoma, S Kuo, M Laroussi, A P Napartovich, S Okazaki
and K H Schoenbach

6.1   Introduction

This chapter treats some more recent developments in the generation of non-equilibrium plasmas. Section 6.2 (Kogelschatz), 6.3 (Kogoma, Okazaki) and 6.4 (Laroussi) are devoted to different aspects of barrier discharges. In addition to the traditional dielectric barrier discharges with a seemingly random distribution of microdischarges, regularly patterned and homogeneous dielectric barrier discharges are also addressed, as well as resistive barrier discharges. The various novel applications in surface treatment, in flat plasma display panels, ozone generation, excimer lamps and high power CO_2 lasers have attracted much interest and have led to a worldwide increase in research activities in all kinds of barrier discharges.

Similar plasma conditions can also be obtained in microhollow cathode discharges (MHCDs) and in a variety of discharges spatially confined in small geometries (section 6.5 (Schoenbach, Becker, Kunhardt)). Of special interest is the capillary plasma electrode discharge (CPED) which uses a perforated dielectric with a large number of equally spaced holes.

Section 6.6 (Akishev, Napartovich) covers recent progress in the generation, modeling and understanding of steady state corona glow discharges. Section 6.7 (Kuo) describes a novel ac torch for the generation of non-equilibrium plasmas.

Many of the discharge types described in this chapter can be used to treat large surfaces or to generate large-volume atmospheric-pressure non-equilibrium plasmas (Kunhardt 2000). Also combinations of different discharge types like the barrier-torch discharge plasma source have been proposed (Hubička *et al* 2002). Current research focuses on dielectric barrier properties (surface structure, electron emission, surface and bulk conductivity) and on micro-structured electrodes, semiconductors or dielectrics to obtain arrays of miniature non-equilibrium plasmas (Miclea *et al* 2001, Park *et al* 2001).

References

Hubička M, Čada, M. Šicha M, Churpita A, Pokorný P, Soukop L and Jastrabík L 2002 *Plasma Sources Sci. Technol.* **11** 195
Kunhardt E E 2000 *IEEE Trans. Plasma Sci.* **28** 189
Miclea M, Kunze K, Musa G, Franzke J and Niemax K 2001 *Spectrochim. Acta B* **56** 37
Park S-J, Chen J, Liu C and Eden J G 2001 *Appl. Phys. Lett.* **78** 419

6.2 Barrier Discharges

Based on experience with ozone research, the major application for many decades, it was believed for a long time that dielectric-barrier discharges always exhibit many discharge filaments or microdischarges. This multifilament discharge with a seemingly random distribution of microdischarges is prevailing in atmospheric-pressure air or oxygen (Samoilovich *et al* 1989, 1997, Eliasson and Kogelschatz 1991, Kogelschatz *et al* 1997, Kogelschatz 2003). Work performed in many different gases under various operating conditions revealed that regularly patterned or diffuse barrier discharges can also exist at atmospheric pressure. The formation of regular discharge patterns, was observed for example by Boyers and Tiller (1982), Breazeal *et al* (1995), Guikema *et al* (2000), Klein *et al* (2001), and Dong *et al* (2003). The physical mechanism of pattern formation has been investigated in a series of papers of the Purwins group at Münster University (Radehaus *et al* 1990, Ammelt *et al* 1993, Brauer *et al* 1999, Müller *et al* 1999a,b). In 1968 Bartnikas reported that ac discharges in helium can also manifest pulse-less 'glow' and 'pseudo-glow' regimes, apparently homogeneous diffuse volume discharges, now often referred to as atmospheric pressure glow discharges (APG/APGD). A few years later this work was extended to discharges in nitrogen and air at atmospheric pressure (Bartnikas 1971). Early work on polymer deposition in pulsed homogeneous barrier discharges in an ethylene/helium mixture was reported by Donohoe and Wydeven (1979). Starting in 1987 the group of S. Okazaki and M. Kogoma at Sophia University in Tokyo (see section 6.3) reported on intensive investigations in homogeneous dielectric-barrier discharges and their applications and proposed the term

APG, short for atmospheric pressure glow discharge. The interesting physical processes in these discharges and their large potential for industrial applications have initiated experimental as well as theoretical studies in many additional groups in France (Massines *et al* 1992, 1998), in the US (see section 6.4), Canada (Nikonov *et al* 2001, Radu *et al* 2003a,b), in Germany (Salge 1995, Kozlov *et al* 2001, Tepper *et al* 2002, Wagner *et al* 2003, Brandenburg *et al* 2003, Foest *et al* 2003), in Russia (Akishev *et al* 2001, Golubovskii *et al* 2002, 2003a,b), and in the Czech Republic (Trunec *et al* 1998, 2001), to name only the most important ones. Much of the work on the physics of filamentary, regularly patterned and diffuse barrier discharges was recently reviewed by Kogelschatz (2002).

6.2.1 Multifilament barrier discharges

The traditional appearance of the barrier discharge used for ozone generation in dry air or oxygen (see section 9.3) or for surface modification of polymer foils in atmospheric air is characterized by the presence of a large number of current filaments or microdischarges (see also section 2.6). Figure 6.2.1 shows a photograph of microdischarges in atmospheric-pressure dry air taken through a transparent electrode.

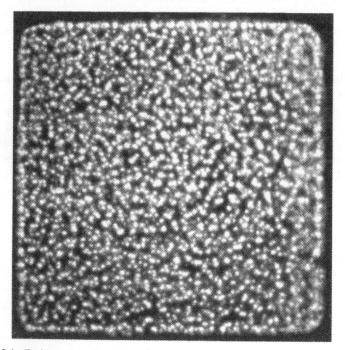

Figure 6.2.1. End-on view of microdischarges in a 1 mm gap with atmospheric-pressure dry air (original size: 6 cm × 6 cm, exposure time: 20 ms).

During the past decades important additional information was collected on the nature of these filaments. Early image converter recordings of microdischarges in air and oxygen were obtained by Tanaka *et al* (1978). Precise current measurements were performed on individual microdischarges (Hirth 1981, Eliasson *et al* 1987, Braun *et al* 1991). The transported charge and its dependence on dielectric properties was determined over a wide parameter range (Dřimal *et al* 1987, 1988, Gibalov *et al* 1991). Typically, many microdischarges are observed per square cm of electrode area. Their number density depends on the power dissipated in the discharge. For a moderate power density of $83 \, mW/cm^2$ about 10^6 microdischarges were counted per cm^2 per second (Coogan and Sappey 1996). The influence of humidity and that of ultraviolet radiation was investigated (Falkenstein 1997). In recent years spectroscopic diagnostics were refined to such a degree that measurements of species concentrations and plasma parameters inside individual microdischarges became feasible (Wendt and Lange 1998, Kozlov *et al* 2001, Lukas *et al* 2001). For a given configuration and fixed operating parameters all microdischarges are of similar nature. They are initiated at a well defined breakdown voltage, and they are terminated after a well defined current flow or charge transfer.

From all these investigations we conclude that each microdischarge consists of a nearly cylindrical filament of high current density and approximately $100 \, \mu m$ radius. At the dielectric surface(s) it spreads into a much wider surface discharge. These are the bright spots shown in figure 6.2.1. The duration of a microdischarge is limited to a few ns, because immediately after ignition local charge build up at the dielectric reduces the electric field at that location to such an extent that the current is choked. Each filament can be considered a self-arresting discharge. It is terminated at an early stage of discharge development, long before thermal effects become important and a spark can form. The properties of the dielectric, together with the gas properties, limit the amount of charge or energy that goes into an individual microdischarge. Typical charges transported by individual microdischarges in a 1 mm gap are of the order $100 \, pC$, typical energies are of the order $1 \, \mu J$. The plasma filament can be characterized as a transient glow discharge with an extremely thin cathode fall region with high electric field and a positive column of quasi-neutral plasma. The degree of ionization in the column is low, typically about 10^{-4}. As a consequence of the minute energy dissipation in a single microdischarge the local transient heating effect of the short current pulse is low, in air typically less than $10 \, °C$ in narrow discharge gaps. The average gas temperature in the discharge gap is determined by the accumulated action of many microdischarges, i.e. the dissipated power, and the heat flow to the wall(s) and from there to the cooling circuit. This way the gas temperature can remain low, even close to room temperature, while the electron energy in the microdischarges is a few eV. Major microdischarge properties of a DBD in a 1 mm air gap are summarized in table 6.2.1.

Table 6.2.1. Characteristic micro-discharge properties in a 1 mm gap in atmospheric-pressure air.

Duration	1–10 ns	Total charge	0.1–1 nC
Filament radius	about 0.1 mm	Electron density	10^{14}–10^{15} cm^{-3}
Peak current	0.1 A	Electron energy	1–10 eV
Current density	100–1000 A cm^{-2}	Gas temperature	Close to average gap temperature

In addition to limiting the amount of charge and energy that goes into an individual microdischarge, the dielectric barrier serves another important function in DBDs. It distributes the microdischarges over the entire electrode area. As a consequence of deposited surface charges the field has collapsed at locations where microdischarges already occurred. As long as the external voltage is rising, additional microdischarges will therefore preferentially ignite in other areas where the field is high. If the peak voltage is high enough, eventually the complete dielectric surface will be evenly covered with footprints of microdischarges (surface charges). This is the ideal situation which leads to the almost perfect voltage charge parallelogram shown in figure 2.6.4. The deposited charges constitute an important memory effect that is an essential feature of all dielectric barrier discharges.

As far as applications are concerned each individual microdischarge can be regarded as a miniature non-equilibrium plasma chemical reactor. Recent research activities have focused on tailoring microdischarge characteristics for a given application by making use of special gas properties, by adjusting pressure and temperature, and by optimizing the electrode geometry as well as the properties of the dielectric(s). Such investigations can be carried out in small laboratory experiments equipped with advanced diagnostics. One of the major advantages of BDBs is that, contrary to most other gas discharges, scaling up presents no major problems. Increasing the electrode area or increasing the power density just means that more microdischarges are initiated per unit of time and per unit of electrode area. In principle, individual microdischarge properties are not altered during up-scaling. Efficient and reliable power supplies are available ranging from a few hundred watts in a plasma display panel, close to 100 kW in an apparatus for high speed surface modification of polymer foils to some MW in large ozone generators.

6.2.2　Modeling of barrier discharges

Numerical modelling efforts have been devoted to describing the physical processes and chemical reactions in a single filament, in adjacent filaments, in a temporal sequence of many filaments and, more recently, in diffuse dielectric-barrier discharges. The problem of modeling the initial phases of a single microdischarge has many similarities with that of treating break-down. Depending on the external voltage, the gap width and the pressure, breakdown can be accomplished either by the Townsend mechanism of successive electron avalanches or by a much faster streamer breakdown (see section 2.4). As soon as a conductive channel is formed and the current in the microdischarge increases, the presence of the dielectric gains a strong influence on further discharge development and on the termination of the current flow. This necessitated the incorporation of additional boundary conditions to adequately treat charge accumulation and distribution on the dielectric surface(s). Early attempts were reported by Gibalov *et al* (1981). With the development of refined numerical algorithms and the availability of faster computers full two-dimensional treatment of a single micro-discharges became possible (Egli and Eliasson 1989, Braun *et al* 1991, 1992, Li and Dhali 1997, Steinle *et al* 1999, Gibalov and Pietsch 2000). In most cases the continuity equations for the major involved species are solved simultaneously with Poisson's equation to determine the electric field due to space charge (see also section 5.3). Secondary effects on the cathode are normally included, in some cases also photo-ionization. Nikonov *et al* (2001) suggested that in gaps wider than 0.02 cm the photo-ionization contribution to the electron density becomes more significant in comparison to the cathode photoemission. In many cases the role of photo-ionization in numerical simulations is approximated by assuming an equivalent density of seed electrons, about 10^7 to $10^8 \, cm^{-3}$, in the background gas (Dhali and Williams 1987). Microdischarge simulations could reproduce measured results about diameter, temporal current variation and transferred charge. They also helped considerably improving our understanding of the physical processes involved.

Steinle *et al* (1999) used a two-dimensional model to predict micro-discharge development in a 0.35 mm wide gap bounded by a metal cathode and a dielectric covered anode in atmospheric pressure air. Their current pulse, reproduced in figure 6.2.2, clearly shows the different phases of the discharge. At 0.54 ns we already have a space charge dominated avalanche phase followed by a streamer phase. The peak current of the microdischarge is preceded by the formation of a cathode fall region, a process that takes only a fraction of a nanosecond. After reaching the peak, within 0.3 ns, the current is already reduced to half of its maximum value. This clearly shows the strong current-choking action of the field reduction caused by charges deposited on the dielectric surface. The development and the radial extension of the cathode fall region was simulated in detail also by Gibalov and Pietsch (2000). Its thickness is less than 20 μm and the maximum field strength, according to this model, reaches over 4000 Td ($1 \, Td = 10^{-21} \, V \, m^2$). Figure 6.2.3 shows the extension of the axial field strength close to the cathode in air at atmospheric pressure. Cathode fall voltage, thickness and current density roughly correspond to values

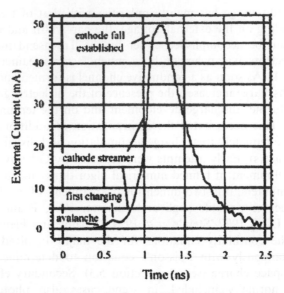

Figure 6.2.2. Computed current pulse for a 0.35 mm gap in atmospheric pressure air (Steinle *et al* 1999).

extrapolated from low-pressure discharges using the similarity laws of the normal glow discharge described in section 2.4. This high current phase of a microdischarge can be regarded as a quasi-stationary high-pressure glow discharge. Such conditions are ideal to induce chemical changes, for example ozone formation or air pollution control. It has also been attempted to model the interaction of adjacent microdischarges (Xu and Kushner 1998).

In many papers the equations treating microdischarge dynamics have been coupled with extensive chemical codes to follow chemical changes.

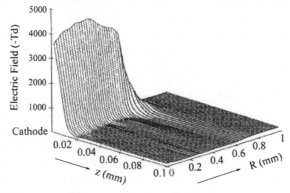

Figure 6.2.3. Numerical simulation of the cathode layer of a microdischarge in a 1 mm atmospheric-pressure air gap (Gibalov and Pietsch 2000).

Since chemical reactions may require longer time to approach equilibrium than the typical duration of a microdischarge, this normally requires the simulation of a large number of microdischarges with a given repetition rate (Eliasson *et al* 1991, 1993, 1994, Gentile and Kushner 1996, Dorai and Kushner 2001). With these tools it became feasible to correlate discharge parameters and volume flow rate to the speed of chemical changes in the gas flow. Recently it has also been attempted to compute the influence of small additives (Niessen *et al* 1998, Dorai and Kushner 2000), of solid particles (Dorai *et al* 2000) and of chemical changes on polymer surfaces (Dorai and Kushner 2003).

With the important and somewhat unexpected experimental advances in the control of diffuse barrier discharges (sections 6.3 and 6.4) one-dimensional numerical modelling of these discharges became an important issue (Massines *et al* 1998, Tochikubo *et al* 1999, Golubovskii *et al* 2002, 2003a). Concentrating mainly on He and N_2 it was soon established that discharge modes resembling a Townsend discharge as well as a glow discharge can be obtained. The Townsend mode is characterized by extremely low current density, negligible influence of space charge and the absence of a quasi-neutral plasma. Typically the ion density is orders of magnitude higher than the electron density, which shows exponential growth from cathode to anode. The glow mode, on the other hand, reaches higher current densities (of the order mA/cm^2). It is influenced by space charge effects leading to a high field region at the cathode, a Faraday dark space with vanishing field and a column of quasi-neutral plasma at current maximum.

These one-dimensional fluid models for atmospheric-pressure discharges bounded by dielectric barriers could produce some of the experimental results, e.g. that the glow-like mode can preferentially be obtained if the gap is sufficiently wide and the barrier is thin or of high dielectric constant. Also the experimental findings of obtaining one current pulse or multiple current pulses per half wave of the feeding voltage can be reproduced by relatively simple models (Akishev *et al* 2001, Golubovskii *et al* 2003a). To exactly reproduce details of measured current pulses it was necessary to introduce additional processes. For example it was found that computations using the ionization coefficient of pure He were not able to reproduce the experimental results. Some low level impurities like Ar (Massines *et al* 1998) or N_2 (Golubovskii *et al* 2003a) had to be introduced to get a better match. Molecular ions He_2^+, He_3^+, N_4^+ had to be considered to get faster recombination. It was also established that there must be a mechanism releasing electrons from the dielectric surface stored in the previous voltage half wave. Models assuming a constant electron desorption rate (Golubovskii *et al* 2003a) or introducing a large γ coefficient ($\gamma = 0.5$) for secondary electron emission by impinging metastables (Khamphan *et al* 2003) achieved better agreement with experimental results. It is apparent that knowledge is still lacking about the fundamental physical processes at dielectric surfaces, namely emission, desorption and recombination of charged particles. Going to two-dimensional models it could be shown that the Townsend discharge in DBDs is immune to filamentation while the glow discharge is inherently unstable (Golubosvkii *et al* 2003b). The situation is comparable to that investigated by Kudryavtsev and Tsendin (2002) between metal electrodes. They could show that a glow discharge operated to the right of the Paschen minimum is inherently unstable. It should be pointed out that the current densities so far reached in diffuse discharges between dielectric barriers are still much lower than those expected for a normal glow discharge at atmospheric pressure (roughly $2\,A/cm^2$ in He and $200\,A/cm^2$ in N_2). To reach those values much thinner dielectrics with higher dielectric constants and/or higher voltage rise times dU/dt are required. With fast pulsing techniques this should be possible.

References

Akishev Yu S, Dem'yanov A V, Karal'nik V B, Pan'kin M V and Trushkin N I 2001 *Plasma Phys. Rep.* **27** 164

Ammelt E, Schweng D and Purwins H-G 1993 *Phys. Lett. A* **179** 348

Bartnikas R 1968 *Brit. J. Appl. Phys. (J. Phys. D) Ser. 2* **1** 659

Bartnikas R 1969 *J. Appl. Phys.* **40** 1974

Bartnikas R 1971 *IEEE Trans. Electr. Insul.* **6** 63

Boyers D G and Tiller W A 1982 *Appl. Phys. Lett.* **41** 28

Brandenburg R, Kozlov K V, Morozov A M, Wagner H-E and Michel P 2003 *Proc. 26th Int. Conf. on Phenomena in Ionized Gases (XXVI ICPIG)* (Greifswald, Germany) http://www.icpig.uni-greifswald.de/

Brauer I, Punset C, Purwins H-G and Boeuf J P 1999 *J. Appl. Phys.* **85** 7569

Braun D, Gibalov V and Pietsch G 1992 *Plasma Sources Sci. Technol.* **1** 166

Braun D, Küchler U and Pietsch G 1991 *J. Phys. D: Appl. Phys.* **24** 564

Breazeal W, Flynn K M and Gwinn E G 1995 *Phys. Rev. E* **52** 1503

Coogan J J and Sappey A D 1996 *IEEE Trans. Plasma Sci.* **24** 91

Dhali S K and Williams P F 1987 *J. Appl. Phys.* **62** 4696

Dong L, Yin Z, Li X and Wang L 2003 *Plasma Sources Sci. Technol.* **12** 380

Donohoe K G and Wydeven T 1979 *J. Appl. Polymer Sci.* **23** 2591

Dorai R and Kushner M J 2000 *J. Appl. Phys.* **88** 3739

Dorai R and Kushner M J 2001 *J. Phys. D: Appl. Phys.* **34** 574

Dorai R and Kushner M J 2003 *J. Phys. D: Appl. Phys.* **36** 666

Dorai R, Hassouni K and Kushner M J 2000 *J. Appl. Phys.* **88** 6060

Dřimal J, Gibalov V I and Samoilovich V G 1987 *Czech. J. Phys.* **B 37** 1248

Dřimal J, Kozlov K V, Gibalov V I and Samoylovich V G 1988 *Czech. J. Phys.* **B 38** 159

Egli W and Eliasson B 1989 *Helvet. Phys. Acta* **62** 302

Eliasson B and Kogelschatz U 1991 *IEEE Trans. Plasma Sci.* **19** 309

Eliasson B, Hirth M and Kogelschatz U 1987 *J. Phys. D: Applied Phys.* **20** 1421

Eliasson B, Simon F-G and Egli W 1993 *Non-Thermal Plasma Techniques for Pollution Control* (Penetrante B M and Schultheis S E, eds), NATO ASI Series G: Ecological Sciences, Vol. 34, Part B (Berlin: Springer) pp 321–337

Eliasson B, Egli W and Kogelschatz U 1994 *Pure Appl. Chem.* **66** 1275

Falkenstein Z 1997 *J. Appl. Phys.* **81** 5975

Foest R, Adler F, Sigeneger F and Schmidt M 2003 *Surf. Coat. Technol.* **163/164** 323

Gentile A C and Kushner M J 1996 *J. Appl. Phys.* **79** 3877

Gibalov V I and Pietsch G J 2000 *J. Phys. D: Appl. Phys.* **33** 2618

Gibalov V I, Dřímal J, Wronski M and Samoilovich V G 1991 *Contrib. Plasma Phys.* **31** 89

Gibalov V I, Samoilovich V G and Filippov Yu V 1981 Russ. *J. Phys. Chem.* **55** 471

Golubovskii Yu B, Maiorov V A, Behnke J and Behnke J F 2002 *J. Phys. D: Appl. Phys.* **35** 751

Golubovskii Yu B, Maiorov V A, Behnke J and Behnke J F 2003a *J. Phys. D: Appl. Phys.* **36** 39

Golubovskii Yu B, Maiorov V A, Behnke J and Behnke J F 2003b *J. Phys. D: Appl. Phys.* **36** 975

Guikema J, Miller N, Niehof J, Klein M and Walhout M 2000 *Phys. Rev. Lett.* **85** 3817

Hirth M 1981 *Beitr. Plasmaphys.* **21** 1 (in German)

Khamphan C, Ségur P, Massines F, Bordage M C, Gherardi N and Cesses Y 2003 *Proc. 16th Int. Symp on Plasma Chem. (ISPC-16)* (Taormina, Italy)

Klein M, Miller N and Walhout M 2001 *Phys. Rev. E* **64** 026402-1

Kogelschatz U 2002 *IEEE Trans. Plasma Sci.* **30** 1400

Kogelschatz U 2003 *Plasma Chem. Plasma Process.* **23** 1

Kogelschatz U, Eliasson B and Egli W 1997 *J. de Phys. IV (France)* **7** C4-47

Kozlov K V, Wagner H-E, Brandenburg R and Michel P 2001 *J. Phys. D: Appl. Phys.* **34** 3164

Kudryavtsev A A and Tsendin L D 2002 *Tech. Phys. Lett.* **28** 1036

Li J and Dhali S K 1997 *J. Appl. Phys.* **82** 4205

Lukas C, Spaan M, Schulz-von der Gathen V, Thomson M, Wegst R, Döbele H F and Neiger M 2001 *Plasma Sources Sci. Technol.* **10** 445

Massines F, Mayoux C, Messaoudi R, Rabehi A and Ségur P 1992 *Proc. 10th Int. Conf. on Gas Discharges and Their Applications (GD-92)* (Swansea) Williams W T Ed 730

Massines F, Rabehi A, Decomps P, Gadri R B, Ségur P and Mayoux C 1998 *J. Appl. Phys.* **83** 2950

Müller I, Punset C, Ammelt E, Purwins H-G and Boeuf J-P 1999a *IEEE Trans. Plasma Sci.* **27** 20

Müller I, Ammelt E and Purwins H-G 1999b *Phys. Rev. Lett.* **82** 3428

Niessen W, Wolf O, Schruft R and Neiger M 1998 *J. Phys. D: Appl. Phys.* **31** 542

Nikonov V, Bartnikas R and Wertheimer M R 2001 *J. Phys. D: Appl. Phys.* **34** 2979

Radehaus C, Dohmen R, Willebrand H and Niedernostheide F-J 1990 *Phys. Rev. A* **42** 7426

Radu I, Bartnikas R and Wertheimer M R 2003a *J. Phys. D: Appl. Phys.* **36** 1284

Radu I, Bartnikas R, Czeremuszkin G and Wertheimer M R 2003b *IEEE Trans. Plasma Sci.* **31** 411

Salge J 1995 *J. de Phys. IV (France)* **5** C5-583

Samoilovich V G, Gibalov V I and Kozlov K V 1997 *Physical Chemistry of the Barrier Discharge* (Düsseldorf: DVS-Verlag) (Conrads J P F and Leipold F eds), Original Russian Edition, Moscow State University 1989

Steinle G, Neundorf D, Hiller W and Pietralla M 1999 *J. Phys. D: Appl. Phys.* **32** 1350

Tanaka M, Yagi S and Tabata N 1978 *Trans. IEE of Japan* **98A** 57

Tepper J, Li P and Lindmayer M 2002 *Proc. 14th Int. Conf on Gas Discharges and their Applications (GD-2002)* vol 1 (Liverpool: 2002) 175
Tochikubo F, Chiba T and Watanabe T 1999 *Jpn. J. Appl. Phys.* **38** Part 1 5244
Trunec D, Brablec A, Šťastný F and Bucha J 1998 *Contrib. Plasma Phys.* **38** 435
Trunec D, Brablec A and Buchta J 2001 *J. Phys. D: Appl. Phys.* **324** 1697
Wagner H-E, Brandenburg R, Kozlov K V, Sonnenfeld A, Michel P and Behnke J F 2003 *Vacuum* **71** 417
Wendt R and Lange H 1998 *J. Phys. D: Appl. Phys.* **31** 3368
Xu X P and Kushner M J 1998 *J. Appl. Phys.* **84** 4153

6.3   Atmospheric Pressure Glow Discharge Plasmas and Atmospheric Pressure Townsend-like Discharge Plasmas

6.3.1   Introduction

In 1987, Okazaki and Kogoma (Kanazawa *et al* 1987) developed a new plasma in He, which they referred to as atmospheric pressure glow (APG) discharge plasma. However, Okazaki and Kogoma did not provide sufficient evidence to prove that the plasma was really a glow discharge. Many researchers had doubts about whether or not the plasma was in fact a glow discharge plasma and have referred to this type of plasma by many other names such as GSD (glow silent discharge), GDBD (glow dielectric barrier discharge) at atmospheric pressure, APGD (atmospheric pressure glow discharge), DBD diffuse barrier discharge and homogeneous barrier discharges at atmospheric pressure (Massines *et al* 2003, Khamphan *et al* 2003, Trunec *et al* 2001, Brandenburg *et al* 2003, Tepper *et al* 2002). Recently, Massines *et al* (2003) demonstrated that the glow-like plasma in He in our discharge configuration was indeed a sub-normal glow discharge, which is very similar to a normal glow discharge.

By contrast, Massines *et al* (2003) found that the same discharge in N_2 is Townsend-like and thus different from a normal glow discharge. Studies of this Townsend-like discharge in nitrogen are continued with fine mesh electrodes (Buchta *et al* 2000, Tepper *et al* 2002). Based on these findings, it is justified to distinguish the discharge plasmas in the two gases, He and N_2, and call one an APG discharge plasma (He) and the other one an atmospheric-pressure Townsend-like (APT) discharge plasma.

Since the semiconductor industry has achieved great success using plasma processing, e.g. in the manufacture of microchips, research into plasma processing has increased significantly worldwide. However, essentially all plasmas used in semiconductor processing are low-pressure plasmas. On the other hand, there are many applications where the vacuum enclosure required for a low-pressure plasma is an obstacle for its technological use.

For instance, the high-speed continuous treatment of sheet-like materials is impossible using a low-pressure plasma. Similarly, materials with a high vapor pressure cannot readily be exposed to a low-pressure plasma or a long soft plastic tube may require plasma treatment of the inner surface, but a low-pressure plasma cannot be generated in the interior of the soft plastic tube. As a consequence, the development of glow discharges at atmospheric pressure has become an urgent need in many areas. At the same time, known discharges at atmospheric pressure (for example sparks, barrier discharges, and arc discharges) could not be used for surface treatment, because they are not homogeneous. The earliest account of a glow discharge at atmospheric pressure is in a paper by von Engel *et al* (1933) where the authors used cooled metal electrodes in hydrogen gas. Thus, atmospheric-pressure glow discharges have been generated for some time, but the principles of their generation and maintenance were never thoroughly researched until recently.

Our group was among the first to develop a stable homogeneous glow discharge at atmospheric pressure and our results are described in the following sections.

6.3.2 Realization of an APG discharge plasma

6.3.2.1 Three conditions for stabilizing APG discharges

Three conditions (Yokoyama *et al* 1990) are generally needed to succeed in producing a stable APG plasma.

(a) The presence of solid dielectric material between discharge electrodes.
(b) A suitable gas passing between the electrodes.
(c) The electric source frequency above 1 kHz.

However, there are situations where not all three conditions are needed.

(a) The first condition: dielectric material

The dielectric material assists pulse formation at low frequencies of the applied voltage in the same way as in an ozone generator (ozonizer), in which many fine filamentary discharges are generated on the dielectric plate. In order to generate an APG discharge, the next two conditions have to be met as well.

Figure 6.3.1 shows a system that has fine metal mesh electrodes. When this mesh size is about 350–400#, a stable discharge, which we believe to be an APT discharge, will be generated even though the other two conditions are not satisfied (Okazaki *et al* 1993). For example, in nitrogen, which is not a gas included in the group of gases that satisfy the second condition (see below), the mesh electrodes can generate a very stable, homogeneous

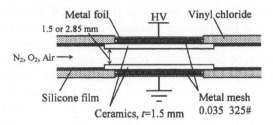

Figure 6.3.1. Parallel plate type plasma generator with fine mesh electrodes.

glow plasma at atmospheric pressure and at 50 Hz applied voltage (which is also outside the range of frequencies that meet the third condition), but the gap distance between the two dielectric plates for stable operation is only about 2–3 mm. When only one electrode is covered with a dielectric plate and when the other electrode consists of many metal needles, an APG discharge will be generated (Kanazawa *et al* 1988). This type of plasma can be used at higher energy than is possible with dielectric plate electrodes on both electrodes. However, the stability of an APG discharge with a multi-needle electrode is lower than in that with conventional electrodes.

If a high frequency excitation source is used, pulse formation caused by charging of the dielectric plate as in the case of a low-frequency source is not important, but the presence of the dielectric plate prevents the build-up of high concentration of discharges.

(b) The second condition: a suitable choice of gas

The use of He as a feed gas, when the first condition is met (i.e. with a dielectric plate inserted between the electrodes) and when the third condition met (i.e. when a high frequency source above 1 kHz is used) will result in the generation of an APG discharge (Yokoyama *et al* 1990, Kanazawa *et al* 1988). Other suitable gases such as Ar+ketone at ppm concentrations or Ar+methane at ppm levels can also be used (Okazaki *et al* 1991). The use of pure Ar gas with the first and third conditions met did not result in a stable APG discharge. The plasma formed was a 'mixture' of a glow-like plasma with a small number of filamentary discharges. However, the addition of an extremely small concentration of any ketone changed this plasma to a stable, uniform glow discharge plasma, whose stability was far higher than that of a He plasma. However, ketones include oxygen atoms, which are often undesirable. Thus, in order to remove oxygen completely from the system, a mixture of methane and Ar was used. The stability of a plasma using a mixture of methane and Ar is, however, lower than that of a ketone–Ar mixture. It has been suggested that plasmas in mixtures containing mostly noble gases are APG discharges (Massines *et al* 2003).

(c)   The third condition: the electric source frequency

In addition to satisfying the first and second conditions, the third condition regarding the frequency of the electric source originally stipulated that the frequency be above 3 kHz. Subsequently, after 1990, we found that the frequency limit could be lowered to 1 kHz. It is only under very special circumstances that a stable APG plasma can be generated at low frequencies, for example around 50–60 Hz, unless very fine mesh electrodes were used.

The APG discharge plasmas are generated in the form of very sharp and narrow discharge current pulses because of the presence of the dielectric. In particular, the APG plasma pulse is generated with a very high frequency, which has no direct relationship to the frequencies of the applied electric source. These discharge current pulses could be observed as a change of charges which passed across the gap between the electrodes.

The use of very high frequency sources, for example a few hundred kHz, can generate a stable APG or APT discharge plasma even in nitrogen without mesh electrodes. The pulse-modulated high frequency discharge can create a homogeneous glow style even from a very high-pressure system. This would be a high-temperature plasma, but its duration is very short.

6.3.2.2   Discharge currents styles and discharge mechanisms

The existence of a dielectric barrier between the electrodes is a common feature in the APG plasma (He), the APT plasma (N_2, perhaps the same for O_2 and air), and in an ozone generator (O_2, air). When a low-frequency source is applied, the form of the discharge current is quite different in terms of the number of pulses per half cycle of the applied voltage and the pulse duration.

Figure 6.3.2 shows the current pulses in an APG discharge in pure Ar and in an Ar–acetone mixture. It is interesting to note that we observed a

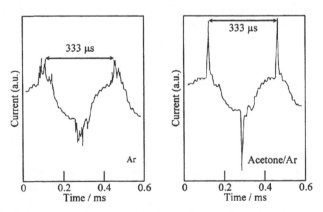

Figure 6.3.2. Pulse current of the APG discharge in pure Ar and acetone–Ar. 3 kHz, Ar 2000 slm, 2.0 kV (Ar), 1.0 kV (acetone–Ar).

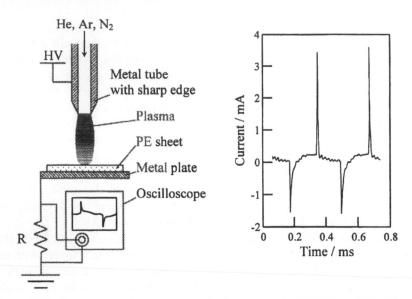

Figure 6.3.3. Downstream plasma at atmospheric pressure: R: 50 Ω, 3 kHz, 1.8 kV, length of plasma; 2 cm.

number of pulses in pure Ar, but only a single pulse per half cycle in the Ar–acetone mixture. We characterize an APG discharge as a discharge having a single pulse per half cycle. Using this criterion, the fine mesh electrode system was shown to have this unique current pulse frequency in all gases at 50 Hz (Okazaki *et al* 1993) and the plasmas generated are thus characterized as APG or APT discharge plasmas. In spray-type plasma treatment, when the outer electrode is located downstream as shown in figure 6.3.3, the current pulse was also observed to be one pulse per half cycle thus classifying the plasma as an APG or APT discharge plasma.

If a very high voltage is applied, the number of pulses per half cycle will increase. If the frequency of the applied voltage is low, e.g. between 50 Hz and 3 kHz, the analysis of the *I–V* characteristics of the discharge using a Lissajous figure on an oscilloscope can be used to establish the nature of the discharge.

The characteristic feature of APG and APT discharges of a single current pulse per half cycle of the applied voltage suggests that these discharges develop in a one shot from the entire surface of the dielectrics in each half cycle. The repetitive formation of filamentary discharges as seen in an ozone generator does not occur. This is a significant difference from the silent electric discharge and allows for the possibility of using the APG and the APT plasma for homogeneous surface treatment. A report of Kekez *et al* (1970) concluded that the transition time from a glow discharge to an arc discharge depends on the kind of gas, the gas pressure, the discharge gap, and the amount of over-voltage applied. Their finding supported our conclusions regarding the formation of an APG discharge, in particular the fact that a very short current pulse can produce a rapid succession of glow discharges.

Discharges using fine mesh electrodes were studied extensively by Trunec *et al* (1998) and Tepper *et al* (2002), but the fundamental mechanisms that generate and sustain the discharges are not yet completely clear and work in this area is continuing (Golubovskii *et al* 2002). Applications of discharges using mesh electrodes, which can generate a homogeneous glow in different gases, are being pursued by many groups and some unexpected and unexplained results have been reported. For example, it has been reported that the mesh has no effect at higher frequencies and, after several hours of operating an APG discharges, the discharge changes to a filamentary discharges. This transition can be reversed by using a new mesh. It seems there is a limit to the useful lifetime of the mesh electrodes (Buchta *et al* 2000).

6.3.3   Applications of APG discharge and APT discharge plasmas

Many technological applications of APG discharge plasmas have been pursued. However, in most applications feed gases and gas mixtures other than air have been used. Thus, these applications are outside the scope of this book and we refer the reader to the original references for more details on applications such as the surface modification of inner surfaces of tubes of polyvinylchloride and surface polymerization applications (Babukutty *et al* 1999, Okazaki and Kogoma 1993, Rzanek-Borocha *et al* 2002, Sawada *et al* 1995, Kojima *et al* 2001, Tanaka *et al* 2001), microwave heating of powders (Sugiyama *et al* 1998, Yamakawa *et al* 2003), exhaust gas treatment (Hong *et al* 2002), adhesive strength control and surface analysis (Nakamura *et al* 1991, Prat *et al* 1998), spray-type plasma applications at atmospheric pressure (Nagata *et al* 1989, Okazaki and Kogoma 1993, Taniguchi *et al* 1997, Tanaka *et al* 1999, Tanaka and Kogoma 2001), powder coating (Mori *et al* 1998, Nakajima *et al* 2001, Ogawa *et al* 2001), sterilization of cavities and surfaces (Japan patent 1994), and surface treatment of woolen fabrics (Okazaki and Kogoma 1999).

Perhaps the only application involving air is a marked improvement in the efficiency of ozone generators using the APG discharge plasma concept. The use of fine mesh metal electrodes in a dielectric barrier discharge produced a glow discharge at atmospheric pressure, even though it showed stability only for a very small gap distance of 2–3 mm, in air, N_2, O_2 and other gases. This gap distance, however, is sufficient for an ozone generator. The ozone formation efficiency in such a reactor was examined (Kogoma *et al* 1994) and an improvement in efficiency of about 20% over that of a conventional ozone generator was found. These results were confirmed by Buchta *et al* 2000 with respect to ozone formation concerning the use of the fine mesh metal electrodes also by Trunec *et al* 1998, Tepper *et al* 1998.

References

Babukutty Y, Prat Y, Endo K, Kogoma M, Okazaki S and Kodama M 1999 *Langmuir* **15** 7055

Brandenburg R, Wagner H-E, Michel P, Trunec D and Stahl D 2003 in *Proc. XXVIth Int. Conference on Phenomena in Ionized Gases*, Greifswald, Germany, vol 4, pp 45–46

Buchta J, Brablec A and Trunec D 2000 *Czech. J. Phys* **50/53** 273. Private discussion with the group

von Engel A, Seelinger R and Steenbeck M 1933 *Z. Phys.* **85** 144

Golubovskii Yu B, Maiorov V A, Behnke J and Behnke J F 2002 in *Proc. VIIIth Int. Symp. on High Pressure Low Temperature Plasma Chem.*, Puhajarve, Estonia, vol 1, pp 53–57

Hong J, Kim S, Lee K, Lee K, Choi J J and Kim Y K 2002 in *Proc. VIIIth Int. Symp. on High Pressure Low Temperature Plasma Chem.*, Puhajarve, Estonia, vol 2, pp 360–363

Japan Patent pending 1994 300911/1994

Japan Patent pending 2002 116459/2002

Kanazawa S, Kogoma M, Moriwaki T and Okazaki S 1987 in *Proc. 8th Int. Symp. on Plasma Chem.*, Tokyo, Japan, vol 3, pp 1839–1844

Kanazawa S, Kogoma M, Moriwaki T and Okazaki S 1988 *J. Phys. D: Appl. Phys.* **21** 838

Kekez M M, Barrault M R and Craggs 1970 *J. Phys. D: Appl. Phys.* **3** 1886

Khamphan C, Segur P, Massines F, Bordage M C, Gherardi N and Cesses Y 2003 in *Proc. 16th Int. Symp. on Plasma Chem.*, Taormina, Italy, p 181

Kogoma M and Okazaki S 1994 *J. Phys. D: Appl. Phys.* **27** 1985

Kojima I, Prat R, Babukutty Y, Kodama M, Kogoma M, Okazaki S and Koh Y J 2001 in *Proc. 15th Int. Symp. on Plasma Chem.*, Orleans, France, vol VI, pp 2391–2396

Massines F, Segur P, Gherardi N, Khamphan C and Ricard A 2003 *Surface and Coating Technology* **174-175C** 8. Private discussion with the group

Mori T, Tanaka K, Inomata T, Takeda A and M Kogoma 1998 *Thin Solid Films* **316** 89

Nagata A, Takehiro S, Sumi S, Kogoma M, Okazaki S and Horiike Y 1989 *Proc. Jpn. Symp. Plasma Chem.* vol 2, pp 109–115

Nakajima T, Tanaka K, Inomata T and Kogoma M 2001 *Thin Solid Films* **386** 208

Nakamura H, Kogoma M, Jinno H and Okazaki S 1991 Proc. *Jpn. Symp. Plasma Chem.*, vol 4, pp 339–344

Ogawa S, Takeda A, Oguchi M, Tanaka K, Inomata T and Kogoma M 2001 *Thin Solid Films* **386** 213

Okazaki S and Kogoma M 1993 *J. Photopolymer Sci. Tech.* **6** 339

Okazaki S and Kogoma M 1999 in *Proc. XXIVth Int. Conference Phenomena in Ionized Gases*, Warsaw, Poland, vol I, pp 123–124

Okazaki S, Kogoma M and Uchiyama H 1991 in *Proc. IIIrd Int. Symp. on High Pressure Low Temperature Plasma Chem.*, Strasburg, France, pp. 101–107

Okazaki S, Kogoma M, Uehara M and Kimura Y 1993 *J. Phys. D: Appl. Phys.* **26** 889

Prat R, Suwa T, Kogoma M and Okazaki S 1998 *J. Adhesion* **66** 163

Rzanek-Borocha Z, Schmidt-Szalowski K, Janowska J, Dudzinski K, Szymanska A and Misiak M 2002 in *Proc. VIIIth Int. Symp. on High Pressure Low Temperature Plasma Chem.*, Puhajarve, Estonia, vol 2, pp 415–419

Sawada Y Ogawa S and Kogoma. M 1995 *J. Phys. D: Appl. Phys.* **28** 1661

Sugiyama K, Kiyokawa K, Matsuoka H, Itoh A, Hasegawa K and Tsutsumi K 1998 *Thin Solid Films* **316** 117

Tanaka K and Kogoma M 2001 *Plasma and Polymers* **6** 27

Tanaka K, Inomata T and Kogoma M 1999 *Plasmas and Polymers* **4** 269

Tanaka K, Inomata T and Kogoma M 2001 *Thin Solid Films* **386** 217

Taniguchi K, Tanaka K, Inomata T and Kogoma M 1997 *J. Photopolymer Sci. Tech.* **10** 113

Tepper J, Li P and Lindmayer M 2002 in *Proc. XIVth Int. Conference on Gas Discharges and their Applications*, Liverpool, vol 1 pp 175–178

Tepper J, Lindmayer M and Salge J 1998 in Proc. *VIth Int. Symp. on High Pressure Low Temperature Plasma Chem.*, Cork, Ireland, pp 123–127

Trunec D, Brablec A and Stastny F 1998 in *Proc. VIth Int. Symp. on High Pressure Low Temperature Plasma Chem.*, Cork, Ireland, pp 313–317

Trunec D, Brablec A and Buchta J 2001 *J. Phys. D: Appl. Phys.* **34** 1697

Yamakawa K, Den S, Katagiri T, Hori M and Goto T 2003 in *Proc. 16th Int. Symp. on Plasma Chem.*, Taormina, Italy, p 832

Yokoyama T, Kogoma M, Moriwaki T and Okazaki S 1990 *J. Phys. D: Appl. Phys.* **23** 1125

6.4 Homogeneous Barrier Discharges

Recently, research on material processing by non-equilibrium atmospheric pressure plasmas witnessed a tremendous growth, both at the experimental and simulation levels. This was motivated by the new technical possibilities in generating relatively large volumes of non-equilibrium plasmas at or near atmospheric pressure, in numerous gases and gas mixtures and at low operating power budgets. Amongst the enabling technologies, the use of 'barrier discharges' has become very prevalent. this started with the use of the 'dielectric barrier discharge' (DBD) which was developed and improved upon over several decades (Bartnikas 1968, Donohoe 1976, Kogelschatz 1990, Kogelschatz *et al* 1997). DBDs use a dielectric material to cover at least one of the electrodes. The electrodes are driven by voltages in the kV range and at frequencies in the audio range (kHz). However, new methods emerged which extended the frequency range down to the dc level. The resistive barrier discharge (RBD) recently developed by Alexeff and Laroussi is such an example (Alexeff *et al* 1999, Laroussi *et al* 2002). The RBD uses a high resistivity material to cover the surface of at least one of the electrodes. It is capable of generating a large volume atmospheric pressure plasma with dc and ac (60 Hz) driving voltages.

The limitations of barrier-based discharges have traditionally been their non-homogeneous nature both in space and time. DBDs, for example, exhibit a filamentary plasma structure, therefore leading to non-uniform material treatment when used in surface modification applications. This situation led some investigators to search for operating regimes under which diffuse and homogeneous discharges can be produced. In the late 1980s and early 1990s, Okazaki's group published a series of papers where they presented their experimental findings regarding the conditions under which a DBD-based reactor can produce homogenous plasma, at atmospheric pressure (Okazaki *et al* 1993, Yokoyama *et al* 1990, Kanazawa *et al* 1988). Their work was soon followed by others (Massines *et al* 1992, 1996, 1998, Gherardi *et al* 2000, Roth *et al* 1992) who validated the fact that non-filamentary plasmas can indeed be produced by DBDs, an outcome not widely accepted by the research community active in this field at that time.

In this section, description of the work of several investigators will be presented. The electrical characteristics, ignition and extinction, stability, and homogeneity of the discharges will be discussed.

6.4.1 DBD-based discharges at atmospheric pressure

6.4.1.1 Experimental set-up

The dielectric barrier discharge (DBD) consists basically of two planar electrodes (sometimes co-axial or adjacent cylinders) made of two metallic plates (or tubes) covered by a dielectric material and separated by a variable gap (see figure 6.4.1). When operated at atmospheric pressure, the electrodes are energized by a high voltage power supply with typical voltages in the 1–20 kV range, at frequencies ranging from a few hundred Hz to a few

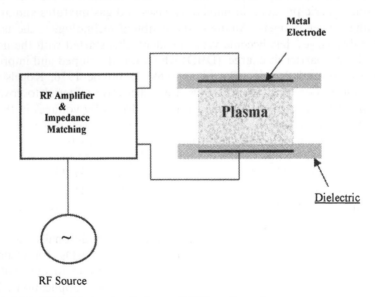

Figure 6.4.1. Dielectric barrier discharge (DBD) configuration.

Figure 6.4.2. Diffuse DBD in a helium/air mixture (photo courtesy: M Laroussi, Old Dominion University).

kHz. To optimize the amount of power deposited in the plasma, an impedance matching network may be introduced between the power supply and the electrodes. The electrode arrangement is generally contained within a vessel or enclosure to allow for the control of the gaseous mixture used. The dielectric material covering the electrodes plays the crucial role in keeping the non-equilibrium nature of the discharge. This is achieved as follows. When a sufficiently high voltage is applied between the electrodes, the gas breaks down (i.e. ionization occurs) and an electrical current starts flowing in the gas. Immediately, electrical charges start accumulating on the surface of the dielectric. These surface charges create an electrical potential, which counteracts the externally applied voltage and therefore limits the flow of current. This process inhibits the glow-to-arc transition. Although traditionally DBDs produce filamentary-type plasmas, under some conditions, which are discussed later in this section, homogeneous plasmas can also be generated. Figure 6.4.2 is a photograph of a diffuse, homogeneous plasma generated by a DBD in an atmosphere of helium with a small admixture of air.

6.4.1.2 *Current–voltage characteristics*

Depending on the operating conditions (gas, gap distance, frequency, voltage), the current waveform can exhibit multiple pulses per half cycle or

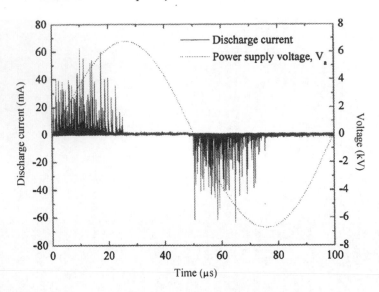

Figure 6.4.3. Current–voltage characteristics of a DBD in N_2 (Gherardi *et al* 2000).

a single wide pulse per half cycle. The presence of multiple current pulses per half cycle is usually taken as an indication that a filamentary discharge is established in the gap between the electrodes. Figure 6.4.3 shows the current and voltage waveforms of a filamentary DBD in nitrogen (Gherardi *et al* 2000). On the other hand, diffuse and homogeneous discharges exhibit a current waveform with a single pulse per half cycle, as shown in figure 6.4.4

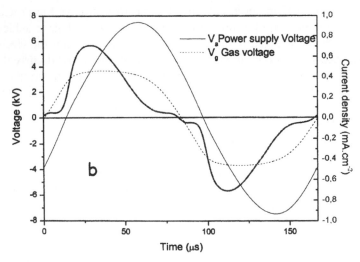

Figure 6.4.4. Current–voltage characteristics of a homogeneous DBD in N_2 (Gherardi *et al* 2000).

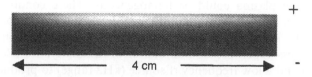

Figure 6.4.5. Ten nanoseconds (10 ns) exposure time photograph of a diffuse DBD in N_2 (Gherardi *et al* 2000).

(Gherardi *et al* 2000). However, a single pulse is not a sufficient test to indicate the presence of homogeneous plasma. Indeed, if a very large number of streamers are generated in a way that they spatially overlap and if the measuring instrument is not capable of resolving the very narrow current pulses, a wide single pulse can be displayed. Gherardi *et al* (2000) used high-speed photography as a second diagnostic method to visually inspect the structure of the discharge channel. Under conditions leading to a homogeneous plasma, photographs taken with exposure times in the order of streamers lifetime (1–10 ns) show a luminous region extending uniformly over the whole electrode surface (see figure 6.4.5). In contrast, when the plasma is filamentary, several localized discharges are clearly visible (figure 6.4.6). Important physical differences between the characteristics of the plasma in a streamer (or microdischarge) and that of a diffuse plasma are to be noted (for details, see section 6.2). Of practical importance are the electron number density, n_e, and kinetic temperature, T_e. In a streamer n_e and T_e are in the 10^{14}–10^{15} cm^{-3} and 1–10 eV range, respectively, while in a diffuse discharge n_e and T_e are in the 10^9–10^{11} cm^{-3} and 0.2–5 eV range, respectively.

6.4.1.3 Discharge homogeneity conditions

The idea of using electrodes covered by a dielectric material to generate a stable non-equilibrium plasma at high pressures is actually an old idea dating from the time Siemens used a discharge to generate ozone (Siemens 1857). However, up until recently the plasma produced by DBDs was filamentary in character, being made of a large number of streamers or microdischarges randomly distributed across the dielectric surface (Kogelschatz *et al* 1997). However, Kanzawa *et al* (1988) showed that, under specific

Figure 6.4.6. Ten nanoseconds (10 ns) exposure time photograph of a filamentary DBD in N_2 (Gherardi *et al* 2000).

conditions, the plasma could be homogeneous. These conditions are (1) helium used as a dilution gas and (2) the frequency of the applied voltage must be in the kHz range. These conclusions were purely empirical, based on more or less experimental trial and error. Similarly, Roth *et al* (1992) used helium and a low frequency rf source (kHz range) to produce a homogeneous discharge in their device, the 'one atmosphere uniform glow discharge plasma' (OAUGDP). The OAUGDP is a DBD-based reactor. They also concluded, based on experimental trials, that helium and the frequency range are the critical parameters, which can lead to a homogeneous plasma at atmospheric pressure. Roth (1995) attempted to explain the frequency range where the homogenous discharge could exist by what he termed the 'ion trapping' mechanism. This idea is based on driving the electrodes by high rf voltages, which induce an electrical field that oscillates at a frequency that is high enough to trap the ions but not the electrons in the space between the electrodes. The electrons ultimately reach the electrodes where they recombine or form a space charge. This theory, however, is different from what has been demonstrated by various modeling results (Kogelschatz 2002). Another argument is the fact that in a highly collisional regime one cannot trap charged particles by a single axially uniform electric field (the axis normal to the plane of the electrodes in this case), even if it is oscillating. Furthermore, collective effects were not taken into account. For example if the ions were trapped and the electrons drifted towards the electrodes, an ambipolar electric field would be established in such a way as to repel the electrons away from the electrodes and towards the ions, a mechanism not taken into account in the proposed analysis.

Massines *et al* (1998) presented a very different theory, which seems to be well supported by experimental and modeling works. The main idea behind Massines' theory is that since the plasma generated by a DBD is actually a self-pulsed plasma, a breakdown of the gas under low electric field between consecutive pulses is possible due to trapped electrons and metastable atoms. These seed particles allow for a Townsend-type breakdown instead of a streamer-type, leading to continued discharge conditions even when the electric field is small. In the case of helium, a density of seed electrons greater than $10^6 \, cm^{-3}$ was found to be sufficient to keep the plasma ignited under low field conditions (Gherardi *et al* 2000). The seed electrons are electrons left over from the previous pulse and those generated via Penning ionization emanating from metastable atoms. In the case of nitrogen, Gherardi reported that the metastables play the dominant role in keeping the discharge ignited between pulses. Their concentration depends strongly on the nature of the surface of the dielectric material, which is a source of metastable-quencher species.

Using a one-dimensional fluid model, Massines calculated the distributions of the electric field, the electron density, and the ion density, and showed that the homogeneous DBD exhibits a structure identical to the

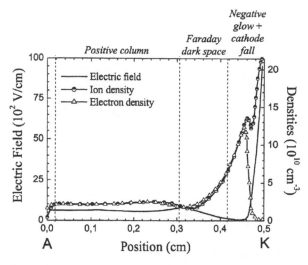

Figure 6.4.7. Electric field, electron density, and ion density spatial distributions between the anode and the cathode of a diffuse DBD in helium (Massines *et al* 1998).

normal glow discharge (positive column, Faraday dark space, negative glow, etc.). Figure 6.4.7 shows such spatial distributions between the anode and cathode of a homogeneous DBD (Massines *et al* 1998).

6.4.2 The resistive barrier discharge (RBD)

To extend the operating frequency range, a few methods were proposed. Okazaki used a dielectric wire mesh electrode in a DBD to generate a glow discharge at a frequency of 50 Hz (Okazaki *et al* 1993). Alexeff and Laroussi (1999, 2002a,b) proposed what came to be known as the resistive barrier discharge (RBD). The RBD can be operated with dc or ac (60 Hz) power supplies. This discharge is based on the dielectric barrier (DB) configuration, but instead of a dielectric material, a high resistivity (few MΩ·cm) sheet is used to cover one or both of the electrodes (see figure 6.4.8). The high resistivity sheet plays the role of a distributed resistive ballast which inhibits the discharge from localizing and the current from reaching a high value, and therefore prevents arcing. It was found that if helium was used as the ambient gas between the electrodes and if the gap distance was not too large (5 cm and below), a spatially diffuse plasma could be maintained for time durations of several tens of minutes. Figure 6.4.9 shows the discharge structure when helium was used. However, if air was mixed with helium (>1%) the discharge formed filaments which randomly appeared within a background of more diffuse plasma. This occurred even when the gap distance was small (Laroussi *et al* 2002a).

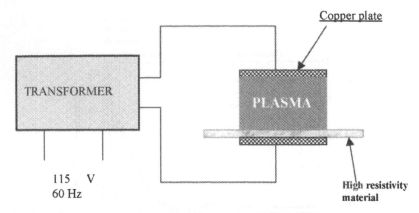

Figure 6.4.8. The resistive barrier discharge (RBD) configuration.

6.4.2.1 Current–voltage characteristics

The RBD can be operated under dc or ac modes. Even when operated in the dc mode, the discharge current was found to be a series of pulses, suggesting that, like the DBD, the RBD is also a self-pulsed discharge. Figure 6.4.10 shows the current waveform and the output signal of a photomultiplier tube (PMT), when a dc voltage of 20 kV was applied. The current pulses are a few microseconds wide and occur at a repetition rate of a few tens of kHz. The PMT signal correlates very well with the current. The pulsed nature of the discharge current can be explained by the combined resistive and capacitive nature of the device. When the gas breaks down and a current of sufficient magnitude flows, the equivalent capacitance of the electrodes becomes charged to the point where most of the applied voltage starts appearing across the resistive layer of the electrodes. The voltage across the gas then becomes too small to maintain a discharge and the plasma extinguishes. At this point, the equivalent capacitor discharges itself through the resistive layer, hence lowering the voltage across the resistive layer and increasing the voltage across the gas until a new breakdown occurs (Wang *et al* 2003).

Figure 6.4.9. Photograph of a diffuse RBD in helium (Laroussi *et al* 2002a).

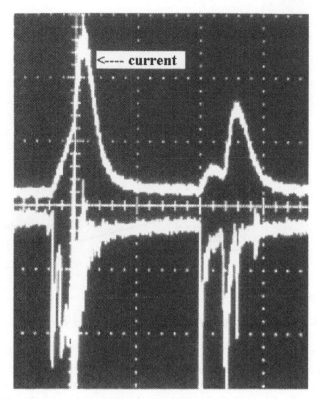

Figure 6.4.10. RBD current waveform under dc excitation. Lower waveform is PMT signal. Horizontal scale is 2 μs/square.

The RBD offers a very practical solution to generate relatively large volumes of low temperature plasma for processing applications and bio-medical applications (Laroussi 2002). For homogeneity purposes, helium was found to be necessary as the main component of the ambient gas mixture between the electrodes. Introduction of air renders the discharge filamentary. If only air is used, plasma can still be initiated for small gaps (millimeters). However, in this case, the structure of the plasma is spatially non-uniform.

6.4.3 Diffuse discharges by means of water electrodes

Although the use of liquid cathodes (such as electrolytes) to generate a discharge has been around for some time (Davies and Hickling 1952), only recently have some investigators applied it to specifically producing diffuse plasmas in air (Andre *et al* 2001, Laroussi *et al* 2002b). Andre used two streams of water as electrodes. A non-equilibrium discharge was ignited between the two streams (few millimeters apart) by means of a dc power supply (applied dc voltage ~3 kV). They reported a current density in the

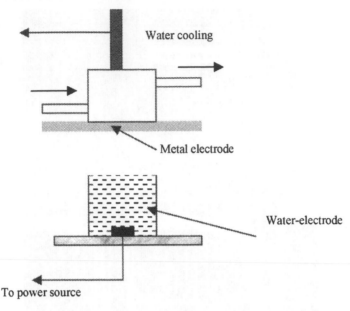

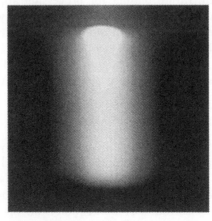

Figure 6.4.11. Discharge configuration with water as a lower electrode (Laroussi *et al* 2002b).

0.2–0.25 A cm^{-2} range. Laroussi used one water electrode (static or flowing water) and as a second electrode a water-cooled metal disk (see figure 6.4.11). The discharge was ignited in the gap between the disk-shaped electrode and the surface of the water by means of an ac power supply (applied voltage ~13 kV, frequency 60 Hz). The plasma generated by this method is diffuse but not necessarily spatially uniform. Figure 6.4.12

Figure 6.4.12. Visual structure of the discharge. Water electrode is at the bottom (Laroussi *et al* 2002b).

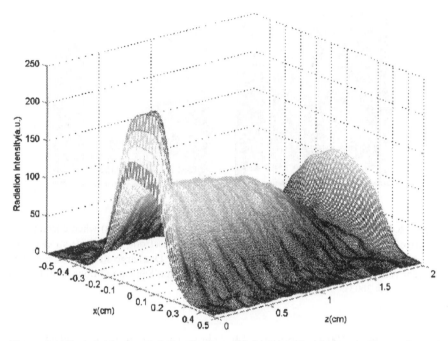

Figure 6.4.13. Axial and radial distribution of light from the discharge (Laroussi *et al* 2003).

shows a typical visual structure of the plasma (Laroussi *et al* 2003). The top, which is the location of the metal disk electrode, exhibits a more intense region, whiter in color than the rest of the column. Next to the surface of the water electrode (bottom), the plasma is more violet in color and rather filamentary. This filamentation is due to the fact that, before breakdown occurs, under the influence of the applied electric field, the water surface develops a number of 'ripples'. These ripples offer sharp curvature points with high electric fields at their tips, which ignite numerous local discharges across the water surface. Figure 6.4.13 shows the axial distribution of light intensity emitted by the discharge. The emission is most intense near the metal electrode (located at $z = 0$ cm), exhibits a nearly constant plateau along most of the gap, then a dark space at about 3 mm from the water surface (located at $z = 2$ cm).

6.4.3.1 Temporal evolution of the plasma structure

In order to characterize the temporal evolution of the plasma structure, a high-speed CCD camera was used to take pictures for different values of the discharge current (Lu and Laroussi 2003). Figures 6.4.14(a), (b) correspond to the positive and negative peaks of the discharge current, respectively. The

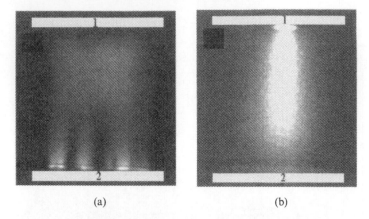

(a)   (b)

Figure 6.4.14. (a) Discharge structure in air (exposure time is 100 μs) when current is at positive peak (water electrode is the cathode). Water electrode is the bottom electrode (2). Gap distance is 1.3 cm (Lu and Laroussi 2003). (b) Discharge structure in air when current is at negative peak (water electrode is the anode). Same conditions as in (a) (Lu and Laroussi 2003).

exposure time is 100 μs. Figure 6.4.14(a) shows that when the water electrode is the cathode the plasma takes the shape of a relatively wide column (about 9 mm wide) but is not visually bright. In contrast, when the water electrode becomes the anode (during the second half cycle of the voltage, figure 6.4.14(b)), the plasma appears as a brighter but narrower column (about 5 mm wide). Structures similar to the dc glow discharge, such as Faraday dark space, negative glow, positive column, and anode dark space, are clearly visible. The 'cathode fall' region is on the metal electrode side. However, when the water electrode is the cathode (figure 6.4.14(a)), the plasma exhibits multi-contact points at the water surface with several localized discharges. These are followed by a dark space, then a single wide bright region, and finally a dark space near the anode (top electrode). The 'cathode fall' region is on the water electrode side. Here, the electric field is high, contributing to the ignition of several local discharges at the rippled surface of the water. It was also found that the discharge always ignites at the water surface and propagates towards the metal electrode at velocities approaching 1 km/s (Lu and Laroussi 2003). This velocity is much smaller than that of streamer heads (~100 km/s) generated in DBDs, suggesting that the breakdown mechanism in this discharge is not similar to the usual electron-driven avalanche.

6.4.3.2   *Electron density and gas temperature measurements*

The electron number density, n_e, was estimated from the electrical parameters of the discharge: the electric field E, the current density j, and electron collision frequency ν_e:

$$j = n_e e^2 E / m_e \nu_e$$

where e and m_e are the electronic charge and mass respectively. Under high pressure and low temperature conditions the electron collision frequency is dominated by electron–neutral collisions. Assuming that the collision cross-section is weakly dependent on temperature, ν_e is related to the electron temperature as $T_e^{1/2}$. For current densities in the range 0.01–$1 \, \text{A/cm}^2$, electron number densities 10^{10}–$10^{12} \, \text{cm}^{-3}$ were calculated.

In order to determine the background gas temperature, the simulated spectra of the 0–0 band of the second positive system of nitrogen were compared with experimentally measured spectra. Because of the low energies needed for rotational excitation and the short transition times, molecules in the rotational states and the neutral gas molecules are in equilibrium. Consequently, the rotational temperature also provides the value of the gas temperature. Using this method, Lu and Laroussi (2003) measured gas temperatures in the 800–900 K range when the water electrode is the cathode, and in the 1400–1500 K range when the water electrode is the anode.

References

Alexeff I and Laroussi M 2002 'The uniform, steady-state atmospheric pressure dc plasma' *IEEE Trans. Plasma Sci.* **30**(1) 174

Alexeff I, Laroussi M, Kang W and Alikafesh A 1999 'A steady-state one atmosphere uniform dc glow discharge plasma' in *Proc. IEEE Int. Conf. Plasma Sci.* p. 208

Andre P, Barinov Y, Faure G, Kaplan V, Lefort A, Shkol'nik S and Vacher D 2001 'Experimental study of discharge with liquid non-metallic (tap-water) electrodes in air at atmospheric pressure' *J. Phys. D: Appl. Phys.* **34** 3456

Bartnikas R 1968 'Note on discharges in helium under ac conditions' *Brit. J. Appl. Phys. (J. Phys. D.) Ser. 2* **1** 659

Davies R A and Hickling A 1952 *J. Chem. Soc. Glow Discharge Electrolysis Part I* 3595

Donohoe K G 1976 'The development and characterization of an atmospheric pressure non-equilibrium plasma chemical reactor' PhD Thesis, California Institute of Technology, Pasadena

Gherardi N, Gouda G, Gat E, Ricard A and Massines A 2000 'Transition from glow silent discharge to micro-discharges in nitrogen gas' *Plasma Sources Sci. Technol.* **9** 340

Kanazawa S, Kogoma M, Moriwaki T and Okazaki S 1988 'Stable glow at atmospheric pressure' *J. Phys. D: Appl. Phys.* **21** 838

Kogelschatz U 1990 'Silent discharges for the generation of ultraviolet and vacuum ultraviolet excimer radiation' *Pure Appl. Chem.* **62** 1667

Kogelschatz U 2002 'Filamentary, patterned and diffuse barrier discharges' *IEEE Trans. Plasma Sci.* **30**(4) 1400

Kogelschatz U, Eliasson B and Egli W 1997 'Dielectric-barrier discharges: principle and applications' *J. Physique IV* **7**(C4) 47

Laroussi M 2002 'Non-thermal decontamination of biological media by atmospheric pressure plasmas: review, analysis and prospects' *IEEE Trans. Plasma Sci.* **30**(4) 1409

Laroussi M, Alexeff A, Richardson J P and Dyer F F 2002a 'The resistive barrier discharge' *IEEE Trans. Plasma Sci.* **30**(1) 158

Laroussi M, Malott C M and Lu X 2002b 'Generation of an atmospheric pressure non-equilibrium diffuse discharge in air by means of a water electrode' in *Proc. Int. Power Modulator Conf.*, Hollywood, CA pp 556–558

Laroussi M, Lu X and Malott C M 2003 'A non-equilibrium diffuse discharge in atmospheric pressure air' *Plasma Sources Sci. Technol.* **12**(1) 53

Lu X and Laroussi M 2003 'Ignition phase and steady-state structures of a non-thermal air plasma' *J. Phys. D: Appl. Phys.* **36** 661

Massines F, Mayoux C, Messaoudi R, Rabehi A and Ségur P 1992 'Experimental study of an atmospheric pressure glow discharge application to polymers surface treatment' in *Proc. GD-92*, Swansea, UK, vol. 2, pp 730–733

Massines F, Gadri R B, Decomps P, Rabehi A, Ségur P and Mayoux C 1996 'Atmospheric pressure dielectric controlled glow discharges: diagnostics and modelling' in *Proc. ICPIG XXII*, Hoboken, NJ 1995, Invited Papers, AIP Conference Proc. vol. 363, pp 306–315

Massines F, Rabehi A, Decomps P, Gadri R B, Ségur P and Mayoux C 1998 'Experimental and theoretical study of a glow discharge at atmospheric pressure controlled by a dielectric barrier' *J. Appl. Phys.* **8** 2950

Okazaki S, Kogoma M, Uehara M and Kimura Y 1993 'Appearance of a stable glow discharge in air, argon, oxygen and nitrogen at atmospheric pressure using a 50 Hz source' *J. Phys. D: Appl. Phys.* **26** 889

Roth J R 1995 *Industrial Plasma Engineering*, vol. 1 (Bristol and Philadelphia, PA: IOP Publishing) pp 453–463

Roth J R, Laroussi M and Liu C 1992 'Experimental generation of a steady-state glow discharge at atmospheric pressure' in *Proc. 27th IEEE ICOPS*, Tampa, FL, paper P21

Siemens W, 1857 *Poggendorfs Ann. Phys. Chem.* **12** 66

Wang X, Li C, Lu M and Pu Y 2003 'Study on Atmospheric Pressure Glow Discharge' *Plasma Source Science and Technology* **12**(3) 358

Yokoyama T, Kogoma M, Moriwaki T and Okazaki S 1990 'The Mechanism of the stabilized glow plasma at atmospheric pressure' *J. Phys. D: Appl. Phys.* **23** 1125

6.5   Discharges Generated and Maintained in Spatially Confined Geometries: Microhollow Cathode (MHC) and Capillary Plasma Electrode (CPE) Discharges

Two discharge types that have been used successfully to generate and maintain atmospheric-pressure plasmas in air are microhollow cathode (MHC) and capillary plasma electrode (CPE) discharges. Common to both discharges is the fact that they are created in spatially confined geometries, whose critical dimensions are in the range 10–500 µm. The MHC discharge is based on the concept of the well-known low-pressure hollow cathode (HC) and, in essence, represents an extension of the HC discharge to atmospheric pressure. The CPE discharge, which uses electrodes with perforated dielectric covers, may be thought of as a variant of the dielectric barrier discharge (DBD). However, the perforated dielectric cover creates an array of capillaries, which critically determine the properties of the discharge and distinguish the CPE discharge properties from those of a DBD. A discharge type which was derived from MHC discharges, but is not based on the hollow cathode effect, has recently been added to the list of spatially confined micro-discharges: the cathode boundary layer (CBL) discharge. Although this discharge has so far only been operated in noble gases, a brief discussion of CBL discharges has been included, because of its potential for the generation of 'two-dimensional' plasmas in atmospheric pressure air.

6.5.1 The microhollow cathode discharge

It is illustrative to start with a brief review of the hollow cathode (HC) discharge. HC discharges have been widely used since the early part of the 20th century, primarily as high-density, low-pressure discharge devices for a variety of applications (Paschen 1916, Güntherschulze 1923, Walsh 1956). An HC discharge device consists of a cathode with hollow structure (hole, aperture, etc.) in it and an arbitrarily shaped anode (figure 6.5.1). Two scaling laws determine the properties of the discharge. The product pd of the pressure p and the anode–cathode separation d obeys the well-known Paschen breakdown law, which applies to all discharges and determines the required breakdown voltage for given values of p, d, and the operating gas (Paschen 1916, White 1959).

   A scaling law that is unique to the HC discharge involves the product pD of the pressure p and cathode opening D. If the product pD is in the range from 0.1 to 10 torr cm, the discharge develops in stages, each with a distinctive I–V characteristics. At low currents, a 'pre-discharge' is observed, which is a glow discharge whose cathode fall region is generally outside the cathode structure. Under these circumstances, there is a single region of positive space charge and the electrons follow a path that is essentially determined by the direction of the vacuum electric field between cathode and anode in the absence of a discharge. As the current increases, the positive space charge region moves closer to the cathode and eventually enters the hollow cathode structure. Now a positive column, which serves as a virtual anode, is formed along the axis of the cathode cavity between two separate cathode sheath regions. This results in a change in the electric field distribution within the hollow cathode. The electric field, which was initially axial, now becomes a radial field and a potential 'trough' is created within the cavity. This trough causes a strong radial acceleration of the electrons towards the

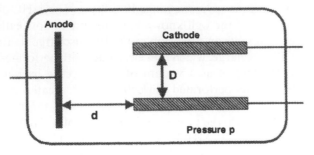

Figure 6.5.1. General hollow cathode geometry.

axis, which may lead to an oscillatory motion of the electrons ('pendulum electrons'; Güntherschulze 1923, Walsh 1956, Helm 1972, Stockhausen and Kock 2001) when they are accelerated into the virtual anode and then repelled at the opposing cathode fall. This may result in an oscillatory motion with ever-decreasing amplitude between the two opposite cathode fall regions. Thus, the path length of the electrons is increased and these pendulum electrons can undergo many ionizing collisions with the background gas. Furthermore, energetic particles within the cathode hole such as ultraviolet photons and metastables have a high probability of producing secondary electrons at the cathode surface, which, in turn, can lead to further ionization and excitation events.

In the transition from an axial pre-discharge to a radial discharge, the sustaining voltage drops as the current increases (Fiala *et al* 1995). The discharge has a 'negative differential' resistance, a mode which is traditionally referred to as the 'hollow cathode' mode. As the current is increased further, the cathode layer expands over the surface of the planar cathode outside the hole. The current–voltage characteristic becomes that of a normal glow discharge with constant voltage at increasing current. Ultimately, when the cathode layer reaches the boundaries of the cathode, any further current increase requires an increase in discharge voltage: the discharge changes into an abnormal glow discharge.

HC discharges are known to have electron energy distributions that are strongly non-Maxwellian and contain a significant amount of very energetic electrons. Most of the earlier diagnostics studies (Gill and Webb 1977) of the electron energy distribution function in HC devices were carried out for low-pressure HC discharges. These studies found copious amounts of electrons with energies well above 10 eV and a tail extending up to the plasma operating voltage. Furthermore, a fraction of high-energy 'beam' electrons was measured with energies near the plasma voltage. These are electrons that were accelerated across the full potential of the cathode fall.

Efforts to increase the operating pressure of HC discharges date back to the late 1950s (White 1959, Sturges and Oskam (1964)). The so-called

'White–Allis' similarity law relates the discharge sustaining voltage V to the product (pD) and the ratio (I/D), where I is the discharge current. As a consequence of this law, operation of a HC discharge at higher pressures can be accomplished by reducing the size D of the hole in the cathode. The lowest value of pD for which the scaling law holds is determined by the condition that the mean free path for ionization cannot exceed the hole diameter (Helm 1972). For argon, the minimum pD value (Güntherschulze 1923) is 0.026 torr cm. Empirically, upper bounds for pD in the rare gases are around 10 torr cm, but less for molecular gases (Gewartkowski and Watson 1965). Physically, the upper limit is determined by the condition that the distance between 'opposite' cathodes cannot exceed the combined lengths of the two cathode fall regions plus the glow region. This would lead to an upper limit (Schoenbach *et al* 1997), in pD for argon of about 1 torr cm, which is almost about a factor of 10 less than the empirically established upper limit. As a result, high-pressure operation of a HC discharge at or near atmospheric pressure in the rare gases is possible, but requires small hole sizes in the cathode. Based on the upper limit of the product pD, atmospheric-pressure operation in the rare gases would require hole sizes of about 10 μm assuming that the gas is at room temperature. Empirically, stable HC operation at atmospheric-pressure in the rare gases has been observed (Schoenbach *et al* 2000) for holes sizes as large as 250 μm. This indicates that physical processes other than pendulum–electron coupling between 'opposite' cathodes must be present to account for the negative differential resistance and the discharge stability at high pD values.

Discharges of this hollow cathode discharge type have been studied by a number of groups and, dependent on particular electrode geometry or on their arrangements in arrays, they have been named differently. In some case, they are just named 'microdischarges' as by the group at the University of Illinois (Frame *et al* 1997) and that at Caltech (Sankaran and Giapis 2001). In another case, where the electrode configuration was designed for parallel operation, the discharges are named by the group at the University of Frankfurt and the University of Dortmund, Germany, 'microstructured electrode arrays' (Penache *et al* 2000). For cylindrical hollow cathode discharges, the term 'microhollow cathode discharge' (MHCD) was coined by the group at Old Dominion University (Schoenbach *et al* 1996). This name is being used by several other groups, who work with microdischarges based on the hollow cathode principle, such as the group at the Steven Institute of Technology (Kurunczi *et al* 1999), the University of Erlangen, Germany (Petzenhauser *et al* 2003), the University of California, Berkeley (Hsu and Graves 2003), the group at Yonsai University, Korea (Park *et al* 2003), the National Cheng Kung University, Taiwan (Guo and Hong 2003), and at the Institute for Low Temperature Plasma Physics in Greifswald, Germany (Adler *et al* 2003).

Most of the experimental studies in high-pressure hollow cathode discharges have so far been performed in rare gases and rare gas halide mixtures. But there is an increasing emphasis on their use in atmospheric pressure air, or at least mixtures of gases containing air. The following sections will give an overview of the various electrode geometries and modes of operation, their plasma parameter range, and some applications.

6.5.1.1   Electrode geometries, materials, and fabrication techniques

Any hollow cathode discharge electrode geometry needs to satisfy the condition that surfaces of the cathode facing each other need to be separated by a distance such that the high-energy electrons generated in one cathode fall can reach the opposite cathode fall. Such cathode geometries can be parallel plates, holes of any shape in a solid cathode, slits in the cathode, and spirals (Schaefer and Schoenbach 1990). For microhollow cathode discharges, initially, cylindrical holes were used to generate the hollow cathode effect (White 1959, Schoenbach *et al* 1996, Frame *et al* 1997). These geometries have been extended to microtubes with the anode at the orifice (Sankaran and Giapis 2002) or inserted through the walls (Adler *et al* 2002), and microslots (Yu *et al* 2003). Paralleling the microholes has resulted in microarrays (Shi *et al* 1999, Park *et al* 2000, Schoenbach *et al* 2003, Allmen *et al* 2003, Penache *et al* 2000, Guo and Hong 2003). Adding microdischarges in series has allowed increasing the light emission (El-Habachi *et al* 2000, Vojak *et al* 2001), and may possibly lead to laser emission (Allmen *et al* 2003). Common to all these geometries are the dimensions of the cathode hollow, which are on the order of 100 μm. Cross-sections of electrode geometries used by the various groups are shown in figure 6.5.2.

Electrode materials range from refractive metals to semiconductors. Whereas mainly molybdenum has been used for high current (>1 mA) discharges (Schoenbach *et al* 2003, Kurunczi *et al* 1999, Petzenhauser *et al* 2003), nickel, platinum, silver and copper were used as the electrode material for microhollow cathode discharges and discharge arrays at lower currents (generally in the sub-mA range for individual holes in microdischarge arrays) (Allmen *et al* 2003, Park *et al* 2003, Penache *et al* 2000). The group at the University of Illinois has early-on concentrated on silicon as material (Frame *et al* 1997, Chen *et al* 2002) a material which allows use of micro-machining techniques. Stainless steel was the choice for microtube cathodes (Sankaran and Giapis 2003). Generally, the choice of electrode materials seems so far to be determined by available fabrication techniques, and the ability to withstand high temperature operation, rather than being guided by the physics of cathode and anode fall.

The dielectric material was mica in initial experiments, but was replaced soon by alumina and other ceramics. In some cases polymers have been used to generate flexible microdischarge arrays (Park *et al* 2000, Penache *et al* 2000). Such materials are well versed for discharges in rare gases or rare gas–halide mixtures, where the gas temperature is relatively low. However,

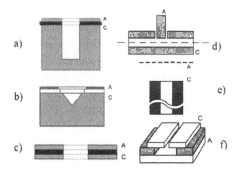

Figure 6.5.2. Various hollow cathode electrode configurations either used for single discharges or as an 'elementary cell' in arrays. (a) Old Dominion University, USA; University of Illinois, Urbana-Champaign, USA; Hyundai Research and Development Center, Korea. (b) University of Illinois, Urbana-Champaign, USA. (c) Old Dominion University, USA; Stevens Institute of Technology, USA; University of Illinois, Urbana-Champaign, USA; University of Erlangen, Germany; University of Frankfurt, Germany; University of Dortmund, Germany; Caltech, USA; University of California, Berkeley, USA; National Cheng Kung University, Taiwan. (d) Institute for Low Temperature Plasma Physics, Greifswald, Germany. (e) Caltech, USA. (f) Colorado State University, USA

for microdischarges in air the material choices are limited. The high gas temperatures (~2000 K) in air microhollow cathode discharges (Block *et al* 1999) require the use of dielectrics and electrode materials with high melting points, such as alumina and molybdenum, respectively.

The microholes in such discharge geometries have initially been drilled mechanically (White 1959, Schoenbach *et al* 1996) or milled ultrasonically (Frame *et al* 1997), with hole diameters of >200 μm. For cylindrical holes with smaller diameter in metals, laser drilling has been the method of choice. For the fabrication of large arrays, silicon bulk micromachining techniques have been successfully used (Chen *et al* 2002).

6.5.1.2 *Array formation of microdischarges*

The application of microdischarges generally requires the arrangement of these discharges in arrays. Such arrays may consist of discharges placed in parallel or in series, or both. Placing the discharges in parallel allows plasma layers to form which could be used as flat plasma sources or as flat light sources. If operated in discharge modes where the current–voltage characteristic has a positive slope, the discharges can be arranged in parallel without individual ballast. This includes operation in the predischarge mode or in an abnormal glow mode.

Parallel operation in the predischarge mode, without individual ballast has been demonstrated by the group at Old Dominion University (Schoenbach *et al* 1996), the University of Illinois (Frame and Eden 1998, Eden *et al* 2003), the University of Frankfurt and University of Dortmund, Germany (Penache *et al* 2000), and the National Cheng Kung University, Taiwan (Guo and Hong 2003). The reference list is by no means exhaustive (only the first published refereed journal publications or papers in conference proceedings for each group are listed), since most of the groups, particularly the group at the University of Illinois, have published extensively on this topic. Because of the relatively low current required for operation in this phase, electrode materials and dielectrics do not need to withstand high thermal loading, and can therefore be fabricated of semiconductor materials (Chen *et al* 2002, Penache *et al* 2000).

Operation in the range of an abnormal glow discharge requires a confined cathode surface. This has been achieved by using a second layer of dielectric material which covers the face of the cathode, and allows the discharge only to develop inside the cathode hole (Miyake *et al* 1999). Another possibility of generating arrays in the abnormal glow mode is to use a geometry as shown in figure 6.5.2a, where the cathode surface is confined to the hole. An example of such an electrode structure with limited cathode area is shown in figure 6.5.2c (Schoenbach *et al* 1997). A series of 30 microholes are placed along a line, with distances of $350\,\mu m$ between hole centers. The cathode area was limited by a dielectric (alumina) to a stripe $250\,\mu m$ wide. The anode was placed on one side on top of the $250\,\mu m$ thick dielectric. The gas was a mixture of 1.5% Xe, 0.03% HCl, 0.06% H_2, and 98.41% Ne. When a voltage of 190 V was applied the microdischarges turned on one after another until the entire set of discharges was ignited. When all discharges were on, the current–voltage characteristic turned positive since all discharges were now operating in an abnormal glow mode.

In the range of operation where the current–voltage characteristic has a negative slope (hollow cathode mode) or is flat (normal glow mode) it is also possible to generate arrays by using distributed resistive ballast. This has been demonstrated by using semi-insulating silicon as the anode material (Shi *et al* 1999). The use of multilayer ceramic structures where each microdischarge has been individually ballasted, with the resistors produced and integrated into the structure by a thick film process, has allowed the generation of arrays 13×13 microdischarges (Allmen *et al* 2003).

Arranging the microdischarges in series, rather than in parallel (as was discussed above) is motivated by the increased radiant excimer emittance. Since the excimer gas does not reabsorb the excimer radiation, the excimer irradiance generated by n discharge plasmas along a common axis should be n times that of a single discharge. A second application for a string of discharges would be its use as an excimer laser medium. A simple estimate of the power density in a string of microdischarges indicates that small signal gains of $>0.1/cm$ are obtainable (El-Habachi *et al* 2000). First experiments with two discharges in series have demonstrated doubling of the studied XeCl excimer irradiance (El-Habachi *et al* 2000). The stable operation of three neon discharges in series in a ceramic discharge device has been demonstrated by Vojak *et al* (2001). Allmen *et al* (2003), have extended such a system to seven sections with an active length of approximately 1 cm, and have found indications of gains for 460.30 nm transition of Xe^+, making this the first example of a microdischarge optical amplifier.

6.5.1.3 Modes of operation

MHCDs are direct current devices, but are not necessarily restricted to dc operation. They have been successfully operated in the pulsed mode as well as in ac and rf modes. Sustaining voltages range from 150 to 500 V, depending on the discharge current, the type of gas, and on the electrode material. Lowest voltages are obtained with rare gases, highest voltages are measured for attaching gases, or mixtures, which contain attaching gases, such as air. The dc voltage–current characteristics of microhollow cathode discharges show distinct regions. An example for such a characteristic is shown in figure 6.5.3 for a discharge in xenon at 760 torr, together with

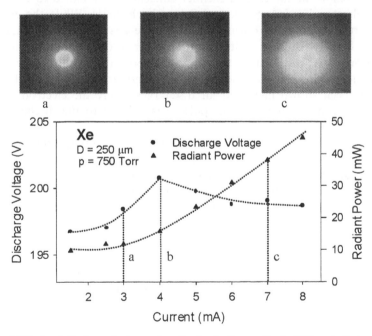

Figure 6.5.3. (a–c) End-on photographs of microhollow cathode (250 μm) discharges in xenon at a pressure of 750 torr for various currents. The photographs were taken through an optical filter, which allowed only the excimer radiation to pass. (d) current–voltage characteristic of the microhollow cathode discharges, and VUV radiant power dependent on current.

images of the discharge obtained in the ultraviolet at 172 nm. In the predischarge mode (lowest current, positive slope in the voltage–current characteristics) and the plasma is confined to the hole. It expands beyond the microhole at the transition from the hollow cathode mode to the normal glow mode. If the cathode surface is limited, the discharge enters an abnormal glow mode, which in the I–V characteristics is indicated as increasing voltage with current.

In order to reduce the thermal loading of microhollow electrodes, but still operate the discharge at high currents, microhollow cathode discharges have been operated in pulsed mode with various duty cycles (such that the average power was kept below a level which causes thermal damage). The pulses were monopolar pulses ranging from milliseconds to nanoseconds. Whereas with millisecond pulses the discharge characteristics was not different from the dc case (Schoenbach *et al* 2000), for microsecond (Adler *et al* 2002, Kurunczi *et al* 2002, Petzenhauser *et al* 2003) and even more for nanosecond pulses (Moselhy *et al* 2001b, 2003), the plasma parameters change strongly. The increase in excimer emission from xenon and argon discharges when pulses of nanosecond duration were applied (and for xenon the increase in excimer efficiency) is assumed to be due to pulsed electron heating (Stark and Schoenbach 2001). While the electron temperature is increased during the pulse, the gas temperature change is small as long as the pulse width is on the order of or less than the electron relaxation time. The shift in the electron energy distribution function to higher energy causes an increase in ionization and excitation rate coefficients. This has been shown in pulsed air discharges where the electron density increased strongly when a 10 ns pulse was applied to the discharge (see chapter 8).

Besides dc and monopolar pulsed operation, radio frequency operation has been explored as a method to generate microplasmas at atmospheric pressure air (Guo and Hong 2003). At frequencies of 13.56 MHz, they could in pure helium (flowed through a microhollow cathode array) generate stable discharges at atmospheric pressure. Recently a group at the Colorado State University has extended this concept to a slotted electrode geometry (Yalin *et al* 2003, Yu *et al* 2003). Stable microdischarges in Ar, Ar–air mixtures, and in open air have been generated when excited with 13.56 MHz with rf voltages of 50–230 V. The slot cathode dimensions are 200 μm by 400–600 μm deep, and 3–35 cm in length.

6.5.1.4   Plasma parameters

(a)   Electron temperature and electron energy distribution
Measurements of the electron temperature in microhollow cathode discharges, in rare gases only, have been performed by means of emission spectroscopy. Based on line intensity measurements in argon an electron

temperature of approximately 1 eV has been determined (Frank *et al* 2001). The electron temperature in pulsed argon discharges was found to be more than twice the dc value. The electron temperature in this case was obtained using information on the temporal development of measured electron densities in plasmas pulsed with 20 ns high-voltage pulses (Moselhy *et al* 2003). This increase in electron temperature, which is correlated to an increase in electron density, is due to pulsed electron heating (Stark and Schoenbach 2001).

Measurements which provide information on average electron energies only give us rather low values. However, from the fact that MHCDs are efficient sources of excimer radiation, large concentrations of high-energy electrons (in excess of the excitation energy of rare gas atoms) must be present. That means that the electron energy distribution must be highly non-thermal. Measurements in the low pressure range confirm this assumption (Badareu and Popescu 1958, Borodin and Kagan 1966). Experiments on plane parallel electrode hollow cathode discharges were performed by Badareu and Popopescu (1958) using Langmuir probes. The electron energy distribution in dry air showed the existence of two groups of electrons, with mean energies of 0.6 and 5 eV. Borodin and Kagan (1966) determined with a similar technique the electron energy distribution in a circular hollow cathode and compared them to that in a positive column. Again, the results indicated a two-electron group distribution with higher concentrations of electrons at high electron energies (>16 eV) compared to that in a positive column.

(b)   Electron density

Electron densities in microhollow cathode discharges in argon have been measured using either Stark broadening and shift of argon lines at 801.699 and 800.838 nm (Penache *et al* 2003), and the hydrogen Balmer-β line at 486.1 nm (Moselhy *et al* 2003). In both cases the measured electron densities were for dc microdischarges on the order of 10^{15} cm^{-3}, showing a slight increase with current. When operated in the pulsed mode, with 10 ns electrical pulses of 600 V applied, the electron densities increased to 5×10^{16} cm^{-3} (Moselhy 2003). Electron densities in microhollow cathode discharges in atmospheric pressure air have been measured using heterodyne infrared interferometry, a method which is described in chapter 8. In a MHCD with a hole diameter of 200 μm, with a current of 12 mA at a voltage of 380 V, the electron density was found to be 10^{16} cm^{-3} (Block *et al* 1999).

(c)   Gas temperature

The MHCD plasma is a non-thermal plasma, that means that the gas temperature is much lower than the electron temperature. Gas temperature measurements have been performed in rare gas MHCDs, as well as in air MHCDs by using optical emission spectroscopy (Block *et al* 1999, Kurunczi *et al* 2003) and by means of absorption spectroscopy (Penache *et al* 2003). The gas temperature in atmospheric-pressure air MHC discharges was measured to be between 1700 and 2000 K for discharge currents between 4 and 12 mA by evaluating the rotational (0,0) band of the second positive N_2 system (Block *et al* 1999). The temperature in a neon MHC discharge (400 torr) was measured to be around 400 K (Kurunzci *et al* 2003) at a current of 1 mA. The temperature was obtained from the analysis of the N_2 band system (using a trace admixture of nitrogen added to the neon). Absorption spectroscopy (Doppler broadening of argon lines) has been used by Penache *et al* (2003) to determine the gas temperature in argon microdischarges. It was found to increase with pressure from 380 K at 50 mbar to 1100 K at 400 mbar. The result indicate that the gas temperature depends on the type of gas. It is highest for molecular gases, such as air (2000 K), and lowest for low atomic weight rare gases (slightly above room temperature). It increases with pressure, but only slightly with current.

6.5.1.5   Applications of microdischarges

(a)   Microdischarges as ultraviolet radiation sources
The electrostatic non-equilibrium resulting from small size (the cathode fall of MHCDs is commensurate with the radial dimensions of the microhole) is the reason for an electron energy distribution with large concentration of high-energy electrons. This, and the stable operation of these discharges at high pressure favors three-body processes, such as ozone generation, and excimer formation. The latter effect has been extensively studied for rare gases such as helium (Kurunzci *et al* 2001), neon (Frame *et al* 1997, Kurunzci *et al* 2002), argon (Schoenbach *et al* 2000, Moselhy and Schoenbach 2003, Petzenhauser *et al* 2003), and xenon (El-Habachi and Schoenbach 1998a,b, Schoenbach *et al* 2003, Adler *et al* 2002, Petzenhauser *et al* 2003) and for some rare gas halide mixtures which generated ArF (Schoenbach *et al* 2000) and XeCl (El-Habachi *et al* 2000) excimer radiation. Internal efficiencies of up to 8% are reported for xenon excimer MHCD sources (El-Habachi 1998b). For rare gas halide mixtures, efficiencies on the order of percent have been measured (Schoenbach *et al* 2000). Ultraviolet/vacuum ultraviolet radiant power densities of several W/cm^2 seem to be obtainable over large areas when MHCDs are operated in parallel.

Applications of excimer light sources, based on microdischarge arrays are flat panel deep ultraviolet sources for a variety of applications, similar to those of barrier discharges (Kogelschatz 2004). One application, which has been pursued at Hyundai Display Advanced Technology R&D Research

Center (Choi 1999, Choi and Tae 1999) and at the University of Illinois (Park *et al* 2001), is their use in flat panel displays. However, applications of microdischarges as light sources go beyond excimer lamps and flat panel displays. First experiments to develop microlasers with a series of microdischarges have been reported (Allmen *et al* 2003).

Besides excimer radiation microhollow cathode discharges have also been shown to emit line radiation at high efficiencies. Kurunczi *et al* (1999) observed intense emission of the atomic hydrogen Lyman-α (121.6 nm) and Lyman-β (102.5 nm) lines from high-pressure microhollow cathode discharges in neon with a small hydrogen admixture. The atomic emission is attributed to near-resonant energy transfer processes between the Neon excimer and H_2. A similar resonant effect in argon microhollow cathode discharges with small admixtures of oxygen has been reported by Moselhy *et al* (2001). The emission of strong oxygen lines at 130.2 and 130.5 nm indicates resonant energy transfer from argon dimers to oxygen atoms.

(c) Microdischarges as plasma-reactors and detectors

The high-energy electrons in high-pressure microdischarges assist in the production of a high-electron density plasma. This is for atmospheric pressure operation desirable for materials processing and surface modification where the microdischarges serve as sources of radicals and ions. Experiments with electrode geometries as shown in figure 6.5.2(c), either in single discharges or in discharge arrays, have been performed in rare gases and mixtures of rare gases with molecular gases. Hsu and Graves (2003) have explored the use of single discharges as flow reactors. Flow of molecular gases was found to induce chemical modifications such as molecular decomposition. Maskless etching of silicon and diamond deposition on a heated Mo substrate has been demonstrated by Sankaran and Giapis (2001, 2003). Surface modifications of polymeric film substrates in a mixture of argon and 10% air (Penache *et al* 2002), and fabrication of amorphous carbon films by adding 1% of hexamethyldisiloxane (HMDSO) to atmospheric pressure helium in a microhollow cathode discharge array with a third biased electrode (Guo and Hong 2003) has been pursued. Microdischarges have also been used as detectors. Due to its high electron density (10^{15} cm^{-3}) and a gas temperature of approximately 2000 K in molecular gases, the plasma has similar plasma parameters as plasmas used in analytical spectroscopy. Based on this concept, high pressure microplasma has e.g. been used as detector of halogenated hydrocarbons (Miclea *et al* 2002). Another interesting application has been explored by Park *et al* (2002). It was found that the photosensitivity of microdischarges is such that microdischarges serve as photodetectors where the photocathode determines the spectral response, and the microplasma serves as an electromultiplier.

(d)   Microdischarges as plasma cathodes

One of the major obstacles in obtaining glow discharge plasmas in gases at
atmospheric pressure are instabilities, particularly glow-to-arc transitions
(GAT), which lead to the filamentation of the glow discharge in times
short compared to the desired lifetime of a homogeneous glow. These
instabilities generally develop in the cathode fall, a region of high electric
field, which in self-sustained discharges are required for the emission of
electrons through ion impact. Eliminating the cathode fall, by supplying
the electrons by means of an external source, is therefore expected to prevent
the onset of GAT. Microhollow cathode discharges have been shown to serve
as electron emitters (plasma cathodes) for direct current glow discharges
between plane parallel electrodes. The stabilizing effect of MHCDs has
been demonstrated for rare gas discharges (Stark and Schoenbach 1999,
Park *et al* 2003, Guo and Hong 2003).

This concept has also been used to generate glow discharges in atmospheric pressure air with dimensions up to cubic centimeters. In a three-
electrode system, as shown in figure 6.5.4, electrons are extracted through
the anode opening at moderate electric fields when the microdischarge was
operated in the hollow cathode discharge mode. These electrons support a
stable plasma between the microhollow anode and a third electrode. The
sustaining voltage of the microhollow cathode discharge in air ranges from
200 to 400 V depending on current, gas pressure and gap distance. The

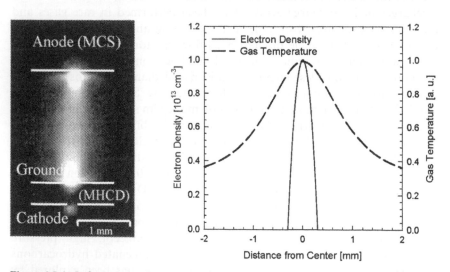

Figure 6.5.4. Left: cross-section of a microhollow electrode geometry with third positively
biased electrode. Superimposed is the photography of a MHCD sustained atmospheric
pressure air plasma. Right: electron density and gas temperature profile of the air
plasma, measured by means of heterodyne infrared interferometry in the middle between
MHCD and the third electrode (anode) (Leipold *et al* 2000).

MHCD current was limited to values of less than 30 mA dc to prevent overheating of the sample. The glow discharges with the MHCVD as plasma cathode were operated at currents of up to 30 mA, corresponding to current densities of $4 \, A/cm^2$ and at average electric fields of 1.25 kV/cm. Electron densities and temperatures have been measured by means of heterodyne laser interferometry and were found to be on the order of $10^{13} \, cm^{-3}$, and 2000 K, respectively (Leipold *et al* 2000). The air plasma can be extended in size by placing MHCD discharges in parallel (Mohamed *et al* 2002).

One of the major obstacles in using such dc glow discharges in atmospheric pressure air is the electrical power density required to sustain these discharges. Operating the discharges in a pulsed mode, with pulses of 10 ns superimposed on a dc MHCD glow discharge in air, has been shown to reduce the required power density for the same average electron density (Stark and Schoenbach 2001). This effect is based on the shift in the electron energy distribution towards higher energies on a timescale shorter than the critical time for the development of a glow-to-arc transition.

6.5.2 The cathode boundary layer discharge

The cathode boundary layer (CBL) discharge is a new type of high-pressure glow discharge between a planar cathode, and a ring-shaped anode separated by a dielectric, with a thickness on the order of 100 μm, and with an opening of the same diameter as the anode (figure 6.5.5) (Schoenbach *et al* 2004). The diameter of the opening is in the range of fractions of millimeters to several millimeters. The discharge is restricted to the cathode fall and negative glow, with the negative glow serving as a virtual anode: the plasma in the negative glow region provides a radial current path to the ring-shaped metal anode. This assumption is supported by the measured thickness of the plasma layer (Moselhy *et al* 2002), which corresponds to the thickness of the cathode fall plus negative glow, but also by the measured sustaining voltage. For high-pressure operation in xenon and argon, the pressure in the normal glow mode was measured as approximately 200 V (Moselhy and Schoenbach 2004), which is on the order of measured cathode fall voltages in noble gases (Cobine 1958).

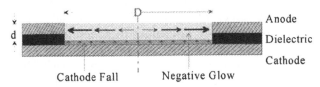

Figure 6.5.5. CBL discharge electrode geometry and estimated current density pattern (Schoenbach *et al* 2004).

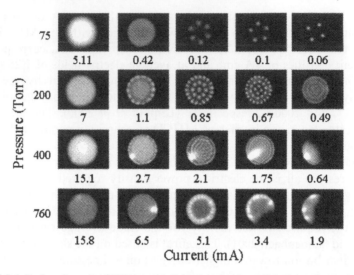

Figure 6.5.6 End-on images of CBL xenon discharges in the visible dependent on pressure and current. The diameter of the cathode is 0.75 mm. The brightness of the images at 75, 200, and 400 torr is for all currents (except the largest one) increased relative to that at 760 torr, in order to better show the pattern structure (Schoenbach *et al* 2004).

The stability of CBL discharges, which allows us to operate them in a dc mode, is assumed to be due to thermal losses through the cathode foil, an effect that is also considered to be the reason for the observed self-organization in xenon discharges (Schoenbach *et al* 2004). The plasma pattern consists of filamentary structures arranged in concentric circles (figure 6.5.6). The self-organization structures are most pronounced at pressures below 200 torr, and become less regular when the pressure is increased.

An important feature of CBL discharges is the positive slope in the voltage–current characteristics over most of the current range, except for low current values (figure 6.5.7). This shows that parallel operation of these discharges is possible without using individual ballast resistors. A consequence of this resistive discharge behavior is the possibility of constructing large-area thin (100 μm) plasma sources.

The experimental studies have so far focused on noble gas operation, because of the importance of such discharges as flat excimer sources. Medium- and high-pressure dc discharges in xenon and argon have been found to emit excimer radiation with efficiencies reaching values of almost 5% in xenon and 2.5% in argon (Moselhy and Schoenbach 2004). However, operation in atmospheric pressure air seems to be feasible, and would allow the generation of ultra-thin (on the order of 100 μm) non-thermal air plasma layers over large surface areas.

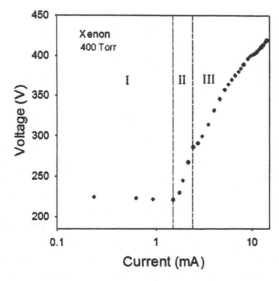

Figure 6.5.7. Voltage–current characteristics of xenon discharges at 400 torr. The characteristics can be divided into three regions. In region I, the discharge behaves as a normal glow; in region II, the self-organized patterns are observed; region III corresponds to abnormal glow (Schoenbach *et al* 2004).

6.5.3 The capillary plasma electrode discharge

The operating principles and basic properties of the capillary plasma electrode (CPE) discharge are much less well understood and the discharge has been much less researched than the MHC discharge. The basis for the atmospheric-pressure operation of the capillary plasma electrode (CPE) discharge is a novel electrode design (Kunhardt and Becker 1999). This design uses dielectric capillaries that cover one or both electrodes of a discharge device, which in many other aspects looks similar to a conventional dielectric barrier discharge (DBD) as shown in figure 6.5.8. However, the CPE discharge exhibits a mode of operation that is not observed in DBDs, the so-called 'capillary jet mode'. Here, the capillaries, with diameters that range from 0.01 to 1 mm and length-to-diameter ratios of the order of

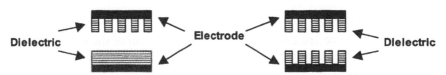

Figure 6.5.8. Schematic diagram of a capillary plasma electrode (CPE) discharge configuration.

10:1, serve as plasma sources, which produce jets of high-intensity plasma at atmospheric pressure under the right operating conditions. The plasma jets emerge from the end of the capillary and form a 'plasma electrode' for the main discharge plasma. Under the right combination of capillary geometry, dielectric material, and exciting electric field, a stable uniform discharge can be achieved. The placement of the tubular dielectric capillary(s) in front of the electrode(s) is crucial for the occurrence of the 'capillary jet mode' of the CPE discharge. In fact, the CPE discharge displays two distinct modes of operation when excited by pulsed dc or ac. When the frequency of the applied voltage pulse is increased above a few kHz, one observes first a diffuse mode similar to the diffuse glow described of a DBD as described by Okazaki *et al* (1993). When the frequency reaches a critical value (which depends strongly on the length-to-diameter value and the feed gas), the capillaries 'turn on' and a bright, intense plasma jet emerges from the capillaries. When many capillaries are placed in close proximity to each other, the emerging plasma jets overlap and the discharge appears uniform. This 'capillary' mode is the preferred mode of operation and has been characterized in a rudimentary way for several laboratory-scale research discharge devices in terms of its characteristic electric and other properties (Kunhardt *et al* 1997a,b, 1998, Panikov *et al* 2002, Moswinski *et al* 2003): peak discharge currents of up to 2 A, current density of up to 80 mA/cm^2, E/p of about 0.25 V/(cm torr), electron density n_e above 10^{12} cm^{-3} (which is about two orders of magnitude higher than the electron density in the diffuse mode of operation), power density of about 1.5 W/cm^3 in He and up to 20 W/cm^3 in air. Using a Monte Carlo modeling code (Amorer 1999), the existence of the threshold frequency, which depends critically on the length-to-diameter ratio of the capillaries and dielectric material, has been verified (Kunhardt *et al* 1997a,b). The model also predicts relatively high average electron energies of 5–6 eV in the 'capillary' mode.

CPE discharges have been operated at atmospheric-pressure in He, Ar, He–N$_2$, He–Air, He–H$_2$O, N$_2$–H$_2$O, and air–H$_2$O gases and gas mixtures and discharge volumes of more than 100 cm^3. The electron density was calculated from the current density, the power input, and an estimate of the electron drift velocity as well as measured using a mm-wave interferometer (Amorer 1999) operating at 110 GHz. Measurements were done in a He discharge in the diffuse mode and in the capillary mode. As can be seen in figure 6.5.9, the transition from the diffuse mode to the capillary mode is accompanied by a drastic increase in the electron density from about 10^{10} to 10^{12} cm^{-3}.

Recently, a spectroscopic analysis of the emission of the unresolved N$_2$ second positive band system from a CPE discharge in atmospheric-pressure air was carried out. Measurements were done for various discharge powers in two geometries. In one case, the emissions arising from inside the capillary, presumably the hottest part of the plasma, were analyzed. In the other

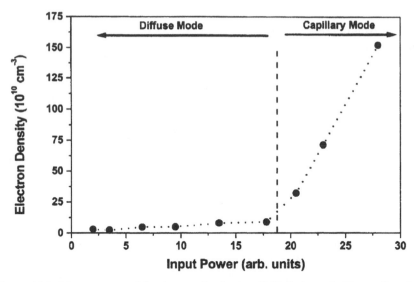

Figure 6.5.9. Measurement of the electron density in a CPE discharge in He as a function of power input. The transition of the discharge from the diffuse mode to the capillary mode with a corresponding drastic increase in the electron density is clearly apparent.

arrangement, the emissions perpendicular to the axis of the capillary, presumably a 'colder' plasma region as the plasma jet emerging from the capillary is beginning to spread out spatially, were studied. The results shown in figure 6.5.10 reveal higher rotational temperatures in the plasma

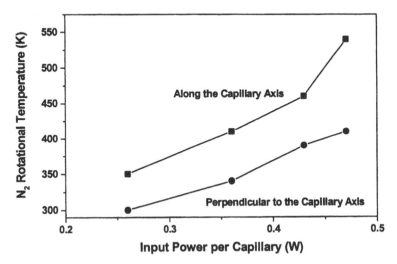

Figure 6.5.10. Rotational temperatures of N_2 in a CPE discharge in atmospheric-pressure air obtained in two geometries from a spectroscopic analysis of the emission of the N_2 second positive band system.

inside the capillary rising from about 350 to 500 K at the highest power level studied (slightly less than 0.5 W input power per capillary). In contrast, the measurements made perpendicular to the capillary axis show a rotational temperature of 300 K (essentially room temperature) at the lowest power setting rising to only about 400 K at the highest power level.

While a full understanding of the fundamental processes in the CPE discharge on a microscopic scale has not been achieved, it seems that the capillaries act as individual high-density plasma sources. The initial step is the formation of a streamer-like discharge inside each capillary, whose properties are critically determined by their interaction with the dielectric walls of the capillaries.

6.5.4 Summary

When the plasma size decreases, plasma–surface interactions gain in importance due to the increase of the surface-area to volume ratio. For microglow discharges, this means that the processes in the cathode fall dominate the discharge even more than in common glow discharges. This allows us to generate plasmas with electron energy distributions which contain large concentrations of high-energy electrons, at low gas temperatures. The energy losses to the surfaces surrounding the plasma seem to be the reason for enhanced plasma stability. Microdischarges have allowed us to generate stable glow discharges in atmospheric-pressure gases. The high-pressure operation, and a relatively large concentration of high-energy electrons from the cathode fall of the discharge, favors three-body reactions, such as excimer formation. Electron densities in dc microdischarges have been found to be on the order of 10^{15} cm^{-3} (rather independent of gas type), gas temperatures range from values close to room temperature to approximately 2000 K (lower for noble gases, higher for molecular gases). For the air plasma community the most attractive feature of these microdischarges seems to be the application as plasma cathodes, which support larger volume dc atmospheric pressure air glows, and the application as plasma reactors for chemical and bacterial decontamination of air. But other applications, such as cold atmospheric air plasma jets, generated by flowing atmospheric pressure air through these microdischarges, are emerging. This research field is still young and promises rewards for researchers in non-equilibrium, high pressure glow discharges.

References

Adler F, Kindel E and Davliatchine E 2002 'Comprehensive parameter study of a microhollow cathode discharge containing xenon' *J. Phys. D: Appl. Phys.* **35** 2291

Allmen P von, McCain S T, Ostrom N P, Vojak B A, Eden J G, Zenhausern F, Jensen C and Oliver M 2003 'Ceramic microdischarge arrays with individually ballasted pixels' *Appl. Phys. Lett.* **82** 2562

Allmen P von, Sadler D J, Jensen C, Ostrom N P, McCain S T, Vojak B A and Eden J G 2003 'Linear, segmented microdischarge array with an active length of 1 cm: continuous wave and pulsed operation in the rare gases and evidence of gain on the 460.30 nm transition of Xe^{+}' *Appl. Phys. Lett.* **82** 4447

Amorer L E 1999 PhD Thesis, Stevens Institute of Technology, unpublished

Badareu E and Popescu I 1958 'Research on the double cathode effect' *J . Electr. Contr.* **4** 503

Becker K, Kurunczi P and Schoenbach K H 2002 'Collisional and radiative processes in high-pressure discharge plasmas' *Phys. Plasmas* **9** 2399

Block R, Laroussi M, Leipold F and Schoenbach K H 1999 'Optical diagnostics for non-thermal high pressure discharges' in *Proc. 14th Intern. Symp. Plasma Chemistry*, Prague, Czech Republic, vol II, p 945

Block R, Toedter O and Schoenbach, K H 1999 'Gas temperature measurements in high pressure glow discharges in air' in *Proc. 30th AIAA Plasma Dynamics and Lasers Conf.*, Norfolk, VA, paper AIAA-99-3434

Borodin V S and Kagan Yu M 1966 'Investigation of hollow-cathode discharge. I. Comparison of the electrical characteristics of a hollow cathode and a positive column' *Sov. Phys.–Tech. Phys.* **11** 131

Chen J, Park S-J, Fan Z, Eden J G and Liu C 2002 'Development and characterization of micromachined hollow cathode plasma display devices' *J. Microelectromechanical Systems* **11** 536

Choi K C 1999 'A new dc plasma display panel using microbridge structure and hollow cathode discharges' *IEEE Trans. Electron Devices* **46** 2256

Choi K C and Tae H-S 1999 'The characteristics of plasma display with the cylindrical hollow cathode' *IEEE Trans. Electron Devices* **46** 2344

Cobine, J D 1958 *Gaseous Conductors: Theory and Engineering Applications* (New York: Dover Publications) pp 218–225

Eden J G, Park S-J, Ostrom N P, McCain S T, Wagner C J, Vojak B A, Chen J, Liu C, von Allmen P, Zenhausern F, Sadler D J, Jensen J, Wilcox D L and Ewing J J 2003 'Microplasma devices fabricated in silicon, ceramic, and metal/polymer structures: arrays, emitters and photodetectors' *J. Phys. D: Appl. Phys.* **36** 2869

El-Habachi A and Schoenbach K H 1998 'Generation of intense excimer radiation from high-pressure hollow cathode discharges' *Appl. Phys. Lett.* **73** 885

El-Habachi A and Schoenbach K H 1998 'Emission of excimer radiation from direct current, high pressure hollow cathode discharges' *Appl. Phys. Lett.* **72** 22

El-Habachi A, Shi W, Moselhy M, Stark R H and Schoenbach K H 2000 'Series operation of direct current xenon chloride excimer sources' *J. Appl. Phys.* **88** 3220

Fiala A, Pitchford L C and Boeuf J P 1995 'Two-dimensional, hybrid model of glow discharge in hollow cathode geometries' in *Proc. 22nd Conf. on Phenomena in Ionized Gases*, Hoboken, NJ, ed. Kurt H Becker and Erich Kunhardt (Hoboken, NJ: Stevens Institute of Technology) p 191

Frame J W and Eden J G 1998 'Planar microdischarge arrays' *Electr. Lett.* **34** 1529

Frame J W, John P C, DeTemple T A and Eden J G 1998 'Continuous-wave emission in the ultraviolet from diatomic excimers in a microdischarge' *Appl. Phys. Lett.* **72** 2634

Frame W, Wheeler D J, DeTemple T A and Eden J G 1997 'Microdischarge devices fabricated in silicon' *Appl. Phys. Lett.* **71** 1165

Frank K, Ernst U, Petzenhauser I and Hartmann W 2001 *Conf. Record IEEE Intern. Conf. Plasma Science*, Las Vegas, NV, p 381

Gill P and Webb C E 1977 *J. Phys. D* **10** 299

Gewartkovski J W and Watson H A 1965 *Principles of Electron Tubes* (Princeton: Van Nostrand-Reinhold)

Güntherschulze A 1923 *Z. Tech. Phys.* **19** 49

Guo Y-B and Hong F C-N 2003 'Radio-frequency microdischarge arrays for large-area cold atmospheric plasma generation' *Appl. Phys. Lett.* **82** 337

Helm H 1972 'Experimenteller Nachweis des Pendel-Effektes in einer zylindrischen Nieder-druck-Hohlkathode-Entladung in Argon' *Z. Naturf.* **A27** 1812

Hsu D D and Graves D B 2003 'Microhollow cathode discharge stability with flow and reaction' *J. Phys. D: Appl. Phys.* **36** 2898

Kogelschatz U 2004 'Excimer lamps: history, discharge physics and industrial applica-tions' in *Atomic and Molecular Pulsed Lasers V*, Tarasenko V F, Mayer G F and Petrash G G (eds) and *Proc., SPIE* **5483** 272

Kunhardt E E and Becker K 1999 US Patents 5872426, 6005349 and 6147452

Kunhardt E E, Becker K and Amorer L E 1997a *Proc. 12th International Conference on Gas Discharges and their Applications*, Greifswald, Germany, p I-374

Kunhardt E E, Becker K, Amorer L E and Palatini L 1997b *Bull. Am. Phys. Soc.* **42** 1716

Kunhardt E E, Korfiatis G P, Becker K and Christodoulatos C 1998 'Non-thermal plasma technology for remediation of air contaminants' in *Proc. 4th International Confer-ence on Protection and Restoration of the Environment*, Halkidiki, Greece 1998 edited by G P Korfiatis

Kurunczi P, Abramzon N, Figus M and Becker K 2003 'Measurement of rotational temperatures in high-pressure microhollow cathode MHC and capillary plasma electrode CPE discharges' to appear in *Acta Physica Slovakia*

Kurunczi P, Shah H and Becker K 1999 'Hydrogen Lyman-α and Lyman-β emissions from high-pressure microhollow cathode discharges in Ne–H_2 mixtures' *J. Phys. B* **32** L651

Kurunczi P, Abramzon N, Figus M and Becker K 2004 *Acta Physica Slovakia* **57** 115

Kurunczi P, Lopez J, Shah H and Becker K 2001 'Excimer formation in high-pressure MHC discharge plasmas in He initiated by low-energy electrons' *Int. J. Mass Spectrom.* **205** 277

Kurunczi P, Martus K and Becker K 2003 'Neon excimer emission from pulsed high-pressure MHC discharge plasmas' *Int. J. Mass. Spectrom.* **223/224** 37

Leipold F, Stark R H, El-Habachi A and Schoenbach K H 2000 'Electron density measurements in an atmospheric pressure air plasma by means of infrared hetero-dyne interferometry' *J. Phys. D: Appl. Phys.* **33** 2268

Miclea M, Kunze K, Franzke J, Leis F, Niemax K, Penache C, Hohn O, Schoessler S, Jahnke T, Braeuning-Demian A and Schmidt-Boecking H 2002 'Decomposition of halogenated molecules in a microstructured electrode glow discharge in atmos-pheric pressure' *Proc. of Hankone VIII*, Puehajaerve, Estonia, vol 1, p 206

Miyake M, Takahaski H, Yasuoka K and Ishii S 1999 *Conference Record, IEEE Intern. Conf. Plasma Science*, Monterey (Piscataway, NJ: CA Institute of Electrical and Electronic Engineers) p 143

Mohamed A-A H, Block R and Schoenbach K H 2002 'Direct current glow discharges in atmospheric air' *IEEE Trans. Plasma Science* **30** 182

Moselhy M and Schoenbach K H 2004 'Excimer emission from cathode boundary layer discharges' *J. Appl. Phys.* **95** 1672

Moselhy M, Petzenhauser I, Frank K and Schoenbach K H 2003 'Excimer emission from microhollow cathode discharges in argon' *J. Phys. D: Appl. Phys.* **36** 2922

Moselhy M, Shi W, Stark R H and Schoenbach K H 2001b 'Xenon excimer emission from pulsed microhollow cathode discharges' *Appl. Phys. Lett.* **79** 1240

Moselhy M, Shi W, Stark R H and Schoenbach K H 2002 *IEEE Trans. Plasma Sci.* **30** 198

Moselhy M, Stark R H, Schoenbach K H and Kogelschatz U 2001a 'Resonant energy transfer from argon dimers to atomic oxygen microhollow cathode discharges' *Appl. Phys. Lett.* **78** 880

Moskwinski L, Ricatto P J, Abramzon N, Becker K, Korfiatis G P and Christodoulatos C 2003 *Proc. XIV Symposium on Applications of Plasma Processes (SAPP)*, Jasna, Slovakia

Okazaki S, Kogoma M, Uehara M and Kimura Y 1993 *J. Phys. D* **26** 889

Panikov N S, Paduraru A, Crowe R, Ricatto P J, Christodoulatos C and Becker K 2002 'Destruction of *Bacillus subtilis* cells using an atmospheric-pressure dielectric capillary electrode discharge plasma' *IEEE Trans. Plasma Sci.* **30** 1424

Park H I, Lee T I, Park K W and Baik H K 2003 'Formation of large-volume, high pressure plasmas in microhollow cathode discharges' *Appl. Phys. Lett.* **82** 3191

Park S-J, Eden G, Chen J and Liu C 2001 'Independently addressable subarrays of silicon microdischarge devices: electrical characteristics of large 30×30 arrays and excitation of a phosphor' *Appl. Phys. Lett.* **13** 2100

Park S-J, Eden J G and Ewing J J 2002 'Photodetection in the visible, ultraviolet, and near-infrared with silicon microdischarge devices' *Appl. Phys. Lett.* **81** 4529

Park S-J, Wagner C J, Herring C M and Eden J G 2000 'Flexible microdischarge arrays: metal/polymer devices' *Appl. Phys. Lett.* **77** 199

Paschen F 1916 *Ann. Phys.* **50** 901

Penache C, Braeuning-Demian A, Spielberger L and Schmidt-Boecking H 2000 'Experimental study of high pressure glow discharges based on MSE arrays' *Proc. of Hakone VII*, Greifswald, Germany, vol 2, p 501

Penache C, Datta S, Mukhopadhyay S, Braeuning-Demian A, Joshi P, Hohn O, Schoessler S, Jahnke T and Schmidt-Boecking H 2002 'Large area surface modification induced by parallel operated MSE sustained glow discharges' *Proc. of Hakone VIII*, Puehajaerve, Estonia, vol 2, p 390

Penache C, Miclea M, Braeuning-Demian A, Hohn O, Schoessler S, Jahnke T, Niemax K and Schmidt-Boecking H 2003 'Characterization of a high-pressure microdischarge using diode laser atomic absorption spectroscopy' *Plasma Sources Science and Technology* **11** 476

Petzenhauser I, Biborosch L D, Ernst U, Frank K and Schoenbach K H 2003 'Comparison between the ultraviolet emission from pulsed microhollow cathode discharges in xenon and argon' *Appl. Phys. Lett.* **83** 4297

Sankaran R M and Giapis K P 2001 'Maskless etching of silicon using patterned microdischarges' *Appl. Phys. Lett.* **79** 593

Sankaran R M and Giapis K P 2002 'Hollow cathode sustained plasma microjets: characterization and application to diamond deposition' *J. Appl. Phys.* **92** 2406

Sankaran R M and Giapis K P 2003 'High-pressure micro-discharges in etching and deposition applications' *J. Phys. D: Appl. Phys.* **36** 2914

Schaefer G and Schoenbach K H 1990 'Basic mechanisms contributing to the hollow cathode effect' in Gundersen M and Schaefer G (eds) *Physics and Applications of Pseudosparks* (New York: Plenum Press) p 55

Schoenbach K H, El-Habachi A, Shi W and Ciocca M 1997 'High-pressure hollow cathode discharges' *Plasma Sources Sci. Techn.* **6** 468

Schoenbach K H, El-Habachi A, Moselhy M M, Shi W and Stark R H 2000 'Microhollow cathode discharge excimer lamps' *Physics of Plasmas* **7** 2186

Schoenbach K H, Moselhy M and Shi W 2004 'Selforganization in cathode boundary layer microdischarges' *Plasma Sources Science and Technology* **13** 177

Schoenbach K H, Moselhy M, Shi W and Bentley R 2003 'Microhollow cathode discharges' *J. Vac. Sci. Technol. A* **21** 1260

Schoenbach, K H, Verhappen R, Tessnow T, Peterkin P F and Byszewski W 1996 'Microhollow cathode discharges' *Appl. Phys. Lett.* **68** 13

Shi W, Stark R H and Schoenbach K H 1999 'Parallel operation of microhollow cathode discharges' *IEEE Trans. Plasma Science* **27** 16

Stark R H and Schoenbach K H 1999 'Direct current high pressure glow discharges' *J. Appl. Phys.* **85** 2075

Stark R H and Schoenbach K H 1999 'Direct current glow discharges in atmospheric air' *Appl. Phys. Lett.* **74** 3770

Stark R H and Schoenbach K H 2001 'Electron heating in atmospheric pressure glow discharges' *J. Appl. Phys.* **89** 3568

Stockhausen G and Kock M 2001 *J. Phys. D* **34** 1683

Sturges D J and Oskam H J 1964 'Studies of the properties of hollow cathode glow discharges in helium and neon' *Appl. Phys.* **35** 2887

Vojak B A, Park S-J, Wagner C J, Eden J G, Koripella R, Burdon J, Zenhausern F and Wilcox D L 2001 'Multistage, monolithic ceramic microdischarge device having an active length of ~0.27 mm' *Appl. Phys. Lett.* **78** 1340

Walsh A 1956 *Spectrochim. Acta* **7** 108

White A D 1959 'New hollow cathode glow discharges' *J. Appl. Phys.* **30** 711

Yalin A P, Yu Z Q, Stan O, Hoshimiya K, Rahman A, Surla V K and Collins G J 2003 'Electrical and optical emission characteristics of radio-frequency-driven hollow slot microplasmas operating in open air' *Appl. Phys. Lett.* **83** 2766

Yu Z, Hoshimiya K, Williams J D, Polvinen S F and Collins G J 2003 'Radio-frequency-driven near atmospheric pressure microplasma in a hollow slot electrode configuration' *Appl. Phys. Lett.* **83** 854

6.6　Corona and Steady State Glow Discharges

6.6.1　Introduction

The negative corona is one of the oldest electrical discharges in ambient air. Usually, its operation is limited towards higher currents by transition to a spark. Recent progress in a special discharge technique resulted in realizing a glow discharge in ambient air (Akishev *et al* 1991), which has a cathode in the form of a pin array and exists at much higher electric current per pin. It is well known that with increase of gas pressure a glow discharge becomes unstable against spark formation. Until now it has been the common opinion that the classical glow discharge may exist only at low gas pressures. Actually, detailed studies on the mechanisms of glow discharge instabilities resulted in a substantial extension of the gas pressure range (up to about 1 atm), where a stable glow discharge can be maintained. Evidently, an alternative way to realize the glow discharge could be stabilization of the traditional corona. Now both these approaches (increase of pressure in the classical glow discharge and increase of current in the traditional negative corona discharge) were pursued, and transitions between negative corona, glow and spark forms of discharge were studied. One of the purposes of this section is to present a modern understanding of relationships between the mentioned forms of discharges at atmospheric pressure.

The material in this section is organized as follows: first, the methods to control negative corona parameters are described, then properties of sub- and atmospheric pressure glow discharge (APGD) in air flow and results of studies on transitions between corona, glow and spark forms are reported, and, finally, pulsed diffused discharge techniques are discussed briefly. Particular attention will be paid to basic physical processes lying behind the observed phenomena.

6.6.2 Methods to control negative corona parameters

Upon applying a step of high voltage (that is, over an inception one), the ignition of a negative corona is accompanied by a sharp peak of discharge current with duration of a pulse of about 10^{-7} s. For electronegative gases (air, O_2, etc.), in which electrons are quickly converted to negative ions, the pulsed regime of the corona with regular spikes of discharge current is established. The current pulses are named Trichel pulses. It is well known (Cross *et al* 1986) that the amplitude of established Trichel pulses is much lower than of the first pulse. The total voltage concentrated around the pin controls the dynamics of the first pulse. The extremely strong electron avalanches create a wave of growing positive charge that moves rapidly to the cathode (Morrow 1985, Černák and Hosokawa 1991, Napartovich *et al* 1997). As a result, the cathode layer and a plasma region are formed in the generation zone. At this moment, the corona current has a maximum value. On the anode this current is closed by the displacement current. In the following pulses, the voltage, which can drop within the pin vicinity, diminishes due to appearance of negative charge in the drift zone. Therefore, these pulses are lower as it is seen in figure 6.6.1 (Akishev *et al* 1999). As expected, the experiments revealed that the amplitude of the first Trichel pulse grows with increasing the step of applied voltage (see figure 6.6.2). In

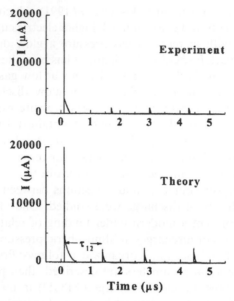

Figure 6.6.1. The establishment in time of Trichel pulses in negative corona in ambient air. Pin to plane distance 7 mm, tip curvature radius 0.057 mm, voltage applied $U_0 = 6$ kV.

contrast, the amplitude of the regular Trichel pulse diminishes with an increase in the applied voltage (figure 6.6.2). The amplitude of the first pulse is greater for shorter spacing, and can reach 0.25 A (figure 6.6.2). Using squared voltage pulses of length shorter than the duration between the first and second pulses τ_{12} and a repetition period long enough to clear

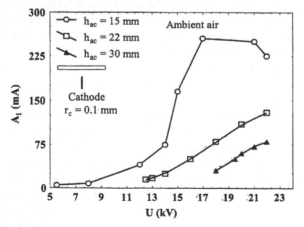

Figure 6.6.2. Amplitude of the first Trichel pulse versus height of applied voltage step for different inter-electrode spacings.

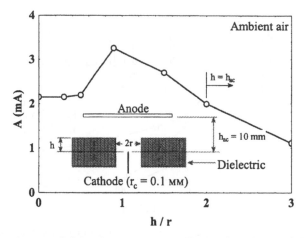

Figure 6.6.3. Influence of dielectric screens on amplitude of regular Trichel pulses at the near-inception applied voltage.

the space from negative charges, one can realize a periodical pulse regime with the height of each pulse of about 1 A.

This auto-pulsing mode of the negative corona in air is observed at low currents of this discharge. This regime can be useful for different practical applications because the current amplitude of a single pulse is far in excess of the average corona current. Experimental studies were carried out on the influence of geometric and gas-dynamic factors and on amplitude and repetition frequency of Trichel pulses, to find out the main experimental parameters controlling them (Akishev *et al* 1996). Generally, it is known that the amplitude of the regular Trichel pulse rises as the radius of pin increases (see, for example, Scott and Haddad 1986). Akishev *et al* (1996) have shown that in dependence on parameter variation a strong increase of the Trichel pulse amplitude, as well as full suppression of them, can be realized for a fixed pin radius.

It was found for ambient and dry air that the amplitude of the regular Trichel pulse depends strongly on the divergence of current lines in the vicinity of the corona pin and on the aperture of the drift region of the corona. To change the geometry of current spreading near the pin, dielectric shields around the pin with variable parameters were employed. Using different shapes of the anode and restriction of the corona cross section modified the geometry of current lines in the drift region. Some results illustrating effects produced by these means on the amplitude of the regular Trichel pulse are shown in figures 6.6.3 and 6.6.4. Restriction of the corona cross-section also influences the repetition frequency of Trichel pulses. Some experimental data are shown in figure 6.6.5. One can see in figures 6.6.4 and 6.6.5 that restriction of the corona space with a dielectric tube results in diminishing the peak corona current and in the rise of the repetition

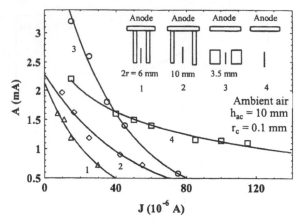

Figure 6.6.4. Amplitude of regular Trichel pulses versus corona current for various corona geometries.

frequency. Figure 6.6.6 demonstrates how the shape of the anode influences the repetition frequency of Trichel pulses. The current profile on the anode can also be broadened by use of a resistive anode. An effect of this resistance-induced current expansion in the drift zone on the amplitude of the regular Trichel pulses is illustrated in figure 6.6.7.

An alternative method to influence the near-pin region of the corona is a powerful jet stream of air directed through a plane mesh anode towards the tip of the pin. The amplitude of the Trichel pulses and repetition period grew with increasing gas stream speed (see figures 6.6.8 and 6.6.9). This effect can be explained by an extension of the generation zone in the vicinity of the corona pin produced by the gas jet stream, which is equivalent, in some degree, to the increase of the pin radius known to enhance pulse amplitudes.

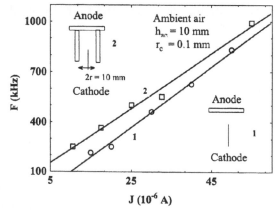

Figure 6.6.5. Frequency of regular Trichel pulses versus corona current for restricted and free-space coronas.

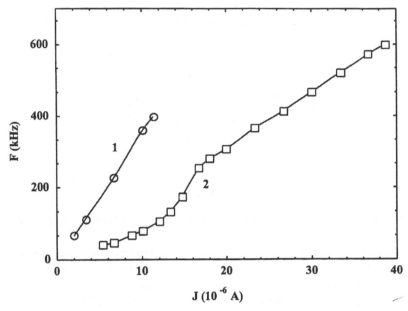

Figure 6.6.6. Repetition frequency of Trichel pulses for different shapes of anode. $h_{ac} = 35\,\text{mm}$, $r_c = 0.08\,\text{mm}$, ambient air. 1, pin–plane geometry; 2, pin inside of semi-sphere.

The presented experimental results demonstrate an opportunity of active control of parameters of regular Trichel pulses by gas-dynamic and geometric factors without changing the pin radius.

Akishev *et al* (1996) reported on a hysteresis in the voltage–current characteristics of the negative corona in the auto-pulsing regime. The

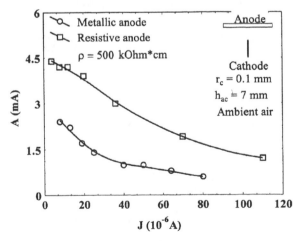

Figure 6.6.7. Amplitude of regular Trichel pulses versus corona current for metal and resistive anodes.

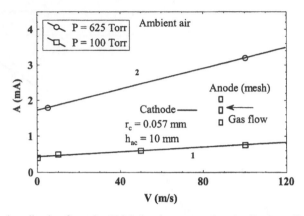

Figure 6.6.8. Amplitude of regular Trichel pulses versus longitudinal gas flow velocity.

experiment showed that the average corona current in this regime depends on the direction of change of the applied voltage (figure 6.6.10). Figure 6.6.11 demonstrates the increase of the upper current of the hysteresis region with gas pressure. While the form and repetition frequency of Trichel pulses can be satisfactorily explained by numerical modeling (Napartovich *et al* 1997, Akishev *et al* 2002b), the phenomenon of hysteresis of this regime reflects complicated physics in the generation zone, which still cannot be described adequately.

6.6.3 DC glow discharge in air flow

The first report on observation of steady glow discharge in transverse air flow at atmospheric pressure (Akishev *et al* 1991) was the result of long-term

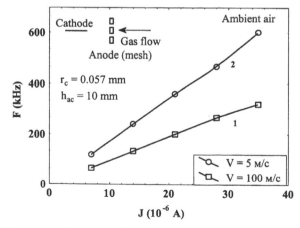

Figure 6.6.9. Frequency of regular Trichel pulses versus longitudinal gas flow velocity.

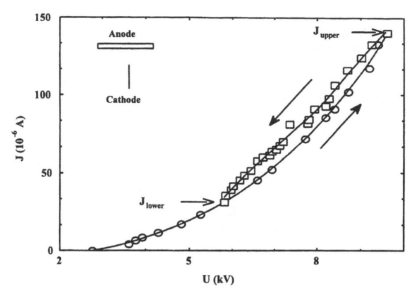

Figure 6.6.10. Hysteresis in voltage–current characteristic (VCC) of negative corona. J_{upper} is the current at pulses disappearance on the growing branch of the VCC, J_{lower} is the current for appearance of Trichel pulses at diminishing voltage, $h_{ac} = 7\,mm$, $r_c = 0.08\,mm$, ambient air.

studies on glow discharge properties at moderate pressures summarized in a paper of Napartovich and Akishev (1993a). The following features were recognized as the most important for approaching the atmospheric pressure range: cathode sectioning with individual ballast resistors for each cathode

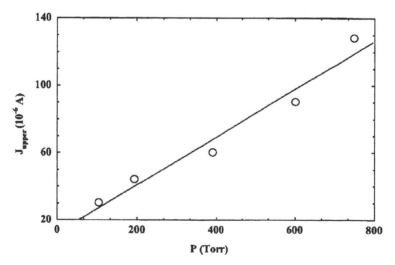

Figure 6.6.11. Current of disappearance of Trichel pulses in pin–plane corona versus gas pressure. $h_{ac} = 10\,mm$, $r_c = 0.06\,mm$, ambient air.

segment and fast gas flow. Cathode sectioning serves to elucidate transition from a high current density at the cathode surface to a required lower current density in the discharge volume. Ballast resistors limiting the current on each segment stabilize the glow discharge against arcing. The gas flow serves to remove heated gas from the discharge gap and additionally stabilizes the discharge by restriction of the residence time of gas in a region with a high electric field.

Sectioning of a cathode makes the spatial structure of a glow discharge near the cathode rather complicated. A transient region appears where the separate current channels originated from different cathode elements are expanding and combining with each other. The cathode is a periodic array of sharp pins, and the anode is a flat plate. In the discharge structure inside one cell of the nearly-periodical array, five regions can be distinguished known from the classical glow discharge at low gas pressures: a cathode layer, a negative glow, a Faraday dark space, a plasma column, and an anode layer.

The well-known dependence of the cathode current density of a normal glow discharge on the gas density ($j_c \approx N^2$) retains its validity up to the pressure of the order of 1 atm. At a fixed cathode area, the current per pin grows with pressure. It was shown by Akishev *et al* (1984) that at higher pressures the amplitude of the current per pin is limited by some instability of the cathode layer resulting in the formation of a cathode spot differing from known low pressures arc spots (Mesyats and Proskurovsky 1989). Non-uniform dielectric films, which are usually present on a metal surface, can trigger this instability. If the current per pin exceeded this critical value, an intermediate cathode spot forms with a current density of the order of $300 \, A/cm^2$. With further current increase, this intermediate spot transforms to the arc spot (Akishev *et al* 1985a) destroying the cathode surface. Existence of the limiting current per pin determines the allowable size of the pin at a given pressure.

The negative glow appears as a result of the relaxation of suprathermal electrons with energies nearly corresponding to the cathode voltage drop, V_c. The thickness of the negative glow region is nearly inversely proportional to pressure, and is on the order of fractions of 1 mm for ambient air. In the Faraday dark space the plasma density decreases from the high value caused by the non-equilibrium ionization by the cathode electron beam to the value corresponding to the balance of ionization, attachment, detachment and recombination processes. The size of this zone is determined by the rate of the plasma decay and by plasma transport processes. For an air plasma, the size of this zone turned out to be on the order of a few centimeters at pressure $p = 100$ torr (Akishev *et al* 1981). With rising pressure the length of this transition region is rapidly decreasing, because a three-body attachment process with the rate proportional to p^2 governs the plasma decay. At atmospheric pressure this length is about of 1–2 mm. Respectively, at

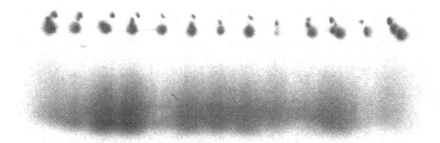

Figure 6.6.12. Photograph (negative) of the discharge in room air; the discharge current per pin is 39 μA.

atmospheric pressure the discharge in the gap of length about 1 cm consists mostly of a plasma column with an electric field strength determined by the local plasma density balance.

Neighboring plasma columns in the multi-pin cathode construction overlap at the distance approximately equal to the pin array period. Provided this period is less than the discharge gap, the major part of the discharge space is occupied by combined plasma columns with weak modulation of its properties. Figure 6.6.12 shows the photograph of this discharge taken in the direction of the air flow. In this device only a single row of pins transverse to gas flow was installed. In general, the multi-pin cathode was arranged in a form of rectangular blocks with some tens of the pins ballasted individually. Parameters of this plasma column in dry and humid air were measured and numerically simulated for fast-flow multi-pin glow discharges (Akishev *et al* 1994a).

Although the anode layer occupies a relatively small fraction of discharge volume, it is of great importance for discharge stability. The voltage–current characteristic of an anode layer at higher gas pressure has a negative slope (Pashkin 1976). As a result, it is unstable to anode spot formation with a high current density and elevated electric field (Dykhne and Napartovich 1979). The plasma layer adjoined to the negatively charged anode sheath plays the role of a distributed ballast resistor that stabilizes the spot-forming instability. Two-dimensional numerical simulations by simultaneous solution of plasma transport equations and the Poisson equation (Dykhne *et al* 1982, 1984) for the glow discharge in nitrogen and air demonstrated that anode spot formation is followed by the contraction of the current channel uniformly through the discharge gap. This model did not include any mechanism of the bulk plasma instability. The formation of anode spots in glow discharges in mixtures of N_2 and O_2 at a very low discharge current was observed experimentally (Akishev *et al* 1982). Because of a low discharge current density it is improbable that any bulk plasma instability may play some role. The time for the appearance of anode spots in the experiment was of the order of that calculated later by Dykhne *et al* (1984). Since the plasma in the plasma column is stable, the formation of the anode spot results in a situation where the high plasma density and the high electric field strength are localized in the same space. Conditions for triggering plasma instability are realized in this region earlier than anywhere else. Then the plasma density will grow further because of the instability and this object will propagate into the bulk of plasmas, forming a bright filament.

Special experiments with plasma perturbations produced by an auxiliary pulsed discharge demonstrated high stability of the bulk plasma (Akishev *et al* 1985b). It turned out that any perturbation created, decayed quickly. However, this perturbation can initiate the formation of the anode spot. Thus this spot serves as an embryo for filament growth.

Akishev *et al* (1987) made special arrangements to study the evolution of a filament under controlled conditions. Filaments propagating from the anode and from the cathode were studied. The influence of a distributed resistance of the anode on the filament evolution was also explored. A simplified theory was formulated which satisfactorily describes the propagation of the filament as a function of its length. The filament growth time was found to be of the order of 100 μs. This indicates that a fast gas flow can prevent its formation.

The knowledge gained in these studies on the glow discharge in air at moderate pressures served as a basis for the development of non-thermal plasma sources in atmospheric pressure air, which were successfully applied for pollutant removal and surface treatments (Akishev *et al* 1993a, 1994a, 2001, 2002a, Napartovich *et al* 1993a,b, Vertriest *et al* 2003). A photograph (negative) of the discharge in the steady-state glow regime is shown in figure 6.6.12. Depending on the gas-flow velocity, the spacing length and the electrode construction (form of the individual pin, shape and resistance of the anode) electric power densities in glow discharge may vary in the interval 30–500 W/ cm^3, which are values that are much higher than those obtained in corona discharges.

6.6.4   Transitions between negative corona, glow and spark discharge forms

To get a clear understanding of how the dc glow discharge in flowing gas relates to known electric discharges at atmospheric pressure, it is important to explore how this form transforms to the known corona and spark discharges under proper variation of parameters. Such studies were performed for single-pin as well as for multi-pin cathode configurations.

6.6.4.1   Single-pin to plane discharge

As a first step, the peculiarities of the voltage–current characteristics (VCC) of the low current discharge between a single cathode pin and an anode plate in air at atmospheric pressure were explored. Contrary to the known experimental studies of other authors (see review article by Chang *et al* 1991), a very large ballast resistor for the cathode pin was taken in the experiments ($R \approx 20\,\mathrm{M\Omega}$) in order to observe the corona-to-glow discharge transition and to avoid the spark discharge. Fast circulation of gas through the inter-electrode gap prevents the local overheating of gas in the vicinity of electrodes and intensifies the turbulent diffusion in the bulk of corona. Therefore, it is a very effective method for stabilization of the diffusive mode of a negative corona. The large ballast resistor is also an effective stabilizer at small gas flow velocities. Use of special procedures for perfecting the shape of electrodes and gas-dynamic stabilization of the near-electrode regions of the corona led to a dramatic increase of the threshold current for sparking, and resulted in a new current mode of discharge, interposed between corona mode and spark mode. The typical reduced VCC of the discharge in transverse flow of air at atmospheric pressure is presented in figure 6.6.13 for metallic and resistive anodes. The reduced electric field in the near-anode region rises with current and reaches a critical value at some critical current I_1. Thereafter the ionization and detachment processes in the drift zone become more intense. This results in the formation of a quasi-neutral plasma. As a consequence, the electrons make a contribution (that will grow more and more with increasing total current) to the charge transfer through the drift zone. In this way, the corona discharge has turned into a glow discharge (Akishev *et al* 1993b).

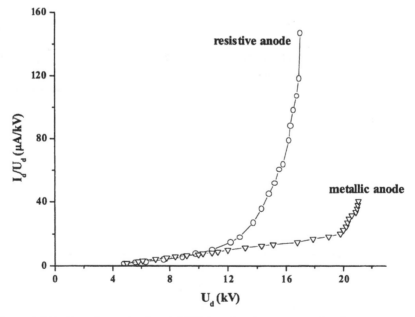

Figure 6.6.13. Experimental reduced VCC for pin–plane construction in transverse air flow, $h = 10.5\,\mathrm{mm}$, pin curvature radius $0.06\,\mathrm{mm}$, $p = 750\,\mathrm{torr}$, gas flow velocity $65\,\mathrm{m/s}$.

Let us designate I_1 as the threshold current for the corona-to-glow discharge transition and I_2 as the threshold current for the glow discharge-to-spark transition. The discharge mode is the classical negative corona, if the current is lower than I_1. The gap between the electrodes is dark in this case. There is a negative space charge in the bulk between electrodes owing to the negative ions. The negative point-to-plane corona at the discharge current lower than nearly 120 µA generates regular Trichel pulses. The typical repetition frequencies for the Trichel pulses were in the range 10–50 kHz. The pulseless corona was observed for currents in excess of 120 µA and lower than I_1. Parametric dependences of I_1 are illustrated in figure 6.6.14 for a metallic plate anode. The critical current grows with gas flow velocity and spacing length.

Once the amplitude of the current has reached the value I_1, the transition from the negative corona to the glow discharge occurs. In this regime, a diffusive glow column is formed near the axis of a pin–plane discharge. The current of the glow discharge is steady and has no pronounced modulation in time. The principal difference between the glow discharge and the negative corona is the existence of quasi-neutral plasma in the bulk of the APGD. The dominant current carriers in the glow mode are free electrons instead of negative ions in the case of the negative corona. If the amplitude of the current surpasses the critical value I_2, the discharge turns into the non-stationary regime. In this regime a lot of irregular bright and fine sparks are observed

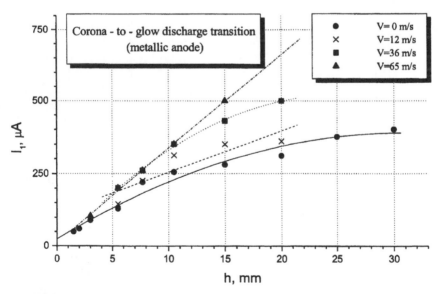

Figure 6.6.14. Critical current of corona discharge, I_1, corresponding to appearance of glow discharge within pin–plane gap versus inter-electrode gap length, h. Ambient air at atmospheric pressure. Anode is metallic plate.

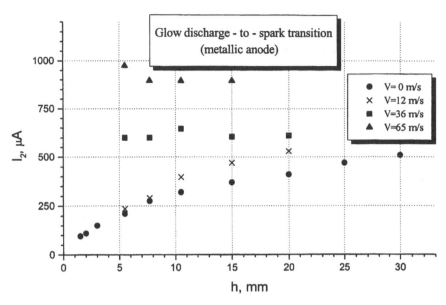

Figure 6.6.15. Critical current of glow discharge, I_2, corresponding to appearance of spark within pin–plane gap versus inter-electrode gap length, h. Ambient air at atmospheric pressure. Anode is metallic plate.

in the gap, and the discharge current exhibits drastic irregular changes in time. Traditionally, spark formation was observed in the corona discharge prior to its transition to the recently revealed glow discharge mode (Akishev *et al* 1993). Special research on the scenarios of corona-to-spark transition is described in this book in section 2.5.2. Parametric dependences of I_2 corresponding to glow discharge-to-spark transition are illustrated in figure 6.6.15 for the metallic anode plate. It is seen that the gas flow velocity is the most important factor efficiently stabilizing the glow discharge. By replacing the metallic anode by a resistive plate the critical current for glow discharge-to-spark transition can be further increased about two to five times. Further studies inspired by an idea to diminish the current density at the anode axis, in order to improve glow discharge stability against sparking, resulted in the development of practical recommendations demonstrating their fruitfulness (Akishev *et al* 2001). Experimental data showed that the local current density on the anode could be decreased by shaping the anode surface, by using a resistive anode material, by using specific-shape cathode pins and by applying a gas flow.

6.6.4.2   *Multi-pin to plane discharge*

Historically, corona and glow forms of the discharge were studied separately: classical glow discharges were observed in low-pressure gases, whereas corona discharges were observed in high-pressure gases, specifically in atmospheric pressure air. The glow discharge is characterized by a high value of the reduced electric field E/N in the inter-electrode gap. This field is sufficiently high for producing intense ionization of a gas resulting in the gap filling with quasi-neutral plasma. In the case of a negative corona, the reduced field in the gap is much lower, and there is a negative space charge in the major part of the gap (ion drift region).

A special electrode system with a multi-pin cathode and a flat metal anode was made to investigate the transition from a negative corona to a glow discharge in air at atmospheric pressure (Akishev *et al* 2000). The pins were stainless-steel needles, 0.5 mm in diameter, tapered to a cone with a tip curvature radius of $R_c = 0.06$ mm. 52 needles were uniformly distributed over an area of 1 cm × 4 cm in four rows of 13 needles in each. The distance d between needles (i.e. the spatial period of the cathode structure) was equal to 3.5 mm and was small compared to the distance between their tips and the anode, $h = 5$–20 mm. In this case, the current density in the negative-corona gap increases substantially (by nearly a factor of $3(h/d)^2$) in comparison with the pin–plane configuration, and the transition from the corona to glow discharge occurs at a relatively low current through each pin.

In order to ensure a stable diffusive regime of the negative corona, the high voltage to each needle was supplied through a high-resistance load: $R = \sim 2$ MΩ. In addition, the anode plate was connected to a high-voltage supply through a 0.2 MΩ resistor. The stability of the corona against its transition to a spark was also ensured by an air flow through the discharge; the cathode unit was oriented with the longer side perpendicular to the air flow. A typical flow velocity was on the order of several tens of meters per second.

Along with recording I–V characteristics, the discharge was photographed in the direction transverse to the air flow. If the discharge is in the corona regime, only the needle ends are luminous, whereas the inter-electrode gap is hardly visible and the anode is dark. The glow discharge, on the other hand, is diffuse and rather uniform, although the discrete structure of the plasma column caused by the discrete structure of the multi-pin cathode is also clearly seen (figure 6.6.12). Figure 6.6.16 shows a typical reduced I–V characteristic of the discharge under study. Here, the ratio I/U (instead of the total discharge current) is plotted versus the discharge voltage U, I being the discharge current per pin.

In the reduced I–V characteristics, we can distinguish two segments (the first in the region of initial corona currents and the second in the region of high currents corresponding to the regime of a developed glow discharge), in which the reduced current is a nearly-linear function of the voltage. It is seen that in the glow discharge the current increases with voltage much more steeply than in the corona regime. This is explained by the increasing role of ionization (which depends strongly on the field) in creating the conductivity in the inter-electrode gap of the glow discharge.

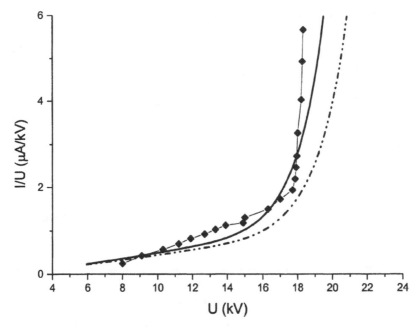

Figure 6.6.16. Reduced *I–V* characteristic of the multi-pin to plane discharge in room air (*I* is the current per pin). The points correspond to the experiment; the solid and dashed-and-dotted lines correspond to the calculations for relative humidity of 30 and 65%, respectively.

The kink point of the reduced *I–V* characteristics can be considered as a critical voltage corresponding to the transition of the corona to a glow discharge. Near this point of the *I–V* characteristic, a luminous thin sheath appears on the anode. This evidences formation of the anode sheath, which is characteristic of a glow discharge. At voltages higher than the critical one, the gap luminosity increases sharply with the current and the discharge exhibits more and more features typical of glow discharges.

Let us define a threshold I_1 for the transition from the corona to a glow discharge as a moment when the luminous anode sheath becomes visible. Figure 6.6.17 shows the dependence of the threshold current on the inter-electrode distance h. A similar dependence of the threshold current I_2 for the transition from the glow discharge to a spark is also shown. Hence, the current range in which a uniform glow discharge at atmospheric pressure can exist is bounded by two curves $I_1(h)$ and $I_2(h)$. Note that this range may be extended substantially by using gas-dynamic effects and anodes of special design.

A 1.5-dimensional numerical model of the discharge described in section 2.5.2 was employed for modeling corona-to-glow discharge transition in multi-pin-to-plane geometry for humid air. The model includes the

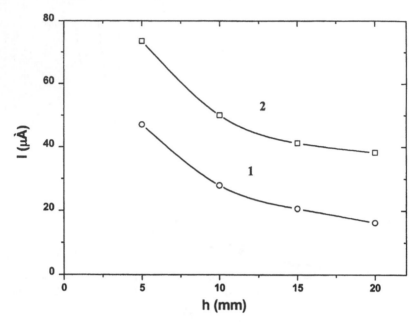

Figure 6.6.17. Threshold currents I_1 (curve 1) and I_2 (curve 2) per pin for the transition from the corona to a glow discharge and from the glow discharge to a spark, respectively, as functions of the inter-electrode distance h.

ionization, three-body attachment of electrons to an oxygen molecule, detachment, and ion–ion recombination. The presence of water vapor in air was taken into account by introducing an additional attachment rate caused by three-body attachment to oxygen with the participation of water molecules acting as a third body. In these calculations, the equivalent radius of the discharge at the anode was determined from the discharge area per pin. The total area was calculated by the formula $S = S_0 + 2\alpha(a + b)h$, where S_0 is the area enveloped by the contour drawn through the edge pins, $2(a + b)$ is the circumference of this contour, and α is a phenomenological parameter ($\alpha = 0.5$). The shape of the current channel was chosen according to visual observations: in a distance of one third of the full distance between the electrodes, the channel rapidly broadens until its radius becomes equal to the anode radius; further, the cross section area remains constant. Possible variations in the shape of the current channel due to variations in the current value were neglected in calculations.

In the calculations, all the parameters were reduced to the conditions referred to one pin. The equivalent ballast resistance in the discharge circuit for one pin was $R = 12.2\,\text{M}\Omega$ (the resistance in the anode circuit was taken into account). Note that a series of calculations of I–V characteristics was performed with various values of the ballast resistance (from $100\,\text{k}\Omega$ to

18 MΩ). These calculations showed that the value of the ballast resistance has little effect on the shape of the I–V characteristics.

Upon calculating the distribution of the reduced electric field across the discharge gap, we calculated the distribution of radiation intensity in the discharge. It was assumed that the first and second positive systems of nitrogen make the main contribution to the radiation and that the total radiation intensity is proportional to the total excitation rate for these levels. The excitation rate constants for these levels were determined numerically by solving the Boltzmann equation for the electron energy distribution function. Densities in the inter-electrode gap were computed based on the numerical 1.5-dimensional code, the I–V characteristics of the discharge, the light emission distribution along the current channel of an individual pin, the longitudinal profile of the electric field, the components of the total current, and the charged-particle (electron, ion, and negative ion).

An example of comparison between the computed reduced I–V characteristics and the experimental ones is shown in figure 6.6.16. It is seen that, for the parameters given, the calculation results are in good qualitative and quantitative agreement with the experimentally observed $I(V)$ dependence. The influence of water vapor on the reduced I–V characteristic is illustrated by calculations for two values of air humidity. The calculated distribution of the radiation intensity across the gap is also in good agreement with the experiment. Figures 6.6.18–6.6.20 show self-consistent variations in the profiles of electric field, relative electron current and charge density in the inter-electrode gap as the discharge current varies. The computation was performed for ambient air (relative humidity 30%) and an inter-electrode distance of 10.5 mm. It is seen in figure 6.6.18 that the electric field within the gap (outside of the cathode sheath) has a maximum near the anode. Hence the ionization rate also has a maximum near the anode. Growth of the field to the anode is explained by the attachment of electrons and the decrease in their contribution to the total current (figure 6.6.19). A specific feature of this discharge is a noticeable space charge even at highest discharge current seen in figure 6.6.20.

For higher discharge voltages the profile of the electron component of the current along the discharge gap becomes non-monotonic: after decrease in the region of low fields near the cathode, the electron flux increases again in the region of high fields far from the cathode. As the voltage increases, the electron current minimum shifts inside the gap, and the contribution of the electron current to the total current increases. It is noteworthy that the electron flux in the gap starts to increase at field values when the ionization rate is still low compared to the attachment rate. This finding indicates that the processes of destruction of negative ions play an important role in the growth of the electron flux and the formation of the anode sheath.

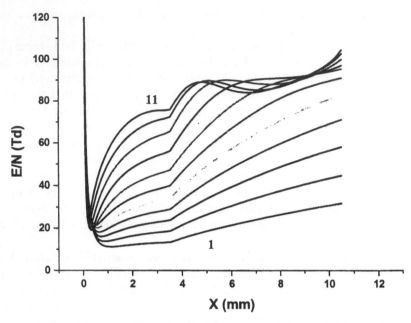

Figure 6.6.18. Axial profile of the reduced electric field for different values of the discharge current listed in table 1 according to numbers 1–11.

Thus, the calculations show that in a multi-pin construction the plasma column in the glow discharge does not form simultaneously along the entire inter-electrode gap. After the anode sheath has formed, the quasi-neutrality conditions are first created near the anode. As the discharge current increases, the region of quasi-neutral plasma extends toward the cathode progressively covering the inter-electrode gap (figure 6.6.20).

It should be noted that the parameters of the plasma column calculated with use of the 1.5-dimensional code are close to that of a glow discharge, which have been computed previously with the zero-dimensional kinetic model (Akishev *et al* 1994a). The results of experimental studies and numerical calculations allow tracing the evolution of the parameters of a

Table 1. Calculated values of current and discharge voltage (U) as a function of power supply voltage (U_0) for ambient air with 30% relative humidity.

	1	2	3	4	5	6	7	8	9	10	11
U_0 (kV)	6	8	10	12	14	16	18	20	22	24	26
U (kV)	5.93	7.90	9.86	11.81	13.75	15.66	17.43	18.92	19.97	20.62	21.08
I (μA)	1.4	2.79	4.76	7.34	10.8	17.5	34.3	75.2	152	261	386

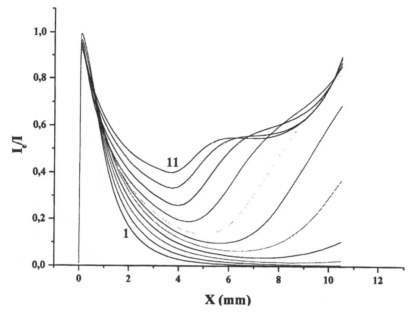

Figure 6.6.19. Axial profile of the electron current contribution to the total current for different values of the discharge current listed in table 1 according to numbers 1–11.

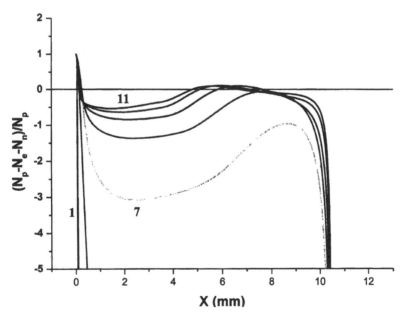

Figure 6.6.20. Axial profile of the normalized space charge for different values of the discharge current listed in table 1 according to numbers 1–11.

multi-pin negative corona during the transition to the glow discharge regime at atmospheric pressure.

6.6.5   Pulsed diffuse glow discharges

At low over-voltages applied to a discharge gap, electron avalanches started near the cathode are weak ones, and the formation of the plasma requires multiple avalanches to proceed with a feedback produced by the secondary-emission processes at the cathode surface (see e.g. Llewellyn-Jones 1966). This is the so-called Townsend mechanism of discharge formation. At high over-voltages in high-pressure gases the discharge gap breakdown usually proceeds in a form of streamers, the number of which depends on many parameters, in particular, on an amount of seed electrons (Korolev and Mesyats 1998). In earlier studies of pulse discharge development in hydrogen at pressures about 1 bar in narrow gaps, Doran and Meyer (1967), Cavenor and Meyer (1969), and Meyer (1969) observed at low over-voltages the formation of a diffuse glow form of the discharge followed by sparking (see also section 2.4).

Applications of high-pressure plasmas for the excitation of gas mixtures for achieving laser action gave a strong impetus to pulse discharge studies. Lasers oscillating on optical transitions of CO_2, excimers, Ar/Xe, N_2 and CO can effectively operate at atmospheric pressures and above, and they have found a wide range of applications (see, for example, Baranov *et al* 1988). For laser applications, it is important to produce uniform non-thermal plasma in a large volume. To solve this problem, a number of methods were proposed for discharge initiation allowing one to avoid streamer and arc formation. In particular, an initial electron number density necessary for overlap of streamers initiated by these electrons was evaluated in works by Koval'chuk *et al* (1970), Baranov *et al* (1972), and Palmer (1974). The criteria derived agreed qualitatively with further more detailed studies. A number of discharge techniques varied by methods of pre-ionization and electrode constructions were developed allowing for pulse glow discharge maintenance in highly electronegative gases like HCl, F_2, and SF_6. An overview of these techniques can be found in Baranov *et al* (1988) and in Korolev and Mesyats (1998).

The pulse-periodical glow discharge is characterized typically by high energy deposition into single pulses, dictated by the necessity to provide sufficiently strong excitation of the laser medium (the almost exclusive application for this discharge type). The pulse periodical mode introduces additional problems of discharge stability (Baranov *et al* 1988): gas-dynamic perturbations from the preceding pulse distort the uniformity of gas flow, resulting in an earlier development of instability in the form of arcs or micro-arcs (also-called filaments). This, in turn, limits repetition frequency, and results in incomplete usage of the gas mixture flow, a serious handicap for some applications of this kind of discharge in industry. However, this problem is important only for high-energy loading in every pulse. Applications not requiring high-energy density can benefit from existing pulse discharge techniques allowing one to achieve homogenous gas excitation with many types of electro-negative additives.

References

Akishev Yu S, Dvurechenskii S V, Zakharchenko A I, Napartovich A P, Pashkin S V and Ponomarenko V V 1981 *Sov. J. Plasma Phys.* **7** 700

Akishev Yu S, Napartovich A P, Pashkin S V and Ponomarenko V V 1982 *Sov. J. Tech. Phys. Lett.* **8** 512

Akishev Yu S, Napartovich A P, Pashkin S V, Ponomarenko V V, Sokolov N A and Trushkin N I 1984 *High Temp.* **22** 157

Akishev Yu S, Napartovich A P, Ponomarenko V V and Trushkin N I 1985a *Sov. Phys. Tech. Phys.* **30** 388

Akishev Yu S, Napartovich A P, Pashkin S V, Ponomarenko V V and Sokolov N A 1985b *High Temp.* **23** 522

Akishev Yu S, Volchek A M, Napartovich A P, Sokolov N A and Trushkin N I 1987 *High Temp.* **25** 465

Akishev Yu S, Levkin V V, Napartovich A P and Trushkin N I 1991 *Proc. XX ICPIG*, Pisa, Italy, vol 4, p 901

Akishev Yu S, Deryugin A A, Kochetov I V, Napartovich A P and Trushkin N I 1993a *J. Phys. D: Appl. Phys.* **26** 1630

Akishev Yu S, Deryugin A A, Karal'nik V B, Kochetov I V, Napartovich A P and Trushkin N I 1993b *Proc. ICPIG XXI*, Bochum, vol. 2, p 293

Akishev Yu S, Deryugin A A, Karal'nik V B, Kochetov I V, Napartovich A P and Trushkin N I 1994a *Plasma Phys. Rep.* **20** 437

Akishev Yu S, Deryugin A A, Elkin N N, Kochetov I V, Napartovich A P and Trushkin N I 1994b *Plasma Phys. Rep.* **20** 511

Akishev Yu S, Deryugin A A, Kochetov I V, Napartovich A P, Pan'kin M V and Trushkin N I 1996 *Hakone V Contr Papers* (Czech Rep.: Milovy) p 122

Akishev Yu S, Grushin M E, Kochetov I V, Napartovich A P and Trushkin N I 1999 *Plasma Phys. Rep.* **25** 922

Akishev Yu S, Grushin M E, Kochetov I V, Napartovich A P, Pan'kin M and Trushkin N I 2000 *Plasma Phys. Rep.* **26** 157

Akishev Yu S, Goossens O, Callebaut T, Leys C, Napartovich A P and Trushkin N I 2001 *J. Phys. D: Appl. Phys.* **34** 2875

Akishev Yu S, Grushin M E, Napartovich A P and Trushkin N I 2002a *Plasmas and Polymers* **7** 261

Akishev Yu S, Kochetov I V, Loboiko A I and Napartovich A P 2002b *Plasma Phys. Rep.* **28** 1049

Baranov V Yu, Borisov V M, Vedenov A A, Drobyazko S V, Knizhnikov V N, Napartovich A P, Niziev V G and Strel'tsov A P 1972 *Preprint of Kurchatov Atomic Energy Inst.* #2248 Moscow (in Russian)

Baranov V Yu, Borisov V M and Stepanov Yu Yu 1988 *Electric Discharge Excimer Noble-Gas Halides Lasers* (Moscow: Energoatomizdat)

Cavenor M C and Meyer J 1969 *Aust. J. Phys.* **22** 155

Černák M and Hosokawa T 1991 *Phys. Rev. A* **43** 1107

Chang J-S, Lawless P A and Yamamoto T 1991 *IEEE Trans. Plasma Science* **19**(8) 1152

Cross J A, Morrow R and Haddad G N 1986 *J. Phys. D: Appl. Phys.* **19** 1007

Doran A A and Meyer J 1967 *Brit. J. Appl. Phys.* **18** 793

Dykhne A M and Napartovich A P 1979 *Sov. Phys. Dokl.* **24** 632

Dykhne A M, Napartovich A P, Taran M D and Taran T V 1982 *Sov. J. Plasma Phys.* **8** 422

Dykhne A M, Elkin N N, Napartovich A P, Taran M D and Taran T V 1984 *Sov. J. Plasma Phys.* **10** 366

Korolev Yu D and Mesyats G A 1998 *Physics of Pulsed Breakdown in Gases* (Yekaterinaburg: URO-PRESS)

Koval'chuk B M, Kremnev V V and Mesyats G A 1970 *Sov.Phys. Dokl.* **191** 76

Llewellyn-Jones F 1966 *Ionization and Breakdown in Gases* (London: John Wiley)

Mesyats G A and Proskurovsky D I 1989 *Pulsed Electrical Discharge in Vacuum* (New York: Springer)

Meyer J 1969 *Brit. J. Appl. Phys.* **20** 221

Morrow R 1985 *Phys. Rev. A* **32** 1799

Napartovich A P and Akishev Yu S 1993a *Proc. XXI ICPIG, Bochum, Germany*, vol 3, pp 207–216

Napartovich A P, Akishev Yu S, Deryugin A A, Kochetov I V and Trushkin N I 1993b in Penetrante B M and Schultheis S E (eds) *Non-Thermal Plasma Techniques for Pollution Control* Part B, NATO ASI Series G, vol 34, pp 355–370

Napartovich A P, Akishev Yu S, Deryugin A A, Kochetov I V and Trushkin N I 1997 *J. Phys. D: Appl. Phys.* **30** 2726

Palmer A J 1974 *Appl. Phys. Lett.* **25** 138

Pashkin S V 1976 *High Temp.* **14** 581

Scott D A and Haddad G N 1986 *J. Phys. D: Appl. Phys.* **19** 1507

Vertriest R, Morent R, Dewulf J, Leys C and van Langenhove H 2003 *Plasma Sources Sci. Technol.* **12** 412

6.7 Operational Characteristics of a Low Temperature AC Plasma Torch

6.7.1 Introduction

Dense atmospheric-pressure plasma can be produced through dc or low frequency discharge operating in the high-current diffused arc mode, such as a plasma torch (Gage 1961, Koretzky and Kuo 1998), which introduces a gas flow to carry the plasma out of the discharge region. Non-transferred dc plasma torches (Boulos *et al* 1994, Zhukov 1994) are usually designed for power levels over 10 kW. In this work, an ac torch for lower power (less than 1 kW) use is described. The volume of a single torch is generally restricted by the gap between the electrodes, which in turn is limited by the available voltage of the power supply. A simple way to enlarge the plasma volume is to light an array of torches simultaneously (Koretzky and Kuo 1998). The torches in an array can be arranged to couple to each other, for example, through capacitors. In doing so, the number of power sources needed to operate the array can be reduced considerably, so that the size of the power supply can be compact—an advantage for practical reasons.

The installation of an array of plasma torches is made easy by introducing a cylindrical-shape plasma torch module (Kuo *et al* 1999, 2001), which has been designed and constructed by remodeling components from two commercially available spark plugs and adding a tungsten wire as the central electrode. A ring-shaped permanent magnet is introduced in the set-up to add a dc magnetic field between the electrodes (Kuo *et al* 2002). Thus each torch module has the size slightly larger than a spark plug and is in the form of a module unit, which screws easily into the base surface of an array. The module as a building block simplifies the design of a large-volume plasma source. It makes the maintenance of the source easy.

The operation and performance of the torch module are described in the following. Power consumption of low frequency discharge for plasma generation is evaluated numerically. The results of numerical simulations for a broad parameter space of plasma species establish a dependence of power consumption on plasma parameters (Koretzky and Kuo 2001), which is useful for minimizing the power budget for each application.

6.7.2 Torch plasma

6.7.2.1 *A magnetized plasma torch module*

A torch module is fabricated by using a surface-gap spark plug (Nippon Denso ND S-29A), which has a concentric electrode pair, as the frame. For torch operation, a gas flow between the electrodes is necessary. Thus, the original electrode insulator, which fills the space between the central and outer electrodes, is replaced with a new one taken from a different spark plug (Champion RN 12YC). This new ceramic insulator has a smaller outer diameter than the original one; hence, an annular gap of 1.81 mm is created for the gas flow. Moreover, the central electrode set in the new ceramic insulator is replaced by a solid 2.4 mm diameter tungsten rod, which is held in place concentrically with the outer electrode, having inner and outer diameters of 6 and 12 mm, by the new insulator and axially by a setscrew in the anode terminal post. The relatively high melting point of tungsten is desirable in the high-temperature environment of the arc. Eight holes of 2 mm diameter each are drilled through the frame (in the section having a screw thread as seen in figure 6.7.1a) of the module to pass gas into the region between the electrodes. The torch is screwed into a plenum chamber (which is not shown) that supplies the feedstock gas and hosts the ring-shaped permanent magnets, one for each torch. The geometry of the

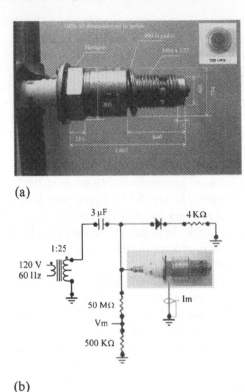

(a)

(b)

Figure 6.7.1. (a) A photo of the plasma torch module, (b) circuit of 60 Hz power supply to run the torch. (Copyright 2002 by AIP.)

electrodes and the dimensions of the parts in the frame of the module are presented in figure 6.7.1a. This torch module has relatively large gap (2 mm) between two electrodes compared to the gaps (usually less than 1 mm) used in the non-transferred dc plasma torches (Boulos *et al* 1994, Zhukov 1994). The discharge is restricted to occur only outside the module by the ceramic insulator inserted between the electrodes. Thus this torch can be operated even with very low gas flow rates. On the other hand, the non-transferred dc plasma torch requires sufficient gas flow to push the arc into the anode nozzle. This electrode feature reduces the power loss to the electrodes considerably.

The ring magnet has outer and inner diameters of 51.8 and 19.6 mm, respectively, and a thickness of 12.2 mm. It produces an axial magnetic field of 0.14 Tesla at the central location of the ring. Each magnet is positioned concentrically around the outer electrode of each module and held inside the plenum chamber. The torch is run by a 60 Hz power supply shown in figure 6.7.1b, which will be described later. Operation of the torch in 60 Hz periodic mode, rather than in dc mode, gives the feedstock

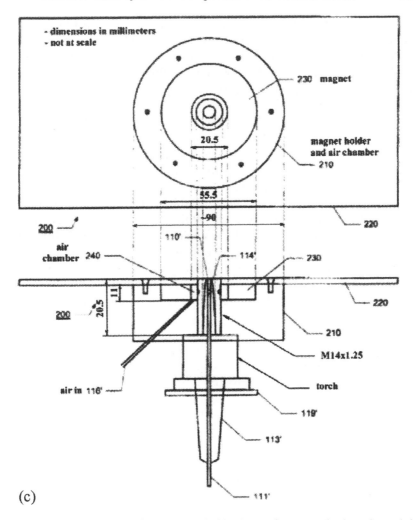

(c)

Figure 6.7.1. (c) Schematics of the top and side views of a magnetized torch module. (Copyright 2002 by AIP.)

gas sufficient time between two consecutive discharges to cool the electrodes. Shown in figure 6.7.1c are schematics of the top and side views of a module. The annular chamber designed for hosting one torch module only is inside the aluminum body indicated in the side view of figure 6.7.1c.

6.7.2.2 Power supply

The power supply and the electrical circuit used to light a single torch module is shown in figure 6.7.1b. As shown, the discharge voltage is provided by a power supply, which includes a power transformer with a turns ratio of

1 : 25 to step up the line voltage of 120 V from a wall outlet to 3 kV, and a 1 μF capacitor in series with the electrodes (i.e. the torch). A branch consists of a diode (15 kV and 750 mA rating) and a resistor (1 kΩ), which is connected in parallel to the torch, is added to the circuit to further step up the peak voltage in half a cycle. During one of the two half cycles when the diode is forward biased, the capacitor is charged to reduce the voltage across the electrodes. During the other half cycle, the diode is reverse biased. The charged capacitor increases the voltage across the electrodes and uses its stored energy to assist the breakdown process and to enhance the discharge.

Using the same circuit for each torch, in general, all of the torches can be connected in parallel to a common power source (i.e. the power transformer) if it has the required power handling capability. The capacitors in the circuit play a crucial role in the discharge. Without them, the torches in the set cannot be lit up simultaneously by a single common source. This is because once one is lit up, it tends to short out the voltage across all of the other electrode pairs connected in parallel. The capacitors work as active ballasting circuit elements. Charging and discharging of each capacitor provides feedback control to the voltage across the corresponding electrode pair.

6.7.2.3   *Plasma torches*

The magnetic field introduced by the ring-shaped permanent magnet is in the (axial) direction perpendicular to the discharge electric field (in the radial direction). It rotates the discharge by the $\mathbf{J} \times \mathbf{B}$ force around the electrodes (in the azimuth direction) and thus enhances the strength and stability of plasma produced by the module, and the lifetime of the electrodes by avoiding discharge at a fixed hot spot. Shown in figure 6.7.2a is a photo of torch plasma produced by this module. Backpressure of air is 17 psia ($\sim$1.156 atm). This torch module can also be run without the ring magnet. A photo of unmagnetized torch plasma is presented in figure 6.7.2b for comparison. The first noticeable difference between these two is their sizes. The volume of magnetized torch plasma is evidently larger. The evenly distributed bright anode spots around the base of magnetized torch demonstrate the rotation of the discharge by the magnetic field, which helps to optimize the torch volume by ballasting the arc constriction and to reduce erosion at hot spots. The disadvantage of adding the magnet to the module is to increase the space between two modules in the array. Use of four magnetized torch modules to enlarge the volume of plasma is demonstrated in figure 6.7.2c.

6.7.2.4   *Voltage and current measurements*

Shown in figure 6.7.3a are the voltage and current waveforms of the discharge in one cycle. During the first half cycle when the diode in the

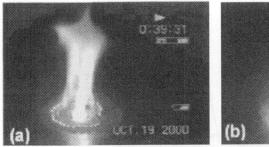

Figure 6.7.2. Torch plasmas produced by (a) a magnetized and (b) an unmagnetized, torch module; the backpressure is 17 psig; (c) a photo of four plasma torches produced by a portable array. (Copyright 2002 by AIP.)

circuit is reversed biased, the discharge is in the low-voltage–high-current arc mode; it evolves to a high-voltage–low-current glow discharge in the other half cycle when the diode becomes forward biased. The product of the voltage and current measurements gives the power function of a single torch, which is shown in figure 6.7.3b. As shown, the peak and average power are about 1.5 kW and 320 W, respectively. The power factor is about 0.62. This may be because the inductance of the transformer is too large. When two torch modules discharge simultaneously by a single power supply, the capacitance of the circuit increases; moreover, the coupling capacitors work as additional dependent sources providing feedback control of the phases of the discharge voltage and current of each torch so that the discharge can stay longer and the system operates with improved power efficiency, as evidenced by the increase of the power factor to 0.96 and the reduction of the total harmonic distortion of the power line to a very low percentage. The results indicate that the electrical performance of the circuit with coupled torches is significantly improved, suggesting that the capacitively coupled plasma torch array be an excellent self-adjusting resistive load to the power line.

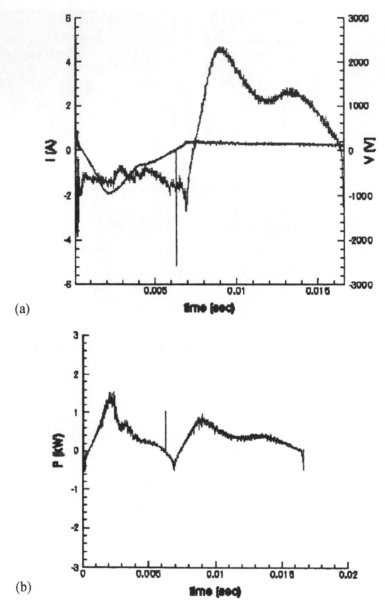

(a)

(b)

Figure 6.7.3. (a) Voltage and current, and (b) power functions of the torch module. (Copyright 2002 by AIP.)

It is noted that the power of this plasma torch depends strongly on the power supply. In an application requiring high power and high temperature torch plasma, the 1 μF capacitor in the power supply is replaced by a 3 μF one and the resistor in series with the diode is increased from 1 to 4 kΩ.

The results (Kuo *et al* 2003) show that the torch plasma has a peak and average power of 3.8 and 1.5 kW, respectively.

6.7.2.5   Temperature and density measurements

A method (Kuo *et al* 1999) based on thermal equilibrium and a detailed analysis of heat loss from a copper wire placed in a torch is applied to measure the temperature of the torch plasma. Consider the model of a long wire with only a portion immersed in the torch. The wire in the torch heats up due to forced convection from the torch and loses energy in the torch via radiation. Outside the torch, the wire acts as a cylindrical pin fin and loses energy via conduction along the wire and natural convection with ambient air. A wire with a small diameter reduces heat loss from the pin fin, which increases the wire temperature in the torch, compared to a larger diameter wire. So systematically reducing the wire diameter placed in the torch eventually results in a critical wire diameter that just melts or shows signs of softening. The wire so determined has a temperature nearly equal to its melting temperature.

Copper wires of different diameters were used in the experiment, because it is easy to assemble a set of different diameter wires with known purity and emissivity $\varepsilon = 0.8$. The diameters of the wires varied from 10 to 33 mil (1 mil = 1/1000 inch, 0.0254 mm) and the burning time of the torch was up to 1 min. It was found that 10 mil wire melted right away and 33 mil wire remained unscathed. By increasing the diameter of the wire from 10 mil graduately, it was found that 16 mil was a critical diameter. For the 16 mil wire, its status (melted or not melted) depended on its surface condition and location in the torch. The hottest burning spot in the torch was identified. With the temperature of the 16 mil wire determined to be about the melting temperature of copper (1083 °C), a power balance equation could be set up, to determine the torch temperature.

In the experiment, the wire was held by a holder placed at $x = l_0 = 27.5$ mm from the center at $x = -10$ mm of the torch. To reach thermal equilibrium, the power q_{in0}, convected from the gas flow in the torch to the wire, must be balanced by the power losses P_{rad0} and P_{cond0} of the wire, via thermal radiation and thermal conduction, respectively. The power balance condition is written as

$$q_{in0} = Ah_c(T - T_{w0}) = P_{rad0} + P_{cond0} = q_{out0}$$

where $A = \pi D D_t = 2.55 \times 10^{-5} \, \text{m}^2$ is the area of the portion of wire immersed in the torch, $D = 406 \, \mu\text{m}$ (16 mil) and $D_t = 20$ mm are the diameters of wire and torch plasma; $h_c = 0.75(k/D)Re^{0.4}Pr^{0.37} \, \text{W} \, \text{m}^{-2} \, \text{K}^{-1}$ is the forced heat convection coefficient; the Prandtl number $Pr \approx 0.7$ and k is the thermal conductivity; T and T_{w0} are the temperatures of the torch and wire. Based on data for air in table A.4 of the reference book by

Incropera and DeWitt (1996), the Reynolds number is calculated for the flow speed $u = 20$ m/s with the air temperature T as a parameter varying from 1350 to 2200 K. Hence, the power input from torch to wire can be evaluated as a function of T.

The temperature gradient of the wire at $x = 0$ (boundary of torch) is determined by the local power balance condition (Siegel and Howell 1992) for the segment of wire outside the plasma flow ($0 < x < l_0$)

$$A_w k_w \, d^2 T_w/dx^2 = (d/dx)(P_{rad} + P_{fin}) = a(T_w^4 - T_a^4) + b(T_w - T_{air}) \quad (6.7.1)$$

where A_w, k_w, and T_w are the cross section area, thermal conductivity, and temperature of the copper wire; P_{rad} and P_{fin} are the thermal radiation and natural convection power of wire; $a = \pi D \varepsilon \sigma$ and $b = \pi D h_{cn}$; $\sigma = 56.7$ nW m^{-2} K^{-4} is the Stefan–Boltzmann constant; h_{cn} and T_{air} are the natural heat convection coefficient, and temperature of air next to the wire.

Collisions keep the plasma flowing with the gas flow. The temperature T_{air} of air outside the plasma is expected to drop quickly to the ambient temperature $T_a \approx 300$ K. Thus, an average value of 86 W m^{-2} K^{-1} is assumed for the natural heat convection coefficient h_{cn}, which is much smaller than h_c.

Equation (6.7.1) can be integrated to be

$$dT_w/dx = -\{(2\alpha/5)[T_w(T_w^4 - T_a^4) - 4T_a^4(T_w - T_a)]$$
$$+ \beta(T_w - T_a)^2 + (P_{holder}/A_w k_w)^2\}^{1/2} \quad (6.7.2)$$

subjected to the boundary conditions $T_w(0) = T_{w0}$ and $T_w(l_0) = T_a$, where $\alpha = a/A_w k_w = 1.324 \times 10^{-6}$ m^{-2} K^{-3}, $\beta = b/A_w k_w = 2.51 \times 10^3$ m^{-2}, and P_{holder} is the conduction power from wire to the holder.

To match the boundary condition $T_w(l_0) = T_a$ at $x = l_0$, $P_{holder} = 1.12$ W is determined self-consistently. The conduction loss of the segment of wire inside of torch can now be evaluated to be $P_{cond0} \approx 3.33$ W. Therefore, the total power loss for the 16 mil wire is $q_{out0} = P_{cond0} + P_{rad0} = 7.16$ W. Set $q_{in0}(T) = q_{out0}$, the time averaged torch temperature T is found (Kuo *et al* 1999) to be about 1760 K.

The electron density of the torch plasma can be deduced, with the aid of temperature information, from the microwave absorption measurements. The experiment (Koretzky and Kuo 1998) was conducted by streaming torch plasma through aligned holes on the bottom and top walls of a rectangular X-band waveguide. This plasma post has a complex dielectric constant $\varepsilon = \varepsilon' - j\varepsilon''$, where $\varepsilon' = 1 - \omega_p^2/(\omega^2 + \nu^2)$ and $\varepsilon'' = \nu \omega_p^2/\omega(\omega^2 + \nu^2)$; ω, ω_p, and ν are the wave, plasma, and electron–neutral collision, frequencies, respectively, and ε'' is determined from the absorption measurement. Since $\omega_p^2 \propto n_e$ and $\nu \propto T_N \cong T$, the time-dependent electron density was found to have a spatially averaged maximum value n_{emax} of about 10^{13} electrons/cm^3.

6.7.3   Power consumption calculation

Plasma growth and decay are governed by the rate equations of plasma species (Zhang and Kuo 1991) in each torch

$$\frac{dn_e}{dt} = -\nu_a n_e + \nu_d n_- - \alpha n_e n_+ + \nu_i n_e$$

$$\frac{dn_+}{dt} = -\alpha n_e n_+ - \beta n_+ n_- + \nu_i n_e \qquad (6.7.3)$$

$$\frac{dn_-}{dt} = \nu_a n_e - \nu_d n_- - \beta n_+ n_-$$

where n_e, n_+, and n_- are the densities of electrons, positive ions, and negative ions, respectively, in cm^{-3}; ν_a is the attachment rate and ν_d is the detachment rate; and α and β are the electron–ion recombination coefficient and ion–ion recombination coefficient, respectively, in cm^3 s^{-1}. The ionization frequency ν_i representing the external driver of the discharge is given by (Lupan 1976, Kuo and Zhang 1990)

$$\nu_i = 383\nu_a [\varepsilon^{3/2} + 3.92\varepsilon^{1/2}] \exp(-7.546/\varepsilon) \qquad (6.7.4)$$

where $\varepsilon = E/E_{cr}$ is the discharge field E normalized to the breakdown threshold field E_{cr}.

By solving (6.7.3), the net electron loss during a number of discharge periods can be evaluated. It turns out that the rate terms on the left hand side of (6.7.3) can be neglected in calculating the electron density decay. It is understandable because the temporal variation of the discharge voltage is, in general, much slower than the transient variations of (6.7.3). The steady state solution of (6.7.3) is given by

$$n_e = -\frac{\nu_d(\beta\nu_a + \alpha\nu_d - \beta\nu_i - \eta)}{\alpha(\beta\nu_a - (\alpha - 2\beta)\nu_d - \beta\nu_i + \eta)}$$

$$n_+ = -\frac{\beta\nu_a - \alpha\nu_d - \beta\nu_i - \eta}{2\alpha\beta} \qquad (6.7.5)$$

$$n_- = \frac{\nu_d(\beta\nu_a + \alpha\nu_d - \beta\nu_i - \eta)}{\beta(\beta\nu_a - (\alpha - 2\beta)\nu_d - \beta\nu_i + \eta)}$$

where $\eta = \sqrt{4\alpha\beta\nu_d\nu_i + (\beta\nu_a + \alpha\nu_d - \beta\nu_i)^2}$ is used to simplify the presentation of (6.7.5). The average power consumption is given by the average electron loss per second times the average ionization energy ($\approx 10\,\text{eV}$) of air (Brown 1967). Shown in figure 6.7.4 is a parametric dependence of the power consumption (W/cm^3) on the average electron density (cm^{-3}) maintained in the plasma, where the electron–ion recombination coefficient α (cm^3 s^{-1}) is used as a variable parameter. It provides a very useful reference for choosing the density regime for the most efficient operation of the plasma torch. The results for two situations are shown. The first is for a completely

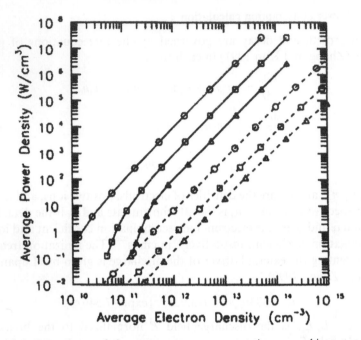

Figure 6.7.4. Dependence of the average power consumption per cubic meter on the average electron density per cubic centimeter with the electron–ion recombination coefficient α ($cm^3 s^{-1}$) as a variable parameter. Solid lines are for transient plasma generation case and the dashed lines are for steady state plasma maintenance case. α is given as (O) 10^{-6}, ($\square$) 10^{-7}, and ($\triangle$) 10^{-8}. ($\nu_a = 4.56 \times 10^7 s^{-1}$, $\nu_d = 1.52 \times 10^7 s^{-1}$, and $\beta = 1.2 \times 10^{-9} cm^3 s^{-1}$). (Copyright 2001 by IEEE.)

transient plasma generation system using equation (6.7.3), where an initial electron density is created and then allowed to recombine. The electron density is averaged over $\Delta T = 1$ ms, which is shorter than the discharge duration of presently reported experiments, but yet very long to demonstrate a significantly different result from that of the second case. The second is for a steady state plasma generation system using equation (6.7.5). The large difference in the average power consumption between the two situations for each α shows the importance of plasma maintenance, which can reduce the power budget considerably. In other words, an increase of the repetition rate of the discharge (i.e. reducing ΔT) works to reduce the power consumption in the transient case. However, the engineering problem of the power supply becomes an issue. The simulation results also show that the power budget is reduced by decreasing the value of α, which can be achieved by increasing the temperature of the plasma (Christophorou 1984, Rowe 1993).

Since the power consumption for plasma maintenance is much less than that for pulse generation, it suggests that a proper trigger mechanism for the start of plasma production may work to reduce the power requirement.

Using the fitting curves of the simulation results, a function giving a parametric dependence of the consumed average power density $\langle P \rangle$ on the normalized average electron density $\langle \underline{n}_e \rangle$ maintained in the plasma is derived (Koretzky and Kuo 2001) to be $\langle P \rangle = 48 \langle \underline{n}_e \rangle^{1.9} \underline{\alpha}^{0.4}$ (W/cm^3), where $\langle \underline{n}_e \rangle$ is normalized to 10^{13} cm^{-3} and where $\underline{\alpha}$, the electron–ion recombination coefficient, normalized to 10^{-7} cm^3 s^{-1}, is used as a variable parameter in the simulation. This relationship provides a useful guide for the choice of the plasma density and temperature to achieve an efficient operation of the plasma torch.

References

Boulos M, Fauhais P and Pfender E 1994 *Thermal Plasmas Fundamentals and Applications* vol 1 (New York: Plenum Press) pp 33–47 and 403–418

Brown S C 1967 *Basic Data of Plasma Physics* (Cambridge, MA: MIT Press)

Christophorou L G 1984 *Electron–Molecule Interactions and Their Applications* vol 2 (Orlando: Academic Press)

Gage R M 1961 *Arc Torch and Process* (United States Patent No. US 2858411)

Incropera F P and DeWitt D P 1996 *Fundamentals of Heat and Mass Transfer* 4th edition (John Wiley)

Koretzky E and Kuo S P 1998 'Characterization of an atmospheric pressure plasma generated by a plasma torch array' *Phys. Plasmas* **5**(10) 3774

Koretzky E and Kuo S P 2001 'Simulation study of a capacitively coupled plasma torch array' *IEEE Trans. Plasma Sci.* **29**(1) 51

Kuo S P and Zhang Y S 1990 'Bragg scattering of electromagnetic waves by microwave produced plasma layers' *Phys. Fluids B* **2**(3) 667

Kuo S P, Bivolaru D and Orlick L 2002 'A magnetized torch module for plasma generation' *Rev. Sci. Instruments* **73**(8) 3119

Kuo, S P, Bivolaru D, Carter C D, Jacobsen L and Williams S 2003 'Operational characteristics of a plasma torch in a supersonic cross flow' *AIAA* Paper 2003-1190 (Washington, DC: American Institute of Aeronautics and Astronautics)

Kuo S P, Koretzky E and Orlick L 1999 'Design and electrical characteristics of a modular plasma torch' *IEEE Trans. Plasma Sci.* **27**(3) 752

Kuo S P, Koretzky E and Vidmar R J 1999 'Temperature measurement of an atmospheric-pressure plasma torch' *Rev. Sci. Instruments* **70**(7) 3032

Kuo S P, Koretzky E and Orlick L 2001 *Methods and Apparatus for Generating a Plasma Torch* (United States Patent No. US 6329628 B1)

Lupan Y A 1976 'Refined theory for an RF discharge in air' *Sov. Phys. Tech. Phys.* **21**(11) 1367

Rowe B R 1993 *Recent Flowing Afterglow Measurements, in Dissociative Recombination: Theory, Experiment and Applications* (New York: Plenum Press)

Siegel R and Howell J R 1992 *Thermal Radiation Heat Transfer* (Hemisphere Publishing)

Zhang Y S and Kuo S P 1991 'Bragg scattering measurement of atmospheric plasma decay' *Int. J. IR & Millimeter Waves* **12**(4) 335

Zhukov M 1994 'Linear direct current plasma torches' in Solonenko O and Zhukov M (eds) *Thermal Plasma and New Material Technology* vol 1: *Investigations of Thermal Plasma Generators* (Cambridge Interscience Publishing) pp 9–43

Chapter 7

High Frequency Air Plasmas

*J Scharer, W Rich, I Adamovich, W Lempert, K Akhtar, C Laux,
S Kuo, C Kruger, R Vidmar and R J Barker*

7.1 Introduction

The use of high-frequency power to produce plasmas in air and high-pressure
gases is a relatively new development. These methods span the regimes of
seed gas ionization via carbon monoxide (CO) and ultraviolet flash tubes
and lasers, seed gas ionization and optical pumping via carbon monoxide
lasers and ionization sustainment by rf plasma torches and microwave
plasma sources. Their advantage is that power can be spatially focused
away from electrodes or wall materials by means of antennas or optical
lenses. In addition, since the focus is adjustable, large, three-dimensional
volumes of plasma can be created in space without the need for electrodes
that can degrade. Historically, rf air plasma torches in air were the first to
be investigated. Then microwave and later flash-tube and laser sources
became of interest. Recently, electron beams propagated through a
vacuum window to protect the cathode and short-pulse high-voltage
plasma sources in air have been investigated. Much of the recent research
presented in this chapter was supported by a Defense Department Research
and Engineering multi-university research initiative (MURI) entitled 'Air
Plasma Ramparts' and AFOSR grants administered by Dr Robert Barker.

This chapter is organized as follows. First, laser and flash-tube ionization and the excitation of gas seeds in air are discussed by Professors William
Rich, Igor V Adamovich and Walter Lempert of Ohio State University in
section 7.2.2. Then laser-formed, seeded, high-pressure gas and air plasma
research is presented by Professor John Scharer and Dr Kamran Akhtar of
the University of Wisconsin in section 7.2.3. This is followed by a presentation on the rf torch in Section 7.3 by Professors Christophe Laux of Ecole
Centrale Paris and Stanford University and Dr Kamran Akhtar and
Professor John Scharer from the University of Wisconsin. Then microwave

air plasma sources are presented in section 7.3.4 by Professor Spencer Kuo of Polytechnic. Thereafter, more complex short-pulse, high-voltage experiments involving rf gas preheating and electrode discharges and laser excitation of electron beam heated air plasmas is presented. This research is described in sections 7.4 and 7.5 by Professors Christophe Laux of the University of Paris and Stanford University, and by Professors William Rich, Igor Adamovich and Walter Lempert of Ohio State University. Finally, section 7.6 presents challenges and new opportunities for research and applications in this field.

Section 7.2 presents an investigation of optically pumped excitation of carbon monoxide (CO) and laser excitation and ionization of organic gas tetrakis-dimethyl-amino-ethelyene (TMAE) seed gases under high pressure and atmospheric air conditions. This is done to create non-equilibrium high-density plasma conditions and maintain low gas kinetic temperatures with a lower power budget. The low power optically pumped CO experiment is augmented with an rf capacitive source and produces air component and air plasma densities in the 10^{10}–10^{11}/cm^3 density range. In addition, detailed optical spectra illustrating the vibrationally excited states are presented. This optically pumped plasma is used together with an electron-beam-produced plasma that is discussed in section 7.5. The ionization of a low ionization energy (6.1 eV) organic seed gas in high-pressure gases and atmospheric air by a short-wavelength (193 nm) high-power excimer laser is then discussed in section 7.2.3. High density (10^{13}/cm^3), large volume (500 cm^3), low temperature plasmas are obtained and millimeter wave interferometry and optical spectra measurements are presented to determine the two- and three-body recombination rates for different cases. Both direct and delayed ionization processes are found to influence the plasma decay process. The high-density and large volume plasma formed in this case provides an excellent load for reduced power rf inductive sustainment that is discussed in section 7.3.3.

Section 7.3.2 presents a review of rf plasma torch experiments that are the most developed of the high-frequency high-pressure plasma sources. They have applications in materials processing and biological decontamination. High density ($>10^{13}$/cm^3), large volume (1000 cm^3) air plasmas in near thermal equilibrium are obtained and electron temperatures and densities in air plasmas as well as the wall plug power density required to sustain the plasma are discussed. This technique is used to increase the neutral air temperature in order to reduce electron attachment to oxygen for the short-pulse high-density experiments discussed in section 7.4. Next, the use of the laser initiated seed gas discussed in section 7.2.3 as a seed plasma load for high-power inductive rf sustainment is presented. It is found that much lower rf power densities for sustainment compared to initiation can be obtained and enhanced rf penetration well away from antenna is observed. Section 7.3.4 discusses the use of higher frequency microwave discharges in air to obtain spatially localized high-density plasmas and can be compared with rf methods.

Section 7.4 discusses a short repetitive pulse, low-duty cycle, high-voltage discharge in air that is used to produce non-equilibrium plasmas with time-averaged densities in the ($10^{12}/cm^3$) range and greatly reduced power consumption and lower neutral temperatures relative to thermal equilibrium. Section 7.5 discusses the reduction in electron attachment to oxygen, one of the major loss processes for air plasmas, for a 60–80 kV electron beam-formed, $10^{11}/cm^3$ density plasma resulting from CO laser pumping of the seed gas that can couple to and detach the electron from the oxygen. Recombination rates and power density estimates are also presented. Section 7.6 concludes with challenges and opportunities for future research.

7.2   Laser Initiated or Sustained, Seeded High-Pressure Plasmas

7.2.1   Introduction

Laser pumping of seed gas and laser ionization of low ionization potential seed organic gas in high-pressure gases and atmospheric air to obtain non-equilibrium, high-density plasmas is presented in this section. These techniques are relatively new and have an objective of high-density, remote plasma creation with substantial reduction in power compared to plasma production in high-pressure gas alone. These experiments are grouped together since they both utilize lower concentration seed gas for which laser power can be efficiently coupled or used for ionization than is the case for the high-pressure gas into which they are injected. They also create non-equilibrium, large volume plasmas that can be sustained remotely from the source region. A key scientific property that is examined is the seed gas and plasma interaction with the background high-pressure gas. The carbon monoxide (CO) laser ($\lambda \approx 5 \mu m$) pumping technique is used to efficiently pump vibrational states of the seed CO gas in the high-pressure background gas. Efficient coupling and transfer to metastable states of high-pressure seed gas and capacitive rf coupling of power to associative ionization of the CO-laser-pumped plasma is discussed. Optical spectra and the associated plasma density are presented.

The use of a low ionization potential seed gas that is ionized and excited by a 193 nm wavelength excimer laser is discussed in section 7.2.3. Both direct ionization and delayed ionization of the seed gas produces a high-density large-volume plasma in high-pressure gases and atmospheric air. This plasma can be produced in space well away from the laser source and can be used as a large volume seed plasma that can be sustained by lower power inductive rf coupling that can be pulsed or continuous. This topic is discussed in section 7.3 on rf and microwave plasmas. Fast Langmuir probe measurements, optical spectroscopy and millimeter wave interferometry are used to determine the plasma density, super-excited neutral states and recombination rates for seed plasma and the properties in high-pressure background gas.

7.2.2 Laser-sustained plasmas with CO seedant

Creating considerable levels of ionization, uniformly distributed in a large-volume high-pressure molecular gas mandates a non-thermal, or non-equilibrium, plasma approach, if relatively low gas kinetic temperatures must be maintained. The first point to be clarified is what is meant by a non-equilibrium, as opposed to an equilibrium, plasma. Figure 7.2.2.1 shows a simple schematic indicating the various modes of motion of diatomic molecules, the dominant species making up the air plasmas which are a principal focus of this book. The plasma can store energy in each of the indicated modes, and each therefore can contribute to the specific heat of the plasma. Note that in addition to the modes shown, the translational motion of the plasma atoms and free electrons are also participating modes. Polyatomic species, if present, would also contribute additional modes. The total energy of each atom, molecule, ion, or free electron in the gas may be written in the form

$$E = E_{\text{trans}} + E_{\text{rot}} + E_{\text{vib}} + E_{\text{electron}} + E_{\text{interaction}} \tag{7.2.2.1}$$

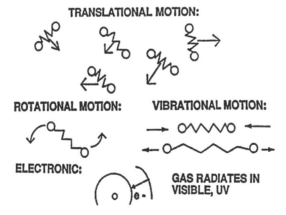

Figure 7.2.2.1. Schematic of the various modes of motion for diatomic molecular species in a plasma.

where each of the energies shown corresponds to one of the modes of motion shown in figure 7.2.2.1. Various other possible energy storage modes (chemical, nuclear) are omitted, both for simplicity and because they are not primarily participating in the processes being described here. $E_{\text{interaction}}$ represents energies associated with the coupling of various modes within a single molecule (vibration with rotation, or vibration with electronic motion, etc.). The 'internal' energy modes (rotation, vibration, electronic) are quantized into discrete energy levels. For engineering systems of macroscopic dimensions, the translational modes of the plasma species are not quantized, and translational motion is described by classical mechanics.

It is convenient to designate the total energy of an atom or molecule corresponding to a particular array of specific quantum energy states as E_i, where the subscript i refers to the collection of quantum numbers for each mode designating the specific energy level. When the plasma is in thermal equilibrium, the distribution of populations of plasma species (electrons, ions, atoms, molecules) among the various energy states E_i is typically governed by Maxwell–Boltzmann statistics. In this equilibrium case, the fractional number of plasma species in the ith energy state, E_i, is

$$\frac{n_i}{N} = \frac{g_i \exp(-E_i/kT)}{Q} \tag{7.2.2.2}$$

where the partition function, Q, is given by:

$$Q = \sum_i g_i \exp(-E_i/kT) \tag{7.2.2.3}$$

where $N = \sum_i n_i$ is the total number of species, g_i is the statistical weight of the ith internal energy state, k is Boltzmann's constant and T is the temperature of the plasma. For this equilibrium case, specification (or measurement) of the single plasma temperature, T, allows the distribution of energy and populations of states to be determined.

In the molecular plasmas of primary interest in this book, one or more modes of motion are not in thermal equilibrium, and some states are not populated according to the simple expressions above. It must be recognized that producing large degrees of such non-equilibrium requires input of considerable work to the plasma, to maintain the non-equilibrium. Thermodynamic laws dictate that this work input must exceed the heat input necessary to maintain a thermal, equilibrium plasma having the same ionization fraction. Non-equilibrium, cool molecular plasmas are easily created in lower pressure gases, usually in small volumes. These are the familiar glow discharge plasmas, that can have near-room gas kinetic (translational mode) temperatures, and which can be readily struck in a gas with electrodes biased with dc or rf electrical potentials. The specific non-equilibrium modes in such glow-type discharges in molecular gases are (1) the free electron gas, whose mean energy or effective temperature is much higher than the translational mode temperature of the molecular and atomic species, (2) at least some of the vibrational modes, whose mean energy is, again, much higher than the mean translational mode energy of atoms and molecules, and, often, (3) some of the electronic modes of the atomic and molecular species, which again may have much higher mean energies than their mean translational mode temperatures. It is this third non-equilibrium that creates the defining 'glow' of the ordinary glow discharge. This often-visible glow arises from radiative decay of the highly energetic electronic states. While such intense radiation is only achieved by heating thermal equilibrium plasmas to thousands of degrees, in radiating glow plasmas, the gas temperature may be only slightly above room temperature. We cite the common examples of 'neon' sign plasmas, or normal fluorescent lighting tubes, which are cool to the touch. In all these glow discharges, it is electrical power that supplies the requisite work for maintaining non-equilibrium. Creating such a cool, non-thermal plasma in any atmospheric pressure gas, and especially in air, is, however, beset with many difficulties, and is exacerbated when a uniform, diffuse ionization is required in a large volume. Chief among these difficulties is the instability that causes the plasma to condense into a thermal arc.

Stability control of large-volume high-pressure non-equilibrium molecular plasmas has long been one of the most challenging problems of gas discharge physics and engineering. At high pressures, the most critical instability, which produces the transition of a diffuse, non-equilibrium, self-sustained discharge into a higher temperature, higher ionization fraction, near-thermal equilibrium arc, is the ionization heating instability. Basically, the transition to an arc develops due to a positive feedback between gas heating and the electron impact ionization rate (Raizer 1991, Velikhov *et al* 1987). In the transition, small electron density perturbations, producing excess Joule heating, result in a more rapid electron generation and eventually lead to runaway ionization. Since the advent of very high-power gas lasers, which require production of extreme disequilibrium in internal molecular energy modes, coupled with low gas kinetic temperature, various approaches to this stabilization problem have been developed. Among a few well known high-pressure plasma stabilization methods are the use of separately ballasted multiple cathodes (Raizer 1991), aerodynamic stabilization (Rich *et al* 1979), rf frequency high-voltage pulse stabilization (Generalov *et al* 1975), and external ionization by a high-energy electron beam (Basov *et al* 1979).

The use of these techniques is tantamount to introducing an additional damping factor into a conditionally stable system, which raises the instability growth threshold and allows the sustainment of a diffuse discharge at higher pressures and/or electron densities. However, they do not affect the original source of the ionization heating instability. For this reason, raising the gas pressure or discharge current eventually results in a glow-to-arc transition.

Even the non-self-sustained dc discharge with external ionization produced by an e-beam is in fact self-sustained in the unstable cathode layer, where ionization is primarily produced by secondary electron emission from the cathode (Velikhov *et al* 1987). Therefore instability growth in the cathode layer of high-power discharges sustained by an e-beam results in the development of high-current density cathode spots extending into the positive column and eventually causing its breakdown.

The cathode layer instability of the e-beam-sustained discharge can be avoided by using an rf instead of a dc electrical field to draw the discharge current between dielectric-covered electrodes. In this case, secondary emission from the electrodes is precluded, the cathode regions do not form, and the current loop is closed by the displacement current in the near-electrode sheaths. This type of discharge remains non-self-sustained in the entire region between the electrodes and is therefore not susceptible to the cathode layer instability (Velikhov *et al* 1987). Indeed, experiments show that an rf beam-driven discharge remains stable at higher E/N and current densities than a dc discharge (Kovalev *et al* 1985). However, at high e-beam currents this type of discharge also becomes unstable since the rate of ionization by the beam is inversely proportional to the gas density, so that gas heating by the beam would eventually produce an ionization instability.

The above discussion shows that even the use of external ionization does not always allow unconditionally stable discharge operation at high currents and pressures. On the other hand, it suggests that a discharge system sustained by an external source with a negative feedback between gas heating and ionization rate, and, if necessary to provide work input to internal modes, using sub-breakdown electric fields to draw the discharge current, might be unconditionally stable (Plonjes *et al* 2000). An ionization process that satisfies this condition is the associative ionization in collisions of two highly vibrationally excited molecules (Plonjes *et al* 2000, Polak *et al* 1977, Adamovich *et al* 1993, 1997, 2000, and Palm *et al* 2000),

$$AB(v) + AB(w) \longrightarrow (AB)_2^+ + e^-, \qquad E_v + E_w > E_{ion}. \qquad (7.2.2.4)$$

In equation (7.2.2.4), AB represents a diatomic molecule, and v and w are vibrational quantum numbers. Basically, ionization is produced in collisions of two highly vibrationally-excited molecules when the sum of their vibrational energies exceeds the ionization energy. This volume ionization method was first detected in nitrogen plasmas, and is the key ion-producing process in many of the well-known CO_2/N_2 high-power gas lasers (Polak *et al* 1977). Of direct relevance for application to air plasmas, ionization by this mechanism has been previously observed in CO–Ar and CO–N_2 gas mixtures optically pumped by resonance absorption of CO laser radiation at pressures of $P = 0.1$–1.0 atm and temperatures of $T = 300$–700 K (Plonjes *et al* 2000, Adamovich *et al* 1993, 1997, 2000, and Palm *et al* 2000). In these optically pumped non-equilibrium plasmas, where high vibrational levels of CO are populated by near-resonance vibration–vibration (V–V) exchange (Treanor *et al* 1968, Rich 1982), a gas temperature rise results in rapid relaxation of the upper vibrational level populations because of the exponential rise of the vibration–translation (V–T) relaxation rates with temperature (Billing 1986). In other words, ionization by mechanism (1) can be limited and even terminated by the heating of the gas.

The present section reviews the work in exciting high-pressure molecular plasmas by such 'optical pumping' of CO. While such plasmas can be created in high-pressure mixtures of pure CO, or CO in an inert (Ar, He) diluent, CO can also be used as a seedant to create other diatomic gas plasmas (N_2, O_2, air). This unconditionally stable high-pressure molecular plasma concept will be reviewed here. To accomplish this, carbon-monoxide-containing gas mixtures are vibrationally excited at high pressures using a combination of a CO laser and a sub-breakdown rf field. More extensive presentations of work with plasmas of this type are given in (Lee *et al* 2000, Plonjes *et al* 2001), from which most of the data given below are obtained.

A schematic of a typical experimental set-up is shown in figure 7.2.2.2. A continuous wave (c.w.) carbon monoxide laser is used to irradiate a high-pressure gas mixture, which is slowly flowing through an optical absorption cell. For purposes of the present discussion consider that gas mixture to be nitrogen containing 1% of carbon monoxide and trace amounts ($\sim$10–100 ppm) of nitric oxide or oxygen, at pressures of $P = 0.4$–1.2 atm. The residence time of the gases in the cell is about 1 s. The CO pump laser is electrically excited, producing continuous wave output on approximately 20 vibrational–rotational lines of the CO fundamental infrared bands, vibrational quantum transitions $\Delta v = 1$. It produces a substantial fraction of its power output on the $v = 1 \longrightarrow 0$ fundamental band component in the infrared. (Note that 50% efficiencies have been demonstrated for these lasers at very high powers.) A typical small-scale laser operates at 10–15 W continuous wave broadband power on the lowest ten fundamental bands. The output on the lowest bands ($1 \longrightarrow 0$ and $2 \longrightarrow 1$) is necessary to start the absorption process in cold CO (initially at 300 K) in the cell. The laser is mildly focused to increase the power loading per CO molecule, providing an excitation region of, typically, $\sim$1–2 mm diameter and up to 10 cm long. The absorbed laser power is of the order of 1 W/cm over the absorption length of about 10 cm, which gives an absorbed power density of $\sim$100 W/cm^3. It is important to note that this technique is not the laser-induced 'breakdown', familiar from the many focused pulsed laser experiments, which create an intense arc-like plasma. In the present technique, up to at least 70% of the laser power is absorbed, but by resonance transitions, initially, into the vibrational mode of the CO seedant only. This use of the CO laser to excite high-pressure gas mixtures is an extension of a technique described numerous times in the literature

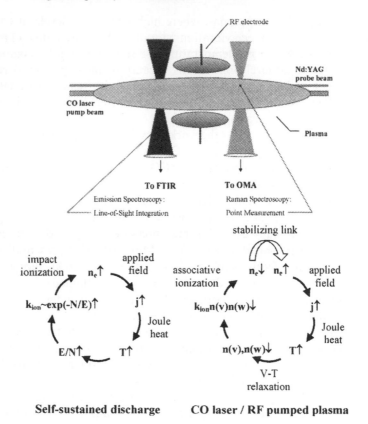

Figure 7.2.2.2. Schematic of the CO laser/rf field pumping experiment.

(Rich *et al* 1979, DeLeon and Rich 1986, Flament *et al* 1992, Wallaart *et al* 1995, Dünnwald *et al* 1985, Saupe *et al* 1993, Plonjes *et al* 2000, Lee *et al* 2000).

The low vibrational states of CO, $v \leq 10$, are populated by direct resonance absorption of CO pump laser radiation in combination with rapid redistribution of the population by vibration–vibration (V–V) exchange processes [14],

$$CO(v) + CO(w) \longrightarrow CO(v-1) + CO(w+1). \qquad (7.2.2.5)$$

The V–V processes then continue to populate higher vibrational levels of CO as well as vibrational levels of N_2, which are not coupled to the laser radiation (Dünnwald *et al* 1985, Saupe *et al* 1993, Plonjes *et al* 2000),

$$CO(v) + N_2(w) \longrightarrow CO(v-1) + N_2(w+1). \qquad (7.2.2.6)$$

The large heat capacity of the gases, as well as conductive and convective cooling of the gas flow, allow the translational/rotational mode temperature in the cell to be controlled. Under steady-state conditions, when the average vibrational mode energy of the CO would correspond to several thousand Kelvin, the temperature never rises above a few hundred degrees (Dünnwald *et al* 1985, Saupe *et al* 1993, Plonjes *et al* 2000). Thus a strong non-equilibrium distribution of mode energies can be maintained in the cell, characterized by a very high energy of the vibrational modes and a low translational–rotational mode temperature. The populations of the vibrational states of N_2 and CO in the cell are monitored by infrared emission and Raman spectroscopy (Plonjes *et al* 2000, Lee *et al* 2000).

Under these highly non-equilibrium conditions, the optically pumped gas mixture becomes ionized by the associative ionization mechanism of equation (7.2.2.4). The ionization of carbon monoxide by this mechanism has been previously observed in CO–Ar and CO–N_2 gas mixtures optically pumped by resonance absorption of CO laser radiation (Plonjes *et al* 2000, Adamovich *et al* 1993, 1997, 2000, and Palm *et al* 2000). The calculated (Adamovich *et al* 1993, 1997, 2000) and measured (Plonjes *et al* 2000, Palm *et al* 2000) steady-state electron density sustained by a 10 W CO laser in these optically pumped plasmas is in the range $n_e \approx 10^{10}$–10^{11} cm^{-3}. Such ionization levels are maintained in CO–Ar and CO–N_2 mixtures by the mechanism of equation (7.2.2.4) with the laser pump only. It is not necessary to do additional work on the plasma. However, an rf field can be imposed, and further energy inputed to the vibrational modes without gas breakdown. For this purpose, two 3 cm diameter brass plate electrodes were placed in the cell as shown in figure 7.2.2.2, so that the laser beam creates a roughly cylindrical excited region between the electrodes, 1–2 mm in diameter. The probe electrodes, 13.5 mm apart, are connected to a 13.56 MHz rf power supply via a tuner used for plasma impedance matching. Typically, the reflected rf power does not exceed 5–10% of the forward power. The applied rf voltage amplitude, measured by a high-voltage probe, is varied in the range of 2–3 kV at $P = 0.8$–1.2 atm, so that the peak reduced electric field does not exceed $E/N \cong 1 \times 10^{-16}$ V cm^2. It should be emphasized that this low value of E/N precludes electron impact ionization by the applied field, so that the associative ionization of equation (7.2.2.6) remains the only mechanism for electron production in the plasma. The applied rf field is used to heat free electrons created by the associative ionization mechanism and to couple additional power to the vibrational modes of the gas mixture molecules by electron impact processes,

$$CO(v) + e^-(\text{hot}) \longrightarrow CO(v + \Delta v) + e^-(\text{cold}) \qquad (7.2.2.7)$$

$$N_2(v) + e^-(\text{hot}) \longrightarrow N_2(v + \Delta v) + e^-(\text{cold}). \qquad (7.2.2.8)$$

It is well known that over a wide range of reduced electric field values ($E/N = (0.5$–5.0$) \times 10^{-16}$ V cm^2) more than 90% of the input electrical power in nitrogen plasmas goes to vibrational excitation of N_2 by electron

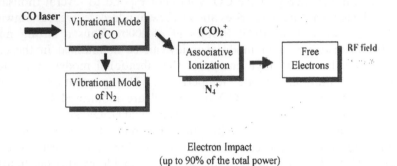

Figure 7.2.2.3. Schematic of the dominant kinetic processes in a CO–N$_2$ plasma pumped by a CO laser and a sub-breakdown rf field.

impact (Raizer 1991). Combined with the high efficiency of the CO laser, this provides a very efficient method of sustaining extreme vibrational disequilibrium in high-pressure molecular gases. In this approach, the laser need only be powerful enough to load one of the molecular vibrational modes to vibrational levels producing significant ionization, in accordance with equation (7.2.2.4). It is not necessary to use a high-power pump laser. However, as shall be seen subsequently, considerably greater laser powers are needed to achieve the same states in air mixtures.

The strong vibrational disequilibrium enhanced by the electron impact processes of equations (7.2.2.7) and (7.2.2.8) results in a faster electron production by the associative ionization mechanism of equation (7.2.2.6). The resultant electron density increase in turn further accelerates the rate of energy addition to the vibrational modes of the molecules. However, this self-accelerating process does not produce an ionization instability such as occurs in other types of high-pressure non-equilibrium plasmas. The reason for this is a built-in self-stabilization mechanism existing in plasmas sustained by associative ionization. In high-pressure self-sustained discharge plasmas, excess Joule heating produced by a local electron density rise accelerates the rate of impact ionization and therefore results in a further increase of electron density (see figure 7.2.2.3). This is the well-known mechanism of ionization-heating instability development (Raizer 1991, Velikhov *et al* 1987). In a plasma sustained by associative ionization, excess Joule heating due to a local electron density rise sharply increases the vibration–translation (V–T) relaxation rates, which results in a rapid depopulation of high vibrational energy levels, slows down the ionization rate, and reduces the electron density (see figure 7.2.2.2). This provides negative feedback between gas heating and the ionization rate and enables the unconditional stability of the plasma at arbitrarily high pressures, for as long as the applied rf field does not produce any impact ionization. Obviously, optically pumped plasmas sustained by the CO laser alone (without the externally applied field) are always unconditionally stable. Indeed, stable and diffuse plasmas of this type have been sustained in CO–Ar mixtures at pressures up to 10 atm (Rich *et al* 1982). Figure 7.2.2.3 shows a schematic of the dominant kinetic processes in the CO laser/rf field sustained CO–N_2 plasma.

Triggering the rf power coupling to the vibrational modes of the cell gases requires the initial electron density, n_e, to exceed a certain threshold value. Recent studies of associative ionization in CO laser pumped plasmas (Plonjes *et al* 2000, Adamovich *et al* 2000, Palm *et al* 2000) showed that the electron density in these plasmas can be significantly increased (from $n_e < 10^{10}\,\text{cm}^{-3}$ to $n_e = (1.5–3.0) \times 10^{11}\,\text{cm}^{-3}$) by adding trace amounts of species such as O_2 and NO to the baseline CO–Ar or CO–N_2 gas mixtures; as discussed in (Lee *et al* 2000), this has the net effect of significantly altering the dissociative recombination rate in the plasma.

Figures 7.2.2.4–7.2.2.7 show the levels of non-equilibrium mode excitation and plasma production with this method. Figure 7.2.2.4 shows the spectrally-resolved emission from the first overtone infrared bands of CO in the CO–N_2 plasma, displayed against the frequency (in wavenumbers) for two plasma pressures, 600 and 720 torr. In this spectrum, each of the peaks displayed is roughly indicative of the population of a CO vibrational quantum level. The large peaks on the left correspond to the lower quantum levels ($v = 2, 3$, and so on) with the highest populated levels (near $v \approx 38$) at the right of the spectrum. The greatly increased populations when the subcritical rf field is turned on are also displayed. Figure 7.2.2.5 shows the corresponding N_2 vibrational populations for one of the same CO–N_2 plasmas, namely for the 600 torr case, from a Raman spectrum. In this figure, each peak is indicative of the vibrational population, starting with $v = 0$ on the right, and increasing to $v = 4$ on the left. Again, the much greater population of the upper states with the rf field on is evident. The Raman measurements are also used to infer the gas kinetic temperature (i.e. the rotational/translational mode temperature) of these plasmas. This temperature is 360 K for excitation with the CO laser alone, not greatly above room temperature, and rises to 540 K when the rf is on for the conditions of the figures. The photograph of figure 7.2.2.6 shows the visual appearance of the plasma, again, with and without the rf field on. The visible emission is from the small amounts of C_2 and CN radicals formed from the reaction of the vibrationally excited CO and N_2. Chemical reaction is not a significant energy absorption channel in the plasmas under these conditions, but the visible electronic emission provides an easy qualitative diagnostic of the plasma size. The substantial increase in volume with the rf is apparent. Again, the electron densities, measured both by probes and microwave attenuation techniques, are in the range $1.5–3.0 \times 10^{11}\,\text{cm}^{-3}$.

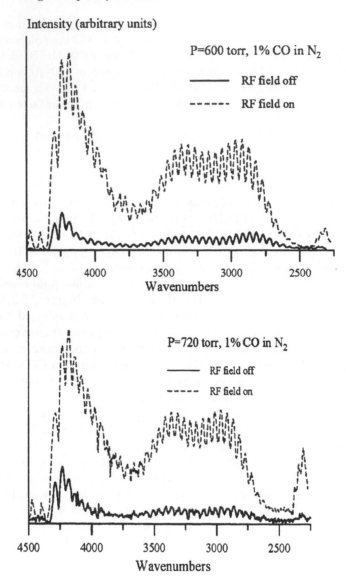

Figure 7.2.2.4. CO first overtone infrared emission spectra in the CO laser/rf field pumped $1\%CO–99\%N_2–150\,ppm$ NO gas mixture at $P = 600\,torr$ (laser power $10\,W$) and $P = 720\,torr$ (laser power $15\,W$).

Figure 7.2.2.7 shows the levels of excitation achieved when pumping atmospheric air, inferred from the Raman spectra of a dry air mixture at one atmosphere, with CO seedant, pumped by the CO laser. Figure 7.2.2.7 is a semilog plot of these experimentally determined relative populations of

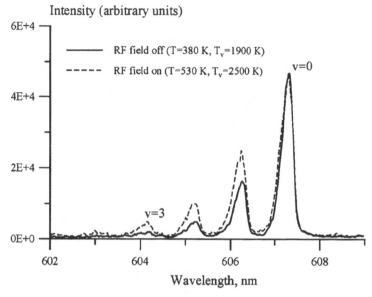

Figure 7.2.2.5. Raman spectra of nitrogen in the CO laser/rf field pumped 1%CO–99%N$_2$–150 ppm NO gas mixture at $P = 600$ torr. The spectra are normalized on the $v = 0$ peak intensity.

each vibrational level for the three species, N$_2$, CO, and O$_2$, plotted against the vibrational quantum level number. The vibrational quantum level number is roughly proportional to the energy of the level. Accordingly, a Boltzmann distribution of populations in such a plot would approximate

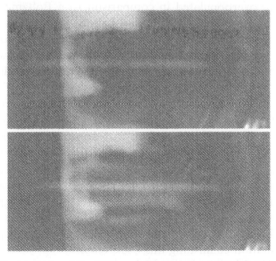

Figure 7.2.2.6. Photographs of the CO laser/rf field pumped 1%CO–99%N$_2$–10 ppm NO gas mixture at $P = 1$ atm. Top, rf field turned off; bottom, rf field turned on.

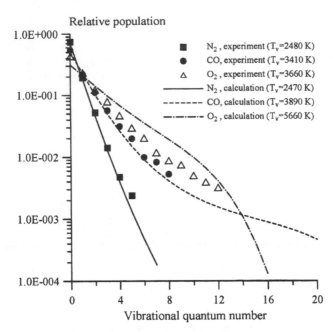

Figure 7.2.2.7. Experimental (symbols) and calculated (lines) vibrational population distribution functions on centerline of optically pumped atmospheric pressure air, for a 580/120/40 torr mixture of $N_2/O_2/CO$.

a straight line; the obvious departure from Boltzmann equilibrium, even within a single species vibrational mode, is evident. The higher level populations are overpopulated in comparison to a Boltzmann plot. Since an equilibrium (Boltzmann) distribution cannot be fitted to these data, a unique 'vibrational mode temperature' cannot be assigned to each species. We can, however, use the populations of only the lowest two vibrational levels in each species to define an approximate vibrational mode temperature. These approximate vibrational temperatures are given on the figure. It can be seen that even these temperatures, which ignore the higher level overpopulations, still are far above the translational mode, or gas kinetic temperature, of the plasma, $T = 540$ K. It can be seen that approximately five vibrational levels of the N_2, eight vibrational levels of the CO, and 12 vibrational levels of the O_2 have significant non-equilibrium populations. The kinetics of such vibrationally excited systems are now well understood, and dictate that in mixtures of species such as in figure 7.2.2.7, the greatest energy loading accumulates in the vibrational mode of the lowest frequency oscillator, in this case, O_2. The figure shows this, and also displays a kinetic modeling calculation confirming this basic result.

The advantages of producing high-pressure low-temperature molecular gas plasmas by the above method are apparent. There are two principal limitations to using the CO seedant optical pumping method as the sole source of volume ionization. One is that associative ionization of the type given by equation (7.2.2.4) is not a particularly efficient volume ionization process, although it is a common ionizing process in conventional glow discharges. It requires that a great deal of the work applied to the plasma must go into vibrational mode excitation; the actual ionization energy supplied to the plasma is only perhaps 0.1% of the total power input. A second limitation is that the laser power requirements rise substantially with more fast-relaxing vibrationally-excited species present. To maintain very high vibrational mode power loadings, the input laser power must be increased. In the dry air case of figure 7.2.2.7, the oxygen is a faster relaxing species than either the N_2 or the CO seedant. With the power density of ~ 1–$10\,W/cm^2$ available from the laser used for these experiments, no molecular species of the 1 atm air case were pumped to levels high enough to give substantial associative ionization. With higher powers, it is possible to achieve this in air mixtures. However, given the relatively inefficient volume ionization obtainable by these means alone, the optical pumping method should be supplemented by more efficient ionization methods if large volume, high electron density plasmas are wanted with minimum work input. When combined with an efficient ionization technique, the vibrationally excited air produced by the optical pumping exhibits striking increases in plasma lifetimes. The means of accomplishing very high levels of ionization in relatively cold air by a combination of optical pumping and an efficient ionizer are presented in a subsequent section (section 7.5).

References

Adamovich I V, 2001 *J. Phys. D: Appl. Phys.* **34** 319

Adamovich I V and Rich J W 1997 *J. Phys. D: Appl. Phys.* **30** 1741

Adamovich I, Saupe S, Grassi M J, Schulz O, Macheret S and Rich J W 1993 *Chem. Phys.* **173** 491

Basov, N G, Babaev, I K and Danilychev, V A *et al* 1979 *Sov. J. Quantum Electronics* **6** 772

Billing, G D 1986 'Vibration-vibration and vibration-translation energy transfer, including multiquantum transitions in atom–diatom and diatom–diatom collisions' in *Nonequilibrium Vibrational Kinetics* (Berlin: Springer) ch 4, pp 85–111

DeLeon R L and Rich J W 1986 *Chem. Phys.* **107** 283

Dünnwald H, Siegel E, Urban W, Rich J W, Homicz G F and Williams M J 1985 *Chem. Phys.* **94** 195

Flament C, George T, Meister K A, Tufts J C, Rich J W, Subramaniam V V, Martin J P, Piar B and Perrin M Y 1992 *Chem. Phys.* **163** 241

Generalov, N A, Zimakov V P, Kosynkin V D, Raizer Yu P and Roitenburg D I 1975 *Tech. Phys. Lett.* **1** 431

Kovalev A S, Muratov E A, Ozerenko A A, Rakhimov A T and Suetin N V 1985 *Sov. J. Plasma Phys.* **11** 515

Lee W, Adamovich I V and Lempert W R 2000 *J. Chem. Phys.* **114** 117

Palm P, Plönjes E, Buoni M, Subramaniam V V and Adamovich I V 2000 'Electron density and recombination measurements in co-seeded optically pumped plasmas', submitted to *J. Appl. Phys.*, December

Plönjes E, Palm P, Chernukho A P, Adamovich I V and Rich J W 2000a *Chem. Phys.* **256** 315

Plönjes E, Palm P, Lee W, Chidley M D, Adamovich I V, Lempert W R and Rich J W 2000b *Chem. Phys.* **260** 353

Plönjes E, Palm P, Adamovich I V and Rich J W 2000c *J. Phys. D: Appl. Phys.* **33**(16) 2049

Plonjes E, Palm P, Lee W, Lempert W R and Adamovich I V 2001 *J. Appl. Phys.* **89** 5911

Polak L S, Sergeev P A and Slovetskii D I 1977 *Sov. High Temp. Phys.* **15** 15

Raizer, Y P 1991 *Gas Discharge Physics* (Berlin: Springer)

Rich, J W 1982 'Relaxation of molecules exchanging vibrational energy,' in Massy H S W, McDaniel E, Bederson B and Nighan W (eds) *Applied Atomic Collision Physics*, vol 3, *Gas Lasers*, ch 4, pp 99–140 (New York: Academic Press)

Rich J W, Bergman R C and Williams M J 1979 'Measurement of kinetic rates for carbon monoxide laser systems', Final Contract Report AFOSR F49620-77-C-0020 (November)

Rich W, Bergman R C and Lordi J A 1975 *AIAA J.* **13** 95

Saupe S, Adamovich I, Grassi M J and Rich J W 1993 *Chem. Phys.* **174** 219

Treanor, C E, Rich, J W and Rehm, R G 1968 *J. Chem. Phys.* **48** 1798

Velikhov E P, Kovalev A S and Rakhimov A T 1987 *Physical Phenomena in Gas Discharge Plasmas* (Nauka: Moscow)

Wallaart H L, Piar B, Perrin M Y and Martin J P 1995 *Chem. Phys.* **196** 149

7.2.3   Ultraviolet Laser Produced TMAE Seed Plasma

Experiments were performed to explore the possibility of creating an initial seed plasma that can be sustained efficiently by the inductive coupling of radiofrequency (rf) power. A large volume (500 cm^3), axially long (100 cm) tetrakis (dimethyl-amino) ethylene (TMAE) seeded plasma in a high-pressure background gas is created by a uniform intensity ultraviolet beam of 193 nm wavelength produced by a Lumonics Pulsemaster (PM-842) excimer laser. The laser runs in the ArF mode (6.4 eV). The long axial extent of the electrodeless laser seed plasma is attractive since it can allow enhanced rf penetration and ionization well away from the 20 cm axial extent of the antenna. A schematic illustrating the initial University of Wisconsin-Madison laser-initiated plasma experiment is shown in figure 7.2.3.1 (Ding *et al* 2001).

The efficiency of the subsequent rf sustainment of the plasma was determined by the plasma density and lifetime that depends on the two- and three-body recombination loss processes in the presence of background gases and electron attachment to oxygen. In this section the laser-produced TMAE plasma is characterized. The first measurement of temporal density and temperature decay of the laser-produced TMAE plasma was carried out using a special fast ($\tau \approx 10$ ns) Langmuir probe whose structure is

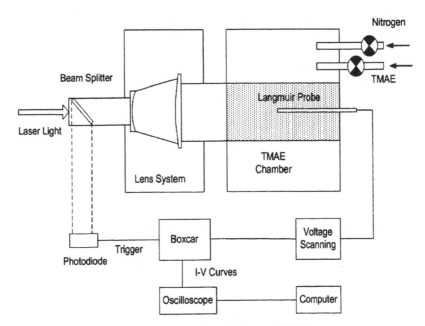

Figure 7.2.3.1. Laser seed plasma experiment. (Ding *et al* 2001.)

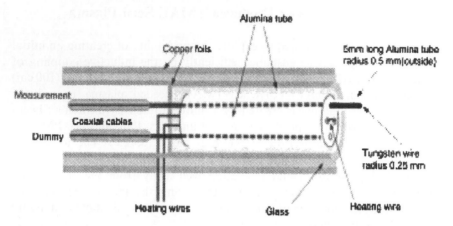

Figure 7.2.3.2. Fast Langmuir probe structure. (Ding *et al* 2001.)

illustrated in figure 7.2.3.2 (Ding *et al* 2001). The instantaneous Langmuir probe (LP) current–voltage characteristic curve is measured by a sampling technique using a boxcar averager triggered by the laser pulse. A heated tungsten wire was used to keep the probe surface very clean and a dummy probe was used differentially to reduce the noise from the laser, the electromagnetic pulse and transient plasma oscillations. The LP current–voltage traces for this plasma were extremely sharp. Very accurate temporal density and temperature data was obtained for the plasma.

The LP temporal plots of electron density and temperature at 20 cm from the Suprasil laser window are shown in figures 7.2.3.3 and 7.2.3.4 (Ding *et al* 2001). The high-density, cold plasma (10^{12}–10^{13} cm^{-3}, ~0.2–0.4 eV) decay was accurately measured 100 ns after the initial 20 ns laser pulse of 4–8 mJ/cm^2 that created the plasma. The electron densities were higher for higher TMAE pressure whereas the electron temperature was higher for lower TMAE pressures. It was also observed that the electron temperature decays sharply for earlier times as compared to the electron density.

Consider the temporal decay of the plasma density. In the absence of an ionizing source, the plasma decay can be described as (Akhtar *et al* 2004, Ding *et al* 2001, Kelly *et al* 2002, Stalder *et al* 1992),

$$\frac{dn_e}{dt} = -D_a \nabla^2 n_e - \alpha_r n_e^2 - \beta_j n_j n_e^2 - \kappa_a n_e n_g^2. \qquad (7.2.3.1)$$

Here, D_a is the ambipolar diffusion term, α_r (cm^3/s) is the two-body (electron–ion) recombination coefficient and $\beta_{j=e,g}$ (cm^6/s) is the three-body (electron–ion) recombination coefficient involving either a neutral

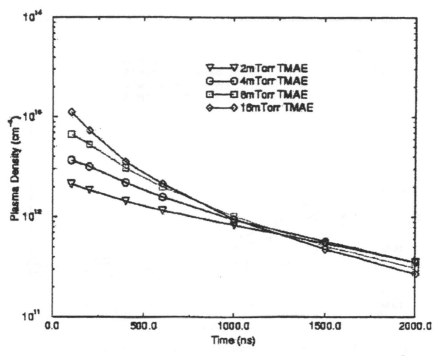

Figure 7.2.3.3. Temporal plots of electron density under conditions of $4\,\text{mJ/cm}^2$ laser fluence. (Ding *et al* 2001.)

atom (β_g; n_g) or an electron (β_e; n_e) as the third species. Here, n_g is the neutral particle density of the neutral background gas. κ_a (cm^6/s) is the three-body electron attachment rate coefficient for the process $e + O_2 + M \longrightarrow O_2^- + M$ (M $= O_2$, N_2). The diffusive loss in the TMAE plasma after the application of the 20 ns laser pulse is small on the microsecond time scale and can be neglected. Since the TMAE molecule is a strong electron donor (Nakato *et al* 1971, Holroyd *et al* 1987), the electron attachment in a pure TMAE plasma is also very small and is neglected.

The rate coefficient for the three-body recombination process where the third body is an electron (i.e. $A^+ + e + e \longrightarrow A^* + e$) is given by $\beta_e \approx 1.64 \times 10^{-9} \{T\,(\text{Kelvin})\}^{-9/2}\,\text{cm}^6/\text{s}$ (Capitelli *et al* 2000). For electron densities, $n_e \approx 10^{13}\,\text{cm}^{-3}$, at room temperature, the loss factor, $\beta_e n_e$, is $1.2 \times 10^{-7}\,\text{cm}^3/\text{s}$. The neutral-stabilized, electron–ion collisional recombination rate for the process, $A^+ + e + B \longrightarrow A^* + B$, where B is a neutral atom, is given as (Capitelli *et al* 2000, Bates 1987).

$$\beta_g \approx 6 \times 10^{-27} \left(\frac{300}{T\,(\text{Kelvin})} \right)^{1.5} (\text{cm}^6/\text{s}). \qquad (7.2.3.2)$$

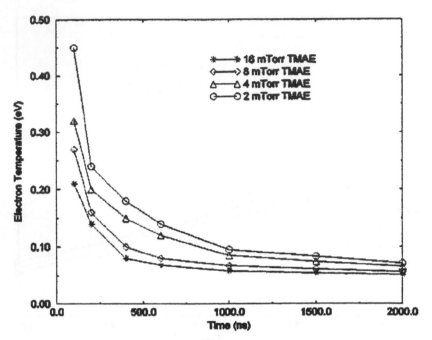

Figure 7.2.3.4. Temporal decay of electron temperature corresponding to the plasma density plots in figure 7.2.3.3. (Ding *et al* 2001.)

For a pure TMAE plasma at a maximum pressure of 50 mtorr at 300 K ($n_g \approx 1.6 \times 10^{15}$ cm^{-3} using Loschmidt's number (NRL Plasma Formulary 2002)), the loss factor, $\beta_g n_g = 9.6 \times 10^{-12}$ cm^3/s, can be neglected. However, at 760 torr where the neutral particle density, $n_g = 2.45 \times 10^{19}$ cm^{-3}, the loss factor, $\beta_g n_g = 1.5 \times 10^{-7}$ cm^3/s, becomes important.

As a result, for a TMAE partial pressures of 4–16 mtorr, the three-body loss processes involving $A^+ + e + e \longrightarrow A^* + e$ and $A^+ + e + B \longrightarrow A^* + B$ can be neglected along with the loss due to electron attachment to oxygen. Therefore, for a temporally decaying TMAE plasma, the continuity equation (7.2.3.1) takes the form

$$\frac{dn_e}{dt} = -\alpha_r n_e^2. \qquad (7.2.3.3)$$

The effective recombination coefficient (α) for a TMAE plasma can be measured from the temporal plot of the plasma density. The numerical solution of equation (7.2.3.3) is obtained by determining the electron densities, n_{e1} and n_{e2}, at two closely-spaced measurement times, t_1 and t_2, respectively. It is given as

$$\alpha_r = \left(\frac{1}{n_{e2}} - \frac{1}{n_{e1}} \right) \bigg/ (t_2 - t_1). \qquad (7.2.3.4)$$

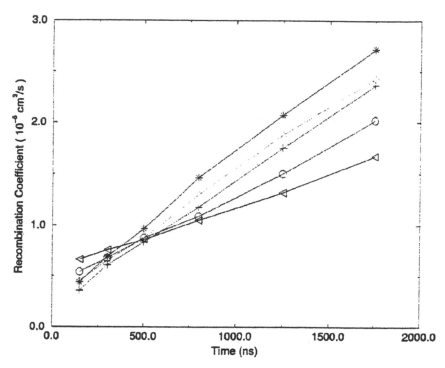

Figure 7.2.3.5. Temporal plot of effective electron–ion recombination coefficient under conditions of $4\,\mathrm{mJ/cm^2}$ laser fluence with TMAE pressures of ($*$) 16 mtorr, ($\square$) 8 mtorr, ($\bigcirc$) 4 mtorr, and ($\triangledown$) 2 mtorr and under $8\,\mathrm{mJ/cm^2}$ laser fluence with a TMAE pressure of ($+$) 8 mtorr. (Ding *et al* 2001.)

Using the data from the temporal density plot in figure 7.2.3.3, a temporal plot of α starting 100 ns after the initial application of the 20 ns laser pulse is shown in figure 7.2.3.5 (Ding *et al* 2001). As shown in the plot, the recombination coefficient increases with time. The experimental result cannot be interpreted by either three-body recombination or multi-ion species effects (Ding *et al* 2001). Since the neutral TMAE density was constant, the three-body process, $A^+ + e + TMAE \longrightarrow A^* + TMAE$, remained constant in time. The recombination process, $A^+ + e + e \longrightarrow A^* + e$, will cause α_r to decrease as the electron density decays. Multi-ion species will also cause α_r to decrease with time. If two components are assumed, the component with a larger α_r decays more rapidly and as a result the global value of α_r must decrease with time. None of these processes can explain the increase of α_r with time.

This increase in α_r with time can be explained in terms of a delayed ionization process in TMAE (Ding *et al* 2001). This has also been observed in large molecules such as metal clusters and C_{60} molecules (Schlag and Levin 1992, Levin 1997). Single photon ionization of large molecules does not necessarily result in prompt ionization, even though the photon energy is above the vertical ionization potential of the molecule (Schlag and Levin 1992). The photons absorbed by molecules AB produce super-excited neutrals AB^{**}. The super-excited AB^{**} molecules store energy in the vibrational states and it is the slow, diffusive-like transfer of this energy to the departing electrons that determines the ionization rate (Levin 1997). This process is known as delayed ionization and plays an important role in the TMAE plasma formation and subsequent decay process. These super-excited $TMAE^{**}$ neutrals decay by electron emission

$$TMAE + h\nu \longrightarrow TMAE^{**} \longrightarrow TMAE^+ + e \text{ (delayed ionization)}.$$

$$(7.2.3.5)$$

The process of delayed ionization can be incorporated in the temporal TMAE plasma density decay as (Ding *et al* 2001)

$$\frac{dn_e}{dt} = -\alpha' n_e^2 + D(t) \qquad (7.2.3.6)$$

where $D(t)$ is the delayed ionization coefficient. Substituting $dn_e/dt_e = -\alpha_r n_e^2$ from equation (7.2.3.3), we obtain $D(t) = (\alpha' - \alpha_r)n_e^2$. This implies that $\alpha' - \alpha_r$ is the change in recombination due to the delayed ionization.

The presence of air components such as oxygen at room temperature has a substantial impact on the TMAE plasma formation and plasma decay process essentially through the electron attachment process. In order to achieve efficient rf sustainment of a laser-preionized TMAE seed plasma, the following scientific issues have to be resolved: (i) The effect of the background gas on the formation and decay characteristics of the TMAE plasma, (ii) the role of delayed ionization, (iii) whether the lifetime of the laser-produced TMAE plasma is long enough such that rf power can be coupled efficiently through inductive wave coupling at lower power levels to sustain the plasma, and (iv) the time scale for modification of TMAE vapor due to its chemical interaction with oxygen, which could reduce its viability as a readily ultraviolet-ionized seed gas in air. In addition, the presence of a background gas makes the plasma very collisional and, therefore, a plasma diagnostic that measures plasma collisionality and recombination losses is also required. Since the previous fast (10 ns) Langmuir probe (LP) measurements (Ding *et al* 2001) could only be carried out 100 ns after the application of the laser pulse when the plasma was in a quiescent decay state, many physical processes, such as delayed ionization, present during the formation and early stages of the decay of the TMAE plasma could not be examined. Recent work (Akhtar *et al* 2003, 2004) with millimeter wave interferometry and fast emission spectroscopy diagnostics have been used to obtain the full temporal decay characteristics of the TMAE plasma.

A 105 GHz (QBY-1A10UW, Quinstar Technology) quadrature-phase, millimeter wave interferometer was used to characterize the temporal development of the plasma during and following the application of the

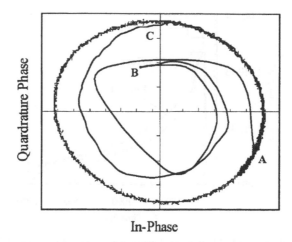

In-Phase

Figure 7.2.3.6. Interferometer trace showing the phase and amplitude variation for 35 mtorr TMAE plasma after the application of a 20 ns laser pulse reaching a maximum line-average plasma density of 4×10^{13} cm^{-3} (A → B), followed by the plasma decay (B → C → A) at a distance of 20 cm from the laser window. The outside circle represents the vacuum phase variation. (Akhtar *et al* 2004 (© 2004 IEEE).)

20 ns laser pulse. The millimeter wave interferometry technique is described in detail in chapter 8 of this book (Akhtar *et al* 2003). The interferometer worked in the Mach–Zehnder configuration, in which the plasma was in one arm of the two-beam interferometer. The interferometer utilized an I-Q (In-phase and Quadrature phase) mixer to obtain the phase and amplitude change of the 105 GHz mm wave signal that passed through the plasma.

The interferometer trace shown in figure 7.2.3.6 is a function of time as the 35 mtorr TMAE plasma formed by the application of 20 ns laser pulse decayed. Since the laser intensity ($I = 6$ mJ/cm^2) was uniform ($\Delta I/I \approx 10\%$) over its 2.8 cm diameter, a uniform radial plasma profile could be assumed. The outside circle represents the phase variation for vacuum conditions without plasma. The onset of plasma followed the path A → B. The line-average plasma density reached its maximum value of 4×10^{13} cm^{-3} at $z = 20$ cm from the Suprasil window. The temporal decay of TMAE plasma was along the path B → C → A. A plane wave model and software were utilized to obtain the plasma density in this collisional regime.

In figure 7.2.3.7, the temporal plot of the TMAE plasma density for 4, 16, and 50 mtorr TMAE vapor pressures is shown. It should be noted that the peak plasma density occurred fairly late in time ($t = 140 \pm 10$ ns) after the application of the laser pulse. Optical emission data also showed the presence of a small (two orders of magnitude lower) direct ionization process during the laser pulse. However, the initial ($\tau \leq 20$ ns) low density plasma ($\sim 10^{11}$ cm^{-3}) produced by direct ionization could not be accurately

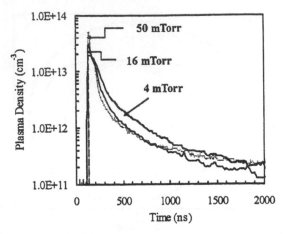

Figure 7.2.3.7. Plots of TMAE plasma density versus time for different TMAE vapor pressures for a laser fluence of $6\,\mathrm{mJ/cm^2}$, $\tau_L = 20\,\mathrm{ns}$. (Akhtar *et al* 2004 (© 2004 IEEE).)

measured by the 105 GHz interferometer. It was also observed that the plasma density increased with vapor pressure, while the plasma density decay was more rapid at higher vapor pressures. The axial plasma density plot in figure 7.2.3.8 reveals a rapid axial plasma density decay for higher vapor pressure plasmas. The fractional peak plasma density at an 80 cm axial location with respect to its value at 20 cm was 40, 30, 14, and 8% for the 4, 10, 30, and 50 mtorr cases, respectively. This was due to the enhanced laser absorption nearer the Suprasil window at higher pressures.

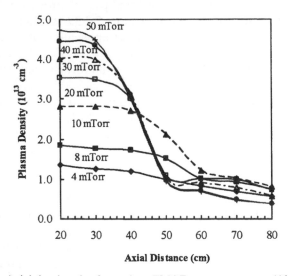

Figure 7.2.3.8. Axial density plot for various TMAE vapor pressures. (Akhtar *et al* 2004 (© 2004 IEEE).)

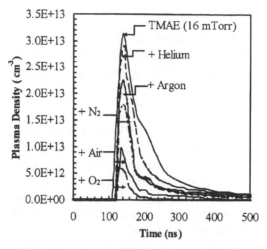

Figure 7.2.3.9. TMAE plasma density versus time plot for different background gases at 760 torr. (Akhtar *et al* 2004 (© 2004 IEEE).)

In order to study the effect of background gases on the TMAE plasma formation and decay characteristics in an evacuated chamber, the TMAE pressure was raised to 16 mtorr and then the background gas pressure was increased slowly to 760 torr. A temporal plot of the TMAE plasma density in the presence of different background gases is shown in figure 7.2.3.9. A laser fluence of 6 mJ/cm^2 was maintained. The temporal variation of 16 mtorr of pure TMAE plasma density is also shown in the plot for reference.

The peak plasma density of pure TMAE was 3.2×10^{13} cm^{-3}. In the presence of 760 torr of noble gases such as helium and argon, the TMAE peak plasma density was reduced to 2.9×10^{13} and 2.3×10^{13} cm^{-3}, respectively. This corresponds to a density reduction of 10% for helium and 30% for argon background gas. It was also observed that a high-density ($>10^{12}$ cm^{-3}) plasma is maintained in the presence of noble background gases for over 2 µs. Since the background gas was at atmospheric pressure with neutral particle densities $\sim 2.5 \times 10^{19}$ cm^{-3}, the effect of three-body recombination involving a neutral as the third particle became an important factor. In the experiment with room temperature air constituents as the background gas, the effect of electron attachment was evident. The peak TMAE plasma densities obtained in the presence of 760 torr of nitrogen, oxygen and air were 1.8×10^{13}, 5.8×10^{12} and 9.8×10^{12} cm^{-3}, respectively. In addition, a TMAE plasma density $\geq 5 \times 10^{11}$ cm^{-3} was maintained in atmospheric air for $t \geq 0.3$ µs. This was long enough so that rf power could be coupled to the seed plasma efficiently (Kelly *et al* 2002). It was also observed that the seed TMAE vapor remained viable for large-volume (~ 500 cm^3) and high-density (10^{13} cm^{-3}) laser ionization in air for $t \leq 10$ minutes.

7.2.3.1 TMAE density decay in the presence of Noble Gases

In the presence of noble gases at 760 torr, three-body recombination involving neutrals as the third particle becomes significant. Neglecting electron attachment, equation (7.2.3.1) can be expressed as

$$\frac{\partial n_{\mathrm{e}}}{\partial t} = -\alpha n_{\mathrm{e}}^2 - \beta_{\mathrm{g}} n_{\mathrm{g}} n_{\mathrm{e}}^2 = -(\alpha + \beta_{\mathrm{g}} n_{\mathrm{g}}) n_{\mathrm{e}}^2. \qquad (7.2.3.7)$$

Here α represents the recombination losses for the pure TMAE plasma described in equation (7.2.3.3) and β_{g} is the loss due to three-body recombination where the third body is a neutral atom. In order to determine β_{g} for TMAE in the presence of helium and argon, a numerical derivative of the TMAE plasma density temporal plot in figure 7.2.3.9 is obtained. Using the recombination coefficients, α, already obtained for pure TMAE (figure 7.2.3.5) along with the neutral particle gas density, n_{g}, equation (7.2.3.7) was numerically solved in time to determine β_{g}. A plot of the resultant three-body recombination coefficient, β_{g}, is presented in figure 7.2.3.10. Since the three-body recombination process depends on the neutral gas density (maintained at 760 torr during this experiment) only a very small temporal variation in β_{g} was observed. The small variation ($\sim$5%) is within statistical error. In this experiment, the three-body recombination rate coefficients for TMAE in the presence of helium and argon were determined to be $\beta_{\mathrm{g}}(\mathrm{He}) = (4.35 \pm 0.7) \times 10^{-26}\,\mathrm{cm}^6\,\mathrm{s}^{-1}$ and $\beta_{\mathrm{g}}(\mathrm{Ar}) = (9.5 \pm 0.8) \times 10^{-26}\,\mathrm{cm}^6\,\mathrm{s}^{-1}$, respectively. The values obtained were comparable to the published collisional three-body recombination rates for singly ionized plasmas (Zel'dovich and Raizer 1966).

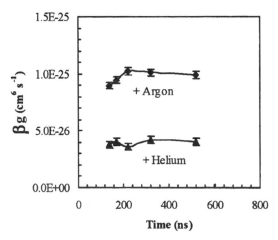

Figure 7.2.3.10. Three-body recombination rate coefficients for a TMAE plasma in the presence of helium and argon at 760 torr. (Akhtar *et al* 2004 (© 2004 IEEE).)

7.2.3.2 *TMAE density decay in the presence of air constituent gases*

In atmospheric pressure air at room temperature, the dominant density loss mechanism in a TMAE plasma in air is electron attachment with oxygen through the process $e + O_2 + M \longrightarrow O_2^- + M$ $(M = O_2, N_2)$. Negative oxygen ions are rapidly removed by ionic recombination and this results in a significant reduction in the plasma density and life-time. The density decay equation (equation (7.2.3.1)) for this case is written as

$$\frac{dn_e}{dt} = -\alpha n_e^2 - \beta_g n_g n_e^2 - \kappa_a n_e n_g^2. \tag{7.2.3.8}$$

Here β_g is the three-body recombination rate coefficient with either oxygen or nitrogen as the third species and κ_a is the electron attachment rate coefficient for oxygen and nitrogen. Based on the classical diffusion model that includes the elastic scattering of electrons by diatomic molecules, the β_g values at room temperature are assumed to be $\cong 10^{-26}\,\mathrm{cm^6\,s^{-1}}$ (Bates 1980, Biberman *et al* 1987) for the present calculation. The differences in β_g values for diatomic molecules with mirror symmetry like oxygen, nitrogen and hydrogen are small due to the absence of permanent dipole moments (Bates 1980).

A numerical solution of equation (7.2.3.8) is obtained for the electron attachment coefficient, κ_a, by using numerical differentiation of the temporal decay of the TMAE plasma density in the presence of air constituents (figure 7.2.3.9) along with the known effective two-body recombination coefficient, α, for TMAE (figure 7.2.3.5). A temporal plot of the electron attachment rate coefficient, κ_a, for nitrogen, oxygen and air when they are individually added to TMAE is shown in figure 7.2.3.11 (Akhtar *et al* 2004). As shown in that

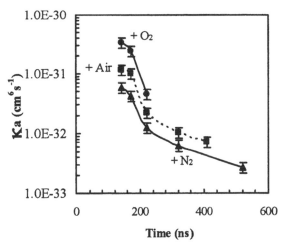

Figure 7.2.3.11. Electron attachment rate coefficients for a TMAE plasma in the presence of nitrogen, oxygen and air at 760 torr. (Akhtar *et al* 2004 (© 2004 IEEE).)

figure, the electron attachment rate decreases temporally with the TMAE plasma density. This illustrates that the probability of electron capture for attachment decreases with a decrease in the plasma density. In the presence of nitrogen, the peak value at the peak plasma density ($t = 140$ ns) for $\kappa_a(N_2)$ is 5.6×10^{-32} cm^6 s^{-1}. As a result, the subsequent nitrogen contribution to the TMAE plasma loss for air is small. This is to be expected since nitrogen does not readily form a negative ion and the dominant plasma loss can be attributed to the presence of the oxygen (Capitelli *et al* 2000). However, for oxygen, the peak electron attachment rate coefficient $\kappa_a(O_2)$ at $t = 140$ ns, when the TMAE density is maximum, is 3.2×10^{-31} cm^6 s^{-1}. This is almost an order of magnitude higher than that for nitrogen. In the presence of atmospheric air, the TMAE plasma electron attachment rate to oxygen is 1.1×10^{-31} cm^6 s^{-1}. These electron attachment rate coefficients for TMAE plasmas in nitrogen, oxygen and air are lower by almost an order of magnitude than the values obtained for the process, $e + O_2 + M \longrightarrow O_2^- + M$ ($M = O_2$, N_2, H_2O) in room temperature air (Raizer 1991). This indicates that the process of delayed ionization of TMAE that has a much longer lifetime ($\tau = 140$ ns) than the direct ionization gradually populates the emissive state and plays an important role in increasing the lifetime of the TMAE plasma for rf sustainment at lower power.

7.2.3.3 Plasma emission spectroscopy

The optical emission spectra of a 193 nm laser-produced TMAE plasma was obtained using a high-resolution spectrometer (Akhtar *et al* 2004). Plasma emission passed through a high-quality ultraviolet (200–800 nm) fiber-optic bundle into a spectrometer, and was then detected by a photomultiplier tube (PMT). An ultraviolet cutoff filter (<300 nm) is used in front of the fiber-optic bundle to eliminate the scattered 193 nm high-power source laser pulse that can saturate the PMT. It utilizes a 500 mm focal length monochromator (Acton Research SpectraPro-500i, Model SP-558) with a 1200 g/mm grating and a high-resolution of 0.05 nm at 435.8 nm. The entrance and exit slit widths were set at 2000 µm to obtain a statistically large number of photon counts per acquisition. A schematic is shown in figure 7.2.3.12.

A wavelength scan of the emission spectrum from 300 to 650 nm, with a step size of 4 nm and averaged over 200 laser pulses was obtained. A user-defined program written in Lab View provided the flexibility of arbitrary integration window size, accurate referencing of the integration window with respect to the laser pulse, and better statistics by averaging over a large number of laser pulses. The emission spectrum of 16 mtorr TMAE plasma alone and in the presence of air constituents, measured for the time window 100 ns $< t < 1100$ ns referenced to the laser pulse turn on with the

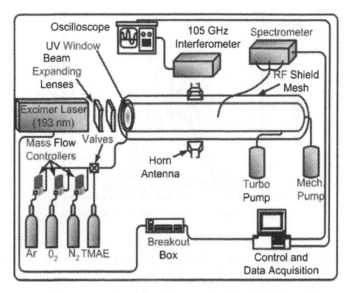

Figure 7.2.3.12. Schematic of the experimental arrangement of the laser-initiated and rf sustained plasma. The lens system is used to modify the laser footprint cross-section to 2.8 cm × 2.8 cm. In this experiment the rf coil has not been energized. (Akhtar *et al* 2004 (© 2004 IEEE).)

laser flux held constant at 6 mJ/cm^2 is shown in figure 7.2.3.13. The spectrum has maxima at 448 and 480 nm. The 480 nm maximum was reported as a peak emission and corresponds to the first Rydberg state TMAE* (R1) with a 20 ns lifetime (Hori *et al* 1968, Nakato *et al* 1972).

The emission spectrum increased in the presence of nitrogen as compared to the pure TMAE spectrum, whereas the peak emission dropped significantly in the presence of pure oxygen and it was only slightly higher than the noise level. The decrease in plasma emission in the presence of oxygen could be explained in terms of the rapid quenching of TMAE plasma through the process of electron attachment to oxygen. This result is in agreement with the interferometric measurements of lower density and a shorter lifetime of the TMAE plasma in the presence of room temperature oxygen. A decrease was observed in the plasma emission with atmospheric pressure air compared to TMAE alone. However, the plasma emission as well as the peak plasma density measurement ($n_e \approx 10^{13}$ cm^{-3}) indicates that a high-density ($>5 \times 10^{11}$ cm^{-3}) TMAE plasma in air can be maintained for $t \leq 0.3$ μs such that efficient coupling at lower rf power for sustainment can occur (Kelly *et al* 2002).

In order to obtain the temporal evolution of the 480 nm line corresponding to the TMAE*(R1) state over $t \leq 800$ ns, a narrow integration window of 10 ns was used. Figure 7.2.3.14 clearly shows that the peak of

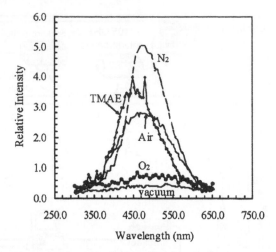

Figure 7.2.3.13. Effect of 760 torr background gases nitrogen, oxygen and air on the emission spectra of a 16 mtorr TMAE plasma measured during the time window 100 ns < t < 1100 ns. (Akhtar *et al* 2004 (© 2004 IEEE).)

480 nm emission for 16 mtorr TMAE occurred fairly late in time ($\tau = 140 \pm 10$ ns) after the application of the 20 ns laser pulse. Small (two orders of magnitude lower) 480 nm emission was also observed due to the direct ionization process during the laser pulse. In order to reference

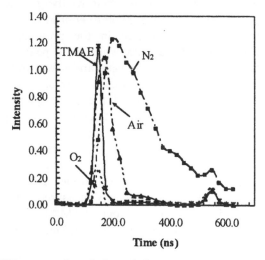

Figure 7.2.3.14. The temporal evolution of the 480 nm line corresponding to TMAE Rydberg states (R1) for 16 mtorr TMAE plasma in the presence of air constituent gases nitrogen, oxygen and air at 760 torr. (Akhtar *et al* 2004 (© 2004 IEEE).)

the plasma temporal emission to the turn-on of the laser pulse, the laser temporal profile was accurately measured by a fast ultraviolet photodiode (Hamamatsu S1226-18BQ with less than 10 ns rise-time) using a 2 GSa/s Lecroy sampling oscilloscope. This late emission of the 480 nm peak was interpreted in terms of the phenomenon of delayed ionization of TMAE.

The absence of direct ionization in TMAE is contrary to the traditional interpretation of the ionization process associated with small molecules. The process of ionization of small molecules is very direct and once the ionization energy is exceeded, free electrons depart on a femtosecond time scale (Platzman 1967). However, for larger molecules such as C_{60} and metal oxide clusters, the ionization is no longer prompt and there is a measurable time delay in the appearance of the electrons (Platzman 1967, Campbell *et al* 1991, Wurz *et al* 1991, Remacle and Levin 1993). Research on photo-ionization of C_{60} (Schlag *et al* 1992, Levin 1997) proposed that even though the photons provide the energy necessary to initiate electron removal, the actual departure of electrons and, hence, ionization is delayed.

Most of the photons absorbed by the TMAE molecules do not contribute to the direct ionization process. Even though the laser photon energy of 6.4 eV was above the TMAE vertical ionization potential (6.1 eV) (Nakato *et al* 1971, 1972), the experiment indicated that the additional energy of 0.3 eV above the ionization potential was not sufficient to produce substantial direct ionization of the large TMAE molecule (molecular weight = 200.3). Instead, these photons excited the neutrals to a super-excited state. These super-excited TMAE neutrals (TMAE**) stored energy in the many degrees of freedom of the molecule and then transfered energy to the departing free electrons on a slower time scale ($\tau = 140$ ns). The delay in the peak 480 nm emission after the application of the laser pulse corresponded to the relaxation time of the super-excited state. From the temporal plot of the 480 nm emission, the relaxation time (the lifetime) of the super-excited state was found to be $\tau \cong 140 \pm 10$ ns. The lifetime of the first Rydberg state of TMAE given by the observed emission spectrum full width at half maximum (FWHM) was 30 ns.

The increase in plasma emission, as shown in figure 7.2.3.13, due to the presence of nitrogen is on the higher wavelength side close to the 480 nm Rydberg line. In addition, figure 7.2.3.13 shows that the peak of the 480 nm line occurs 200 ns after the laser pulse and that the full-width at half-maximum of the Rydberg emission process increased to 170 ns. Since nitrogen does not react with TMAE and also does not absorb 193 nm photons, the enhancement of the emission intensity implies that the nitrogen molecules enhanced the excitation of the TMAE** state, where energy was stored, during the application of the laser pulse (Ding *et al* 2001). These highly excited TMAE** states gradually decayed by electron emission and populated the first Rydberg state through the process, $TMAE^{**} + N_2 \rightarrow TMAE^*(R1) + N_2$. This gradual population of the $TMAE^*(R1)$ state and subsequent emission resulted in a broad temporal profile of 480 nm emission.

The experiment showed that it is possible to create a large-volume ($\sim$500 cm^3), high-density ($\sim$10^{13} cm^{-3}) TMAE plasma in 760 torr air. The density decay was such that $n_e \geq 5 \times 10^{11}$ cm^{-3} for $t \geq 0.3$ µs. In addition, the long axial extent (100 cm) of the laser seed plasma allowed enhanced rf penetration and ionization well away from the 20 cm antenna axial extent. This suggests an optimum electrodeless scenario where TMAE is pulse-injected into heated air at 2000 K, thus reducing the electron attachment and enhancing plasma lifetime in air. The plasma could be formed by ultraviolet flash tube optical means that facilitates the efficient coupling of high-power pulsed rf power to the plasma and substantially reduces rf power requirements for high-density (10^{13} cm^{-3}), large volume air plasma for a variety of applications.

References

Akhtar K, Scharer J, Tysk S and Denning C M 2004 *IEEE Trans. Plasma Sci.* **32**(2) 813

Akhtar K, Scharer J, Tysk S and Kho E 2003 *Rev. Sci. Instrum.* **74** 996

Bates D R 1980 *J. Phys. B* **13** 2587

Biberman L M, Vorob'ev V S and Yakubov I T 1987 *Kinetics of Nonequilibrium Low-Temperature Plasmas* (New York: Consultants Bureau) p 412

Campbell G E E B, Ulmer G and Hertel I V 1991 *Phys. Rev. Lett.* **67** 1986

Capitelli M, Ferreira C M, Gordiets B F and Osipove A I 2000 *Plasma Kinetics in Atmospheric Gases* (Berlin: Springer) p 140

Ding G, Scharer J E and Kelly K 2001 *Phys. Plasmas* **8** 334

Holroyd R A, Preses J M, Woody C L and Johnson R A 1987 *Nucl. Instr. and Meth. Phys. Res. A* **261** 440

Hori M, Kimura K and Tsubomura H 1968 *Spectrochimica Acta A* **24** 1397

Kelly K L, Scharer J E, Paller E S and Ding G 2002 *J. Appl. Phys.* **92** 698

Levin R D 1997 *Adv. Chem. Phys.* **101** 625

Nakato Y, Ozaki M, Egawa A and Tsubomura H 1971 *Chem. Phys. Lett.* **9**(6), 615

Nakato Y, Ozaki M and Tsubomura H 1972 *J. Phys. Chem.* **76** 2105

NRL Plasma Formulary, revised edition 2002

Platzman R L 1967 in Silini G (ed) *Radiation Research* (Amsterdam: North-Holland)

Raizer Y P 1991 *Gas Discharge Physics* (Berlin Heidelberg: Springer) p 62

Remacle F and Levin R D 1993 *Phys. Lett. A* **173** 284

Schlag E W and Levin R D 1992 *J. Phys. Chem.* **96** 10608

Stalder K R and Eckstrom D J 1992 *J. Appl. Phys.* **72** 3917

Stalder K R, Vidmar R J and Eckstrom D J 1992 *J. Appl. Phys.* **72** 5098

Wurz P, Lykke K R, Pellin M J and Gruen D M 1991 *J. App. Phys.* **70** 6647

Zel'dovich Y B and Raizer Y P 1966 *Physics of Shock Waves and High-Temperature Hydrodynamic Phenomena* (New York: Academic Press) vol 1, p 407

7.3 Radiofrequency and Microwave Sustained High-Pressure Plasmas

7.3.1 Introduction

Radiofrequency and microwave sources for plasma production at low pressures in the milli-torr range are highly developed and used in applications for materials processing and surface modification. In this section, we describe their characteristics for high density plasma production at high pressure and in atmospheric air. The properties of near thermal equilibrium air plasmas produced by a rf inductive source or plasma torch are discussed in section 7.3.2. Optical spectroscopy is used to measure the plasma density and electron temperature. Radiofrequency plate power is used to determine power balance and efficiency characteristics for the air plasma in steady-state. These results serve as a benchmark for air plasmas and illustrate the power densities required to sustain air plasmas near thermal equilibrium at high density.

Section 7.3.3 discusses rf sustainment of a flashtube or laser initiated plasma. This can be accomplished at much lower power levels than is required for breakdown and ionization in high-pressure air or other gas. It should be noted that power levels for initial ionization of atmospheric air are substantially higher that those discussed for steady-state in section 7.3.2. The laser-formed, large volume, high density plasma provides an ideal plasma load that can be efficiently sustained at lower power levels by short pulse or steady-state rf power. Detailed characteristics of the temporal density characteristics of these plasmas are discussed using millimeter wave interferometry, optical spectroscopy and detailed rf coupled power measurements.

Section 7.3.4 discusses the use of microwaves to produce breakdown and high density in air. Intersecting microwave beams can produce spatial localization and microwaves can be used in a microwave cavity for highly localized plasmas. They can also be beamed to space for plasma ionization for use as a microwave mirror reflector in the atmosphere.

7.3.2 Review of rf plasma torch experiments

7.3.2.1 Introduction

Thermal plasma devices, such as rf or microwave torches, represent a convenient way to produce relatively large volumes of atmospheric pressure air plasma with electron number densities up to $10^{15} \, \text{cm}^{-3}$. However, the plasmas generated with such devices are generally near local thermodynamic equilibrium (LTE), which implies that the gas temperature increases with the electron number density as shown in figure 7.3.2.1. From that plot, one can

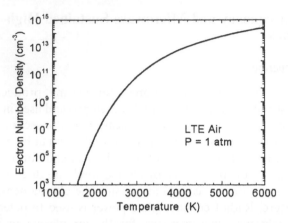

Figure 7.3.2.1. Electron number density in atmospheric pressure air under LTE conditions.

see that the equilibrium electron density in atmospheric pressure air is approximately $3.3 \times 10^6 \, cm^{-3}$ at 2000 K, $6.5 \times 10^{10} \, cm^{-3}$ at 3000 K, $6.1 \times 10^{12} \, cm^{-3}$ at 4000 K, and $6.2 \times 10^{13} \, cm^{-3}$ at 5000 K. Once produced, the thermal plasma can be sustained for an indefinite duration if placed in a perfectly insulated container. In this ideal situation, no power would be needed to sustain the plasma and therefore the power budget could be infinitesimally small. In practice, however, the thermal plasma is flowing into a non-perfectly insulated container or into ambient air, where it undergoes recombination by conductive and radiative cooling and by mixing with entrained air. The power required to sustain the plasma depends on the geometry of the device, the environment into which the plasma flows, and the flow velocity. In this section, the goal is to determine the minimum power required to produce and sustain an open-air plasma volume by means of a typical, industrial-scale rf, inductively coupled plasma torch. First, a baseline experiment was performed to determine the 'brute force' unoptimized power necessary to produce a plasma with an electron number density greater than $10^{13} \, cm^{-3}$, and with dimensions greater than 5 cm in all directions. Section 7.3.2.2 describes the rf torch facility that was used and the set-up for the optical diagnostics. Section 7.3.2.3 presents measurements of the gas and electron density profiles produced by the torch for various gas injection modes. Finally section 7.3.2.4 presents measurements of the power required to sustain the plasma.

7.3.2.2   Radiofrequency plasma torch facility

The measurements presented here were obtained in the rf torch facility of the High Temperature Gas dynamics Laboratory at Stanford University. This facility is centered around a 50 kW inductively coupled plasma torch (TAFA

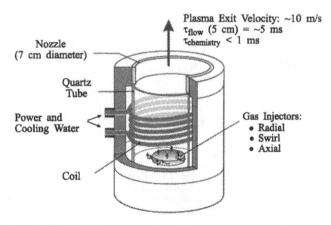

Figure 7.3.2.2. Schematic cross-section of torch head with 7 cm diameter nozzle.

Model 66) powered by an rf LEPEL Model T-50-3 power supply operating at 4 MHz. The power supply delivers up to 120 kVA of line power to the oscillator plates with a maximum of 12 kV dc and 7.5 A. The oscillator plates have a maximum rf power output of 50 kW. The basic design for inductively coupled plasma torches has not changed much since their introduction by Reed (1961). A schematic drawing of the plasma torch head is shown in figure 7.3.2.2. The feed gas is injected at the bottom of a quartz tube (inner diameter 7.6 cm, thickness 3 mm) surrounded by a coaxial five-turn copper induction coil (mean diameter 8.6 cm) traversed by an rf current. The outer Teflon body acts as an electrical insulator and electromagnetic screen. The coil is cooled with deionized water to prevent arcing between its turns. The rf current produces an oscillating axial magnetic field that forces the free electrons to spin in a radial plane and thereby generates eddy currents. The energetic free electrons produced by rf excitation can then ionize and dissociate heavy particles through collisions. Further details on inductively coupled plasma torches can be found in Eckert *et al* (1968), Dresvin *et al* (1972), Davies and Simpson (1979), and Boulos (1985), and advanced numerical models in Mostaghimi *et al* (1987, 1989) and van den Abeele *et al* (1999).

The plasma torch can operate with a variety of gases (air, hydrogen, nitrogen, oxygen, methane, argon, or mixtures thereof). For the baseline experiments described here, the feed gas was primarily air with a small amount of hydrogen (less than 2% mole fraction) added for purposes of electron number density measurements from the Stark-broadened H_β line shape.

The feed gas can be injected in axial, radial or swirl modes through a manifold located at the bottom of the torch. Axial injection provides bulk movement to the gas during the start-up phase. In normal operation, only swirl and radial injectors are used. As will be seen below, the swirl-to-radial feed ratio has a large impact on the temperature and concentration profiles of the plasmas produced by the torch.

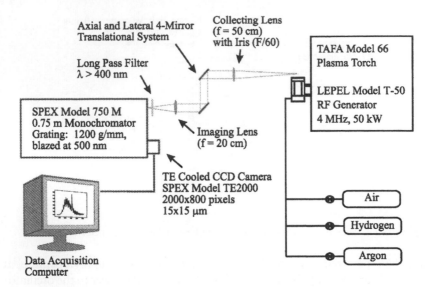

Figure 7.3.2.3. Experimental set-up for emission diagnostics. (Laux *et al* 2003.)

The plasma generated in the coil region expands into ambient air through a converging copper nozzle, 7 cm in diameter. At the nozzle exit plane, the maximum axial velocity is estimated to be 10 m/s, the maximum temperature is measured at about 7000 K, the density $\rho \cong 5.04 \times 10^{-2}$ kg m^{-3} and the dynamic viscosity $\mu \cong 1.6 \times 10^{-4}$ kg m^{-1} s^{-1}. Based on the nozzle diameter of 7 cm, the Reynolds number at the nozzle exit is about 220. The plasma jet is therefore laminar at locations of 1 and 5 cm downstream of the nozzle exit where our measurements were made. A few nozzle diameters downstream of the nozzle exit, the plasma plume becomes turbulent as a result of mixing with ambient air.

The radial profiles of temperature and electron number density were measured by optical emission spectroscopy. The experimental set-up, shown in figure 7.3.2.3, includes a 0.75 m monochromator (SPEX model 750M) fitted with a 1200 lines/mm grating blazed at 500 nm and a backthinned, ultraviolet-coated SPEX Model TE-2000 Spectrum One thermoelectrically cooled CCD camera. The CCD chip measures 30×12 mm and contains 2000×800 square pixels of dimension 15×15 µm. Absolute intensity calibrations were obtained with an Optronics model OL550 radiance standard traceable to NIST standards.

7.3.2.3 Plasma characterization

Figure 7.3.2.4 shows photographs of the plasma plume for three different swirl/radial injection ratios. In the 'low swirl case', the flow rates were 67 slpm (standard liter per minute) in the radial mode and 33 slpm in the

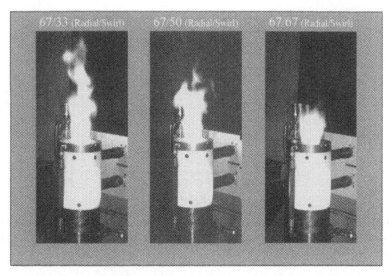

Figure 7.3.2.4. Air plasma plume for three conditions of the radial/swirl flowrates.

swirl mode. The 'medium' and 'high' swirl cases correspond to radial/swirl flow rates of 67/50 and 67/67, respectively. In all three cases, the plate power was kept constant at approximately 41.2 kW, and a small quantity of hydrogen (2.3 slpm) was premixed prior to injection into the torch. To a good approximation the flow injected into the torch was thermodynamically equivalent to humid air with 2.3 slpm of water vapor.

As can be seen from figure 7.3.2.4, the swirl/radial injection ratio had a noticeable influence on the physical aspect of the plasma. The length of the plume was approximately 35, 20, and 10 cm for the low (67/33), medium (67/50) and high (67/67) swirl cases, respectively. In the low swirl case the plasma luminosity exhibited a strong radial gradient, but in contrast it was almost radially uniform in the high swirl case (it is not possible to observe radial variations of the luminosity in figure 7.3.2.4 because the photographs are intensity-saturated). Thus the plasma properties (temperature, electron number density) were more uniform radially in the case with highest swirl injection.

Measurements were made of temperature and electron number density radial profiles at locations 1 and 5 cm downstream of the nozzle exit. Temperature profiles were determined from the absolute intensity of the atomic line of oxygen at 777.3 nm, using an Abel-inversion technique. The temperature profiles measured at 1 and 5 cm downstream of the nozzle exit for a plate power of 41.2 kW are shown in figures 7.3.2.5 and 7.3.2.6 for both the low and high swirl cases. The radial profiles were found to be flatter in the high swirl case (67/67) than in the low swirl case (67/33), in accordance with the visual aspect of the plume.

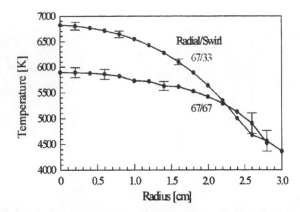

Figure 7.3.2.5. 1 cm downstream of the nozzle exit. Measured temperature profiles from Abel-inverted absolute intensity profiles of the atomic oxygen triplet at 777.3 nm. Plate power = 41.2 kW. Gas: air + 2.3 slpm H_2.

Previous studies conducted at Stanford University (Laux 1993) had shown that air plasmas generated by this torch under similar conditions of temperature and velocity were close to local thermodynamic equilibrium (LTE). This was because the characteristic chemical relaxation time was about 10 times faster than the characteristic flow time between the coil region, where the plasma was in a state of non-equilibrium, and the nozzle exit where the measurements were made. Thus the relatively slow plasma flowing through the 7 cm diameter nozzle was close to LTE both at 1 and 5 cm downstream of the nozzle exit. Under LTE conditions, electron number densities were determined from the knowledge of the plasma

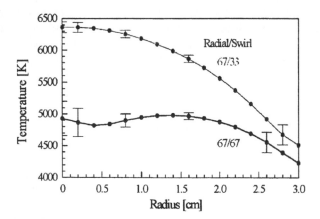

Figure 7.3.2.6. 5 cm downstream of the nozzle exit. Measured temperature profiles from Abel-inverted absolute intensity profiles of the atomic oxygen triplet at 777.3 nm. Plate power = 41.2 kW. Gas: air + 2.3 slpm H_2.

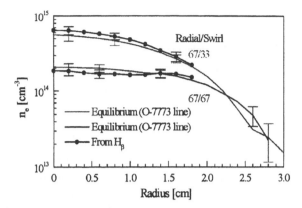

Figure 7.3.2.7. 1 cm downstream of the nozzle exit. Measured electron number density profiles from Abel-inverted H_β line shapes and equilibrium electron number density profiles based on the temperature profiles of figure 7.3.2.5. Plate power $= 41.2$ kW. Gas: air $+ 2.3$ slpm H_2.

temperature using chemical equilibrium relations (Saha equation). Figures 7.3.2.7 and 7.3.2.8 show the equilibrium electron number density profiles based on the temperature profiles of figures 7.3.2.5 and 7.3.2.6. In order to verify the LTE, direct electron number density measurements were also made from the Stark-broadened atomic hydrogen Balmer β line at 486 nm, using the spectroscopic technique detailed in chapter 8, section 8.3.

In air plasmas, the H_β line sits on top of an intense emission background that is mainly composed of bands of the second positive system of molecular nitrogen. In order to extract the H_β lineshape, spectral measurements were

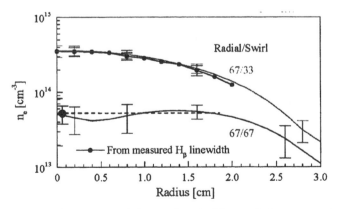

Figure 7.3.2.8. 5 cm downstream of the nozzle exit. Measured electron number density profiles from Abel-inverted H_β lineshapes and equilibrium electron number density profiles based on the temperature profiles of figure 7.3.2.6. Plate power $= 41.2$ kW. Gas: air $+ 2.3$ slpm H_2.

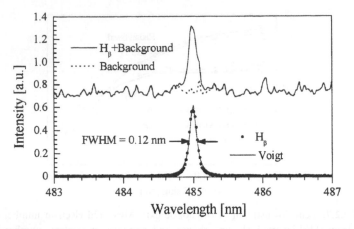

Figure 7.3.2.9. H_β lineshape extraction procedure. 1 cm downstream of the nozzle exit. Low swirl case (67/33 radial/swirl). Plate power = 41.2 kW. Gas: air + 2.3 slpm H_2. The two spectra in the figure are those obtained after Abel-inversion at $r = 10$ mm from the plasma centerline.

made both with the mixture of air/hydrogen (spectrum labeled 'H_β + background' in figure 7.3.2.9) and with pure air (spectrum labeled 'Background'). Spectra measured at several lateral locations along chords of the plasma were then Abel-inverted to provide local emission spectra as a function of the radial location. At each radial location, the H_β lineshapes were recovered by subtracting the background from the total signal. The H_β lineshapes were then fitted with Voigt profiles, which represent the convolution of several broadening mechanisms including pressure (van der Waals, resonance), Doppler, instrumental, and Stark broadening (see figure 7.3.2.10). Pressure and Doppler broadening widths only depend on the pressure and temperature of the gas. Instrumental broadening was minimized by using a very small entrance slit on the monochromator (30 μm). Radial electron number density profiles were determined with the aid of the curves of figure 7.3.2.10. These curves were obtained as discussed in chapter 8, section 8.5. The resulting electron density profiles are shown in figures 7.3.2.7 and 7.3.2.8.

For the high swirl case shown in figure 7.3.2.8, the H_β line intensity was so weak relative to the nitrogen background (see figure 7.3.2.11) that it was not possible to obtain a reliable series of Abel-inverted H_β lineshapes. Nevertheless, since the plasma temperature did not vary significantly over the central part of the plasma, the electron number density determined from the H_β lineshape measured along the diameter of the plasma provided an estimate of the average electron density in the central region. As can be seen from figure 7.3.2.8, the measured line-of-sight-averaged electron density agreed well with the expected equilibrium value in the central region of the plasma.

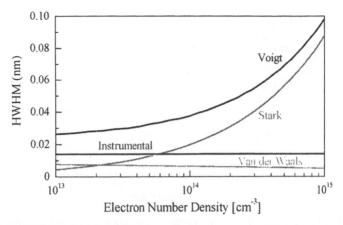

Figure 7.3.2.10. H_β line broadening in atmospheric pressure, equilibrium air. Instrumental broadening is well approximated by a Gaussian of half width at half maximum of 0.014 nm.

The foregoing measurements demonstrated that the rf plasma torch could generate steady-state open air plasmas with electron number densities greater than 10^{13} cm^{-3} over volumes with dimensions greater than 5 cm in all directions. The shape of the electron number density profiles could be controlled by modifying the ratio of radial-to-swirl injection. The measurements presented up to this point were obtained with a mixture of air and hydrogen. Measurements were also made for dry air, in which case the electron number density could only be determined by assuming chemical equilibrium at the local temperature measured from the oxygen triplet at 777.3 nm. Results of this series of experiments are shown in figures 7.3.2.12

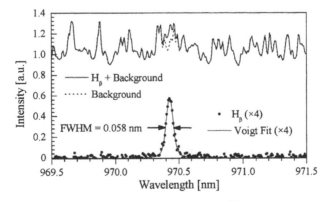

Figure 7.3.2.11. H_β lineshape extraction procedure. Line-of-sight H_β lineshape at location 5 cm downstream of the nozzle exit. High swirl case (67/67 radial/swirl). Plate power = 41.2 kW. Gas: air + 2.3 slpm H_2. Here the H_β lineshape was measured in second order so as to reduce instrumental broadening to approximately 0.007 nm. (Laux *et al* 2003.)

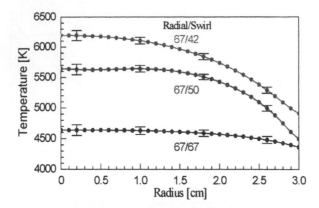

Figure 7.3.2.12. Dry air. Temperature profiles 5 cm downstream of the nozzle exit as a function of radial/swirl injection ratio. Plate power = 40.1 kW.

and 7.3.2.13. It can be seen that the profiles of temperature and electron density are very similar to those measured in humid air.

7.3.2.4 The power budget

Power requirements can be defined either as the total 'wall plug' power (which depends on the efficiency of the specific device utilized to generate the plasma) or as the net power deposited into the plasma. In this work, both measurements were made. The total wall plug power was determined by directly measuring the current in each power lead of the 440 V, triphase power supply, by means of a Fluke model 33 ammeter. The average measured rms current was approximately 72 A (70, 72, and 74 A in each phase). The rms voltage was $V_{rms} = 440$ V. The total wall power is then

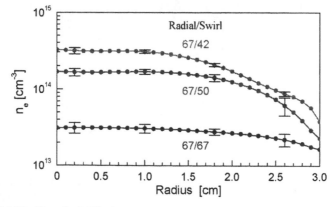

Figure 7.3.2.13. Dry air. LTE electron number density profiles at location 5 cm downstream of the nozzle exit as a function of radial/swirl injection ratio. Plate power = 40.1 kW.

given by the following expression:

$$P_{\text{wall}} = \sqrt{3} I_{\text{rms}} V_{\text{rms}} \cos \phi \cong 48 \,\text{kW} \qquad (7.3.2.1)$$

where $\cos \phi$, the power factor, was approximately 0.9 according to power supply specifications. The volume of plasma probed was well approximated by a cylinder of diameter 6 cm and length 5 cm ($140 \,\text{cm}^3$). Thus the total volumetric wall power was about $340 \,\text{W/cm}^3$.

To determine the power actually deposited into the plasma and, thereby, the torch efficiency, a calorimetric balance was done on the cooling-water circuit. To this end, the cooling circuit was instrumented with thermocouples at the inlet and outlet of the generator, and turbine flow meters in the flow lines. The total power removed by the cooling water was measured to be

$$P_{\text{cooling water}} = \dot{m} C_{\text{p}} \, \Delta T \cong 33 \,\text{kW}. \qquad (7.3.2.2)$$

The power deposited into the plasma is given by

$$P_{\text{plasma}} = P_{\text{wall}} - P_{\text{cooling water}} \cong 15 \,\text{kW}. \qquad (7.3.2.3)$$

The torch efficiency, defined as $\eta = P_{\text{plasma}}/P_{\text{wall}}$, was therefore about 31%. Thus, the minimum power required to sustain the thermal plasma volume was $105 \,\text{W/cm}^3$.

The schematic plasma torch diagram presented in figure 7.3.2.14 shows the power inputs and losses measured with the techniques described in the

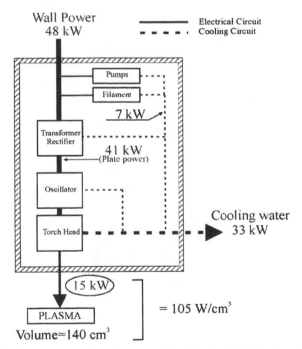

Figure 7.3.2.14. Typical power flow diagram of the Stanford 50 kW rf torch.

foregoing paragraphs. Approximately 15% of the total wall power was dissipated as heat by the pumps and the filament, and about 54% by the transformer/rectifier, oscillator, and torch head.

The total volume of plasma generated in the low swirl case (67/33) depicted in figure 7.3.2.4 was actually larger than the probed volume of 140 cm^3. The total volume with electron number densities greater than 10^{13} cm^{-3} was estimated to be on the order of 1000 cm^3. This estimate included the volume of plasma generated inside the torch head and the volume extending 10 cm downstream of the nozzle exit. Basing the power requirements on this larger volume, the wall-plug power was about 48 W/cm^3, and the minimum volumetric power for an ideal (100% efficient) generator would be 15 W/cm^3.

7.3.3 Conclusions

Steady-state air plasmas with electron number densities greater than 10^{13} cm^{-3} and volumes with dimensions greater than 5 cm in all directions were generated in both dry and humid air. The ratio of radial-to-swirl injection controlled the shape of the electron number density profile, and the power injected controlled the magnitude of the electron density profile.

The wall-plug power required to sustain an open volume of thermal air plasma with electron density greater than 10^{13} cm^{-3} with a typical rf torch was measured to be about 340 W/cm^3, or $105/\eta$ W/cm^3, where η represents the efficiency of the specific device used to produce the plasma. Additional experiments were conducted in the Stanford University High Temperature Gas Dynamics Laboratory with an atmospheric pressure microwave torch (model Litmas Red). This torch operated at a frequency of 2.45 GHz and nominal power 5 kW, with up to 3.5 kW of microwave power deposited

Figure 7.3.2.15. Atmospheric pressure air plasma produced with a Litmas Red 5 kW microwave torch. The nozzle exit diameter is 1 cm.

into the plasma. The torch could produce thermal air plasmas with temperatures up to 5000 K, with a volume in open air of about 10 cm^3. A photograph of the plasma plume is shown in figure 7.3.2.15. The wall-plug power required to produce electron densities greater than 10^{13} cm^{-3} with the microwave torch was about 200 W/cm^3, which is comparable to the power requirements of the rf torch. It is important to emphasize again that the rf and microwave torches produce plasmas that are thermal (i.e. in a state of LTE) and accordingly that the gas temperature tends to be relatively high (e.g. 4200 K for 10^{13} electrons/cm^3). Reducing the plasma temperature while maintaining a high electron number density requires the use of non-equilibrium plasmas. This motivates the work presented later on dc and repetitively pulsed plasma discharges in section 7.4.

References

Boulos M I 1985 *Pure Appl. Chem.* **57**(9) 1321

Davies J and Simpson P 1979 Induction Heating Handbook (London, New York: McGraw-Hill)

Dresvin S V *et al* 1972 in Dresvin S V (ed) *Physics and Technology of Low-Temperature Plasmas* (Moscow: Atomizdat)

Eckert H U, Kelly F L and Olsen H N 1968 *J. Appl. Phys.* **39**(3) 1846

Laux C O 1993 'Optical diagnostics and radiative emission of air plasmas' PhD Thesis in Mechanical Engineering, Stanford University, Stanford, CA

Laux C O, Spence T G, Kruger C H and Zare R N 2003 *PSST* **12** 1

Mostaghimi J and Boulos M I 1989 *Plasma Chem. Plasma Proc.* **9**(1) 25

Mostaghimi J, Proulx P and Boulos M I 1987 *J. Appl. Phys.* **61**(5) 1753

Reed T B 1961 *J. Appl. Phys.* **32** 821

van den Abeele D *et al* 1999 *Heat and Mass Transfer under Plasma Conditions* **891** 340

7.3.3 Laser initiated and rf sustained experiments

7.3.3.1 Introduction

Near atmospheric pressure plasmas of higher densities (10^{13} cm^{-3}) and larger volumes ($\sim$2000 cm^3) have a variety of applications. At higher pressures, however, there is a decrease in the mean electron temperature at constant rf power and fewer high-energy electrons are present. This effect, in addition to the increasing collision frequency due to high gas pressures, makes the energy cost per electron–ion pair created prohibitively high. A model based on electron-beam delta function excitation and electric field sustainment estimates a power density of 9 kW/cm^3 for an air plasma density of $\sim$$10^{13}$/cm^3 at sea level (Vidmar and Stalder 2003). In a classic experiment, Eckert *et al* (1968) created an atmospheric pressure plasma in both argon and air to study the emission spectrum given off by a high-pressure plasma. Following the work of Babat (1947), he created a plasma using an inductive coil at a lower pressure of ~1 torr, and slowly increased the neutral pressure and rf power until he could open the plasma chamber to the atmosphere. To protect the quartz chamber from heat damage and to help stabilize the discharge, the gas was injected in a vortex, essentially forming a thermal gas barrier between the hot plasma and the chamber wall. The coupled power required to maintain the discharge was 18–50 kW at 4 MHz to create the plasma at lower pressure and sustain it up to atmospheric pressure with a volume of about 2500 cm^3 (7–10 W/cm^3). Moreover, the time scale for creating high-pressure plasma from the low pressure discharge is several minutes and there is a great interest in the instantaneous creation of large volume (>1000 cm^3), high density (10^{12}–10^{13}/cm^3) discharges at atmospheric pressures with minimum power.

In addition, the inductively coupled rf power required to ionize high-pressure air is much higher than the rf power level (~9 kW/cm^3) needed to sustain the plasma at sea level. In an atmospheric pressure plasma torch, a 300 kV potential was required to initiate a discharge, whereas only 100 V was needed to maintain the discharge with operating currents of 200–600 A (Ramakrishnan and Rogozinski 1986, Schutze *et al* 1998). Therefore, there is a need for an alternative scheme to reduce the power budget required to initiate and sustain the discharge at higher gas pressures. We envisioned that if we could ionize a low ionization energy seed gas such as tetrakis (dimethyl-amino) ethylene (TMAE) by (193 nm) ultraviolet laser or flashtube photon absorption, then we could efficiently couple rf power to the plasma at higher gas pressures and sustain the plasma at a much reduced rf power level. A seed plasma can also be created by placing electrodes inside the chamber where a small plasma formed by the spark is localized between the electrodes. If the electrode is located close to the rf antenna so as to provide the required plasma load, arcing from the rf source to the electrode can occur. In addition, plasma bombardment of the electrode will result in deterioration and plasma impurities over time.

Therefore, an electrodeless method for creating a large volume (500 cm^3) seed plasma using ultraviolet photo-ionization is sought that will provide a good plasma load for efficient rf coupling at lower power level via pulsed inductively coupled sources. Previous experiments (Akhtar *et al* 2004, Kelly *et al* 2002, Ding *et al* 2001) described in section 7.2.3, have shown that a high initial density (~10^{13} cm^{-3}), long axial extent (~100 cm) TMAE plasma can be efficiently created by a 193 nm laser in 760 torr of nitrogen, air or argon. In addition, the long axial extent (100 cm) of the laser seed plasma can allow enhanced rf penetration and ionization well away from the 15–20 cm axial extent of the antenna. The possibility of initiating a discharge by 193 nm laser photo-ionization of TMAE seeded in high-pressure background argon gas that was later sustained by inductive coupling of an rf wave has been demonstrated by Kelly *et al* (2002). This section describes the experiments where laser-initiated seed discharge in high-pressure background gas is sustained by the efficient coupling of rf power with a reduced power budget.

7.3.3.2   Experimental set-up

A schematic of the experimental set-up is shown in figure 7.2.3.12. A uniform intensity ultraviolet beam of 193 nm wavelength is produced using an excimer laser (Lumonics Pulsemaster PM-842) that runs in the ArF (6.4 eV per photon) mode. The half-width of the laser pulse is 20 ± 2 ns with a 2 ns rise/fall time and a maximum laser energy of 300 mJ. The laser output cross-section of 2.8 cm $\times$ 1.2 cm is increased to 12.8 cm $\times$ 2.8 cm using a lens system of fused silica cylindrical plano-convex and plano-concave lenses in order to increase the plasma filling fraction of the vacuum chamber. The laser beam enters a 5.4 cm diameter by 80 cm long alumina plasma chamber through a 2.8 cm diameter Suprasil quartz window (98% transparency at 193 nm) at one end. Laser energy passing through the ultraviolet window is measured using an energy meter Scientech (Astral AD30). In order to account for the laser attenuation by the ultraviolet window, the ultraviolet window is placed in front of the energy meter. A laser fluence of 6 mJ/cm^2 is maintained. Gas mass flow controllers along with a swirl gas injection system are also located at the laser window end as shown in figure 7.2.3.12. The plasma chamber is pumped down to a base pressure of 10^{-6} torr using a turbo-molecular pump. In the evacuated chamber, the TMAE is either introduced by slowly raising the pressure to the optimum values of 4–50 mtorr or by raising the chamber pressure to 5 torr with argon pressurized TMAE admixture and then the air or noble gas is added slowly over a minute to a pressure of 760 torr while ensuring a laser-produced TMAE plasma density $>10^{12}$ cm^{-3}. The gas flow condition here is similar to the static case and is used as a reference to measure the comparable efficiency of the scheme.

The rf source is a 13.56 MHz single frequency generator and a maximum output power of 10 kW (Comdel CX-10000S) with variable duty cycle (90–10%) and variable pulse repetition frequency (100 Hz–1 kHz) and very fast (microsecond) turn-on/off time. Power is transmitted through a 50 Ω cable to an efficient capacitive matching network and to the antenna which surrounds the plasma chamber. The rf power is coupled to the plasma using a five-turn water-cooled helical antenna in conjunction with a capacitive matching network. The equivalent series resistance of the antenna and the capacitive match box are 1.5 Ω and 300–400 mΩ, respectively. We have experimentally determined that a five-turn helical coil is the most effective antenna for initiating and maintaining the plasma which excites the $m = 0$ TE mode field distribution. One interesting aspect that this antenna has over the other antennas studied is that the dominant electric field lines, which accelerate the electrons, have primarily an azimuthal component, and close on themselves. This eliminates the radial component of the current density thought to be a major loss mechanism in the type-III antennas which also excites $m = 1$ modes (Kelly *et al* 2002). The chamber and antenna are enclosed in a screen shield at a 10 cm radius. The capacitive network consists of two high-voltage vacuum variable capacitors and is shielded from the plasma chamber by enclosing it in a conducting box. The lower plasma radiation resistance $(1–5\,\Omega)$ mandates special care required to reduce ohmic losses in the impedance matching network and connections.

7.3.3.3   *Experimental results*

The hypotheses was confirmed in an earlier experiment (Kelly *et al* 2002) where a laser-initiated seed discharge of 2–5 mtorr of TMAE in 150 torr of argon is sustained by an rf coupling power of 2.8 kW, whereas with rf power alone the maximum pressure at which plasma could be created was 80 torr. The line average plasma density scan versus pressure with different rf sustaining power level is shown in figure 7.3.3.1.

We have recently improved the rf system by redesigning the capacitive matching network and reduced the ohmic losses in the rf connections. A very accurate, computer-controlled timing circuit sequences the seed gas injection, laser firing, the rf turn-on and data acquisition. This exact timing

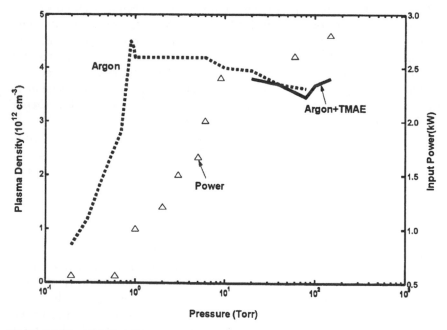

Figure 7.3.3.1. Collisionally corrected plasma density versus pressure for the five-turn helical antenna in argon and a TMAE/argon mixture (Kelly *et al* 2002).

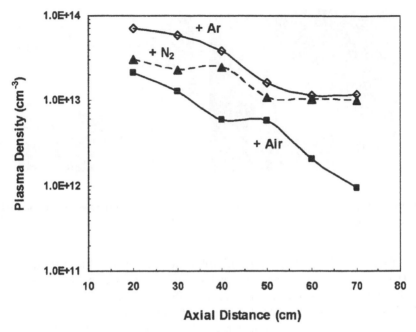

Figure 7.3.3.2 Axial plot of the laser-initiated plasma density of 5 torr of argon-pressurized TMAE with the addition of 760 torr of background gas.

sequence is very critical since the rf pulse must be enabled during the TMAE plasma lifetime ($\tau \approx 1\,\mu s$) where the seed plasma density is sufficiently large ($n > 10^{12}/cm^3$) to provide sufficient plasma radiation resistance load ($R_{pl} > 1\,\Omega$) for efficient rf coupling. Figure 7.3.3.2 shows the axial plot of laser-initiated line-average plasma density of 5 torr argon pressurized TMAE admixture to which 760 torr of background gas is slowly added. The line-average plasma density is measured by the collisional plasma inter-ferometry technique (Akhtar *et al* 2003). The long axial extent ($\sim$100 cm) of high-density seed plasma acts as a good plasma load for efficient rf coupling.

Figure 7.3.3.3 shows photographs of argon plasmas at 760 torr. Part (a) shows the plasma created by inductive coupling of 3.0 kW of rf power in a Pyrex plasma chamber where the chamber pressure was raised to 760 torr. In this case plasma is localized under the antenna. In contrast, as shown in part (b), an axially uniform ($\sim$80 cm), high-density argon plasma ($10^{13}/cm^3$) is produced using rf sustainment of laser initiated discharge at a substantially reduced rf power level of 700 W. A large volume plasma of about 2000 cm^3 is maintained at a density of $\sim$10$^{13}/cm^3$. The photograph illustrates that the long axial extent of the seed plasma allows increased axial penetration of inductive waves and helps maintain a plasma away from the source region.

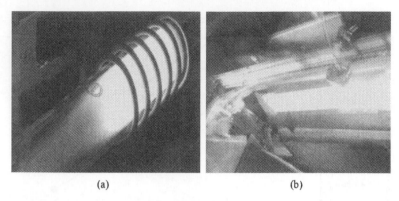

(a)   (b)

Figure 7.3.3.3. *760 torr argon plasmas produced by (a) 3.0 kW of rf power alone and (b) by rf sustainment of a laser initiated-discharge at 700 W.*

Figure 7.3.3.4 shows a photograph of a laser-initiated and rf sustained 300 torr nitrogen plasma at a power level of 4.0 kW. The pressure variation of the time-averaged plasma density and effective collision frequency of the nitrogen plasma at a constant power of 4.0 kW is shown in figure 7.3.3.5. A very bright plasma of high density ($>10^{12}$ cm^{-3}) fills the entire plasma chamber. As can be seen from the photograph, the laser preionization has a noticeable influence on the final rf sustained plasma density. It was also observed that in the absence of a seed plasma or a low density seed plasma, the background plasma could not be sustained even at higher rf power levels. These results show that the laser initiation substantially enhances the rf penetration and reduces the sustainment rf power levels. Future research will examine air plasmas and higher rf power short pulses for reduction of power densities for large-volumes high-density air plasmas.

Figure 7.3.3.4. *Laser-initiated and rf sustained 300 torr nitrogen plasma at a coupled power level of 4.0 kW.*

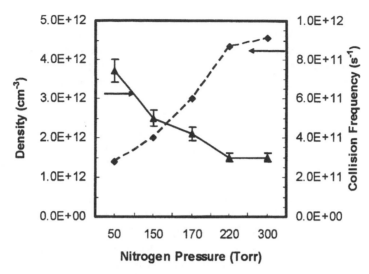

Figure 7.3.3.5 Line average ($d = 5$ cm) plasma density and effective collision frequency for a laser-initiated and rf sustained nitrogen plasma measured 10 cm from the antenna.

References

Akhtar K, Scharer J, Tysk S and Denning C M 2004 *IEEE Trans. Plasma Sci.* **32**(2) 813

Akhtar K, Scharer J, Tysk S and Kho E 2003 *Rev. Sci. Instrum.* **74** 996

Babat G 1947 *J. Inst. Elec. Engineers (London)* **94** 27

Ding G, Scharer J E and Kelly K 2001 *Phys. Plasmas* **8** 334

Eckert H U, Kelly F L and Olsen H N 1968 *J. Appl. Phys.* **3** 1846

Kelly K L, Scharer J E, Paller E S and Ding G 2002 *J. Appl. Phys.* **92** 698

Ramakrishnan S and Rogozinski, M W 1986 *J. Appl. Phys. D* **60** 2771

Schutze A, Young J Y, Babayan S E, Park J, Selwyn G S and Hicks R F 1998 *IEEE Trans. Plasma Sci.* **26** 1685

Vidmar R J and Stalder K R 2003 'Air chemistry and power to generate and sustain plasmas: plasma lifetime calculations', in *Proc. AIAA 2003*, pp. 1–8

7.3.4 Methods for spatial localization of a microwave discharge

7.3.4.1 *Characteristics of microwave discharge*

As discussed in section 1.2, at sea level, the molecular composition of air is roughly 80% nitrogen (N_2) and 20% oxygen (O_2). The ionization energies ε_i of O_2 and N_2 are 12.1 and 15.6 eV, respectively. These molecules can be ionized by ultraviolet radiation, for example, the earth's ionospheric plasma is principally generated by solar ultraviolet radiation (see also section 1.3). Photon ionization requires that the wavelength λ_0 of the radiation be less than $\lambda_c = hc/\varepsilon_i$. Thus the wavelengths of the ultraviolet radiation for ionizing O_2 and N_2 must be less than 102.6 and 79.6 nm, respectively. Therefore, microwave wavelengths are too long to cause photon ionization. On the other hand, the microwave electric field can accelerate background charge particles. When the electric field intensity, E, of a high-power microwave beam propagating in air exceeds the breakdown threshold field, E_{cr}, of the background air, avalanche ionization can occur through the impact process (i.e. some of charge particles' (mainly electrons') kinetic energies can exceed the ionization energies of O_2 and N_2). In each elastic collision with a neutral molecule, an electron loses only a very small fraction of its total kinetic energy, thus electrons can easily build up the thermal energy through multiple collisions in the microwave field. However, as the electron energy increases, the cross sections of inelastic collisions also increase. For electron energies between 2 and 4 eV, the cross section for the excitation of vibrational levels experiences a very large nearly step-like leap. This vibrational excitation process hinders the continuous acceleration of electrons by the microwave field toward the ionization energy level. It increases the required field intensity for the microwave discharge, which occurs when the quiver speed v_q of 'seed' electrons exceeds a critical value, $v_{qc} = eE_{cr}/m\nu_c$, where ν_c is the electron–neutral particle collision frequency. Then a significant fraction of seed electrons can bypass the vibrational excitation loss band and are accelerated continuously by the microwave field to the ionization energy level. The breakdown threshold field, E_{cr}, for a continuous wave or long pulse microwave beam is given by (Lupan 1976)

$$E_{cr} = 3.684p(1 + \omega^2/\nu_c^2)^{1/2} \text{ kV/m} \qquad (7.3.4.1)$$

where p is the background air pressure measured in torr; and ω and ν_c are the microwave frequency and electron–neutral particle collision frequency, respectively.

The density, n, of the microwave plasma is normally limited by the microwave frequency. In the density range of $n \leq 10^{17} \text{ m}^{-3}$, the dominant loss mechanism of free electrons in air is through their attachment with neutral molecules. Avalanche breakdown occurs when the ionization rate, ν_i, is larger than the attachment rate, ν_a. The ionization frequency, ν_i, is given by (Yu 1976, Kuo and Zhang 1991)

$$\nu_i = 2.5 \times 10^7 p[8.8\varepsilon^{1/2} + 2.236\varepsilon^{3/2}] \exp(-7.546/\varepsilon) \quad \text{s}^{-1} \qquad (7.3.4.2)$$

where $\varepsilon = E/E_{cr}$. Equation (7.3.4.2) can be reduced to $\nu_i/\nu_a \approx \varepsilon^{5.3}$ for $1.3 < \varepsilon < 3.5$ (Gurevich 1980).

This microwave-generated plasma attenuates the microwave beam spatially, which in turn affects the volume and uniformity of plasma generation. If the background is uniform, ionization tends to occur near the source, which hinders the propagation of the microwave beam. Therefore, the electric field intensity of a high-power microwave beam cannot be increased indefinitely. Its power density has an upper bound set by the avalanche breakdown of air.

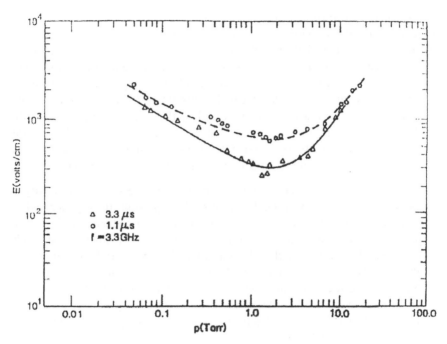

Figure 7.3.4.1. Dependence of air breakdown threshold fields on the pressure for microwave pulse lengths of 1.1 and 3.3 μs. (Kuo *et al* 1990.)

It is noted that the breakdown process requires an initiation time interval that depends on the number of seed electrons pre-existing in the background. Normally, the breakdown threshold field increases as the pulse length, τ, of the microwave radiation decreases. This tendency is demonstrated (Kuo and Zhang 1990) by the two Paschen breakdown curves shown in figure 7.3.4.1, which show the dependence of the air breakdown threshold field on the air pressure for the cases of 1.1 and 3.3 μs pulses. In both cases, the minimum of the breakdown threshold appears at about the same pressure, where $\nu_c \cong \omega$ consistent with equation (7.3.4.1). In the pressure region having $\nu_c \gg \omega$, the breakdown threshold field, E_{cr}, is essentially independent of the pulse length and microwave frequency. Thus $E_{cr} = 3.684p$ kV/m, the same as in the dc discharge case. In this pressure region, it should be noted that the thermal ionization instability (Gildenburg and Kim 1978) might become dominant in the discharge. This instability arises due to mutual enhancements of the electron density and gas temperature. It evolves the discharge into filaments parallel to the wave electric field, which form a fishbone structure (Vikharev *et al* 1988) as can be seen from the luminescence of the discharge.

To use microwaves to produce atmospheric pressure air plasma in a designated region away from the source, it is necessary to avoid the undesirable ionization along the propagation path before reaching the preferred ionization region. Such undesirable ionization causes attenuation of the microwave radiation, which could then be left with insufficient power density to cause air breakdown in the designated region. The maximum power density of microwave radiation that propagates in the air at atmospheric pressure without causing air breakdown is about 10 GW/ m^2. This thus determines an upper bound of the microwave power for the application of air plasma generation at atmospheric pressure. Therefore, additional arrangements are needed to achieve spatial localization of the discharge. Several prominent approaches are discussed in the following.

1. Use a microwave resonant cavity to enhance the electric field intensity at localized resonant peak-field regions. The plasma thus generated is confined inside the cavity. An air jet can be introduced to blow the plasma out of the cavity through a nozzle such as a microwave torch; however, the volume of the plasma is usually small, and the generation efficiency is low because most of the plasma is lost inside the cavity.
2. Add a lens to focus the microwave beam so that the electric field intensity of the microwave beam in the region around the focal point can exceed the air breakdown threshold. Again, the volume of the plasma is limited by the size of the focal spot.
3. Add a seeding source to produce preliminary plasma, which can lower the breakdown threshold field considerably in the region of space containing the seed. The possible seeding sources include ultraviolet and x-ray radiation, laser and electron beams, and dc and low frequency discharges (e.g. plasma torches).
4. Use two intersecting microwave beams with parallel polarization. The field intensity of each beam is below the breakdown threshold (Vikharev *et al* 1984, Kuo and Zhang 1990). However, in the intersection region of the two beams, the field intensity can be doubled and can exceed the breakdown threshold. This approach makes it possible to achieve better spatial localization of the discharge and yet to produce plasma in a large region (determined by the size of the intersection region). In fact, this approach was first (Gurevich 1980) suggested to generate an artificial ionospheric mirror in the lower ionosphere by ground-transmitted high-power microwave beams for over-the-horizon (OTH) radar applications (Kuo *et al* 1992). This is the approach to be described in detail in the next subsection.

7.3.4.2 *Plasma generated by two intersecting microwave beams*

In the experiments discussed here, microwave power at a frequency of 3.27 GHz was generated by a single magnetron driven by a pulse forming network, which had a pulse length of 1.1 μs and a repetition rate of 60 Hz.

The peak output power of the tube was 1 MW. Since the power density of the microwave radiation was too low to cause air breakdown at atmospheric pressure, the experiment was conducted in a Plexiglas cube chamber, 2 ft (61 cm) on a side, which was pumped down to a pressure of about 1 torr. First, it was found that using a single pulse it was possible to generate a localized plasma only near the chamber walls. Therefore, a second pulse provided by the same magnetron was fed into the cube through a second S-band microwave horn placed at a right angle to the first. With such an arrangement, the power of each pulse was reduced to below the breakdown threshold for the low-pressure air inside the chamber. Hence, air breakdown could only occur in the central region of the chamber, where the two pulses intersected. The wave fields added to form a standing-wave pattern in the intersecting region in a direction perpendicular to the bisecting line of the angle between the two intersecting pulses. Thus parallel plasma layers with a separation $d = \lambda/\sqrt{2}$ were generated, where λ was the wavelength of the wave. This is shown in figure 7.3.4.2(a), in which seven such layers can be seen. The spatial distribution of the plasma layers was measured with a Langmuir double probe. This was accomplished by using a microwave phase shifter to move the plasma layers across the probe. The peak density distribution for one half of a spatial period was thus obtained and is presented in figure 7.3.4.2(b). It is shown that the plasma layers produced are well confined with very good spatial periodicity.

Using the same approach but with much higher microwave power, a plasma having similar characteristics to those presented in figure 7.3.4.2 can be generated in the open air. The volume of plasma generated by this approach would depend on the dimensions, a and b, of the cross section of the (rectangular) waveguide used (i.e. on the frequency band of the microwave). Because the maximum field intensity has to be lower than the breakdown threshold field inside the waveguide and slightly higher than half of the breakdown threshold field in the intersecting region, the volume of the resulting intersecting region could be estimated to be $8a^2b$ and the volume of the region containing plasma would be about $4a^2b$. Using S-band microwave radiation and a standard rectangular waveguide having a cross section of 7.2 cm × 3.4 cm, the cross section of the horn should not exceed 14.4 cm × 6.8 cm. Therefore, the volume of microwave plasma layers is estimated to be about $2a^2b = 350\,\text{cm}^3$.

Air plasma is very collisional and thus quite different from the more widely investigated plasmas having low background gas pressures. The collision frequency is much larger than the plasma frequency for plasma densities less than $5 \times 10^{13}\,\text{cm}^{-3}$. In this density regime, the real part of the index of refraction is positive for all wave frequencies, and there is no cutoff to the wave propagation. Thus the applicable microwave field, rather than the microwave frequency, limits the plasma density, which is estimated to have a maximum at about $10^{13}\,\text{cm}^{-3}$. Inelastic collision processes dominate the microwave plasma

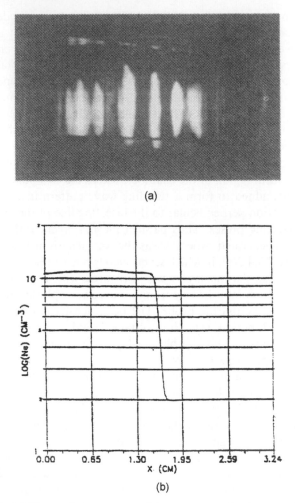

Figure 7.3.4.2. (a) Plasma layers generated by two crossed microwave pulses having parallel polarization, (Kuo *et al* 1990) and (b) the plasma peak density distribution across the plasma layers, from the central point at $x = 0$ of one layer to the midpoint at $x = 3.24$ cm of the next layer. (Kuo *et al* 1990.)

produced. Thus the electron temperature is usually limited to about 2 eV by the vibrational excitation loss.

The power required to maintain such a microwave discharge depends on the electron–ion recombination rate and on the heating rate of the neutral gas by the plasma (mainly through electron–neutral inelastic collisions). The electron–ion recombination rate decreases with the temperature of the plasma (Christophorou 1984, Rowe 1993). The electron–neutral inelastic collision rate can be significantly reduced either by elevating the electron temperature to exceed 4 eV or by limiting it to be well below 2 eV. An auxiliary plasma heating mechanism, such as could be provided by an auxiliary dc or low frequency field, may be used to maintain a non-equilibrium microwave plasma and to reduce the microwave power budget. However, it is not clear if the overall power budget can be thus reduced.

References

Christophorou L G 1984 *Electron–Molecule Interactions and Their Applications*, vol 2 (Orlando: Academic Press)

Gildenburg V B and Kim A V 1978 *Sov. Phys. JETP* **47** 72

Gurevich A V 1980 *Sov. Phys. Usp.* (Engl. Transl.) **23** 862

Kuo S P 1990 *Phys. Rev. Lett.* **65**(8) 1000

Kuo S P and Zhang Y S 1990 *Phys. Fluids* **2**(3) 667

Kuo S P and Zhang Y S 1991 *Phys. Fluids B* **3**(10) 2906

Kuo S P, Zhang Y S, Lee M C, Kossey P A and Barker R J 1992 *Radio Sci.* **27**(6) 851

Lupan Y A 1976 *Sov. Phys. Tech. Phys.* **21**(11) 1367

Rowe B R 1993 *Recent Flowing Afterglow Measurements, in Dissociative Recombination: Theory, Experiment and Applications* (New York: Plenum Press)

Vikharev A L, Gildenburg V B, Golubev S V *et al* 1988 *Sov. Phys. JETP* **67** 724

Vikharev A L, Gildenburg V B, Ivanov O A and Stepanov A N 1984 *Sov. J. Plasma Phys.* **10** 96

7.4 Repetitively Pulsed Discharges in Air

7.4.1 Introduction

As we have seen in chapter 5, the power required to sustain elevated electron densities with dc discharges is extremely large. Therefore, we have explored a power reduction strategy based on pulsed electron heating. This strategy is illustrated in figure 7.4.1. Short voltage pulses are applied to increase the electron number density. After each pulse, n_e decreases according to electron recombination processes. When n_e reaches the minimum desired value, a second pulse is applied. The average electron density obtained with this method depends on the pulse duration, pulse voltage, and the interval between pulses.

As seen in chapter 5, dc discharges can maintain $n_e \geq 10^{12} \, \text{cm}^{-3}$ in atmospheric pressure air with electric fields producing an electron temperature on the order of 1 eV. To produce the same average electron density with short (1–10 ns) pulsed discharges, a higher electron temperature of about 3–5 eV is required. Although the corresponding field is higher than for a dc discharge, the ionization efficiency is much larger in the pulsed

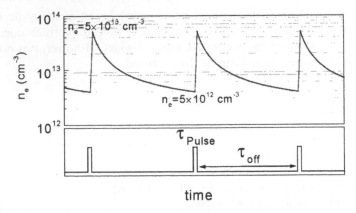

Figure 7.4.1. Repetitively pulsed strategy. (Kruger *et al* 2002.)

case than in the dc case because the energy lost to nitrogen molecules, per electron created, is several orders of magnitude smaller at $T_e = 3\text{--}5\,\text{eV}$ than at 1 eV. This increase in efficiency allows the power budget to be dramatically reduced with pulsed discharges. It may be shown (Nagulapally *et al* 2000) that the power reduction afforded by the repetitively pulsed approach relative to dc is given by

$$R \cong \frac{k_{\text{ion}}(T_{e,\text{pulse}})N}{k_{\text{DR}}n_e^*} \times \alpha^2 e^\alpha (e^\alpha - 1)^{-2} \times \left(\frac{T_{e,\text{dc}}}{T_{e,\text{pulse}}}\right)^{3/2}$$

where $\alpha \equiv k_{\text{ion}}N\tau_1$, and where k_{ion} is the species-weighed rate coefficient for electron impact ionization of O_2, N_2, and O, k_{DR} is the rate coefficient for dissociative recombination of NO^+, N is the total number density of species, τ_1 is the pulse length, n_e^* is the average electron number density produced by the repetitively pulsed discharge, and $T_{e,\text{dc}}$ and $T_{e,\text{pulse}}$ represent the electron temperatures produced by the dc and pulsed discharges, respectively.

Figure 7.4.2 shows the predicted inelastic energy losses of electrons by collisions with N_2, per unit number density of N_2 and electrons. The losses to nitrogen represent the main fraction of the losses in air. This is because at electron temperatures below $\sim$20 000 K the resonant $e\text{--}V$ transfer in ground state N_2 is by far the dominant loss channel (there is no such resonant channel for O_2 or NO). At electron temperatures above 20 000 K, the inelastic losses are dominated by electron-impact electronic excitation, dissociation, and ionization, and the total losses per unit number density of N_2 and electrons are about the same as the total losses per unit number density of O_2 and electrons. Nitrogen losses dominate at $T_e > 20\,000\,\text{K}$ because the density of N_2 is much higher than the density of O_2. Figure 7.4.2 shows that the (useful) power into ionization represents an increasingly large fraction of the total power as the electron temperature is increased. This explains why pulsed discharges with electron temperatures of several eV are

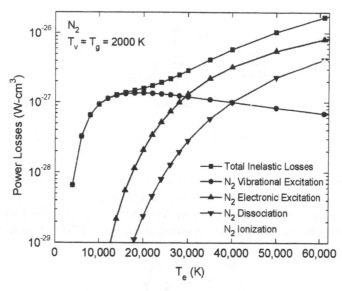

Figure 7.4.2. Inelastic power losses by electron-impact vibrational excitation, electronic excitation, dissociation, and ionization of N_2. In these calculations, the vibrational temperature is fixed equal to the gas temperature ($T_v = T_g = 2000$ K), and the electronic temperature of internal energy levels is fixed equal to the electron temperature.

more efficient in terms of ionization than the dc discharges which operate at about 1 eV.

7.4.2 Experiments with a single pulse

To test the pulsing scheme, experiments were conducted (Nagulapally *et al* 2000) in atmospheric pressure 2000 K air using a pulse forming line capable of generating a 10 ns rectangular pulse with peak voltage up to 16 kV. To experimentally simulate the conditions of a repetitively pulsed discharge, the initial elevated electron number density generated by the 'previous' pulse is created by means of a dc discharge in parallel with the pulser. The circuit schematic is shown in figure 7.4.3. With a dc voltage of 2 kV and current of 150 mA, the initial electron density is 6.5×10^{11} cm^{-3}. A 10 kV, 10 ns pulse is superimposed to further increase the electron density. The measured discharge diameter of about 3 mm is comparable with the diameter of the dc discharge (figure 7.4.4). The temporal variation of plasma conductivity was measured from the voltage across the electrodes and the current density through the plasma. The electron density increases from 6.5×10^{11} to 9×10^{12} cm^{-3} during the pulse, then decays to 10^{12} cm^{-3} in about 12 μs (figure 7.4.5). The average measured electron density over the 12 μs duration is 2.8×10^{12} cm^{-3}.

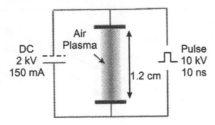

Figure 7.4.3. Schematic of the combined pulsed and dc discharge experiments. (Kruger *et al* 2002.)

Figure 7.4.5 shows a comparison of the measured electron number density with the predictions of our two-temperature model. The predictions agree well with the measured electron decay time of 12 μs. This decay time is consistent with the dissociative recombination time of NO^+ predicted to be

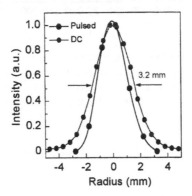

Figure 7.4.4. Spatial extent of the plasma produced with pulsed and dc discharges. (Kruger *et al* 2002.)

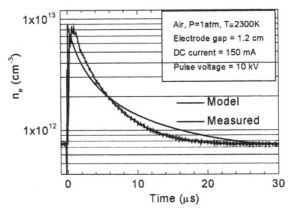

Figure 7.4.5. Temporal electron density profile in the 10 ns pulsed discharge. (Kruger *et al* 2002.)

8.7 µs without the dc background. Thus these results provide validation of our chemical kinetic model of the recombination phase.

7.4.3   Experiments with 100 kHz repetitive discharge

The success of the proof-of-concept experiments conducted with the single pulse discharge led us to investigate the generation of air plasmas with a repetitively pulsed discharge. A repetitive pulser capable of generating 10 ns pulses, with peak voltages of 3–12 kV and pulse repetition frequencies up to 100 kHz, was acquired from Moose-Hill/FID Technologies. This pulser operates with a solid-state opening switch or drift-step recovery diode (DSRD). The experimental set-up is shown in figure 7.4.6 and the electrical circuit in figure 7.4.7. The discharge is applied to preheated, LTE air at atmospheric pressure and about 2000 K. The dc circuit in parallel with the pulser was used only to determine the electron number density from the plasma conductivity. In regular operation, the dc circuit is disconnected and the discharge operated with the pulser only.

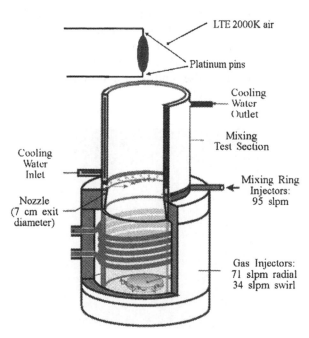

Figure. 7.4.6. Set-up for repetitive pulse discharge in air at 2000 K, 1 atm. (Kruger *et al* 2002.)

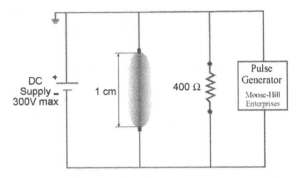

Figure. 7.4.7. Repetitive pulse discharge circuit schematic (dc circuit applied only for conductivity measurements). (Kruger *et al* 2002.)

A photograph of the repetitively pulsed discharge in operation in atmospheric pressure preheated (2000 K) air is shown in figure 7.4.8. The diffuse character of the discharge was confirmed with time-resolved (1.5 ns frames every 2 ns) measurements of plasma emission during the pulse (see figure 7.4.9). These measurements were made with a high-speed intensified camera, Roper Scientific PI-MAX1024. The diameter of the discharge is approximately 3.3 mm. Additional time- and spectrally-resolved measurements of emission during the pulse and the recombination phase show that the pulse excites the C state of N_2 and the A state of NO. After the pulse, emission from the C state of N_2 decays to a constant value within 30 ns, and emission from the A state of NO shows a two-step decay, with first an abrupt decrease by over four orders of magnitude from the end of the pulse until 320 ns after the pulse, and then a slower decrease by one order of magnitude until the next pulse.

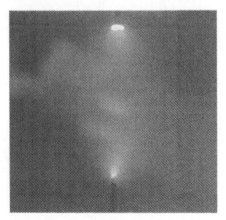

Figure 7.4.8. Photograph of 10 ns, 100 kHz repetitive pulse discharge in air at 2000 K, 1 atm. (Kruger *et al* 2002.)

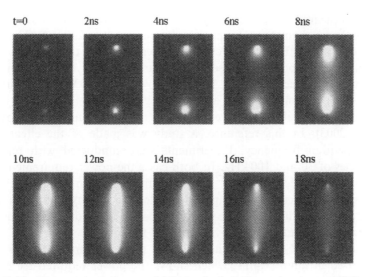

Figure 7.4.9. Time-resolved images of 10 ns pulsed discharge in air at 2000 K, 1 atm. (Dulen *et al* 2002 (© 2002 IEEE).)

Figure 7.4.10 shows the measured temporal variations of the electron density during three cycles of the pulsed discharge. The electron number density varies from 7×10^{11} to 1.7×10^{12} cm^{-3}, with an average value of about 10^{12} cm^{-3}. The power deposited into the plasma by the repetitive discharge was determined from the pulse current (measured with a Rogowski coil), the voltage between the electrodes (6 kV peak) minus the cathode fall

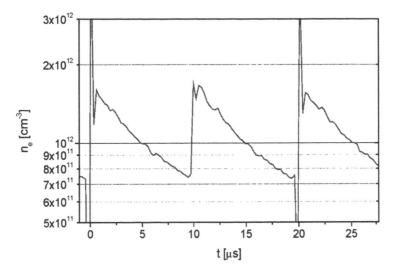

Figure 7.4.10. Electron number density measurements in the repetitive pulse discharge in air at 1 atm, 2000 K.

voltage (measured to be 1525 V by varying the gap distance), and the measured discharge diameter. The peak pulse current was 240 mA. The power deposited is found to be 12 W/cm^3, consistent with the theoretical value of 9 W/cm^3 for an optimized pulsed discharge producing 10^{12} electrons/cm^3. It is lower, by a factor of 250, than the power of 3000 W/cm^3 required to sustain 10^{12} electrons/cm^3 with a dc discharge.

More details about these experiments and modeling can be found in (Packan 2003). In this reference, a study was made of the effect of the pulse repetition frequency. Experiments were conducted with repetition frequencies of 30 and 100 kHz. In both cases the power requirements were close to 10 W/cm^3 for about 10^{12} electrons/cm^3. The main difference between the plasmas produced is the amplitude of electron density variations. In the 30 kHz discharge, the amplitude varies by about a factor of 10, whereas in the 100 kHz the amplitude varies by a factor of two only.

The results of our research on dc and pulsed electrical discharges are summarized in figure 7.4.11, which shows the power required to generate elevated electron number density in 2000 K, atmospheric pressure air, with dc and pulsed discharges. The experimental point represents the measured power requirement of our repetitively pulsed discharge experiment. Power budget reductions by an additional factor of about 5 are possible with repetitive pulses of 1 ns duration. Such repetitive pulsers are already commercially available. Therefore, power budget reductions by a factor of 1000 relative to the dc case at 10^{12} electrons/cm^3 can be readily obtained with a repetitively pulsed technique.

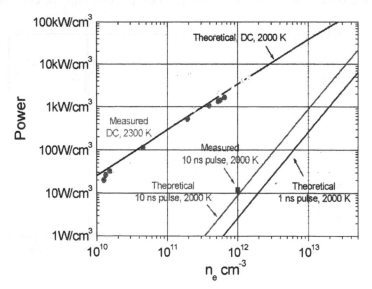

Figure 7.4.11. Power budget requirements versus electron number density for dc and pulsed discharges in air at 1 atm, 2000 K.

7.4.4 Conclusions

We have described a plasma generation technique using a repetitively pulsed discharge in which electron number densities of more than $10^{12}\,cm^{-3}$ in air are produced with approximately $12\,W/cm^3$, more than two orders of magnitude lower than the power required for a dc discharge. The basis of the technique is to apply short (10 ns), high voltage ($\sim$10 kV) electric pulses with a repetition frequency tailored to match the recombination time of electrons. Both single-shot and repetitively pulsed diffuse discharges at 100 kHz have been demonstrated, with power reductions of over two orders of magnitude for average electron densities greater than $10^{12}\,cm^{-3}$. Power reductions of approximately three orders of magnitude are possible with a 1 ns repetitive pulsing technique.

References

Duten X, Packan D, Yu L, Laux C O and Kruge C H 2002 *IEEE Trans. Plasma Sci.* **30**(1) 178

Kruger C H, Laux C O, Yu L, Packan D and Pierrot L 2002 *Pure and Applied Chemistry* **74**(3) 337

Nagulapally M, Candler G V, Laux C O, Yu L, Packan D, Kruger C H, Stark R and Schoenbach K H 2000 'Experiments and simulations of dc and pulsed discharges in air plasmas' in *31st AIAA Plasmadynamics and Lasers Conference*, Denver, CO

Packan D M 2003 'Repetitively pulsed glow discharge in atmospheric pressure air' PhD Thesis in Mechanical Engineering, Stanford University, Stanford, CA

7.5 Electron-Beam Experiment with Laser Excitation

7.5.1 Introduction

In this section, we present a method of sustaining large-volume plasmas in cold, atmospheric pressure air, using the optical pumping technique reviewed in section 7.2.2.1 above, combined with an electron beam ionizer. The combination of these techniques was adopted in an effort to mitigate the most critical problems of creating such plasmas: reducing the required power budget and insuring stability. The techniques described in this section are examples of 'non-self-sustained' electric discharges, in contrast to 'self-sustained' discharges, in which ionization is provided by applying high voltage to the electrodes maintaining the plasma. Typically, self-sustained discharges, lacking an external ionization source, are usually only struck at low gas pressures, well below even 0.1 atm, if a low temperature, diffuse, glow-type plasma is required. As gas pressure is increased, higher voltages are required to strike. Operation at such higher voltages and pressures usually leads to a marked transition, in which the plasma changes from a diffuse, cool column of weakly ionized gas, a 'glow discharge', to a much higher-temperature higher-conductivity plasma between the electrodes. This transition is sometimes termed the 'glow-to-arc transition', and is described in standard plasma references, e.g. [Rai91]. After transition, very high temperatures are reached, with a large fraction of the gas becoming ionized, the resistivity of the plasma greatly decreasing, and the electron temperature coming into near thermal equilibrium with the gas temperature. Such discharges do not normally provide the relatively cold, large-volume diffuse plasmas desired here. To circumvent this problem, various methods have been used to extend the range of self-sustained glow-type discharges to near atmospheric pressures, such as the use of individually ballasted multiple cathodes, short duration rf high-voltage pulse stabilization, or aerodynamic stabilization [Rai91, Vel87, Gen75, Ric75, Zhd90]. The energy efficiency of such discharges is, however, much lower than desirable for large-volume plasmas, since the fraction of the input electrical power going into ionization is often quite small.

An alternative approach is the use of non-self-sustained glow discharges, in which some or all of the required volume ionization is provided by an external source, such as an electron beam [Bas79, Kov 85]. Electron beams are identified as having by far the lowest power budget among all non-equilibrium ionization methods [Ada00, Mac00, Mac99]. Further, reliance on an external ionization source mitigates the glow-to-arc break-down problem. The glow-to-arc transition, with subsequent plasma thermalization, can be significantly delayed or avoided altogether. Even when using this efficient ionization source, however, the power budget required to sustain a relatively cold, large-volume air plasma remains huge, greater than $1\,\mathrm{GW/m^3}$. This is predominantly due to the rapid attachment of electrons to oxygen molecules. Consequently, reduction of the air plasma power budget mandates mitigation of electron attachment, and, for further power reduction, lowering of the electron–ion recombination rate. The method of the present section uses an electron beam to produce electrons efficiently, and uses the optical pumping technique reviewed previously to mitigate electron loss. In brief, we use the approach of section 7.2.2.1, i.e. optical pumping by a CO laser, to modify the electron removal rates in an electron beam sustained, CO-seeded high-pressure air plasma.

7.5.2   Electron loss reduction

There is recent experimental evidence that vibrational excitation of diatomic species produced by a CO laser may reduce the rates of electron removal (dissociative recombination and attachment to oxygen) in non-equilibrium plasmas [Pal01a]. We give a brief discussion of this effect.

First, electron impact ionization of vibrationally excited molecules produced by a CO laser can create vibrationally excited molecular ions such as N_2^+ and O_2^+,

$$N_2(v) + e_{beam}^- \longrightarrow N_2^+(v) + e_{beam}^- + e_{secondary}^-. \qquad (7.5.1)$$

Vibrationally excited ions can also be created by a rapid resonance charge transfer from vibrationally excited parent molecules, such as

$$O_2^+ + O_2(v) \longrightarrow O_2^+(v) + O_2. \qquad (7.5.2)$$

Recent experimental data [Mos99] show that vibrational excitation of molecular ions such as NO^+ or O_2^+ can considerably reduce the rate of their dissociative recombination, such as

$$O_2^+(v) + e^- \longrightarrow O + O. \qquad (7.5.3)$$

Secondly, three-body attachment of secondary electrons produced by the electron beam to vibrationally-excited oxygen molecules created by a CO laser,

$$O_2(v) + e^- + M \longrightarrow O_2^-(v) + M \qquad (7.5.4)$$

will produce vibrationally excited ions $O_2^-(v)$. Since the electron affinity of this ion is only about 0.4 eV [Rai91], vibrational excitation of oxygen molecules to vibrational levels $v \geq 2$ can provide enough energy for auto-detachment of an electron,

$$O_2^-(v \geq 2) \longrightarrow O_2 + e^-. \qquad (7.5.5)$$

Since the three-body electron attachment to oxygen molecules is by far the most rapid mechanism of electron removal in cold, high-pressure air plasmas, reduction of the attachment rate greatly reduces the plasma power budget. We now proceed to the details of an experimental demonstration of the use of this vibrational excitation technique, together with an electron-beam ionizer, which produces cool, atmospheric pressure air plasma with markedly improved efficiency.

7.5.3 Experimental discharge; electron beam ionizer

Figures 7.5.1 and 7.5.2 show schematics of the experimental set-up. An electron gun (Kimball Physics EGH-8101) generates an electron beam with energy of up to 80 keV and a beam current of up to 20 mA. The electron gun can be operated continuously or pulsed. From the vacuum inside the electron gun the electron beam passes through an aluminum foil window into a plasma cell that can be pressurized up to atmospheric pressure. The foil window with a thickness of 0.018 mm is glued onto a vacuum flange with an aperture of 6.4 mm. About 30 keV of the electron beam energy is lost in the window, which results in heating of the window. Pulsed operation

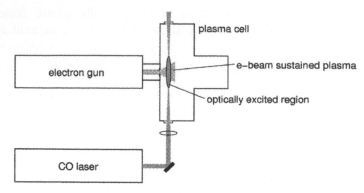

Figure 7.5.1. Schematic of the electron beam and laser set-up [Pal01b].

of the electron gun at a low duty cycle prevents overheating and failure of the window. In the electron gun the electron beam has a relatively small divergence that increases significantly (~90° full angle) due to scattering in the foil window. A 12.7 mm diameter brass electrode faces the window at a distance of 10 mm. This defines a volume of the e-beam excited plasma of ~1 cm^3 between the beam window and the electrode. The beam window together with the entire chamber is grounded. For the current experiments the electrode was usually also grounded. The electron gun was typically operated at beam energies between 60 and 80 keV and different beam currents measured using an unbiased Faraday cup placed behind the beam window. The plasma cell is pressurized with air at pressures between 100 torr and 1 atm. A slow gas flow is maintained in the cell to provide flow convective cooling and to remove chemical products. The residence time of the gas mixture in the cell is of the order of a few seconds.

Perpendicular to the e-beam axis a CO laser beam is directed into the e-beam excited plasma. The laser is used to vibrationally excite the diatomic

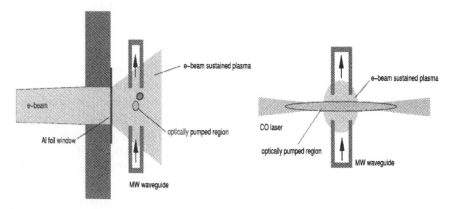

Figure 7.5.2. Schematic of the plasma cell [Pal01b].

plasma constituents. The liquid nitrogen-cooled continuous wave CO laser [Plo00a] produces a substantial fraction of its power output on the $v = 1$–0 fundamental band component in the infrared. In the present experiment, the laser is typically operated at $\sim$10 W continuous wave broadband power on the lowest ten vibrational bands. The output on the lowest bands (1–0 and 2–1) is necessary to start the optical absorption process in cold CO at 300 K, 1–5% of which is seeded into the cell gases. The laser beam is focused ($f = 250$ mm) to a focal area of $\sim$0.5 mm diameter to increase the power loading per CO molecule, producing an excited region $\sim$5 cm long. As indicated in figures 7.5.1 and 7.5.2, the vibrationally excited region is only a part of the total e-beam ionized plasma. Typically, the laser pump maintains the gas molecules in this region with high energies in the CO, O_2, and N_2 vibrational modes. The energies in each mode would correspond to a few thousand Kelvin if the gas were in equilibrium. These mode energies are maintained in steady state in the plasma by the laser. The external gas kinetic modes of translational and rotation molecular motion remain relatively cold in this steady state. This gas kinetic temperature is easily measured by monitoring the spontaneous infrared emission from the fundamental vibrational bands of the vibrationally excited CO. From the relative intensity of the spectrally-resolved vibrational–rotational lines, the rotational temperature can be inferred from a Boltzmann plot. The rotational temperature is equal to the translational mode temperature in these high-pressure collision-dominated plasmas. Since the emission arises only from the laser-excited region of the plasma, this temperature inference is not compromised by the surrounding e-beam-only excited region, and only reflects the temperature of the laser-excited part of the plasma. Figure 7.5.3 shows such an emission spectrum, from which a gas kinetic temperature of $T = 560$ K is inferred.

The electron density in the e-beam/optically sustained plasma is measured by microwave attenuation. The microwave experimental apparatus consists of a $\nu = 40$ GHz oscillator, a transmitting and receiving antenna/waveguide system, oriented perpendicular to the e-beam axis and to the laser axis (figure 7.5.2), and a transmitted microwave power detector. The receiving waveguide is positioned directly opposite the transmitting waveguide, with the plasma located between them (figure 7.5.2). The microwave detector produces a dc voltage proportional to the received microwave power. From the relative difference of the transmitted power with and without a plasma the attenuation of the microwave signal across the plasma was determined.

7.5.4 Results and analysis of discharge operation

A reduction of the electron removal rates (i.e. the electron–ion recombination rate and/or the electron attachment rate) in the vibrationally excited

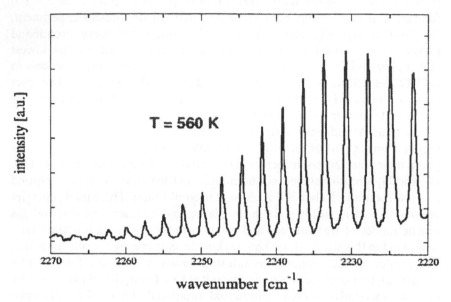

CO 1->0 R-Branch Emission

T = 560 K

Figure 7.5.3. Translational temperature in vibrationally-excited air at 1 atm measured by Fourier transform emission spectroscopy.

region should manifest itself in two experimental observations: (i) the steady-state electron density reached after an electron beam pulse is turned on should rise, and (ii) the electron density decay after the beam is turned off should become slower. In the present experiment, the average electron density in the e-beam sustained plasma, n_e^{baseline}, is inferred from microwave attenuation measurements using the relationship [Pal01a]

$$n_e^{\text{baseline}} = (m_e c \varepsilon_0 / e^2) \nu_{\text{coll}} \left(\frac{\delta V}{V} \right) \frac{1}{D} \qquad (7.5.6)$$

where ν_{coll} is the electron–neutral collision frequency, $\delta V / V = (V_{\text{trans}} - V_{\text{inc}}) / V_{\text{inc}}$ is the relative attenuation factor in terms of the forward power detector voltage proportional to the incident and the transmitted microwave power, and $D \cong 0.8\,\text{cm}$ is the size of the ionized region along the microwave signal propagation (see figure 7.5.2). Note that equation (7.5.6) assumes a uniform ionization across the plasma.

A CO laser beam propagating across the electron beam sustained plasma creates a cylindrically shaped vibrationally excited region of $d \cong 2\,\text{mm}$ diameter (see figure 7.5.2). The analysis of the microwave absorption measurements in electron beam sustained plasmas enhanced by laser excitation is somewhat complicated by the fact that the plasma volume affected by a focused CO laser is considerably smaller than the volume ionized by the electron beam. For this reason, equation (7.5.6) should be modified to take this effect into account. If one assumes that the electron removal rate modification due to vibrational excitation is significant, and that consequently the electron density in the optically pumped region, n_e^{modified}, is much higher than in the e-beam ionized region, n_e^{baseline}, equation (7.5.6) becomes

$$n_e^{\text{modified}} = (m_e c \varepsilon_0 / e^2) \nu_{\text{coll}} \left(\frac{\delta V}{V} \right) \frac{W}{\pi d^2 / 4} \tag{7.5.7}$$

where $W \cong 0.33\,\text{cm}$ is the width of the waveguide perpendicular to the laser beam axis. In addition, inference of the electron density should account for the change of the electron–neutral collision frequency, ν_{coll}, in the vibrationally excited plasma, which primarily depends on the electron temperature. In the present paper, the dependence of the collision frequency on the average electron energy is calculated by solving the coupled master equation for the vibrational level populations of CO, N_2, and O_2, and Boltzmann equation for the secondary (low-energy) plasma electrons [Ada98]. In the laser-excited plasma, the electron temperature is strongly coupled to the vibrational temperatures of the diatomic species due to rapid energy transfer from vibrationally excited molecules to electrons in superelastic collisions [Ale78, Ale79, Ada97]. The average electron energy in the optically pumped plasma is about 5000 K, as determined by the modeling calculations and recent Langmuir probe measurement in these laser pumped plasmas [Plo02]. This gives a collision frequency of $\nu_{\text{coll}} = 6.1 \times 10^{11}\,\text{s}^{-1}$ in air at $p = 1\,\text{atm}$ and $T = 560\,\text{K}$. In the purely e-beam sustained plasma, the average electron energy is $\sim 300\,\text{K}$, and the collision frequency is $\nu_{\text{coll}} = 1.1 \times 10^{11}\,\text{s}^{-1}$ at $p = 1\,\text{atm}$ and $T = 300\,\text{K}$. Summarizing, the electron densities in the electron beam sustained plasma and in the laser-enhanced region are evaluated from equations (7.5.6) and (7.5.7), respectively.

The experimental results are compared with a kinetic model of the electron production, electron removal, and charge transfer processes in the investigated air plasmas. The model takes into account rates for electron production by the e-beam, S, electron–ion recombination, β, three-body ion–ion recombination k_R, electron attachment in three-body collisions to O_2, $k_a^{O_2}$, and to N_2, $k_a^{N_2}$, electron detachment from O_2^- in collisions with O_2, $k_d^{O_2}$, and in collisions with N_2, $k_d^{N_2}$. Electron densities, n_e, positive ion densities, n^+, and O_2^- densities are calculated integrating the differential equations

$$dn_e/dt = S - k_a^{O_2} n_e [O_2]^2 - k_a^{N_2} n_e [O_2][N_2] - \beta n_e n^+$$

$$+ k_d^{O_2} n_e [O_2^-][O_2] + k_d^{N_2} n_e [O_2^-][N_2] \tag{7.5.8}$$

$$d[O_2^-]/dt = k_a^{O_2} n_e [O_2]^2 + k_a^{N_2} n_e [O_2][N_2] - k_R [O_2^-] n^+ N$$

$$- k_d^{O_2} n_e [O_2^-][O_2] - k_d^{N_2} n_e [O_2^-][N_2] \tag{7.5.9}$$

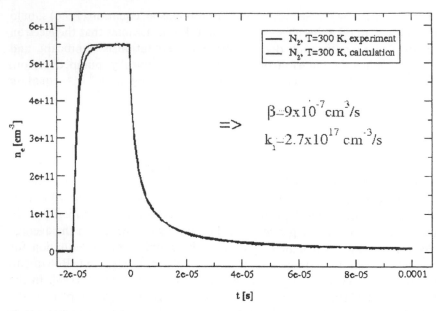

Figure 7.5.4. Measured and calculated electron densities during and after a 20 μs e-beam pulse in 1 atm N_2. In the calculation the electron production rate k_i and recombination rate β were chosen to best fit the measurement.

$$dn^+/dt = S - \beta n_e n^+ - k_R[O_2^-]n^+. \qquad (7.5.10)$$

In a first step, modeling results are fitted to the time resolved electron density in 1 atm of pure N_2 after a 20 μs e-beam pulse. Figure 7.5.4 shows the electron density measurement and the calculated electron density that best agrees in peak electron density and electron density decay. From the fit we obtain an electron production rate of $S = 2.7 \times 10^{12}\,\mathrm{cm^{-3}\,s^{-1}}$ and, since the decay in N_2 is dominated by electron–ion recombination, the effective dissociative electron–ion recombination rate for our e-beam ionized N_2 plasma. The determined recombination rate $\beta = 0.9 \times 10^{-6}\,\mathrm{cm^3\,s^{-1}}$ lies between the known recombination rates for the expected dominant ions N_4^+ $(\beta = 2 \times 10^{-7}\,\mathrm{cm^3\,s^{-1}})$ and N_2^+ $(\beta = 2 \times 10^{-6}\,\mathrm{cm^3\,s^{-1}})$. Consequently, the measurement suggests that about 50% of the positive ions in the plasma are the faster recombining N_4^+ that is produced in a conversion reaction.

Electron density measurements in the laser excited part of the e-beam plasma are somewhat more uncertain than measurements in purely e-beam sustained plasmas. This is due to (i) the uncertainty in the diameter d of the laser excited region (see equation 7.5.7), (ii) the uncertainty in the translational temperature in the laser excited region, and (iii) the uncertainty in the electron temperature T_e in the laser excited region. From the size of the visible glow of a laser-excited N_2/CO plasma at $p = 1$ atm and, critically, from Raman spectroscopic measurements [Lem00] the diameter of the

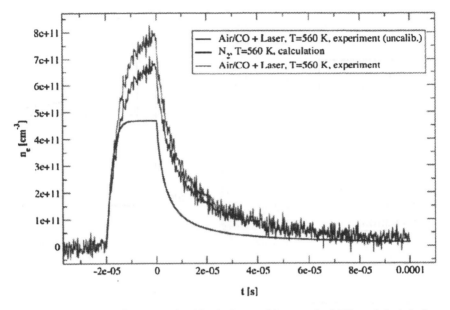

Figure 7.5.5. Measured electron densities in 1 atm of laser-excited CO-seeded air before and after calibration by comparison with N_2. Assuming identical electron production rates in 1 atm of air and 1 atm of N_2 the slope of the initial electron density rise should be identical for air and N_2. Very good agreement is achieved by changing the diameter of the laser-excited region from $d = 0.2$ cm to $d = 0.185$ cm (equation 7.5.6).

laser excited region was estimated to be $d = 0.2$ cm. As noted previously translational temperature in the laser region was measured spectroscopically from a Boltzmann plot of the infrared emission intensities of CO $1 \to 0$ R-branch lines (figure 7.5.3). For 1 atm of air seeded with 5% CO and optically excited by a 10 W CO laser, the temperature was found to be 560 K.

Figure 7.5.5 shows the measured electron density assuming $d = 0.2$ cm and $T_e = 5000$ K and a calculated electron density pulse in N_2 at $T = 560$ K. Assuming identical electron production rates in 1 atm of air and 1 atm of N_2, the slope of the initial electron density rise in laser excited air should be identical to the slope in N_2. Very good agreement is achieved by changing the diameter of the laser excited region in equation 7.5.6 from $d = 0.2$ cm to $d = 0.185$ cm, also shown in figure 7.5.5. The signal-to-noise ratios for electron density measurements in the laser excited region are much lower than purely e-beam excited plasmas. This is caused by the much smaller size of the laser excited region and the consequently lower MW attenuation. In fact, a microwave attenuation measurement in the laser excited region is always accompanied by a measurement in the surrounding, purely e-beam excited region. The illustrations of figure 7.5.6 show how the net signal is combined.

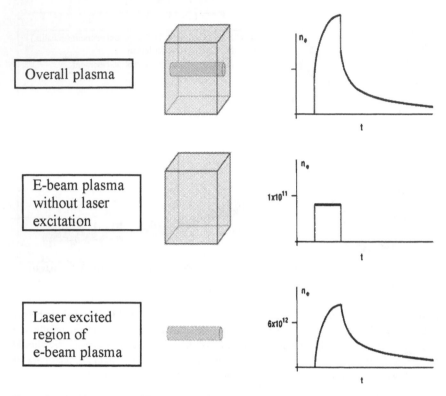

Figure 7.5.6. Illustration of how the e-beam-ionized region and the e-beam-ionized/laser excited region contribute to the overall electron density signal recorded by the microwave system. Note the different scales on the n_e-axes [Pal01b].

Figure 7.5.7 shows the electron density pulse in the vibrationally excited air plasma (the same data as figure 7.5.5, now calibrated), together with a calculated pulse in N_2 at $T = 560$ K and the assumed $T_e = 5000$ K. Both traces appear to be in very good agreement. Most notably, the decay of the electron density in laser excited air is equally slow as in N_2, i.e. attachment of electrons to O_2 does not seem to be a relevant process in vibrationally excited air. The importance of attachment to oxygen in cold equilibrium air can be seen in the dashed trace in figure 7.5.7 showing the corresponding electron density measurement without laser excitation. This experiment shows the markedly low level of ionization maintained by the e-beam only. Note the 200-fold higher peak electron density in laser excited air [$(7.9 \times 10^{11}\,\mathrm{cm}^{-3})/(4.4 \times 10^9\,\mathrm{cm}^{-3})$].

As mentioned before, the CO laser excited air plasma is in a very strong vibrational non-equilibrium. The vibrational temperature of the diatomic species exceeds the translational temperature by at least a factor of 4.

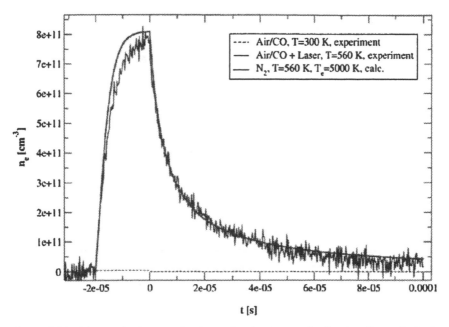

Figure 7.5.7. Measured electron density pulse in 1 atm of vibrationally excited air compared with calculated electron density in N_2. In strong contrast to a plasma in cold equilibrium air (dashed line) vibrationally-excited air does not seem to exhibit any electron attachment to O_2, i.e. peak electron density and plasma decay in vibrationally-excited air seem to be purely caused by electron–ion recombination.

Nevertheless, the fraction of molecules in excited vibrational states is still small compared to the population of the vibrational ground state. Therefore, the apparent complete mitigation of electron attachment to oxygen in vibrationally excited air cannot be caused by a vibrationally induced modification of the attachment rate itself. This is because the ground-state O_2 molecules (>50%) would still be exhibiting the full attachment rate, i.e. the total attachment rate could only be reduced by less than 50%. Consequently, the vibrational excitation has to be acting on the electron detachment side. On the other hand, the detachment rate shows a strong temperature dependence that raises the question of whether the observed effect might be due to the temperature rise (from 300 to 560 K) associated with the optical excitation of our air plasma.

Figure 7.5.8 shows calculated and experimental electron densities in a 30 µs e-beam pulse in $p = 1$ atm air at slightly higher beam current than in figure 7.5.7, there is no laser excitation of vibration in this experiment. The modeling calculation, assuming $S = 0.5 \times 10^{18}\,\mathrm{cm}^{-3}\,\mathrm{s}^{-1}$, $\beta = 2 \times 10^{-6}\,\mathrm{cm}^3\,\mathrm{s}^{-1}$, $k_a^{O_2} = 2.5 \times 10^{-30}\,\mathrm{cm}^6\,\mathrm{s}^{-1}$, $k_a^{N_2} = 0.16 \times 10^{-30}\,\mathrm{cm}^6\,\mathrm{s}^{-1}$, $k_d^{O_2} = 2.2 \times 10^{-18}\,\mathrm{cm}^3\,\mathrm{s}^{-1}$, $k_d^{N_2} = 1.8 \times 10^{-20}\,\mathrm{cm}^3\,\mathrm{s}^{-1}$, $k_R = 1.55 \times 10^{-25}\,\mathrm{cm}^3\,\mathrm{s}^{-1}$ [Rai91] and $T = 300\,\mathrm{K}$ shown in figure 7.5.8 agrees well with

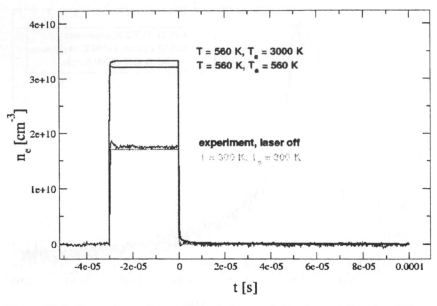

Figure 7.5.8. Comparison of experimental data and kinetic modeling for different translational and electron temperatures.

the experimental data. The two other traces in figure 7.5.8 show the calculated electron densities taking into account modified electron detachment and electron–ion recombination rates due to increased electron and translational temperatures. The modified rates for increased translational temperature only and increased translational and electron temperature used are $\beta = 1.5 \times 10^{-6}\,\mathrm{cm^3\,s^{-1}}$, $k_d^{O_2} = 2.2 \times 10^{-14}\,\mathrm{cm^3\,s^{-1}}$, $k_d^{N_2} = 1.8 \times 10^{-16}\,\mathrm{cm^3\,s^{-1}}$, and $\beta = 6.3 \times 10^{-7}\,\mathrm{cm^3\,s^{-1}}$, $k_d^{O_2} = 2.2 \times 10^{-14}\,\mathrm{cm^3\,s^{-1}}$, $k_d^{N_2} = 1.8 \times 10^{-16}\,\mathrm{cm^3\,s^{-1}}$, respectively [Rai91]. It can be seen that the change of electron and translational temperatures associated with the laser excitation would not produce a very strong effect on the electron density (×2) and the plasma decay time. Therefore, the strong effect observed in the experimental data with laser excitation can be attributed to the vibrational excitation, not temperature effects.

Figure 7.5.9 shows the measured electron densities for the conditions of figures 7.5.4, 7.5.5, 7.5.7, and calculations for these conditions, using hugely increased electron detachment rates. The experimental electron density shown in this figure represents the best performance achieved in the 1 atm air plasma, reaching high electron density with greatly increased plasma lifetime. Increase of the detachment rates by five orders of magnitude fully mitigates the effect of attachment and the calculated trace for laser excited air practically coincides with the calculated trace for N_2. The change of the

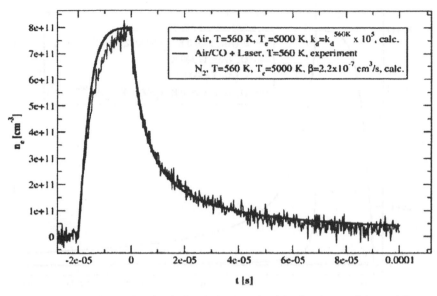

Figure 7.5.9. Experimental and calculated electron densities for the conditions of figures 7.5.4, 7.5.5 and 7.5.7 using hugely increased electron detachment rates. Increase of the detachment rates by five orders of magnitude mitigates the effect of attachment and the calculated trace for laser excited air practically coincides with the calculated trace for N_2.

electron–ion recombination rate from $\beta = 0.9 \times 10^{-6}\,\mathrm{cm}^3\,\mathrm{s}^{-1}$ to $\beta = 2.2 \times 10^{-7}$ is due to the increase of the electron temperature from $T_e = 300\,\mathrm{K}$ in cold gas to $T_e = 5000\,\mathrm{K}$ in the vibrationally excited gas.

Finally, figure 7.5.10 shows the number densities for the negatively charged species e^- and O_2^-, calculated from the rates determined from the experiment using the analysis reviewed above. Due to attachment, the dominant negative species in cold air is O_2^-, whereas in vibrationally excited air the O_2^- population is insignificant ($<2 \times 10^9\,\mathrm{cm}^{-3}$) and the dominant negative species is e^-. Note the higher total number density of charged species in vibrationally-excited air that is due to the reduced ion–ion recombination channel. The experimental results and modeling calculations are consistent with the hypothesis given in section 7.5.2, equations (7.5.1)–(7.5.5) for the effect of vibrational excitation on electron attachment to oxygen and electron–ion recombination in electron beam sustained atmospheric pressure air plasmas: (i) since the electron affinity of O_2^- is only about $0.4\,\mathrm{eV}$ [Rai91], vibrational excitation of O_2^- to vibrational levels $v \geq 2$ can provide sufficient energy for the detachment of the attached electron

$$O_2^-(v \geq 2)[+M] \longrightarrow O_2 + e^-[+M] \tag{7.5.11}$$

while charge transfer from O_2^- to vibrationally excited oxygen is sufficiently rapid to make this process very efficient and (ii) superelastic collisions of

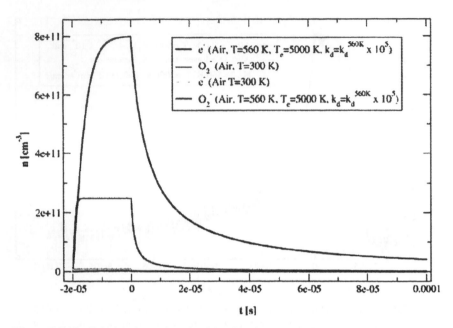

Figure 7.5.10. Calculated number densities for the negatively charged species e^- and O_2^-. Due to attachment, the dominant negative species in cold air is O_2^-, whereas in vibrationally excited air the O_2^- population is insignificant ($<2 \times 10^9 \, cm^{-3}$) and the dominant negative species is e^-. Note the higher total number density.

the initially cold secondary electrons produced by the electron beam with highly vibrationally excited molecules increase the electron temperature significantly to $T_e \approx 5000 \, K$, which reduces the electron–ion recombination rate.

7.5.5 Summary; appraisal of the technique

These time-resolved electron density measurements in electron beam sustained cold atmospheric pressure air plasmas demonstrate the effect of vibrational excitation of the diatomic air species on electron removal processes, notably dissociative recombination and attachment to O_2. Vibrational excitation of the diatomics is produced by laser excitation of CO seeded into the air and subsequent vibration–vibration energy transfer within the CO vibrational mode and from the CO to O_2 and N_2. The experimental results are consistent with a model that assumes rapid vibrationally induced detachment of electrons from O_2^- and vibrationally induced heating of the free electrons to temperatures on the order of $T_e \approx 5000 \, K$, thus effectively mitigating the effect of electron attachment and electron–ion recombination, respectively.

What is the overall influence of these electron loss mitigation effects on the overall plasma power budget? This can be estimated as follows: In cold air plasmas the dominant electron removal process is attachment to oxygen. The minimum power budget (assuming 100% ionization efficiency) to sustain a cold air plasma with an electron density of $n_e = 10^{13}\,\mathrm{cm}^{-3}$ is therefore given by $P_a = E_{ion}k_a[O_2]^2$. For an average ionization energy in air of $E_{ion} \approx 14\,\mathrm{eV}$ this gives $P_a = 1.4\,\mathrm{kW/cm^3} = 1.4\,\mathrm{GW/m^3}$. In the case of vibrationally-excited air, the electron loss by attachment is replenished by detachment of electrons from O_2^- instead of O_2 in the case of cold air. With an electron affinity of $E_{det} \approx 0.4\,\mathrm{eV}$ the minimum power budget to overcome attachment decreases to $P_a = E_{det}k_a[O_2]^2 = 40\,\mathrm{W/cm^3}$ at $T = 300\,\mathrm{K}$ or $P_a = 10\,\mathrm{W/cm^3}$ at the reduced gas density at $T = 560\,\mathrm{K}$. In the case of mitigated attachment the main electron removal process in an electron-beam-sustained air plasma is dissociative electron–ion recombination. The minimum power budget to overcome recombination is given by $P_{rec} = E_{ion}\beta n_e^2$. With an electron–ion recombination rate of $\beta \approx 1 \times 10^{-6}\,\mathrm{cm^3\,s^{-1}}$ we obtain $P_{rec} = 225\,\mathrm{W/cm^3}$. With the measured recombination rate in vibrationally excited air, $\beta \approx 2 \times 10^{-7}\,\mathrm{cm^3\,s^{-1}}$, the minimum power budget to overcome recombination decreases to $P_{rec} = 45\,\mathrm{W/cm^3}$.

In summary, the theoretical minimum power budget to overcome attachment and recombination in our vibrationally excited air plasmas is approximately $50\,\mathrm{W/cm^3}$, which represents a significant reduction compared to almost $2000\,\mathrm{W/cm^3}$ in cold equilibrium air.

The $45\,\mathrm{W/cm^3}$ power budget estimate does not, however, include the efficiency of the laser excitation process and the efficiency of the electron beam ionization process. The laser power required is approximately $1\,\mathrm{W/cm^3}$. The laser used in the experimental demonstration is a continuous wave, electrically-excited CO laser, which is the most efficient laser known with demonstrated very high continuous wave powers. Several hundred kW lasers of this type have been built, with 50% conversion of the input electric power into the beam. It is possible to project other means of achieving the required vibrational mode excitation. For example, use of other lasers with molecular seedants other than CO could be possible. Auxiliary electrodes, producing reduced electric fields operating at values to optimize vibrational mode power loading are conceivable. These alternatives all have their own problems. At the time of writing, the vibrational mode loading method used here seems the most effective.

The electron beam as an ionization source is efficient, with perhaps 50% of the beam energy going into ionization of the air. There are not major losses in producing the beam. We estimate that perhaps total beam power requirements increase the power budget by another $1–2\,\mathrm{W/cm^3}$.

A feature of this method of plasma generation is its exclusive reliance on beamed energy (laser, electron beam) to produce the plasma. This feature would be useful in applications in which it electrodeless plasma, or one created at a distance from the power source, is desirable.

The principal limitations of the method should be noted, however:

1. The performance achieved here is only achieved in dry air. Moisture or the presence of hydrocarbons in the air rapidly increases the rate of energy loss from the excited vibrational modes, mandating higher laser powers, and increasing plasma heating.
2. The system complexity and the attendant costs accompanying the electron beam. The foil window is fragile, and vulnerable to heating from the high pressure plasma; window failure leads to the air plasma contaminating the electron gun. Improvement in window materials, window cooling, and, even, electrodeless window development are subjects of on-going research, but this remains a key problem in the use of large electron beams for high pressure plasmas.
3. The systems complexity and the attendant costs accompanying the laser. The CO laser achieves its high efficiencies when cooled to near cryogenic temperatures. Large CO lasers have elaborate circulating gas systems with heat exchangers, or use fast, even supersonic flows for convective cooling. Research and development is also on-going in these laser systems.

References

[Ada97] Adamovich I V and Rich J W 1997 *J. Phys. D: Appl. Phys.* **30**(12) 1741

[Ada98] Adamovich I V, Rich J W and Nelson G L 1998 *AIAA J.* **36**(4) 590

[Ada00] Adamovich I V, Rich J W, Chernukho A P and Zhdanok S A 2000 'Analysis of the power budget and stability of high-pressure non-equilibrium air plasmas' Paper 00-2418, *31st Plasmadynamics and Lasers Conference*, Denver, CO, 19–22 June

[Ale78] Aleksandrov N L, Konchakov A M and Son E E 1978 *Sov. J. Plasma Phys.* **4** 169

[Ale79] Aleksandrov N L, Konchakov A M and Son E E 1979 *Sov. Phys. Tech. Phys.* **49** 661

[Bas79] Basov N G, Babaev I K, Danilychev V A *et al* 1979 *Sov. J. Quantum Electronics* **6** 772

[Gen75] Generalov N A, Zimakov V P, Kosynkin V D, Raizer Yu P and Roitenburg D I 1975 *Technical Phys. Lett.* **1** 431

[Kov85] Kovalev A S, Muratov E A, Ozerenko A A, Rakhimov A T and Suetin N V 1985 *Sov. J. Plasma Phys.* **11** 515

[Lee01] Lee W, Adamovich I V and Lempert W R 2001 *J. Chemical Phys.* **114**(3) 1178

[Lem00] Lempert W R, Lee W, Leiweke R and Adamovich I V 2000 'Spectroscopic measurements of temperature and vibrational distribution function in weakly ionized gases', Paper 00-2451, *21st AIAA Aerodynamic Measurement Technology and Ground Testing Conference*, Denver, CO, 19–22 June

[Mac99] Macheret S O, Shneider M N and Miles R B 1999 AIAA Paper 99-3721, *30th AIAA Plasmadynamics and Lasers Conference*, Norfolk, VA, 28 June–1 July

[Mac00] Macheret S O, Shneider M N and Miles R B 2000 'Modeling of air plasma generation by electron beams and high-voltage pulses', AIAA Paper 2000-2569, *31st AIAA Plasmadynamics and Lasers Conference*, Denver, CO, 19–22 June

[Mae91] Maetzing H 1991 *Adv. Chem. Phys.* **80** 315

[Mos99] Mostefaoui T, Laube S, Gautier G, Ebrion-Rowe C, Rowe B R and Mitchell J B A 1999 *J. Phys. B: At. Mol. Opt. Phys.* **32** 5247

[Pal01a] Palm P, Plönjes E, Buoni M, Subramaniam V V and Adamovich I V 2001 *J. Appl. Phys.* **89** 5903

[Pal01b] Palm P, Plönjes E, Adamovich I V, Subramaniam V V, Lempert W R and Rich J W 2001 'High pressure air plasmas sustained by an electron beam and enhanced by optical pumping', AIAA-Paper 2001-2937, *32nd AIAA Plasmadynamics and Lasers Conference*, 11–14 June, Anaheim, CA

[Plo00a] Plönjes E, Palm P, Chernukho A P, Adamovich I V and Rich J W 2000 *Chem. Phys.* **256** 315

[Plo00b] Ploenjes E, Palm P, Lee W, Chidley M D, Adamovich I V, Lempert W R and Rich J W 2000 *Chem. Phys.* **260** 353

[Plo01] Ploenjes E, Palm P, Lee W, Lempert W R and Adamovich I V 2001 *J. Appl. Phys.* **89**(11) 5911

[Plo02] Plönjes E, Palm P, Adamovich I V and Rich J W 2002 'Characterization of electron-mediated vibration-electronic (V-E) energy transfer in optically pumped plasmas using Langmuir probe measurements', AIAA-Paper 2002-2243, *33rd AIAA Plasmadynamics and Lasers Conference* 20–23 May, Maui, Hawaii

[Rai91] Raizer Y P 1991 *Gas Discharge Physics* (Berlin: Springer)

[Ric75] Rich W, Bergman R C and Lordi J A 1975 *AIAA J.* **13** 95

[Vel87] Velikhov E P, Kovalev A S and Rakhimov A T 1987 *Physical Phenomena in Gas Discharge Plasmas* (Moscow: Nauka)

[Zhd90] Zhdanok, S A, Vasilieva, E M and Sergeeva, L A 1990 *Sov. J. Engineering Phys.* **58**(1) 101

7.6 Research Challenges and Opportunities

The air plasma research techniques discussed in this chapter have yielded several important results and concepts that need further development. The use of lasers producing optically pumped or low ionization energy seed gases in atmospheric air to provide seed plasmas of high density (10^{12}–$10^{13}/cm^3$) of small size ($20\,cm^3$) to larger ($500\,cm^3$) volume should be pursued further. These techniques can overcome the high power densities required to ionize atmospheric air and provide an initial condition for lower power plasma sustainment by inductive rf waves or other techniques. The use of a laser allows plasma production well away from material surfaces which can be attractive for certain applications. Although some of these techniques were examined utilizing lasers to concentrate on the air plasma chemistry issues, less expensive focused flash tubes with reflectors could also be considered for these techniques.

An important issue in sustaining high density air plasmas is the formation of negative oxygen ions, O_2^-, at room temperature. By preheating the air to provide a higher neutral temperature of 2000 K by means such as rf heating, this process can be greatly reduced and plasma lifetimes and power sustainment densities required to provide average plasma densities in the $10^{13}/cm^3$ range and larger volumes substantially reduced. Another important experimental technique is to carry out individual air component experiments where the nitrogen, oxygen and other components of air including residual water vapor concentrations, H_2O, are isolated. Due to the complexity of air plasma chemistry, the role of the individual and collective processes can be examined in a more systematic way. Important optical spectroscopy and millimeter wave interferometery techniques and associated analytic codes that have been developed by the researchers in this area will make important contributions to this field.

The use of inductive rf waves to provide a plasma torch in near local thermodynamic equilibrium provides an analysis of baseline condition for steady-state, high density ($>10^{13}/cm^3$), large volume (1000 cm^3), atmospheric air plasma wall plug power density that is quite high ($P = 48\,W/cm^3$). The use of gas flow enhances these discharges, cools the source region and allows plasma production remote from the material source region. Microwave plasma torch power densities for smaller plasmas require power densities in region of 200 W/cm^3. In both cases the gas temperature is fairly high, at 4200 K with electron plasma temperatures of 5000 K. These parameters are deleterious for materials in the plasma region and illustrate the need for pulsed, non-equilibrium plasma that can reduce the plasma temperature, yet maintain high plasma densities and large volumes (1000 cm^3). In addition, further research on pulsed plasmas should be carried out in the microwave range to obtain high plasma density remote from the microwave source region. The use of short, repetitive pulsed power, high voltage plasmas in preheated (2000 K) air has been demonstrated to produce high average density, non-equilibrium plasmas with a higher ionization efficiency, with 100 times lower time-average power densities than in the steady-state case. The decaying plasma provides a seed for the next pulse when the repetition rate matches the electron recombination rate. The volumes of initial experiments were quite small (0.3 cm^3) and arrays and methods for creation of these lower time averaged power density air plasmas and creating plasmas remotely from electrodes should be pursued further.

The use of pulsed, moderate energy (60–80 keV) electron beams can also be used to provide plasmas with lower power density and optical pumping to reduce electron attachment to oxygen is an interesting technique. Initial experiments show that due to the increased electron and gas temperatures of 5000 K, electron attachment to oxygen could be reduced so that minimum power densities of 50 W/cm^3 could be obtained to offset electron recombination processes. Scaling of this technique to larger volumes and improvement of electron beam window are areas that need to be developed further. The use of pulsed dc, rf, microwave and electron beam power with seed gas and techniques used to reduce electron recombination with oxygen as well as more advanced aspects of air plasma chemistry are areas that need to be explored further to obtain non-equilibrium plasmas with lower power density in the atmospheric air for a variety of applications.

Chapter 8

Plasma Diagnostics

*B N Ganguly, W R Lempert, K Akhtar, J E Scharer, F Leipold,
C O Laux, R N Zare and A P Yalin*

8.1 Introduction

Measurements of plasma parameters in high-pressure plasma environment offer challenges and opportunities which usually have to satisfy requirements that are different compared to both partially ionized and highly ionized low-pressure plasmas. The highly collisional nature of atmospheric pressure plasma, compared to lower (<10 torr) pressure plasmas, can significantly modify the data analysis procedure and, more importantly, sometimes even the applicability of methods used to measure plasma characteristics in diffusion-dominated lower pressure plasmas. Also, the scaling laws of collisionally dominated *self-sustained* plasmas are usually bounded by ionization and thermal instabilities, which impose different operating requirements for maintaining self-sustained non-equilibrium plasmas at atmospheric pressure compared to low-pressure plasmas. Well developed low-pressure plasma diagnostics methods for both partially ionized (Auciello and Flamm 1989, Herman 1996) and highly ionized plasmas (Fonck and den Hartog 2002, Hutchinson 2002) can be adopted for collisionally dominated plasmas. The examples of applicability of electron density measurement by millimeter wave and mid infrared interferometric methods, with appropriate modifications for collisionally dominated plasmas, are discussed in this chapter in sections 8.3 and 8.4, respectively. Also, elastic and inelastic laser light diagnostic methods which are better suited for characterizing plasmas at elevated gas density are described in section 8.2. In section 8.2, both theoretical and experimental descriptions of Rayleigh scattering, pure rotational and ro-vibrational Raman scattering and Thomson scattering measurements in air plasma are described.

In this section filtered (by resonance absorption of atomic optical transition) laser light scattering techniques are discussed in detail which permit measurement of gas temperature from Doppler broadening of Rayleigh scattering under conditions where stray light scattering is significantly greater than the Rayleigh scattering intensity. Similarly, examples of filtered pure rotational Raman and Thomson scattering in plasmas at elevated pressure are also described in this section. Some of the Thomson scattering results discussed in section 8.2 are more applicable to the conditions for near equilibrium plasmas than highly non-equilibrium plasmas. The incoherent Thomson scattering data are fitted to a Gaussian-shape intensity distribution (Hutchinson 2002), which is appropriate if the EEDF is Maxwellian. The EEDF in many molecular gas non-equilibrium plasmas are not Maxwellian (see chapter 3). The procedure for obtaining non-Maxwellian EEDF measurement by incoherent Thomson scattering in atmospheric pressure plasmas has been discussed in a recent publication by Huang *et al* (2000).

Pure rotational Raman scattering of N_2 can permit gas temperatures measurement with high accuracy (± 10 K). Such a measurement technique can be very useful to quantify the operating conditions of a short pulse excited, low average power DBD where the gas temperature rise may be only be 100–200 K above the ambient gas temperature.

Electron density measurement by millimeter wave interferometry is described in section 8.3. In atmospheric pressure plasmas, the 105 GHz probe frequency is smaller than the electron–neutral collision frequency. Under such a measurement condition both probe beam intensity attenuation and phase shift need to be measured to estimate electron density. The details of such measurement and data analysis procedures are described in section 8.3. The choice of microwave or millimeter wave probe frequency is determined by the required resolution of the electron density measurement. For most non-equilibrium atmospheric pressure plasmas the electron density is $n_e \leq 10^{13}$ cm^{-3}. The 105 GHz probe frequency permits electron line density measurement with resolution $n_e l \leq 10^{14}$ cm^{-2}, where l is the linear plasma dimension.

The spatially resolved electron density measurement using mid-infrared CO_2 laser interferometry is described in section 8.4. This interferometric approach is ideally suited for electron density measurement in microhollow-cathode and other atmospheric pressure boundary dominated discharges with $PD \leq 10$ torr cm, where P is the gas pressure and D is the inter-electrode gap (Stark and Schoenbach 1999).

In low-pressure plasmas, Langmuir probes are used to measure electron density and EEDF (Auciello and Flamm 1989). Probes always perturb the local plasma surrounding. The extent of such perturbation depends on some characteristic lengths in plasma, namely, electron Debye length λ_D, ionization mean free path λ_e, and charge exchange mean free path λ_{ex}. If the probe dimension is larger than these characteristic lengths, the probe perturbs the local plasma properties and the validity of probe measurement becomes questionable (Auciello and Flamm 1989).

Plasma emission based measurements of rotational temperatures from electronically excited states are widely used to infer gas temperature in plasmas (Auciello and Flamm 1989, Herman 1996, Ochkin 2002). Measurements of rotational temperatures in atmospheric pressure air plasmas are described in chapter 8.5. It should be noted that such measurements would be a valid indicator of gas temperature only if the excited states are produced by direct electron impact excitation from the ground state. Since the electron collision with molecules cannot impart any significant amount of angular momentum, the rotational population distribution of the excited state should replicate the ground state rotational population distribution. Other factors which can impact such measurements include self-absorption of radiation and rotational quantum number dependent collisional quenching. If the excited states are formed through dissociative excitation or other processes where a significant amount of internal energy can be deposited, plasma emission from those excited molecular states cannot be used for estimating the rotational temperature of the ground state. Even when these conditions are met, in discharges where the EEDF is time modulated, such as in rf discharge, additional complications can arise where time modulated radiative cascade can modify the population distribution of the electronically excited rotational states. A comparison of time resolved rotational temperature measurements from H_2 Fulcher-α band and N_2^+ *B-X* (0,0) plasma emission showed a radiative cascade can influence the estimate of 'rotational temperature' measurement from the H_2 Fulcher-α band (Gans *et al* 2001). The accuracy of this relatively simple measurement technique can be compromised if all the necessary conditions are not met. In view of this, the plasma emission based rotational temperature measurement should be calibrated with rotational Raman or Doppler broadening of diode laser absorption measurements (Penache *et al* 2002). Although Doppler broadening measurement permits measurement of gas temperature with high accuracy in low-pressure plasmas, it may have limited accuracy in high-pressure plasmas, since the Doppler broadening scale is $\Delta_D = 7.16 \times 10^{-7} \nu_0 \sqrt{T/M}$, where ν_0 is the line-center transition frequency and M is the mass of the absorbing species in atomic mass units, whereas pressure broadening increases linearly with gas pressure (Demtroder 1981). For atmospheric pressure plasmas with a gas temperature rise $\leq 200\,K$ from ambient, pressure broadening may dominate over Doppler broadening. Under this condition, the diode laser absorption line shape becomes a Voigt profile, which is a convolution of Gaussian (Doppler broadened) and a Lorentzian (pressure broadened) line shape (Demtroder 1981). The Voigt, Gaussian, and Lorentzian linewidths (FWHM) are approximately given by (Penache *et al* 2002):

$$\Delta\lambda_G^2 = \Delta\lambda_V^2 - \Delta\lambda_V \cdot \Delta\lambda_L \tag{1}$$

where $\Delta\lambda_G$ is the Gaussian component width, $\Delta\lambda_V$ is the Voigt linewidth, and $\Delta\lambda_L$ is the Lorenztian component width. The Lorentzian component width can be de-convolved from the total Voigt linewidth in the wings of the absorption line, since the Lorentzian predominates in the wing, and the Gaussian component width is then determined from equation (1). If the pressure broadening becomes the dominating contributor to the Voigt profile, the accuracy of the Doppler broadening estimate from equation (1) becomes limited.

Plasma emission based measurement of electron density in air plasma from Stark broadening H_β is described in section 8.5. More details of the electron temperature and the electron density dependent H_β line shape fitting information can also be found in a recent review of spectroscopic measurements at or near atmospheric pressure plasma (Ochkin 2002).

The N_2^+ and NO^+ ion density measurements in atmospheric pressure air plasmas by ring-down spectroscopy are described in section 8.6.

The diagnostics methods presented in this chapter allows quantification of the fundamental plasma characteristics, which can be used to either validate model calculations and/or experimentally demonstrate scaling properties of high-pressure plasmas. Application specific diagnostics, such as measurements of O, H, or N atom or other radical densities in plasmas, have not been included in this chapter since the end use of atmospheric pressure non-equilibrium plasmas covers a wide scope, such as high flux radicals for materials processing, surface properties modification, detoxification, plasma display panel, and VUV/UV photon source. Some of the optical spectroscopic based measurements of process control and optimization are described in a recently published proceeding of the International Society for Optical Engineering (Ochkin 2002). It should be noted that commonly used one- or two-photon allowed laser-induced fluorescence (LIF) measurement of absolute radical densities in low pressure plasmas (Dreyfus *et al* 1985) may not be readily applicable to similar absolute density measurement of radical species, at atmospheric pressure, which have high collisional quenching rates, e.g. the H atom (Preppernau *et al* 1995). The LIF measurement can still be used to measure radical production efficiency in atmospheric pressure discharges, using methods similar to the combustion diagnostics of reactive species (Eckbreth 1996). Under some conditions, where spatial resolution is not required, ring-down spectroscopic measurement is very well suited for sensitive laser spectroscopic measurement of line integrated absolute density of radical (McIlroy 1998, Staicu *et al* 2002) and ionic species (see section 8.6) formed in an atmospheric pressure plasma.

References

Aucillo O and Flamm D L (eds) 1989 *Plasma Diagnostics* vols 1 and 2 (New York: Academic)

Demtroder W 1981 *Laser Spectroscopy* (Berlin: Springer)

Dreyfus R W, Jasinski J M, Walkup R E and Selwyn G S 1985 *Pure and Appl. Chem.* **57** 1265

Eckbreth A C 1996 *Laser Diagnostics for Combustion Temperature and Species* (Amsterdam: Gordon and Breach)

Fonck R J and Den Hartog D J (eds) 2003 Proceedings of the 14th Topical Conference on High Temperature Plasma Diagnostics, *Rev. Sci. Instrum.* **74**(3). And other previous conference proceedings published in *Rev. Sci. Instrum.*

Gans T, Schulz-von der Gathen V and Dobele H F 2001 *Plasma Sources Sci. Technol.* **10** 17

Herman I P 1996 *Optical Diagnostics for Thin Film Processing* (New York: Academic)

Huang M, Warner K, Lehn S and Hieftje G M 2000 *Spectrochimica Acta B* **55** 1397

Hutchinson I H 2002 *Principles of Plasma Diagnostics* (Cambridge: Cambridge University Press)

McIlroy A 1998 *Chem. Phys. Lett.* **296** 151

Ochkin V N (ed) 2002 'Spectroscopy of nonequilibrium plasma at elevated pressure', *Proceedings of SPIE*, vol 4460

Penache C, Micelea M, Brauning-Demian A, Hohn O, Schossler S, Jahnke T, Niemax K and Schmidt-Bocking H 2002 *Plasma Sources Sci. Technol.* **11** 476

Preppernau B L, Pearce K, Tserpi A, Wurzburg E and Miller T A 1995 *Chem. Phys.* **196** 371

Staicu A, Stolk R L and ter Meulen J J 2002 *J. Appl. Phys.* **91** 969

Stark R H and Schoenbach K H 1999 *Appl. Phys. Lett.* **74** 3770

8.2   Elastic and Inelastic Laser Scattering in Air Plasmas

8.2.1   Background and basic theory

8.2.1.1   Scattering intensities

Laser scattering is a relatively simple yet powerful optical diagnostic tool for high pressure molecular plasmas, capable of quantitative determination of heavy species rotational/translational temperature, vibrational distribution function of all major species, and electron number density and electron temperature. We begin this section by providing a brief overview of spontaneous scattering theory, emphasizing the essential elements relevant to measurements in molecular, non-equilibrium plasmas. More detail can be found in Long (2002), Eckbreth (1996), and Weber (1979). The discussion assumes knowledge of the fundamentals of diatomic spectroscopy such as Dunham expansions for calculating individual rotational and vibrational transition frequencies, nuclear spin degeneracy, and the Boltzmann distribution for equilibrium partitioning of internal energy, from which rotational temperature can be determined. If necessary a compact summary can be found in chapter 6 of Long (2002).

Scattering can be explained, classically, as the result of an incident electromagnetic wave inducing an oscillating electric dipole moment $p(t)$

which is given by the product of the polarizability, α, of the medium and the time-varying incident electric field, $E(t)$.

$$p(t) = \alpha \cdot E(t). \tag{1}$$

The polarizability, which has units of volume, is a measure of the distortion of the electron charge cloud in response to the applied electric field and is a function of the relative coordinates of the nuclei. It is customarily expanded with respect to the vibrational normal coordinates (or 'normal modes') (Q) of the molecule as

$$\alpha = \alpha_0 + \left(\frac{\partial \alpha}{\partial Q}\right)_0 Q + \cdots \tag{2}$$

where α_0 and $(\partial \alpha / \partial Q)_0$ are evaluated at the equilibrium internuclear displacement. Note that for diatomic molecules, which dominate air plasmas, there is only a single vibrational normal mode, corresponding to relative motion parallel to the axis connecting the nuclei. Assuming harmonic oscillation with natural frequency ω_k, so that $Q = Q_0 \cos(\omega_k t)$, and sinusoidal applied electric field, E, with frequency ω_1 and amplitude E_0, the induced electric dipole moment is given by

$$p(t) = \left[\alpha_0 + \left(\frac{\partial \alpha}{\partial Q}\right)_0 Q_0 \cos(\omega_k t)\right] E_0 \cos(\omega_1 t)$$

$$= \alpha_0 E_0 \cos(\omega_1 t) + \left(\frac{\partial \alpha}{\partial Q}\right)_0 \frac{Q_0 E_0}{2} [\cos(\omega_1 - \omega_k)t + \cos(\omega_1 + \omega_k)t]. \tag{3}$$

The first term in equation (3) contributes to two well known scattering phenomena. The first is the quasi-elastic scattering from bound electrons, commonly referred to as Rayleigh scattering, which can be used to extract heavy species translational temperature and number density. As will be discussed in section 8.2.4, the analogous quasi-elastic scattering from free electrons is termed Thomson scattering, which can be used for determination of electron density and temperature. The first term is also responsible for pure rotational Raman scattering, an inelastic scattering process corresponding to quantized molecular rotation which, as will be shown, can be used to extract extremely accurate values of rotational temperature. The second term represents vibrational Raman scattering, which can be used to measure the vibrational distribution functions of all major species.

Raman scattering requires a change in the polarizability with respect to motion of internal degrees-of-freedom. For pure rotational Raman scattering, this requires the polarizability to vary with molecular orientation, so that there must exist an anisotropic component to the molecular polarizability, generally expressed as $\alpha_\parallel - \alpha_\perp$. A spherically symmetric molecule, such as CH_4, yields no pure rotational Raman effect. For a vibrational Raman transition to occur, the polarizability must change as the molecule oscillates or as part of it bends.

Since Raman scattering does not require a permanent dipole moment, it is an excellent diagnostic for air plasmas, which are dominated by the homonuclear diatomic molecules N_2 and O_2. In general, the polarizability increases as the number of electrons increases so that heavier molecules tend to have inherently larger Rayleigh scattering intensities.

For quantized transitions between rotational–vibrational quantum states, the quantum mechanical expression for the polarizability matrix element, analogous to the classical expression given by equation (2), is

$$\alpha_{J''v'',J'v'} = \langle J''v''|\alpha|J'v'\rangle = \langle J''v''|\alpha_0|J'v'\rangle + \left(\frac{\partial\alpha}{\partial Q}\right)_0 \langle J''v''|Q|J'v'\rangle \quad (4)$$

where $J''v''$ and $J'v'$ are rotational–vibrational quantum numbers labeling the initial and final states, respectively, and the brackets indicate integration. In equation (4), the first term represents Rayleigh and pure rotational Raman scattering, which vanish unless $v' = v''$ due to the orthogonality of the vibrational wave functions, and the second term is responsible for vibrational Raman scattering. Assuming separation of the rotational and vibrational parts of the wave functions, evaluation of the matrix elements in equation (4) leads to the well known selection rules, which for linear molecules are

$$\Delta J = 0, \pm 2 \quad (5)$$

for pure rotational Raman scattering (where $\Delta J = 0$ corresponds to Rayleigh scattering) and

$$\Delta v = \pm 1, \qquad \Delta J = 0, \pm 2 \quad (6)$$

for vibrational transitions between harmonic oscillators. Transitions with $\Delta J = -2, 0, +2$ are called O, Q and S branches, respectively. Overtone transitions ($\Delta v = \pm 2, \pm 3, \ldots$) are allowed for anharmonic oscillators, but their intensities are very weak.

Figure 8.2.1 shows the basic geometry employed in most scattering measurements. The incident laser beam is linearly polarized with the polarization vector orthogonal to the plane defined by the propagation directions of the incident and detected scattered radiation, commonly referred to as the z axis. For such a geometry the detector, by definition, is located in the scattering plane so that the angle θ_z (see equation (49)) is equal to $90°$.

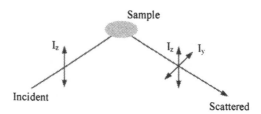

Figure 8.2.1. Basic scattering geometry for polarized light.

For this case the scattering intensity, I, or power (P) per unit solid angle (Ω), from an ensemble of scatterers in rotational level J, is given by (Long 2002)

$$I_{\|} = \frac{\pi^2}{\varepsilon_0^2} \tilde{\nu}_s^4 \left[(a_{00})^2 + b_{J,J} \frac{4(\gamma_{00})^2}{45} \right] N_J I_L, \qquad \Delta J = 0 \qquad (7)$$

$$I_{\perp} = \frac{\pi^2}{\varepsilon_0^2} \tilde{\nu}_s^4 \left[b_{J,J} \frac{(\gamma_{00})^2}{15} \right] N_J I_L, \qquad \Delta J = 0 \qquad (8)$$

for Rayleigh scattering

$$I_{\|} = \frac{\pi^2}{\varepsilon_0^2} \tilde{\nu}_s^4 \left[b_{J\pm2,J} \frac{4(\gamma_{00})^2}{45} \right] N_J I_L, \qquad \Delta J = \pm 2 \qquad (9)$$

$$I_{\perp} = \frac{\pi^2}{\varepsilon_0^2} \tilde{\nu}_s^4 \left[b_{J\pm2,J} \frac{(\gamma_{00})^2}{15} \right] N_J I_L, \qquad \Delta J = \pm 2 \qquad (10)$$

for pure rotational Raman scattering and, assuming harmonic oscillator wave functions,

$$I_{\|} = \frac{\pi^2}{\varepsilon_0^2} \tilde{\nu}_s^4 \left[(a_{10})^2 + b_{J,J} \frac{4(\gamma_{10})^2}{45} \right] N_J I_L, \qquad \Delta v = 1, \Delta J = 0 \qquad (11)$$

$$I_{\perp} = \frac{\pi^2}{\varepsilon_0^2} \tilde{\nu}_s^4 \left[b_{J,J} \frac{(\gamma_{10})^2}{15} \right] N_J I_L, \qquad \Delta v = 1, \Delta J = 0 \qquad (12)$$

$$I_{\|} = \frac{\pi^2}{\varepsilon_0^2} \tilde{\nu}_s^4 \left[b_{J\pm2,J} \frac{4(\gamma_{10})^2}{45} \right] N_J I_L, \qquad \Delta v = 1, \Delta J = \pm 2 \qquad (13)$$

$$I_{\perp} = \frac{\pi^2}{\varepsilon_0^2} \tilde{\nu}_s^4 \left[b_{J\pm2,J} \frac{(\gamma_{10})^2}{15} \right] N_J I_L, \qquad \Delta v = 1, \Delta J = \pm 2 \qquad (14)$$

for vibrational Raman scattering. In equations (7)–(13) the symbols $\|$ and $\perp$ correspond to scattering polarized parallel and perpendicular, respectively, to the incident laser polarization direction, N_J is the number density of scatterers in the level J, I_L the irradiance (power/area) of the incident laser beam, and a_{00} and γ_{00} represent the matrix elements for the mean and anisotropic parts of the polarizability, respectively, given by

$$a_{00} = \tfrac{1}{3}(\alpha_{xx} + \alpha_{yy} + \alpha_{zz}) \qquad (15)$$

$$\gamma_{00} = \tfrac{1}{2}[(\alpha_{xx} - \alpha_{yy})^2 + (\alpha_{yy} - \alpha_{zz})^2 + (\alpha_{zz} - \alpha_{xx})^2 + 6(\alpha_{xy}^2 + \alpha_{yz}^2 + \alpha_{zx}^2)]^{1/2}. \qquad (16)$$

Similarly, a_{10}/γ_{10} represent the corresponding polarizability derivative components.

In equations (7)–(14) the symbols $b_{J',J}$, known as the Plazeck–Teller factors (or rotational line strengths), represent the part of the polarizability matrix elements in equation (4) which arise from summation over the magnetic sublevels, m_J. For linear molecules which behave as rigid rotors (or more precisely, for symmetric top wave functions with the 'K' quantum number equal to 0), $b_{J',J''}$ have the following form (Long 2002).

$$b_{J+2,J} = \frac{3(J+1)(J+2)}{2(2J+1)(2J+3)} \tag{17}$$

$$b_{J-2,J} = \frac{3J(J-1)}{2(2J+1)(2J-1)} \tag{18}$$

$$b_{J,J} = \frac{J(J+1)}{(2J-1)(2J+3)} \tag{19}$$

8.2.1.2   Cross sections

Scattering intensities are most commonly tabulated by combining the constants and molecule dependent matrix elements that occur in equations (7)–(14) to form what is known as the differential scattering cross section, $(d\sigma/d\Omega)$, which is defined as

$$\left(\frac{d\sigma}{d\Omega}\right)_{\|/\perp} = \frac{I_{\|/\perp}}{NI_L} \tag{20}$$

where $\|$ and $\perp$ again refer to polarization of scattered light which is parallel or perpendicular, respectively, to the incident z axis polarization. Note that the cross sections scale as the scattering frequency, ν, to the fourth power (with the exception of Thomson scattering) and are independent of both the incident laser intensity and the scatterer number density. Some selected Rayleigh and Raman cross sections are given in table 8.2.1. More extensive tables can be found in (Eckbreth 1996, Shardanand and Rao 1977, Weber 1979).

While not essential to the primary purpose of this chapter, it is worth pointing out that the differential Rayleigh cross section is typically cast in a form different than equations (7) and (8). First, since the scattering originating from particles with different values of J spectrally overlaps, N_J can be replaced by N, the total number density, and the b_{JJ} sector can be set to 1. More significantly, it is traditional to express a and γ in terms of n, the index of refraction, and ρ_0, the natural light depolarization ratio, so that the cross sections become (Miles *et al* 2001)

$$\left(\frac{d\sigma}{d\Omega}\right)_{\|} = \left(\frac{3\sigma}{8\pi}\right)\left(\frac{2-\rho_0}{2+\rho_0}\right) \tag{21}$$

$$\left(\frac{d\sigma}{d\Omega}\right)_{\perp} = \left(\frac{3\sigma}{8\pi}\right)\left(\frac{\rho_0}{2+\rho_0}\right) \tag{22}$$

Table 8.2.1. Selected Rayleigh, rotational and vibrational Raman differential cross sections are listed. Excitation wavelength is 532 nm. Values for Rayleigh and vibrational Raman correspond to the sum of $\|+\perp$ contributions. Values for rotational Raman correspond to the $\|$ contribution from specified values of J'' and J'.

Molecule	Rayleigh differential cross section $(\times 10^{28}\,cm^2\,sr^{-1})$ (from Shardanand 1977)	Rotational Raman differential cross section $(J'' \to J')$ at 488 nm $(\times 10^{30}\,cm^2\,sr^{-1})$ (from Eckbreth 1996, Penney 1974)	Vibrational Raman differential cross section, Q-branch $(\times 10^{31}\,cm^2\,sr^{-1})$ (from Weber 1979)
N_2	3.9	5.4 $(6 \to 8)$	3.7
O_2	3.4	14 $(7 \to 9)$	4.4
CO		0.61 $(6 \to 8)$	3.5
He	0.080	—	—
H_2	0.81	2.2 $(1 \to 3)$	8.0
CO_2	12	53 $(16 \to 18)$	5.3 (v_1 mode)
CH_4	8.6		29 (v_1 mode)

where σ is the integrated cross section given by

$$\sigma = \frac{32\pi^3 (n-1)^2}{3\lambda^4 N^2} \left(\frac{6 + 3\rho_0}{6 - 7\rho_0} \right). \tag{23}$$

Note that ρ_0 is equal to zero for isotropic molecules and is of the order 0.01–0.05 for typical diatomic gases. The relatively small term in parentheses in equation (23) is known as the 'King correction factor' (Miles *et al* 2001).

8.2.1.3 Anharmonicity effects

In the previous sections we have ignored the vibrational level dependence of the Raman scattering cross sections. However, since non-equilibrium plasmas are characterized by very substantial vibrational mode disequilibrium it is important to assess the influence of anharmonicity and rotation/vibration coupling on the matrix elements, defined by equation (4). In particular, it is important to note that for harmonic oscillator wavefunctions, the polarizability derivative matrix elements scale as $(v+1)^{1/2}$ for $\Delta v = +1$ and $v^{1/2}$ for $\Delta v = -1$ so that the vibrational scattering cross sections are predicted to scale as $v'' + 1$ for Stokes scattering and v'' for anti-Stokes (Eckbreth 1996). Real molecules, however, exhibit anharmonicity which needs to be taken into consideration, particularly at high v. One approach is to substitute Morse potential wave functions in equation (4). Assuming vibrational transitions originating in level v with $\Delta v = \pm 1$, the result is (Gallas 1980)

$$\langle w|r|v\rangle = \frac{1}{a_M(k-2v-2)}\left[(v+1)\frac{(k-2v-1)(k-2v-3)}{(k-v-1)}\right]^{1/2},$$

$$\Delta v = +1 \qquad (24)$$

$$\langle w|r|v\rangle = \frac{1}{a_M(k-2v)}\left[v\frac{(k-2v-1)(k-2v+1)}{(k-v)}\right]^{1/2}, \qquad \Delta v = -1 \qquad (25)$$

where

$$a_M = \left(\frac{2\mu\omega_e x_e}{\hbar}\right)^{1/2}, \qquad k = \omega_e/\omega_e x_e$$

and v and w are vibrational quantum numbers, and ω_e and $\omega_e x_e$ are the first two terms in standard Dunham expansions for vibrational frequency. If it is assumed that $(\partial\alpha/\partial Q)_0$ is constant with respect to vibrational quantum number, then the vibrational Raman scattering cross sections will scale as

$$I_v \propto (v+1)\frac{(k-2v-1)(k-2v-3)}{(k-2v-2)^2(k-v-1)} \qquad \text{(Stokes)} \qquad (26)$$

$$I_v \propto v\frac{(k-2v-1)(k-2v+1)}{(k-2v)^2(k-v)} \qquad \text{(anti-Stokes)}. \qquad (27)$$

The influence of anharmonicity can be seen in figure 8.2.2 which plots the relative scattering cross section as a function of vibrational quantum number assuming harmonic oscillators (filled circles), and equation (26) for carbon monoxide (squares) and hydrogen (triangles). It can be seen that

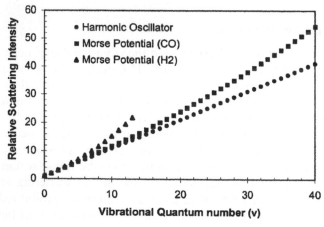

Figure 8.2.2. Relative scattering cross section as a function of v for harmonic oscillator (●), CO (■), and H_2 (▲).

for v less than ~ 5, the anharmonicity correction is quite small, even for hydrogen which has the largest anharmonicity of any diatomic molecule. For carbon monoxide, which is representative of other common diatomic species, such as nitrogen and oxygen, the correction becomes appreciable ($\sim 7\%$) for vibrational levels exceeding ~ 10, and reaches $\sim 33\%$ at level $v \approx 40$. As will be seen below, such high levels of CO have been observed in optically pumped as well as certain electric discharge plasmas. In such cases, the anharmonicity correction to the vibrational cross sections cannot be ignored.

In addition to anharmonicity effects, it is important to consider the effect of rotation-vibration interaction, particularly for pure rotational Raman scattering (Drake 1982). As stated previously, the Plazeck–Teller factors given in equations (17)–(19) assume rigid rotor wave functions, which while an excellent approximation at low J, can introduce significant uncertainty at high J, even in $v = 0$. The effect becomes even larger at high v, due to the increase in the average internuclear separation and corresponding increase in the polarizability anisotropy. Following the notation of Drake (1982) and Asawaroengchai and Rosenblatt (1980), the matrix element for the polarizability anisotropy can be expressed by

$$\langle J''v''|\alpha_{\parallel} - \alpha_{\perp}|J'v'\rangle \equiv \langle\alpha\rangle_v^2 = CS(J)f(J)\beta_v^2 \tag{28}$$

where C is a constant, $S(J)$ are the rigid rotor Plazeck–Teller factors, $f(J)$ is a centrifugal distortion correction, and β_v is the change in the polarizability accompanying rotation, which is a function of v and can be expressed as

$$\beta_v = \beta_e + \beta_{e'}\langle r - r_e\rangle_v + \tfrac{1}{2}\beta_{e''}\langle r - r_e\rangle_v^2 + \cdots \tag{29}$$

where $\beta_{e'} = (\partial\beta/\partial r)_e$, $\beta_{e''} = (\partial^2\beta/\partial r^2)_e$, and the average value of internuclear displacement, from first-order perturbation theory (Wolniewicz 1966), is given by

$$\langle r - r_e\rangle_v = r_e\left[\left(\frac{3B_e}{\omega_e}\right) + \left(\frac{\alpha_e}{2B_e}\right)\right]\left(v + \frac{1}{2}\right). \tag{30}$$

In equation (30) B_e is the first term in the standard Dunham expansion for rotational frequency and α_e is the rotation–vibration spectroscopic coupling constant. The significance of equations (28)–(30) is that they provide a method for correcting pure rotational Raman cross sections, which are tabulated for $v = 0$, for use in vibrationally non-equilibrium environments. Figure 8.2.3, which is a plot of the square of β_v/β_0 (which is proportional to the cross section) for H_2, CO, and NO, illustrates the magnitude of the effect. This can also be important for high resolution measurements in high temperature equilibrium systems, such as flames, in which temperature is determined by the ratio of intensities for fixed J and different v.

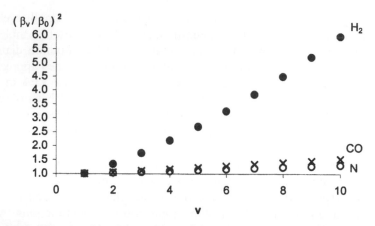

Figure 8.2.3. Vibrational level dependence of the square of the polarizability anisotropy for H_2, CO, and NO.

Finally, for highly non-rigid rotors, such as hydrogen, the $f(J)$ factor, while less significant than β_v needs to be considered, since it can impact spectra of molecules in the $v = 0$ level. Again, from Asawaroengchai and Rosenblatt (1980), $f(J)$ for pure rotational Stokes scattering is given by

$$f(J)_{00} = [1 + (4/\chi)(B_e/\omega_e)^2(J^2 + 3J + 3)]^2 \tag{31}$$

where χ is defined as $\beta_e/r_e\beta_e'$. (Note that for anti-Stokes scattering, J is replaced by $J - 2$). Similarly, for Stokes rotation–vibration scattering $f(J)$ is given by

$$f(J)_{01}^S = [1 - 3(B_e/\omega_e)^2 J(J + 1) - 4\chi(B_e/\omega_e)(2J + 3)]^2, \qquad (\Delta J = +2) \tag{32}$$

$$f(J)_{01}^O = [1 - 3(B_e/\omega_e)^2 J(J + 1) + 4\chi(B_e/\omega_e)(2J - 1)]^2, \qquad (\Delta J = -2). \tag{33}$$

Figure 8.2.4 plots $f(J)_{00}$ for H_2 and N_2 pure rotational Stokes transitions. It can be seen that for N_2 the correction is essentially negligible, where as for H_2 the correction factor is approximately 15% for $J = 4$, which corresponds to a rotational energy of $1200\,cm^{-1}$ (or characteristic temperature of $\sim 1750\,K$).

8.2.1.4 Spectral line shapes

For most, albeit not for all, diagnostic measurements extraction of quantitative information requires accurate knowledge of the appropriate spectral line shape function. We provide here a brief introduction to the subject which will serve as a basic foundation. Additional details can be found in the cited references.

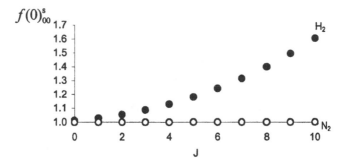

Figure 8.2.4. Centrifugal distortion correction to the pure rotational Raman cross section as a function of J for H_2 and N_2.

We begin with a slight digression, pointing out an important distinction between Raman and Rayleigh/Thomson scattering. For Raman scattering from an ensemble of gas phase scatterers what is known as the 'random phase approximation' is generally assumed valid since it is reasonable that the relative phases of internal oscillation or rotation of individual 'particles' are randomly distributed. The result is that the total scattering intensity seen at the detector is the simple incoherent sum of the intensity from each scattering particle. This leads directly to a total intensity which is proportional to N, the particle number density, as per equations (7)–(14). However, Rayleigh and Thomson scattering are inherently coherent so that the relative phases seen at the detector are dictated by the differences in the total propagation path, which depends upon the positions of the individual particles within the scattering sample volume, as well as the scattering geometry. For the idealized case of a perfectly ordered array of stationary scattering particles the vector sum of the Rayleigh scattered electric fields at the detector is identically zero, except for the trivial case of zero degree scattering angle (or 'forward' scattering). In the gas phase, however, the random motion of particles gives rise to instantaneous fluctuations in the local scattering number density such that phase cancellation at the detector is not perfect. This 'dynamic' light scattering mechanism was first described by Einstein and is discussed in more detail in many standard textbooks on the subject (Chu 1991, Berne and Pecora 1976). Without going through the details we simply state that for almost all cases the total Rayleigh scattering intensity is also proportional to the number density of scatterers. Exceptions occur at very small scattering angle and/or long wavelength light (Gresillon *et al* 1990) or in the vicinity of critical points (Ornstein and Zernike 1926).

For Raman scattering in gases, therefore, we can ignore collective motion and focus our discussion of spectral line shapes on mechanisms which affect individual molecules. A central consideration, for both Raman and Rayleigh/Thomson scattering, is the scattering wave-vector, **k**,

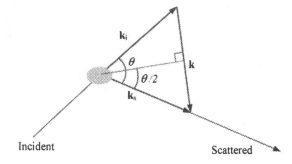

Figure 8.2.5. Scattering diagram illustrating magnitude and direction of wave-vector.

defined as

$$\mathbf{k} = \mathbf{k}_s - \mathbf{k}_i \tag{34}$$

where the subscripts i and s refer to the incident and scattered propagation directions, respectively (see figure 8.2.5). It can be seen that the direction of $\mathbf{k}$ is perpendicular to the bisector of the angle formed by the incident and scattering directions, referred to a common origin. From simple geometry (the law of cosines) it is easy to show that the magnitude of the vector $\mathbf{k}$, Δk, is given, in general, by

$$\Delta k = |\mathbf{k}_s - \mathbf{k}_i| = [k_s^2 + k_i^2 - 2k_s k_i \cos(\theta)]^{1/2} \tag{35}$$

where k_s and k_i are equal to $2\pi/\lambda_s$ and $2\pi/\lambda_i$, respectively, λ is the radiation wavelength and θ is the scattering angle.

Note that for Rayleigh and Thomson scattering it is easy to show (using $2\sin^2[\theta/2] = 1 - \cos[\theta]$), that equation (35) reduces to

$$\Delta k \cong 2|\mathbf{k}_0|\sin\left(\frac{\theta}{2}\right) = \frac{4\pi}{\lambda}\sin\left(\frac{\theta}{2}\right). \tag{36}$$

From the perspective of spectral line shapes, the $\mathbf{k}$ vector dictates the contribution to the phase of the detected scattering due to the position, $\mathbf{r}$, of individual scatterers, through the expression

$$E_{det}(t) = E_0 \exp[-i(\omega_s t + \mathbf{k} \cdot \mathbf{r}(t)] \tag{37}$$

where ω_s represents the central scattering frequency and the $\mathbf{k} \cdot \mathbf{r}(t)$ term has units of phase angle. If $\mathbf{r}(t) = \mathbf{v}t$, where $\mathbf{v}$ is the vector velocity, then equation (37) becomes

$$E_{det}(t) = E_0 \exp[-i(\omega_s + \mathbf{k} \cdot \mathbf{v})t] = E_0 \exp[-i(\omega_s + \omega_{Dop})] \tag{38}$$

where the $\mathbf{k} \cdot \mathbf{v}$ term is the well known Doppler shift due to the vector velocity $\mathbf{v}$.

We now introduce the parameter often given the symbol Y, defined as

$$Y = \frac{1}{\Delta kl} \propto \frac{\lambda}{l} \tag{39}$$

where l is the collision mean free path (Miles 2001). If $Y \ll 1$, then the scattering particles, on average, traverse a distance such that the $\mathbf{k} \cdot \mathbf{v}$ term in equation (38) oscillates through many cycles of 2π in the time interval between collisions. This condition, which typically corresponds to low density, results in the well known 'Doppler' spectral profile, is representative of scattering from an ensemble of particles with Maxwellian velocity distribution. The spectral profile, $S(\omega)$, is given by

$$S(\omega) = \frac{1}{\gamma_{\mathrm{Dop}}} \left[\frac{\ln(2)}{\pi} \right]^{1/2} \exp\left[\frac{-\ln(2)(\omega - \omega_s)^2}{\gamma_{\mathrm{Dop}}^2} \right] \tag{40}$$

where γ_{Dop}, the half width at half maximum (HWHM), is given by

$$\gamma_{\mathrm{Dop}} = \left[\frac{\Delta k}{2\pi} \right] \left[\frac{2\ln(2)k_{\mathrm{B}}T}{m} \right]^{1/2} \tag{41}$$

and k_{B} is Boltzmann's constant. Note that equation (40) is valid for Raman, Rayleigh, and Thomson scattering so long as $Y \ll 1$. As we shall see in section 8.2.4, however, the definition of Y is different than equation (39) for Thomson scattering. Equation (41), in combination with equation (35), represents the general expression for the Doppler scattering line width, taking into account scattering geometry as well as, in the case of Raman scattering, Stokes or anti-Stokes frequency shifts.

For $Y \gg 1$, the $\mathbf{k} \cdot \mathbf{v}(t)$ term in equation (38) evolves by much less than 2π in the time interval between collisions. In this limit, a spectral phenomena known as Dicke narrowing (Dicke 1953) occurs, in which the Doppler contribution to the line width goes to zero and is replaced by a Lorentzian component due to mass diffusion given by

$$S_{\mathrm{L}}(\omega) = \frac{1}{\pi \gamma_{\mathrm{Diff}}} \left[\frac{\gamma^2}{(\omega_s - \omega)^2 + \gamma^2} \right] \tag{42}$$

where $\gamma_{\mathrm{Diff}} \propto D_{\mathrm{m}}$, D_{m} being the coefficient of mass diffusion which scales as the inverse of pressure. In the case of Raman scattering an additional 'Lorentzian' contribution to the spectral line width also develops due to collisions which limit the 'lifetime' of the oscillation at ω_s. This phenomenon, known as 'pressure broadening', has HWHM given by

$$\gamma_P = a(T)P \tag{43}$$

where $a(T)$ is the temperature dependent pressure broadening coefficient, which is most commonly given in units of $\mathrm{cm}^{-1}\,\mathrm{bar}^{-1}$. A full conceptual treatment of the determination of $a(T)$ is beyond the scope of this chapter, but we will simply state that it is typically of order $0.1\,\mathrm{cm}^{-1}\,\mathrm{bar}^{-1}$ at room temperature and is generally determined experimentally (Bonamy *et al* 1988, Rosasco *et al* 1983).

In the intermediate Y regime, the most commonly employed approach for Raman scattering is to utilize the Voigt profile, $S_V(\omega)$, given by

$$S_V(\omega) = \left(\frac{\ln 2}{\pi}\right)^{1/2} \frac{1}{\gamma_{\text{Dop}}} \left(\frac{B}{\pi}\right) \int dy \left[\frac{e^{-y^2}}{(D-y)^2 + B^2}\right] \tag{44}$$

with

$$B = (\ln 2)^{1/2} \left(\frac{\gamma_L}{\gamma_{\text{Dop}}}\right), \qquad D = (\ln 2)^{1/2} \frac{(\omega - \omega_s)}{\gamma_{\text{Dop}}}$$

which treats the simultaneous Doppler and Lorentz components as a convolution integral (Demtroder 1998). In some cases, where very high resolution data are available, more complex treatments employing line shape functions such as the Galatry profile are employed (Galatry 1961).

For Rayleigh and/or Thomson scattering, collective motion begins to influence the line shape as Y approaches ~ 1. As will be described in some detail in section 8.2.4, in this regime acoustic modes begin to propagate in the fluid, inducing correlated density fluctuations scattering from which results in the development of frequency shifted side bands, symmetrically displaced from the 'narrowed' central component. For molecular scattering this phenomenon is known as Rayleigh–Brillouin (or Mandelstem) scattering.

For completeness we note briefly one additional spectral effect that occurs at elevated pressures. In the previous discussion we have assumed that the intensities from a set of individual spectral transitions, for example O/S branch Raman transitions, are independent of one another. However, in cases where lines begin to spectrally overlap it is often the case that this so-called 'isolated line' hypothesis fails so that the total intensity at any wavelength is not equal to the simple sum of contributions from adjacent lines. In particular, individual Q-branch Raman transitions overlap significantly for pressures of order 1 bar or greater and it is well known that 'line-mixing' techniques must be used to accurately fit experimental spectra. While the details are beyond the scope of this chapter, the basic approach requires incorporation of state-to-state J-dependent rotational energy exchange, which constitutes the primary mechanism of pressure broadening in most diatomic systems. As molecules begin to experience J changing collisions with a frequency exceeding the difference frequency between adjacent transitions, the individual lines begin to merge together. This 'rotational narrowing' is analogous to the Dicke narrowing of the Doppler profile described previously, and is well recognized in spectral models of coherent anti-Stokes Raman spectroscopy (CARS) (Hall *et al* 1979).

8.2.2 Practical considerations

Figure 8.2.6 illustrates a somewhat generic scattering apparatus, typical of that which would be employed for single spatial point scattering measurements.

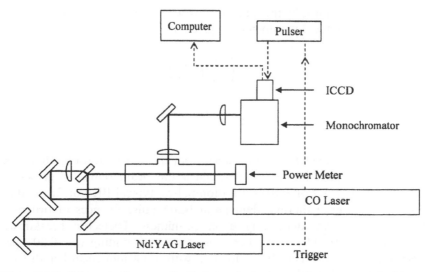

Figure 8.2.6. Schematic diagram of typical spontaneous scattering apparatus, in this case used for CO laser optical pumping studies described in section 7.2.

The most common laser source for application to luminous environments is the 'Q-switched' Nd:YAG laser, which is readily available from several commercial vendors. Typical single pulse output energy at the second harmonic wavelength of 532 nm ranges from ~0.3 to 1.0 J with pulse duration and repetition rate equal to ~10 ns and 10–30 Hz, respectively. While 532 nm systems are most common, it can be useful, in some cases, to employ the third (355 nm) or fourth (266 nm) harmonic or to use KrF (248 nm) or ArF (193 nm) excimer lasers. Such systems take advantage of the fourth power of frequency dependence of the cross section (equations (7)–(14)), but require more expensive, and somewhat less robust, optics. For non-equilibrium air plasmas, strong interferences from O_2 laser induced fluorescence must also be considered, particularly at 193 and 248 nm. This generally requires the use of line-narrowed, tunable sources, which are readily available but considerably more expensive. Nonetheless, if ultimate sensitivity is essential, for example to capture instantaneous 'single laser shot' data, shorter wavelength systems are often a necessity. It should always be recalled, however, that in photon units the scattering cross section scales as frequency to the third power, since the photon energy is proportional to frequency.

Laser focusing into the scattering medium is straightforward but subject to the dual constraints of dielectric breakdown, which limits the intensity at the 'waist' of the focused laser beam, and damage to the scattering medium access windows. For what are known as 'Gaussian' laser beams, these two constraints are coupled by the following expressions for the 'beam waist', w_0, and the beam confocal parameter, z_0, given by

$$w_0 = \left(\frac{\lambda}{2\pi}\right)\left(\frac{f}{D}\right) \tag{45}$$

$$z_0 = \frac{\pi w_0^2}{\lambda}. \tag{46}$$

The confocal parameter is the distance from the waist location, along the laser beam propagation axis, after which the beam diameter grows to $\sqrt{2}w_0$ (Yariv 1975). For $\sim$10 ns duration pulses, typical BK7 or fused silica windows experience thermal damage at beam pulse fluences of order 1–10 J/cm^2 at 532 nm, depending upon cleanliness. Dielectric breakdown occurs at $\sim$5 $\times$ 10^3 J/cm^2 at 1 bar pressure, corresponding to $\sim$0.20 J per pulse, for a typical $\sim$1 cm beam focused with a 300 mm focal length lens. Note that this value is based on experience and assumes a focal spot which is substantially greater ($\sim$50 μm) than that calculated from equation (45). None the less, as can be seen from equations (45) and (46), if w_0 is increased in order to avoid breakdown, the accompanying increase in the confocal beam parameter can lead to window damage.

For Raman scattering, signal is typically collected at 90° with respect to the laser beam propagation direction. The 'étendue' of the resolving instrument (in this case a spectrometer), which, for fixed resolution, is the maximum product of the collection solid angle and 'source' (which in this case is the scattering volume) cross sectional area, places some additional constraints on the collection optics (Vaughan 1989). For moderate resolution Raman spectra, the sampling volume is typically 1:1 imaged onto the entrance slit of an $\sim$0.25–0.3 m focal length grating spectrometer with slits set to 100 μm, or $\sim$2–4 w_0 of the focused laser beam. The solid collection angle is matched to that of the spectrometer optics, typically $\sim f/4$, where f is the ratio of the collection lens focal length to clear aperture, and the cylindrical scattering volume is aligned with its long axis parallel to the entrance slit. Faster collection can be performed, but only with accompanying loss of spectral resolution. For example if an $f/2$ collection lens is used with an $f/4$ imaging lens the accompanying magnification would require an increased slit size to pass all the collected light into the spectrometer.

In general, the detector of choice for Raman measurements in air plasmas is the microchannel plate intensified CCD (ICCD) camera, which combines high quantum efficiency with fast gating capability. This is essential in highly luminous environments, typical of such plasmas, where interference due to spontaneous emission can be far larger than the desired scattering signal, often by eight orders of magnitude or more.

Anticipated scattering signal levels can be estimated by considering the following simple expression

$$S = \left(\frac{E_L}{h\nu A}\right) N \left(\frac{d\sigma}{d\Omega}\right) (d\Omega) V \eta \phi \qquad (47)$$

where $E_L/h\nu A$ is the fluence of a single laser pulse (in photons/cm^2), N is the number density of scatterers, $d\sigma/d\Omega$ is the scattering differential cross-section, $d\Omega$ is the collection solid angle, V is the object plane measurement volume, η is the detector quantum efficiency, and ϕ is an optical collection efficiency which accounts for window losses, spectrometer grating efficiency, filters, etc.

As an example, we consider the vibrational Q-branch spectrum to be given in the next section (figure 8.2.7). For N_2, $d\sigma/d\Omega = \sim 5 \times 10^{-31}$ cm^2/sr for $v = 0$. If we assume that all of the molecules are in level $v = 0$ then $N = \sim 1.6 \times 10^{19}$ cm^{-3}, corresponding to 1 bar pressure and 500 K temperature. If we further assume $E = 0.20$ J/pulse, the V, the object plane cylindrical volume, is 0.5 cm in length × the focused beam cross sectional area, $d\Omega = 0.049$ sr$(f/4)$, $\eta = 0.06$, and $\phi = 0.1$, then substitution into equation (11) yields $S \approx 600$ photoelectrons/laser pulse, or 3.6×10^5 photoelectrons/min (at 10 Hz laser repetition rate). The actual N_2 data in figure 8.2.7 was obtained by integrating for ~ 30 s, whereas the CO data, for which the number density is lower, was integrated for 5 min.

8.2.3 Measurements of vibrational distribution function

As alluded to in the previous section, figure 8.2.7 shows a Q-branch vibrational Raman spectrum obtained in a weakly ionized CO seeded N_2 plasma, which has been created using the CO laser optical pumping technique discussed in section 7.2. The total pressure is 410 torr and the seed fraction is 4%. Each peak represents an unresolved Q-branch Stokes Raman shift from a vibrational level with different vibrational quantum number. The left part of the spectrum shows vibrational levels of CO up to $v = 37$ while the right part shows nitrogen vibrational levels 0–5. The spectrum is obtained using a standard spontaneous Raman scattering instrument, similar to that shown in figure 8.2.6, with Nd:YAG pulse energy of ~ 0.20 J at 532 nm and a 0.25 m grating spectrometer equipped with an ICCD detector. The cylindrical measurement volume had dimensions of ~ 0.5 cm length and 0.01 cm diameter. Since at the employed spectrometer resolution the ICCD detector can capture ~ 10 nm, the spectrum displayed is actually a composite of multiple spectra which were obtained in immediate succession. As mentioned in the previous section, the N_2 signal was averaged for approximately 30 s at a laser repetition rate of 10 Hz whereas the CO spectra were averaged for 5 min. More experimental details can be found in (Lee *et al* 2001).

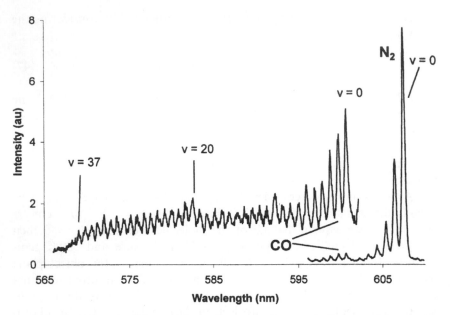

Figure 8.2.7. Q-branch vibrational Raman spectrum from optically pumped (see section 7.2) 4% CO seeded N_2 plasma at 410 torr total pressure.

Figure 8.2.8 shows the corresponding vibrational distribution functions (VDFs) of CO and N_2, which are obtained by dividing the integrated individual Q-branch intensities by the relative v-dependent cross sections given by equation (26). Also included in figure 8.2.8 is the result of master equation modeling, as discussed in section 7.2. In this regard it is important to point out that Raman scattering, unlike infrared emission spectroscopy, provides absolute population fractions for all observed levels, including $v = 0$.

When comparing VDFs of multi-component mixtures, it is sometimes useful to define a 'first level' vibrational temperature by

$$T_v = \frac{1.44(\tilde{\nu}_1 - \tilde{\nu}_0)}{\ln(P_0/P_1)} \tag{48}$$

where $\tilde{\nu}_0$ and $\tilde{\nu}_1$ are the vibrational energies of vibrational levels $v = 0$ and $v = 1$ (in wavenumber units), and P_0 and P_1 are their fractional populations. Predicted and measured first level vibrational temperatures, defined by equation (48), are shown in figure 8.2.8.

As a second example, figure 8.2.9 shows a Q-branch Raman spectrum obtained from an optically pumped mixture similar to that of figure 8.2.7 except that 15 torr of oxygen has been added and the total pressure increased to 755 torr. The CO seed fraction is also increased somewhat, to ~5%. It can be seen that the energy previously accumulated in the vibrational mode of CO has been substantially transferred to O_2, due, as discussed in section

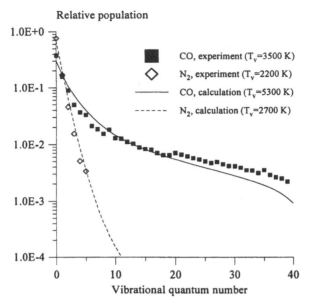

Figure 8.2.8. VDFs extracted from data of figure 8.2.7, along with master equation modeling predictions (see section 7.2).

7.2, to the lower vibrational mode spacing of O_2, relative to CO. The top spectrum shows six vibrational levels ($v = 0$–5) of nitrogen with corresponding first level vibrational temperature $T_v = 2500 \pm 100$ K. The middle spectrum shows nine vibrational levels ($v = 0$–8) of CO with $T_v = 3400 \pm 250$ K. The bottom spectrum contains 13 vibrational levels ($v = 0$–12) of O_2 with $T_v = 3660 \pm 400$ K. The vibrational distributions are, again, non-Boltzmann.

As a final example, figure 8.2.10 shows a pair of pure H_2 rotational Raman spectra obtained from a recent pump/probe study of V–V transfer rates (Ahn 2004). The system was initially prepared, via stimulated Raman pumping, to a state in which about one third of the H_2 molecules in the $v = 0$, $J = 1$ rotation–vibration level were excited to the $v = 1$, $J = 1$ level. The displayed spectrum was obtained 1 µs after application of the pump pulse, and shows that detectable population has been V–V transferred to vibrational levels, in $J = 1$, as high as $v = 6$. As can be seen in figure 8.2.3, ignoring rotation–vibration coupling effects on the value of β_v would result in an overestimate of the $v = 3$, $J = 1$ level population by a factor of approximately two and the $v = 6$ level population by a factor of approximately three.

We end this section by noting that vibrational Q-branch spectra have also been widely utilized for temperature measurements, particularly in combustion environments. In particular, N_2 CARS thermometry is a well established temperature diagnostic which can yield rotational and/or

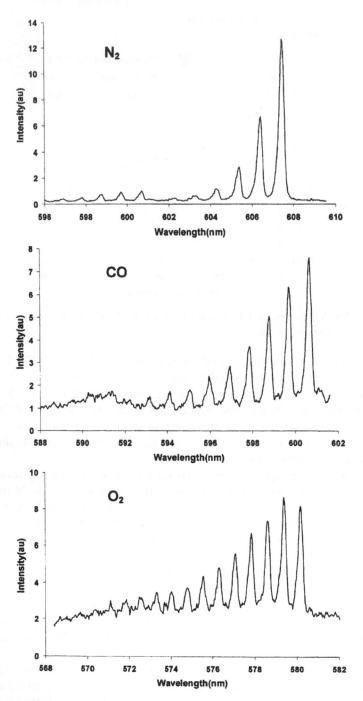

Figure 8.2.9. Q-branch Raman spectrum from optically pumped synthetic air mixture with 2% O_2. Total pressure is 1 bar.

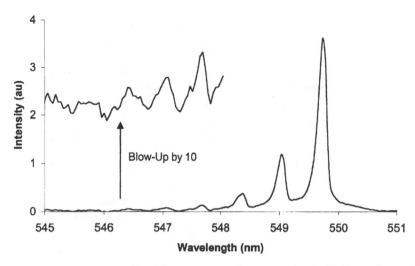

Figure 8.2.10. H_2 pure rotational Raman spectra from pump/probe V–V transfer study. The spectrum corresponds to vibrational distribution 1 μs after initial excitation of ~33% of molecules to $v = 1$. Pressure/temperature is 1 bar/300 K, respectively.

vibrational temperature (Regnier and Taran 1973). As stated previously, rotational temperature determination at pressures of order 1 bar or higher requires incorporation of rotational narrowing phenomena into the spectral model. In principal, high resolution nonlinear Raman 'gain/loss' techniques can be used to obtain complete rotationally resolved spectra (Rahn and Palmer 1986, Lempert *et al* 1984) but the approach is, in general, somewhat impractical as a diagnostic method due to the required slow spectral tuning of a very narrow spectral line width single longitudinal mode (SLM) laser.

8.2.4 Filtered scattering

8.2.4.1 Basic concept

Small wave-number shift scattering diagnostics, such as Rayleigh/Thomson or pure rotational Raman, have traditionally suffered from large interferences due to elastic scattering from window and/or wall surfaces, or, in the case of Thomson scattering in weakly ionized plasmas, from molecular Rayleigh scattering. Such interferences, which are typically orders of magnitude more intense than the desired signal, can completely overwhelm the measurement when performed with traditional instrumentation such as grating spectrometers. In recent years, however, several optical diagnostic techniques based on the use of atomic/molecular vapor filters as narrow bandwidth filters and/or as spectral discriminators have been developed. The basic idea, illustrated in figure 8.2.11, is to utilize a narrow spectral line width laser which is tuned to a strong absorption resonance of the

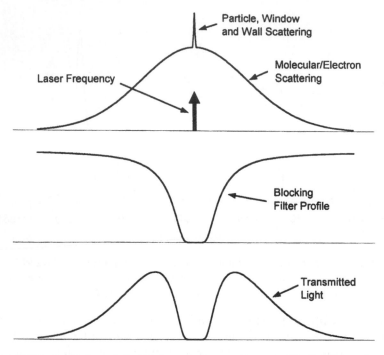

Figure 8.2.11. Basic filtered Rayleigh scattering concept, specifically illustrating thermometry diagnostic.

vapor. If a cell filled with the vapor is then inserted into the path between the scattering volume and the detector, elastic scattering can be attenuated while Doppler shifted and/or broadened scattering can be transmitted. In fact, the use of such vapor filters for Raman scattering dates to near the discovery of the Raman effect itself (Rasetti 1930), although it is only with recent advances in laser technology that their true utility has been realized. In addition to continuous wave (cw) Raman instruments incorporating mercury vapor (Pelletier 1992) and rubidium vapor (Indralingan *et al* 1991, Clops *et al* 2000), the availability of high power, narrow spectral line width pulsed laser sources as common laboratory tools has enabled a wide range of new vapor filter-based scattering techniques. Most of these have utilized iodine vapor, which is particularly convenient because of strong absorption resonances within the tuning range of injection-seeded, pulsed Nd:YAG lasers, as well as the relative ease of filter construction, and availability of high quantum efficiency detectors, both for point measurements and for imaging. A recent special issue of the journal *Measurement Science and Technology* (2001) contains a variety of molecular filter-based diagnostics, including velocity imaging, in which Doppler-shifted Rayleigh or Mie scattering is converted to velocity by determination of the fractional transmission through a vapor filter, and temperature imaging, which is similar to velocity imaging but is based on Doppler broadening of molecular Rayleigh scattering, as opposed to Doppler shift. Other examples include: high spectral resolution light detection and ranging (HSRL) (Shimizu *et al* 1983) and, most recently, Thomson and pure rotational Raman scattering.

A comprehensive discussion of filtered scattering-based diagnostics is beyond the scope of this book. Instead, we will focus on three techniques, ultraviolet filtered Rayleigh scattering, which has been used for temperature field mapping in a glow discharge plasma, filtered rotational Raman scattering, which can give extremely accurate rotational temperature, and filtered Thomson scattering, for which electron density sensitivity as low as order $5 \times 10^{11}\,\mathrm{cm}^{-3}$ and electron temperature sensitivity of $\sim 0.10\,\mathrm{eV}$ has been demonstrated (Bakker and Kroesen 2000).

8.2.4.2 *Filtered Rayleigh scattering temperature diagnostic*

As can be seen from consideration of equations (7), (9), and (11), Rayleigh scattering has the advantage that the signal depends, principally, upon the isotropic part of the polarizability, a_{00}, as opposed to the anisotropic part γ_{00} (rotational Raman scattering), or the polarizability derivatives, a_{10} and/or γ_{10} (vibrational Raman scattering). Since the anisotropic part of the polarizability is typically of order a few percent of the isotropic, and the polarizability derivatives are only $\sim 0.1\%$ of the static polarizability, Rayleigh scattering is inherently more intense, by two to three orders of magnitude, than Raman scattering.

The traditional difficulty with Rayleigh scattering as a general quantitative diagnostic technique has been, as stated above, the interference due to stray scattered light. This has now been largely overcome through the use of vapor filters, which enables the high inherent sensitivity of Rayleigh scattering to be utilized in a variety of traditionally harsh environments. For example, iodine vapor based filtered Rayleigh scattering has recently been utilized for two-dimensional temperature field imaging in hydrogen–air and methane–air flames (Elliott *et al* 2001). Sensitivity was sufficiently high that instantaneous 'single laser shot' images were obtained, in addition to mean field data. The H_2–air data were found to agree with coherent anti-Stokes Raman (CARS) profiles to within $\sim 2\%$.

A particularly novel ultraviolet filtered Rayleigh temperature instrument utilizes the third harmonic output of a single frequency, injection-seeded titanium:sapphire laser at 253.7 nm in combination with an atomic mercury vapor filter (Miles *et al* 2001). This system, while somewhat more complex than Nd:YAG–iodine systems, takes advantage of the sensitivity enhancement realizable by shifting the measurement to shorter wavelengths. In addition to the 4th power of frequency scaling of the scattering cross section, this system takes advantage of the nearly ideal behavior of filters constructed from atomic mercury vapor. In particular, exceedingly high extinction can be

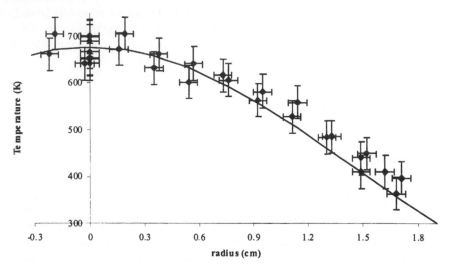

Figure 8.2.12. Radial temperature profile from 50 torr argon glow discharge plasma obtained by ultraviolet filtered Rayleigh scattering (UV FRS).

realized from filters which are quite simple to fabricate. Ultraviolet Rayleigh scattering also offers the somewhat subtle advantage that the Y parameter, for fixed collision mean free path, is smaller than for visible or near infrared radiation (see equation (39)). In other words, the spectrum remains Gaussian at relatively high pressure, up to several atmospheres, depending upon scattering angle, which simplifies its use as a temperature diagnostic.

Figure 8.2.12 shows radial temperature profiles obtained with the ultraviolet-FRS system in an Ar glow discharge plasma at 50 torr pressure (Miles *et al* 2001). The data were obtained as part of a program to study shock-wave propagation through weakly ionized plasmas. Prior to performing the measurements the laser is tuned to the center of an absorption resonance of the filter, as illustrated by the bold arrow in figure 8.2.11. At this fixed laser frequency, since the Y parameter is of order 0.02, the fractional transmission of Doppler broadened scattering is a function of the temperature only. The data in figure 8.2.12 were used to confirm the prediction that the significant temperature gradient, characteristic of the wall-bounded flow discharge plasma, would result in the development of substantial shock curvature during its propagation (Macheret *et al* 2001).

8.2.4.3 *Filtered rotational Raman scattering*

As discussed in the previous section, filtered Rayleigh scattering provides significant temperature measurement potential, even in relatively low density plasmas. However, the temperature accuracy is somewhat limited by the resulting relatively weak dependence of the filter transmission on temperature due to the inherent $\sqrt{T}$ scaling of the Doppler line width (see equation (41)). An alternative approach is based on rotational Raman scattering, which has the advantage that the complete rotational distribution function is determined so that the inherent temperature sensitivity is higher. The disadvantage, as seen in table 8.2.1, is that the cross section for pure rotational Raman scattering is a factor of ~100 weaker than that for Rayleigh scattering, and this lower integrated signal is also distributed amongst the individual populated rotational levels. Fortunately laser and CCD-based detector technology has developed substantially over the past ten years so that high signal-to-noise spectra can be readily obtained. For most practical purposes, however, the measurement is constrained to single spatial points.

Figure 8.2.13 illustrates the enabling capability provided by filtered scattering. Figure 8.2.13(a) is a scattering spectrum from a static cell of 500 torr of nitrogen at room temperature. The spectrum was obtained in an apparatus similar to that illustrated in figure 8.2.6 except that an injection-seeded, single frequency titantium:sapphire laser was employed, in place of the Nd:YAG laser. The output energy was ~50 mJ per pulse at 780 nm and the signal was integrated on a near-infrared sensitive ICCD detector for 1 min. The cylindrical scattering volume has dimensions of ~0.5 cm (length) × 50 μ (diameter). The spectrum appears as a single central component with apparent spectral line width of 0.20 nm FWHM, completely determined by the resolution of the grating spectrometer, since the line width of the laser is ~30 MHz (or ~6 × 10^{-5} nm). Figure 8.2.13(b) shows the spectrum obtained under identical conditions except that a 5 cm path length rubidium vapor filter, heated to 320 °C, has been inserted into the detection path. Note that the intensity axis for the filtered spectrum is the same as that for the unfiltered, so that the relative intensities are directly comparable. It can be seen that the peak rotational Raman intensity is a factor of ~800 weaker than the original elastic and Rayleigh scattering. While difficult to determine directly from the figure, it has been shown that the peak residual fractional intensity of the central components is ~6 × 10^{-6} so that the peak rotational Raman intensity is now ~200 times greater (rather than ~800 times weaker) than the peak central component (Lee and Lempert 2002).

Figure 8.2.14 shows a spectrum (Stokes side only) similar to that of figure 8.2.13(b) except that it was obtained in a CO laser optically pumped N_2/CO mixture at ~1 bar total pressure, as described in section 7.2. Also shown is a least squares fit to a simple sum of pressure broadened transitions spectral model, including convolution with the instrumental spectral response function. The inferred rotational temperature is 355 K with 2σ statistical uncertainty of ±7 K (Lee 2003).

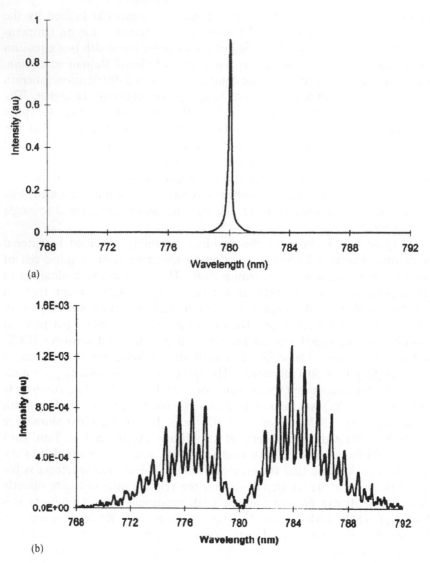

Figure 8.2.13. Illustration of rubidium vapor filtered pure rotational Raman spectra. (a) From the static cell of pure N_2 at 500 torr and 300 K, obtained without filtering; (b) is identical except that a vapor filter was employed.

8.2.4.4   Filtered Thomson scattering

Thomson scattering is a well known technique for determination of spatially resolved electron density and electron temperature (Hutchinson 1990, Evans and Katzenstein 1969). Similar to Rayleigh scattering, Thomson scattering results from laser-induced polarization of charged species, principally, at

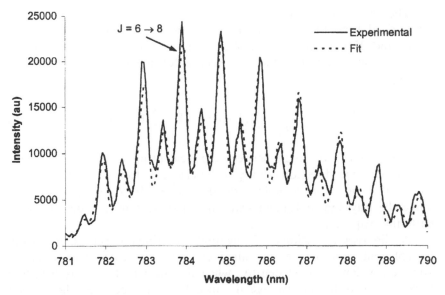

Figure 8.2.14. Filtered pure rotational Raman spectrum of optically pumped N_2/CO mixture at 1 bar pressure and least squares spectral fit. Inferred temperature is 355 ± 7 K.

least in weakly ionized plasmas, from free electrons. While the cross section for free electron scattering is approximately one hundred times greater than that for Rayleigh scattering of common air species, the typically low free electron number density in weakly ionized plasmas ($\sim 10^{10}$–10^{13} cm^{-3}) results in extremely low scattering signals. Further aggravating this problem is the fact that the electron temperature of molecular plasmas is typically quite low (a few eV). The corresponding relatively low Doppler broadened linewidth complicates the use of grating-based instruments for spectral rejection of stray scattering, although the reader is referred to a recently reported triple grating instrument incorporating a physical central component blocking mask in place of the normal slit separating the first two gratings (Noguchi *et al* 2001).

Recently vapor filter-based Thomson scattering instruments, similar to the filtered Rayleigh and Raman instruments discussed above, have been developed and demonstrated in weakly ionized plasmas. The first reported system utilized a commercial Nd:YAG pumped dye laser in combination with a sodium vapor filter at ~ 580 nm (Bakker *et al* 2000) and, shortly thereafter, independently developed rubidium vapor systems were also reported (Miles *et al* 2001, Lee 2003). Compared to rubidium-based systems, sodium systems have the advantage that the laser is relatively simple and is readily available commercially. The sodium vapor filter, however, is somewhat more complex to fabricate.

The theory of Thomson scattering is well known and will only be summarized here. More detail can be found in Hutchinson (2000) and Evans and Katzenstein (1969). We begin with the expression for the Thomson scattering differential cross section for linearly polarized photons given by

$$\frac{\mathrm{d}\sigma}{\mathrm{d}\Omega} = \left(\frac{e^2}{4\pi\varepsilon_0 m_e c_0^2}\right)^2 (1 - \cos^2\theta_z) = r_e^2(1 - \cos^2\theta_z) \tag{49}$$

where θ_z, again, is the angle between the incident light polarization vector and the detection direction, and r_e is the classical electron radius equal to 2.818×10^{-15} m. For $\theta_z = 90°$, $\mathrm{d}\sigma/\mathrm{d}\Omega = r_e^2 = 7.94 \times 10^{-26}$ cm^2/sr and, unlike the Rayleigh or Raman scattering cross section, is independent of scattering frequency.

As discussed in section 8.2.2, the total scattering intensity is, in general, the coherent sum of the individual contributions from each electron. However, at low electron density, when the incident laser wavelength is short compared to the average distance between electrons, the photon 'sees' the moving electrons as individual particles, randomly distributed in the plasma. In this case, the phase from each scattering 'particle', as seen at the detector, is completely uncorrelated from that of all other particles and the total scattering intensity is just the summation of intensities from each electron. This is called incoherent Thomson scattering. However, if the average distance is short compared to the laser wavelength, the phase differences are no longer random and individual scattering intensities add in a coherent manner. Analogous to equation (39) we define a parameter α as

$$\alpha = \frac{1}{\Delta k \lambda_D} \tag{50}$$

where Δk, again, is the magnitude of the scattering wave vector (see equation (36)), and λ_D is the Debye length given by

$$\lambda_D = \left(\frac{\varepsilon_0 k_B T_e}{e^2 n_e}\right)^{1/2} \cong 743 \left(\frac{T(\mathrm{eV})}{n_e(\mathrm{cm}^{-3})}\right)^{1/2} (\mathrm{cm}) \tag{51}$$

where n_e and T_e are the electron number density and temperature, respectively. When $\alpha \ll 1$, the effects of Coulomb interactions on the scattering spectrum are negligible since the scattering length scale, $1/\Delta k$, is much smaller than the Debye length, which is the characteristic length scale over which significant net charge separation can exist. In this case, the scattering is completely incoherent, provided that the electrons are randomly distributed in space. In the limit of $\alpha \to 0$, the scattering line shape is Gaussian, corresponding to a Maxwellian velocity distribution of electrons, with γ, the half width at half maximum, for $\lambda = 780$ nm, given by

$$\gamma(\text{nm}) = \frac{\lambda_{780}}{c} \sqrt{\frac{2\ln(2)kT_e}{m_e}} \sin\frac{\theta}{2} = 2.57\sqrt{T_e}\sin\frac{\theta}{2} \tag{52}$$

where T_e is in eV units and θ is the scattering angle.

For $\alpha > 1$, the incident wave interacts with the Debye-shielded charges and the scattered spectrum depends on the collective behavior of groups of charges. The Gaussian shape becomes distorted and a distinct symmetric side-band peaks arise. Physically this coherent Thomson scattering is analogous to Rayleigh/Brillouin scattering, mentioned in section 8.2.1, except that the scattering originates from correlated fluctuations in charge density due to what are known as 'ion-acoustic' waves. When the correlation length for these fluctuations exceeds the reciprocal of the scattering wave vector, the side-bands begin to appear.

A full treatment of coherent scattering is beyond the scope of this book. However, a common approximation to the scattering spectrum is that given by Salpeter (1960), which, strictly speaking, applies when T_e/T_i, the ratio of electron to ion temperatures, is approximately 1. In this case the total scattering is the sum of components originating from correlated electron motion and that from correlated ion motion, given by

$$S(k,\omega) = \frac{2\pi^{1/2}}{ka}\Gamma_\alpha(x_e) + \frac{2\pi^{1/2}}{kb}Z\left(\frac{\alpha^2}{1+\alpha^2}\right)^2 \Gamma_\beta(x_i) \tag{53}$$

where $x_e = \omega/ka$, $a = (2kT_e/m_e)^{1/2}$, $x_i = \omega/kb$, $b = (2kT_i/m_i)^{1/2}$, and $\Gamma_\alpha(x_e)$, $\Gamma_\beta(x_i)$ are identical line shape functions which are plotted in figure 8.2.15 as a function of the non-dimensional parameter x. Note, however, that for $T_e/T_i \approx 1$ the ion average velocity b is much smaller than a. This implies that, in frequency units, the ion scattering contribution is located much closer to the unshifted laser frequency than the electron contribution.

The significance of figure 8.2.15 is that, for α of order 1 or greater, electron density can be determined from the shape of the Thomson scattering spectrum, without the need for absolute scattering intensity calibration. Figure 8.2.16 illustrates an example filtered Thomson spectrum of an atmospheric pressure argon lamp, obtained with a rubidium vapor–titanium: sapphire system very similar to that used to obtain the filtered rotational Raman spectra in the previous section, except that a scanning monochrometer and photomultiplier tube detector were used rather than an ICCD (Miles 2001). This measurement is complicated by the limited optical access, which was solved by employing a $180°$ backscattering geometry. As can be seen by comparison of the shape of the experimental and fit spectra with those given in figure 8.2.15, the measurement clearly corresponds to the onset of the incoherent scattering regime. The inferred electron density

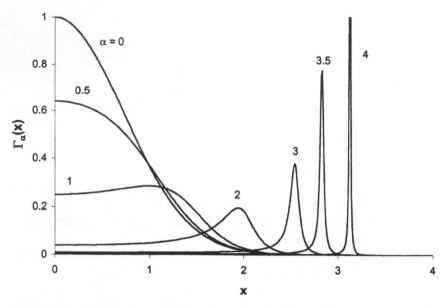

Figure 8.2.15. Saltpeter approximation to Thomson scattering profile as a function of the non-dimensional parameter α.

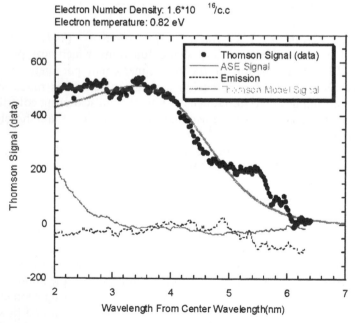

Figure 8.2.16. Rubidium vapor filtered Thomson scattering spectrum from atmospheric pressure argon lamp.

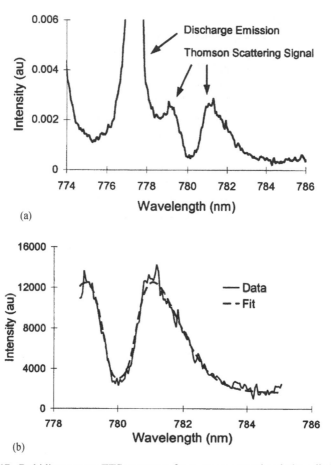

Figure 8.2.17. Rubidium vapor FTS spectrum from argon constricted glow discharge. (a) Spectrum illustrates scattering signal relative to spontaneous emission detected despite utilization of gated ICCD detector. (b) Least squares fit to incoherent scattering model.

and temperature are $1.61 \times 10^{16}\,\mathrm{cm}^{-3}$ and 0.82 eV, respectfully, which from equations (36), (50), and (51) corresponds to a value of $\alpha \approx 1.2$.

As an example of filtered Thomson scattering at lower electron density (Lee 2002), figure 8.2.17 shows a spectrum from a dc argon 'constricted' glow discharge, obtained using the same instrument employed for the filtered rotational Raman spectra in figures 8.2.13 and 8.2.14. The argon pressure is 30 torr and the discharge current is 100 mA. The constricted glow is ~1–2 mm in diameter and is stabilized by incorporation of a $500\,\Omega$ current limiting ballast resistor in series with the dc discharge. Figure 8.2.17(a) shows the Thomson scattering signal superimposed upon the relatively large argon spontaneous emission, which is many orders of magnitude more intense despite employing a gated ICCD camera. Figure 8.2.17(b) is a least squares fit of the experimental spectrum in figure 8.2.17(b) to a simple incoherent Thomson scattering model. The absolute intensity is calibrated using a N_2 pure rotational Raman spectrum similar to that of figure 8.2.13, taking advantage of the accurately known differential rotational Raman cross section of $5.4 \times 10^{-30}\,\mathrm{cm}^2/\mathrm{sr}$ for the $J = 6 \to 8$ transition of nitrogen at 488.0 nm (Penney *et al* 1974) (see table 8.2.1). From this procedure the inferred values of electron number density and temperature are $(2.0 \times 10^{13}) \pm (6 \times 10^{11})\,\mathrm{cm}^{-3}$ and $0.67 \pm 0.03\,\mathrm{eV}$, respectfully, corresponding to α equal to ~ 0.06. We note, however, that the inferred value of electron temperature seems somewhat low for this plasma and may reflect systematic error associated with spatial non-uniformity and/or temporal unsteadiness, which was observed by the authors.

As noted previously, sensitivity of $\sim 10^{11}\,\mathrm{cm}^{-3}$ has been reported for the conceptually similar sodium vapor filter system (Bakker and Kroesen 2000).

8.2.5 Conclusions

Recent years have seen very significant advances in laser and detector technology which has allowed spontaneous scattering-based methods to evolve into routine diagnostic tools for molecular, non-equilibrium plasmas. The emergence of novel diagnostic approaches, such as those based on narrow pass band atomic and molecular vapor filters, has enabled several orders of magnitude improvement in sensitivity, so that the techniques can now be applied to weakly ionized plasmas, a feat which was previously considered all but impossible. Clearly the future looks bright for the use of elastic and inelastic laser light scattering techniques for plasma diagnostics.

References

Ahn T and Lempert W 2004 to be published
Asawaroengchai C and Rosenblatt G M 1980 *J. Chem. Phys.* **72** 2664
Bakker L P and Kroesen G M 2000 *J. Appl. Phys.* **88** 3899
Bakker L P, Freriks J M, deGroog F J and Kroesen G M W 2000 *Rev. Sci. Instrum.* **71** 2007
Berne B J and Pecora R 1976 *Dynamic Light Scattering* (New York: Wiley)
Bonamy L, Bonamy J, Robert D, Lavorel B, Saint-Loup R, Chaux J, Santos J and Berger H 1988 *J. Chem. Phys.* **89** 5568
Chu B 1991 *Laser Light Scattering—Basic Principles and Practice* 2nd edition (Boston: Academic Press)
Clops R, Fink M, Varghese P L and Young D 2000 *Appl. Spectroscopy* **54** 1391
Demtroder W 1998 *Laser Spectroscopy* 2nd edition (Berlin: Springer)
Dicke R H 1953 *Phys. Rev.* **89** 472

Drake M 1982 *Optics Lett.* **7** 440

Eckbreth, A C 1996 *Laser Diagnostics for Combustion Temperature and Species* 2nd edition (Amsterdam: Gordon and Breach)

Elliott G S, Glumac N and Carter C D 2001 *Measurement Science and Technology* **12** 452

Evans D K and Katzenstein J 1969 *Rep. Progress in Phys.* **32** 207

Galatry L 1961 *Phys. Rev.* **122** 1281

Gallas J A 1980 *Phys. Rev. A* **21** 1829

Gresillon D, Gemaux G, Cabrit B and Bonnet J P 1990 *European J. Mechanics B* **9** 415

Hall R J, Verdieck J F and Eckbreth A C 1979 *Optics Commun.* **35** 69

Hutchinson I H 1990 *Principles of Plasma Diagnostics* (Cambridge: Cambridge University Press)

Indralingan R, Simeonsson J B, Petrucci G A, Smith B W and Winefordner J D W 1991 *Analytical Chem.* **64** 964

Lee W 2003 'Development of Raman and Thomson scattering diagnostics for study of energy transfer in nonequilibrium molecular plasmas' PhD thesis, Ohio State University, June

Lee W and Lempert W R 2002 *AIAA J.* **40** 2504

Lee W, Adamovich I V and Lempert W R 2001 *J. Chem. Phys.* **114** 1178

Lempert W R, Rosasco G J and Hurst W S 1984 *J. Chem. Phys.* **81** 4241

Long D A 2002 *The Raman Effect* (London: Wiley)

Macheret S O, Ionikh Y Z, Chernysheva N V, Yalin A P, Martinelli L and Miles R B 2001 *Phys. of Fluids* **13** 2693

Measurement Science and Technology 2001 **12**(4)

Miles R B, Lempert W R and Forkey J N 2001a *Measurement Science and Technology* **12**

Miles R B, Yalin A P, Tang Zhen, Zaidi S H and Forkey J N 2001b *Measurement Science and Technology* **12** 442

Noguchi Y, Matsuoka A, Bowden M D, Uchino K and Muraoka K 2001 *Japanese J. Appl. Phys.* **40** 326

Ornstein L S and Zernike F 1926 *Phys. Z.* **27** 761

Pelletier M J 1992 *Appl. Spectroscopy* **46** 395

Penney C M, St Peters R L and Lapp M 1974 *J. Opt. Soc. America* **64** 712

Rahn L A and Palmer R E 1986 *J. Opt. Soc. America B* **3** 1165

Rasetti F 1930 *Nuovo Cimento* **7** 261

Regnier P R and Taran J P E 1973 *Appl. Phys. Letters* **23** 240

Rosasco G J, Lempert W, Hurst W S and Fein A 1983 in *Spectral Line Shapes*, vol 2 (Berlin: Walter de Gruyter) p 635

Salpeter E E 1960 *Phys. Rev.* **120** 1528

Shardanand and Rao A D P 1977 'Absolute Rayleigh scattering cross sections of gases and freons of stratospheric interest in the visible and ultraviolet regions', NASA Technical Note, TN D-8442

Shimizu, H, Lee, S A and She, C Y 1983 *Appl. Optics* **22** 1373

Vaughan, J M 1989 *The Fabry–Pérot Interferometer* [Adam Hilger Series on Optics] (Bristol: Institute of Physics Publishing)

Weber A 1979 *Raman Spectroscopy of Gases and Liquids* (Berlin: Springer)

Wolniewicz L 1966 *J. Chem. Phys.* **45** 515

Yariv A 1975 *Quantum Electrodynamics* 2nd edition (New York: Wiley)

8.3   Electron Density Measurements by Millimeter Wave Interferometry

8.3.1   Introduction

Interferometry is primarily a non-perturbing plasma density diagnostic technique through the interaction of electromagnetic waves with plasma. It measures the refractive and dissipative properties of the plasma which in turn depend on the plasma properties including the plasma density and the collision frequency. The interferometer works on the Mach–Zehnder principle (Hutchinson 2002) in which the plasma is in one arm of the two-beam interferometer. Phase and amplitude differences between the two arms are the measures of the electron plasma density and the effective collision frequency. However, the specific interferometric measurement technique depends on the choice of the wave frequency (ω), relative to the plasma (ω_p) and the effective collision (ν_{eff}) frequencies. If the probing wave frequency is much greater than the plasma frequency and collision frequencies ($\omega \gg \omega_p \gg \nu_{eff}$), the electromagnetic wave suffers almost little or no attenuation as it travels through the plasma. Therefore, only phase change data are needed for a density measurement. In this case a linear relationship exists between the line-average plasma density and the phase shift for a radially uniform plasma column (Wharton 1965). Also, if $\omega_p \geq \omega$, in low collisionality plasmas the ordinary wave mode (O-mode, $\mathbf{E} \| \mathbf{B_O}$) is in cutoff (Wharton 1965, Stix 1992) and interferometry data cannot be obtained. On the other hand, for high-pressure discharges, where the collision frequency can be higher than both the plasma and the millimeter wave frequency ($\nu_{eff} \geq \omega \approx \omega_p$), an electromagnetic wave propagating through the plasma arm undergoes phase change as well as strong attenuation. The wave attenuation is caused by the presence of high collisionality. In this situation the plasma density has a complex dependence on phase change as well as on amplitude change and, therefore, the correct evaluation of plasma density can only be obtained if both phase-change and amplitude-change data are used (Akhtar *et al* 2003). In addition, we experimentally observe O-mode transmission for $\omega_p \geq \omega$ as predicted by the theory (Wharton 1965).

For atmospheric pressure air pressure discharges, the diagnostic technique will depend on the choice of the probing wave frequency. Choice of a higher wave frequency such as a CO_2 laser ($\omega = 1.78 \times 10^{14}$ Hz) satisfies the condition ($\omega \gg \omega_p \gg \nu_{eff}$), where plasma density is linearly related to phase change. However, the contribution of neutral particle density to the refractive index and to the phase change which can be neglected for microwave diagnostics becomes very important for infrared diagnostics (Podgornyi 1971). A technique to infer phase contributions of the electrons and those of heavy particles is described in detail in the section 8.4.

In this section we present a measurement and analysis technique where both amplitude and phase change data are used simultaneously to uniquely determine both plasma density and effective collision frequency. This treatment does not limit the application of interferometry to the relative values of collision frequency and hence can be used for measurements at both low gas pressure ($\omega \gg \omega_p \gg \nu_{eff}$) and high gas pressure ($\nu_{eff} \geq \omega, \omega_p$). The analysis does not assume, *ab initio*, a particular value of the collision frequency; rather, it calculates the collision frequency along with density using the phase and amplitude change data.

8.3.2   Electromagnetic wave propagation in plasma

In order to calculate the refractive and dissipative properties of a collisional plasma, we consider an electromagnetic wave propagating in an infinite, uniform, collisional plasma. In this model, electron motion is induced by the electromagnetic wave and the ions are assumed to form a stationary background. The equation of motion for plasma electrons in the absence of a magnetic field is written as (Wharton 1965)

$$m\ddot{\mathbf{r}} = -e\mathbf{E} - \nu_{eff}\, m\dot{\mathbf{r}} \tag{1}$$

where $\mathbf{r}$ is the electron displacement vector, $\mathbf{E}$ is the electromagnetic field and ν_{eff} is the effective collision frequency for momentum transfer. If the electric field varies as $\exp(j\omega t)$, the displacement vector $\mathbf{r}$ is given as

$$\mathbf{r} = \frac{e\mathbf{E}}{m\omega(\omega - j\nu_{eff})}. \tag{2}$$

Using the current density equation $\mathbf{J} = -en_e\mathbf{v} = \boldsymbol{\sigma} \cdot \mathbf{E}$, the complex conductivity σ is given as

$$\sigma \equiv \sigma_r + j\sigma_i = -\frac{n_e e\dot{\mathbf{r}}}{E} = \frac{n_e e^2}{m}\frac{(\nu_{eff} - j\omega)}{(\omega^2 + \nu_{eff}^2)}. \tag{3}$$

The complex relative dielectric constant for a linear medium is given by (Wharton 1965)

$$\frac{\varepsilon}{\varepsilon_0} = \kappa = \kappa_r + j\kappa_i = 1 - j\frac{\sigma}{\varepsilon_0\omega} = 1 - \frac{\omega_p^2}{\omega^2 + \nu_{eff}^2}\left(1 + j\frac{\nu_{eff}}{\omega}\right) \tag{4}$$

where ω_p is the plasma frequency and ε_0 is the free space permittivity. The complex refractive index (n) and the complex propagation (γ) constants are

$$n = \frac{c}{v} = \mu_r - j\chi = \kappa^{1/2}, \qquad \gamma = \alpha + j\beta = \frac{\omega}{c}(j\mu) = \frac{\omega}{c}\sqrt{\kappa} \tag{5}$$

where ω/c is the phase velocity, $\alpha = \chi\omega/c$ is the attenuation constant in Np/m and $\beta = \mu_r\omega/c$ is the phase constant in rad/m. The solution for the plane wave phase and attenuation constants in the plasma yields

$$\beta_p = \frac{\omega}{c}\left\{\frac{1}{2}\left(1 - \frac{\omega_p^2}{\omega^2 + \nu_{eff}^2}\right)\right.$$
$$\left. + \frac{1}{2}\left[\left(1 - \frac{\omega_p^2}{\omega^2 + \nu_{eff}^2}\right)^2 + \left(\frac{\omega_p^2}{\omega^2 + \nu_{eff}^2}\frac{\nu_{eff}}{\omega}\right)^2\right]^{1/2}\right\}^{1/2} \qquad (6)$$

$$\alpha_p = \frac{\omega}{c}\left\{-\frac{1}{2}\left(1 - \frac{\omega_p^2}{\omega^2 + \nu_{eff}^2}\right)\right.$$
$$\left. + \frac{1}{2}\left[\left(1 - \frac{\omega_p^2}{\omega^2 + \nu_{eff}^2}\right)^2 + \left(\frac{\omega_p^2}{\omega^2 + \nu_{eff}^2}\frac{\nu_{eff}}{\omega}\right)^2\right]^{1/2}\right\}^{1/2}. \qquad (7)$$

Assuming a plasma slab of uniform average density profile, the total change in phase and amplitude for interferometric signal are given as

$$\Delta\phi = \int_0^d (\beta_0 - \beta_p)\,dr, \qquad \Delta A = \int_0^d (\alpha_0 - \alpha_p)\,dr. \qquad (8)$$

Here β_0 and α_0 are the free space values and β_p and α_p are the plasma values. Simultaneous solution of plasma density and ν_{eff} are obtained from experimentally measured $\Delta\phi$ and ΔA values.

The relative frequency condition $\omega \gg \omega_p \gg \nu_{eff}$ is usually satisfied in low pressure discharges ($p \leq 10\,\text{mtorr}$), where most interferometry operates. In this limiting case the phase constant and attenuation constant are given as

$$\beta_p = \frac{\omega}{c}\left(1 - \frac{\omega_p^2}{\omega^2}\right)^{1/2} \approx \frac{\omega}{c}\left(1 - \frac{\omega_p^2}{2\omega^2}\right), \qquad \alpha_p = \frac{\nu_{eff}\omega_p^2}{2\omega^2 c}\left(1 - \frac{\omega_p^2}{\omega^2}\right)^{-1/2}. \qquad (9)$$

Therefore, in such low pressure discharges the electromagnetic wave suffers almost little or no attenuation as it travels through the plasma and the phase difference between the two arms with the plasma present to that without the plasma is a measure of the plasma density. The plasma density can be expressed in this limit for a uniform density profile using equation (7) as

$$n_e = \left(\frac{4\pi c\varepsilon_0 m_e}{e^2}\right)\frac{f\,\Delta\phi}{d} = 2.073\frac{f\,\Delta\phi}{d}\,\text{cm}^{-3}. \qquad (10)$$

Here the phase change is in degrees, the diameter in centimeters and wave frequency is in s^{-1}. It can be seen that a linear relationship exists between the line-average plasma density and the phase shift for a radially uniform plasma column. Also if $\omega_p \geq \omega$ and $\omega \gg \nu_{eff}$, the ordinary wave mode (O-mode) is cut off and interferometry data cannot be obtained as shown in the normalized plot (figure 8.3.1) of $\beta_p c/\omega$ and $\alpha_p c/\omega$ versus ω_p/ω using equations (6) and (7). Also shown is the propagation of wave even when $\omega_p > \omega$, when the collision frequency is equal to the wave frequency. It

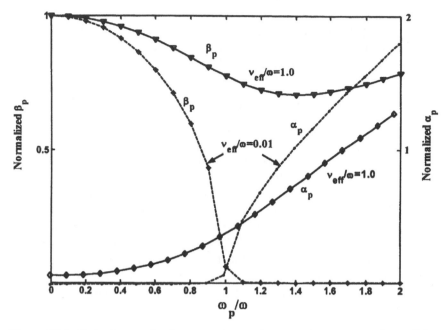

Figure 8.3.1. Plot of normalized propagation and attenuation constant for collision frequency relative to the wave frequency.

should be noted here that this approximation depends on the values of wave frequency relative to the collision and plasma frequencies. As described in section 8.4, this approximation for highly collisional atmospheric pressure air plasmas is obtained by choosing a CO_2 laser wave frequency of $\omega = 1.78 \times 10^{14}$ Hz.

For highly collisional plasmas at high gas pressures where the condition $\nu_{eff} \gg \omega \geq \omega_p$ is satisfied, the effect of collisions can be accounted for through the phase function (Laroussi 1999). However, these approximations are valid only for limiting cases. The propagation phase constant and the corresponding density terms in this limiting case for a uniform plasma profile are

$$\beta_p = \frac{\omega}{c}\left\{\frac{1}{2} + \frac{1}{2}\left[1 + \frac{\omega_p^4}{\omega^2 \nu_{eff}^2}\right]^{1/2}\right\}^{1/2} \approx \frac{\omega}{c}\left[1 + \frac{\omega_p^4}{8\omega^2 \nu_{eff}^2}\right] \tag{11}$$

$$n_e = 38.6\nu_{eff}\left(\frac{f\,\Delta\phi}{d}\right)^{1/2} = 5.09 \times 10^{-5}\nu_{eff}\left(\frac{f\,\Delta\phi}{d}\right)^{1/2} \text{ cm}^{-3}. \tag{12}$$

However, for moderately to highly collisional plasma where relative frequency condition $\nu_{eff} \gg \omega \geq \omega_p$ is satisfied, wave undergoes a phase change as well as amplitude change. Therefore, it is instructive to use both phase and amplitude change data from equations (8) and (9) simultaneously to solve for both plasma density and effective collision frequency accurately. This treatment does not limit the application of interferometry to the relative values of the collision frequency and, hence, can be used for both low pressure discharges ($\omega \gg \omega_p \gg \nu_{eff}$) and high pressure discharges ($\nu_{eff} \gg \omega, \omega_p$).

8.3.3 Plasma density determination

A 105 GHz quadrature mm wave interferometry system (QBY-1A10UW, Quinstar Technology) is used to measure the plasma density and the effective collision frequency of an rf produced plasma. The rf source is a 10 kW solid-state unit (Comdel Inc.) with variable duty cycle (90–10%), variable pulse repetition frequency (100 Hz–1 kHz) and very fast (μs) turn-on/off time and a 25 kW unit (Comdel Inc.). The rf power is coupled through a helical antenna that excites the $m = 0$ TE (transverse electric) mode very efficiently using a capacitive matching network. The helical antenna is a five-turn coil of $\frac{1}{4}$ inch (6.35 mm) copper tube wound tightly over the 5 cm diameter Pyrex plasma chamber. The coil is 10.0 cm long axially and has a 6 cm internal diameter. Figure 8.3.2 shows the schematic of the experimental system.

The interferometer works by using an I-Q (in-phase and quadrature phase) mixer to determine the phase and amplitude change of the 105 GHz mm wave signal going through the plasma. The two outputs are transferred to the computer through an oscilloscope with a GPIB interface and stored

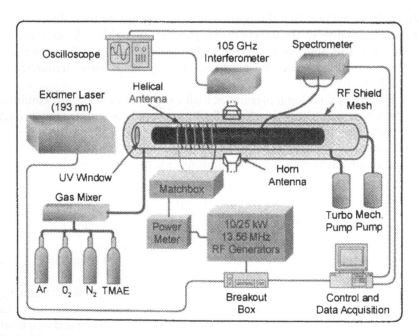

Figure 8.3.2. Schematic of the laser-initiated and rf sustained plasma experiment.

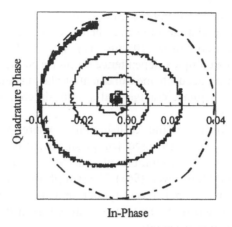

Figure 8.3.3. Interferometer trace showing a nearly cut-off density of $9 \times 10^{13} \, \text{cm}^{-3}$ in 10 torr argon plasma at 1.0 kW using a five turn helical antenna. The vacuum circle is represented by the dotted line.

using a Labview program. In order to shield rf-sensitive Gunn and detector diodes, the interferometer assembly is housed in a Faraday shielded conducting box. In addition, cables with very high shielding ($\geq$90 dB, Times Microwave Systems) have been used to reduce the noise level on the interferometer signal. The interferometric trace shown in figure 8.3.3 illustrates that electromagnetic wave attenuates significantly for high-density plasma even at low neutral pressures.

The results for the plasma density computation using equations (9), (10) and (12) are presented in table 8.3.1, for typical phase change and collision frequency data in an rf-produced air plasma at 10, 100, and 760 torr maintained at different rf power. From the experimentally determined phase and attenuation data, the plasma density and effective collision

Table 8.3.1. Air plasma density using a 105 GHz ($\omega = 6.59 \times 10^{11} \, \text{s}^{-1}$) interferometer for 5 cm diameter tube.

Air pressure (torr)	$\Delta\phi$ (degrees)	Attenuation (dB)	ν_{eff} (s^{-1})	n_e (cm^{-3}) Using phase and amplitude data, equation (8)	n_e (cm^{-3}) Collisionless limit, equation (10)	n_e (cm^{-3}) Highly collisional plasma, equation (12)
10	200	0.94	2.1×10^{10}	8.5×10^{12}	8.7×10^{12}	2.1×10^{12}
100	239.2	16.31	2.91×10^{11}	1.2×10^{13}	1.03×10^{13}	3.3×10^{13}
760	16.7	5.8	1.6×10^{12}	4.5×10^{12}	7.25×10^{11}	4.5×10^{13}
760	25.1	14.79	2.5×10^{12}	1.7×10^{13}	1.1×10^{12}	9.25×10^{13}
760	50.4	35.3	2.8×10^{12}	4.5×10^{13}	2.2×10^{12}	1.46×10^{14}

frequency are determined using the analysis presented above. The collision frequency and phase change data are then used to calculate the limiting plasma density using equations (10) and (12). The result clearly shows that plasma density has a complex dependence on phase change and attenuation data and, therefore, an accurate measurement of plasma density must involve measurement of both phase change and amplitude change of the probing electromagnetic wave.

At high gas pressure and collisionality, where optical diagnostics including the Stark effect are used for plasma density and temperature, characterizations require a minimum plasma density ($n_e \geq 10^{14}$–$10^{15}/\text{cm}^3$) (Griem 1997, Lochte-Holtgreven 1968). This simple diagnostic is particularly valuable for collisional air plasmas of moderate densities ($n_e < 10^{14}\,\text{cm}^{-3}$) at higher gas pressures where probe and optical emission diagnostics are not suitable for density measurements.

References

Akhtar K, Scharer J, Tysk S and Kho E 2003 'Plasma interferometry at high pressures' *Rev. Sci. Instrum.* **74** 996

Griem H R 1997 in *Principles of Plasma Spectroscopy* (Cambridge: Cambridge University Press) p 258

Hutchinson I H 2002 *Principles of Plasma Diagnostics* (Cambridge: Cambridge University Press) p 114

Laroussi M 1999 *Int. J. Infrared and Millimeter Waves* **20** 1501

Lochte-Holtgreven W 1968 in *Plasma Diagnostics* ed. Lochte-Holtgreven W (Amsterdam: North-Holland) p 186

Podgornyi I M 1971 in *Topics in Plasma Diagnostics* (New York: Plenum Press) p 141

Stix T H 1992 in *Waves in Plasmas* (New York: AIP Press, Springer)

Wharton C B 1965 in *Plasma Diagnostic Techniques* ed. Huddlestone R H and Leonard S L (Academic Press, New York) p 477

8.4   Electron Density Measurement by Infrared Heterodyne Interferometry

8.4.1   Introduction

The electron density, n_e, determines to a large extent the refractive index of a plasma. The complex refractive index in turn determines the phase shift and the attenuation of electromagnetic waves of frequency ω passing through the plasma. Phase shift and attenuation can be measured by using interferometric techniques, and consequently allow us to obtain information on the electron density.

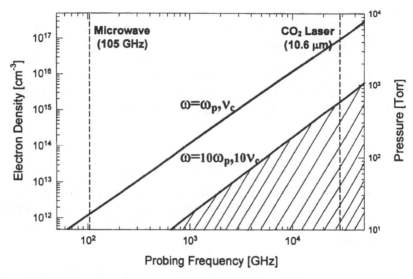

Figure 8.4.1. Probe frequency range (hashed area) for which an air plasma at room temperature can be considered as transparent ($\omega \gg \omega_p$) and lossless ($\omega \gg \nu_c$). The collision frequency, ν_c, for air was obtained from Raizer (1991).

The index of refraction of a plasma as shown in the next section is a nonlinear function of probe wave frequency, ω, the plasma frequency, ω_p, which contains information on the electron density, and the collision frequency, ν. However, if the probing frequency is large compared to the collision frequency ($\omega \gg \nu$) the attenuation of the probing beam can be neglected. If, in addition, the probing frequency is large compared to the plasma frequency ($\omega \gg \omega_p$) the relation between phase shift and electron density becomes linear, and the electron density can be obtained directly from the phase shift. Figure 8.4.1 shows the frequency-dependent range of electron densities and gas pressures for which the two conditions hold.

According to these conditions (probing frequency at least an order of magnitude higher than plasma and collision frequency, respectively), a microwave interferometer operating at a frequency of 105 GHz allows us to measure electron densities up to 2×10^{12} cm^{-3} in a plasma with a heavy particle density equivalent to 20 torr or less at room temperature. For high-pressure plasmas, such as atmospheric pressure plasmas, the heavy particle density in plasmas and consequently the electron collision frequency increases. Furthermore, the plasma frequency in a high-pressure discharge may exceed the probing frequency due to higher electron densities. To use an interferometric technique in this case, and still stay in the range where the plasma can be considered collisionless and transparent, requires an increase in probing frequency, e.g. using a laser in the infrared range. For the interferometer operating at 10.6 μm described in this chapter, an electron density of up to 10^{17} cm^{-3} can be measured and up to a heavy particle density equivalent to 6 atm at room temperature, still satisfying the condition ($\omega \gg \omega_p, \nu$). Changes of the heavy particle density in the plasma (caused by heating) contribute also to the phase shift of the probing beam. While this contribution can be neglected compared to the contribution of electrons for microwaves, it has a considerable contribution in the infrared and must be taken into account. A technique how this can be accomplished is outlined in section 8.5.3.

8.4.2   Index of refraction

The index of refraction, N, for an optical thin plasma with a low degree of ionization, contains contributions from electrons, ions and neutrals. The refractive index can be obtained from the dispersion relation for a monochromatic electromagnetic wave with an electric field $\mathbf{E} = \mathbf{E}_0 \exp \mathrm{i}(\mathbf{kr} - \omega t)$ in a conducting medium:

$$k^2 = \varepsilon_0 \varepsilon_r \mu_0 \mu_r \omega^2 + \mathrm{i} \mu_0 \mu_r \omega \sigma = \frac{\omega^2}{c^2} \mu_r \varepsilon_r \left(1 + \frac{\mathrm{i} \sigma(\omega)}{\omega \varepsilon_0 \varepsilon_r} \right) \tag{1}$$

with k being the wave number, ω the angular frequency, ε_0 and ε_r the absolute and relative permittivity, μ_0 and μ_r, the absolute and relative permeability, and c is the speed of light in vacuum. The conductivity σ depends on the wave frequency ω and on the plasma frequency ω_p and is given by the equation (Greiner 1986)

$$\sigma(\omega) = \frac{\omega_p^2 \varepsilon_0}{(\nu - \mathrm{i}\omega)} = \frac{e^2 n_e}{m_e(\nu - \mathrm{i}\omega)} \tag{2}$$

where ν is the electron collision frequency, and e and m_e the electron charge and mass, respectively. Substituting the conductivity in equation (1) by $\sigma(\omega)$ (equation (2)) the dispersion relation yields

$$k^2 = \frac{\omega^2}{c^2} \mu_r \varepsilon_r \left(1 + \frac{\mathrm{i} e^2 n_e}{\omega \varepsilon_0 \varepsilon_r m_e (\nu - \mathrm{i}\omega)} \right). \tag{3}$$

If the probe wave frequency ω is much higher than the collision frequency ν in the plasma, equation (3) simplifies to

$$k^2 = \frac{\omega^2}{c^2} N^2 = \frac{\omega^2}{c^2} \mu_r \varepsilon_r \left(1 - \frac{e^2 n_e}{\omega^2 \varepsilon_0 \varepsilon_r m_e} \right) = \frac{\omega^2}{c^2} \mu_r \varepsilon_r \left(1 - \frac{\omega_p^2}{\omega^2 \varepsilon_r} \right) \tag{4}$$

where N is the refractive index. The contribution of the heavy particles to the refractive index can be expressed by the susceptibilities κ ($\varepsilon = 1 + \kappa$) of the particles. The dispersion relation then reads

$$k^2 = \frac{\omega^2}{c^2} N^2 = \frac{\omega^2}{c^2} \mu_r \left(1 + \kappa_{\mathrm{ion}} + \kappa_{\mathrm{neutral}} - \frac{\omega_p^2}{\omega^2} \right) \tag{5}$$

where κ_{ion} and κ_{neutral} are the contributions from ions and neutral particles, respectively. They are small compared to 1. With $\mu_r = 1$, and for ω_p/ω small compared to 1, the square root of the expression for N^2 can be written as

$$N = 1 - \frac{\omega_p^2}{2\omega^2} + \frac{\kappa_{\text{ion}}}{2} + \frac{\kappa_{\text{neutral}}}{2}. \tag{6}$$

The contribution to the refractive index from neutrals and ions, respectively, can be described by (Duschin and Pawlitschenko 1973)

$$\frac{\kappa_{\text{ion}}}{2} = N_{\text{ion}} - 1 = \left(A_{\text{ion}} + \frac{B_{\text{ion}}}{\lambda^2}\right)\frac{n_{\text{ion}}}{n_{\text{ion0}}} \tag{7a}$$

$$\frac{\kappa_{\text{neutral}}}{2} = N_{\text{neutral}} - 1 = \left(A_{\text{neutral}} + \frac{B_{\text{neutral}}}{\lambda^2}\right)\frac{n_{\text{neutral}}}{n_{\text{neutral0}}} \tag{7b}$$

where A and B are specific values for a specie, n is the density, n_0 is the density under standard temperature and pressure (STP) condition ($T = 273\,\text{K}$, $p = 1\,\text{bar}$) and λ is the wavelength. If A and B for ions are not available, the values for neutrals can be used as a reasonably good approximation. In the following, the notation for ions and neutrals are combined into one expression for heavy particles. A and B for selected species are listed in Duschin and Pawlitschenko (1973). Substituting all terms in equation (6) yields

$$N = 1 - \frac{e^2}{2(c^2 m_e \varepsilon_0 4\pi^2)}\lambda^2 n_e + \left(A + \frac{B}{\lambda^2}\right)\frac{n_{\text{heavy}}}{n_{\text{heavy0}}} \tag{8}$$

where n_{heavy} and n_{heavy0} are the heavy particle density (ions and neutrals) at a given pressure and temperature and under STP condition ($T = 273\,\text{K}$, $p = 1\,\text{bar}$), respectively. A and B are constants. The first term describes the contribution of the electrons; the second term that of neutrals and ions.

Interferometry can be used to measure changes in the refractive index and consequently provides information on changes of particle densities. The phase shift $\Delta\Phi$ (rad) of a laser beam with a wavelength λ passing through a non-homogeneous plasma of length L caused by changes in the electron density and heavy particle density is

$$\Delta\Phi = \frac{2\pi}{\lambda}\int_0^L \Delta N(l)\,dl \tag{9a}$$

$$\Delta\Phi = \Delta\Phi_{\text{el}} + \Delta\Phi_{\text{heavy}}$$

$$= -\frac{e^2}{4\pi\varepsilon_0 m_e c^2}\lambda\int_0^L \Delta n_e(l)\,dl + \left(A + \frac{B}{\lambda^2}\right)\frac{2\pi}{\lambda n_{\text{heavy0}}}\int_0^L \Delta n_{\text{heavy}}(l)\,dl. \tag{9b}$$

8.4.3 The infrared heterodyne interferometer

In order to measure the phase shift and consequently the refractive index, a Mach–Zehnder heterodyne interferometer operating at a wavelength of $\lambda = 10.6\,\mu m$ (CO_2 laser) has been used (figure 8.4.2). The laser beam is separated into two equal intensity beams by means of a beam splitter (ZnSe). One beam passes through the plasma. In order to provide the required spatial resolution it has been focused into the plasma, with a waist width of less than $50\,\mu m$. The plasma can be shifted transverse to the beam direction, allowing us to scan the plasma column. The second beam bypasses the plasma and is frequency shifted by means of a 40 MHz acousto-optic modulator. The beat frequency of 40 MHz, obtained by superimposing both beams, is recorded by an infrared detector, which operates at room temperature, and the signal is compared to the driver signal of the acousto-optic modulator. The phase shift of the laser beam is transferred to the high-frequency signal and is recorded by a phase detector, which converts the phase shift into a voltage signal. The resolution of the interferometer is about $0.01°$.

The characteristic of the phase detector is sinusoidal. In order to calibrate the phase detector, the interferometer is tuned (manually) to a phase of $\Phi = \pi/2$. The corresponding voltage $V_{\pi/2}$ at the phase detector is recorded. For measurements, the interferometer is tuned to a phase of $\Phi = 0$. This is the preferred operation point Φ_0 of the interferometer. The relation between measured phase detector signal $V(\Phi)$ and the phase shift $\Delta\Phi$ (rad) is given by the equation

$$\Delta\Phi = \Phi - \Phi_0 = \arcsin\frac{V(\Phi)}{V_{\pi/2}} - \Phi_0. \qquad (10)$$

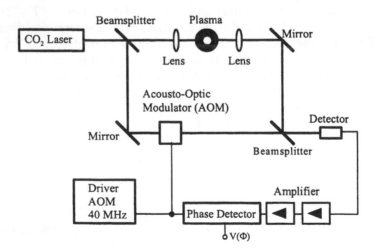

Figure 8.4.2. Schematics of the infrared heterodyne interferometer.

The correlation between phase detector signal and particle densities is obtained by substituting the phase shift $\Delta\Phi$ in equation (9).

8.4.4 Application to atmospheric pressure air microplasmas

The conditions for the validity of equation (10) are that (a) the electron collision frequency is small compared to the probing wave frequency and (b) the plasma frequency is small compared to the probing wave frequency. The electron collision frequency for air, which is the gas of choice in our experiments, at atmospheric pressure and 2000 K is 4.4×10^{11} Hz (Raizer 1991). For a probing frequency of $\omega = 1.78 \times 10^{14}$ Hz (CO_2 laser), the expression for N (equation (8)) can be used to get information on the electron density in air plasmas with heavy particle densities up to 1.4×10^{20} cm^{-3} ($\nu_c/\omega < 0.1$). The plasma frequency is determined by the electron density. Assuming that ω_p/ω needs to be less than 0.1 allows us to use equation (9) to determine the index of refraction in ionized gases with electron densities up to 10^{17} cm^{-3}.

Since interferometry provides the total phase shift ($\Delta\Phi$) of a plasma, the contributions of electrons ($\Delta\Phi_{el}$) and heavy particles ($\Delta\Phi_{heavy}$) need to be separated. In general, separation of electron and heavy particle contribution can be achieved by using a second wavelength, since the contribution of electrons and heavy particles to the phase shift are frequency dependent (equation (8)). This technique provides information on both the electron density and the heavy particle density. However, under certain conditions it is possible to separate the contribution due to electrons (in which we are interested) using a single-wavelength interferometer. By using light sources which provide long-wavelength radiation, the contribution of the heavy particles to the refractive index can be disregarded compared to the contribution of electrons. These requirements are met for conditions of gas pressure of several tens of torr (condition (a)) and an electron density corresponding to a plasma frequency exceeding the probing frequency by a factor of 10 (condition (b)), using microwave interferometry. In this case, the measured phase shift provides the electron density without the need to use a separate diagnostic technique.

However, in order to probe microplasmas with characteristic dimensions in the 100 μm range, light sources with wavelengths on the order of, or less than, the characteristic dimensions need to be used, in order to provide sufficient spatial resolution. This condition requires, for microplasma studies, the use of infrared light sources. For infrared illumination and with electron densities on the order of 10^{13} cm^{-3} in an atmospheric pressure gas, the contribution to the phase shift caused by changes of the heavy particles may exceed the one for electrons by more than one order of magnitude. In this case, the different response time for electrons and heavy particles, when a pulsed voltage is applied to the plasma, can be used to separate the phase shift signals $\Delta\Phi_{el}$

and $\Delta\Phi_{\text{heavy}}$. As discussed in the following, using a microplasma in atmospheric air as an example, this method, which is based on the difference in time constants, can be used in diagnosing dc plasmas (Leipold *et al* 2000) and pulsed plasmas (Leipold *et al* 2002).

8.4.5   Measurement of the electron density in dc plasmas

The plasma that was studied is a cylindrically symmetric atmospheric pressure air glow discharge column with a diameter of less than 1 mm and a column length of 2 mm (figure 8.4.3) (Stark and Schoenbach 1999). The spatial resolution requires a wavelength in the infrared range. For this application a CO_2 laser with an operation wavelength of $\lambda = 10.6\,\mu m$ has been chosen. According to Raizer (1991), the collision frequency of an atmospheric pressure air plasma for a heavy particle density of $3.6 \times 10^{18}\,\text{cm}^{-3}$ is $\nu = 4.4 \times 10^{11}\,\text{Hz}$. Since this frequency is small compared to the laser frequency of $\omega = 1.78 \times 10^{14}\,\text{Hz}$, the simplified equation (4) can be used for the evaluation of the refractive index. An electron density of $10^{17}\,\text{cm}^{-3}$ corresponds to a plasma frequency of $1.78 \times 10^{13}\,\text{Hz}$. Consequently, the ratio ω_p^2/ω^2 is approximately 1%.

The electrode system consists of a microhollow cathode electrode system (MHCD) and an additional (third) electrode with a variable distance from the MHCD. The electrode configuration and the plasma are shown in figure 8.4.3. The MHCD geometry consists of two plane-parallel electrodes with a centered hole in each electrode. The electrodes are made of 100 μm thick molybdenum foils, and the cathode and anode hole size of the plasma cathode is also 100 μm. The dielectric between the electrodes is

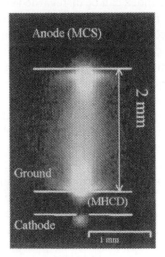

Figure 8.4.3. Atmospheric pressure air discharge.

alumina (Al_2O_3, 96% purity) of 250 µm thickness. The anode of the micro-hollow cathode geometry is connected to ground. The third electrode, placed at a distance of 2 mm in front of the plasma cathode, is also made of molybdenum and biased positively. The MHdc sustained glow discharge (MCS) is operated in dc mode, optional with a superimposed high voltage pulse (1600 V) of 10 ns duration. The time between pulses was on the order of 100 ms. The discharge dc current was limited by means of a ballast resistor of 300 kΩ to 16 mA. The measurements were performed in air at a pressure of 1000 mbar and a humidity of 30%.

For a wavelength of $\lambda = 10.6$ µm and in air plasma ($A = 2.871 \times 10^{-4}$, $B = 1.63 \times 10^{-18}$ m² (Duschin and Pawlitschenko 1973), the ratio of the contributions to the phase shift due to electrons and heavy particles is given by

$$\frac{\Delta\Phi_{el}}{\Delta\Phi_{heavy}} = 4.5 \times 10^3 \frac{\Delta n_e}{\Delta n_{heavy}}. \tag{11}$$

The change of the heavy particle density Δn_{heavy} after switching the discharge on is estimated using the ideal gas law. The gas temperature varies between room temperature when the plasma is off and a temperature of 2000 K when the plasma is on (Leipold *et al* 2000). For a pressure of 1 atm, $\Delta n_{heavy} = 2.3 \times 10^{19}$ cm^{-3} at room temperature. With electron densities at ignition of 10^{13} to 10^{15} cm^{-3}, the ratio $\Delta\Phi_{el}/\Delta\Phi_{heavy}$ varies between 0.002 and 0.2. This means that the major phase shift during the switching transient is still determined by the change of the heavy particle density. In spite of this difficulty, the phase signal can be separated due to the different response times for electrons and heavy particles to rapid changes in voltage (ignition of the plasma) (Leipold *et al* 2000). Figure 8.4.3 shows the phase shift signal through the center of the discharge. The fast rising part of the phase shift signal is assumed to be due to the change of the electron density; the slowly rising part is assumed to be due to the change of the heavy particle density caused by gas heating.

The electron density decays to the dc value after breakdown, while the gas heats up causing a change in the heavy particle density. At ignition, the electron density provides a significant fraction of the total phase shift $d\Phi_{el}/(d\Phi_{el} + d\Phi_{heavy})$ (at $t = 5$ ms in figure 8.4.4). When the plasma approaches steady state conditions, the fraction decreases to approximately 0.2% (at $t > 10$ ms in figure 8.4.4). Therefore, the total amplitude of the phase shift for $t > 10$ ms after ignition can be considered the change in the heavy particle density with an error of less than 1%. In order to obtain information on the electron density during this steady-state phase, where the electron density is identical to that for a dc plasma, the plasma was operated in a pulsed mode with time intervals between pulses continuously decreasing. The electron density can be measured during the re-ignition phase of each pulse. By reducing the time between pulses towards zero, the

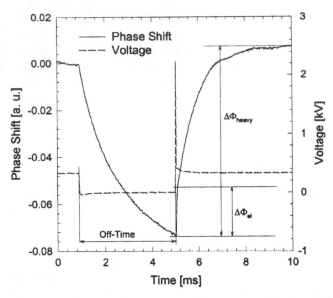

Figure 8.4.4. Phase shift signal through the center of the discharge for an off-time of 4 ms.

electron densities measured for the re-ignition transients approach that of the dc plasma.

This method has been applied to the discharge in atmospheric pressure air. The discharge was operated in the dc mode and was switched off for a specific time (off-time) (figure 8.4.4). The electron density in the center of the discharge at ignition calculated from the phase shift $\Delta\Phi_{el}$ and the change of the heavy particle density in the center of the discharge calculated from the phase shift $\Delta\Phi_{heavy}$ were recorded and plotted versus various off-times (figure 8.4.5). Shortening the off-time allowed us to approach the dc mode (off-time $= 0$). The extrapolation of the curve in figure 8.4.5 towards zero change in heavy particle density provides the electron density in the dc case.

In order to obtain absolute electron densities, the radial profile of the electron density needs to be known. In side-on measurements the plasma was shifted in the z direction (insert, figure 8.4.6) through the laser beam, providing the spatial phase shift distribution. In order to obtain the radial phase shift distribution, a parabolic radial profile was assumed and the corresponding spatial profile was calculated. The parameters for the parabolic profile were varied for best fit of measured and calculated relative spatial profiles. The results are shown in figure 8.4.6. This relative radial profile was used for calculating the electron density from the spatially resolved phase shift. The same procedure was applied for the relative radial heavy particle density profile. The gas temperature was obtained by

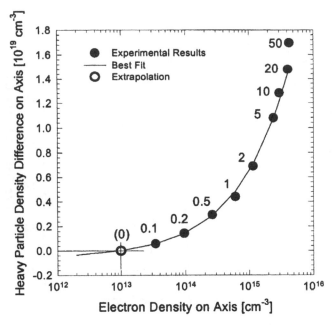

Figure 8.4.5. Electron density in the center of the plasma column after breakdown versus the change in heavy particle density. The numbers along the curve indicate the corresponding off-times.

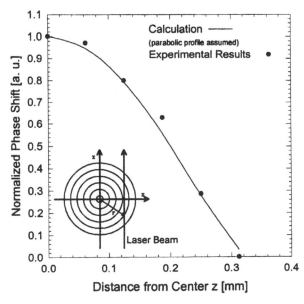

Figure 8.4.6. Spatial distribution of the measured and computed relative phase shift $\Delta\Phi_{el}$.

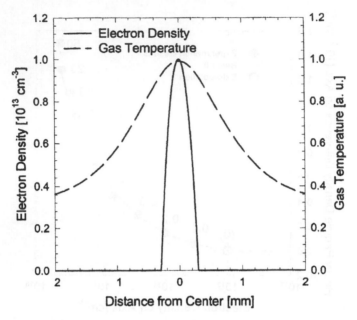

Figure 8.4.7. Radial distribution of electron density and relative gas temperature distribution.

using the information on the heavy particle density and assuming that the ideal gas law holds. The electron density distribution and the relative radial temperature profile are shown in figure 8.4.7.

8.4.5 Measurement of the electron density in pulsed operation

A strong increase in electron density can be obtained by applying a voltage pulse with a duration on the order of, or less than, the dielectric relaxation time of the electrons to a dc plasma. The application of such a pulsed voltage causes a shift in the electron energy distribution function to higher energies, with negligible gas heating, thus reducing the probability for glow-to-arc transition. The shift in electron energy causes a temporary increase of the ionization rate and consequently an increase in electron density (Stark and Schoenbach 2001).

The same atmospheric pressure air plasma, which was studied in the dc mode, was pulsed with a 10 ns pulse of 1.6 kV amplitude (superimposed to the dc voltage), and the electron density was measured by means of infrared heterodyne interferometry. The change of the electron density caused by the high voltage pulse can, in this case, be obtained directly from the phase shift signal. The spatially resolved relative phase shift $\Delta\Phi(z)$ for various times after pulse application is shown in figure 8.4.8. The spatial profiles could be fit to a Gaussian profile with a width of

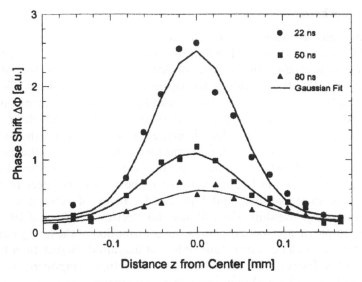

Figure 8.4.8. Spatially resolved relative phase shift $\Delta\Phi(z)$ for various times after pulse application.

$\sigma = 0.056$ mm. This means that the radial profile is also Gaussian with the same width. Figure 8.4.8 shows the temporally resolved electron density in the center of the discharge obtained from the measured phase shift signal. The voltage pulse causes an increase in electron density to at least 2.8×10^{15} cm^{-3}. The electron density decays hyperbolically to its dc value. The temporal resolution of this diagnostic method, with the currently used experimental set-up, is 20 ns.

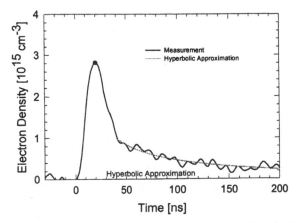

Figure 8.4.9. Temporally resolved electron density in the center ($z = 0$) of the discharge.

8.4.6 Conclusions

Interferometry is widely used for measurements of the electron density in partially ionized plasmas (Hutchinson 1991). The choice of the probing frequency is determined by the range of electron density, by the gas pressure, and the desired spatial resolution. Increasing the probing frequency allows us to increase the range of electron densities and gas pressures, utilizing only the measured phase shift of the probe radiation passing through the plasma. Also, the spatial resolution, which is limited to dimensions on the order of the probe radiation wavelength, is improved by increasing the probing frequency, The drawback of moving from e.g. the microwave into the infrared or even visible frequency range is the increasing effect of heavy particles, atom, molecules, and ions on the index of refraction, which determines the phase shift. For instance, for electron densities of 10^{13} cm^{-3} in an atmospheric pressure air plasma the contribution of the heavy particles to the measured phase shift is four orders of magnitude higher than that of the electrons. Extracting information on the electron component therefore requires phase shift measurements at two wavelengths.

A method which does not require a second probing radiation source but still allows us to obtain electron density distributions and gas temperature distributions in atmospheric pressure air plasmas with a spatial resolution of better than 100 μm (using a CO_2 laser) makes use of the different time constant for ionizing and for heating of the weakly ionized plasma (Leipold *et al* 2000). This concept is not only applicable to pulsed plasmas, but also to dc plasmas. In the second case, the dc electron density is obtained by a process where the dc discharge is turned on and off with increasingly smaller intervals between the on-state. Extrapolating the electron densities to the case of diminishing time between off- and on-states allows us to obtain the steady-state (dc) value of the electron density and the gas temperature. Although the diagnostic procedure for obtaining electron densities with this method in weakly ionized atmospheric pressure air or other high-pressure plasmas is rather complex, the high spatial resolution makes this diagnostic technique attractive for the study of microdischarges or micro-structures in large-volume high-pressure discharges.

References

Duschin L A and Pawlitschenko O S 1973 *Plasmadiagnostik mit Lasern* (Berlin: Akademie-Verlag) p 8

Greiner W 1986 *Theoretische Physik* (Frankfurt am Main: Verlag Harri Deutsch)

Hutchinson I H 1991 *Principles of plasma diagnostics* (Cambridge: Cambridge University Press)

Leipold F, Mohamed A-A and Schoenbach K H 2001 'Electron temperature measurements in pulsed atmospheric pressure plasmas' *Bull. APS GEC* **46**(6) 22

Leipold F, Mohamed A-A H and Schoenbach K H 2002 'High electron density, atmospheric pressure air glow discharges' Conf. Record, 25th Int. Power Modulator Symp. and 2002 High Voltage Workshop, Hollywood, CA, June, p 130

Leipold F, Stark R H, El-Habachi A and Schoenbach K H 2000 'Electron density measurements in an atmospheric pressure air plasma by means of IR heterodyne interferometry' *J. Phys. D: Appl. Phys.* **33** 2268

Raizer Y P 1991 *Gas Discharge Physics* 2nd edition (Berlin: Springer)

Stark R H and Schoenbach K H 1999 'Direct current glow discharges in atmospheric air' *Appl. Phys. Lett.* **74** 3770

Stark R H and Schoenbach K H 2001 'Electron heating in atmospheric pressure glow discharges' *J. Appl. Phys.* **89** 3568

8.5 Plasma Emission Spectroscopy in Atmospheric Pressure Air Plasmas

8.5.1 Temperature measurement

Atmospheric pressure air plasmas are often thought to be in local thermodynamic equilibrium (LTE) owing to fast interspecies collisional exchange at high pressure. This assumption cannot be relied upon, particularly with respect to optical diagnostics. Velocity gradients in flowing plasmas, or elevated electron temperatures created by electrical discharges, or both can result in significant departures from chemical and thermal equilibrium. This section reviews diagnostic techniques based on optical emission spectroscopy (OES) that we have found useful for making temperature measurements in atmospheric pressure air plasmas, under conditions ranging from thermal and chemical equilibrium to thermochemical non-equilibrium.

8.5.1.1 Temperature measurements in LTE air plasmas

For plasmas in LTE, a single temperature characterizes all internal energy modes (vibrational, rotational, and electronic). This temperature can be determined from the absolute intensity of any atomic or molecular feature, or from Boltzmann plots of vibrational or rotational population distributions. Such measurements were made (Laux 1993) at 1 cm downstream of the exit of a 50 kW, rf (4 MHz), inductively coupled plasma torch operating with atmospheric pressure air (Figure 8.5.1). Because the plasma flows at relatively low velocity (10 m/s) in the field-free region between the induction coil and the nozzle exit where the measurements are made, all chemical reactions equilibrate well before reaching the nozzle exit and therefore the plasma is close to LTE. The experimental set-up for OES measurements, shown in Figure 8.5.2, comprises a 0.75 m monochromator fitted with

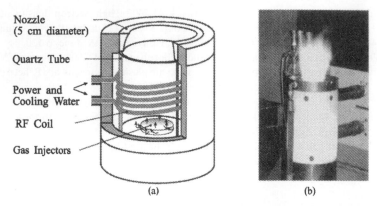

(a)   (b)

Figure 8.5.1. (a) Schematic of 50 kW plasma torch head. The distance from the top of the induction coil to the nozzle exit is about 10 cm. (b) Torch head and LTE air plasma plume.

either a 2000×800 pixel CCD camera (SPEX TE2000) or a photomultiplier tube (Hamamatsu R1104). Absolute calibrations of the spectral intensities between 200 and 800 nm are made with radiance standards including a calibrated tungsten strip lamp for the range 350–800 nm and a 1 kW dc argon arc-jet in the range 200–400 nm. The optical train is constructed with spherical mirrors or MgF_2 lenses to minimize chromatic aberrations in the ultraviolet. Long-pass filters inserted in the optical train eliminate second- and higher-order light. Figure 8.5.3 shows the radial temperature profiles obtained after the emission measurements are inverted with the

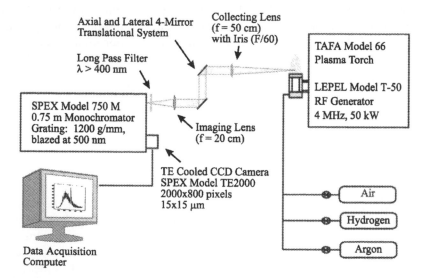

Figure 8.5.2. Experimental set-up for emission diagnostics.

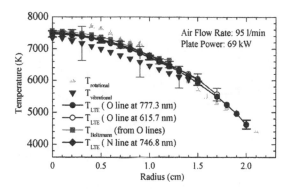

Figure 8.5.3. Measured electronic, vibrational, and rotational temperature profiles in LTE air.

help of the Abel transform. The 'LTE' and Boltzmann temperatures shown in figure 8.5.3 are based on the absolute and relative intensities, respectively, of various atomic lines of oxygen and nitrogen. The rotational temperature profiles are obtained from measurements of the NOγ (0,1) band shape, using the technique proposed by Gomès *et al* (1992). The vibrational temperature profile is measured from the relative intensities of the (0,0) and (2,1) bandheads of N_2^+ B-X (first negative band system) at 391.4 nm and 356.4 nm, respectively. As can be seen from figure 8.5.3, the measured vibrational, rotational, and electronic temperature profiles are to within experimental uncertainty in good agreement with one another, as expected because the plasma is close to LTE.

8.5.1.2 *Temperature measurements in non-equilibrium air plasmas*

In non-equilibrium plasmas, the techniques described in the foregoing paragraph may not provide reliable information about the gas temperature because the population distribution of internal energy states tends to depart from Boltzmann distributions at the gas temperature. This behavior is especially the case for the electronic and vibrational population distributions, but the rotational populations tend to follow a Boltzmann distribution at the gas temperature owing to fast rotational relaxation at atmospheric pressure. Thus, the gas temperature can often be inferred from the intensity distribution of rotational lines. Various transitions of O_2, N_2, N_2^+, and NO (dry air) and OH (humid air) can be used, depending on the level of plasma excitation. To illustrate the variety of emission bands available for OES in air plasmas, figure 8.5.4 shows the ultraviolet emission spectra of equilibrium, atmospheric pressure air with a water vapor mole fraction of 1.3%, for temperatures in the range 3000–8000 K. Below about 5000 K, bands of NO, OH, and O_2 dominate the spectrum. The second positive

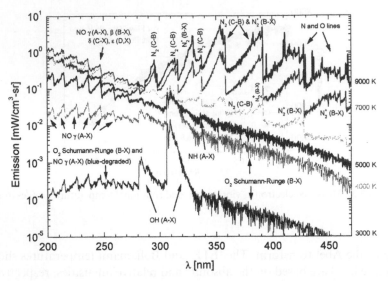

Figure 8.5.4. Ultraviolet emission spectra of LTE air at atmospheric pressure with 1.3% mole fraction of water vapor. These simulations were performed with SPECAIR, using a trapezoidal instrumental broadening function of base 0.66 nm and top 0.22 nm.

band system of N_2 (C-B), the first negative band system of N_2^+ (B-X), and atomic lines of O and N appear at higher temperatures. Emission features similar to those of figure 8.5.4 can also be observed in low-temperature non-equilibrium air plasmas such as those produced by electrical discharges.

All spectral simulations presented here have been made with the SPECAIR code (Laux 2002), which was developed on the basis of the NonE-Quilibrium Air Radiation code (NEQAIR) of Park (1985). The current version of SPECAIR models 37 molecular transitions of NO, N_2, N_2^+, O_2, CN, OH, NH, C_2, and CO, as well as atomic lines of N, O, and C. The model provides accurate simulations of the absolute spectral emission and absorption of air from 80 nm to 5.5 μm. As an illustration of the capabilities of the model, figure 8.5.5 shows a comparison between *absolute* intensity emission spectra measured in LTE air and SPECAIR predictions. The plasma conditions are those corresponding to the temperature profile of figure 8.5.3, with a peak centerline temperature of approximately 7500 K. As can be seen in figure 8.5.5, the model is able to reproduce the line positions and intensities of the experimental spectra.

We now turn our attention to techniques best suited for quantitative temperature measurements in discharges. The rotational temperature can be measured from N_2 C-B rotational lines. At even higher temperatures or higher electric field excitation, many molecular transitions appear in the spectrum and an accurate spectroscopic model is required to extract individual lines of a particular system. For these conditions, we recently proposed a

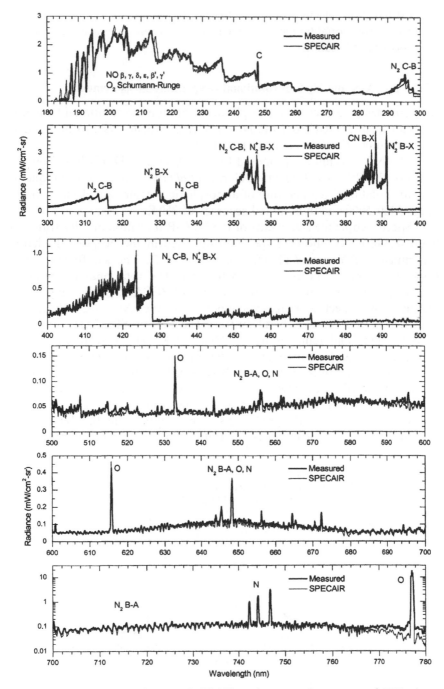

Figure 8.5.5. Comparison between SPECAIR and measured spectrum of LTE air at ~7500 K.

method based on selected rotational lines of N_2^+ B-X (Laux *et al* 2001). The N_2 and N_2^+ rotational temperature measurement techniques are described in the following subsections.

8.5.2 NO A-X and N_2 C-B rotational temperature measurements

At higher temperatures or higher plasma excitation the rotational temperature can be measured from the NO A-X (NO γ-band system) or N_2 C-B (N_2 second positive band system) transitions.

The NO γ technique proposed by Gomès *et al* (1992) is based on the width of the NO A-X (0,1) band. Gomès *et al* (1992) used the technique to measure rotational temperature of atmospheric pressure air plasmas in the range 3000–5000 K with a quoted accuracy of 250 K.

Spectroscopic measurements of the N_2 C-B transition are illustrated in figure 8.5.6, which shows a spectrum obtained in the dc glow discharge experiments of Yu *et al* (2002) (see figure 8.5.7). The slit function is a trapezoid of base 0.66 nm and top 0.22 nm. The rotational temperature was determined by fitting the spectrum with SPECAIR in the range 260–382 nm. This spectral range corresponds to the $\Delta\nu = -2$ vibrational sequence of the N_2 C-B band system. The best-fit SPECAIR spectrum yields a rotational temperature of 2200 ± 50 K. The best-fit vibrational temperature, based on the relative intensities of the (0,2), (1,3), (2,4), and (3,5) vibrational bands of the N_2 C-B system, is 3400 ± 50 K. It should be noted that the vibrational temperature of the C state of N_2 is not necessarily the same as the vibrational temperature of the ground state of N_2.

Figure 8.5.8 shows the predicted spectral width of the (0,2) band of N_2 C-B at 20 and 40% of the peak intensity, as a function of the rotational

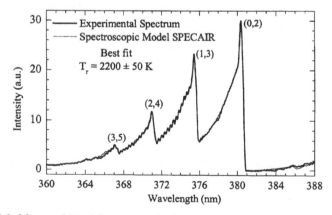

Figure 8.5.6. Measured N_2 C-B spectrum in the atmospheric pressure air glow discharge (conditions of figure 8.5.7). SPECAIR best-fit provides a rotational temperature of 2200 ± 50 K.

Figure 8.5.7. DC glow discharge experiments in air at 2200 K and 1 atm. The glow discharge is created by applying a dc electric field (1.4 kV/cm 200 mA) in fast flowing (~450 m/s) low-temperature (2200 K) atmospheric pressure air. Interelectrode distance = 3.5 cm. The measured electron number density in the bright central region of the discharge is approximately 10^{12} cm^{-3}.

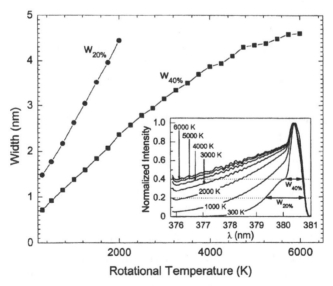

Figure 8.5.8. Spectral widths of the N_2 C-B (0,2) band at 20 and 40% of the peak's height. These calculations were made with SPECAIR assuming a trapezoidal slit function of base 0.66 and top 0.22 nm. The inset shows the (0,2) band spectra at various rotational temperatures, normalized to the intensity of the peak at 380.4 nm.

temperature. These simulations were made with SPECAIR, assuming a trapezoidal slit function of base 0.66 and top 0.22 nm. The width curves provide a quick way to estimate the rotational temperature if a full spectral model is not available.

8.5.3 N_2^+ B-X rotational temperature measurements

At higher excitation levels, the NO γ and N_2 second positive band systems suffer from increasing overlap by transitions from higher NO states (NO δ, ε), and by the O_2 Schumann–Runge, CN violet, and N_2^+ first negative band systems. The N_2^+ first negative band system (B-X transition) can be used to measure the rotational temperature, provided that an accurate spectroscopic model is available to extract N_2^+ lines from the encroaching lines of CN and N_2 that emit in the same spectral range. The modeling is complicated by perturbations that affect the positions, intensities, and splittings of the N_2^+ lines. Recent spectroscopic analyses by Michaud *et al* (2000) have provided accurate spectroscopic constants, incorporated in SPECAIR, that enable the precise identification of high rotational lines of N_2^+ B-X up to rotational quantum numbers of about 100. We showed in (Laux *et al* 2001) that the group of rotational lines R(70) and P(97) at 375.95 nm is well isolated from lines of other transitions, and that the intensity of these two lines relative to the bandhead of the N_2^+ B-X (0,0) band at 391.55 nm is a very sensitive function of the rotational temperature. This technique was successfully applied to rotational temperature measurements in a non-equilibrium recombining nitrogen/argon plasma (Laux *et al* 2001). The rotational temperature was measured to be 4850 ± 100 K, an accuracy far superior to that of other N_2^+ rotational temperature measurement techniques (see for instance the review by Scott *et al* (1998)).

8.5.4   Measurements of electron number density by optical emission spectroscopy

In plasmas with electron number densities greater than $\sim 5 \times 10^{13}$ cm^{-3}, spatially and temporally resolved electron number density measurements can be made by emission spectroscopy from the lineshape of the Balmer β transition (4-2) of atomic hydrogen at 486.1 nm. This technique requires the addition to the plasma of a small amount (typically 1 or 2% mole fraction) of hydrogen, which may come either from dissociated water vapor in humid air or from premixing H_2 into the air stream. For detection by emission spectroscopy, the population of the $n = 4$ electronic state of atomic hydrogen must be high enough for the H_β line to be distinguishable from underlying air plasma emission (mostly coming from the B-A or second positive band system of N_2). This condition is usually fulfilled in equilibrium air plasmas with temperatures greater than 4000 K, or in non-equilibrium plasmas with sufficient excitation of hydrogen electronic states.

8.5.4.1 Broadening coefficients of the H_β lineshape

The lineshape of the H_β transition is determined by Lorentzian (Stark, van der Waals, resonance, natural) and Gaussian (Doppler, instrumental) broadening mechanisms that result in a Voigt profile. The Lorentzian half-width at half-maximum (HWHM) is the sum of the Lorentzian HWHMs. The Gaussian HWHM is the square root of the sum of the squared Gaussian HWHMs. If monochromator slits of equal width are used, the instrumental slit function is well approximated by a Gaussian profile. Numerical expressions for the Stark, van der Waals, resonance, Doppler, and natural HWHMs are derived below for the case of an air plasma with a small amount (a few percent) of hydrogen.

8.5.4.2 Stark broadening

Stark broadening results from Coulomb interactions between the radiating species (here the hydrogen atom) and the charged particles present in the plasma. Both ions and electrons induce Stark broadening, but electrons are responsible for the major part because of their higher relative velocities. The lineshape can be approximated by a Lorentzian function except at the linecenter where electrostatic interactions with ions cause a dip. The Stark broadening width is mostly a function of the free electron concentration, and a weak function of the temperature. The Stark HWHM expression given in table 8.5.1 corresponds to a fit of the widths listed by Gigosos and Cardeñoso (1996) for electron densities between 10^{14} and $4 \times 10^{17}\,\text{cm}^{-3}$ and for reduced masses between 0.9 and 1.0, which covers all perturbers present in the air plasma except hydrogen. (The Stark broadening of hydrogen by hydrogen ions is neglected here because we assume that the mole fraction of hydrogen is less than a few percent.) The fit is within $\pm 5\%$ of the values of Gigosos and Cardeñoso for temperatures up to 10 000 K, $\pm 13\%$ up to 20 000 K, and $\pm 20\%$ up to 40 000 K. If better precision is needed, the actual values of Gigosos and Cardeñoso can be substituted for the present fit.

Table 8.5.1. Half widths at half maximum (in nm) for the H_β line at 486.132 nm. P is the pressure in atm, T the gas temperature in Kelvin, n_e the electron number density in cm^{-3}, and X_H the mole fraction of hydrogen atoms.

$\Delta\lambda_{\text{Stark}}$	$\Delta\lambda_{\text{resonance}}$	$\Delta\lambda_{\text{van der Waals}}$	$\Delta\lambda_{\text{natural}}$	$\Delta\lambda_{\text{Doppler}}$
$1.0 \times 10^{-11}(n_e)^{0.668}$	$30.2 X_H(P/T)$	$1.8 P/T^{0.7}$	3.1×10^{-5}	$1.74 \times 10^{-4} T^{0.5}$

8.5.4.3   Reonance broadening

Resonance broadening is caused by collisions between 'like' particles (e.g. two hydrogen atoms) where the perturber's initial state is connected by an allowed transition to the upper or lower state of the radiative transition under consideration. Typically, the three perturbing transitions that must be considered are $g \to l$, $g \to u$, and $l \to u$, where g stands for the ground electronic state, and l and u for the lower and upper states of the radiative transition. Using the expression given by Griem (1964, p 97), we obtain

$$\Delta\lambda_{\text{resonance}} = \underbrace{\frac{3e^2}{16\pi^2\varepsilon_0 m_e c^2}}_{6.72 \times 10^{-16}\,\text{m}^{-2}}$$

$$\times \lambda_{ul}^2 \left[\lambda_{lg} f_{gl} \sqrt{\frac{g_g}{g_l}} n_g + \lambda_{ug} f_{gu} \sqrt{\frac{g_g}{g_u}} n_g + \lambda_{ul} f_{lu} \sqrt{\frac{g_l}{g_u}} n_l \right]. \tag{1}$$

Using the constants of Wiese *et al* (1966) ($\lambda_{ul} = 486.132\,\text{nm}$, $\lambda_{lg} = 121.567\,\text{nm}$, $\lambda_{ug} = 97.2537\,\text{nm}$, $g_u = 32$, $g_g = 2$, $g_l = 8$, $f_{gl} = 0.4162$, $f_{gu} = 0.02899$, $f_{lu} = 0.1193$), we obtain the resonance HWHM listed in table 8.5.1.

8.5.4.4   Van der Waals broadening

Van der Waals broadening is caused by collisions with neutral perturbers that do not share a resonant transition with the radiating particle. Griem (1964, p 99) gives the following expression for a radiating species r colliding with a perturber p:

$$\Delta\lambda_{\text{van der Waals}} \approx \frac{\lambda_{ul}^2}{2c} \left(\frac{9\pi\hbar^5 \overline{R_\alpha^2}}{16m_e^3 E_p^2} \right)^{2/5} v_{rp}^{3/5} N_p \tag{2}$$

where v_{rp} is the relative speed of the radiating atom and the perturber, E_p is the energy of the first excited state of the perturber connected with its ground state by an allowed transition, N_p is the number density of the perturber, and the matrix element $\overline{R_\alpha^2}$ is equal to

$$\overline{R_\alpha^2} \approx \frac{1}{2} \frac{E_{\text{H}}}{E_\infty - E_\alpha} \left[5 \frac{z^2 E_{\text{H}}}{E_\infty - E_\alpha} + 1 - 3l_\alpha(l_\alpha + 1) \right]. \tag{3}$$

In equation (3), E_{H} and E_∞ are the ionization energies of the hydrogen atom and of the radiating atom, respectively, E_α is the term energy of the upper state of the line, l_α its orbital quantum number, and z is the number of effective charges ($z = 1$ for a neutral emitter, $z = 2$ for a singly ionized emitter,...). For H_β, we have $E_{\text{H}} = E_\infty = 13.6\,\text{eV}$, $E_\alpha = 12.75\,\text{eV}$, and $z = 1$. The H_β transition is a multiplet of seven lines (see table 8.5.2)

Table 8.5.2. Components of the H_β transition multiplet and their properties.

Wavelength air (nm)	A_{ul} (s^{-1})	Upper level configuration	Lower level configuration	g_u	g_l	Relative intensity (% of total H_β emission)
486.12785	1.718×10^7	$4d\,^2D_{3/2}$	$2p\,^2P^0_{1/2}$	4	2	25.5
486.12869	9.668×10^6	$4p\,^2P^0_{3/2}$	$2s\,^2S_{1/2}$	4	2	14.4
486.12883	8.593×10^5	$4s\,^2S_{1/2}$	$2p\,^2P^0_{1/2}$	2	2	0.6
486.12977	9.668×10^6	$4p\,^2P^0_{1/2}$	$2s\,^2S_{1/2}$	2	2	7.2
486.13614	2.062×10^7	$4d\,^2D_{5/2}$	$2p\,^2P^0_{3/2}$	6	4	45.9
486.13650	3.437×10^6	$4d\,^2D_{3/2}$	$2p\,^2P^0_{3/2}$	4	4	5.1
486.13748	1.719×10^6	$4s\,^2S_{1/2}$	$2p\,^2P^0_{3/2}$	2	4	1.3

originating from upper states $4s$, $4p$, and $4d$ of orbital angular momenta $l_\alpha = 0$, 1, and 2. For $l_\alpha = 0$, 1, and 2, $(\overline{R_\alpha^2})^{2/5}$ takes the values 13.3, 12.9, and 12.0, respectively. As listed in table 8.5.2, the components issued from the $4s$, $4p$, and $4d$ states represent 1.9, 21.6, and 76.5% of the total H_β emission, respectively. We use these percentages as weighting factors to determine an average value of $(\overline{R_\alpha^2})^{2/5} = 12.2$.

The relative velocity term $\overline{v_{rp}^{3/5}}$ of equation can be related to the mean speed as follows:

$$\overline{v_{rp}^{3/5}} = (4/\pi)^{2/10}\Gamma(9/5)(\overline{v_{rp}})^{3/5} \cong 0.98(\overline{v_{rp}})^{3/5} = 0.98(8kT/\pi m_{rp}^*)^{3/10} \quad (4)$$

where m_{rp}^* is the reduced mass of the radiating species and its perturber.

Summing over all perturbers present in the plasma, and introducing the mole fraction X_p of perturber p, equation becomes

$$\Delta\lambda_{\text{van der Waals}} \approx 0.98\,\frac{\lambda_{ul}^2}{2c}\left(\frac{9\pi\hbar^5 R_\alpha^2}{16m_e^3 E_p^2}\right)^{2/5}(8kT/\pi)^{3/10}\frac{P}{kT}\sum_p\left[\frac{X_p}{E_p^{4/5}(m_{rp}^*)^{3/10}}\right].$$
$$(5)$$

In air plasmas, O, N, N_2, O_2, and NO represent 98% of the chemical equilibrium composition for temperatures up to 10 000 K. We computed the equilibrium mole fractions of these five species up to 10 000 K and combined them with the E_p and m_{rp}^* values listed in table 8.5.3 in order to evaluate the summation term in equation (5). The value of this term is found to be approximately constant over the entire temperature range and equal to 0.151 ± 0.007. The final expression for the van der Waals HWHM of H_β in air plasmas with a small amount of hydrogen added is given in table 8.5.1.

Table 8.5.3. Constants needed in equation (5) when the radiating species is a hydrogen atom.

Perturber	M_{rp}^* (g/mole)	Transition issued from the first excited state optically connected to the ground state	E_p (eV)	$M_{rp}^{*-0.3} E_p^{-0.8}$ (g/mol)$^{-0.3}$ eV$^{-0.8}$
O	0.94	$^3S^0 \longrightarrow {}^3P$	9.5	0.17
N	0.93	$^4P \longrightarrow {}^4S^0$	10.3	0.16
O_2	0.97	$B^3\Sigma_u^- \longrightarrow X^3\Sigma_g^-$ (Schumann–Runge)	6.2	0.23
N_2	0.97	$b^1\Pi_g \longrightarrow X^1\Sigma_g^+$ (Birge–Hopfield 1)	12.6	0.13
NO	0.97	$A^2\Sigma^+ \longrightarrow X^2\Pi$ (gamma)	5.5	0.26

8.5.4.5 Doppler broadening

For a collection of emitters with a Maxwellian velocity distribution (characterized by a temperature T_h), Doppler broadening results in a Gaussian lineshape with HWHM given by Griem (1964, p. 101):

$$\Delta\lambda_{\text{Doppler}} = \frac{1}{2} \lambda_{ul} \sqrt{\frac{8kT_h \ln 2}{m_h c^2}} = 3.58 \times 10^{-7} \lambda_{ul} \sqrt{\frac{T_h[\text{K}]}{\hat{M}_h[\text{g/mol}]}}. \tag{6}$$

The Doppler HWHM of H_β is given in table 8.5.1.

8.5.4.6 Natural broadening

Natural broadening gives a Lorentzian line profile of HWHM:

$$\Delta\lambda_{\text{natural}} = \frac{\lambda_{ul}^2}{4\pi c} \left(\sum_{n<u} A_{un} + \sum_{n<l} A_{ln} \right) \tag{7}$$

where the two summation terms represent the inverses of the transition's upper and lower level lifetimes, which can be calculated using the Einstein A coefficients tabulated by Wiese *et al* (1966). As can be seen from table 8.5.1, natural broadening is negligible in comparison with the other mechanisms.

8.5.4.7 Fine structure effects

Because of fine structure spin–orbit splitting in the upper and lower levels, the H_β transition is in fact a multiplet of seven lines (see table 8.5.2). The resulting lineshape is the sum of these lines, each of which can be calculated with the broadening widths listed in table 8.5.1. The resulting lineshape will be close to a Voigt profile only if the HWHM of each line is much greater than 0.005 nm, half the separation between the extreme lines. The technique presented below should only be used when the measured HWHM is much greater than 0.005 nm. This condition is fulfilled in most situations of practical interest.

8.5.4.8   *Electron density measurements in equilibrium air plasmas*

We have applied the H_β lineshape technique to the measurement of electron densities in the LTE air plasma characterized in section 8.5.1.1. Application of the technique to non-equilibrium air and nitrogen plasmas (Gessman *et al* 1997) will be discussed in section 8.5.2.3. We used a 0.75 m monochromator with a 1200-groove/mm grating, and entrance and exit slits of 20 μm. The instrumental slit function was approximately Gaussian with HWHM of 0.011 nm. A small amount of H_2 (1.7% mole fraction) was premixed with air before injection into the plasma torch. The spatial resolution of the measurements, determined by the width of the entrance slit and the magnification of the optical train, was approximately 0.13 mm.

Figure 8.5.9 shows the line-of-sight emission spectrum measured along the plasma diameter and the 'background' emission spectrum, mainly due to the N_2 $B^3\Pi_g - A^3\Sigma_u^+$ first positive system, which was measured after switching off the hydrogen flow. Without hydrogen, the measured plasma temperature is lower by approximately 200 K. The torch power was slightly readjusted in order to return to the same temperature conditions by matching the intensity of the background spectral features away from the H_β line-center. Figure 8.5.10 shows the H_β lineshape obtained by subtracting the background signal from the total spectrum. The measured lineshape is well fitted with a Voigt profile of HWHM = 0.11 nm. From the HWHMs of the various broadening mechanisms shown in figure 8.5.11, we infer an electron number density of approximately 1.0×10^{15} cm^{-3}. Because the intensity of the H_β line is proportional to the population of hydrogen in excited state $n = 4$, which is a strong function of the temperature

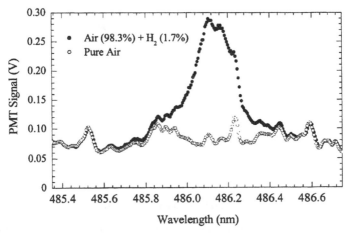

Figure 8.5.9. Typical emission scan of the H_β line. The underlying emission features are mostly from the N_2 first positive band system.

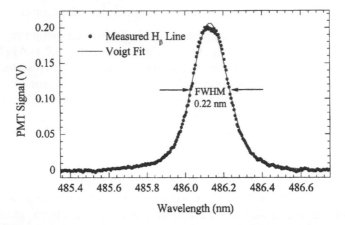

Figure 8.5.10. H_β lineshape obtained from the difference of the two signals shown in figure 8.5.9, and Voigt fit.

($n_4 \sim \exp[-150\,000/T]$), the line-of-sight-integrated emission scan is dominated by emission from the hot central plasma core and therefore provides a good approximation of the electron density at the plasma center.

Radial profiles of electron density were obtained by an Abel inversion. To this end, we scanned the H_β lineshape at 25 lateral locations along chords of the 5 cm diameter plasma. Figure 8.5.12 shows the radial profile of electron densities determined from the Abel-inverted lineshapes. Figure 8.5.12 also shows the radial profile of chemical equilibrium electron densities

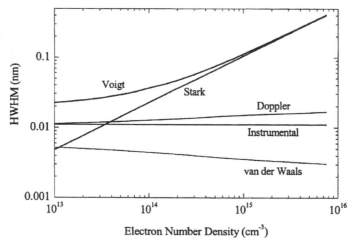

Figure 8.5.11. H_β lineshape broadening as a function of the electron number density in LTE air at atmospheric pressure. (Instrumental HWHM = 0.011 nm.) The resonance broadening HWHM is less than 2×10^{-4} nm for present experimental conditions.

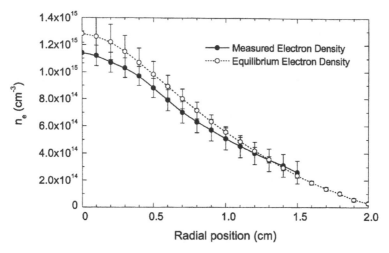

Figure 8.5.12. Measured (solid line) and equilibrium (dashes) electron number density profiles.

based on the measured LTE temperature of the oxygen line at 777.3 nm. Because the plasma is expected to be in LTE, the excellent agreement between the two profiles provides validation of the technique. Note that the electron density of 1.0×10^{15} cm^{-3} determined from the line-of-sight integrated lineshape is consistent with the Abel-inverted electron densities in the central core of the plasma.

Additional examples of electron density measurements based on the Stark-broadened H_β lineshape can be found in chapter 7. In the latter case (figure 8.5.13), the instrumental broadening was minimized using the narrowest possible slit width (HWHM = 0.015 nm), but still was not

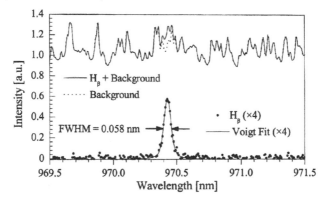

Figure 8.5.13. H_β lineshape in a low temperature ($\sim$4500 K) LTE air plasma. Here the H_β lineshape was measured in second order to reduce the instrumental broadening width by a factor of 2 relative to other broadening widths.

negligible relative to Stark broadening. To improve the sensitivity we measured the spectrum in the second order of the grating. This had the effect of reducing instrumental broadening by a factor of 2 with respect to the other broadening widths. CCD averaging times of 10 s were employed. The inferred number density of 5×10^{13} cm^{-3} represents a lower detection limit in equilibrium air plasmas because the intensity of the H_β line becomes very weak relative to the underlying N_2 first positive signal.

Acknowledgments

The authors acknowledge Richard G. Gessman, Denis Packan and Lan Yu for their contributions to the work presented here.

References

Copeland R A and Crosley D R 1984 'Rotational level dependence of electronic quenching of hydroxyl OH ($A\,^2\Sigma^+$, $v' = 0$)' *Chem. Phys. Lett.* **107**(3) 295–300

Gessman R J 2000 'An experimental investigation of the effects of chemical and ionizational nonequilibrium in recombining air plasmas' Mechanical Engineering Dept., Stanford University, Stanford, CA

Gessman R J , Laux C O and Kruger C H 1997 'Experimental study of kinetic mechanisms of recombining atmospheric pressure air plasmas' 28th AIAA Plasmadynamics and Lasers Conference, Atlanta, GA

Gigosos M A and Cardeñoso V 1996 'New plasma diagnosis tables of hydrogen Stark broadening including ion dynamics' *J. Phys. B: At. Mol. Opt. Phys.* **29** 4795–4838

Gomès A M, Bacri J, Sarrette J P and Salon J 1992 'Measurement of heavy particle temperature in a rf air discharge at atmospheric pressure from the numerical simulation of the NOγ system' *J. Analytical Atomic Spectroscopy* **7** 1103–1109

Griem H R 1964 *Plasma Spectroscopy* (New York: McGraw-Hill)

Laux C O 1993 'Optical diagnostics and radiative emission of air plasmas' PhD thesis, HTGL Report 288, Mechanical Engineering, Stanford University, Stanford, CA

Laux C O 2002 'Radiation and nonequilibrium collisional-radiative models' *Special Course on Physico-Chemical Modeling of High Enthalpy and Plasma Flows* ed. Fletcher T M D and Sharma S (Rhode-Saint-Genèse, Belgium: von Karman Institute)

Laux, C O, Gessman R J, Kruger C H, Roux F, Michaud F and Davis S P 2001 'Rotational temperature measurements in air and nitrogen plasmas using the first negative system of N$_2^+$' *JQSRT* **68**(4) 473–482

Levin D A, Laux C O and Kruger C H 1999 'A general model for the spectral radiation calculation of OH in the ultraviolet' *JQSRT* **61**(3) 377–392

Michaud F, Roux F, Davis S P, Nguyen A-D and Laux C O 2000 'High resolution Fourier spectrometry of the ^{14}N$_2^+$ ion' *J. Molec. Spectrosc.* **203** 1–8

Park C 1985 *Nonequilibrium Air Radiation (NEQAIR) Program: User's Manual* (Moffett Field, CA: NASA-Ames Research Center)

Scott C D, Blackwell H E, Arepalli S and Akundi M A 1998 'Techniques for estimating rotational and vibrational temperatures in nitrogen arcjet flow' *J. Thermophys. Heat Transfer* **12**(4) 457–464

Wiese W L, Smith M W and Glennon B M 1966 *Atomic Transition Probabilities* vol 1. *Hydrogen through Neon* (Washington, DC: US National Bureau of Standards, National Standard Reference Series **1** 153.

Yu L, Laux C O, Packan D M and Kruger C H 2002 'Direct-current glow discharges in atmospheric pressure air plasmas' *J. Appl. Phys.* **91**(5) 2678–2686.

8.6 Ion Concentration Measurements by Cavity Ring-Down Spectroscopy

8.6.1 Introduction

Measurements of ion and/or electron number density are needed to characterize experiments and validate models for atmospheric pressure air and nitrogen plasmas. As discussed in the previous section (on Stark broadening), the electron density can be measured from the H_β Stark-broadened lineshape down to densities of about 5×10^{13} cm^{-3}. Below this value, more sensitive measurement techniques are required. Physical probes tend to disturb the plasma (see section 8.1), and techniques such as EM wave interferometry (see section 8.3) does not readily provide results with high spatial resolution. Optical techniques that measure ion concentrations are widely used. Of these, emission provides information only on excited species, fluorescence suffers from quenching effects, predissociation, and optical interference that complicate interpretation, and absorption often lacks sensitivity.

Cavity ring-down spectroscopy (CRDS), on the other hand, is a sensitive line-of-sight averaged laser absorption technique that has been used to measure species concentrations in low-pressure plasmas (Grangeon *et al* 1999, Quandt *et al* 1999, Booth *et al* 2000, Kessels *et al* 2001, Schwabedissen *et al* 2001). The CRDS is additionally attractive as it enables measurements of the speciation of the ion density. In particular, the N_2^+ ion has been studied in low-pressure hollow cathode sources (Kotterer *et al* 1996, Aldener *et al* 2000). In this section, we describe the use of CRDS to measure ion concentrations in atmospheric pressure discharges. By implementing CRDS in its 'standard' (i.e. not temporally-resolved) form, we perform spatially resolved (by Abel inversion) ion concentration measurements. We also develop a temporally resolved variant of CRDS, which we used to study ion recombination in pulsed plasmas. Measurements have been performed in both air and nitrogen plasmas. In nitrogen plasmas, N_2^+ tends to the dominant ion at temperatures below ~6000 K, and under these conditions the CRDS measurement of N_2^+ enables one of the most direct measurements of electron number density. In LTE air, the concentration of N_2^+ may be linked to that of electrons through chemical equilibrium relations. In non-equilibrium plasmas, a collisional-radiative model may be used to relate the ion and electron concentrations. Alternatively, CRDS measurements of the NO^+ ion can be performed.

An overview of the CRDS technique is provided in section 8.6.2, including a discussion of temporally resolved CRDS. Section 8.6.3 presents the experimental schemes used for spatially resolved ion concentration measurements of the N_2^+ ion in dc discharges, as well as temporally resolved N_2^+ ion concentration measurements in pulsed discharges. Measurement results, and discussion, are provided. To aid in interpreting results, a collisional radiative (CR) model is used to compute population fractions and to relate the measured ion concentrations to electron number densities. The inferred electron number density profiles are compared with electrical measurements, and the non-equilibrium nature of the plasma is discussed. Section 8.6.4 discusses CRDS measurements of the NO^+ ion in air plasmas. The experimental scheme and a discussion of results are presented. Conclusions are provided in section 8.6.5.

8.6.2   Cavity ring-down spectroscopy

Cavity ring-down spectroscopy (CRDS) has become a widely used method in absorption spectroscopy owing primarily to its high sensitivity. Detailed reviews of the technique may be found in Busch and Busch (1999) and Berden *et al* (2000). Essentially, a laser beam is coupled into a high-finesse optical cavity containing a sample, where it passes many times between the mirrors. As the light bounces back and forth inside the cavity, its intensity decays (rings down) owing to sample absorption, particle scattering loss (generally negligible), and mirror transmission loss. A photodetector is used to measure the ring-down signal, which is fitted to yield the sample loss. The technique affords high sensitivity owing to a combination of long effective path length and insensitivity to laser energy fluctuations. Therefore, CRDS is well suited to the detection of trace species in plasmas. Under appropriate conditions, the laser lineshape may be neglected, and the ring-down signal $S(t)$ decays exponentially (Zalicki and Zare 1995, Yalin *et al* 2002) as:

$$S(t) = S_0 \exp[-t/\tau], \qquad 1/\tau = \frac{c}{l}\left[l_{abs}k(\nu_L) + (1-R)\right] \qquad (1)$$

where τ is the $1/e$ time of the decay (termed the ring-down time), c is the speed of light, l is the cavity length, l_{abs} is the absorber column length, $k(\nu)$ is the absorption coefficient, ν_L is the laser frequency, and $1 - R$ is the effective mirror loss (including scattering and all other empty-cavity losses). Generally, the measured ring-down signal is fit with an exponential, and the ring-down time τ is extracted. Combining τ with the ring-down time τ_0 measured with the laser detuned from the absorption feature allows a determination of the sample absorbance, and hence absorption coefficient:

$$\text{abs} \equiv l_{\text{abs}} k(\nu_L) = \frac{l}{c} \left[\frac{1}{\tau} - \frac{1}{\tau_0} \right]. \tag{2}$$

We minimize any potential laser lineshape dependence by tuning the laser frequency across an absorption line, and measuring the frequency-integrated absorption coefficient (Yalin *et al* 2002). This approach is equivalent to assuming that laser broadening causes an effective absorption lineshape, found as the convolution of the symmetric laser lineshape with the actual absorption lineshape (Yalin *et al* 2002).

The prior discussion of CRDS has implicitly assumed that the sample concentration (and associated absorption loss) is independent of time, as would be the case in a dc discharge. In pulsed discharges, however, the ion concentration varies over the duration of the optical ring-down (decay of light in the cavity). A more complex approach is then required. It might be tempting to consider using lower reflectivity mirrors with shorter ring-down times so that the losses may be treated as constant over the ring-down, but the sensitivity of such an approach is inferior (Zalicki *et al* 1995). Although a number of kinetics studies have been performed with CRDS, nearly all of these experiments study processes that are slow compared to experimental ring-down times. An exception is the work of Brown *et al* (2000), who perform gas-phase measurements in cases where the populations do change over the duration of the ring-down. We follow a related approach to measure ion recombination in a plasma over time-scales comparable to the ring-down time (microseconds). For the case of a time-dependent absorption, the ring-down signal $S(t)$ may be written as (Brown *et al* 2000)

$$S(t) = S_0 \exp \left[-\frac{c}{l} \left[\int_0^t k(\nu, t) l_{\text{abs}} \, dt + (1 - R)t \right] \right] \tag{3}$$

where the absorption coefficient now has a time dependence. Rearranging equation leads to an expression for the absorbance as a function of time:

$$\text{abs}(t) \equiv k(\nu, t) l_{\text{abs}} = -\frac{l}{c} \frac{d}{dt} \left[\ln \left(\frac{S(t)}{S_0} \right) \right] - (1 - R). \tag{4}$$

The derivative (local slope) of the logarithm of the ring-down signal is proportional to the loss (sample plus empty cavity) at that time. To obtain time-dependent concentrations directly, Brown *et al* analyzed their data with this method. We choose to follow an alternative approach in which the ring-down signal is divided into a series of time-windows, each of which is fit to an exponential decay (ring-down time). We believe that this approach is less noisy because it avoids differentiation.

8.6.3 N_2^+ measurements

8.6.3.1 *Atmospheric pressure discharge*

We have developed a compact atmospheric pressure plasma source for diagnostic development. The discharge may be operated with both nitrogen and air. A photograph of the nitrogen discharge with a schematic representation of the ring-down cavity is shown in figure 8.6.1. Nitrogen is injected through a flow straightener and passes through the discharge region with a velocity of about 20 cm/s. The discharge is formed between a pair of platinum pins (separation 0.85 cm) that are vertically mounted on water-cooled stainless-steel tubes. The discharge is maintained by a dc current supply ($i_{max} = 250$ mA) in a ballasted circuit ($R_b = 9.35$ kΩ). The pins are brought together to ignite the discharge, and are then separated using a translation stage. The position of the discharge is observed to be stable and reproducible. The discharge is contained within a Plexiglas cylinder (diameter 12 inches, 30.5 cm) that isolates it from room air disturbances. Small holes allow weak ventilation by a fan through the top to avoid accumulation of undesirable by-products of the discharge (such as ozone or oxides of nitrogen), and enable passage of the laser beam through the discharge. A second translation stage is used to displace the entire discharge cylinder relative to the optical axis in order to obtain spatial profiles.

To explore the repetitively pulsed approaches, we connect a high-voltage pulser in parallel to the dc discharge circuit. The pulser is capacitively coupled to the discharge so that it is isolated from the dc supply. The dc field serves to give a baseline of ionization and to heat the gas. We operate the high-voltage pulser (pulse width ~10 ns, pulse voltage ~8 kV) at 10 Hz so that it may be synchronized relative to the laser. At this repetition rate the plasma equilibrates between high-voltage pulses so that the behavior during and following each pulse is not affected by the presence of other pulses.

8.6.3.2 *CRDS measurements*

We study the N_2^+ ion by probing the (0,0) band of its first negative band system ($B\,^2\Sigma_u^+ - X\,^2\Sigma_g^+$) in the vicinity of 391 nm. We select this spectral

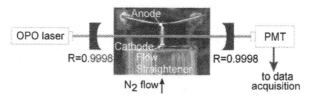

Figure 8.6.1. Photograph of the atmospheric pressure nitrogen discharge and schematic diagram of the ring-down cavity. Electrode separation: 0.85 cm. Discharge current: 187 mA.

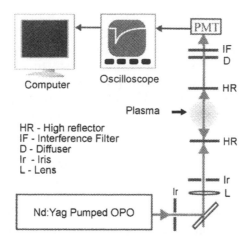

Figure 8.6.2. Schematic diagram of CRDS set-up. The ring-down cavity has a length of 0.75 m and uses 0.5 m radius of curvature mirrors. An OPO is used as the light source, and a photomultiplier tube (PMT) detects the light exiting the cavity.

feature because it is comparatively strong and optically accessible. The optical layout is shown in figure 8.6.2. An OPO system (doubled idler) is used as the light source (repetition rate = 10 Hz, pulse width ~7 ns, pulse energy ~3 mJ, linewidth ~0.14 cm^{-1}). The output from the OPO passes through a Glan–Taylor polarizer to attenuate the energy, and several irises. Typically, about 100 μJ per pulse is incident on the back face of the entrance ring-down mirror. The irises serve to select a relatively uniform portion of the beam and to reduce the beam diameter prior to cavity injection (from ~6 mm to less than < ~1 mm). Because the OPO laser light is multi-mode, and spectrally broad (~4 GHz) compared to the cavity free spectral range (~400 MHz), we operate in a continuum-mode (meaning that many transverse cavity-modes are active) rather than attempting to mode-match the beam to the cavity. Exciting many transverse cavity-modes has the advantage that mode-beating effects (interferences between different cavity modes causing temporal 'beats' in the ring-down signals) are minimized. Also, by averaging multiple ring-down signals, the mode-beating effects are further reduced (averaged away) and a near single-exponential decay is obtained (see Berden *et al* 2000 and references therein). We use a linear cavity of 75 cm length with 50 cm radius-of-curvature (ROC) mirrors from Research Electro-Optics. The spatial resolution and selection of cavity geometry are discussed below. The ring-down signal is collected behind the output mirror with a fast photomultiplier tube (Hamamatsu-R1104), which we filter against the pump laser and other luminosity with two narrow-band interference filters (CVI-F10-390-4-1). The PMT signals are passed to a digitizing oscilloscope (HP 54510A, 250 MHz analog bandwidth,

8-bit vertical resolution) and are read to computer with custom data acquisition software. In a typical ring-down spectrum, 16 or 32 decay curves are averaged at each wavelength (to minimize mode-beating effects), and the resulting waveform is fitted with an exponential to yield the ring-down time τ. For the dc measurements, the portion of the ring-down signal used in the fit is that in between 90 and 10% of the peak (initial) signal amplitude. The detuned ring-down time τ_0 is determined with the laser tuned off the absorption features. Spectral scans use a step-size of 0.001 nm. When performing spatial scans, we use step-sizes of 0.2 mm.

The CRDS set-up used for the pulsed measurements differs only in terms of data fitting and timing. Because sample absorption loss is no longer constant during the measurement, the time dependent equation (4) is used to fit the data. Rather than computing derivatives we fit a series of line segments to the logarithm of the ring-down signal. The fitting windows are 1 µs in length. This time interval represents a good compromise in making the window short compared to the timescale of the process studied yet affording an acceptable signal-to-noise level. Because our time windows are long compared to the pulse length ($\sim$10 ns), we do not resolve the build-up of ionization that occurs during the time the pulse is on; yet we are able to resolve the subsequent recombination. We synchronize the laser relative to the firing of the discharge with an external timing circuit. In order to obtain concentration information at different times relative to the firing of the high voltage pulse, we vary the delay time between the firing of the laser and the firing of the high voltage pulser. Delay times ($T_{pulse}-T_{laser}$) of -10, -8, -6, -4, -2, -1, 0, 1, and 2 µs are used. Jitter from the external timing circuit and triggering scheme are negligible compared to the measurement temporal resolution. Again, to minimize mode-beating effects, we average 16 or 32 ring-down decay signals.

Implementing CRDS in the atmospheric plasma requires special care in the choice of cavity geometry. For a linear cavity formed with mirrors of equal radius of curvature (ROC), the cavity geometry is determined by the dimensionless g-parameter, defined as unity minus the cavity length divided by the mirror radius of curvature (Siegman 1986). Initial attempts to form a ring-down cavity with $g = 0.875$ (length 75 cm, 6 m ROC mirrors), resulted in distorted and irreproducible profiles, owing to beam steering from index-of-refraction gradients (similar to a mirage). Recent work by Spuler and Linne (2002) simulates the effect of cavity geometry on beam propagation in CRDS experiments. Their results indicate that a g-parameter of about -0.5 represents a good compromise between beam waist and beam walk in environments where beam steering may be present. Accordingly, we form a cavity of length 75 cm, with 50 cm ROC mirrors (Research Electro Optics). No beam steering is detected with this geometry. Qualitatively, this geometry tends to recenter deviated beams, whereas the more planar geometry is not as effective.

8.6.3.3 Conductivity measurements

We also determine the electron number density in the discharge by an electrical conductivity approach. For the dc discharge, we measure the time-independent discharge current and electric field, and use Ohm's law to compute the product of the average electron number density and column area. The electric field is found as the slope of the discharge voltage versus electrode separation. We write Ohm's law as $j = i/\text{area} = (n_e e^2/m_e \sum \nu_{eh})E$ where ν_{eh} is the average collision frequency between electrons and heavy particles. Because of the low ionization fraction ($< \sim 10^{-5}$), ν_{eh} is dominated by collisions with neutrals, so we can write $\nu_{eh} \cong n_n g_e Q_{en}$, where $n_n = P/kT_g$ is the number density of neutrals, $g_e = (8kT_e/\pi m_e)^{0.5}$ is the thermal electron velocity, and Q_{en} is the average momentum transfer cross-section for electron–nitrogen collisions. We determine T_e by means of a collisional radiative model (Pierrot *et al* 1999), with input parameters (vibronic ground state population and rotational temperature) obtained from the CRDS measurements. Q_{en} is determined as a function of T_e from tabulated values (Shkarofsky *et al* 1966).

We follow an analogous approach to determine the time-varying electron number density in the pulsed discharge. In this case we measure the time-dependent current and electric field, and use these to determine temporally resolved electron number densities. The time-dependent electric field is found (after subtracting the cathode fall) from the discharge voltage, which we measure using fast probes (time response $\sim$3 ns). The pulsed discharge has the complication that the shape of the profile is altered following the high-voltage pulse, because higher concentrations near the center recombine more quickly. We model these effects based on chemical kinetic considerations (see Yalin *et al* 2002).

8.6.3.4 N_2^+ ring-down spectra

N_2^+ ring-down spectra are recorded as a function of discharge current and position. We discuss the spectra in terms of measurement accuracy and detection sensitivity.

Figure 8.6.3 shows measured and simulated absorption spectra in the vicinity of the (0,0) bandhead of the N_2^+ first negative band system. Rotationally resolved lines from the P and R branches are visible. The lines are identified using tabulated line locations (Michaud *et al* 2000, Laux *et al* 2001), and are labeled with the angular momentum quantum number N'' of the lower state. The displayed spectrum is recorded along the discharge centerline, at a current of 187 mA, and averages 16 shots at each wavelength. The experimental spectrum is plotted in terms of single-pass cavity loss and illustrates the high sensitivity attainable with the CRDS technique. The cavity loss is the sum of mirror reflective loss and sample absorptive loss.

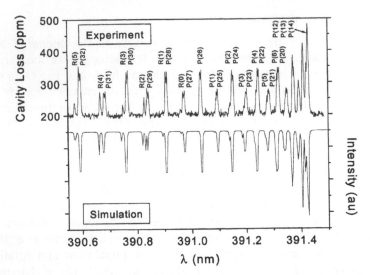

Figure 8.6.3. Measured and simulated N_2^+ absorption spectra near the (0,0) bandhead of the first negative band system. Lines from the P and R branches are identified.

(The Rayleigh scattering losses are computed to be negligible.) Near the bandhead, the signal is ~280 ppm/pass, the baseline reflective loss is ~200 ppm/pass, and the baseline noise is ~5 ppm/pass, so that the signal-to-noise ratio is ~56. This ratio suggests an absorbance-per-pass sensitivity of about 1 ppm, which corresponds to a detection limit of about 7×10^{10} cm^{-3} for N_2^+ ions at our experimental conditions. The baseline reflective loss of ~200 ppm/pass corresponds to a mirror reflectivity of ~0.9998 which is in accord with the manufacturer's specifications.

Figure 8.6.4 shows an expanded view of the P(28) and R(1) lines after baseline subtraction. Fitted Voigt peaks (constrained to have the same shape and width) are shown with solid lines, and their sum is shown with a dotted line. For the P(28) lines, the doublet structure arising from the unpaired electron is apparent. The fit yields a doublet spacing of 0.005 nm, in good agreement with the literature (Michaud *et al* 2000, Laux *et al* 2001). The R(1) lines are close to the detection limit and correspond to a N_2^+ X state population of about 10^{10} cm^{-3}. Their splitting is much less than the linewidths and is not resolved. The fitted FWHM of each peak is 0.0042 nm, or 7.9 GHz, which is consistent with an expected thermally broadened linewidth of ~7 GHz and a measured laser linewidth of ~4 GHz. Similar to conventional laser-based absorption, the use of an effective lineshape in CRDS is only rigorously valid in the optically thin (weakly absorbing) limit. However, our calculations indicate that for the linewidths and absorption parameters used in these experiments, the laser lineshape will have a negligible effect on the area of the measured absorption features (Yalin *et al* 2002).

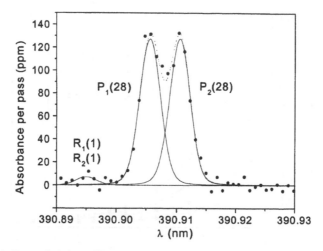

Figure 8.6.4. Expanded view of P(28) and R(1) lines from figure 8.6.3. Background absorption has been subtracted. Voigt profiles fitted to the doublet are shown with solid lines, while their sum is shown with a dotted line.

8.6.3.5 Spatial profiles of ion concentration and electron number density

We obtain spatial profiles of the N_2^+ concentration by displacing the discharge perpendicularly to the optical axis. CRDS is a path-integrated technique and the discharge has axial symmetry. We verify the symmetry of the discharge by performing measurements with the plasma rotated by 90°, and find that the cases have <2% deviation. We use an Abel inversion to recover the radial N_2^+ concentration profile. The concentration measurements are based on the (frequency integrated) area of the lines P(9)–P(17) in the (0,0) band head vicinity. We use tabulated line strengths from Michaud *et al* (2000) and Laux *et al* (2001).

Figure 8.6.5 shows concentration profiles determined for different values of current ($i = 52$, 97, 142, and 187 mA). We find peak (centerline) N_2^+ concentrations of 7.8×10^{11}, 1.5×10^{12}, 2.4×10^{12}, and 3.6×10^{12} cm^{-3} for $i = 52$, 97, 142, and 187 mA respectively. The shape of the concentration profile remains approximately uniform at the different conditions, though we observe that the radial half-maximum values increase slightly with current. We find radial half-maximums of 0.80, 0.82, 0.93, and 1.05 mm for $i = 52$, 97, 142, and 187 mA respectively.

The error bars on the N_2^+ concentrations represent one standard deviation (1σ). They primarily arise from the uncertainties in relating the measured population of several rotational levels in the ground vibronic state, to the overall population of N_2^+. Because the discharge is out of equilibrium, this relationship depends on how the rotational, vibrational, and electronic energy levels are populated. The rotational levels are equilibrated at the

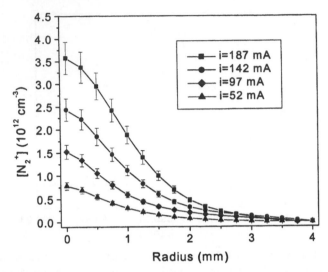

Figure 8.6.5 Radial concentration profiles of N_2^+ measured by CRDS in an atmospheric pressure glow discharge. Experimental data points are joined with line segments for visual clarity.

gas temperature owing to fast collisional relaxation. The rotational temperatures used in the analysis are obtained from Boltzmann plots, and are $T_r = 3100$, 3600, 4150, and 4700 K for currents of $i = 52$, 97, 142, and 187 mA, respectively. The vibrational and electronic energy levels are out of equilibrium, and a collisional–radiative (C-R) model (Pierrot *et al* 1999) is used to determine the fraction of the population in the ground vibronic state, and predicts 0.37 ± 0.02, 0.35 ± 0.02, 0.33 ± 0.02, and 0.31 ± 0.02 for $i = 52$, 97, 142, and 187 mA, respectively. Combining the rotational temperature uncertainties with those from the C-R model and those from the Abel inversion ($\sim$4%) results in an overall experimental uncertainty in concentration of $\sim$10%.

The spatial resolution of our measurements is determined by the spatial step-size (0.2 mm). To justify this claim, we need to verify that the dimension of our laser beam waist does not influence the measured spatial profiles. The simulations by Spuler and Linne (2002) indicate that our expected beam waist is approximately 160–320 μm, depending on the level of mode matching achieved. Deconvoluting the broader case has an effect of only about 1% (0.02 mm) on the measured profiles, which is negligible compared to the spatial step-size. Therefore the resulting spatial resolution is about 0.2 mm.

We incorporate the electrical measurements by comparing the electron number density inferred from the CRDS ion measurements, to the electron number density from the electrical conductivity approach. To infer electron number densities from the CRDS, we need to know the fraction of positive ions that are N_2^+. At our conditions, the C-R model predicts

Table 8.6.1. Comparison of electron number densities (at the radial half-maximum) inferred by CRDS to those found by electrical measurement, for the dc discharges. The last column is the ratio of the electron number density inferred by CRDS to that found from electrical measurement.

i (mA)	$n_{e\text{-CRDS}}$ (cm^{-3})	$n_{e\text{-Elec}}$ (cm^{-3})	CRDS/electrical
52	$4.1 \pm 0.4 \times 10^{11}$	$3.8 \pm 0.4 \times 10^{11}$	1.08 ± 0.16
97	$8.2 \pm 0.8 \times 10^{11}$	$7.8 \pm 0.8 \times 10^{11}$	1.05 ± 0.16
142	$1.4 \pm 0.1 \times 10^{12}$	$1.4 \pm 0.1 \times 10^{12}$	0.96 ± 0.14
187	$2.1 \pm 0.2 \times 10^{12}$	$2.0 \pm 0.2 \times 10^{12}$	1.06 ± 0.16

that 96, 93, 89, and 85% of ions are N_2^+, and the remainder is N^+, for $i = 52$, 97, 142, and 187 mA, respectively. By charge balance, the sum of the N_2^+ and N^+ concentrations equals the electron number density. We convert the N_2^+ concentration profiles (found by CRDS) to electron number density profiles using these percentages. In order to determine electron number densities from the conductivity measurements (which yield the product of average electron number density with area), we assume that the shape of the electron number density profile is the same as that for the ions. The electron number densities (at the radial half-maximum) found in this way from electrical measurements are compared with those inferred from the CRDS ion measurements in table 8.6.1. The values are plotted in figure 8.6.6. The

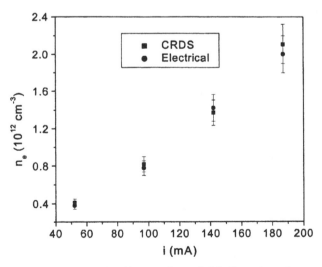

Figure 8.6.6. Electron number densities (at the radial half-maximum) as a function of discharge current. Number densities are derived from CRDS ion measurements (squares), and from electrical measurement (circles).

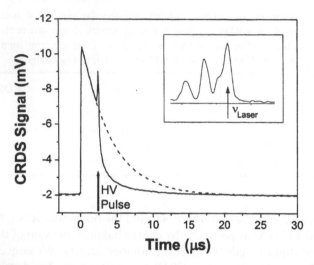

Figure 8.6.7. Experimental ring-down traces with the laser tuned to the N_2^+ absorption bandhead (inset) with the high-voltage pulse (solid line) and without the high-voltage pulse (dashed line).

uncertainty in the electrical measurement (10%) is primarily from uncertainties in the momentum transfer cross-section (5%), the discharge area (4%), and the average gas temperature (8%). Column 4 of table 8.6.1 shows that the electron number densities found from optical and electrical measurements overlap within their error bars. This excellent agreement gives us confidence in our results for the electron number density.

8.6.3.6 Temporal profiles of N_2^+ concentration and electron number density

Figure 8.6.7 shows ring-down traces obtained with and without firing the high voltage pulse, and with the laser tuned to the N_2^+ B-X (0,0) bandhead. In the absence of the high-voltage pulse (dashed line) the absorption losses are constant in time, and the signal decays as a single-exponential. In the trace with the pulse (solid line), the light decays more steeply after the pulse, reflecting an increased concentration of N_2^+. The spike in the latter trace coincides with the firing of the pulse, and is caused by rf interference generated by the pulser. To verify that we are observing changes in the N_2^+ concentration, we examine the analogous traces but with the laser detuned from the absorption band (see figure 8.6.8). These traces confirm that the only effect of the high voltage pulse on the ring-down system is to generate the interference spike. We analyze these traces to determine over what region the interference spike affects the data. We vary the delay of the high-voltage pulse relative to the laser shot so that we can obtain ion concentrations at different times.

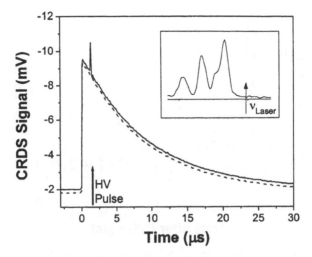

Figure 8.6.8. Experimental ring-down traces with the laser tuned away from the N_2^+ absorption (inset). We slightly scale (<5%) the amplitude of the traces for visual clarity. The detuned trace (dashed line) is offset by 0.2 mV to make it more visible.

We quantify the time-varying N_2^+ concentration using equation (4) with a 1 μs window. This time interval represents a good compromise in making the window short compared to the timescale of the process studied yet affording an acceptable signal-to-noise level. The empty-cavity losses (mirror reflectivity) are found from the ring-down signals with the laser detuned, and these losses are subtracted in the analysis. Using tabulated line strengths and the discharge dimensions, we find the absolute N_2^+ center-line concentrations as a function of time. Figure 8.6.9 presents the time-varying concentrations (symbols). The error bars reflect uncertainties in the population fractions, as well as uncertainty associated with a possible change in shape of the concentration profile. The latter uncertainty is estimated by chemical kinetic considerations (see Yalin *et al* 2002). One microsecond after the pulse, the N_2^+ concentration is ~1.5×10^{13} cm^{-3}, and then N_2^+ recombines to the dc level in about 10 μs. The dc level is found by analyzing the pulsed data at sufficiently long time delays after the pulse, and its value is consistent with that found in the dc plasma without the pulser.

For the pulsed discharge, we also determine the electron concentration by measuring the electrical conductivity. The temporally resolved electron concentrations are shown with a swath in figure 8.6.9. The uncertainty in the dc electron concentration reflects uncertainties in the profile shape, the momentum transfer cross-section, and the gas temperature. The collisional-radiative model predicts that N_2^+ is the dominant ion produced by the pulse. Thus, the agreement between the time-dependent electron and N_2^+ concentrations during plasma recombination verifies the temporally

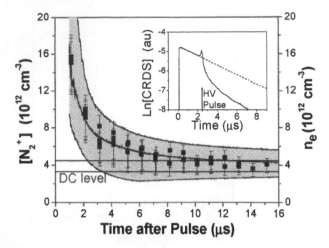

Figure 8.6.9. CRDS measurements of N_2^+ concentrations (circles) and conductivity measurements of electron densities (swath) versus time following the firing of a high-voltage pulse in an atmospheric pressure nitrogen dc plasma. The dc level of N_2^+ concentration found by CRDS is shown with a hatched bar. The inset shows the ring-down signals (plotted on a semi-log scale) with the HV pulse (solid), and without the HV pulse (dotted).

resolved CRDS measurement. The measured recombination time is consistent with reported (Park 1989) dissociative recombination rate coefficients for N_2^+ (approximately 5×10^{-8} cm^3/s).

8.6.3.7 Non-equilibrium discharge

To have a measure of the degree of non-equilibrium in the dc discharges, we examine the ratio of the measured electron number density (at the radial half-maximum) to the LTE electron number density at the corresponding gas temperature. These ratios are given in column 3 of table 8.6.2 for the four conditions studied in the dc discharge. The measured ion and electron concentrations in the discharge are significantly higher than those

Table 8.6.2. Ratio of the measured dc electron number density to the concentration corresponding to a LTE plasma at the same gas temperature.

i (mA)	T_g (K)	$n_{e\text{-CRDS}}/n_{e\text{-LTE}}$
52	3100	2.8×10^4
97	3600	980
142	4200	48
187	4700	5.6

corresponding to LTE conditions at the same gas temperature. The results quantify the degree of ionization non-equilibrium in the discharges. At higher values of discharge current the LTE concentration of charged species rises steeply, so that the ratio of measured concentration to LTE concentration reduces. Related work in our laboratory has shown that by more rapidly flowing the gas, comparable electron densities may be achieved with lower gas temperatures. Clearly, additional non-equilibrium is generated in the pulsed discharge. The high voltage pulse has a negligible effect on the gas temperature (and hence corresponding LTE number density) yet the measured electron number density in the discharge increases by a factor of at least 4 immediately following the high voltage pulse.

8.6.4 NO$^+$ measurements

8.6.4.1 *RF air plasma*

The experimental set-up is shown schematically in figure 8.6.10. Atmospheric pressure air plasmas are generated with a 50 kW rf inductively coupled plasma torch operating at a frequency of 4 MHz. The torch is operated with a voltage of 8.9 kV and a current of 4.6 A. The torch has been extensively characterized at similar conditions, and the plasma is known to be near LTE with a temperature of about 7000 K (Laux 1993).

8.6.4.2 *CRDS measurements*

Unlike the N$_2^+$ ion, the NO$^+$ ion does not have optically accessible electronic transitions. To perform CRDS measurements, the ion must be probed by accessing its infrared vibrational transitions. The strongest vibrational transitions are the fundamental bands, and for these transitions one finds that the

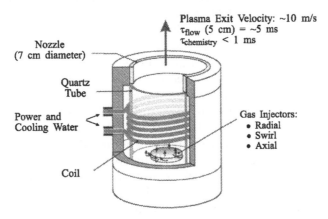

Figure 8.6.10. Schematic cross-section of torch head with 7 cm diameter nozzle.

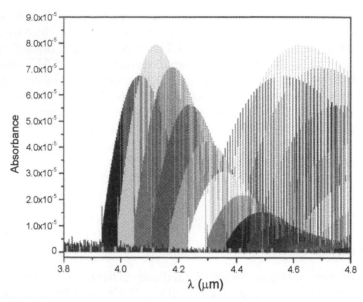

Figure 8.6.11. Modeled absorbance of the air plasma at LTE temperature of 7000 K over pathlength of 5 cm. Absorption by NO, OH, and NO$^+$ are included. Rotationally resolved lines of the vibrational transitions are shown.

absorbance per NO$^+$ ion is about 20 000 times less than that of the electronic transitions of the N$_2^+$ ion. Figure 8.6.11 shows the modeled absorbance, as a function of wavelength, for the air plasma at the conditions used. The simulation is performed with SPECAIR and assumes a pathlength of 5 cm, and LTE conditions at a temperature of 7000 K ($T_g = T_r = T_v = T_{electronic} = 7000$ K). The simulation includes the infrared absorption features of NO, OH, and NO$^+$. The absorption by NO and OH is relatively weak, while the various fundamental bands of NO$^+$ have stronger predicted absorbances.

It is evident that the NO$^+$ absorption begins at a wavelength of about 3950 nm, and is a maximum at about 4100 nm. Accessing these infrared wavelengths is challenging in terms of available laser sources. The current measurements have been performed using a Continuum-Mirage OPO system. The Mirage laser is designed to operate at a maximum wavelength of 4000 nm; however, we optimized the alignment in a manner that enabled operation in the vicinity of 4100 nm, in order to be nearer to the peak NO$^+$ absorption. Ring-down cavity alignment at these wavelengths is challenging, since the beam (and its back-reflections) are not readily observable. The ring-down cavity was aligned using a combination of LCD (liquid crystal display) paper to locate the beam, and a helium–neon laser to act as a reference. With the plasma off, ring-down times of about 1.2 μs were obtained, corresponding to mirror reflectivities of about 0.998 (approximately an order of magnitude worse than the mirrors used for the N$_2^+$ experiments).

Our initial attempts to perform CRDS measurements in the plasma torch used the same cavity-geometry as was used in the N_2^+ experiments— a g-parameter of 0.5. With the plasma off, this geometry yielded excellent stability in the ring-down times: 1% standard deviation in ring-down time for single shot ring-down signals. However, with the plasma on, the beam steering reduced the stability significantly. In the rf plasma, as compared to the smaller nitrogen plasma, the cavity-geometry considerations are different. In the smaller nitrogen plasma, we wanted to minimize simultaneously the cavity beam-waist and the beam-walk, leading to a g-parameter of -0.5 (see discussion above). On the other hand in the rf plasma, the plasma dimension (about 5 cm) is significantly larger than the beam dimension (about 1 mm). Therefore, the exact beam dimension is not critical, and the cavity-geometry may be selected solely to minimize beam-walk. The numerical modeling of Spuler and Linne (2002) indicates that minimizing the beam-walk may be accomplished with a g-parameter of about 0.25, which we implemented by using a cavity of length 75 cm, and mirrors of radius-of-curvature of 1 m. This geometry did indeed reduce the beam-walk and enabled improved stability (about 2% standard deviation in empty cavity ring-down times).

As will be discussed, the identification of spectral lines in the analysis of the air plasma spectra is challenging. In order to assist in identifying NO^+ spectral features, we also collected CRDS spectra with the plasma running with argon and nitrogen (as opposed to air), conditions that are not expected to have any significant NO^+ concentration.

8.6.4.3 Results and discussion

Figure 8.6.12 shows a measured absorbance spectrum along the centerline of the air plasma. The experimental data were obtained by averaging 16 laser shots at each spectral position. The plotted CRDS data have been converted to absorbance, and fitted with a peak-fitting program. (Fitted peaks are shown in black, while raw data are shown with blue symbols.) Also shown is the modeled NO absorbance assuming the expected plasma conditions of path length 5 cm, and LTE at 7000 K. The modeled contributions from OH and NO absorption are negligible on this scale. Comparing the CRD spectrum in the air plasma to the CRD spectrum in the argon/nitrogen plasma provides information as to line identities. The largest spectral feature (at $\sim$4127.7 nm) is present in both spectra, and therefore is presumed not to be NO^+. Comparing the other observed spectral features with the model does not yield good agreement. To the best of our knowledge, the spectroscopic constants used in our modeling are the most recent and accurate ones available (Jarvis *et al* 1999). The exact locations of the rotationally resolved lines are largely determined by the rotational constants B, which have a quoted uncertainty of $\pm 0.005\,\mathrm{cm}^{-1}$ (or about 0.25%). Based on the quoted

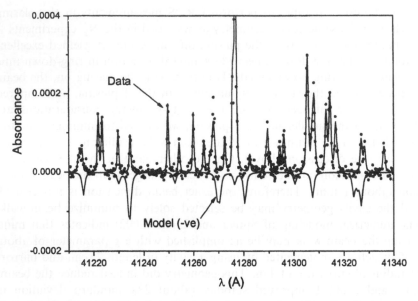

Figure 8.6.12. Experimental and modeled absorbance spectrum from the air plasma near 4100 nm. Raw data (blue symbols) as well as fitted peaks (top black line) are shown, as well as the modeled NO$^+$ lines (plotted negative for visual clarity). The precision of the spectroscopic constants used in the model is insufficient to predict accurately the locations of the rotational lines.

uncertainty we performed an uncertainty analysis, and found that with this level of precision it is not possible to accurately predict the locations of the rotational lines. Therefore, any match between the experimental data and model would be fortuitous. Our experimental features are repeatable (to within experimental uncertainty) and have approximately the correct integrated area, so we do believe they belong to NO$^+$.

8.6.5 Conclusions

Spatial and temporal profiles of N$_2^+$ concentration have been measured in dc and pulsed atmospheric pressure nitrogen glow discharges by cavity ringdown spectroscopy. Special care in the selection of cavity geometry is needed in the atmospheric pressure plasma environment. Sub-millimeter spatial resolution, microsecond temporal resolution, and sub-ppm concentration sensitivity have been achieved. The signal-to-noise ratio suggests a dc detection limit of about $7 \times 10^{10} \, \text{cm}^{-3}$ for N$_2^+$ ions at our experimental conditions (corresponding to an uncertainty in column density of about $1.4 \times 10^{10} \, \text{cm}^{-2}$). Using a collisional–radiative model we infer electron number densities from the measured ion profiles. The values of electron number density found in this way are consistent with those found from spatially integrated electrical conductivity measurements. The spectroscopic technique is clearly favorable, because it offers spatial resolution and does not require knowledge of other discharge parameters. Furthermore, the spectroscopic technique enables measurements of the speciation of the ion density, information not available from direct electrical measurements.

Measurements of the NO^+ ion in air plasmas have also been demonstrated. The accessible spectral features of NO^+ are vibrational transitions, considerably weaker than the ultraviolet electronic transitions used to probe N_2^+. Nevertheless, CRDS data from air plasmas were obtained, and spectral features attributed to NO^+ were observed. This technique shows promise for the measurement of NO^+ concentrations once more accurate spectroscopic constants of NO^+ become available.

References

Aldener M, Lindgren B, Pettersson A and Sassenberg U 2000 'Cavity ringdown laser absorption spectroscopy: nitrogen cation' *Physica Scripta* **61**(1) 62–65

Berden G, Peeters R and Meijer G 2000 'Cavity ring-down spectroscopy: experimental schemes and applications' *Int. Rev. Phys. Chem.* **19**(4) 565–607

Booth J P, Cunge G, Biennier L, Romanini D and Kachanov A 2000 'Ultraviolet cavity ring-down spectroscopy of free radicals in etching plasmas' *Chem. Phys. Lett.* **317**(6) 631–636

Brown S S, Ravishankra A R and Stark H 2000 'Simultaneous kinetics and ring-down: rate coefficients from single cavity loss temporal profiles' *J. Chem. Phys. A* **104** 7044–7052

Busch K W and Busch A M (eds) 1999 *Cavity-Ringdown Spectroscopy* (acS Symposium Series) (Oxford: Oxford University Press)

Grangeon F, Monard C, Dorier J-L, Howling A A, Hollenstein C, Romanini D and Sadeghi N 1999 'Applications of the cavity ring-down technique to a large-area RF-plasma reactor' *Plasma Sources Sci. Technol.* **8** 448–456

Jarvis G K, Evans M, Ng C Y and Mitsuke K 1999 'Rotational-resolved pulsed field ionization photoelectron study of NO^+ $X\,^1\Sigma^+$, $v^+ = 0$–32) in the energy range of 9.24–16.80 eV' *JCP* **111**(7) 3058–3069

Kessels W M M, Leroux A, Boogaarts M G H, Hoefnagels J P M, van de Sanden M C M and Schram D C 2001 'Cavity ring down detection of SiH_3 in a remote SiH_4 plasma and comparison with model calculations and mass spectrometry' *J. Vac. Sci. Technol. A* **19**(2) 467–476

Kotterer M, Conceicao J and Maier J P 1996 'Cavity ringdown spectroscopy of molecular ions: $A\,^2\Pi_u X\,^2\Sigma_g^+$ (6–0) transition of N_2^+' *Chem. Phys. Lett.* **259**(1–2) 233–236

Laux C O 1993 'Optical diagnostics and radiative emission of air plasmas' Mechanical Engineering. Stanford University, Stanford, CA, p 232

Laux C O, Gessman R J, Kruger C H, Roux F, Michaud F and Davis S P 2001 'Rotational temperature measurements in air and nitrogen plasmas using the first negative system of N_2^+' *JQSRT* **68**(4) 473–482

Michaud F, Roux F, Davis S P, Nguyen A-D and Laux C O 2000 'High resolution Fourier spectrometry of the $^{14}N_2^+$ ion' *J. Molec. Spectrosc.* **203** 1–8

Park C 1989 *Nonequilibrium Hypersonic Aerothermodynamics* (New York: Wiley)

Pierrot L, Yu L, Gessman R J, Laux C O and Kruger C H 1999 'Collisional-radiative modeling of non-equilibrium effects in nitrogen plasmas' in 30th AIAA Plasma-dynamics and Lasers Conference, Norfolk, VA

Quandt E, Kraemer I and Dobele H F 1999 'Measurements of Negative-Ion Densities by Cavity Ringdown Spectroscopy' *Europhysics Lett.* **45** 32–37

Schwabedissen A, Brockhaus A, Georg A and Engemann J 2001 'Determination of the gas-phase Si atom density in radio frequency discharges by means of cavity ring-down spectroscopy' *J. Phys. D: Appl. Phys.* **34**(7) 1116–1121

Shkarofsky I P, Johnston T W and Bachynski M P 1966 *The Particle Kinetics of Plasmas* (Addison-Wesley)

Siegman A E 1986 *Lasers* (Mill Valley: University Science Books)

Spuler S and Linne M 2002 'Numerical analysis of beam propagation in pulsed cavity ring-down spectroscopy' *Appl. Optics* **41**(15) 2858–2868

Yalin, A P and Zare R N 2002 'Effect of laser lineshape on the quantitative analysis of cavity ring-down signals' *Laser Physics* **12**(8) 1065–1072

Yalin A P, Zare R N, Laux C O and Kruger C H 2002 'Temporally resolved cavity ring-down spectroscopy in a pulsed nitrogen plasma' *Appl. Phys. Lett.* **81**(8) 1408–1410

Zalicki P and Zare R N 1995 'Cavity ring-down spectroscopy for quantitative absorption measurements' *J. Chem. Phys.* **102**(7) 2708–2717

Chapter 9

Current Applications of Atmospheric Pressure Air Plasmas

M Laroussi, K H Schoenbach, U Kogelschatz, R J Vidmar, S Kuo, M Schmidt, J F Behnke, K Yukimura and E Stoffels

9.1 Introduction

High-pressure non-equilibrium plasmas possess unique features and characteristics which have provided the basis for a host of applications. Being non-equilibrium, these plasmas exhibit electron energies much higher than that of the ions and the neutral species. The energetic electrons enter into collision with the background gas causing enhanced level of dissociation, excitation and ionization. Unlike the case of thermal plasmas, these reactions occur without an increase in the gas enthalpy. Because the ions and the neutrals remain relatively cold, the plasma does not cause any thermal damage to articles they may come in contact with. This characteristic opens up the possibility of using these plasmas for the treatment of heat-sensitive materials including biological tissues. In addition, operation in the high-pressure regime lends itself to the utilization of three-body processes to generate useful species such as ozone and excimers (excited dimers and trimers).

Low-temperature high-pressure non-equilibrium plasmas are already routinely used in material processing applications. Etching and deposition, where low-pressure plasmas have historically been dominant, are examples of such applications. In the past two decades, non-equilibrium high-pressure plasmas have also played an enabling role in the development of excimer VUV and ultraviolet sources (Eliasson and Kogelschatz 1991, El-Habachi and Schoenbach 1998), plasma-based surface treatment devices (Dorai and Kushner 2003), and in environmental technology such as air pollution control (Smulders *et al* 1998). More recently, research on the biological and medical applications of these types of plasmas have witnessed a great

537

interest from the plasma and medical research communities. This is due to newly found applications in promising medical research such as electro-surgery (Stoffels *et al* 2003, Stalder 2003), tissue engineering (Blakely *et al* 2002), surface modification of bio-compatible materials (Sanchez-Estrada *et al* 2002), and the sterilization of heat-sensitive medical instruments (Laroussi 2002). These exciting applications would not have been possible were it not for the extensive basic research on the generation and sustainment of relatively large volumes of 'cold' plasmas at high pressures and with rela-tively small input power. However, as seen in the previous chapters of this book, in the case of air several challenges still remain to be overcome to arrive at an optimal generation scheme that is capable of producing large volume of air plasmas without a prohibitive level of applied power. Nonethe-less, as will be shown in this chapter, success in this research endeavor will potentially bring with it substantial economical and societal benefits. In particular, the semiconductor industry, chemical industry, food industry, and health and environmental industries, as well as the military stand to be great beneficiaries from the novel applications of 'cold' air plasmas.

In this chapter, several applications of non-equilibrium air plasma are covered in details by experts who have extensively contributed to this research. The selected applications are of the kind that have had or poten-tially will have a significant impact on industrial, health, environmental, or military sectors. The first two sections (9.2 and 9.3) discuss electrostatic pre-cipitation and ozone generation. This choice is motivated by the fact that historically these two applications of electrical discharges were the first to have been applied on a large industrial scale: electrostatic precipitation for the cleaning of air from fumes and particulates, and ozone generation for the disinfection of water supplies. Section 9.4 discusses the reflection and absorption of electromagnetic waves by air plasmas. This has direct applica-tions in military radar communications, and opens the possibility of using plasmas as a protective shield from radar and high power microwave weapons. Section 9.5 introduces the concept of using air plasmas to mitigate the effects of shock waves in supersonic/hypersonic flights. Plasma has been shown to reduce drag, which leads to lower thermal loading and higher fuel efficiency. Section 9.6 discusses the use of air plasma to enhance combustion. Ignition delays can be reduced and the combustion of hydrocarbon fuels can be increased by the presence of radicals generated by the plasma. Section 9.7 gives an extensive coverage of material processing by high-pressure non-equilibrium plasmas. The cleaning of surfaces, functionalization (such as for better adhesion), etching, and deposition of films are discussed and prac-tical examples are presented. Section 9.8 explores on the use of plasma discharges for the decomposition of NO_x and VOCs. All practical aspects of the decomposition processes are discussed in detail. Sections 9.9 and 9.10 introduce the reader to the biological and medical applications of 'cold' plasmas. The emphasis of section 9.9 is on the use of air plasma to inactivate bacteria efficiently and rapidly. The sterilization of heat-sensitive medical tools and food packaging and the decontamination of biologically contaminated surfaces are particularly attractive applications. The emphasis of section 9.10 is the use of 'bio-compatible' plasmas for *in vivo* treatment such as in electrosurgery. Cell detachment without damage using the 'plasma needle' is discussed. Wound healing is one example where 'bio-compatible' plasma sources can be used.

Research on non-equilibrium air plasmas has been to a large extent application-driven. Inter-disciplinary and cross-disciplinary efforts are necessary to drive plasma-based technology forward and into new fields and applications where air plasma has not been traditionally a component, but its use can substantially improve the established conventional processes.

References

Blakely E A, Bjornstad K A, Galvin J E, Montero O R and Brown I G 2002 'Selective neutron growth on ion implanted and plasma deposited surfaces' in *Proc. IEEE Int. Conf. Plasma Sci.*, Banff, Canada, p 253

Dorai R and Kushner M 2003 'A model for plasma modification of polypropylene using atmospheric pressure discharges', *J. Phys. D: Appl. Phys.* **36** 666

El-Habachi A and Schoenbach K H 1998 'Emission of excimer radiation from direct current, high pressure hollow cathode discharges' *Appl. Phys. Lett.* **72** 22

Eliasson B and Kogelschatz U 1991 'Non-equilibrium volume plasma processing' *IEEE Trans. Plasma Sci.* **19**(6) 1063

Laroussi M 2002 'Non-thermal decontamination of biological media by atmospheric pressure plasmas: Review, analysis and prospects', *IEEE Trans. Plasma Sci.* **30**(4) 1409, 1415

Sanchez-Estrada F S, Qiu H and Timmons R B 2002 'Molecular tailoring of surfaces via rf pulsed plasma polymerizations: Biochemical and other applications' in *Proc. IEEE Int. Conf. Plasma Sci.*, Banff, Canada, p 254,

Smulders E H W M, Van Heesch B E J M and Van Paasen B S V B 1998 'Pulsed power corona discharges for air pollution control' *IEEE Trans. Plasma Sci.* **26**(5) 1476

Stadler K 2003 'Plasma characteristics of electrosurgical discharges' in *Proc. Gaseous Electronics Conf.*, San Fransisco, CA, p 16

Stoffels E, Kieft I E and Sladek R E J 2003 'Superficial treatment of mammalian cells using plasma needle' *J. Phys. D: Appl. Phys.* **36** 1908

9.2   Electrostatic Precipitation

9.2.1   Historical development and current applications

The influence of electric discharges on smoke, fumes and suspended particles was described by William Gilbert as early as 1600. Gilbert acted as the president of the British Royal College of Physicians and also as physician to Queen Elizabeth I of England. His famous work *De Magnete* (on the magnet) was a comprehensive review of what was then known about electrical and magnetic phenomena. In 1824 Hohlfeld in Leipzig reported an experiment of clearing smoke in a jar by applying a high voltage to a corona wire electrode. Similar experiments were later repeated in Britain by Guitard in 1850 and by Lodge in 1884. Sir Oliver Lodge was the first to systematically investigate this effect and to put it to test on large scale in lead smelters at Bagillt in Flintshire, UK, to suppress the white lead fume escaping from the chimney (Hutchings 1885, Lodge 1886). To supply the corona current special electrostatic induction machines of the Wimshurst type were designed, with rotating glass plates of 1.5 m diameter. This can be considered the first, although not totally successful, commercial application of electrostatic precipitation for pollution control. The importance of this new 'electrical process of condensation for a possible purification of the atmosphere' was clearly recognized, and international patent coverage was obtained (Walker 1884). Practically simultaneously and independently a German patent was issued for a cylindrical precipitator (Möller 1884).

A number of important industrial applications followed the pioneering work of Frederick Gardner Cottrell, a professor of physical chemistry at the University of California-Berkeley. Starting in 1906 he conducted research on air pollution control, responding to growing nuisance caused by factories in his native San Francisco. The result was an improved precipitator, an electrical device, which could collect dusts and fumes as well as acid mists and fogs. Cottrell was the first to realize that for precipitation the negative corona discharge was superior to the positive corona, and who took advantage of the newly developed synchronous mechanical rectifier (Lemp 1904) and better high voltage step-up transformers. Within a few years commercial applications evolved for collecting sulfuric acid mists, for zinc and lead fumes, for cement kiln dust, for gold and silver recovery from electrolytic copper slimes, and for alkali salt recovery from waste liquors in paper-pulp plants (Cottrell 1911). In 1923 the first use of electrostatic precipitators (ESPs) collecting fly ash from a pulverized coal-fired power plant was reported. This process became by far the largest single application of ESPs. The fine wire corona discharge electrode, as it is used in many precipitators today, one of the most important advances in precipitator technology, was introduced and patented W A Schmidt (1920), a former student of Cottrell. In the following years investigations by Deutsch (1922, 1925a,b) and Seeliger (1926) brought new insight in the physical processes involved in electrostatic precipitation and a first quantitative formulation of precipitator performance. The Deutsch equation has been used ever since for sizing precipitators. For further details the reader is referred to the classical comprehensive treatment of industrial electrostatic precipitation by H J White (1963), to some more recent books (Oglesby and Nicholls 1978,

Cross 1987, Parker 1997) and to well written review articles (White 1957, 1977/78 1984, McLean 1988, Lawless *et al* 1995, Lawless and Altman 1999).

The main advantages of electrostatic precipitators are that various types of dust, mist, droplets etc. can be collected under both dry and wet conditions, and also that submicron size particles can be collected with high efficiency. ESPs can handle very large air or flue gas streams, typically at atmospheric pressure, with low power consumption and low pressure drop.

These properties have led to a number of large-scale commercial applications in the following industries: steel mills, non-ferrous metal processing, cement kilns, pulp/paper plants, power plants and waste incinerators, sulfuric acid plants, and in petroleum refineries for powder catalyst recovery. Much smaller ESPs of different design are used for indoor air cleaning in homes and offices.

9.2.2   Main physical processes involved in electrostatic precipitation

Electrostatic precipitation is a physical process in which particles suspended in a gas flow are charged electrically by ions produced in a corona discharge, are separated from the gas stream under the influence of an electric field, and are driven to collecting plates, from which they can be removed periodically by mechanical rapping (dry ESP) or continuously by washing (wet ESP). Typical configurations are corona wires centered in cylinders or wires mounted at the center plane between parallel plates forming ducts (figure 9.2.1).

The discharge electrodes can be simple weighted wires, barbed wires, helical wires, or rods, serrated strips and many other kinds. They all have in common that they have parts with a small radius of curvature or sharp edges to facilitate corona formation (see also chapters 2 and 6). The particle laden gas flow is channeled to pass through many cylinders or ducts either in

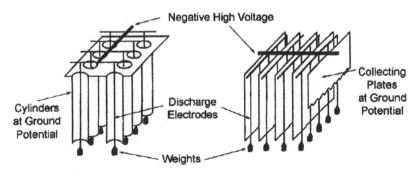

Figure 9.2.1. Cylindrical and planar precipitator configurations with weighted wire corona discharge electrodes.

the vertical (cylinders) or horizontal direction (ducts). In large precipitators negative coronas are used almost exclusively because they have a larger stability range and can be operated at higher voltages. For these devices electrode plate distances of 0.2–0.4 m and voltages in the range 50–110 kV are common. Small ESPs for indoor air cleaning normally use positive coronas, because they produce less ozone, a matter of great concern for indoor applications.

9.2.2.1 Generation of electrons and ions

The active corona region in which electrons as well as positive and negative ions are generated is restricted to a very thin layer around the corona electrodes. Typically ionization occurs only in a layer extending a fraction of 1 mm into the gas volume. Positive ions travel only a short distance to the negative electrode, while electrons and negative ions start moving towards the collecting surface at ground potential. In air or flue gas mixtures at atmospheric pressure electrons rapidly attach to O_2, CO_2 or H_2O molecules, thus forming negative ions. As a consequence, most of the space in the duct is filled with negative ions. They are utilized to charge dust particles so that these can be subjected to electrical forces in order to separate the dust from the gas stream. With modern computational tools it is possible to calculate the ion charge density distribution for complicated electrode structures. An example is given in figure 9.2.2 for one helical electrode (left part) and for three helical electrodes in a duct formed by specially shaped collecting plates (right part).

It is interesting to note that practically no ions are produced on the inner side of the helical discharge electrode (dark zone) because of shielding effects. The shape and orientation of the ion clouds in the duct depends very much on

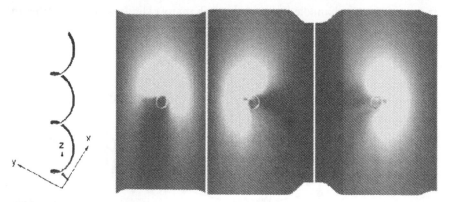

Figure 9.2.2. Ion charge density on a helical corona electrode and in three different horizontal planes of an ESP duct formed by specially shaped collecting plates (maximum charge density: 10^{-4} As m^{-3}).

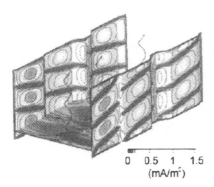
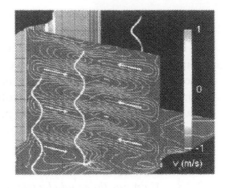

Figure 9.2.3. Current density on collecting plates and ion-induced secondary flow in an ESP duct with helical corona electrodes and specially formed collecting plates.

where the horizontal plane used in the visualization cuts the helix as well as on the location and shape of the closest collecting plane and on the distance to the neighboring electrodes. The complicated ion flow leads to a very inhomogeneous current density distribution on the collecting plates including zones of zero current density (figure 9.2.3). Such inhomogeneous current distributions were measured as well. They also show up in the deposited dust patterns.

9.2.2.2   Space charge limitations and saturation current

For practical purposes the active corona layer where ionization takes place can be regarded as very thin and as a copious source of charge carriers, in this case negative ions. The amount of current that is drawn depends on the characteristics of the ion drift region, which again depends on the applied voltage. The maximum current scales linearly with the ion mobility μ and with U^2, when U is the applied voltage. The current is limited by the space charge accumulated in the duct. A unipolar ion drift region can be described by the following set of equations:

$$\mathbf{E} = -\nabla\Phi = -\mathrm{grad}\,\Phi \tag{9.2.1}$$

$$\nabla^2\Phi = \mathrm{div}\,\mathrm{grad}\,\Phi = -\rho/\varepsilon_0 \tag{9.2.2}$$

$$\mathbf{j} = \rho\mu\mathbf{E} \tag{9.2.3}$$

$$\nabla\cdot\mathbf{j} = \mathrm{div}\,\mathbf{j} = 0. \tag{9.2.4}$$

In these equations E stands for the electric field, Φ for the potential, ρ for the ion space charge density, ε_0 for the vacuum permittivity (8.85×10^{-12} As/V m), and j for the current density. Poisson's equation (9.2.2) enforces a strong coupling between the ion space charge and the electric field. Adequate boundary conditions have to be formulated at the rim of the active corona region and at the collecting plane.

Because of this strong dependence on the voltage, ESPs operate at the maximum possible voltage stable corona discharge operation will allow. Since the highest possible voltage is beneficial both for charging and precipitation, ESPs are automatically controlled to run close to the sparking limit by allowing a certain number of sparks per unit of time to occur (up to 60 sparks per minute). Modern ESPs utilize all-solid-state high voltage rectifiers and microcomputer controls.

9.2.2.3   Main gas flow and electric wind

Ions, traveling in the duct at a speed of the order 100 m/s, move perpendicularly to the gas stream flowing at a speed of about 1 m/s. Since they have practically the same mass as the neutral components of the gas flow there is an efficient collisional momentum transfer. As a result strong secondary flows are induced. This phenomenon, referred to as the ion wind or electric wind, has been known for a long time and has been reviewed by Robinson (1962). At high applied voltages the magnitude of the ion-induced secondary flow component in an ESP becomes comparable to the main flow velocity. In a complicated electrode duct geometry like the helical discharge electrodes discussed earlier, this leads to stationary or oscillating vortex structures (Egli *et al* 1997), as demonstrated in the right-hand part of figure 9.2.3. The computed cross flow velocity distribution is shown in a vertical plane perpendicular to the main flow, located between the second and third helical discharge electrode.

As already suspected by Ladenburg and Tietze (1930) the electric wind can have a major adverse influence on particle collection. Recent 3D computations of corona charging, particle transport in the flow field and particle collection show that this is indeed the dominating effect at certain operating conditions (Egli *et al* 1997, Lowke *et al* 1998).

9.2.2.4   Particle charging

The physical processes involved in corona charging of powders and droplets have been studied in great detail. Apart from precipitators these phenomena are utilized in electrophotography (Crowley 1998), copying machines, printers, liquid spray guns, and in powder coating (Mazumder 1998). Solid particles or droplets entering a precipitator pass many corona zones, undergo collisions with ions resulting in charge accumulating, and are subjected to Coulomb forces in the electric field and to drag forces in the viscous flow. The charging process of solid particles or droplets has two main contributions, the relative importance of which depends on particle size. Field charging is the dominating process for particles of diameter of about 2 µm or more. It is described by the following differential equation:

$$\frac{dq_f}{dt} = p\pi r_p^2 \mu \rho E \left(1 - \frac{q_f}{q_s}\right)^2 \tag{9.2.5}$$

in which q_f is the accumulated particle charge due to field charging, $p = 3\varepsilon_r/(2 + \varepsilon_r)$, r_p is the particle radius, and q_s is the saturation charge. The parameter p depends on the relative dielectric constant ε_r of the particle and varies only moderately between the value $p = 1$ for $\varepsilon_r = 1$ and $p = 3$ for a metallic particle ($\varepsilon_r = \infty$). Charging stops when the saturation charge q_s is reached. At this point additional approaching ions will be deflected in the electric field of the previously accumulated charges on the particle and will no longer be able to impact.

$$q_s = 4\pi p \varepsilon_0 r_p^2 E. \tag{9.2.6}$$

At the ion densities and electric fields encountered in ESPs, field charging is a fast process. Its rate is proportional to the ion density, the cross section of the particle and to the electric field strength. Also the maximum attainable charge is proportional to the particle cross section and the electric field. Under typical precipitator conditions a 5 μm particle may accumulate several thousand elementary charges.

For very small particles with $r_p \leq 1$ μm, field charging gets very slow and another charging process depending on the Brownian motion of ions takes over (Fuchs 1964). This process is referred to as diffusion charging and follows a different law:

$$\frac{dq_d}{dt} = \frac{\mu \rho}{\varepsilon_0} \frac{q_d}{\exp\left(\dfrac{q_d \cdot e}{4\pi\varepsilon_0 r_p kT}\right) - 1} \tag{9.2.7}$$

where q_d is the particle charge accumulated due to diffusion charging, e is the elementary charge, k is the Boltzmann constant (1.38×10^{-23} J/K), and T is the gas temperature.

Diffusion charging is a much slower process than field charging. It does not depend on the electric field and does not reach a saturation charge. At the exit of a precipitator, after 10–15 s transit time, a 0.3 μm particle has accumulated about 100 elementary charges. The theoretical limit is reached (if ever) when the field at the particle surface has reached a value where gas breakdown is initiated. In the intermediate particle size rage 0.1 μm $< r_p <$ 10 μm both charging mechanisms are of comparable speed and occur simultaneously. The charging equations (9.2.5) and (9.2.7) have to be integrated along the particle trajectories, simultaneously with solving the coupled codes describing the corona discharge and the fluid phenomena (Choi and Fletcher 1997, Egli *et al* 1997, Meroth 1997, Gallimberti 1998, Medlin *et al* 1998). Instead of integrating (9.2.7) often a useful approximate relation for the charge q_d reached at time t is used:

$$q_d(t) = \frac{3r_p kT}{e} \ln(A\mu\rho t). \qquad (9.2.8)$$

In this relation, suggested by Kirsch and Zagnit'ko (1990), A is a constant. It shows that the charge obtained by diffusion charging is proportional to the gas temperature and that it grows with the logarithm of the time t.

9.2.3 Large industrial electrostatic precipitators

Industrial precipitators can be very large installations. As an example the precipitator at the exit of a pulverized-coal fired utility boiler of a 500 MW power plant is described. Coal consumption is about 200 tons per hour resulting in fly ash quantities of 20–80 tons per hour, depending on the origin and quality of coal. Fly ash particles range from 0.1 to 10 μm size. At the exit of the boiler they are dispersed in a flue gas stream of about 2.5 million m^3 per hour with a mass concentration of about $20\,g/m^3$. To meet tolerable output concentrations of $20\,mg/m^3$ the precipitator has to reach a weight collection efficiency of 99.9%. With modern technology this can be achieved. In extreme cases even 99.99% efficiency has been obtained. It is one of the major achievements of modern precipitator technology that these goals can be reached with an almost negligible power consumption of 0.1% of the generated power and a pressure drop of only 1 mbar.

9.2.3.1 *Structural design*

To handle such a large gas flow the flue gas is slowed down to about 1 m/s, channeled into many parallel ducts of 15 m height, up to 15 m length, and 0.3–0.4 m width. Such large ESPs are subdivided into fields of about 5 m length. About 110–150 such ducts add up to a total width of 45 m, being typically sectionalized into 3×15 m. In total $60\,000\,m^2$ of collecting area are provided. At the center plane of each duct the discharge electrodes are mounted. (See figure 9.2.4.) The helical electrodes shown in figures 9.2.2 and 9.2.3 have the advantage that, mounted under tension in metal frames at the center plane of each duct, they are always self centered. In addition, rapping of the metal frames induces vibration of the discharge electrodes, thus efficiently cleaning them of deposited fly ash. The charged particles impinging on the collecting plates, usually made of mild steel, and kept at ground potential, form a dust cake, which is held in position by electric forces. It is removed periodically by mechanical rapping using either side- or top-mounted hammers. Upon rapping the collected material is dislodged and slides down into hoppers at the bottom from where it is removed by conveyor belts. The special shape of the collecting plates indicated in figures 9.2.2 and 9.2.3 is chosen to give them mechanical strength and to reduce rapping-induced re-entrainment of already collected material.

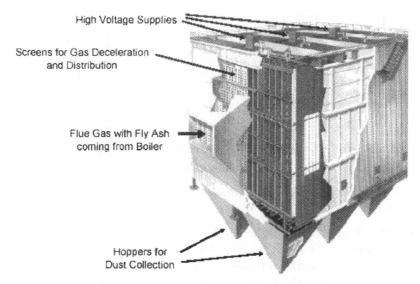

High Voltage Supplies

Screens for Gas Deceleration
and Distribution

Flue Gas with Fly Ash
coming from Boiler

Hoppers for
Dust Collection

Figure 9.2.4. Structure of a large precipitator behind a coal-fired utility boiler (Fläkt design).

9.2.3.2 Numerical modeling

For many years ESPs have been sized according to the Deutsch equation which was derived in 1922 and which, for the first time, established a quantitative relation between the collection efficiency η of a precipitator and some operational and geometry parameters:

$$\eta = 1 - (c_{exit}/c_0) = 1 - \exp(-wA/Q). \qquad (9.2.9)$$

The quantities c_{exit} and c_0 are the dust concentrations at the exit and entrance of the precipitator, respectively, A is the total collection area and Q is the volumetric gas flow. The parameter w has the dimension of a velocity and is called the migration velocity. For ESP sizing this parameter was determined empirically and contained all the pertinent information about precipitator design, dust properties and corona operation.

With a better understanding of all the physical processes involved, and taking advantage of fast computers and advanced computational tools, individual particle paths can now be followed through a large industrial precipitator. This approach requires that sufficiently accurate computational models are available for the field distribution and ion production, the charging process, the flow field and the particle motion. Since there is a strong interaction between the different processes involved the differential equations describing the different processes have to be solved simultaneously with appropriate boundary conditions. As an example some results are given of numerical studies in which individual particle paths where followed through a 12 m long ESP duct in which they passed 45 helical corona

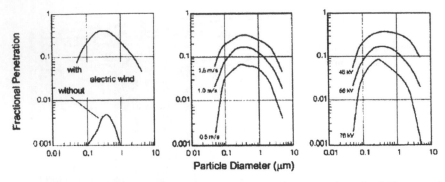

Figure 9.2.5. Fractional particle penetration curves demonstrating the influence of different parameters.

electrodes (Kogelschatz *et al* 1999). For each size class 2000 particles with different initial positions at the entrance were traced.

The plots, referred to as penetration curves, show the fraction of particles that are able to pass the whole precipitator without getting collected, as a function of particle size. The left-hand part of figure 9.2.5 demonstrates the overwhelming influence of the electric wind. If it were not present, collection would improve by more than 2 orders of magnitude. In the model computation this was simulated by switching off the electric volume forces on the flow. These computations were performed for the specially formed collecting plates (figures 9.2.2, 9.2.3), a 0.4 m duct, an initial flow velocity of 1 m/s and a corona voltage of 56 kV. The middle graph of figure 9.2.5 shows results for different flow velocities at a fixed voltage of 56 kV in a 0.4 m duct with planar walls. Clearly, slower transport velocity, and consequently longer residence time, results in better particle collection. The right-hand part shows the influence of the applied voltage for a fixed initial flow velocity of 1 m/s. All computations show that there is a particle size range between 0.1 and 1 μm diameter that is difficult to collect. Larger particles are more efficiently collected because they accumulate sufficient charge in the corona zones and are subjected to strong electric forces. Very small particles are also easily collected despite the reduced electric forces. The reason is that they experience less flow resistance when particle diameters approach the mean free path of the gas molecules (Cunningham slip). Measurements of particle size distributions at the entrance and exit of large industrial precipitators yield the same form of the penetration curves. Such numerical simulations, based on the fundamental physical processes and validated in real situations, have become a powerful tool for optimizing ESP performance.

9.2.3.4 *Limitations by corona quenching and dust cake resistivity*

The practical performance of electrostatic precipitators can be limited by additional effects not mentioned so far. If large amounts of fine dust enter the precipitator, the corona current in the entrance sections can drop to a small fraction of what it had been without dust. This very pronounced effect is called corona quenching. The reason is that the properties of the corona discharge that were originally determined by ion mobility and ion space charge are now determined by the much smaller dust mobility and the dust space charge. Fortunately, after collecting most of this fine dust, the corona recovers to its original current density, typically after a few meters in the duct.

The collected material on the collecting plates can also pose limitations on electrostatic precipitation. If particles have a very low electrical resistivity, for example metal particles, they do not adhere to the collecting plates, thus preventing collection. On the other hand, if dust resistivity is very high, one might expect that the deposited dust layer would finally limit the current flow and stop the corona. Normally a different phenomenon, called back corona, occurs instead. Since the deposited dust forms a porous layer of growing thickness and voltage drop, gas breakdown in interstices and on particle surfaces can occur. When this happens, the corona current suddenly increases and collection is severely effected. Now positive ions, generated by back corona inside the dust cake, travel towards the center electrodes and counteract the charging process with negative ions. This results in what is called a bipolar corona. Obviously, for optimum charging conditions we depend on a unipolar ion flow.

Back corona is observed in precipitators serving boilers using low sulfur coal and also in powder coating, where high resistivity polymer particles and pigments are deposited. It was first observed by Eschholz in 1919. The described effects limit the useful range of electrostatic precipitators to material with resistivity in the range of about $10^8 \, \Omega \cdot cm$ to less than $10^{13} \, \Omega \cdot cm$. The resistivity range for optimum ESP performance is 10^8 to $10^{10} \, \Omega \cdot cm$. In many cases high dust resistivity can be reduced by raising the temperature or by conditioning, which means by using additives like H_2O or SO_3. The cohesive properties of the dust cake can be influenced by adding NH_3 to the gas stream. It is also possible to detect malfunctioning of a precipitator section as a consequence of corona quenching or back corona and counteract by modifying the electrical feeding of the corona.

9.2.4 Intermittent and pulsed energization

In many cases pronounced improvement of ESP performance has been obtained by abandoning the classical dc high voltage on the discharge electrodes. Microprocessor control of the supply voltage allows simple variations in the way the corona discharge in ESPs is fed. Intermittent energization can be achieved by suppressing voltage half cycles or even several cycles in the rectifier circuit. This way, peak voltages higher than those achievable with dc energization, and lower average voltages and average currents are obtained. In addition to energy savings this can result in improved performance if back corona is a problem.

Even better results can be obtained if pulsed energization is used. This technique originated about 1950 following pioneering research and development by Hall and White (Hall 1990). We speak of a pulsed corona if the duration of the applied voltage pulse is shorter than the ion transit time from the discharge electrode to the collecting plate. In a large ESP this is typically of the order 1 ms. Using this technique, periodic short high-voltage pulses are superimposed on a dc high voltage. Typical pulse widths of <1 μs to about 300 μs and repetition rates of about 30 to 300 per second are used. Pulsed energization introduces a number of new parameters that can be optimized: pulse duration, pulse repetition frequency, base dc voltage. It increases the uniformity of the corona along the discharge electrodes and on the collecting plates. It helps to suppress back corona in the collection of high resistivity dust. Experience shows that application of short HV pulses to high resistivity dusts of 10^{10}–10^{13} Ω·cm results in significant performance improvement over that achievable with dc energization.

In conclusion it can be stated that electrostatic precipitation is the leading and most versatile procedure for high-efficiency collection of solid particles, fumes and mists escaping from industrial processes. It presents by far the most important application of industrial air pollution control. About one hundred years of practical experience with various kinds of dust, a growing understanding of the physical processes involved, and more recently, the use of advanced computational tools simulating the whole particle charging, particle motion and collection process have led to its present supremacy.

References

Choi B S and Fletcher C A J 1997 *J. Electrost.* **40/41** 413–418
Cottrell F C 1911 *J. Ind. Eng. Chem.* **3** 542–550
Cross J A 1987 *Electrostatics: Principles, Problems and Applications* (Bristol: Adam Hilger)
Crowley J M 1998 'Electrophotography' in *Wiley Encyclopedia of Electrical and Electronic Engineering* Webster J G (ed) (New York: Wiley-Interscience) vol 6, pp 719–734
Deutsch W 1922 *Ann. Phys.* **68** 335–344
Deutsch W 1925a *Z. Techn. Phys.* **6** 423–437
Deutsch W 1925b *Ann. Phys.* **76** 729–736
Egli W, Kogelschatz U, Gerteisen E A and Gruber R 1997 *J. Electrostat.* **40/41** 425–439
Eschholz O H 1919 *Trans. Am. Inst. Mining Metall. Eng.* **LX** 243–279
Fuchs N A 1964 *The Mechanics of Aerosols* (Oxford: Pergamon)
Gallimberti I 1998 *J. Electrostat.* **43** 219–247
Gilbert W 1600 *Tractatus, sive Physiologia de Magnete, Magnetisque corporibus magno Magnete tellure, sex libris comprehensus* (London: Excudebat Petrus Short)
Guitard C F 1850 *Mech. Mag. (London)* **53** 346

Hall H J 1990 *J. Electrostat.* **25** 1–22

Hohlfeld M 1824 *Arch. f. d. ges. Naturl.* **2** 205–206

Hutchings W M 1885 *Berg- u Hüttenmänn Zeitschr.* **44** 253–254

Kirsch A A and Zagnit'ko A V 1990 *Aerosol Sci. Technol.* **12** 465–470

Kogelschatz U, Egli W and Gerteisen E A 1999 *ABB Rev.* **4/1999** 33–42

Ladenburg R and Tietze W 1930 *Ann. Phys.* **6** 581–621

Lawless P A and Altman R F 1999 'Electrostatic precipitators' in *Wiley Encyclopedia of Electrical and Electronic Engineering*, Webster J G (ed) (New York: Wiley-Interscience) vol 7 pp 1–15

Lawless P A, Yamamoto T and Oshani 1995 'Modeling of electrostatic precipitators and filters' in *Handbook of Electrostatic Processes*, Chang J S, Kelly A J and Crowley J M (eds) (New York: Marcel Dekker) pp 481–507

Lemp H 1904 Alternating current selector, US Pat No. 774,090

Lodge O J 1886 *J. Soc. Chem. Ind.* **5** 572–576

Lowke J J, Morrow R and Medlin A J 1998 *Proc. 7th Int. Conf. on Electrostatic Precipitation (ICESP VII)*, Kyonju, Korea 1998, pp 69–75

Mazumder M K 1999 'Electrostatic processes' in *Wiley Encyclopedia of Electrical and Electronic Engineering*, Webster J G (ed) (New York: Wiley-Interscience) vol 7 pp 15–39

McLean K J 1988 *IEE Proc.* **135** 347–361

Medlin A J, Fletcher C A J and Morrow R 1998 *J. Electrostat.* **43** 39–60

Meroth A M 1997 *Numerical Electrohydrodynamics in Electrostatic Precipitators* (Berlin: Logos-Verlag)

Möller K 1884 *Röhrenförmiges Gas und Dampffilter*, German Pat. No. 31911

Oglesby S and Nichols G 1978 *Electrostatic Precipitation* (New York: Decker)

Parker K R (ed) 1997 *Applied Electrostatic Precipitation* (London: Blackie)

Robinson M 1962 *Am. J. Phys.* **30** 366–372

Schmidt W A 1920 *Means for separating suspended matter from gases*, US Pat. No. 1,343,285

Seeliger R 1926 *Z. Techn. Phys.* **7** 49–71

Walker A O 1884 *A process for separating and collecting particles of metals or metallic compounds applicable for condensing fumes from smelting furnaces and for other purposes*, Brit Pat No. 11,120

White H J 1957 *J. Air Poll. Contr. Ass.* **7** 167–177

White H J 1963 *Industrial Electrostatic Precipitation* (Reading: Addison-Wesley)

White H J 1977/78 *J. Electrostat.* **4** 1–34

White H J 1984 *J. Air Poll. Contr. Ass.* **34** 1163–1167

9.3 Ozone Generation

9.3.1 Introduction: Historical development

In 1785 the natural scientist Martinus van Marum described a characteristic odor forming close to an electrostatic machine, and in 1801 Cruikshank, performing water electrolysis, noticed the same odor at the anode. Only in 1839 Schönbein, professor at the University of Basel, also working on electrolysis, established that this very pronounced smell was due to a new chemical compound which he named ozone after the Greek word οζειν for to reek or smell. It took another 25 years of scientific vehement dispute before J L Soret could establish in 1865 that this new compound was made up of three oxygen atoms.

Industrial ozone generation is the classical application of non-equilibrium air plasmas at atmospheric pressure. Low temperature is mandatory because ozone molecules decay fast at elevated temperature. At the same time a relatively high pressure is required because ozone formation is a three-body reaction involving an oxygen atom, an O_2 molecule and a third collision partner, O_2 or N_2. The dielectric barrier discharge (silent discharge) originally proposed by Siemens (1857) for 'ozonizing air' is ideally suited for this purpose. Siemens' invention came at the right time. The foundations of bacteriology had been laid through the work of the French microbiologist Louis Pasteur and the German district surgeon Robert Koch. It had been established that infectious diseases like cholera and typhoid fever were caused by living micro-organisms, which were dispersed by contaminated drinking water, food and clothing. Cholera epidemics like the ones reported in Hamburg (1892) and in St Petersburg (1908) caused hundreds of casualties per day. Occasional typhoid fever epidemics were common in many cities.

Ozone is an extremely effective oxidant, surpassed in its oxidizing power only by fluorine or radicals like OH or O atoms. Siemens succeeded in persuading Ohlmüller, professor at the Imperial Prussian Department of Health, to test the effect of ozone exposure on cholera, typhus and coli bacteria. The result was complete sterility after ozone treatment. Soon after the first official documentation of these bactericidal properties (Ohlmüller 1891), industrial ozone production started for applications in small water treatment plants in Oudshoorn, Holland (1893) and in Wiesbaden and Paderborn, Germany (1901/2). Within the following years major drinking water plants using ozone disinfection were built in Russia (St Petersburg 1905), in France (Nice 1907, Chartres 1908, Paris 1909) and in Spain (Madrid 1910). The water works at St Petersburg already treated $50\,000\,m^3$ of drinking water per day with ozone, those of Paris $90\,000\,m^3$. Thus, historically speaking, ozonation was the first successful attempt of disinfecting drinking water on a large scale. Ever since, ozone generating technology has been closely linked to the development of water purification processes. In many countries ozonation in water treatment was later replaced by more cost-effective processes using chlorine or chlorine compounds, which are not only cheaper but also more soluble in water than ozone. Recent concerns about potentially harmful disinfection by-products have reversed this, tending towards the use of ozone again. Many European cities and some Canadian cities have abandoned chlorination in favor of ozone technology to disinfect water. Water works in the US as well as in Japan are increasingly turning to ozone, in order to be able to meet more stringent legislation about disinfection by-products like trihalomethanes (THMs) and haloacetic acids. These compounds can be formed when chlorine is added to the raw water containing organic water pollutants or humic materials. Some THMs are suspected to cause cancer. For this reason many experts consider ozone treatment the technology of choice for potable water treatment. In the United States more than 250 operating plants use ozone. For many years the Los Angeles Aqueduct Filtration Plant treating two million m^3/day (600 mgd) of drinking water with ozone generating capacity of close to 10 000 kg per day, was the largest US plant. Very recently larger ozone generating facilities have been installed at the Alfred Merrit Water Treatment Plant in Las Vegas, the East Side Water Treatment Plant in Dallas, Texas, and the Metropolitan Water District in Southern California. In Europe, more than 3000 cities use ozone to disinfect their municipal water supplies.

9.3.2 Ozone properties and ozone applications

O_3 is a triangle shaped molecule with a bond angle of 117° and equal bond lengths of 0.128 nm. Ozone is a practically colorless gas with a characteristic pungent odor (Horváth *et al* 1985, Wojtowicz 1996). At -112 °C it condenses to an indigo blue liquid which is highly explosive. Below -193 °C ozone forms a deep blue-violet solid. Because of explosion hazards ozone is used only in diluted form in gas or water streams. Its solubility is about 1 kg per m^3 of water. Due to its oxidizing power it finds applications as a potent germicide and viricide as well as a bleaching agent. In many applications ozone is increasingly used to replace other oxidants such as chlorine that present more environmental problems and safety hazards. Strong oxidants are chemically active species. Their storage, handling and transportation involve substantial hazards. An important issue is also the question of residues and side reactions. In all respects ozone represents a superior choice due to its innocuous side product, oxygen. As a consequence of its inherent instability ozone is neither stored nor shipped. It is always generated on the site at a rate controlled by its consumption in the process.

The most important application of ozone is still for the treatment of water. It is capable of oxidizing many organic and inorganic compounds in water. Ozone chemistry in water involves the generation of hydroxyl free radicals, very reactive species approaching diffusion controlled reaction rates for many solutes such as aromatic hydrocarbons, unsaturated compounds, aliphatic alcohols, and formic acid (Glaze and Kang 1988, Hoigné 1998). Besides applications in drinking water, ultra-pure process water, swimming pools, and cooling towers, ozone also finds applications in municipal waste water treatment plants and in industrial processes. Very large amounts of ozone are also used for pulp bleaching.

9.3.3   Ozone formation in electrical discharges

Ozone can be generated in different types of gas discharges in which the electron energy is high enough to dissociate O_2 molecules and in which the gas temperature can be kept low enough for the O_3 molecules to survive without undergoing thermal decomposition. Mainly non-equilibrium discharges can meet these requirements, above all corona discharges and dielectric barrier discharges.

9.3.3.1   *Ozone formation in corona discharges*

Ozone formation in both positive and negative corona discharges has been extensively investigated and is reasonably well understood. Ozone formation is restricted to the thin active corona region where ionization takes place. Since it is rarely used on an industrial scale it will not be treated in detail. The reader is referred to the following references: Peyrous (1986, 1990), Peyrous *et al* (1989), Boelter and Davidsen (1997), Held and Peyrous (1999), Yehia *et al* (2000), Chen (2002), and Chen and Davidson (2002, 2003a,b).

9.3.3.2   *Ozone formation in dielectric barrier discharges*

The preferred discharge type for technical ozone generators has always been the dielectric barrier discharge (silent discharge) as originally proposed by Siemens. In recent years industrial ozone generation profited substantially from a better understanding of the discharge properties and of the ozone formation process (Filippov *et al* 1987, Kogelschatz 1988, 1999, Samoilovich *et al* 1989, Braun *et al* 1991, Kogelschatz and Eliasson 1995, Pietsch and Gibalov 1998). Operating in air or oxygen at pressures between 1 and 3 bar, at frequencies between 0.5 kHz and 5 kHz, and using gap spacings in the mm range the discharge is always of the filamentary type. Major improvements were obtained by tailoring microdischarge properties in air or in oxygen in such a way that recombination of oxygen atoms is mimimized and ozone formation is optimized. This can be achieved by adjusting the width of the discharge space, the operating pressure, the properties of the dielectric barrier, and the temperature of the cooling medium. Changing the operating frequency has little influence on individual microdischarge properties. The power dissipated in the discharge is determined by the amplitude and frequency of the operating voltage. In connection with the cooling circuit, it determines the average temperature in the discharge gap. Cylindrical as well as planar electrode configurations have been used. The majority of commercial ozone generators use cylindrical electrodes forming narrow annular discharge spaces of 0.5–1 mm radial width. The outer electrode is normally a stainless steel tube, which is at ground potential and which is

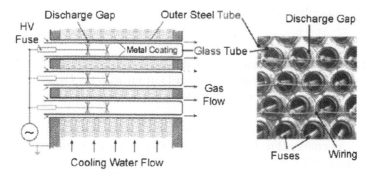

Figure 9.3.1. Configuration of water-cooled discharge tubes in an ozone generator.

water-cooled. These tubes have a length of 1–4 m. The coaxial inner electrode is a glass or ceramic tube, closed at one side, and having an inner metal coating as a high voltage electrode (figure 9.3.1), or a closed steel cylinder which is covered by a dielectric layer (ceramic, enamel). The feed gas is streaming in the axial direction through the annular discharge region between the inner and outer tube. Each volume element of the flowing gas is subjected to the action of many microdischarges and leaves enriched with ozone.

9.3.4   Kinetics of ozone and nitrogen oxide formation

Any electric discharge in air or oxygen causes chemical changes induced by reactions electrons or ions with N_2, O_2 or trace elements like H_2O and CO_2 and subsequent free radical reactions. Extensive lists of possible reactions have been collected, and reliable sets of rate coefficients have been established (Krivosonova *et al* 1991, Kossyi *et al* 1992, Herron 1999, 2001, Herron and Green 2001, Sieck *et al* 2001). As far as ozone formation is concerned, extensive reaction schemes also exist (Yagi and Tanaka 1979, Samoilovich and Gibalov 1986, Eliasson and Kogelschatz 1986a,b, Eliasson *et al* 1987, Braun *et al* 1988, Peyrous 1990, Kitayama and Kuzumoto 1997, 1999). It turns out that ion reactions play only a minor role and that the main trends can be described by tracing the reactions of the atoms generated by electron impact dissociation of O_2 and N_2 and those of a few excited molecular states.

9.3.4.1   Ozone formation in oxygen

In pure oxygen, which is actually used in many large ozone generation facilities, ozone formation is a fairly straightforward process. Ozone always originates from a three body reaction of oxygen atoms reacting with $2O_2$ molecules:

$$O + O_2 + O_2 \longrightarrow O_3^* + O_2 \longrightarrow O_3 + O_2 \qquad (9.3.1)$$

where O_3^* stands for a transient excited state in which the ozone molecule is initially formed after the reaction of an O atom with an O_2 molecule. The time scale for ozone formation in atmospheric pressure oxygen is a few microseconds.

O is formed in reaction of electrons with O_2 after excitation to the $A\,^3\Sigma_u^+$ state with an energy threshold of about 6 eV and via excitation of the $B\,^3\Sigma_u^-$ state starting at 8.4 eV.

Fast side reactions, also using O atoms or destroying O_3 molecules, compete with ozone formation.

$$O + O + O_2 \longrightarrow 2O_2 \qquad (9.3.2)$$

$$O + O_3 + O_2 \longrightarrow 3O_2 \qquad (9.3.3)$$

$$O(^1D) + O_3 \longrightarrow 2O_2 \qquad (9.3.4)$$

$$O + O_3^* + O_2 \longrightarrow 3O_2. \qquad (9.3.5)$$

The undesired side reactions (9.3.2)–(9.3.5) pose an upper limit on the atom concentration, or the degree of dissociation, tolerable in the microdischarges. Since equation (9.3.2) is quadratic in atom concentration while the ozone formation equation (9.3.1) is linear one would expect that extremely low atom concentrations are preferable. Computations with large reactions schemes show that complete conversion of O to O_3 can only be expected if the relative atom concentration $[O]/[O_2]$ stays below 10^{-4}. There are other considerations, however, that exclude the use of extremely weak microdischarges. If the energy density in a microdischarge and consequently also the degree of dissociation is too low, a considerable fraction of the deposited energy is dissipated by ions (up to 50%). Since ions do not appreciably contribute to ozone formation this situation has to be avoided. A reasonable compromise between excessive energy losses due to ions and best use of O atoms for ozone formation is found when the relative oxygen atom concentration in a microdischarge reaches about 2×10^{-3} in the microdischarge channel. This concentration can be obtained at an energy density of about $20\,\text{mJ/cm}^{-3}$ (Eliasson and Kogelschatz 1987). In this case energy losses to ions are negligible and 80% of the oxygen atoms are utilized for ozone formation. At zero ozone background concentration this leads to a maximum energy efficiency of ozone formation corresponding to roughly 25%. The efficiency of ozone formation is normally related to the enthalpy of formation, which is 1.48 eV/O_3 molecule or 0.82 kWh/kg. Thus 100% efficiency corresponds to the formation of 0.68 O_3 molecules per eV or 1.22 kg ozone per kWh. The indicated reaction paths requiring dissociation of O_2 first (dissociation energy: 5.16 eV) puts an upper limit at 0.7 kg/kWh.

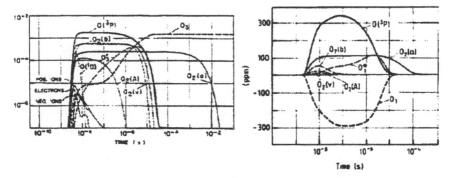

Figure 9.3.2. Evolution of particle species after a short current pulse: with zero ozone background concentration (left) and at the saturation limit (right) ($p = 1$ bar, $T = 300$ K).

If the electron energy distribution in oxygen is considered, and the combined actual dissociation processes at 6 and 8.4 eV, this value is further reduced to 0.4 kg/kWh. The best experimental laboratory values obtained at vanishing O_3 background concentration are in the range 0.25–0.3 kg/kWh.

The ozone concentration in the gas stream passing through the ozone generator is built up due to the accumulated action of a large number of microdischarges. With increasing ozone concentration back reactions gain importance. In addition to the already mentioned reactions equations (9.3.2)–(9.3.5), O_3 reactions with electrons and excited O_2 molecules have to be considered. This finally leads to a situation where each additional microdischarge destroys as much ozone as it generates (figure 9.3.2, right-hand section). The attainable saturation concentration defined by this equilibrium depends strongly on pressure and on gas temperature.

9.3.4.2 Ozone formation in dry air

In air the situation is more complicated. The presence of nitrogen atoms and excited atomic and molecular species as well as the nitrogen ions N^+, N_2^+, N_4^+ add to the complexity of the reaction system. Again, ions are of minor importance for ozone formation. Excitation and dissociation of nitrogen molecules, however, lead to a number of additional reaction paths involving nitrogen atoms and the excited molecular states $N_2(A\,^3\Sigma_u^+)$ and $N_2(B\,^3\Pi_g)$, that can produce additional oxygen atoms for ozone generation.

$$N + O_2 \longrightarrow NO + O \tag{9.3.6}$$

$$N + NO \longrightarrow N_2 + O \tag{9.3.7}$$

$$N + NO_2 \longrightarrow N_2O + O \tag{9.3.8}$$

$$N_2(A, B) + O_2 \longrightarrow N_2 + 2O \tag{9.3.9}$$

$$N_2(A) + O_2 \longrightarrow N_2O + O. \tag{9.3.10}$$

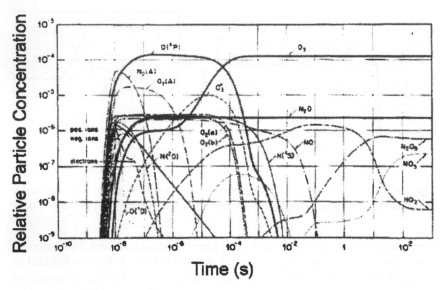

Figure 9.3.3. Evolution of particle species after a short current pulse in a mixture of 80% N_2 and 20% O_2 simulating dry air ($p = 1$ bar, $T = 300$ K).

These oxygen atoms, generated in addition to those obtained from direct electron impact dissociation of O_2, contribute about 50% of the ozone formed in air, which now takes longer, roughly about 100 µs. The result is that a substantial fraction of the electron energy initially lost in collisions with nitrogen molecules can be recovered and utilized for ozone generation through reactions (9.3.6)–(9.3.10). In addition to ozone a variety of nitrogen oxide species are generated: NO, N_2O, NO_2, NO_3, and N_2O_5. All these species have been measured at realistic ozone generating conditions (Eliasson and Kogelschatz 1987, Kogelschatz and Baessler 1987). In the presence of ozone only the highest oxidation stage N_2O_5 is detected in addition to the rather stable molecule N_2O (nitrous oxide, laughing gas). Figure 9.3.3 shows results of a numerical simulation using a fairly extended reaction scheme in dry air (20% O_2, 80% N_2). The formation of ozone and different NO_x species due to a single short discharge pulse is followed for a reasonably long time.

A few results demonstrating special characteristics of ozone generation in air are added. The maximum attainable energy efficiency is reduced to about to 0.2 kg/kWh and it shifted to higher reduced electric field values (200–300 Td). This has to be expected because dissociation of N_2 requires higher electron energies.

The maximum attainable ozone concentration is lower and, surprisingly enough, no saturation concentration exists. When the power is increased or the air flow is reduced, the ozone concentration passes through a maximum and then decreases again until it drops to zero. This effect, referred to as discharge poisoning, was reported by Andrews and Tait (1860), only a few years after Siemens had presented his ozone discharge tube. The poisoning effect was correctly associated with the presence of nitrogen oxides. Today we know that catalytic processes involving the presence of NO and NO_2 can use up O atoms at a fast rate thus preventing O_3 formation and can also destroy already formed ozone. This is a phenomenon that involves only fast chemical reactions between neutral particles and has little influence on electrical discharge parameters. Addition of 0.1% NO or NO_2 to the feed gas of an ozone generator can completely suppress ozone formation. In the absence of ozone only NO, NO_2 and N_2O can be detected at the exit. In dry air the catalytic reactions leading to enhanced removal of O and O_3 are as follows:

$$O + NO + M \longrightarrow NO_2 + M \qquad (9.3.11)$$

$$\underline{O + NO_2 \qquad \longrightarrow NO + O_2} \qquad (9.3.12)$$

$$O + O \qquad \longrightarrow O_2 \qquad (9.3.13)$$

and

$$O + NO_3 \longrightarrow NO_2 + O_2 \qquad (9.3.14)$$

$$\underline{O + NO_2 \longrightarrow NO + O_2} \qquad (9.3.15)$$

$$O + O_3 \qquad \longrightarrow 2O_2 \qquad (9.3.16)$$

These NO_x reactions also play a dominant role in atmospheric chemistry (Crutzen 1970, Johnston 1992).

9.3.4.3   Ozone formation in humid oxygen and air

The situation is further complicated if water vapor is present in the feed gas. Even traces of humidity drastically change the surface conductivity of the dielectric. At the same electrical operating conditions fewer and more intense microdischarges result. In addition, a strong influence on major reaction paths results from the presence of OH and HO_2. The hydroxyl radical OH is formed by electron impact dissociation of H_2O and, in most cases more importantly, by fast reactions of electronically excited oxygen atoms and nitrogen molecules:

$$e + H_2O \longrightarrow e + OH + H \qquad (9.3.17)$$

$$O(^1D) + H_2O \longrightarrow 2OH \qquad (9.3.18)$$

$$N_2(A\,^3\Sigma_u^+) + H_2O \longrightarrow N_2 + OH + H. \qquad (9.3.19)$$

HO_2 is then formed in a reaction of OH radicals with ozone:

$$OH + O_3 \longrightarrow HO_2 + O_2. \qquad (9.3.20)$$

The presence of OH and HO_2 can limit ozone production in oxygen by introducing a further catalytic ozone destruction cycle:

$$OH + O_3 \longrightarrow HO_2 + O_2 \tag{9.3.21}$$

$$\underline{HO_2 + O_3 \longrightarrow OH + 2O_2} \tag{9.3.22}$$

$$2O_3 \qquad \longrightarrow 3O_2 \tag{9.3.23}$$

In air an additional fast NO oxidation reaction occurs:

$$NO + HO_2 \longrightarrow NO_2 + OH. \tag{9.3.24}$$

The main paths for NO removal in wet air are oxidation to NO_2 and fast conversion to HNO_2 and HNO_3.

$$NO + OH + M \longrightarrow HNO_2 + M \tag{9.3.25}$$

$$NO_2 + OH + M \longrightarrow HNO_3 + M. \tag{9.3.26}$$

9.3.5 Technical aspects of large ozone generators

Large ozone generators use several hundred discharge tubes and now produce up to 100 kg ozone per hour. In most water works several ozone generators are installed. Figure 9.3.4 shows a photograph of the entrance section of a large ozone generator. One can see the glass tubes mounted in slightly wider steel tubes, the high voltage fuses at the center of each tube

Figure 9.3.4. Large ozone generator at the Los Angeles Aqueduct Filtration Plant.

and the electric wires connecting them. Depending on the feed gas, ozone concentration up 5 wt% (from air) or up to 18 wt% (from oxygen) can be obtained. Advanced water treatment processes utilize ozone at concentrations up to 12 wt%. Depending on the desired ozone concentration the energy required to produce 1 kg of O_3 ranges from 7.5 to 10 kWh in oxygen and from about 15 to 20 kWh in air. Information on the technical aspects of ozone generation and ozone applications can be found in Rice and Netzer (1982, 1984) or in Wojtowicz (1996).

9.3.5.1 *Design aspects and tolerances*

To obtain such performance several design criteria and operating conditions have to be met. The desired small width of the discharge gap in the range 0.5–1 mm puts severe tolerance limits on the diameters and on the straightness of the cylindrical dielectric and steel tubes. It is essential that the inner dielectric tube is perfectly centered inside the outer steel tube. Even a small displacement results in a drastic drop of performance. Microdischarge efficiency, heat removal and axial flow velocity depend strongly on the width of the discharge gap, which must be kept in tight tolerances. Also the pressure has to be kept close to the design value, about 2 bar in O_2 and closer to 3 bar in air. For a given dielectric tube its optimum value depends on the desired ozone concentration, the gap width, the temperature of the cooling fluid, and the power density the ozone generator is operated at.

9.3.5.2 *Feed gas preparation*

The feed gas for most ozone generators is air or oxygen. In large installations operating at high ozone concentrations and power density also O_2 with a small admixture of N_2 is used. It is essential that the feed gas contains only a few ppm H_2O (dew point below −60 °C). As mentioned above, humidity has a strong influence on the surface conductivity of the dielectric and on the properties of the microdischarges. In addition, we observe the changes in the chemical reaction scheme as described in section 9.3.4.3. Also traces of other impurities like H_2, NO_x and hydrocarbons have an adverse influence on ozone formation. Some of them lead to a catalytically enhanced recombination of O atoms, others to catalytic ozone destruction cycles.

These requirements necessitate a feed gas preparation unit to remove humidity even if air is used. For this reason many large ozone installations use oxygen as a feed gas. If cryogenic oxygen is used one has to be aware of the fact that in polluted areas hydrocarbons may accumulate in the liquid oxygen. Oxygen prepared by pressure swing or vacuum swing adsorption–desorption techniques, on the other hand, is practically free of hydrocarbons (<1 ppm).

9.3.5.3   Heat balance and cooling circuit

The ozone formation efficiency and the stability of the O_3 molecule deteriorate at elevated temperature. As a consequence only non-equilibrium discharges are suited for ozone generation and efficient cooling of the discharge gap is mandatory. This is the reason why ozone generators are essentially built like heat exchangers. The average temperature increase due to discharge heating in the narrow annular discharge gap can be approximated by a simple formula. After a few cm of entrance length stationary radial profiles of velocity and temperature are established. The radial temperature profile is a half parabola with its maximum at the inner uncooled dielectric tube if a uniform power deposition in the discharge is assumed. The average temperature increase in the gap ΔT_g is then determined by the power dissipated in the discharge and the heat removed through the cooled steel electrode and kept at the wall temperature T_w. Unfortunately, only a minor fraction of the energy is used for ozone formation (efficiency: η).

$$\Delta T_g = \frac{1}{3}\frac{d}{\lambda}\frac{P}{F}(1-\eta). \qquad (9.3.26)$$

In this formula d is the gap width, λ is the heat conductivity of the feed gas (discharge plasma) and P/F is the power density referred to the electrode area F. For efficient ozone generation, especially at higher O_3 concentrations, the temperature has to kept as low as possible, definitely below 100 °C. If a second cooling circuit is used to additionally cool the inner tube, the average temperature increase ΔT_g is reduced by a factor of four. This allows for a considerable increase of power density. However, it is rarely done in commercial ozone generators, because it requires cooling of the high voltage electrodes and introduces additional sealing problems.

9.3.5.4   Power supply units

Originally ozone generators were run at line frequency or were fed by motor generators operating at rather low frequencies. Step-up transformers are required to reach the desired voltage level. To achieve reasonable power densities, high voltages (up to 50 kV) had to be used. Dielectric failure was a common problem. Since all tubes are connected in parallel, high voltage fuses were used to disconnect faulty elements. Modern high-power ozone generators take advantage of solid state power semiconductors. They utilize thyristor or transistor controlled frequency converters to impress square-wave currents or special pulse trains in the frequency range 500 Hz to 5 kHz. Using this technology, applied voltages can be reduced to the range of about 5 kV. Dielectric failure is no longer a problem. With large ozone generators power factor compensation has become an important issue.

Typical power densities now reach $1-10\,kW/m^2$ of electrode area. Using semiconductors at higher frequencies brought several advantages: increased power at lower voltage, fast shut off and improved process control.

9.3.6 Future prospects of industrial ozone generation

A better understanding of microdischarge properties in non-equilibrium dielectric barrier discharges and advances in power semiconductors resulted in improved performance and reliability of industrial ozone generation in recent years. Raised ozone generating efficiency and drastically reduced size of the ozone generators helped to lower the cost. Today, ozone can be produced at a total cost of about 2 US$/kg. Further progress can be expected. Engineering efforts for superior dielectric properties, better flow control and improved thermal management will continue. Rapid advances in power semiconductor design resulting in improved GTOs (gate turnoff thyristors) and IGBTs (insulated gate bipolar transistors) will have a major impact. Encapsulated IGBT modules now switch 1000 A at 5 kV. It is foreseeable that soon bulky step-up transformers will be no longer required and that almost arbitrary wave forms can be generated. Investigations into homogeneous self-sustained volume discharges may even lead to more favorable plasma condition for ozone formation (Zakharov *et al* 1988, Kogoma and Okazaki 1994, Nilsson and Eninger 1997).

References

Andrews T and Tait P G 1860 *Phi. Trans. Roy. Soc. (London)* **150** 113
Boelter K and Davidsen J H 1997 *Aerosol Sci. Technol.* **27** 689–708
Braun D, Küchler U and Pietsch G 1988 *Pure Appl. Chem.* **60** 741–746
Braun D, Küchler U and Pietsch G 1991 *J. Phys. D: Appl. Phys.* **24** 564–572
Chen J 2002 *Direct current corona-enhanced chemical reactions* PhD thesis, Minneapolis, University of Minnesota
Chen J and Davidson J H 2002 *Plasma Chem. Plasma Process* **22** 199–224
Chen J and Davidson J H 2003a *Plasma Chem. Plasma Process* **23** 83–102
Chen J and Davidson J H 2003b *Plasma Chem. Plasma Process* **23** 501–518
Crutzen P J 1970 *Quart. J. Roy. Meteor. Soc.* **96** 320–325
Eliasson B and Kogelschatz U 1986a *J. Chim. Phys.* **83** 279–282
Eliasson B and Kogelschatz U 1986b *J. Phys. B: At. Mol. Phys.* **19** 1241–1247
Eliasson B and Kogelschatz U 1987 *Proc 8th Int Symp on Plasma Chemistry (ISPC–8)*, Tokyo 1987, vol 2, pp 736–741
Eliasson B, Hirth M and Kogelschatz U 1987 *J. Phys. D: Appl. Phys.* **20** 1421–1437
Filippov Yu V, Boblikova V A and Panteleev V I 1987 *Electrosynthesis of Ozone* (in Russian), (Moscow: Moscow State University Press).
Glaze W H and Kang J W 1988 *J. AWWA* **88** 57–63
Held B and Peyrous R 1999 *Eur. Phys. J AP* **7** 151–166
Herron J T 1999 *J. Phys. Chem. Ref. Data* **28** 1453–1483

Herron J T 2001 *Plasma Chem. Plasma Proc.* **21** 581–609

Herron J T and Green D S 2001 *Plasma Chem. Plasma Process* **21** 459–481

Hoigné J 1998 'Chemistry of aqueous ozone and transformation of pollutants by ozonation and advanced oxidation processes' in *Handbook of Environmental Chemistry*, Vol 5, Part C: Quality and Treatment of Drinking Water II, Hrubec J (ed) (Berlin: Springer) pp 83–141

Horváth M, Bilitzky L and Hüttner J 1985 *Ozone* (New York: Elsevier Science Publishing)

Johnston H S 1992 *Ann. Rev. Phys. Chem.* **43** 1–32

Kitayama J and Kuzumoto M 1997 *J. Phys. D: Appl. Phys.* **30** 2453–2461

Kitayama J and Kuzumoto M 1999 *J. Phys. D: Appl. Phys.* **32** 3032–3040

Kogelschatz U 1988 'Advanced ozone generation' in *Process Technologies for Water Treatment* Stucki S ed (New York: Plenum Press) pp 87–120

Kogelschatz U and Baessler P 1987 *Ozone Sc. Eng.* **9** 195–206

Kogelschatz U 1999 *Proc. Int. Ozone Symp.*, Basel, pp 253–265

Kogelschatz U 2000 'Ozone generation and dust collection' in *Electrical Discharges for Environmental Purposes: Fundamentals and Applications* van Veldhuizen E M (ed) (Huntington, NY: Nova Science Publishers) pp 315–344

Kogelschatz U and Eliasson B 1995 'Ozone generation and applications' in *Handbook of Electrostatic Processes*, Chang J S, Kelly A J and Crowley J M (eds) (New York: Marcel Dekker) pp 581–605

Kogoma M and Okazaki S 1994 *J. Phys. D: Appl. Phys.* **27** 1985–1987

Kossyi I A, Kostinsky A Yu, Matveyev A A and Silakov V P 1992 *Plasma Sources Sci. Technol.* **1** 207–220

Krivosonova O E, Losev S A, Nalivaiko V P, Mukoseev Yu K and Shatolov O P 1991 'Recommended data on the rate constants of chemical reactions among molecules consisting of N and O atoms' in *Reviews of Plasma Chemistry*, Smirnov B M Ed (New York: Consultants Bureau) vol 1, 1–29

Nilsson J O and Eninger J E 1997 *IEEE Trans. Plasma Sci.* **25** 73–82

Ohlmüller W 1891 *Ueber die Einwirkung des Ozons auf Bakterien* (Berlin: Springer)

Peyrous R 1986 *Simulation de l'évolution temporelle de diverses espèces gazeuses creées par l'impact d'une impulsion électronique dans l'oxygène ou de l'air, sec ou humide* PhD Thesis, Université de Pau

Peyrous R 1990 *Ozone Sci. Eng.* **12** 19–64

Peyrous R, Pignolet P and Held B 1989 *J. Phys. D: Appl. Phys.* **22** 1658–1667

Pietsch G and Gibalov V I 1998 *Pure Appl. Chem.* **70** 1169–1174.

Rice R G and Netzer A 1982 and 1984 (eds) *Handbook of Ozone Technology and Applications* vol 1 and 2 (Ann Arbor: Ann Arbor Science Publishers)

Samoilovich V G and Gibalov V I 1986 *Russ. J. Phys. Chem.* **60** 1107–1116

Samoilovich V G, Gibalov V I and Kozlov K V 1989 *Physical Chemistry of the Barrier Discharge* (in Russian) (Moscow: Moscow State University Press) (English translation: Düsseldorf: DVS-Verlag 1997, Conrads J P F and Leipold F (eds))

Schönbein C F 1840 *Compt. Rend. Hebd. Séances Acad. Sci.* **10** 706–710

Sieck L W, Herron J T and Green D S 2001 *Plasma Chem. Plasma Process* **20** 235–258

Siemens W 1857 *Poggendorfs Ann. Phys. Chem.* **102** 66–122

Soret J L 1865 *Ann. Chim. Phys. (Paris)* **7** 113–118

Wojtowicz J A 1996 'Ozone' in *Kirk-Othmer Encyclopedia of Chemical Technology*, (John Wiley) 4th edition, vol 17, pp 953–994

Yagi S and Tanaka M 1979 *J. Phys. D: Appl. Phys.* **12** 1509–1520

Yehia A, Abdel-Salam M and Mizuno A 2000 *J. Phys. D: Appl. Phys.* **33** 831–835

Zakharov A I, Klopovskii K S, Opsipov A P, Popov A M, Popovicheva O B, Rakhimova T V, Samarodov V A and Sokolov A P 1988 *Sov. J. Plasma Phys.* **14** 191–195

9.4 Electromagnetic Reflection, Absorption, and Phase Shift

9.4.1 Introduction

The effect of plasma on electromagnetic (EM), wave propagation in the ionosphere is well known and documented by Budden (1985) and Gurevich (1978). A particularly striking example of plasma in air is the EM black out and fluctuation of radar cross section (RCS), associated with re-entry vehicles reported by Gunar and Mennella (1965) and discussed by Ruck *et al* (1970, pp 874–875). A shock wave and resulting plasma develop around a vehicle because of the increasing gas pressure and friction as it descends from space. At an altitude of 200 000 ft (60.9 km) and higher, a 5 GHz radar frequency is greater than the plasma frequency and the momentum-transfer collision rate between electrons and the bulk gas, the RCS corresponds to the bare skin value. At $\sim$180 000 ft (55 km), however, the plasma frequency increases to approximately the radar frequency and the RCS decreases up to 10 dB because of refraction from the plasma enclosing the re-entry vehicle. At 150 000 ft (45.7 km) the plasma frequency is significantly greater than the radar frequency and an enhanced reflection produces a net increase in RCS of 5–10 dB. At 60 000 ft (18.3 km) the atmosphere is significantly thicker, and the momentum-transfer collision rate is $\sim$9 $\times$ 10^9 s^{-1}, which is roughly equal to the plasma frequency with both exceeding the radar frequency. In this collision dominated plasma, absorption dominates and the RCS decreases approximately 15 dB. At lower altitudes the re-entry vehicle slows, the plasma dissipates, and the RCS returns to its bare skin value.

Another example is an artificial ionospheric mirror. Borisov and Gurevich (1980) and Gurevich (1980) suggest that a reflective plasma layer below the D-layer could be generated at the intersection of two high-power EM pulses. The utility of such a mirror is the ability to reflect radio waves at frequencies above those supported by the ionosphere to great distances. This would permit long range high-frequency point-to-point communication and may even permit some radar to extend their range by bouncing their signals off such mirrors.

In this section, EM effects based on a cold collisional plasma with a spatially varying plasma density are discussed. The dispersion relation and density profile theory is quantified, summary formulas for reflection, transmission, absorption, and phase shift provided, air-plasma characteristics quantified, electron-beam produced plasmas discussed, and typical applications described.

9.4.2   Electromagnetic theory

The theory of an EM wave propagating in air plasma is that of a wave propagation in a cold collisional plasma. In this approximation ions are assumed to be at rest compared to electrons. In the presence of a strong electric field it is possible for a non-equilibrium system to develop with an electron, ion, and bulk gas temperatures that are all different. The following material describes a cold system where the contribution to electrical conductivity by ions is small and has been neglected.

9.4.2.1   Cold collisional dispersion relationship

For wave propagation in an air plasma the effect of collisions between electrons and the bulk gas is important. The Langevin equation of motion for electrons includes the damping of electron motion due to momentum-transfer collisions (Tanenbaum 1967),

$$m_e \frac{du}{dt} = -e(E + u \times B) - m_e \nu u \qquad (9.4.1)$$

where m_e is electron mass, u is electron velocity, e is electron charge, E is electric field strength, B is magnetic field density, ν is momentum-transfer collision rate, and MKS units are used throughout. For propagation of a transverse EM wave at frequency f through a collisional plasma, the dispersion relation provides a succinct relation between angular frequency $\omega = 2\pi f$ and a complex wavenumber k,

$$k(\omega) = \frac{\omega}{c} \sqrt{1 - \frac{\omega_p^2}{\omega(\omega - i\nu)}} \qquad (9.4.2)$$

where c is the speed of light, $\omega_p = (n_e e^2 / \varepsilon_0 m_e)^{1/2}$ is the plasma frequency, $i = +(-1)^{1/2}$, n_e is electron density, e is electron charge, and ε_0 is the free-space permittivity. Wave propagation is proportional to $\exp[+i(\omega t - kz)]$, where t is time and z is distance. For $\nu = 0$ in (9.4.2) the dispersion relation reduces to a cold lossless dispersion relation with a cutoff frequency at $\omega = \omega_p$.

For a lightly ionized collisional plasma with $|\omega_p^2 / \omega(\omega - i\nu)| \ll 1$ equation (9.4.2) can be expanded and factored into real and complex parts,

$$k_r(\omega) = \frac{\omega}{c}\left[1 - \frac{\omega_p^2}{2(\omega^2 + \nu^2)}\right], \qquad k_i(\omega) = -\frac{\omega_p^2 \nu}{2c(\omega^2 + \nu^2)}. \qquad (9.4.3)$$

The value of k_r is directly proportional to frequency. The leading term of k_r can be interpreted as $k_0 = \omega/c$, which is the free-space wavenumber, but the first-order plasma term is proportional to both ω and n_e. Consequently, an EM wave will encounter an impedance that depends on n_e. If n_e exhibits a step-like change in number density there will be a coherent reflection. If the change in n_e is smooth and extends over several free-space wavelengths, reflections along the smooth profile add incoherently and can be quite small. The value of k_i for $\omega < \nu$ is effectively independent of frequency and so implies that a collisional plasma is a broadband EM wave absorber. These two processes of reflection and absorption are present in collisional plasmas with the dominant effect depending on the profile for n_e and the frequency of observation.

9.4.2.3 Electron density profiles

The exact profile for n_e depends on the plasma source and the intended application. Large changes in n_e over a distance of less than one free-space wavelength generally result in a strong coherent reflection (often modeled as a slab discontinuity), whereas the same change in n_e over several wavelengths produces an incoherent reflection. Ruck *et al* (1970, pp 473–484) describe a layered-media matrix approach that takes internal reflections into account and can be applied to an arbitrary plasma distribution. The values of n_e and the momentum-transfer collision rate can change for each layer and equation (9.4.2) is used to generate a complex wavenumber for each frequency of interest. A few distribution functions for n_e yield analytic results. Budden (1985) provides analytic expressions for the reflection and transmission coefficients for linear, piecewise linear, parabolic, Epstein, and sech^2 electron distributions. The Epstein distribution is used to model a variety of plasma sources that generates a high electron density near the source, which diminishes with distance from the source. The Epstein distribution is particularly useful in modeling the plasmas generated by a high-energy electron beam or beta rays and photo processes that adhere to the Beer–Lambert law such as photo-ionization.

10.3.2.3 Epstein distributions

Epstein (1930) discussed a general electron density distribution with three arbitrary constants and wave propagation in absorbing media. Specific wave solutions are discussed by Budden (1985). Vidmar (1990) adapt the Epstein distribution to one suitable for modeling ionization sources. The electron number density utilized is

$$n(z) = \frac{n_0}{1 + \exp(-z/z_0)} \tag{9.4.4}$$

where z_0 is a dimensional scale factor and n_0 is the maximum electron concentration for $z \to +\infty$. Equation (9.4.4) varies from $n(z = -\infty) = n_0$

to $n(z = +\infty) = 0$. For a source that deposits energy over a finite distance, it is possible to match $n(z)$ at the 95% ($z/z_0 = +2.944$), 50% ($z/z_0 = 0$), and 5% ($z/z_0 = -2.944$) values and so determine an approximate value for z_0.

9.4.2.3   Epstein's power reflection and transmission coefficients

Using the Epstein distribution in (9.4.4) for a wave incident at an angle θ, where $\theta = 0$ implies backscatter and $\theta = 90°$ implies grazing incidence. The power reflection, R, and transmission, T, coefficients are

$$R = \left|\frac{C-q}{C+q}\right|^2 \left|\frac{\Gamma[1+ik_0z_0(q+C)]}{\Gamma[1+ik_0z_0(q-C)]}\right|^4 \tag{9.4.5}$$

$$T = \frac{4C^2}{|C+q|^2}\left|\frac{\Gamma^2[1+ik_0z_0(q+C)]}{\Gamma[1+2ik_0z_0q]\Gamma[1+2ik_0z_0C]}\right|^2 \exp[+2\,\mathrm{Im}(k_0q)z] \tag{9.4.6}$$

$$q^2 = C^2 - \frac{\omega_\mathrm{p}^2}{\omega(\omega - i\nu)} \tag{9.4.7}$$

where $C = \cos\theta$ and q is a solution of the Booker quartic. For some atmospheric plasma the arguments of the gamma functions become large, complex, and produce an overflow condition. Lanczos (1964) provides an asymptotic expansion for Γ and evaluation of $\ln\Gamma$ avoids overflow.

9.4.2.4   Attenuation and phase-shift coefficients for an Epstein profile

In some applications, such as those relating to radar, the effects on signal attenuation and phase path-length for round trip propagation through plasma with reflection from a good conductor are of interest. Analytic expressions are evaluated using the approximate values for k in (9.4.3) and evaluating $\exp(+2i\int k\,dz)$, where the integral is from $z = -\infty$ to the reflective surface. For reference the integration of ω_p^2 is proportional to n_e, (9.4.4), and the integral of n_e from $z = -\chi$ to $z = +\chi$ is $n_0\chi$. By noting that the ionization source plasma was modeled by (9.4.4) from the 95–5% values, the integration of $\int k\,dz$ is from free space for $z = -2.944z_0$ to the conductive body and ionization source at $z = +2.944z_0$. For $|\omega_\mathrm{p}^2/\omega(\omega - i\nu)| \ll 1|$ the round trip attenuation, A in dB, and the net phase change, $\Delta\Phi$ in radians, compared to free space propagation for a lightly ionized collisional plasma simplify to

$$A(\mathrm{dB}) = 4.343\left(\frac{e^2h}{\varepsilon_0m_\mathrm{e}c}\right)\left(\frac{n_0\nu}{\omega^2+\nu^2}\right) \tag{9.4.8}$$

$$\Delta\Phi(\mathrm{radians}) = \Delta\Phi_\mathrm{plasma} - \Delta\Phi_\mathrm{air} = -\pi\left(\frac{e^2h}{\varepsilon_0m_\mathrm{e}c}\right)\left(\frac{n_0f}{\omega^2+\nu^2}\right) \tag{9.4.9}$$

where $h = 5.888z_0$ is the thickness of the plasma distribution from its 5–95% values. These analytic formulas are useful in generating estimates of absorption and phase shifts and provides insight on the functional dependencies of A and $\Delta\Phi$ on n_0, h, f, and ν.

9.4.3 Air plasma characteristics

The air chemistry for a plasma depends on many factors such as air density determined from altitude, moisture content, electron density, present populations of excited states, electron temperature, bulk gas temperature, magnitude of electric field, and method of ionization. For production of plasma without any external wire electrodes, a high-energy electron-beam source is proposed. A 250 kV electron beam source, for example, is capable of producing a plasma cloud that extends 1.5 m from its source at 30 000 ft (~9.14 km) altitude. Macheret *et al* (2001) investigated electron beam produced air plasmas and quantified a return current from free space to the source, due to charge transport by fast electrons. Their electric field varies spatially, being most intense near the source. Consequently, the plasma generated by an electron beam varies spatially in electron concentration, electric field, and electron temperature. The air chemistry production–deionization solution must also treat these variations. Analytic air-chemistry approaches are tedious due to the complexity and nonlinear aspects of the air chemistry. Numerical approaches can easily involve hundreds of reactions to model the air chemistry but provide useful estimates of plasma lifetime for pulsed systems and estimates of power expenditure with curves of species as a function of time for a variety of excitation waveforms.

9.4.3.1 Momentum-transfer collision rate

For an electron beam source an electric field may be present with sufficient magnitude to elevate the electron temperature above thermal. Lowke (1992) has investigated free electrons in air as a function of water-vapor content and the reduced electric field E/N, where N is the bulk gas density. The curves Lowke generated explicitly treat the effects of N_2, O_2, CO_2, and H_2O as a gas mixture on the electron energy as a function of E/N. The momentum-transfer collision rates in table 9.4.1 were deduced from Lowke and appear as a function of altitude from sea level to 300 000 ft (~91.4 km). Atmospheric parameters of pressure, bulk gas density, and temperature appear below each altitude.

9.4.3.2 Major attachment mechanisms

Electrons attach primarily to oxygen molecules in a three-body process, Bortner and Baurer (1979) and Vidmar and Stalder (2003) for E/N

Table 9.4.1. Momentum transfer collision rate and atmospheric parameters.[†]

E/N	Momentum transfer collision rate (s^{-1})					
	Sea level	30 000 ft (9.14 km)	60 000 ft (18.3 km)	100 000 ft (30.5 km)	200 000 ft (60.9 km)	300 000 ft (91.4 km)
	764 torr	228 torr	54.8 torr	8.45 torr	149×10^{-3} torr	1.31×10^{-3} torr
	2.55×10^{19} cm^{-3}	9.58×10^{18}	2.43×10^{18}	3.58×10^{17}	5.86×10^{15}	5.91×10^{13}
$V - cm^{-2}$	288.1 K	228.8 K	216.6 K	226.9 K	244.6 K	214.2 K
0.0	9.53×10^{10} s^{-1}	3.58×10^{10}	9.09×10^{9}	1.34×10^{9}	2.19×10^{7}	2.21×10^{5}
1.0×10^{-19}	9.53	3.58	9.09	1.34	2.19	2.21
5.0×10^{-19}	1.31×10^{11}	4.92	1.25×10^{10}	1.84	3.01	3.03
1.0×10^{-18}	1.75	6.60	1.67	2.47	4.04	4.07
1.5×10^{-18}	6.75	1.91×10^{11}	4.90	7.18	1.11×10^{8}	1.19×10^{6}
1.0×10^{-17}	7.25	2.63	6.72	9.86	1.54	1.63
1.5×10^{-17}	1.11×10^{12}	5.60	1.42×10^{11}	2.09×10^{10}	3.41	3.45
1.0×10^{-16}	1.94	8.41	2.13	3.14	5.14	5.18
1.5×10^{-16}	3.31	1.24×10^{12}	3.16	4.66	7.62	7.68
1.0×10^{-15}	3.92	1.47	3.74	5.52	9.03	9.10×10^{7}

[†] 1962 US standard atmosphere

dependencies. The resulting O_2^- ion undergoes numerous charge-transfer reactions, hydration, and eventually becomes NO_3^- and $NO_3^- \cdot H_2O$ prior to negative-ion/positive-ion recombination. The rate for three-body attachment of electrons to O_2 depends on the altitude-dependent O_2 concentration and the E/N-dependent electron temperature. The extent to which O_2^- or NO_3^- is the dominant ion depends on how long the plasma is generated. A typical time scale for generation and deionization for an aircraft flying near the speed of sound that generates then flies through a plasma cloud $\sim 1.5\,\text{m}$ in extent is ~ 5 ms. A time scale of several hundred microseconds to several milliseconds typifies many plasma applications for aircraft.

9.4.3.3   1/e plasma lifetime

The $1/e$ plasma lifetime is the time for plasma that has been suddenly ionized to an electron density of n_0 to deionize to a value of n_0/e. A set of curves that quantifies plasma lifetime as a function of altitude and electron density appears in Vidmar (1990) and quantifies electron densities, where the dominant process for electron loss is three-body attachment to O_2 with an electron as the third body, three-body attachment with O_2 as the third body, and electron–positive ion recombination. These curves have been extended to include an E/N dependency in Vidmar and Stalder (2003). Plasma lifetime is shown to increase by approximately an order of magnitude for $10^{-17}\,\text{V cm}^{-2} < E/N < 10^{-16}\,\text{V cm}^{-2}$. The increase in lifetime corresponds to a decrease in the rate of three-body attachment for $E/N > 10^{-17}\,\text{V cm}^{-2}$ predicted by Aleksandrov (1993). This trend towards longer lifetime reverses for $E/N \approx 10^{-16}\,\text{V cm}^{-2}$, when the reaction rate for dissociative attachment to oxygen increases significantly and dominates the attachment process.

9.4.4   Plasma power

The energy deposited by an electron beam to generate an electron–ion pair in dry air, E_i, is 33.7 eV. For a pulsed source a lower estimate of the power per unit volume, P/V, is approximated by using E_i, the electron number density, and the plasma lifetime:

$$\frac{P}{V} = \frac{n_0 E_i}{\tau} \qquad (9.4.10)$$

where n_0 is the peak electron concentration and τ is plasma lifetime. The value of τ as a function of altitude is quantified in Vidmar (1990) and Vidmar and Stalder (2003). For example, an electron density of $10^{10}\,\text{cm}^{-3}$ at 30 000 ft (9.14 km) with $E/N = 0$ has a plasma lifetime of 157 ns with $P/V = 343\,\text{mW/cm}^3$ or $343\,\text{kW/m}^3$. Plasma lifetime is effectively independent of electron number density below $10^{10}\,\text{cm}^{-3}$, because the dominant electron loss mechanism is three-body attachment to O_2, which is linear with respect to electron concentration. Consequently, power is proportional to n_0, and the total power is the integral of P/V over the electron distribution.

For plasma generated by an electron beam and sustained by an electric field, the expression for power includes a term to account for Joule heating,

$$\frac{P}{V} = \frac{n_0 E_i}{\tau} + J \cdot E \tag{9.4.11}$$

where $J = \sigma E$ is current density and σ is the plasma conductivity. Vidmar and Stalder (2003) calculated plasma lifetime as a function of E/N for a continuous electric field and quantified total power at 30 000 ft (9.14 km). Although Joule heating increases as the square of electric field strength, the increase in plasma lifetime for $10^{-17}\,\mathrm{V\,cm}^{-2} < E/N < 10^{-16}\,\mathrm{V\,cm}^{-2}$ results in a net decrease in total power from 343 to 230 mW/cm^3 for a plasma density of $10^{10}\,\mathrm{cm}^{-3}$. This decrease in net power is also accompanied by an increase in excited states with $O_2(^1\Delta_g)$ reaching $8 \times 10^9\,\mathrm{cm}^{-3}$.

Additional research on power in air plasma involves continuous and pulsed ionization to quantify the concentrations and effect of excited states as a function of time. Because the energy deposited in plasma eventually heats the bulk gas, the concentration of all species will decrease due to volumetric expansion. Over short intervals such as those for an aircraft in flight, the generation of excited states under some conditions can significantly reduce the concentration of ground state species. These two effects slow the attachment process. The reaction rates for all the excited states on the major attachment, detachment, and deionization processes are not well known. Consequently, additional research, both theoretical and experimental, is necessary to quantify total power deposition in air plasma as a function of electron concentration, E/N, and altitude.

9.4.5 Applications

The application of collisional plasma for reflection, absorption, and phase shift has been motivated by early investigations of the ionosphere (Epstein 1930). Reflection from plasma slabs with sharp discontinuities is well understood and application to a surface radar for beam steering has been investigated (Manheimer 1991). Reflections from an ionospheric mirror have been advanced by Borisov and Gurevich (1980) and Gurevich (1980). A set of curves that apply to an ionospheric mirror at 230 000 ft (70.1 km) appears in Vidmar (1990) based on the Epstein distribution and the profile for $n(z)$ in equation (9.4.4). These curves quantify the power reflection coefficient at a shallow angle of 75° off broadside for an electron density of $10^7\,\mathrm{cm}^{-3}$ and $\nu = 7.4 \times 10^7\,\mathrm{s}^{-1}$. It was found that the power reflection coefficient was 0.80 or greater for frequencies below 100 MHz and $z_0 < 10\,\mathrm{m}$. At higher frequencies or for $z_0 > 10\,\mathrm{m}$ the power reflection coefficient decreased substantially. In terms of the profile in (9.4.4) the value of $z_0 = 10$ m implies the means of ionization must transition the air at 230 000 ft (70.1 km) from 5–95% of the maximum electron concentration over a distance of $h = 5.888z_0 = 58.88$ m.

The use of microwave absorption as a diagnostic technique to determine electron concentration is well known. Spencer *et al* (1987) experimentally measured the amplitude and phase in a microwave cavity to quantify the plasma lifetime, complex conductivity, and momentum-transfer collision rate of an electron-beam generated plasma.

The application of the Epstein distribution to model collisional plasma as a broadband absorber by Vidmar (1990) has curves of absorption versus frequency and z_0. These curves quantify total reduction, which refers to the sum of the reflected power, R in equation (9.4.5), the round-trip absorption, A in equation (9.4.9), and points out the power advantage of generating plasma in a noble gas rather than air. The total reduction curves that appear in Vidmar (1990) imply 10–40 dB signal reduction at frequencies that extend from $f_{\text{low}} > c/(4z_0)$ and extends to $f_{\text{high}} < \nu/5$. Physically, the broadband reduction requires approximately five collisions per cycle and the 5–95% gradient of the Epstein distribution, $h = 5.888z_0$ must be one to two wavelengths at the lowest frequency. The total reduction noted transfers of EM energy from a wave to heat via momentum-transfer collisions with the bulk gas. This reduction in reflected power reduces the RCS for the surface directly behind the plasma. The results of Santoru and Gregoire (1993) provide an experimental link between the Epstein theory for reflection and absorption with laboratory measurements.

Some radar systems utilize coherent integration over many cycles to improve their signal-to-noise ratio. For such radars a sudden change in phase interferes with the coherent integration and so degrades radar performance. The phase change $\Delta\Phi$ in (9.4.9) can be used to quantify such effects in terms of radar frequency, electron number density, collision rate, and Epstein gradient.

For all of these applications the EM effects of plasma on reflectivity and RCS are approximated by the Epstein distribution and the derived expressions for reflectivity, transmission, absorption, and phase shift. The means to achieve a man-made Epstein distribution in air all require power. The means of plasma generation for a particular application that minimizes net power required is not known at this time. Electron-beam generated air plasma is a candidate system for some applications because it has a unique excited-state air chemistry, the advantage that no wires are necessary in the plasma, and that the beam energy controls the Epstein gradient. A detractor on the use of electron beams is the problem of window heating that limits beam current and duty cycle. This problem is addressed by liquid cooling around the window or within the window (Vidmar and Barker 1998), or by propagation from vacuum to air through a small opening. Additional research on power required as a function of a continuous or pulsed source, altitude, and electron concentration is necessary to prove the utility of the electron beam approach.

References

Aleksandrov N L 1993 *Chem. Phys. Lett.* **212** 409–412
Borisov N D and Gurevich A V 1980 *Geomagn. Aeronomy* **20** 587–591
Bortner M H and Baurer T 1979 *Defense Nuclear Agency Reaction Rate Handbook*, 2nd edition, NTIS AD–763699 ch 22
Budden K G 1985 *The Propagation of Radio Waves, The Theory of Radio Waves of Low Power in the Ionosphere and Magnetosphere* (New York: Cambridge University Press) 438–479
Epstein P S 1930 *Proc. Nat. Acad. Sci.* **16** 627–637
Gunar M and Mennella R 1965 *Proceedings of the 2nd Space Congress—New Dimensions in Space Technology, Canaveral Council of Technical Societies* 515–548
Gurevich A V 1978 *Nonlinear Phenomena in the Ionosphere, Physics and Chemistry in Space* vol 10 (New York: Springer) p 370
Gurevich A V 1980 *Sov. Phy. Usp.* **23** 862–865
Lanczos C 1964 *J. SIAM Numer. Anal. Ser. B* **1** 86–96
Lowke J J 1992 *J. Phys D: Appl. Phys.* **25** 202–210
Macheret S O, Shneider M N and Miles R B 2001 *Physics of Plasmas* **8** 1518–1528
Manheimer W M 1991 *IEEE Trans. Plasma Sci.* **PS-19** 1228–1234
Ruck G T, Barrick D E, Stuart W D and Krichbaum C K 1970 *Radar Cross Section Handbook* vol 2 (New York: Plenum) 473–484 and 874–875
Santoru J and Gregoire D J 1993 *J. Appl. Phys.* **74** 3736–3743
Spencer M N, Dickinson J S and Eckstrom D J 1987 *J. Phys D: Appl. Phys.* **20** 923–932
Tanenbaum B S 1967 *Plasma Physics* (New York: McGraw-Hill) 62–86
Vidmar R J 1990 *IEEE Trans. Plasma Sci.* **PS-18** 733–741
Vidmar R J and Barker R J 1998 *IEEE Trans. Plasma Sci.* **PS-26** 1031–1043
Vidmar R J and Stalder K R 2003 AIAA 2003–1189

9.5   Plasma Torch for Enhancing Hydrocarbon–Air Combustion in the Scramjet Engine

9.5.1   Introduction

The development of the scramjet propulsion system [1–3] is an essential part of the development of hypersonic aircraft and long-range (greater than 750 miles (1207 km)) scramjet-powered air-to-surface missiles with Mach-8 cruise capability [4]. This propulsion system has a simple structure as required by the hypersonic aerodynamics. Basically, the combustor has the shape of a flat rectangular box with both sides open. Air taken in through the frontal opening mixes with fuel for combustion and the heated exhaust gas at the open end is ejected through a MGD accelerator and a nozzle to produce the engine thrust.

For the hydrocarbon-fueled scramjet in a typical startup scenario, cold liquid JP-7 is injected into a Mach-2 air crossflow (having a static temperature of $\sim$500 K); under these conditions, the fuel–air mixture will not auto-ignite. Instead, some ignition aid—for example a cavity flameholder in conjunction with some mechanism to achieve a downstream pressure rise—is necessary to initiate main-duct combustion. With sufficient downstream pressure rise, a shock front will propagate upstream of the region for heat release. The heat release from combustion will maintain the pre-combustion shock front, while subsonic conditions in the mixing and combustion region favor stable combustion and flameholding.

Of course, even though the device operates as a ramjet under startup conditions (i.e. subsonic flow downstream of the pre-combustion shock) the residence time through the combustion region is short, of order 1 ms. Within scramjet test facilities, the typical mechanisms for achieving the required downstream pressure rise (and stable combustion) are the so-called aero-throttle, where a 'slug' of gas is injected in the downstream region, and the heat is released from the pyrophoric gas silane (SiH_4). Indeed, silane injection into the combustor is the current mechanism by which the X43A scramjet vehicle is started. Both of these approaches, however, have their disadvantages: for example, the aero-throttle approach may not allow re-lighting attempts and silane poses obvious safety risks. Thus, an alternative approach is desired.

For the purpose of developing techniques to reduce the ignition delay time and increase the rate of combustion of hydrocarbon fuels, Williams *et al* [5] have carried out kinetics computations to study the effect of ionization on hydrocarbon–air combustion chemistry. The models being developed—which include both the normal neutral–neutral reactions and ion–neutral reactions—focus primarily on the development of plasma-based ignition and combustion enhancement techniques for scramjet combustors. The results computed over the 900–1500 K temperature range show that the ignition delay time can be reduced significantly (three order of magnitude over the 900–1500 K temperature range) by increasing the initial temperature of fuel–air mixture.

Moreover, detailed kinetics modeling also shows a significant decrease in ignition delay in the presence of initial ionization—in the form of a $H_3O^+/NO^+/e^-$ plasma—at levels of ionization mole fractions greater than 10^{-6}. The ignition delay time is decreased most significantly at low temperatures. Indeed, the computational results suggest that even larger effects may be observed at the low temperatures encountered under engine startup.

Plasma torches can deliver enough heat to replace silane for ignition purpose. Moreover, use of a torch as a fuel injector also introduces an initial ionization in the fuel. The significant decrease in the ignition delay time and

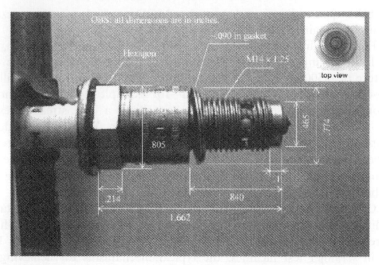

Figure 9.5.1. A photo of the plasma torch module. (Copyright 2004 by IEEE.)

the initial energy carried by plasma may elevate the heat release from combustion to exceed a threshold level for flameholding. These are the primary reasons that plasma torches [6–8] are being developed for the application.

Nevertheless, to make use of the high-temperature torch effluent, which may include quantities of radicals, ions, and electrons, it is necessary to project this gas into the engine in such a way that it readily mixes with a fuel–air stream. Poor penetration of the torch plume into the combustor, and/or improper placement of each torch—that is, more than one torch may be required—will limit its effectiveness. Shown in figure 9.5.1 is a photo of a plasma torch module, which is developed [9–10] in the present effort for the generation of torch plasma. The unique features of this plasma torch make it well suited for the purpose of ignition in a scramjet engine. These features include the following.

1. The compact size. It can be easily mounted to the combustor wall and requires no water cooling.
2. Flexible design. It can deliver high peak powers (and pulse/cycle energy) in 60 Hz or pulsed modes. Furthermore, it can deliver high mass flow rates due to the large annular flow area.
3. High mass flow operation. It can be configured to deliver 10 g of feedstock (which can be the fuel) per second.
4. Durability. It can be run for long periods with an air feedstock.
5. High-voltage operation. Rather than running at high current, the torch runs at high voltage, which allows greater penetration of the arc into the combustor and reduces the power loss to the electrodes (leading to

longer electrode life); higher E/N also enhances dissociations in fuel and air by direct electron impact.

9.5.2 Plasma for combustion enhancement

In the combustion, fuel–air mixing is critical. Without oxygen, fuel will not burn by itself. The hydrocarbon fuel provides hydrogen and carbon to react with oxygen in the combustion process. The reaction rate increases with the temperature of the mixture, which changes the ratios of the components in the composition of the mixture. In low temperature, the gas mixture contains mainly neutral molecules, and neutral–neutral reactions are often immeasurably slow. For example, the rate coefficient for the reaction between H_2 and O_2 is $6 \times 10^{-23} \, cm^3 \, s^{-1}$. As temperature increases, some radicals such as atomic species are produced. Neutral–radical reactions have rates in the range of 10^{-16}–$10^{-11} \, cm^3 \, s^{-1}$. For example, the reaction between H and O_2 has a rate coefficient equal to $1 \times 10^{-13} \, cm^3 \, s^{-1}$. Reactions also occur between radicals, which in fact have higher rates in the range 10^{-13}–$10^{-10} \, cm^3 \, s^{-1}$. Hence, the combustion rate is increased as the percentage of radicals in the mixture becomes significant by the temperature increase. If the temperature of the mixture is high enough to cause significant ionizations, the combustion rate is further enhanced. This is because ion–neutral and ion–electron reactions have rates larger than 10^{-9} and $10^{-7} \, cm^3 \, s^{-1}$, respectively. For instance, the reaction $H_2^+ + O_2$ has a rate coefficient of $8 \times 10^{-9} \, cm^3 \, s^{-1}$. It turns out only long-range ion–electron and ion–dipole reactions are fast enough to react on hypersonic flow time scales in the microsecond range. Therefore, it is desirable to use energy to heat the mixture and also to introduce ionized species to the mixture. Usually, thermal plasma is not very energy efficient to introduce ionized species to the mixture. Nonequilibrium plasmas produced by corona, streamer, pulsed glow and microwave discharges have been suggested, as alternatives to the torch plasma, for aiding the ignition. These discharges run at high E/N can potentially enhance dissociations in fuel and air by direct electron impact [11], where E is the electric field and N is the gas density. However, the practical issue of the research efforts is the combustion efficiency, rather than the energy efficiency of the igniter. The combustion efficiency depends not only on the chemical processes but also on the spatial distribution of the plasma energy, in particular, in a supersonic combustor. If the igniter can only start the ignition locally, for instance, near the wall, a considerable percentage of injected fuel will not be ignited before exiting the combustor. The plasma torch presented in the following demonstrates that it can produce high enthalpy supersonic plasma jet to penetrate the supersonic cross flow, as required to be a practical igniter of a supersonic combustor.

Two types of power supply are applied to operate the torch module shown in figure 9.5.1. One is a 60 Hz source, which sustains the discharge periodically. Such produced plasma will be termed '60 Hz torch plasma' in the following. This power source [12] includes (1) a power transformer with a turn ratio of 1:25 to step up the line voltage of 120 V from a wall outlet to 3 kV, (2) capacitors of $C = 3\,\mu F$ in series with the electrodes, and (3) a serially connected diode (made of four diodes, connected in parallel and each having 15 kV and 750 mA rating) and resistor ($R = 4\,k\Omega$) placed in parallel to the electrodes to further step up the peak voltage. The series resistor is used to protect the diode by preventing the charging current of the capacitor from exceeding the specification (750 mA) of each diode and to regulate the time constant of discharge. In one half cycle when the diode is forward biased, the capacitor is charging, which reduces the available voltage for the discharge in the torch module. However, since the time constant $RC = 12\,ms$ is longer than the half period 8.5 ms of the ac input, the discharge can still be initiated during this half cycle (even though the discharge has lower current and voltage than the corresponding ones in the other half cycle). During this other half period, the diode is reversed biased and the charged capacitor increases considerably the available voltage and current for the discharge in the torch module. The torch energy (i.e. the thermal energy carried by torch plasma) in each cycle varies with the gas supply pressure p_0. The dependence measured in the pressure range from 1.36 to 7.82 atm is presented in figure 9.5.2(a).

As shown, the dependence has a maximum at the gas supply pressure $p_0 = 6.12$ atm, where the plasma energy is 25.6 J. The increasing dependence of the plasma energy on the flow rate in the region of low gas supply pressure (i.e. $p_0 < 6.12$ atm) is realizable because the supplied gas flow works to increase the transit time of charge particles by keeping the discharge away from the shortest (direct) path between two electrodes. As the flow rate increases, the transit time loss of charge particles is reduced and thus the plasma energy increases. However, when the flow rate becomes too high (i.e. $p_0 > 6.12$ atm), the mobilities of charge particles crossing the flow becomes significantly affected by the flow. In such a way that the torch energy decreases with increasing pressure. It is noted in figure 9.5.2(a) that there is a significant plasma energy drop at $p_0 = 4.08$ atm. This unexpected result may be explained as follows. Schlieren images indicate that a transition from subsonic to supersonic flow at the exit of the module occurs near $p_0 = 3.4$ atm, which was identified by the sudden appearance of the shock structure at the exit of the torch nozzle in the schlieren image of the flowfield. After the transition, the flow becomes underexpanded. At $p_0 = 4.08$ atm, the low pressure region in the flow that favors gas breakdown is narrow in the flow direction and close to the exit of the module. Thus the discharge channel is narrow and the transit times of charge particles are small. Consequently, the plasma energy is reduced. As the pressure is further increased, this low-pressure region extends rapidly outward from the exit of the module so that the discharge can again appear in a larger region.

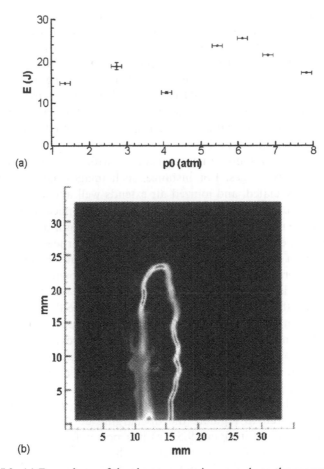

(a)

(b)

Figure 9.5.2. (a) Dependence of the plasma energy in one cycle on the gas supply pressure and (b) a planar image of torch plasma taken by an ultra-fast CCD camera with 10 ns exposure to laser-induced fluorescence from NO molecules. (Copyright 2004 by IEEE.)

As a consequence of the high-voltage nature of the discharge, the arc loop can be many times the distance between the anode and cathode. The arc loop structure is illustrated in the image (typical of those recorded) shown in figure 9.5.2(b), which was recorded through a 239 nm interference filter, 10 nm FWHM, with an intensified CCD camera (Roper Scientific PIMAX) set for an 80 ns exposure time. The current loop is coincident with the thin, intense emission loop shown in the figure. For this measurement, pure nitrogen with a pressure of 1.7 atm was supplied to the torch module. The horizontal extent of the arc loop is *ca.* 3.2 mm, whereas the vertical extent is about 2.5 cm. Such an extended arc loop increases the path length of the charged particles in the discharge by more than 15 times the direct path length from the cathode to the anode. Also shown in figure

9.5.2(b) is laser-induced fluorescence (LIF) from nitric oxide, NO, obtained using a Nd:YAG-pumped dye laser system to generate laser radiation at 226 nm probing the overlapped $Q_1(12.5)$ and $Q_2(19.5)$ transitions in the $\delta(0,0)$ band of NO. The LIF image appears as the diffuse, less intense background and is best seen on the left-hand side of the figure towards the outer portion of the arc loop. NO is produced within the torch plume in the region where the hot torch gas (pure N_2), i.e. the gas near the arc, mixes with quiescent laboratory air. Thus, NO is formed primarily near the outer portion of the arc loop.

The extended arc loop structure produced with this torch module has several distinct advantages. For instance, such images indicate that high temperature, dissociated, and ionized air extends well above the surface of the torch module, which is important for ignition applications. The long electrode lifetime may in part be due to extended arc length since the charged particles' kinetic energy is reduced before hitting electrodes. Furthermore, the conversion of electrical energy to plasma energy may be enhanced due to the longer interaction region. Images such as that shown in figure 9.5.2(b) indicate that the length of the arc loop is not strongly sensitive to the flow rate, but the width of the loop becomes narrower as the flow rate increases, which is consistent with the change in the flowfield structure as the jet becomes underexpanded and supersonic with increased supply pressure.

The other power supply applied is a dc pulsed discharge source, which uses a RC circuit for charging and discharging, where a 28 µF capacitor is used. A very energetic torch plasma, albeit one with a low repetition rate, can be generated. In the circuit, a ballasting resistor R_2 is connected in series with the torch to regulate the discharging current and adjust the pulse duration. Shown in figure 9.5.3(a) is a power function obtained by connecting a resistor of $R_2 = 26\ \Omega$ in series with the torch. This power function has a peak of about 300 kW and a pulse length of about 800 µs, which is very close to the time constant $R_2C = 728$ µs. The difference is accountable from the effective resistance of the discharge. As R_2 is increased to 250 Ω, now the power function shown in figure 9.5.3(b) consists of two parts: an initial part with a large peak of about 20 kW for the ignition of the discharge and a subsequent near-constant low-power part keeping at about 2.5 kW for 10 ms, which maintains the discharge. The energy contained in the pulse is about 50 J.

Because torch plasma delivers adequate energy, it can be an ignition aid and combustion enhancer within a scramjet engine.

9.5.3 Plasma torch for the application

The performances of plasmas produced by the torch module in a Mach-2.5 supersonic crossflow are discussed in the following. Measurements consist of video images of the torch emission and of the flowfield schlieren. We

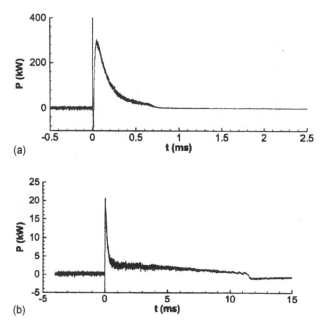

Figure 9.5.3. Power functions of pulsed dc discharges with no flow in the background; gas supply pressure of the torch module is 2.72 atm. (a) $R_2 = 26\ \Omega$ and (b) $R_2 = 250\ \Omega$.

note that due to the limited framing rate, 30 frames per second, these images represent a temporal average during the frame time. Thus, one does not freeze the arc-loop structure as was done with the intensified CCD (figure 9.5.2(b)). This is true regardless of whether one is viewing the 60 Hz or pulsed discharge.

Experiments [13, 14] were conducted in the test section, measuring 38 cm × 38 cm, of a supersonic blow-down wind tunnel. The upstream flow had a flow speed of 570 m/s, a static temperature $T_1 = 135\ K$, and a pressure $P_1 = 1.8 \times 10^4\ N/m^2$ (about 0.20 atm). These conditions approximate the scramjet startup conditions listed earlier, though the temperature and pressure are somewhat low (e.g. the static temperature for engine startup is about 500 K). The torch plume is injected normally into the supersonic flow, and the performance of torch plasma in terms of its height and shape in the supersonic flow is studied. In experiments, the air supply pressure is varied from 1.7 to 9.2 atm.

We first investigate the 60 Hz torch plasma in the wind tunnel. Presented in figure 9.5.4(a) is an airglow image of the plasma torch produced in the Mach-2.5 crossflow with 4.1 atm of air pressure supplied to the gas chamber of the torch module. This image shows the typical shape of the plasma torch in each half cycle; clearly, the supersonic flow causes significant deformation in the shape of the plasma torch. The penetration height of

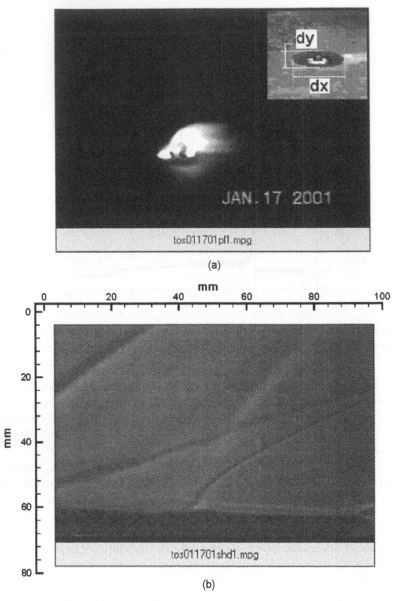

(a)

(b)

Figure 9.5.4. (a) Sideview of the airglow image of ac torch plasma in each half cycle in the Mach-2.5 crossflow. The gas supply pressure of the torch module is 4.1 atm. In the insert, $d_x = d_y = 11.4$ mm define the horizontal and vertical scales of the image. (b) Shadow image of the flow; an oblique shock wave is generated in front of the torch. (c) Airglow image of pulsed dc torch plasma in a supersonic crossflow (about $10°$ off the sideview line); the field of view is estimated to be 9.5 cm × 6 cm; the gas supply pressure of the torch module is 2.72 atm. (d) Schlieren image of pulsed dc torch plasma; the backpressure of the torch is 9.2 atm. (Copyright 2004 by IEEE.)

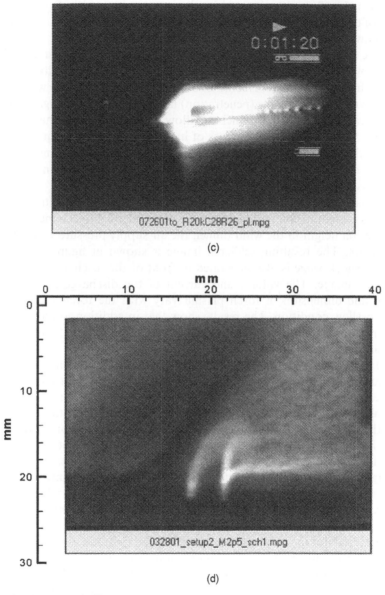

(c)

(d)

Figure 9.5.4. *(Continued)*

the torch is reduced significantly as the plume is swept downstream by the high-speed flow; nevertheless, the torch plume can still penetrate into the supersonic crossflow by more than 1 cm and also extends downstream about 1 cm, based on these emission images. A bow shock wave is also generated in front of the torch (since the torch acts as an obstruction to the oncoming flow), as observed by the image presented in figure 9.5.4(b). This, of course, is typical behavior for a jet injected normally in a supersonic crossflow.

We next study the torch operation in the supersonic flow using the high-power pulsed power supply. Shown in figure 9.5.4(c) is an airglow image of the torch plasma in the supersonic crossflow; the supply pressure was 2.7 atm. As shown in the figure, the (penetration) height of the torch is again reduced considerably by the wind tunnel crossflow. Comparing with that shown in figure 9.5.4(a), obtained in the case of higher gas supply pressure but lower power, the one shown in figure 9.5.4(c) extends about five times as far in the downstream direction and has a slightly larger penetration depth into the crossflow.

Clearly, the increased discharge power produces larger volume plasma, which is evident in comparing figures 9.5.4(a) and 9.5.4(c). To increase torch penetration height in the wind tunnel, the air supply pressure was increased to 9.2 atm. The resulting schlieren image is shown in figure 9.5.4(d). An oblique shock wave is also generated in front of the torch as shown in this schlieren image. The voltage and current of the discharge as well as the shape and dimension of torch plasma vary with the torch flow rate and the crossflow condition. The results show that in addition to increasing the flow rate, one can increase the torch power to improve the penetration of the plasma into the crossflow.

Initial evaluation of plasma-assisted ignition of hydrocarbon fuel was conducted in a supersonic, Mach-2 flow facility, at Wright-Patterson Air Force Research Laboratory, with heated air at a total temperature and pressure of 590 K and 5.4 atm, respectively. The resulting static temperature was thus ~330 K, still a relatively low value insofar as ignition is concerned. This facility allows testing of an individual concept with both gaseous and liquid hydrocarbon fuels without a cavity based flame-holder. In the tested configuration, a 15.2 cm × 30.5 cm test section floor plate fits into a simulated scramjet combustor duct with an initial duct height of 5.1 cm. At the upstream edge of the test section insert, the simulated combustor section diverges on the injector side by 2.5°. This particular hardware was intention-ally designed not to study main-duct combustion (ignition of the entire duct), but to reduce the chance of causing main-duct combustion by limiting the equivalence ratio of the tunnel below 0.1. In particular, this was accom-plished by placing the fuel injector at the centerline of the tunnel and not adding any flame-holding mechanisms such as a cavity or backwards-facing step. This approach allows the interactions of the fuel plume with the plasma torch to be studied by itself, and any flame produced is strictly created by this interaction, hence decoupling the ignition and flameholding problems as much as possible from the combustor geometry. Tests have been conducted using gaseous ethylene fuel, with the 15° downstream-angled single hole.

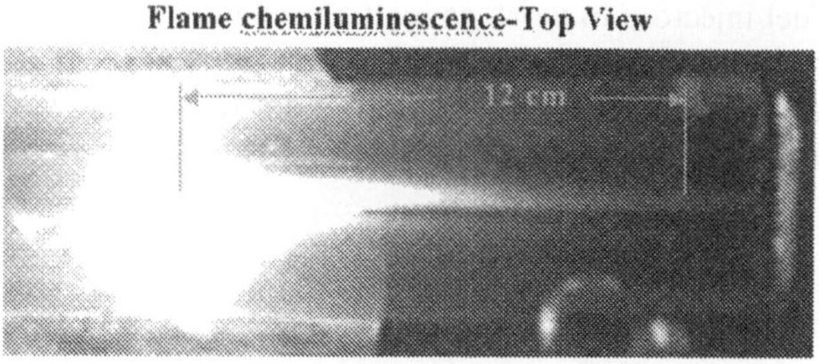

Flame chemiluminescence-Top View

Figure 9.5.5. Flame plume ignited by 60 Hz torch plasma with fuel injected by a single-hole injector. (Copyright 2004 by IEEE.)

The 60 Hz plasma torch module was evaluated and was found to produce a substantial flame plume as observed both from flame chemiluminescence and OH planar laser-induced fluorescence [14]. The flame chemiluminescence (blue emission in the tail of the plume) is illustrated in figure 9.5.5, which shows a single frame taken from video recordings of a flame plume ignited by the 60 Hz plasma torch in operation 5 cm downstream of the ethylene-fueled single-hole injector. Several feedstock flowrates were tried over the torch module operational range and a flowrate of ~500 SLPM was determined to produce the largest visible flame for the current electrode configuration. Air produced a larger flame when compared to nitrogen as the torch feedstock. This difference in flame size indicates that this type of flame is very sensitive to the local equivalence ratio and coupling of the ignition source with the mixture.

Shown in figure 9.5.6 is a schematic of a conceptual Ajax vehicle and its engine. The engine is located at the bottom of the vehicle. Plasma torch modules are installed on the top wall of the box-shaped combustor right

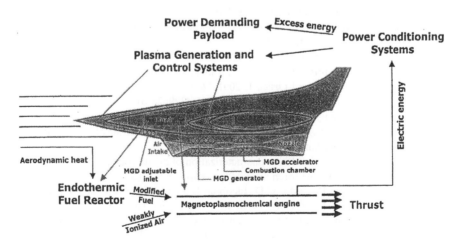

Figure 9.5.6. Schematic of a conceptual Ajax vehicle and the engine.

behind the fuel injectors to work as igniters. The torch modules can also be used as injectors to directly introduce ionizations and heat in the fuel for reducing ignition delay. It is worth pointing out that shock waves generated in front of torch plasma can help for holding flame and increasing its spread to achieve thorough combustion.

References

[1] Gruber M, Jackson, K Mathur T, Jackson T and Billig F 1998 'A cavity-based fuel injector/flameholder for scramjet applications' *35th JANNAF Airbreathing Propulsion Subcommittee and Combustion Subcommittee Meeting*, Tucson, AZ, p 383

[2] Mathur T, Streby G, Gruber M, Jackson K, Donbar J, Donaldson W, Jackson T, Smith C and Billig F 1999 'Supersonic combustion experiments with a cavity-based fuel injector' *AIAA Paper 99–2102*, American Institute of Aeronautics and Astronautics, Washington, DC, June 1999

[3] Gruber M, Jackson K, Mathur T and Billig F 1999 'Experiments with a cavity-based fuel injector for scramjet application' *ISABE Paper IS-7154*

[4] Mercier R A and Weber J W 1998 'Status of the US Air Force Hypersonic Technology Program' *35th JANNAF Airbreathing Propulsion subcommittee and Combustion Subcommittee Meeting*, Tucson, AZ, p 17

[5] Williams S, Bench P M, Midey A J, Arnold S T, Viggiano A A, Morris R A, Maurice L Q and Carter C D 2000 *Detailed Ion Kinetic Mechanisms For Hydrocarbon/Air Combustion Chemistry, AFRL report 2000*, Hanscom AFB, MA 01731–3010, p 1

[6] Wagner T, O'Brien W, Northam G and Eggers J 1989 'Plasma torch igniter for scramjets' *J. Propulsion and Power* 5(5)

[7] Masuya G, Kudou K, Komuro T, Tani K, Kanda T, Wakamatsu Y, Chinzei N, Sayama M, Ohwaki K and Kimura I 1993 'Some governing parameters of plasma torch igniter/flameholder in a scramjet combustor' *J. Propulsion and Power* 9(2) 176–181

[8] Jacobsen L S, Carter C D and Jackson T A 2003 'Toward plasma-assisted ignition in scramjets' *AIAA Paper 2003–0871*, American Institute of Aeronautics and Astronautics, Washington, DC

[9] Kuo S P, Koretzky E and Orlick L 1999 'Design and electrical characteristics of a modular plasma torch' *IEEE Trans. Plasma Sci.* 27(3) 752

[10] Kuo S P, Koretzky E and Orlick L 2001 *Methods and Apparatus for Generating a Plasma Torch* (United States Patent No. US 6329628 B1)

[11] Parish J and Ganguly B 2004 'Absolute H atom density measurement in short pulse methane discharge' *AIAA Paper 2004–0182*, American Institute of Aeronautics and Astronautics, Washington, DC

[12] Koretzky E and Kuo S P 1998 'Characterization of an atmospheric pressure plasma generated by a plasma torch array' *Phys. Plasmas* 5(10) 3774

[13] Kuo S P, Bivolaru D, Carter C D, Jacobsen L S and Williams S 2003 'Operational Characteristics of a Plasma Torch in a Supersonic Cross Flow', *AIAA Paper 2003–1190*, American Institute of Aeronautics and Astronautics, Washington, DC

[14] Kuo S P, Bivolaru D, Carter C D, Jacobsen L S and Williams S 2004 'Operational characteristics of a periodic plasma torch', *IEEE Trans. Plasma Sci.*, February issue

9.6   The Plasma Mitigation of the Shock Waves in Supersonic/Hypersonic Flights

9.6.1   Introduction

A flying object agitates the background air; the produced disturbances propagate, through molecule collisions, at the speed of sound. When the object flight approaches the speed of sound (roughly 760 mph in level flight), those disturbances deflected forward from the object move too slowly to get away from the object and form a sound barrier in front of the flying object. Ever since Chuck Yeager and his Bell X-1 first broke the sound barrier in 1947, aircraft designers have dreamed of building a passenger airplane that is supersonic, fuel efficient and economical. However, the agitated flow disturbances by the flying object at supersonic/hypersonic speed coalesce into a shock appearing in front of the object. The shock wave appears in the form of a steep pressure gradient. It introduces a discontinuity in the flow properties at the shock front location, at the reachable edge of the flow perturbations made by the object. The background pressure behind the shock front increases considerably, leading to significant enhancement of the flow drag and friction on the object.

Shock waves have been a detriment to the development of supersonic/hypersonic aircraft, which have to overcome high wave drag and surface heating from the additional friction. The design of high-speed aircraft tends to choose slender shapes to reduce the drag and cooling requirements. While that profile is fine for fighter planes and missiles, it has long dampened dreams to build a wide-bodied airplane capable of carrying hundreds of people at speeds exceeding 760 mph. This is an engineering tradeoff between volumetric and fuel consumption efficiencies and this tradeoff significantly increases the operating cost of commercial supersonic aircraft. Moreover, shock wave produces a sonic boom on the ground. This occurs when flight conditions change, making the shock wave unstable. The faster the aircraft flies, the louder the boom. The noise issue raises environmental concerns, which have precluded for, example, the Concorde supersonic jetliner from flying overland at supersonic speeds.

A physical spike [1] is currently used in the supersonic/hypersonic object to move the original bow shock upstream from the blunt-body nose location to its tip location in the new form of a conical oblique shock. It improves the body aspect ratio of a blunt-body and significantly reduces the wave drag. However, the additional frictional drag occurring on the spike structure and related cooling requirements limit the performance of a physical spike. Also another drawback of a physical spike is its sensitivity to off-design operation of the vehicle, i.e. flight Mach number and vehicle angle of attack. A failure regime at aspect ratios less than one also prohibits the practical uses of these physical spikes alone for shock wave modification.

Therefore, the development of new technologies for the attenuation or ideal elimination of shock wave formation around a supersonic/hypersonic vehicle has attracted considerable attention. The anticipated results of reduced fuel consumption and having smaller propulsion system requirements, for the same cruise speed, will lead to the obvious commercial gains that include larger payloads at smaller take-off gross weights and broadband shock noise suppression during supersonic/hypersonic flight. These gains can make commercial supersonic flight a reality for the average traveler.

9.6.2 Methods for flow control

Considerable theoretical and experimental efforts have been devoted to the understanding of shock waves in supersonic/hypersonic flows. Various approaches to develop wave drag-reduction technologies have been explored since the beginning of high-speed aerodynamics. In the following, a few of these are discussed.

Buseman [2] suggested that geometrical destructive interference of shock waves and expansion waves from two different bodies could work to reduce the wave drag. However, the interference approach is effective only for one Mach number and one angle of attack, which makes the design for practical implementation difficult.

Using electromagnetic forces for the boundary layer flow control have been suggested as possible means to ease the negative effect of shock wave formation upon flight [3]. However, an ionized component in the flow has to be generated so that the fluid motion can be controlled by, for instance, an introduced $j \times B$ force density, where j and B are the applied current density and magnetic field in the flow.

Thermal energy deposition in front of the flying body to perturb the incoming flow and shock wave formation has been studied numerically [4, 5]. Heating of the supersonic incoming flow results in a local reduction of the Mach number. This in turn causes the shock front to move upstream and thus in this process the stronger bow shock is modified to a weaker oblique shock with significantly lower wave drag to the object and much less shock noise. Although this heating effect is an effectual means of reducing the wave drag and shock noise in supersonic and hypersonic flows, it requires a large power density to significantly elevate the gas temperature [5]. It is known that using the thermal effect to achieve drag reduction in supersonic and hypersonic flight does not, in general, lead to energy gain in the overall process. Thus this is not an efficient approach for drag reduction purposes, but it can be a relatively easy approach for sonic boom attenuation.

Direct energy approaches have also been applied to explore the non-thermal/non-local effect on shock waves. Katzen and Kaattari [6] investigated aerodynamic effects arising from gas injection from the subsonic region of the shock layer around a blunt body in a hypersonic flow. In one particular case, when helium was injected at supersonic speed, the injected flow penetrated the central area of the bow shock front, modifying the shock front in that area to a conical shape with the vertex much farther from the body (at about one body diameter). Laser pulses [7, 8] could easily deposit energy in front of a flying object. However, plasma generated at a focal point in front of the model had a bow radius much smaller than the size of the shock layer around the model, and its non-local effect on the flow was found to be insignificant.

Plasma can effectively convert electrical energy to thermal energy for gas heating. Moreover, it has the potential to possibly offer a non-thermal modification effect on the structure of shock waves. The results from early and recent experiments conducted in shock tubes exhibited an increased velocity and dispersion on shock waves propagating in the glow discharge region [9, 10]. Measurements using laser beam photodeflection concluded that the dispersion and velocity increase of shock wave were attributed to the inhomogeneous plasma heating by the local electric field [11]. Plasma experiments were also conducted in wind tunnels. When plasma was generated ahead of a model either by the off-board or on-board electrical discharge [12–15] or microwave pulses [16, 17] the experimental results showed that the shock front had increased dispersion in its structure as well as increased standoff distance from the model. One of the non-thermal plasma effects was evidenced by an experiment [18] investigating the relaxation time of the shock structure modification in decaying discharge plasma. The observed long-lasting effect on the shock structure was attributed to the existence of long-lived excited states of atoms and molecules in the gas.

The study of the plasma effect on shock waves was further inspired by a wind tunnel experiment conducted by Gordeev *et al* [19]. High-pressure metal vapor (high Z) plasma, produced inside the chamber of a cone–cylinder model by exploding wire by electrical short circuit, is injected into the supersonic flow through a nozzle. A significant drag reduction was measured [19]. A brief history of the development in this subject area was reported in an article published in *Jane's Defence Weekly* [20].

The research in plasma mitigation of the shock waves has two primary goals:

1. to improve the effective aerodynamic shape of an aircraft, but without the cooling requirements of a physical spike, and
2. to reduce the shock noise and possibly make net energy savings.

9.6.3 Plasma spikes for the mitigation of shock waves: experiments and results

To further study plasma effects on shock waves, Kuo *et al* [21] have carried out experiments in a Mach-2.5 wind tunnel. A cone-shaped model having a

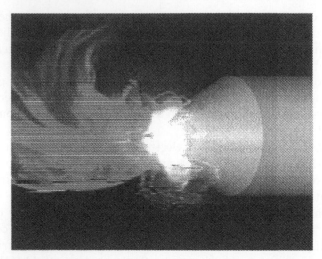

Figure 9.6.1. Plasma produced in front of the model, which is moving around the tip in spray-like forms. (Copyright 2000 by AIP.)

60° cone angle was placed in the test section of the wind tunnel. The tip and the body of the model were designed as two electrodes with the tip of the model designated as the cathode for gaseous discharge. A 60 Hz power supply was used in the discharge for plasma generation. The peak and average powers of the discharge during the wind tunnel runs were measured to be about 1.2 kW and 100 W, respectively. Shown in figure 9.6.1 is the airglow image of a spray-like plasma generated by the 60 Hz self-sustained diffused arc discharge, at the nose region of the model, where the usual attached conical shock is formed in the supersonic flow. The plasma density and temperature of the discharge were not measured. However, the electrode arrangement and the power supply were similar to those used in producing a 60 Hz torch plasma, which was measured [22] to have peak electron density and temperature exceeding 10^{13} electrons/cm^3 and 5000 K (time averaged temperature [23] is less than 2000 K), respectively. During the run, the background pressure drops, thus the plasma density is expected to increase slightly. On the other hand, the electron plasma is cooled considerably by the supersonic flow. The produced spray-like plasma acted as a spatially distributed spike, which could deflect the incoming flow before the flow reached the original shock front location. The effect of this plasma spike on the shock wave formation was explored by examining a sequence of shadowgraphs taken during typical wind tunnel runs.

The shadowgraph technique is briefly described as follows. A uniform collimated light beam is introduced to illuminate the flow. The second derivative of the flow density deflects the light rays to a direction perpendicular to the light beam, which results in light intensity variation on a projection

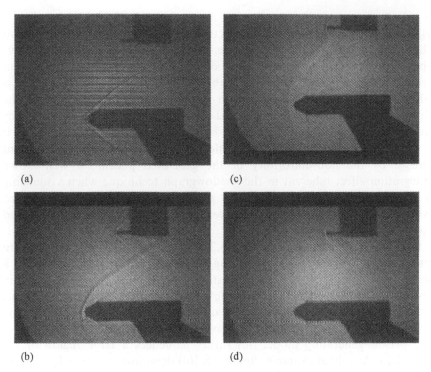

(a)   (c)

(b)   (d)

Figure 9.6.2. A sequence of shadowgraphs taken during a wind tunnel run at Mach-2.5 in the presence of plasma. (a) At the instant close to initiating plasma, (b) at a later time during the run, (c) at a later time during the same run, and (d) at the time when the discharge is around the peak and the shock wave is eliminated. (Copyright 2000 by AIP.)

screen showing the shadow image of the flow field. Thus the location of a stationary shock front in the flow, where the second derivative of the density distribution is very large, is revealed in the shadowgraph as a dark curve because the light transmitted through that region is reduced to a minimum.

In the shadowgraphs shown in figure 9.6.2 the flow is from left to right. The upstream flow has a flow speed $v = 570\,\text{m/s}$, temperature $T_1 = 135\,\text{K}$, and a pressure $p_1 = 0.175\,\text{atm}$. Figure 9.6.2(a) is a snapshot of the flow at the instant close to initiating the plasma. As shown, an undisturbed conical shock is formed in front of the plasma-producing model. To further examine the flow structure, a Pitot tube was installed in the tunnel, which can be seen on the top portion of the shadowgraph with its usual detached shock front. Figure 9.6.2(b) taken at a later time during the run, on the other hand, clearly demonstrates the pronounced influence of plasma on the shock structure. Comparison of figures 9.6.2(a) and (b) clearly indicates an upstream displacement of the shock front along with a larger shock angle, indicating a transformation of the shock from a well defined attached shock into a classic highly curved bow shock structure. It is also interesting to note that the shock in front of the Pitot probe, which is placed at a distance above the plasma-producing model, has been noticeably altered as is evident from the larger shock angle. A highly diffused detached shock front is observed in figure 9.6.2(c) taken at a later time during the same run. The diffused form of the shock front could be the result of less spatial coherency in the flow perturbations introduced by the spatially distributed plasma; it could also be ascribed to a visual effect from an asymmetric shock front caused by the non-uniformity of the generated plasma, a well-known integration effect inherent in the shadowgraph technique when visualizing a three-dimensional flow field. This phenomenon is commonly observed when the spatial extent of the region leading to the shock is small compared to the test section dimensions.

Closer examination of figure 9.6.2(c) demonstrates a further upstream propagation of the bow shock, having an even more dispersed shape and a larger shock angle. It is also interesting to note that the shock wave in front of the Pitot probe has also moved upstream and some evidence of flow expansion may be seen near the tip of the probe. This is an interesting result indicating that the effect of plasma is not confined to the vicinity of the plasma-generating model but rather influences a large region of the flow field. As a final example, figure 9.6.2(d) demonstrates the effectiveness of the plasma in eliminating the shock near the model, an encouraging result, which may have significant consequences in the effectiveness of this scheme in minimizing wave drag and shock noise at supersonic speeds.

In summary, the experimental results represented by the shadowgraphs (figures 9.6.2(b)–(d)) of the flowfield show that the spray-like plasma has strong effect on the structure of the shock wave. It causes the shock front to move upstream toward the plasma front and to become more and more dispersed in the process (figures 9.6.2(b) and (c)). A shock-free state (figure 9.6.2(d)) is observed as the discharge is intensified.

A follow up experiment by Bivolaru and Kuo [24] further demonstrated the plasma effect on shock wave mitigation. The experiment used a similar truncated cone model except that the nose of the model has a 9 mm protruding central spike, which also served as the discharge cathode. Moreover, the power supply was a dc pulse discharge source using RC circuits for charging ($R_c = 10\,k\Omega$) and discharging ($R_d = 150\,\Omega$ to ballast the discharging current) and a $5\,kV/400\,mA$ dc power supply to charge the capacitor ($C = 150\,\mu F$). It produced very energetic plasma with a low repetition rate. The peak power exceeded 40 kW and the energy in each discharge pulse was about 150 J. Again, the plasma density and temperature were not measured during the runs. However, from the current measurement, the peak electron density is estimated to exceed 10^{14} electrons/cm^3. Without the spike, a detached curved shock would be generated in front of the

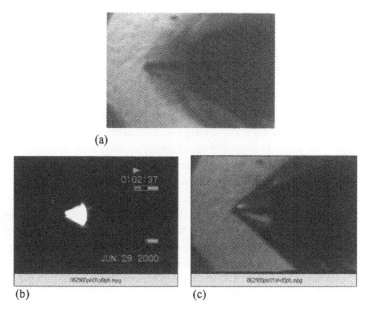

(a)

(b)   (c)

Figure 9.6.3. (a) A baseline schlieren image of a Mach-2.5 flow over 60° truncated cone (pin hole knife-edge of 0.2 mm in diameter); the aspect ratio of the spike length l to the spike diameter d, $l/d = 6$, (b) video graph of the plasma airglow showing a cone-shaped plasma around the spike of the model; and (c) schlieren image of the flowfield modified by the cone-shaped plasma shown in (b). (Copyright 2002 by AIP.)

truncated cone model. The added spike with the selected length modified the structure of the curved shock (which is the one intended to be modified by the plasma) only in the central region around the spike, where the shock front becomes conical and attached to the tip. This is seen in figure 9.6.3(a), a baseline schlieren image of the flow field around the spike and the nose of the cone; the flow is from left to right. The use of this design facilitates the discharge (starting at the base of the truncated cone model) to move upstream through the subsonic region of the boundary layer, along the spike/electrode surface, so that plasma can always be generated in the region upstream of the curved shock front (but it will appear behind the oblique part of the shock front as shown later).

In the schlieren method, again, a uniform collimated light beam is introduced to illuminate the flow. In addition, an obstruction (i.e. a light ray selecting device) is introduced in the light path (e.g. a knife-edge placed at the focal point of the image-forming lens). It uniformly decreases the image illumination in the absence of any disturbance; however, when a density gradient exists in the flow, only some rays will pass the obstruction with a specific variation in the image illumination. The contrast of the image will be proportional to the density gradient in the flow. When rays are deflected toward the knife-edge, the image field becomes darker (negative contrast) and vice-versa. The images can be recorded directly by a CCD camera, without going through an image projection screen. It is noted that if too many rays are stopped, the image quality will deteriorate. Therefore, the knife-edge must be adjusted with a compromise between image quality and contrast.

Much more energetic plasma was generated by this pulsed dc discharge than that generated by 60 Hz discharge in the other experiment. This spike also guided the pulsed electrical discharge to move upstream such that plasma was easily generated in the region upstream of the curved shock front. As plasma was generated, it was found that the schlieren image of the flowfield became quite different from that shown in figure 9.6.3(a). The discharge was symmetric; it produced a cone-shaped plasma around the spike of the model, as shown by the video graph in figure 9.6.3(b). Comparing the corresponding schlieren image of the flowfield presented in figure 9.6.3(c), again the flow is from left to right, with the baseline schlieren image shown in figure 9.6.3(a), it is found that the original curved shock structure in front of the truncated cone is not there any more. The complicated shock structure in figure 9.6.3(a) is now modified to a simple one displaying a single attached conical (oblique) shock similar to the one generated by a perfect cone in the absence of plasma. In other words, it seems that plasma has reinstated the model to a perfect cone configuration. The wave drag to the model caused by oblique shock is much smaller than that caused by the original bow shock.

This experiment has demonstrated that the performance of a small physical spike on the body aerodynamics can be greatly improved by generating plasma around it to form a plasma aero-spike, without increasing the cooling requirement to that for a large physical spike. A change of the shock wave pattern from bow shock dominated structure to oblique shock structure is equivalent to an effective increase in the body aspect ratio (fineness), from $L/D = 0.5$ (blunt conical body) to $L/D = 0.85$ (conical body), by 1.7 times (70%). Although the modification on the shock wave structure by this plasma aero-spike is characteristically different from that by a spread-shaped plasma that causes the shock front to have increased dispersion in its structure as well as standoff distance from the model, both are effective in the mitigation of shock waves. Moreover, it was found, in both experiments, that significant plasma effect on the shock wave was observed only when two criteria were met: (1) plasma is generated in the region upstream of the baseline shock front and (2) plasma has a symmetrical spatial distribution with respect to the axis of the model.

Although experiments have clearly demonstrated that plasmas can significantly modify the shock structure and reduce the wave drag to the object, neither the physical mechanism nor a net energy saving from the drag reduction were confirmed. More experiments are needed to resolve these issues. Some of the facts deduced from the experimental results, however, suggest that deflection of the incoming flow by a symmetrically distributed plasma spike in front of the shock may prove to be a useful process against shock formation.

The effect of plasma aerodynamics on the shock wave observed in experiments may be understood physically. A shock wave is formed by coherent aggregation of flow perturbations from an object. In the steady state, a sharp shock front signified by a step pressure jump is formed to separate the flow into regions of distinct entropies. The shock wave angle β depends on the Mach number M and the deflection angle θ of the flow through a θ–β–M relation, where β increases with θ. Since the shock front is at the far reachable edge of the flow perturbations deflected forward from an object, flow is unperturbed before reaching the shock front. In order to move the shock wave upstream, the flow perturbations have to move upstream beyond the original shock front. An easy way is to start the flow perturbation in front of the location of the original one by, for instance, introducing a longer physical spike. The added plasma spike serves the same purpose; it encounters the flow in the region upstream of the location of the original shock front. It increases the deflection angle θ of the incoming flow as well as the oblique angle β of the tip-attached shock. As the discharge is intensified, the induced flow perturbations from the plasma spike can be large enough to coalesce into a new shock front, which replaces the original one located behind it. This is also realized by the θ–β–M relation. When the deflection angle of the flow exceeds the maximum deflection angle in the θ–β–M relation, then the oblique shock in this region does not exist any more. Instead, the shock structure in this region becomes curved and detached (figure 9.6.2(c)). The deflection mechanism is also applicable for explaining the plasma effect shown in figure 9.6.3(c). As shown in figure 9.6.3(b), on-board generated plasma filled the truncated part of the model. It deflected the incoming flow as effectively as a perfect cone. Because much less flow could reach and be deflected by the frontal surface of the truncated cone, the original bow shock was replaced by an oblique shock attached to the tip of this 'virtually perfect cone'.

The shock front is also expected to appear in a dispersed form because the effective plasma spike is distributed spatially and is not as rigid as the tip of the model or a physical spike. In other words, the flow perturbations by the plasma spike are less coherent as they coalesce into a shock and consequently form a weaker new shock.

References

[1] Chang P K 1970 *Separation of Flow* (Pergamon Press)
[2] Buseman A 1935 'Atti del V Convegna "Volta"' Reale Accademia d'Italia, Rome

[3] Kantrowitz A 1960 *Flight Magnetohydrodynamics* (Addison-Wesley) pp 221–232

[4] Levin V A and Taranteva L V 1993 'Supersonic flow over cone with heat release in the neighborhood of the apex' *Fluid Dynamics* **28**(2) 244–247

[5] Riggins D, Nelson H F and Johnson E 1999 'Blunt-body wave drag reduction using focused energy deposition' *AIAA J.* **37**(4)

[6] Katzen E D and Kaattari G E 1965 'Inviscid hypersonic flow around blunt bodies' *AIAA J.* **3**(7) 1230–1237

[7] Myrabo L N and Raizer Yu P 1994 'Laser induced air-spike for advanced transatmospheric vehicles' *AIAA Paper 94-2451*, 25th AIAA Plasmadynamics and Laser Conference, Colorado Springs, CO, June

[8] Manucci M A S, Toro P G P, Chanes Jr J B, Ramos A G, Pereira A L, Nagamatsu H T and Myrabo L N 2000 'Experimental investigation of a laser-supported directed-energy air spike in hypersonic flow' 7th International Workshop on Shock Tube Technology, hosted by GASL, Inc., Port Jefferson, New York, September

[9] Klimov A N, Koblov A N, Mishin G I, Serov Yu L, Khodataev K V and Yavov I P 1982 'Shock wave propagation in a decaying plasma' *Sov. Tech. Phys. Lett.* **8** 240

[10] Voinovich P A, Ershov A P, Ponomareva S E and Shibkov V M 1990 'Propagation of weak shock waves in plasma of longitudinal flow discharge in air' *High Temp.* **29**(3) 468–475

[11] Bletzinger P, Ganguly B N and Garscadden A 2000 'Electric field and plasma emission responses in a low pressure positive column discharge exposed to a low Mach number shock wave' *Phys. Plasmas* **7**(7) 4341–4346

[12] Mishin G I, Serov Yu. L and Yavor I P 1991 *Sov. Tech. Phys. Lett.* **17** 413

[13] Bedin A P and Mishin,G I 1995 *Sov. Tech. Phys. Lett.* **21** 14

[14] Serov Yu L and Yavor I P 1995 *Sov. Tech. Phys.* **40** 248

[15] Kuo S P and Bivolaru D 2001 'Plasma effect on shock waves in a supersonic flow' *Phys. Plasmas* **8**(7) 3258–3264

[16] Beaulieu W, Brovkin V, Goldberg I *et al* 1998 'Microwave plasma influence on aerodynamic characteristics of body in airflow' in *Proceedings of the 2nd Workshop on Weakly Ionized Gases*, American Institute of Aeronautics and Astronautics, Washington, DC, p 193

[17] Exton R J 1997 'On-board generation of a "precursor" microwave plasma at Mach 6: experiment design' in *Proceedings of the 1st Workshop on Weakly Ionized Gases*, vol 2, pp EE3–12, Wright Lab. Aero Propulsion and Power Directorate, Wright-Patterson AFB, OH

[18] Baryshnikov A S, Basargin I V, Dubinina E V and Fedotov D A 1997 'Rearrangement of the shock wave structure in a decaying discharge plasma' *Tech. Phys. Lett.* **23**(4) 259–260

[19] Gordeev V P, Krasilnikov A V, Lagutin V I and Otmennikov V N 1996 'Plasma technology for reduction of flying vehicle drag' *Fluid Dynamics* **31**(2) 313

[20] 'Drag Factor' 1998 *Jane's Defence Weekly* (ISSN 0265–3818) **29**(24) 23–26

[21] Kuo S P, Kalkhoran I M, Bivolaru D and Orlick L 2000 'Observation of shock wave elimination by a plasma in a Mach 2.5 flow' *Phys. Plasmas* **7**(5) 1345

[22] Kuo S P, Bivolaru D and Orlick L 2003 'A magnetized torch module for plasma generation and plasma diagnostic with microwave', *AIAA Paper 2003-135*, American Institute of Aeronautics and Astronautics, Washington, DC

[23] Kuo S P, Koretzky E and Vidmar R J 1999 'Temperature measurement of an atmospheric-pressure plasma torch' *Rev. Sci. Instruments* **70**(7) 3032–3034

[24] Bivolaru D and Kuo S P 2002 'Observation of supersonic shock wave mitigation by a plasma aero-spike' *Phys. Plasmas* **9**(2) 721–723

9.7 Surface Treatment

9.7.1 Introduction

Low-temperature non-equilibrium plasmas are effective tools for the surface treatment of various materials in micro-electronics, manufacturing and other industrial applications. The application of atmospheric pressure discharges presents advantages such as plasma treatment with cheap gas mixtures, low specific energy consumption and short processing time. Plasma procedures in chemically reactive gases are easy to control and, as dry processes with low material insert, they are environmentally friendly.

The interaction of plasmas with surfaces can be systematized according to the following definitions:

1. Etching means the removal of bulk material. The process includes chemical reactions which produce volatile compounds containing atoms of the bulk material. Sputtering is a physical process which removes bulk atoms by collisions of energetic ions with the surface. Applications are, for example, structuring in micro-electronics and micro-mechanics. These processes are connected with a loss of a weighable amount of the bulk substance.
2. Cleaning is the removal of material located on the surface which is not necessarily connected with the removal of bulk material. This process is applied, for example, in assembly lines as a preparation step for subsequent procedures.
3. Functionalization leads to the formation of functional groups and/or of cross links on the surface by chemical reactions between gas-phase species and surface species/reactive sites and/or surface species (Chan 1994). Grafting is a surface reaction between gas phase and polymer material. The mass yield or loss in these processes is very small. Functionalization changes, but mostly improves the wettability, the adhesion, lamination to other films, the printability, and other coating applications. Biological properties may be influenced too, for example, the probability of settlements of cells or bacteria.
4. Interstitial modifications occur, for example, by ion implantation for the hardening of metal surfaces.
5. Deposition of films of non-substrate material change the mechanical (tribology), chemical (corrosion protection), and optical (reflecting and

Table 9.7.1. Plasma components and their efficiency in surface treatment (Meichsner 2001).

Plasma component	Kinetic energy	Processes and effects in the material	Depth of interaction
Ions, neutrals	~10 eV	Adsorbate sputtering, chemical reactions	Monolayer
Electrons	5–10 eV	Inelastic collisions, surface dissociation, surface ionization	~1 nm
Reactive neutrals	Thermal 0.05 eV	Adsorption, chemical surface reactions, formation of functional groups, low molecular (volatile) products	Monolayer
		Diffusion and chemical reactions	Bulk
Photons	>5 eV (VUV)	Photochemical processes	10–50 nm
	<5 eV (UV)	Secondary processes	μm range

decorative) properties of materials. For films that are not too thin the mass yield is weighable. Systems of thin films with different electrical properties are the basic essentials of micro electronics.

6. The depth scale of the different processes are as follows: etching 10–100 nm, functionalization 1 nm, coating 10–1000 nm (Behnisch 1994).

In reality these different processes are not strongly separated, e.g. cleaning may include sputtering or functionalization. The efficiency of the various plasma components in surface treatment is presented in table 9.7.1 (Meichsner 2001).

The dielectric barrier discharge (DBD) seems to be the most promising plasma source for a plasma-assisted treatment of both large-area metallic and polymer surfaces at atmospheric pressure. Investigations of the homogeneous DBD commonly known as 'atmospheric pressure glow discharge' (APGD) (Kogoma *et al* 1998), and of the filamentary or disperse DBD (Behnke 1996, Schmidt-Szalowski *et al* 2000, Massines *et al* 2000, Sonnenfeld 2001b) proved the applicability of DBD for surface treatment techniques.

Special applications of DBD under atmospheric pressure exist in the modification of large-area surfaces for the purpose of the corrosion protection of metals and of an improvement of e.g. the wetting behavior of polymers.

This modification of surfaces usually consists of three steps:

1. the cleaning of the bulk material of hydrocarbon containing lubricants and other fatty contaminants,
2. especially for metals, the deposition of a stable oxide layer of a thickness of some 10 nm as a diffusion barrier of the metallic bulk material, and

3. the deposition of a surface protecting thin layer (thickness of some hundreds of nm) with a good adhesive characteristics of a primer coating.

The surface functionalization of polymers takes place after the cleaning procedure.

The advantage of the surface treatment of metals by means of the DBD plasma consists in the fact that all three sub-processes can run off successively in the same plasma equipment (Behnke *et al* 2002).

The effect of plasma treatment depends on the energy input into the process. For the energy flow on the mostly moving substrate, the dosage D is used (Softal Report 151 E Part 2/3)

$$D = \frac{P}{sv} \left[\frac{J}{m^2} \right]$$

where P is the power introduced into the discharge [W], s is the electrode width [m], and v is the substrate velocity [m/s].

The power density L in the discharge volume is given by

$$L = Ej = \frac{P}{Aa} \qquad [W/m^3]$$

where E is the averaged voltage gradient inside the plasma [V/m], j is the current density [A/m^2], A is the electrode surface [m^2], and a is the gap distance of the discharge [m].

The power density O on the electrode surface is defined by

$$O = \frac{P}{A} \qquad [W/m^2].$$

D is an important parameter to achieve desired surface properties, L characterizes the plasma properties, O is a measure of the electrode strain. For a resting substrate the dosage is given by the product of O and the treatment time.

This section is organized as follows: it first deals with experimental questions mainly oriented to the dielectric barrier discharge. The next part is devoted to cleaning by atmospheric pressure discharges. Then oxidation and functionalization are discussed, followed by plasma etching. The final topic deals with coating of substrates by deposition of a thin film. Closing remarks outline the advantages and limits of surface treatment by atmospheric pressure discharges in air.

9.7.2 Experimental

Here are presented special investigations with typical parameters which are used for surface cleaning, oxidation and thin film deposition (Behnke 2002). The DBD apparatus consists of two dielectric high-voltage electrodes of rectangular cross section. The ceramic shell (Al$_2$O$_3$) of this hollow block is

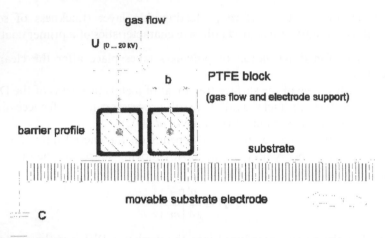

Figure 9.7.1. Scheme of the DBD equipment for surface treatment with a dynamic electrode arrangement.

about 0.1 cm thick, 2 cm wide and 15–50 cm long, and coated inside with a silver layer for the electrical contacts.

The DBD operates within the region between the electrodes and the substrate (grounded electrode) with a gap of 0.05–0.1 cm. The electrodes are moved periodically along the substrate by a step motor. The effective treatment time t_P depends on the relative speed between the substrate and dielectric electrodes v_s, the length b and the number n of the electrodes and the number of the moving periods p during the plasma process $t_P = pnb/v_s$. The slit between the rectangular profiles is used to introduce a laminar flow of the process gas mixture (air, vapors of silicon organic compounds as hexamethyldisiloxane (HMDSO, $(CH_3)_3SiOSi(CH_3)_3$) and tetraethoxysilane (TEOS, $(CH_3CH_2O)_4Si$)) into the discharge zone. To reduce excess heating the electrode system as well as the substrate holder are cooled by a flowing liquid. The DBD is driven by a sinusoidal voltage of some 10 kV in a continuous or pulsed mode of frequencies between 5 and 50 kHz. For characterization of the experimental conditions the electrical power absorbed in the discharge is measured.

A schematic view of the experimental set-up is given in figure 9.7.1. The typical operating conditions during plasma treatment are represented in table 9.7.2. The cleaning and coating experiments are carried out with aluminum plates (80 mm × 150 mm) and Si wafers for ellipsometric measurements of the layer properties.

For the investigation of the cleaning process the substrates were covered with defined quantities of oil (80–300 nm). For the deposition experiments the substrates are chemically pre-cleaned and cleaned in the DBD in air under atmospheric pressure with effective treatment times of about 100 s.

Table 9.7.2. Typical operation conditions during DBD-plasma treatment.

	Cleaning	Oxidation	Deposition	Functionalization
Frequency (kHz)	10–25	10–25	6.6	0.050–125
Voltage (kV)	<15	<15	<15	3–50
Power (W)	60–80	80	45	
Power density (W cm^{-2})	2.2–3.0	3	1–1.6	
Volume power density (W cm^{-3})	20–60	30–60	10–30	
Dosage (J/cm^2)	5–10	5–10	50–80	1–300
Discharge gap (mm)	0.5–1.0	0.5–1.0	0.5–1.0	1–5
Process gas	dry air	dry air	N$_2$ or dry air	Air
Reactive gas	O$_2$20%	O$_2$20%	TEOS 0.1 % HMDSO 0.1%	
Gasflow (slm)	1.6	1.6	1	1–10
Effect. treatment time (s)	<120	<600	<90	10–100
Mean residence time (s)	0.06	0.06	0.1	

The time dependence of the oil removal and of the mass increase during the oxidation phase as well as the deposition of SiO$_x$C$_y$H$_z$ coatings are measured gravimetrically by weighing the samples with a micro-scale. The contaminated and cleaned substrates are quasi *in-situ* characterized ellipsometrically by a spectroscopic polarization modulation ellipsometer (633 nm). The thickness of the deposited Si organic layer is also measured gravimetrically.

The chemical composition of the substrate surface before and after plasma treatment is studied by x-ray photoelectron spectroscopy (XPS) and Fourier transform infrared (FTIR) spectroscopy. The surface morphological properties are investigated by scanning electron microscopy (SEM) and contact angle measurements.

9.7.3 Cleaning

Metal surfaces are frequently covered with fats and oils in order to protect them temporarily against corrosion and to improve their manufacturing properties. For the following surface treatments this contamination must be removed by wet-chemical cleaning procedures or by vapor cleaning techniques using chlorinated and chloro-fluoro compounds. These processes are critically estimated to be environmentally undesirable. A plasma-assisted treatment operating at atmospheric pressure without greenhouse gases represents an environmentally friendly economical alternative. Since for such procedures no vacuum equipment is needed, they can be easily integrated in process lines (Klages 2002).

Non-thermal atmospheric pressure air plasmas generate reactive oxygen atoms and ozone, which easily react with organic compounds and produce

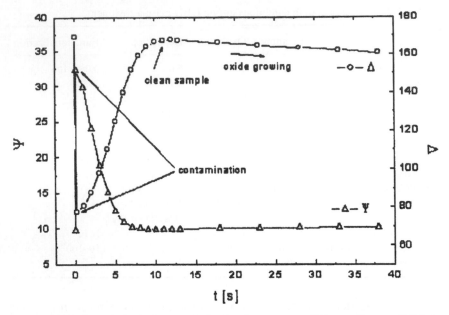

Figure 9.7.2. Ψ and Δ during a whole cleaning process (633 nm, $P_{DBD} = 80\,W$) in dependence on the effective treatment time in seconds.

volatile reaction products like CO, CO_2 and H_2O. Air plasmas have been tested for surface cleaning, especially of contaminated metal.

In order to understand the cleaning procedure in a DBD in air, the erosion of oil contamination on silicon surfaces was investigated by ellipsometry and fluorescence microscopy (Behnke *et al* 1996a,b, Thyen *et al* 2000, Behnke *et al* 2002).

Figure 9.7.2 shows a typical plot of the ellipsometric angles Ψ and Δ versus treatment time, which was monitored during the whole cleaning procedure ($\lambda = 633\,nm$, DBD power 80 W) of a contaminated Si wafer. The ellipsometric angles were measured before and after the oil contamination (Wisura Akamin) (Behnke *et al* 2002).

The angles Ψ (decreases) and Δ (increases) change considerably during the surface treatment. In a short time they approach the values of pure silicon. That means the purification process runs very fast ($<10\,s$). However, the initial values before the contamination are not reached, because the Si surface properties were changed by oxidation.

More information about cleaning and the following oxidation process is elucidated by spectroscopic ellipsometrical investigations. The layer thickness $d(t)$ and therefore the etching rate $r(t)$ are also evaluated from the ellipsometrical data of the wavelengths between 1.5 and 4.5 eV by means of the dispersion formula of Cauchy using model approximations. The contamination thickness and etching rate decrease nearly exponentially.

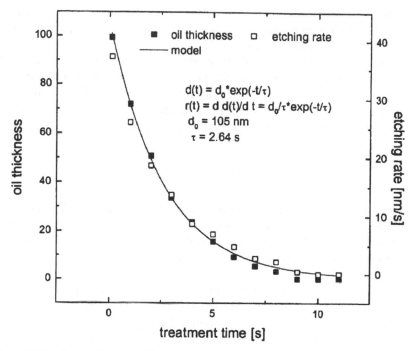

Figure 9.7.3. Contamination thickness d and etching rate r versus treatment time for the discharge power of 80 W. Substrate: Si wafer.

The etching rate reaches values up to 40 nm s^{-1}. It decreases linearly with the contamination thickness. An example for the exponential decay of thickness and etching rate is given in figure 9.7.3. The following relations are valid:

$$d(t) = d_0 \, e^{-t/\tau}$$

$$r(t) = \left| \frac{\partial \, dt}{\partial t} \right| = \frac{dt}{\tau}$$

$$\frac{d(t)}{r(t)} = \frac{1}{\tau} = \text{const}$$

where τ is a time constant which characterizes the cleaning process in dependence on the discharge power and of the initial contamination d_0. The same functional correlation is described by (Thyen *et al* 2000) for the cleaning of contaminated Si wafers. A similar exponential temporal behavior of the erosion of the contamination was determined from gravimetric measurements on aluminum substrates (Behnke *et al* 2002) as well as from fluorescence microscopic measurements on steel substrates (Thyen *et al* 2000).

   In contrast to these results, cleaning investigations in rf oxygen low-pressure discharges show a linear reduction of the contamination thickness and thus a constant etching rate during the entire plasma process. Hence it follows that in low-pressure discharges each sub-layer of the contamination is removed with a constant rate.

One reason for the exponential behavior may be the statistical character of the cleaning process. A single filament removes nearly all the contamination from the sample within the relevant area. The temporal sequence of the filaments is statistically distributed on the substrate. That means that removed mass dm in the time interval dt is proportional to the mass m of the contamination.

$$dm = -m\frac{dt}{\tau}.$$

The second reason is the polymerization of the lubricant for higher initial thickness. That is clearly seen from the increase of the optical constants n and k of the layer which is related to higher layer density. Also Thyen *et al* (2000) explained the exponential decline of $d(t)$ by initiation of polymerization reactions of the oil.

An improved understanding will be achieved by studying the etching process in the remote plasma outside the DBD. There the contaminated metallic plate is not touched by filaments. Etching takes place only due to active species which are produced by the discharge. Under these conditions the etching rates are much lower and the process stops if the contamination reaches about 20% of the initial thickness. That means the filaments are essential for the cleaning process. Without filaments the polymerization of the lubricant becomes the most preferred mechanism. In case of small contamination thickness (100–150 nm) substrates can be completely cleaned using any tested values of power. The time constants for the removal of the contamination decrease approximately linearly with discharge power. Contamination above $6\,g\,m^{-2}$ could not be removed by a barrier discharge.

The cleaning rate r depends strongly on the oxygen content in the process gas. Thyen *et al* (2000) found that in pure nitrogen the rate is over ten times lower than in the air mixture. An admixture of 0.5% oxygen to the process gas raises the rate in relation to that in pure nitrogen by a factor around 3, but in pure oxygen this factor again decreases to 1.3. On the other hand the removal rate increases in dependence on the gas flow. A saturation is reached at a gas throughput of around 5 slm (Thyen *et al* 2000). With increasing flow rate more dismantling products of the hydrocarbons in the exhaust gas stream are removed, because a higher flow counteracts a reassembly of these products on the surface. The saturation of the rate is achieved if the flux of broken hydrocarbon chains equals the products removed by the gas flow (Behnke 1996b). Concerning the chemical reactions of an air plasma with hydrocarbons the reader will be referred to the discussion of the plasma-functionalization of polypropylene as an example of hydrocarbons in section 9.7.5.

Becker and coworkers (Korfiatis *et al* 2002, Moskwinski *et al* 2002) have been using a non-thermal atmospheric-pressure plasma generated in a capillary plasma electrode configuration (Kunhardt 2000; see also chapter 2 of this book) to clean Al surfaces contaminated with hydrocarbons. Efficient hydrocarbon removal of essentially 100% of the contaminants in this discharge type was reported for plasma exposure times of only a few seconds and contaminant films of up to 300 nm. Specifically, these researchers have studied the utility of a plasma-based cleaning process in removing oils and grease from Al surfaces both during manufacturing and prior to the use of the Al in a specific application.

All these experimental investigations show that hydrocarbons can be removed completely from metallic substrates by using an atmospheric plasma in air. From the ellipsometric measurements on a silicon wafer it was found that the residual contamination is in the order of one atomic layer.

One important parameter for the characterization of the surface cleanness is the specific surface energy, which is determined by means of contact angle measurements of several liquids (Owen plot). After the plasma cleaning procedure the total surface tension (67 mN/m) is very high. For further treatment procedures the time behavior of the surface tension is important. While the dispersive fraction does not change (27 mN/m) the polar fraction decreases exponentially in time (time constant: 166 h). A high wettability of the cleaned surface remains stable for 24 h if the energy dosage of the DBD plasma process is between 50 and 100 J cm^{-2}.

9.7.4 Oxidation

Metallic substrates (e.g. Al, Si, Cu) are usually covered with a native, mostly fragile oxide coating with a thickness of some nm during long storage in air. This layer must be conventionally chemically eliminated in order to treat the surface for corrosion protection. Afterwards the deposition of a stable thicker oxide coating follows (e.g. Al_2O_3 on aluminum surfaces) which is produced conventionally by a galvanic anodization. The plasma-supported treatment will also win extra relevance in the future because of the polluting disposal of galvanic baths.

In the example given in figure 9.7.2 the values of the initial ellipsometric angles Ψ and Δ of a silicon wafer without contamination cannot be reached completely after the air plasma cleaning in a DBD. Moreover Δ decreases again after reaching a maximum. The main reason for this is the oxide growth on the substrate. This result is also confirmed by the XPS measurements. The XPS spectra of an Al layer were measured before and after the plasma treatment. Before treatment the intensity of the Al 2p peak reaches 20% of the oxide peak. After the treatment the oxide peak remarkably increases and the Al 2p peak almost disappears (figure 9.7.4). An increase of the oxide

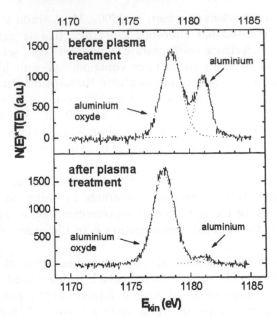

Figure 9.7.4. XPS spectra of an aluminum layer deposited on a silicon wafer before and after the air plasma treatment.

thickness from 3.2 to 8.6 nm is shown by angle resolved measurements. Figure 9.7.5 shows the increase of the weight of an Al-substrate in dependence on the plasma treatment time (Behnke *et al* 2002). In both cases the thickness of the oxide increases approximately proportional to $\sqrt{t}$.

Therefore oxide growth of the oxide is diffusion determined. Diffusion coefficients of about $2–7 \times 10^{-16}\,\mathrm{cm}^2\,\mathrm{s}^{-1}$ are estimated. These are typical

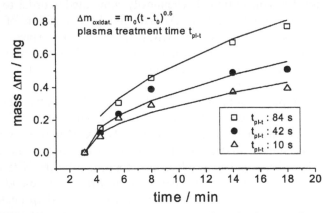

Figure 9.7.5. Increase of the weight after treatment of an aluminum surface with a DBD ($P = 80\,\mathrm{W}$), parameter: plasma treatment time.

values for grain boundary diffusion (Wulff and Steffen 2001). The quality of this oxide depends on the treatment time. If the samples are treated continuously for some minutes the oxide layer is rough. If the samples are treated intermittently only for some seconds with breaks, no roughness can be observed. For aluminum samples the thickness of the oxide reaches about 10 nm after some minutes.

The formation of an oxide layer (Al_2O_3, SiO_2) starts if the DBD is filamented. The high local energy input by the individual filaments leads to a restructuring of the natural oxide coating and to a local evaporation of the bulk material (Al, Si).

The evaporated aluminum or silicon atoms are oxidized by the oxygen atoms inside the DBD plasma and deposited as oxide on the surface. The high current densities between 10^2 and 10^3 A cm^{-2} of an individual micro-discharge causes a compaction of the deposited oxide coating. The local evaporation of the bulk atoms is prevented by increasing oxide thickness and the layer growth is finished. The oxide coating in filamentary air discharge reaches a layer thickness of up to 10–20 nm. This process was monitored by the time-dependent measurement of the aluminum resonance line in a ferro-electrical barrier discharge. The relative line intensity decreased exponentially with the treatment time (Behnke *et al* 1996b). In summary it can be asserted that the DBD supported oxide coating is of a high quality. It has a high density with small roughness.

9.7.5 Functionalization

One important task of functionalization is the improvement of adhesion properties, e.g. for better printing and easier coating. Plastic foils, fibers and other polymer materials are mostly characterized by non-polar chemically inert surfaces with surface energies in the 20–40 mN/m range (polyamide 43.0 mN/m, polyethylene 31.0 mN/m, polytetrafluorethylene 18.5 mN/m). In general polymers are wetted by liquids when the surface energy of the polymer exceeds the surface energy of the liquid. The surface energy of common organic solvents is lower (toluene 28.4 mN/m, carbon tetrachloride 27 mN/m, ethanol 22.1 mN/m) than that of the polymers, therefore paint and inks based on organic solvents are successfully applied to polymers. Environmental requirements call for a replacement by water-based paints, inks, or bonding agents. Because of the high surface strength of water (72.1 mN/m) a treatment of polymer surfaces is necessary to improve their surface energy (Softal Report 102 E).

On the one hand low surface energy impedes surface contamination and allows easy cleaning, but on the other hand it complicates printing, coating, sticking, etc. The surface properties are determined by a thin layer of molecular dimensions and can be changed without influencing the bulk properties of the polymer. Various processes have been developed for surface treatment to enhance adhesion, such as mechanical treatment, wet-chemical treatments, exposure to flames, and plasma treatments in corona and glow discharge plasmas. What is meant by corona discharge is explained in chapter 6. In most cases the corona discharge for the polymer treatment is a dielectric barrier discharge because the non-conductive, dielectric plastic film inside the discharge gap is the barrier. Corona treatment is a well established method. High-capacity systems have been developed and offered by various manufacturers, and are applied to various synthetics. The principles of the action of an air plasma on a polymeric material will be exemplified by the case of polypropylene (PP). After this some characteristic examples for recent activities in surface functionalization will be presented.

Dorai and Kushner (2002a,b, 2003) investigated in detail the processes associated with surface functionalization of an isotactic polypropylene film (0.05 mm thick) in an atmospheric pressure discharge in humid air. Industrial equipment (Pillar Technologies, Hartland, WI) was used for the corona treatment. The discharge is operated at a frequency of 9.6 kHz between a ceramic coated steel ground roll and stainless steel 'shoes' as the powered electrode, separated by a gap of 1.5 mm. The corona energy varied from 0.1 to 17 W s/cm^2. The relative humidity of the air flow in the discharge region was either 2–5% or 95–100% at 25 °C. The treated surface was analyzed to determine its chemical composition by ESCA, its surface energy by contact-angle measurements and its topology by AFM. Additionally the molecular weight of water-soluble low-molecular-weight oxidized material (LMWOM) was investigated. These materials can be separated by washing of the surface in polar solvents like water and alcohols.

The untreated polypropylene surface is free of oxygen. The oxygen content grows with increasing discharge energy. A significant decrease of oxygen is observed after washing. A careful investigation of the LMWOM shows an averaged molecular weight of 400 amu. These oligomers originate from cleavage of the PP chain and contain oxidized groups such as COOH, CHO, or CH$_2$OH. The molecular weight is independent on the discharge energy and the humidity of air. Agglomerates of LMWOM are visible by AFM.

The increase of the discharge energy is associated with a decrease as well as of the advancing and receding water contact angle, that means increasing wettability. The decrease of the advanced contact angle is much smaller for washed samples than for unwashed.

For the treatment of PP in humid air plasma a model was developed (Dorai and Kushner 2003). It includes gas phase chemistry with the formation of O, H, OH radicals and O$_3$ as important active species. Excited O$_2^*$ molecules, N atoms and HO$_2$ need not to be taken into account because of their lower reactivity towards PP. The reactivity of radicals with the PP is different for the position of the C atom where the reaction occurs. Primary C atoms are bound with only one C atom, secondary with two and tertiary with three C atoms inside the polymer. The reaction probability is maximum for the primary C atoms, decreases for secondary and is minimum for tertiary C atoms. The surface reactions can be classified in analogy to polymerization processes in initiation, propagation, and termination.

The initiation reaction is the abstraction of an H atom from the polypropylene surface by an O radical

$$O_{(g)} + \sim CH_2{-}\overset{\overset{\displaystyle H}{|}}{C}{-}CH_2\sim \longrightarrow \sim CH_2{-}\overset{\overset{\displaystyle \cdot}{}}{C}{-}CH_2\sim + OH_{(g)}$$
$$\underset{CH_3}{|} \qquad\qquad \underset{CH_3}{|}$$

or by an OH radical

$$OH_{(g)} + \sim CH_2{-}\overset{\overset{\displaystyle H}{|}}{C}{-}CH_2\sim \longrightarrow \sim CH_2{-}\overset{\overset{\displaystyle \cdot}{}}{C}{-}CH_2\sim + H_2O_{(g)}$$
$$\underset{CH_3}{|} \qquad\qquad \underset{CH_3}{|}$$

associated with the generation of an alkyl radical.

The propagation leads to peroxy radicals on the PP surface in a reaction of the alkyl radical with O_2:

$$O_2 + \sim CH_2{-}\overset{\overset{\displaystyle \cdot}{}}{C}{-}CH_2\sim \longrightarrow \sim CH_2{-}\overset{\overset{\displaystyle O{-}O\cdot}{|}}{C}{-}CH_2\sim$$
$$\underset{CH_3}{|} \qquad\qquad \underset{CH_3}{|}$$

Alkoxy radicals are formed by the reaction of O atoms with the PP alkyl radicals:

$$O_{(g)} + \sim CH_2{-}\overset{\overset{\displaystyle \cdot}{}}{C}{-}CH_2\sim \longrightarrow \sim CH_2{-}\overset{\overset{\displaystyle O\cdot}{|}}{C}{-}CH_2\sim$$
$$\underset{CH_3}{|} \qquad\qquad \underset{CH_3}{|}$$

Also reaction with ozone results in alkoxy radical formation:

$$O_3 + \sim CH_2{-}\overset{\overset{\displaystyle \cdot}{}}{C}{-}CH_2\sim \longrightarrow \sim CH_2{-}\overset{\overset{\displaystyle O\cdot}{|}}{C}{-}CH_2\sim + O_{2(g)}$$
$$\underset{CH_3}{|} \qquad\qquad \underset{CH_3}{|}$$

The abstraction of a neighboring H atom of the PP surface by a peroxy radical produces hydroperoxide:

$$\sim CH_2{-}\overset{\overset{\displaystyle O{-}O\cdot}{|}}{C}{-}CH_2\sim + \sim CH_2{-}\overset{\overset{\displaystyle H}{|}}{C}{-}CH_2\sim \longrightarrow \sim CH_2{-}\overset{\overset{\displaystyle O{-}O{-}H}{|}}{C}{-}CH_2\sim + \sim CH_2{-}\overset{\overset{\displaystyle \cdot}{}}{C}{-}CH_2\sim$$
$$\underset{CH_3}{|} \qquad\qquad \underset{CH_3}{|} \qquad\qquad \underset{CH_3}{|} \qquad\qquad \underset{CH_3}{|}$$

The reaction of the alkyl radical with O_2 may generate, as shown, new peroxy radicals.

A scission of the carbon chain occurs via alkoxy radicals and leads to the formation of ketones

$$\sim CH_2-\underset{\underset{CH_3}{|}}{\overset{\overset{O\cdot}{|}}{C}}-CH_2\sim \quad \left\{ \begin{array}{l} \longrightarrow \quad \sim CH_2-\underset{\underset{O}{\|}}{C}-CH_2\sim \quad + \quad CH_3 \\[2em] \longrightarrow \quad \sim CH_2-C\underset{O}{\overset{CH_3}{\diagdown}} \quad + \quad \cdot CH_2\sim \end{array} \right.$$

or aldehydes:

$$\sim CH_2-\underset{\underset{CH_3}{|}}{\overset{\overset{H}{|}}{C}}-\underset{\underset{H}{|}}{\overset{\overset{O\cdot}{|}}{C}}-\underset{\underset{CH_3}{|}}{\overset{\overset{H}{|}}{C}}-CH_2\sim \quad \longrightarrow \quad \sim CH_2-\underset{\underset{CH_3}{|}}{\overset{\overset{H}{|}}{C}}\cdot \quad + \quad \underset{\underset{H}{|} \underset{CH_3}{|}}{\overset{\overset{O}{\|}}{C}}-C-CH_2\sim$$

Alcohol groups are formed in reactions of alkoxy radicals with the polypropylene:

$$\sim CH_2-\underset{\underset{CH_3}{|}}{\overset{\overset{O\cdot}{|}}{C}}-CH_2\sim \quad + \quad \sim CH_2-\underset{\underset{CH_3}{|}}{\overset{\overset{H}{|}}{C}}-CH_2\sim \quad \longrightarrow$$

$$\sim CH_2-\underset{\underset{CH_3}{|}}{\overset{\overset{OH}{|}}{C}}-CH_2\sim \quad + \quad \sim CH_2-\underset{\underset{CH_3}{|}}{\overset{\overset{\bullet}{|}}{C}}-CH_2\sim$$

Alkoxy radicals are generated by reactions of O and OH radicals:

$$O_{(g)} + \sim CH_2-\underset{\underset{CH_3}{|}}{\overset{\overset{OH}{|}}{C}}-CH_2\sim \quad \longrightarrow \quad \sim CH_2-\underset{\underset{CH_3}{|}}{\overset{\overset{O\cdot}{|}}{C}}-CH_2\sim \quad + \quad OH_{(g)}$$

$$OH_{(g)} + \sim CH_2-\underset{\underset{CH_3}{|}}{\overset{\overset{OH}{|}}{C}}-CH_2\sim \quad \longrightarrow \quad \sim CH_2-\underset{\underset{CH_3}{|}}{\overset{\overset{O\cdot}{|}}{C}}-CH_2\sim \quad + \quad H_2O_{(g)}$$

Termination reactions are

$$H_{(g)} + \sim CH_2-\underset{\underset{CH_3}{|}}{\overset{\overset{\bullet}{|}}{C}}-CH_2\sim \quad \longrightarrow \quad \sim CH_2-\underset{\underset{CH_3}{|}}{\overset{\overset{H}{|}}{C}}-CH_2\sim$$

$$OH_{(g)} + \sim CH_2-\underset{\underset{CH_3}{|}}{\overset{\overset{\bullet}{|}}{C}}-CH_2\sim \quad \longrightarrow \quad \sim CH_2-\underset{\underset{CH_3}{|}}{\overset{\overset{OH}{|}}{C}}-CH_2\sim$$

$$OH_{(g)} + \sim CH_2-\underset{\underset{CH_3}{|}}{\overset{\overset{H}{|}}{C}}-\overset{\bullet}{C}=O \quad \longrightarrow \quad \sim CH_2-\underset{\underset{CH_3}{|}}{\overset{\overset{H}{|}}{C}}-\overset{\overset{OH}{|}}{C}=O$$

The reactions with OH result in the formation of alcohols and acids, respectively.

These reactions illustrate some possibilities of radical production by plasma reactions with a polypropylene surface. Reactions leading to cross linking of the polypropylene matrix must also be taken into account in a detailed description of the plasma–polymer interaction. The probabilities of surface reactions of ultraviolet radiation and ions are supposed to be small.

The surface reaction processes together with the reaction probabilities or reaction rate coefficients are listed in table 9.7.3 (Dorai and Kushner 2003). The calculated values for the percentage coverage of the polypropylene surface by alcohol (−C−OH), peroxy (−C−OO) and acid (−COOH) groups accord well with experimental results (O'Hare *et al* 2002). This successful approach indicates that in spite of the complexity the essential processes of this plasma–surface interaction were comprehensible.

Table 9.7.3. Surface reaction mechanism for polypropylene (Dorai and Kushner 2003).

Reaction[a]	Probabilities or reaction rate coefficients[b]	Comment[c]
Initiation		
$O_g + PP-H \longrightarrow PP^* + OH_g$	$10^{-3}, 10^{-4}, 10^{-5}$	C
$OH_g + PP-H \longrightarrow PP^* + H_2O_g$	$0.25, 0.05, 0.0025$	C
Propagation		C
$PP^* + O_g \longrightarrow PP-O^*$	$10^{-1}, 10^{-2}, 10^{-2}$	C
$PP^* + O_{2,g} \longrightarrow PP-OO^*$	$1.0 \times 10^{-3}, 2.3 \times 10^{-4}, 5.0 \times 10^{-4}$	C
$PP^* + O_{3,g} \longrightarrow PP-O^* + O_{2,g}$	$1.0, 0.5, 0.5$	
$PP-OO^* + PP-H \longrightarrow PP-OOH + PP^*$	$5.5 \times 10^{-16}\,cm^2\,s^{-1}$	
$PP-O^* \longrightarrow$ aldehydes $+ PP^*$	$10\,s^{-1}$	
$PP-O^* \longrightarrow$ ketones $+ PP^*$	$500\,s^{-1}$	
$O_g + PP=O \longrightarrow OH_g +^* PP=O$	0.04	
$OH_g + PP=O \longrightarrow H_2O_g +^* PP=O$	0.4	
$O_g +^* PP=O \longrightarrow CO_{2,g} + PP-H$	0.4	
$OH_g +^* PP=O \longrightarrow (OH)PP=O$	0.12	
$PP-O^* + PP-H \longrightarrow PP-OH + PP^*$	$8.0 \times 10^{-14}\,cm^2\,s^{-1}$	
$O_g + PP-OH \longrightarrow PP-O + OH_g$	7.5×10^{-4}	
$OH_g + PP-OH \longrightarrow PP-O + H_2O_g$	9.2×10^{-3}	
Termination		
$H_g + PP^* \longrightarrow PP-H$	$0.2, 0.2, 0.2$	C
$OH_g + PP^* \longrightarrow PP-OH$	$0.2, 0.2, 0.2$	C

[a] Subscript g denotes gas phase species, PP–H denotes PP.
[b] Those coefficients without units are reaction probabilities.
[c] C = reaction probabilities for tertiary, secondary, and primary radicals, respectively.

The atmospheric plasma surface treatment of polypropylene was a subject of various studies.

A comparison of the action of a homogenous N_2 barrier discharge and a filamentary air discharge (Guimond *et al* 2002) shows that the maximum surface energy γ is higher in the first than in the second one (N_2: $\gamma = 57\,mN/m$, E: $2.8\,W\,s/cm^2$, air: $\gamma = 39\,mN/m$, E: $0.6\,W\,s/cm^2$), but requires a higher specific energy input E. A rapid decrease of the surface energy is observed during the first week of storage, but then the surface energy is fairly stable for more than three months (N_2: $\gamma = 49\,mN/m$, untreated film: $\gamma = 27\,mN/m$).

The action of homogenous and filamentary DBD in various gases, including air, on polypropylene was studied by (Massines *et al* 2001). Cui and Brown (2002) studied the chemical composition of a polypropylene surface during the air plasma treatment. Changes appear to terminate after about 25% of the surface carbon is oxidized. Oxidation produces polar groups like acetals, ketones and carboxyl groups which enhance the surface energy.

A comparison of the treatment of several hydrocarbon polymers (polyethylene PE, polypropylene PP, polystyrene PS and polyisobutylene PIB) by air plasmas at atmospheric pressure of a silent or dielectric barrier discharge and at low pressure (0.2 torr) of an inductively coupled 13.56 MHz discharge was presented by Greenwood *et al* (1995). The dielectric barrier discharge between two plane Al electrodes with a gap of 3 mm was driven by an operating voltage of 11 kV at 3 kHz. The samples on the lower grounded electrode were treated for 30 s and investigated by x-ray photoelectron spectroscopy and atomic force microscopy. Carbon singly bonded to oxygen was found to be the predominant oxidized carbon functionality for all polymers and discharges. The maximum amount of oxygen is incorporated into polystyrene with its π bonds. DBD modification increases the surface roughness of PP, PIB, and PS more than the low pressure discharge. For PE a smoothing is observed. Atmospheric pressure plasma treatment of polyethylene was studied also by Lynch *et al* (1998) and Akishev *et al* (2002). The latter compare the results with polypropylene and polyethylene terephthalate. The surface properties of polypropylene and tetrafluoroethylene perfluorovinyl ether copolymer were investigated after treatment in an atmospheric plasma pretreatment system with a discharge distance of up to 40 cm, which is suitable for a large plastic molding, e.g. an automobile bumper (Tsuchiya *et al* 1998). The increase of the water contact angle with storage time after plasma treatment is explained by a migration of oxygen from a very thin surface area into the inner layer.

Polyimide is an interesting material in the electronics industry for flexible chip carriers. It is characterized by low costs, outstanding properties such as flame resistance, high upper working temperature (250–320 °C), high tensile strength (70–150 MPa), and high dielectric strength (22 kV/cm). The application as a chip carrier demands a metallization with copper. The low surface energy must be enhanced to improve the adhesion of copper. The modification of polyimide surface in a DBD in air is studied by Seeböck *et al* (2000, 2001) and Charbonnier *et al* (2001). The DBD operates at 125 kHz between two plane copper or stainless steel electrodes which have diameters between 0.6 and 2 cm and are separated by a gap of 0.1 mm. There, the dielectric barrier is the polyimide film (thickness 50 or 38 μm). The dielectric barrier discharge with a specific energy input of 3×10^3 W s/ cm^2 leads to an increase of the surface roughness. For a polyimide foil filled with small alumina grains (to improve thermal conductivity) a roughness between 50 and 100 nm is measured. Microscopic inspection shows an increasing number of alumina grains visible at the surface as a consequence of the etching of the polymer. On the surface of the plasma-treated pure polyimide foil, crater-like structures are observed. The DBD in air at atmospheric pressure is filamentary with ignition of the filaments at random spatial positions. The crater formation is assumed as a consequence of repeated ignition of a filament at the same site. This surface roughness enables a metallization with good adhesion (Seeböck *et al* 2001). An obvious enhancement of the surface energy is observed after air plasma treatment. This is caused by the formation of oxygen containing polar groups at the polyimide surface (Seeböck *et al* 2000). XPS investigations demonstrate the increase of oxygen concentration at the surface and show the opening of the aromatic ring under the action of the plasma (Charbonnier *et al* 2001). This bond scission in the imide rings is an important step in the plasma surface reaction with aromatic polymers. For aliphatic polymers H atom abstraction is an essential reaction step, as has been discussed for polypropylene above.

An example for air plasma treatment of a natural material refers to the felt-resistant finishing of wool. By means of an atmospheric pressure barrier discharge in air the content of carboxyl-, hydroxyl- and primary amino-groups on the wool surface is increased. The resulting improved adhesion to special resins enables a uniform and complete coating that leads to a felt-resistance comparable with the results of the environmentally polluting traditional procedures (VDI-TZ 2001, Rott *et al* 1999, Jansen *et al* 1999, Softal Report 152 E).

Non-woven fabrics of synthetic material were successfully treated to increase the surface energy by an air plasma at atmospheric pressure (Roth *et al* 2001a). The treatment of metals was also reported. The removal of mono-layers of contaminants is supposed to be the dominant process of surface energy improvement (Roth *et al* 2001b).

9.7.6 Etching

Concerning the chemical processes, etching is closely related to cleaning, especially if the removal of hydrocarbons or similar materials is studied.

Here examples will be presented of the plasma etching of photo-resists supplemented by one example of plasma etching of Si-based materials and the decomposition of soot in the diesel engine exhaust.

The etch rate of photo-resist on a silicon wafer in a He/O_2 mixture placed on the powered electrode is investigated in an atmospheric pressure dielectric barrier discharge (20–100 kHz, air gap 5–15 mm) (Lee *et al* 2001). Both electrodes are coated with 50 µm polyimide. The grounded electrode is additionally covered with a dielectric plate (thickness 8 mm) furnished with capillaries to induce glow discharges. For a He/O_2 mixture (2.5 or 0.2 slm) 20.7 kHz, 10 mm air gap, and an aspect ratio of 10 average etch rates up to 200 nm/min were obtained. In front of the capillaries an etch rate >3 µm/min was observed.

The photo-resist etching in a dielectric barrier discharge in pure oxygen is studied in dependence on the specific energy input (J/cm^2 and J/cm^3) with the result that the DBD at atmospheric pressure is an alternative to low-pressure plasma processing (Falkenstein and Coogan 1997).

To overcome the difficulties in surface treatment of thick samples or samples with a complicated shape, spray-type reactors were developed (Tanaka *et al* 1999). In a reaction gas Ar/O_2 (100:1) ashing rates of organic photo-resist of up to 1 µm/min were achieved.

The application of a barrier discharge in air (5–7 kHz, 8.5–11 kV, gap width up to 1.5 cm) leads to etching rates of 270 nm/min (Roth *et al* 2001b). The appearance of vertical etching structures under such conditions is observed.

The remote and active plasma generated in a pulsed corona (400 Hz 20 ns rise time, 30 kV) is tested for etching of a photo-resist coating on a silicon wafer (air plasma, remote, 9 nm/min) and the removal of organic films. Etching of the latter is more effective in the active plasma than under remote conditions (Yamamoto *et al* 1995).

An increase of the etch rate of Si-based materials (SiO_2: 1 µm/min; SiN: 2 µm/min; poly Si: 2 µm/min) by more than one order of magnitude in relation to low-pressure plasma etching is observed in an atmospheric pressure of 40.68 MHz discharge in an O_2/CF_4 (up to 1:1) mixture (Kataoka *et al* 2000).

An interesting application of plasma etching in an air discharge concerns the soot decomposition in diesel engine exhaust (Müller *et al* 2000). The reactor operates with a dielectric barrier discharge ($10 kV_{pp}$, ~10 kHz, power on/power off: 3:7, 1:1, 3:7) with an outer tube like porous SiC ceramics electrode (width of the honeycomb channel 5.6 mm) and an inner dielectric barrier electrode (4.2 mm diameter). The flue gas from the diesel engine flows across the discharge gap and is afterwards filtered by the porous outer electrode, leaving the soot particles on its surface. They were decomposed either in the continuous mode or by a regeneration procedure from time to time. More than 95% of the soot particles are removed by the reactor and due to the soot decomposition on the surface a continuous gas flow is achieved across the reactor.

9.7.7 Deposition

Investigations about plasma deposition with DBD have been performed on a broad variety of films in the past ten years. The spectrum ranges from coatings on plastic materials (e.g. polypropylene) for the improvement of the long-term behavior of the wetting ability (Meiners *et al* 1998, Massines *et al* 2000) and hard carbon-based films (Klages *et al* 2003) up to layer systems for the corrosion protection on metal surfaces (Behnke *et al* 2002, 2003, 2004, Foest *et al* 2003, 2004). The kind of precursor used determines the functionality of the deposited layer. The precursors hexamethyldisiloxane (HMDSO, $(CH_3)_3SiOSi(CH_3)_3$) and tetraethoxysilane (TEOS, $(CH_3CH_2O)_4Si$) are frequently studied in atmospheric plasmas concerning their applicability for plasma-supported chemical vapor deposition of silicon–organic thin films (Sonnenfeld *et al* 2001b, Schmidt-Szalowski *et al* 2000, Behnke *et al* 2002, Klages *et al* 2003).

The decomposition of HMDSO and TEOS in the plasma of DBD is controlled by electron impacts (Sonnenfeld *et al* 2001a,b, Basner *et al* 2000). The electron impact induced scission of Si–CH_3 and/or the Si–O bond of the HMDSO monomer is important for the layer deposition via this precursor. The cleavage of the Si–O bond is the main reaction path of the plasma chemical conversion of TEOS with the separation of CH_3–CH_2–O– radicals. In further reaction sequences ethanol and water are produced.

The silicon–organic polymer film is mostly deposited from nitrogen or air DBD with an admixture of the silicon–organic precursor in the order of 0.1% (see table 9.7.2).

The deposition occurs on the basis of small fragments of the silicon–organic precursor. These radicals are adsorbed on the substrate surface. For high energy dosage the gas phase reactions of the precursor and the interaction of the plasma with the surface leads to highly cross-linked films. The films have good adhesion to the substrate surface, they are visually uniform, and transparent. The films are chemically resistant and protect the substrate against corrosive liquids (e.g. NaOH, NaCl, water). SEM images show that damages of the substrate surface ($<350\,m$) are uniformly covered by the films.

The thickness of the deposited silicon–organic polymer films are estimated by gravimetric measurements under the assumption of a film mass density of $1\,g\,cm^{-3}$, also by XPS, SEM and interferometric measurements. The average deposition rate strongly depends on the discharge power density and on the structure of the DBD plasma. One example is presented in figure 9.7.6. Up to the maximum of the deposition rate the DBD appears

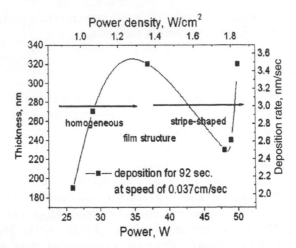

Figure 9.7.6. Thickness and deposition rate of SiO_x polymer films versus discharge power, effective deposition time 92 s, N_2 DBD with admixture of 0.1% TEOS.

quasi-homogeneous: in this discharge range the deposition is quasi-homogeneously dispersed across the substrate. As long as the discharge changes to the mode of stronger filamentation with higher power densities the deposition rate describes a minimum. The film morphology alters to a stripe-shaped structure on the substrate, possibly due to some turbulent convection processes connected with the non-homogeneity of the discharge.

FTIR measurements were carried out on the substrate after plasma treatment. Figure 9.7.7 shows spectra of films produced by an air plasma,

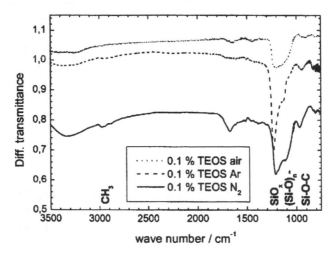

Figure 9.7.7. FTIR spectra of SiO_x polymer films deposited in air, N_2 and Ar DBD with precursor admixture of 0.1% TEOS ($P = 50$ W).

a N_2 plasma and an Ar plasma with 0.1% TEOS admixture. The spectra are dominated by a broad peak in the region of 1000–1250 cm^{-1} which denotes a macromolecular structure of the form $(Si-O_x)_n$. The feature is more broadened for the films produced with air- and N_2-containing plasmas, indicating a slightly enhanced cross linking as compared to the Ar-based film. The specific energy per precursor molecule is comparable in all three cases, hence the effect is presumably caused by increased oxidation of the film.

The $(Si-O_x)_n$ structure is overlapped by the prominent SiO_x peak at 1240 cm^{-1}. Both features along with the very low carbon content (e.g. CH_3 at 2950 cm^{-1}) reveal the pronounced inorganic chemical nature of the film indicative for rather high specific energies per precursor molecule. With increasing specific energy the inorganic character of the film increases—a common effect proven for several silicon–organic precursors, such as HMDSO (Behnke *et al* 2002).

Different technical test procedures for the estimation of the adhesion of a primer on the polymer layer and the determination of the corrosion protection properties of the coating show a sufficient effect only for substrate temperatures above 40 °C. With the dissociation of TEOS in the DBD, ethanol and water are formed, which are linked into the layer without a chemical bonding. The stoichiometric relationship of SiO_x ($x \approx 2$) thereby is never reached and the layer does not become leak-proof. The water stored in the layer withdraws with time and the residual hole-like laminated structure decreases both the adhesion and the anti-corrosion properties of the coating. The water entering the layer is avoided if the coating process is performed at higher substrate temperatures. Layers, which are deposited in a filamentary air DBD plasma, show a better adhesion and corrosion protection effect in contrast to those which are coated by quasi-homogeneous nitrogen DBDs. There will be also an improvement of these layer characteristics, if the layer is deposited only by a one-cycle procedure as a 'mono' layer in relation to a deposition in a multi-cycles procedure (Behnke *et al* 2003).

9.7.8 Conclusions

Atmospheric pressure plasmas are successfully implemented for various surface treatment tasks. When comparing atmospheric-pressure plasma processing with the well established low-pressure plasma processes, one has to consider that the latter methods have been continuously developed for more than 50 years. In contrast, the study of plasma processing at atmospheric pressure on a broader scale has just begun.

The main advantage of atmospheric pressure plasma processing is that it requires much lower investment costs, because no vacuum devices are needed—in the case of ambient air, not even a housing. Hence, the implementation of devices into assembly lines with renouncement of batch procedures is greatly facilitated. The majority of atmospheric plasmas such as DBD and corona discharges are easily scaled up.

The low level of maturity is one of the disadvantages of atmospheric pressure plasma processing in our day. Tailored plasma diagnostic techniques have to be developed for an effective process control.

The state of the art atmospheric pressure plasma technology is holding promising prospects from the economical and environmental point of view. Therefore it is encouraging further research and development activities.

Acknowledgments

The financial support of our activities in the field of atmospheric pressure discharges by the BMBF of Germany Project no. 13N7350/0 and 13N7351/0 is gratefully acknowledged.

References

Akishev Y, Grushin M, Narpatovich A and Trushkin N 2002 'Novel ac and dc non-thermal plasma sources for cold surface treatment of polymer films and fabrics at atmospheric pressure' *Plasma and Polymers* **7** 261–289

Basner R, Schmidt M, Becker K and Deutsch H 2000 'Electron impact ionization of organic silicon compounds' *Adv. Atomic, Molecular and Optical Phys.* **43** 147–185

Behnisch J 1994 'Plasmachemische Modifizierung von Cellulose—Möglichkeiten und Grenzen', Das Papier no. 12 780–783

Behnke J F, Lange H, Michel P, Opalinski T, Steffen H and Wagner H-E 1996b 'The cleaning process of metallic surfaces in barrier discharges' Proc. 5th Int. Symp. on High Pressure Low Temperature Plasma Chemistry (HAKONE V) Janca J *et al* (eds) Milovy/Czech Rep. pp 138–142

Behnke J F, Sonnenfeld A, Ivanova O, Hippler R, To R T X H, Pham G V, Vu K O and Nguyen T D 2003 'Study of corrosion protection of aluminium by siliconoxide-polymer coatings deposited by a dielectric barrier discharge under atmospheric pressure' 56th Gaseous Electronics Conference, 21–24 October 2003, San Francisco, CA. Poster GTP.015 *http://www.aps.org/meet/GEC03/baps/abs/S110015.html*

Behnke J F, Sonnenfeld A, Ivanova O, To T X H, Pham G V, Vu K O, Nguyen T D, Foest R, Schmidt M and Hippler R 2004 'Study of corrosion protection of aluminium by siliconoxide-polymer coatings deposited by a dielectric barrier discharge at atmospheric pressure' Proc. 9th Int. Symp. on High Pressure Low Temperature Plasma Chemistry (HAKONE IX) M. Rea *et al* (eds) 23–26 August 2004, Padova (Italy) in print

Behnke J F, Steffen H and Lange H 1996a 'Elipsometric investigations during plasma cleaning: Comparison between low pressure rf-plasma and barrier discharge at atmospheric pressure' Proc. 5th Int. Symp. on High Pressure Low Temperature Plasma Chemistry (HAKONE V) Janca J *et al* (eds) Milovy/Czech Rep. pp 133–137

Behnke J F, Steffen H, Sonnenfeld A, Foest R, Lebedev V and Hippler R 2002 'Surface modification of aluminium by dielectric barrier discharges under atmospheric pressure' Proc. 8th Int. Symp. on High Pressure Low Temperature Plasma Chemistry (HAKONE VIII) Haljaste A and Planck T (eds), Tartu/Estonia **2** 410

Chan C-M 1994 *Polymer Surface Modification and Characterization* (Munich: Carl Hauser)

Charbonnier M, Romand M, Esrom H and Seeböck R 2001 'Functionalization of polymer surfaces using excimer UV systems and silent discharges. Application to electroless metallization' *J. Adhesion* **75** 381–404

Cui N-Y and Brown N M D 2002 'Modification of the surface properties of a propylene (PP) film using an air dielectric barrier discharge plasma' *Appl. Surf. Sci.* **189** 31–38

Dorai R and Kushner M 2002a 'Atmospheric pressure plasma processing of polypropylene' 49th Int. Symp. Am. Vac. Soc. Banff, Canada, Nov. 2002

Dorai R and Kushner M 2002b 'Plasma surface modification of polymers using atmospheric pressure discharges' 29th ICOPS Banff, Canada

Dorai R and Kushner M 2003 'A model for plasma modification of polypropylene using atmospheric pressure discharges' *J. Phys. D: Appl. Phys.* **36** 666–685

Falkenstein Z and Coogan J J 1997 'Photoresist etching with dielectric barrier discharges in oxygen' *J. Appl. Phys.* **82** 6273–6280

Foest R, Adler F, Sigeneger F and Schmidt M 2003 'Study of an atmospheric pressure glow discharge (APG) for thin film deposition' *Surf. Coat. Technol.* **163/164** 323–330

Foest R, Schmidt M and Behnke J 2004 'Plasma polymerization in an atmospheric pressure dielectric barrier discharge in a flowing gas' in *Gaseous Dielectrics* vol X, ed. Christophorou L G (New York: Kluwer Academic/Plenum Publisher) in print

Greenwood O D, Boyd R D, Hopkins J and Badyal J P S 1995 'Atmospheric silent discharge versus low pressure plasma treatment of polyethylene, polypropylene, polyisobutylene, and polystyrene' *J. Adhesion Sci. Technol.* **9** 311–326

Guimond S, Radu I, Czeremuszkin G, Carlsson D J and Wertheimer M R 2002 'Biaxially orientated polypropylene (BOPP) surface modification by nitrogen atmospheric pressure glow discharge (APGD) and by air corona' *Plasma and Polymers* **7** 71–88

Jansen B, Kümmeler F, Müller H B and Thomas H 1999 'Einfluß der Plasma- und Harzbehandlung auf die Eigenschaften der Wolle' Proc. Workshop Plasmaanwendungen in der Textilindustrie Stuttgart, Germany, 17–23

Kataoka Y, Kanoh M, Makino N, Suzuki K, Saitoh S, Miyajima H and Mori Y 2000 'Dry etching characteristics of Si-based materials used CF4/O2 atmospheric-pressure glow discharge plasmas' *Jpn. J. Appl. Phys.* **39** 294–298

Kersten H, Behnke J F and Eggs C 1994 'Investigations on plasma-assisted surface cleaning of aluminium in an oxygen glow-discharge' *Contr. Plasma Phys.* **34** 563

Klages C P and Eichler M 2002 'Coating and cleaning of surfaces with atmospheric pressure plasmas' (in German) *Vakuum in Forschung und Praxis* **14** 149–155

Klages C P, Eichler M and Thyen R 2003 'Atmospheric pressure PA-CVD of silicon- and carbon-based coatings using dielectric barrier discharges' *New Diamond Front C Tec* **13** 175–189

Kogoma M, Okazaki S, Tanaka K and Inomata T 1998 'Surface treatment of powder in atmospheric pressure glow plasma using ultra-sonic dispersal technique' Proc. 6th Int. Symp. on High Pressure Low Temperature Plasma Chemistry (HAKONE VI), Cork, Ireland, 83–87

Korfiatis G, Moskwinski L, Abramzon N, Becker K, Christodoulatos C, Kunhardt E, Crowe R and Wieserman L 2002 'Investigation of Al surface cleaning using a novel capillary non-thermal ambient-pressure plasma' in *Atomic and Surface Processes* eds Scheier P and Märk T D, University of Innsbruck Press (2002)

Kunhardt E E 2000 'Generation of large-volume atmospheric-pressure, non-equilibrium plasmas' *IEEE Trans. Plasma Sci.* **28** 189–200

Lee Y-H, Yi C-H, Chung M-J and Yeom G-Y 2001 'Characteristics of He/O$_2$ atmospheric pressure glow discharge and its dry etching properties of organic materials' *Surface and Coatings Technology* **146/147** 474–479

Lynch J B, Spence P D, Baker D E and Postlethwaite T A 1999 'Atmospheric pressure plasma treatment of polyethylene via a pulse dielectric barrier discharge: Comparison using various gas composition versus corona discharge in air' *J. Appl. Polym. Sci.* **71** 319–331

Massines F, Gherardi N and Sommer F 2000 'Silane based coatings on propylene. Deposited by atmospheric pressure glow discharge' *Plasmas and Polymers* **5** 151–172

Massines F, Gouda G, Gherardi N, Duran M and Croquesel E 2001 'The role of dielectric barrier discharge atmosphere and physics on polypropylene surface treatment' *Plasma and Polymers* **6** 35–49

Meichsner J 2001 'Low-temperature plasmas for polymer surface modification' in *Low Temperature Plasma Physics* Hippler R, Pfau S, Schmidt M and Schönbach K (eds) (Berlin: Wiley-VCH) 453–472

Meiners S, Salge J G H, Prinz E and Foerster F 1998 'Surface modifications of polymer materials by transient gas discharges at atmospheric pressure' *Surf. Coat. Technol.* **98** 1112–1127

Moskwinski L, Ricatto P J, Babko-Malyi S, Crowe R, Abramzon N, Christodoulatos C and Becker K 2002 'Al surface cleaning using a novel capillary plasma electrode discharge' GEC 2002, Minneapolis, MN (USA), *Bull. APS* **47**(7) 67

Müller S, Conrads J and Best W 2000 'Reactor for decomposing soot and other harmful substances contained in flue gas' International Symposium on High Pressure Low Temperature Plasma Chemistry, (Hakone VII), Greifswald, Germany, Contr. Papers **2** 340–344

O'Hare L A, Leadley S and Parbhoo B 2002 'Surface physicochemistry of corona-discharge-treated polypropylene film' *Surface and Interface Analysis* **33** 335–342

Roth J R, Chen Z, Sherman D M, Karakaya F, Tsai P P-Y, Kelly-Wintenberg K and Montie T C 2001a 'Increasing the surface energy and sterilization of nonwoven fabrics by exposure to a one atmosphere uniform glow discharge plasma (OAUGDP)' *Int. Nonwoven J.* **10** 34–47

Roth J R, Chen Z Y and Tsai P P-Y 2001b 'Treatment of metals, polymer films, and fabrics with a one atmosphere uniform glow discharge plasma (OAUGDP) for increased surface energy and directional etching' *Acta Metallurgica Sinica* (English Letters) **14** 391–407

Rott U, Müller-Reich C, Prinz E, Salge J, Wolf M and Zahn R-J 1999 'Plasmagestützte Antifilzausrüstung von Wolle—Auf der Suche nach einer umweltfreundlichen' Alternative Proc. Workshop Plasmaanwendungen in der Textilindustrie Stuttgart, Germany, 7–16

Schmidt-Szalowski K, Rzanek-Boroch Z, Sentek J, Rymuza Z, Kusznierewicz Z and Misiak M 2000 'Thin film deposition from hexamethyldisiloxane and hexamethyldisilazane under dielectric barrier discharge (DBD) conditions' *Plasmas and Polymers* **5** 173

Seeböck R, Esrom H, Charbonnier M and Romand M 2000 'Modification of polyimide in barrier discharge air-plasma: Chemical and morphological effects' *Plasma and Polymers* **5** 103–118

Seeböck R, Esrom H, Charbonnier M, Romand M and Kogelschatz U 2001 'Modification of polyimide using dielectric barrier discharge treatment' *Surf. Coating Technol.* **142/144** 455–459

Softal Report 102 E 'Corona pretreatment to obtain wettability and adhesion' Softal Electronic GmbH, D21107 Hamburg, Germany

Softal Report 151 E Part 2/3 'New trends in corona technology for stable adhesion' Softal Electronic GmbH, D21107 Hamburg, Germany

Softal Report 152 E Part 3/3 'New trends in corona technology for stable adhesion' Softal Electronic GmbH, D21107 Hamburg, Germany

Sonnenfeld A, Kozlov K V and Behnke J F 2001a 'Influence of noble gas on the reaction of plasma chemical decomposition of silicon organic compounds in the dielectric barrier discharge' Proc. 15th Int. Symp. on Plasma Chem. Contr. Orléans/France 9–13 July 2001 Bouchoule A *et al* (eds) vol 5, pp 1829–1834

Sonnenfeld A, Tun T M, Zajickova L, Wagner H-E, Behnke J F and Hippler R 2001 'The deposition process based on silicon organic compounds in two different types of an atmospheric barrier discharge' in Proc.15th Int. Symp. on Plasma Chem. Contr. Orléans/France 9–13 July 2001, Bouchoule A *et al* (eds) vol 5, pp 1835–1840

Sonnenfeld A, Tun T M, Zajickova M, Kozlov K V, Wagner H E, Behnke J F and Hippler R 2001b 'Deposition process based organosilicon precursors in dielectric barrier discharges at atmospheric pressure' *Plasma and Polymers* **6** 237

Steffen H, Schwarz J, Kersten H, Behnke J F and Eggs C 1996 'Process control of rf plasma assisted surface cleaning' *Thin Solid Films* **283** 158

Tanaka K, Inomata T and Kogoma M 1999 'Ashing of organic compounds with spray-type plasma reactor at atmospheric pressure' *Plasma and Polymers* **4** 269–281

Thyen R, Höpfner K, Kläke N, and Klages C-P 2000 'Cleaning of silicon and steel surfaces using dielectric barrier discharges' *Plasma and Polymers* **5** 91–102

Tsuchiya Y, Akutu K and Iwata A 1998 'Surface modification of polymeric materials by atmospheric plasma treatment' *Progress in Organic Coatings* **34** 100–107

VDI-TZ Physikalische Technologien, Düsseldorf, Germany (Ed.) 2001 Plasmagestützte Filzausrüstung von Wolle Info. Phys. Tech. No. 32

Wulff H and Steffen H 2001 'Characterization of thin solid films' in *Low Temperature Plasma Physics* Hippler R, Pfau S, Schmidt M and Schoenbach K H (eds) (Wiley-VCH)

Yamamoto T, Newsome J R and Ensor D S 1995 'Modification of surface energy, dry etching, and organic film removal using atmospheric-pressure pulsed-corona plasma' *IEEE Transactions Ind. Applications* **31** 494–495

9.8 Chemical Decontamination

9.8.1 Introduction

NO_x gases are emitted from coal burning electric power plant, boilers in factories, co-generation system and diesel vehicles. Some liquids and gases such as trichloroethylene, acetone and fluorocarbon are useful for clean-up of materials used in the semiconductor industry, for refrigerants, and so on. However, recently, it has been noticed that these are harmful to human health. These must be processed for global environmental problems.

Concerning NO_x processing, selective catalytic reductions (SCRs) have been used. Soot and SO_2 exhausted from diesel engines prevent the conventional SCR from removing NO_x. Non-thermal plasmas (NTP) are attractive for decomposing these gases because the majority of the electrical energy goes into the production of energetic electrons with kinetic energies much higher than those of the ions or molecules. Energetic electron impact brings about the decomposition of the harmful gases or induced radicals facilitate the decompositions.

In this section, removal of the harmful gases by NTPs is discussed. In sections 9.8.2–9.8.4, mainly de-NO_x processes and kinetics, instrumentation and influencing parameters for de-NO_x will be treated. In section 9.8.5, processing of environmentally harmful gases such as halogen gases, hydrocarbons, and chlorofluorocarbon removed by NTPs will be presented.

9.8.2   de-NO_x process

Decomposition of NO_x to their molecular elements (N_2 and O_2) is the most attractive method. However, it is seen that the major mechanism of NO_x removal is oxidation to convert NO into NO_2 as shown in figure 9.8.1 for $NO/N_2/O_2$ without water vapor. First, N_2 and O_2 collide with energetic electrons in the NTP to generate ions, excited species and radicals, in which oxygen related species such as O, O_2 and O_3 mainly contribute to convert NO into NO_2. In the case of exhaust gases, including air with water vapor, not only oxygen related radicals but also hydroxyl radicals (OH radicals) are produced and contribute to oxidize NO to NO_2. However, in these systems, NO is only oxidized to NO_2, directly or indirectly, by these radicals. As a result, the net reduction of NO_x ($NO + NO_2$) remains unchanged. Gases such as ammonia, H_2O_2, hydrocarbon, N_2H_4, hydrogen and catalyst as additives are used to dissolve NO_2. The case that ammonia is added into the NO stream field is shown in figure 9.8.2. NO is converted into NO_2 by hydroxyl and peroxy radicals as well as oxygen radicals. NO_2

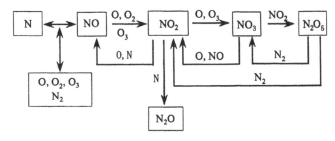

Figure 9.8.1. $NO/N_2/O_2$ system without H_2O.

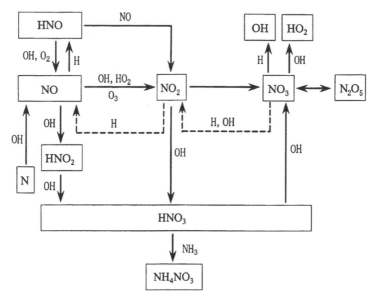

Figure 9.8.2. NO/N$_2$/O$_2$ system with H$_2$O and NH$_3$ as an additive.

reacts with OH to form HNO$_3$ and, further, NH$_4$NO$_3$ is produced by the reaction between HNO$_3$ and ammonia. When ammonia is subjected to electron impact in NTP, ammonia radicals are generated. This reaction scheme is shown in figure 9.8.3. NO reacts with ammonia radicals (NH$_3$, NH$_2$ and NH)

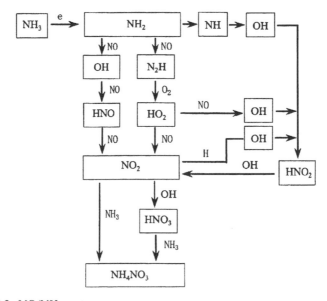

Figure 9.8.3. NO/NH$_3$ system.

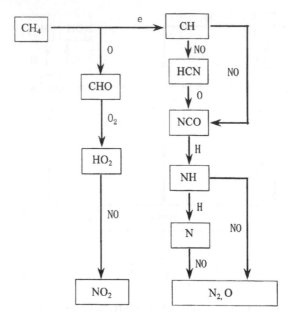

Figure 9.8.4. Hydrocarbon system.

produced by electron impact. NH_2 radicals are a major contributor to oxidize NO to NO_2, through which NH_4NO_3 that is used for fertilizer is produced.

NO decomposition by hydrocarbons is shown schematically in figure 9.8.4. When hydrocarbons are added, the reaction by peroxy radicals (R–OO) is a major pathway to decompose NO [1–4], although the reactions are complicated. CH_i $(i = 1$–$3)$ radicals (CH_3, CH_2, CH etc.) are also produced by electron impact in NTPs to decompose NO through HCN, NCO and HCO radicals [5].

There are many kinds of hydrocarbons such as CH_4, C_2H_2, C_2H_4, C_3H_6 and C_3H_8. However, reactions generated are commonly used to produce peroxy radicals R–OO. HO_2 is an example of a peroxy radical [3], i.e.

$$R + O + O + M \longrightarrow R\text{–}OO + M. \qquad (9.8.2.1)$$

R–OO strongly oxidizes NO into NO_2 as shown in equation (9.8.2.2) [6].

$$R\text{–}OO + NO \longrightarrow R\text{–}O + NO_2. \qquad (9.8.2.2)$$

The detailed R–OO species of C_3H_6 is described in references [2] and [6] and C_3H_8 in reference [6].

NO_2 reacts with OH radicals to make HNO_3. A part of NO_2 is changed into CO_2, where NO_2 is reacted with deposited soot at the proper temperature. Oxygen radicals preferably react with hydrocarbon molecules thereby initiating a reaction chain forming several oxidizing radicals [7].

Carbon dioxide, CO_2, is also included in exhaust gases [8]. CO_2 hardly contributes to the decomposition of NO, because the majority of the energy deposited from the non-thermal plasma may be lost to the vibrational and rotational excitations of CO_2. Although it is thought that electrons impact CO_2 to make CO, NO can be reduced only at very high temperatures as shown in equations (9.8.2.3) and (9.8.2.4) [9].

$$e + CO_2 \longrightarrow CO + O + e \qquad (9.8.2.3)$$

$$CO + NO \longrightarrow CO_2 + \tfrac{1}{2}N_2. \qquad (9.8.2.4)$$

NO is reproduced by the reaction between CO_2 and nitrogen radicals as shown in equation (9.8.2.5) [10].

$$N + CO_2 \longrightarrow NO + CO. \qquad (9.8.2.5)$$

NOs are reproduced by NO_2 reduction by oxygen and hydrogen radicals, and reactions between nitrogen and OH radicals as shown in equations (9.8.2.6)–(9.8.2.8).

$$NO_2 + O \longrightarrow NO + O_2 \qquad (9.8.2.6)$$

$$NO_2 + H \longrightarrow NO + OH \qquad (9.8.2.7)$$

$$N + OH \longrightarrow NO + H. \qquad (9.8.2.8)$$

In summary, NO is converted into final products through the production of NO_2 by additives in a NO stream field. The energetic electron impact is the origin of these reactions. Electrons directly impact to NO or produce radicals to convert NO into NO_2. NO_2 further changes to NH_4NO_3 when ammonia is added. NO is also reproduced by oxygen and hydroxyl radicals.

9.8.3 Non-thermal plasmas for de-NO_x

Plasma reactors that have been utilized for NO_x remediation are: (1) dielectric barrier discharge, (2) corona discharge, (3) surface discharge, (4) glow discharge and (5) microwave discharge. Reactor groups are subdivided according to their power source: dc and pulsed. Electrode configurations in corona discharge and dielectric barrier discharge are (1) plate, (2) needle or multi-needle, (3) thin wire and (4) nozzle. A grounded electrode is placed in parallel or coaxial form near these electrodes.

Hybrid systems combining plasma with electron beam [11, 12] or catalysts were also developed [13–15]. As indirect decomposition systems, radical shower systems were developed using ammonia gases [16, 17] and methane gases [18].

9.8.3.1 Efficiency

The efficiency of NO_x reduction using pulsed or stationary NTPs is a complex function of parameters that include pulse width, pulse polarity, current density, repetition rate and reactor size. For de-NO$_x$, removal efficiency η_{NO_x} and energy efficiency η_E are often used to evaluate the decomposition system. These are defined as equations (9.8.3.1) and (9.8.3.2).

$$\eta_{NO_x} = \frac{[NO]_{before} - [NO]_{after}}{[NO]_{before}} \times 100 \quad (\%) \tag{9.8.3.1}$$

$$\eta_E = L \times \frac{[NO]_{before}}{10^6} \times \frac{\eta_{NO_x}}{100} \times \frac{30}{22.4} \times \frac{1}{P} \quad (g/kWh) \tag{9.8.3.2}$$

where [NO]$_{before}$ and [NO]$_{after}$ are NO concentrations before and after the process in units of ppm. L is NO flow rate in units of l/min, the molecular weight of NO is 30 g, and P is consumed energy in units of kWh. The electrical conversion efficiency that refers to the efficiency for converting wall plug electrical power into the plasma is important in the evaluation of the total efficiency for the decomposition of NO$_x$.

9.8.3.2   Plasma reactors

Figures 9.8.5(a)–(f) show schematics of fundamental plasma reactors for NO decomposition. Figure 9.8.5(a) shows a DBD reactor. The electrode is coated with dielectric materials. To prevent charging-up of the dielectric materials, the power source is ac or burst ac signals with a frequency of 50 Hz to several tens to hundreds of kHz. For the electrode arrangement, parallel plate, multipoint [19] and coaxial types [16] are used. A series of filamentary discharges are produced at the gap. Figure 9.8.5(b) shows a coaxial electrode configuration [20] for generating corona discharge. The central electrode consists of a thin wire. By applying a high voltage, corona discharges are produced around the wire by stationary (ac and dc) and pulsed discharges [21–24]. For dc corona discharge, a polar effect appears (positive and negative corona discharges). The electrode configurations are a wire [20], pipe and

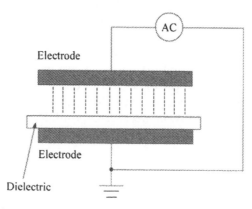

Figure 9.8.5. (a) Dielectric barrier discharge reactor.

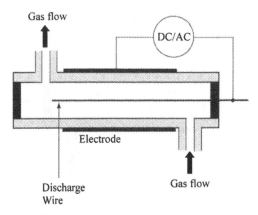

Figure 9.8.5. (b) Corona discharge reactor.

nozzle electrodes [17]. For generating pulsed corona discharges, there are several types of electrode arrangement, i.e. point-to-plate [25], wire-to-plate [26, 27], wire-to-cylinder [28, 29], nozzle-to-plate [30] and pin-to-plate [31, 32]. For power sources, dc/ac superimposed source [33] and bi-polar polarity of pulsed source [28, 34] are also used. Streamer corona discharge, which is generated with a voltage rise time of 10–50 ns and a duration of 50–500 ns FWHM (full-width at half-maximum), can decompose pollutant gases.

The catalyst coated-electrode configuration to facilitate de-NO_x is shown in figure 9.8.5(c). NO_x gases flow in the plasma and the catalyst to undergo decomposition.

Figure 9.8.5(d) shows a tubular packed-bed corona reactor. The pellets of dielectric materials are coated with or without catalyst. The catalyst is activated by energetic particles, i.e. electrons, photons, excited molecules, ions etc. [14]. By applying a high ac voltage to pellets filled in a chamber,

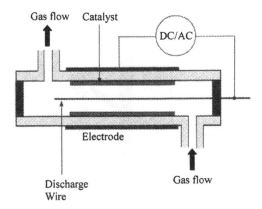

Figure 9.8.5. (c) Corona discharge–catalyst reactor.

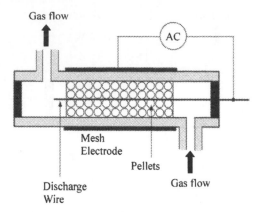

Figure 9.8.5. (d) Packed-bed corona discharge reactor.

micro-discharges in the gap and/or on the surface are generated. This is called a packed bed discharge, which is also expected to have a catalytic effect at the surface of pellets [35].

Figure 9.8.5(e) shows a radical injection NTP system: a pipe electrode with nozzle pipes from which gas additives flow, that are spouted to generate

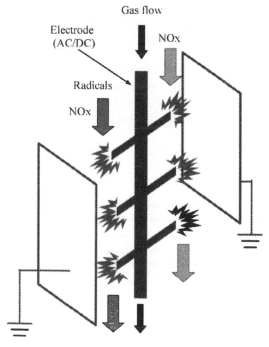

Figure 9.8.5. (e) Radical injection reactor.

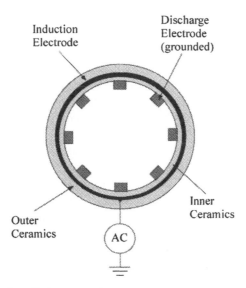

Figure 9.8.5. (f) Surface discharge reactor.

streamer corona discharges in the NO_x stream field. Thus, the NO_x is directly exposed to the corona discharge [30]. On the other hand, radicals are supplied to the NO_x stream field by DBD generated in a separate chamber from the NO_x stream field. In this case, NO_x is not exposed to the plasma. DBD is generated by an intermittent power source so as to control the discharge power. Ammonia radicals are injected into the NO stream field [16]. Remediation by radical shower systems is achieved using dielectric barrier discharges and corona discharges. Plasma-induced radicals from ammonia [16, 17, 36, 37], methane [18, 36] and hydrogen [36], are injected into the NO_x stream region or via the corona zone.

Figure 9.8.5(f) shows a reactor of surface discharge. One of the electrodes is inside the ceramics. By applying a high ac voltage, surface discharge (a kind of dielectric barrier discharge) is generated at a surface of the inner ceramics [38].

Microwave discharges at atmospheric pressure are also used for NO_x removal [39, 40] and are effective to decompose N_2/NO and $N_2/O_2/NO$ mixtures [40]. Because the gas temperature becomes high when operating stationary discharges, a pulsed mode operation is employed [39]. NO is also decomposed into N_2 and O_2 by a microwave discharge in a NO/He mixture [41]. Micro-structured electrode arrays allow generation of a large-area glow discharge, which removes two nitrogen oxides (NO and N_2O). DC or rf power is applied to the arrays [42].

A hybrid system using NTP and an electron beam is effective in simultaneous removal of NO and SO_2 [12]. An electron beam is used together with a corona discharge ammonia radical injection system.

9.8.4   Parametric investigation for de-NO$_x$

In the de-NO$_x$ process by NTPs, optimization of the following parameters is desired: (1) energy efficiency, (2) removal efficiency, (3) process cost, (4) controllability, (5) by-products and (6) lifetime of the system and maintenance. These parameters are directly influenced by: (1) power source (output voltage, pulse width and polarity etc.), (2) electrode configuration, (3) catalyst, (4) radical species, (5) additives, (6) reactor size etc.

In addition to the conventional electrode configurations mentioned above, pyramid [19, 43] and multi-needle geometry [44] have been employed to lower the operating voltage. In the pyramid type, tip angle and height were varied [19]. In the multi-needle type, gap length was varied [44]. These parameters of gap length and height have a close relationship to the plasma initiation voltage leading to the reduced electric field strength and the consumed energy in the plasma. When the angle of the tip point becomes small, energy efficiency decreases due to larger energy consumption. The lower reduced electric field strength was obtained for a shorter gap length to lead to a lower rate of ozone production for the multi-needle type. As a result, the de-NO$_x$ rate becomes low.

The influence of height of the pyramid-shaped electrode was also investigated [43]. It was shown that NO removal rate increases with decreasing heights, in other words, depth of the groove, at the same gas residence time. This change of the removal rate may be related to the change of the discharge modes in DBD and surface discharge.

A heated wire is used for corona discharge generation and energetic electrons are emitted [20]. A heated corona wire is able to produce energetic electrons and activate the oxidation by the generated ozone. It was shown that the average corona currents increased and the corona starting voltages decreased with an increase in the wire temperature. The relation between de-NO$_x$ rate and wire temperature was investigated. For generating corona discharge, metallic wires are often used to make a high electric field. The dependence of de-NO$_x$ rate on the wire materials, tungsten and copper, was examined by a pulsed corona discharge with a wire-to-plate electrode system. A higher de-NO$_x$ rate is obtained by tungsten wire covered with WO$_3$ because a streamer corona discharge is easily generated, while a dc stationary corona is only generated in the case of copper wire [26].

A pair of reticulated vitreous carbon (10 pores per inch) is used for generating streamer corona discharge to convert NO into NO$_2$. This electrode configuration is advantageous in scaling-up the system and gives rise to large total NO$_x$ removal. At the surface of the carbon electrodes, NO$_2$ oxidizes carbon surfaces and finally nitrous acid is formed [9].

Reactor size and power sources are also parameters that influence the de-NO$_x$ characteristics. Instead of the conventional ac and dc power sources to generate corona discharges, a high voltage (60 kV) and large current (approximately 200 A) pulsed power unit was used to generate a 100 ns-duration streamer corona discharge. The output voltage is from a Blumlein line generator. The short-duration pulsed power produces high-energy electrons while the temperature of the ions and the neutrals remains unchanged, and thus the energy consumed is reduced. The maximum energy efficiency was 62.4 g/kWh [45]. A similar test is carried out using the Blumlein line system with an output voltage of 40 kV and a current of 170 A [23]. Actual flue gas from a thermal power plant was used. It was shown that about 90% of the NO was removed at a flow rate of 0.8 liters/min and a repetition rate of 7 pps [23]. Using a traveling wave transmission in a coaxial cable, a series of alternative discharge pulses generate pulsed corona discharge. Filament streamer discharges were generated at an applied reciprocal voltage with an output of 40 kV. The NO gas with a concentration of 170 ppm was reduced to one fourth of the original concentration in a time of 0.6 s [46]. The influence of the reactor diameter for pulsed positive corona discharges on the de-NO_x rate is discussed for a concentric coaxial cylindrical configuration of the electrode. As a result, the increase of inner diameter of the reactor from 10 to 22 mm could be a way to minimize energy losses in the process of NO_x removal from flue gas [47]. Generally, the current through the plasma increases with increasing an applied voltage. In an ammonia radical injection system, the corona current shows a hysteresis characteristic against the applied voltage. This might be based on the NH_4NO_3 aerosol production. The deposition of aerosol particles also affects the NO_x removal rate [30].

The main pathway for NO_x removal in catalyst-based technology is reduction. Selective catalytic reduction (SCR) has been studied using either ammonia (NH_3) or hydrocarbons (HCs) as additional reducing agents. The combination of NTP, catalyst and the additives are effective to significantly reduce nitric oxides (NO and NO_2) synergistically to molecular nitrogen. For example, NO_x is converted into N_2 and H_2O through electron impact in NTP, gas-phase oxidation and catalytic reduction as shown in equations (9.8.4.1)–(9.8.4.3). This is called plasma-enhanced NH_3-SCR [48]. When HCs are used, this is called HC-SCR.

NTP: $$e + O_2 \longrightarrow e + 2O \qquad (9.8.4.1)$$

Gas phase oxidation: $$O + NO + M \longrightarrow NO_2 + M \qquad (9.8.4.2)$$

Catalytic reduction: $$NO + NO_2 + 2NH_3 \longrightarrow 2N_2 + 3H_2O. \qquad (9.8.4.3)$$

As catalysts, Pd-Al_2O_3, TiO_2, aluminosilicate, Ag/mordenite, γ-Al_2O_3 and ZrO_2 were examined for plasma-enhanced HC-SCR [48, 49].

The pulsed corona plasma reactor was followed by a Co-ZSM5 catalyst bed of honeycomb type [14]. NO is converted into NO_2 in the plasma reactor and then NO_2 is reduced in the Co-ZSM5 catalyst bed. No formation of NH_4NO_3 occurs. In the plasma-enhaced SCR system, plasma-treated NO_2 was reduced effectively with NH_3 over the Co-ZSM catalyst at a relatively low temperature of 150 °C [14]. TiO$_2$ [50] as catalyst is also effective to de-NO$_x$. NTP improves the de-NO$_x$ rate with an appropriate content of water vapor and Na-ZSM-5 catalyst at any temperature [13].

9.8.5   Pilot plant and on-site tests

The de-NO$_x$ exhausted from pilot plants and diesel engines can be directly processed by NTP. A diesel engine exhaust of a vehicle with a 3 liter exhaust output is used as a stationary NO$_x$ source with the engine speed set at 1200 rpm, where the plasma reactor consisting of a coaxial DBD with a screw-type electrodes is mounted on the vehicle [51]. The DBD deNO$_x$ system is applied to an actual vehicle with an exhaust output of 2.5 liters and the oxidation of hydrocarbon is recognized, where geometric and electric parameters such as dielectric surface roughness and gap width of the coaxial reactor are investigated [52]. A pulsed corona discharge process is applied to simultaneously remove SO$_2$ and NO$_x$ from industrial flue gas of an iron-ore sintering plant. The corona reactor is connected to the power source consisting of a magnetic pulse compression modulator with a system supplying chemical additives such as ammonia and propylene. The problem regarding the lifetime of the closing switch can be solved by using magnetic pulse compression technology [53]. Propylene used as the chemical additive was very effective in the enhancement of NO$_x$ removal. The increase in C$_3$H$_6$ concentration gives rise to an enhancement of NO$_x$ [53].

NO$_x$ and SO$_2$ from coal burning boiler flue gases are simultaneously removed by dc corona discharge ammonia radical shower systems in pilot scale tests, where multiple-nozzle electrodes are utilized for generating a corona discharge. Tests were conducted for the flue gas rate from 1000 to 1500 N m^3/h, the gas temperature from 62 to 80 °C, the ammonia-to-total acid gas molecule ratio from 0.88 to 1.3, applied voltage from 0 to 25 kV and NO initial concentration from 53 to 93 ppm for a fixed SO$_2$ of 800 ppm. As a result, approximately 125 g of NO$_x$ was removed by 1 kWh of energy input with 75% of removal efficiency [54]. A plasma/catalyst continuously regenerative hybrid system is introduced to reduce diesel particulate matter (DPM), NO$_x$, Co etc., contained in diesel exhaust gas from a passenger diesel car (2500 cm^3). A corona discharge is generated in front of a nozzle-type hollow electrode, where ammonia, hydrocarbon, steam, oxygen, nitrogen etc. are injected. The hybrid system test shows that DPM and CO were almost removed and NO$_x$ reduced to 30% simultaneously by the system [25].

9.8.6   Effects of gas mixtures

It is known that, in addition to NO$_x$, exhaust systems also release varying concentrations of N$_2$, O$_2$, CO$_2$, H$_2$O etc. In coal burning electric plant, sulfur oxide (SO_2) and fly ash are also contained. In diesel exhaust gas, soot is included. One must consider the effect of these mixtures with NO_x. These are molecules and therefore, when present together with NO_x in a plasma, the plasma energy is partly consumed in these mixtures and is expended as vibrational and rotational energies. This energy expense may not contribute to the reaction. Thus, de-NO_x efficiency can be enhanced using chemicals like H_2O, H_2O_2, O_3, NH_3, or hydrocarbons that are introduced into NTPs as an additive. As a result, NO and SO_2 are finally converted into NH_3NO_4 and $(NH_4)_2SO_4$, respectively, where ammonia is used as an additive.

9.8.6.1 Particulate matter, soot, and fly ash

Fly ash is contained in the exhaust gas from coal-burning thermal electrical power plants. Diesel particulate matter, NO_x, CO_2, etc., contained in diesel exhaust gas emitted from a passenger car, were reduced using a dc corona discharge plasma/catalyst regenerative hybrid system. The effects of repetitive pulses and soot chemistry on the plasma remediation of NO_x are computationally investigated [55]. It was pointed out that NO_2 reacts with deposited soot in the plasma reactor at the proper temperature [25]. An outer porous electrode made of SiC ceramics is used for decomposition of soot-containing exhaust gas and acts as both electrode for dielectric barrier discharge and particulate filter. Toxic and soot containing harmful substances from exhaust gas are subjected to plasma processing. The flue gas is let out through the porous electrode which is gas-permeable but filters hold back the soot particles. Reaction products were CO and CO_2. The soot decomposition was achieved by a cold oxidation process. Thus, the soot is constantly oxidized during all engine operating conditions [56].

Fly ash including NO_x gas was removed using pulsed streamer discharges, generated by the configuration of wire and cylinder electrodes. Fly ash with particle sizes from 0.08 to 3000 µm was injected into the discharge region. The removal rate of NO and NO_x including the fly ash was increased in the presence of moisture. It was explained that the presence of H_2O generates the OH radicals by dissociation [57].

9.8.6.2 SO_2

SO_2 is often processed using ammonia as an additional gas. The reaction is shown as

$$2SO_2 + 4OH \longrightarrow 2H_2SO_4 \tag{9.8.6.1}$$

$$H_2SO_4 + 2NH_3 \longrightarrow (NH_4)_2SO_4. \tag{9.8.6.2}$$

When SO_2 reacts with oxygen atoms to form SO_3, SO_3 is converted into H_2SO_4 as

$$SO_3 + H_2O \longrightarrow H_2SO_4. \tag{9.8.6.3}$$

SO_2 was simultaneously removed with NO_x using dc corona discharge ammonia radical shower systems as pilot plant tests. Both removal and energy efficiencies for SO_2 decomposition increase with increasing ammonia-to-acid gas ratio and decrease with increasing flue gas temperature. The maximum removal efficiency exists at an applied power of about 300 W. Approximately 9 kg of SO_2 were removed by an energy input of 1 kWh with 99% of SO_2 removal [54].

SO_2 and NO_x from industrial flue gas of iron-ore sintering plant were processed using pilot-scale pulsed streamer corona discharges generated by magnetic pulse compression technology. The sulfuric acid was neutralized by ammonia in the discharges to finally obtain ammonia sulfate. The removal of SO_2 was greatly enhanced when ammonia was added to the flue gas. The high removal efficiency may be caused by chemical reaction between SO_2 and NH_3 in the presence of water vapor as well as the heterogeneous chemical reaction among SO_2, NH_3 and H_2O [53].

Flue gas from a heavy oil-fired boiler contains 200–1000 ppm of SO_2 and about 50–200 ppm of NO_x. When processed at a hybrid gas cleaning test plant using a corona discharge–electron beam hybrid system, up to 5–22% of NO_x and 90–99% of SO_2 could be removed by operating the corona discharge with an ammonia radical injection system. It was found that total NO_x and SO_2 reduction rates increase non-monotonically with increasing applied voltage, hence, corona current or discharge input power [12].

9.8.6.3 O_2

When oxygen molecules are mixed with a mixture of N_2 and NO, oxygen atoms are generated by electron impact, followed by formation of ozone by a reaction with oxygen molecules as shown in equations (9.8.6.4) and (9.8.6.5),

$$e + O_2 \longrightarrow O + O + e \qquad\qquad (9.8.6.4)$$

$$O + O_2 + M \longrightarrow O_3 + M. \qquad\qquad (9.8.6.5)$$

Ozone oxidizes NO to form NO_2 as shown in equation (9.8.6.6),

$$NO + O_3 \longrightarrow NO_2 + O_2. \qquad\qquad (9.8.6.6)$$

When the NO_2 with ammonia as additive is used, NH_4NO_3 is formed as shown in figure 9.8.3. However, because of the excessive concentration of oxygen molecules, NO_2 is reduced to NO. In this case, oxygen atoms do not contribute to remove NO_x, but reproduce NO as shown in equation (9.8.6.7),

$$NO_2 + O \longrightarrow NO + O_2. \qquad\qquad (9.8.6.7)$$

Using dielectric barrier discharge with multipoint electrodes [44], NO removal was carried out. NO removal rate and NO conversion into NO_2

were discussed in $NO/N_2/O_2$ mixed gas, where the oxygen concentration was varied from 1 to 4%. Removal rates of NO and NO_x increase with increasing concentration of O_2 in gas mixture, but conversion into NO_3 via NO_2 from NO is limited in low NO concentration.

9.8.6.6 *H_2O*

Water vapor H_2O leads to production of OH and HO_2 radicals. As H_2O vapor concentration increases, more OH and HO_2 radicals can be generated to oxide NO to form NO_2 and further HNO_3 [7]. Therefore, NO and NO_x ($NO + NO_2$) are removed with increasing H_2O vapor concentration being in a range of 1100–32 000 ppm [5]. Increase in the de-NO_x rate was also seen in humid (10% H_2O) gas mixture [58], and in dc corona discharge over a water surface [59].

9.8.6.5 *Hydrocarbon radical injection*

Hydrocarbons were used as an additive. NO/NO_x is removed with acetylene (C_2H_2) as an additive using a coaxial wire-tube reactor with dielectric barrier discharge, where the feeding gases include N_2, O_2, NO and C_2H_2. The effect of oxygen with concentrations of 0–10% is discussed for de-NO_x. The rate of NO converted into NO_2 increases with increasing oxygen concentration. Thus, NO to NO_2 oxidation is largely enhanced as the amount of hydrocarbon increases. The hydrocarbon acts as a getter of O and OH radicals, with the products reacting with O_2 to yield peroxy radicals (HO_2) which efficiently convert NO to NO_2. The conversion of NO into N_2 by NH and N radicals produced via HCN, NCO and HCO radicals is shown in figure 9.8.4. The de-NO_x rate decreases with increasing the oxygen concentration from 2.5–10%. This is due to the oxidation to CO or CO_3 by the reaction between CH_x and oxygen radicals. In low oxygen concentration, acetylene C_2H_2 reacts with oxygen radicals to form hydrocarbon radicals that facilitate to form HCN, NCO and HCO radicals. Thus, oxygen strongly influences the de-NO_x process [5].

9.8.6.6 *Ammonia radical injection*

An ammonia radical injection system for converting NO into harmless products was developed [60], where the radicals are generated in a separate chamber from the NO stream chamber. NO gas is not in the plasma. In order to confirm the energy efficiency of de-NO_x using an intermittent one-cycle sinusoidal source for generating DBDs, the NO concentration is increased to 3000 ppm by varying the oxygen concentration from 2–5.6%. For containing oxygen gas in the NO stream field, lower NO temperature operation is possible to obtain a higher de-NO_x rate. At an applied voltage slightly higher than the threshold voltage for plasma initiation, the removal amount of NO reaches maximum, presenting maximum energy efficiency. In particular, for an oxygen concentration of 5.6% and a duty cycle of 5–10%, a high energy efficiency is obtained to be 98 g/kWh. This means that the appropriate electrical power is deposited in the DBD plasma at this duty cycle. In the system, NO is mainly reduced by NH_2 radicals for NO to convert into NH_4NO_3 through HO_2 radicals as shown in figure 9.8.3.

9.8.7 Environmentally harmful gas treatments

Volatile organic compounds (VOCs) are converted into CO_2 and H_2O and other by-products (e.g. HCl and H_2) in the desired reaction stoichiometry by oxygen and hydroxyl radicals. This stoichiometry is difficult to achieve by NTPs, because other intermediate products are produced. According to the process conditions, not only CO and nitric oxide such as N_2O but also phosgene ($COCl_2$) may be produced, which may require a second-stage treatment. The end products include poisonous materials such as phosgene which must be separated from the gas stream and/or be processed in a second-stage treatment [61].

The mechanism of decomposition is based on the electron impact on the harmful gases [62, 63]. Therefore, the simulation model includes a solution of Boltzmann's equation for the electron energy distribution [61]. It was reported that more N_2O was generated for higher concentration of water vapor and decomposition energy efficiency. Power sources with frequencies such as 50 and 60 Hz are often used. In this case, the metal catalyst is contained in the dielectric barrier discharge to remove the by-product by facilitating the decomposition of the harmful gases. NTPs are effective to decompose VOCs and the increase of the decomposition rate is desirable for a practical flue gas process system.

The parameters influencing their decompositions are (1) electrical characteristics of plasmas (power, energy, applied voltage, frequency, repetition rates and rise time), (2) water, (3) carrier gases and flow rate, (4) ionization potential of the target gases, and (5) gas temperatures. These parameters are closely related to bring high selectivity of the target products [64].

A parametric study for decomposing VOC will be introduced below.

9.8.7.1 *Plasma sources*

Plasma chemical processes have been known to be highly effective in promoting oxidation, enhancing molecular dissociation, and producing free radicals to enhance chemical reactions [65]. VOCs are also processed using NTPs, in the same way as NO is used. Four types of plasma reactor have been mainly used for the application of VOC destruction: surface discharge [66], dielectric barrier discharge [67], ferroelectric packed-bed discharge [68], and pulsed corona discharge. Most of the power source frequency is 50–60 Hz [62, 68, 69]. The destruction is also carried out by dc discharge [65], capillary tube discharge [65] and microwave discharge processes as well as electron beam. In order to improve energy efficiency and control of undesirable by-products, hybrid systems in which NTPs are combined with catalysts are used [67]. Synergetic effects are expected. Deposition of by-products is not desirable during the process. Pevovskite oxides such as barium titanate ($BaTiO_3$) act as a highly dielectric compound [68]. The perovskite oxides can be catalytically activated by free radicals of ultraviolet irradiation from the plasma [68].

Uniform generation of the corona discharge contribute to reduce toluene. The higher destruction efficiency of toluene is attributed to more uniform corona-induced plasma activities throughout the reactor volume. The size of the pellets contributes to the plasma uniformity [62].

9.8.7.2 Processes

Halogen gases such as chlorine and fluorine are finally converted into CO_2 and halogenated hydrogen, respectively. It was found that the destruction efficiency decreases in the order of toluene, methylene chloride and trichlorotrifluoroethane (CFC-113: $CF_2ClCFCl_2$). CFC113 has the strongest bonding and is stable [62]. Toluene ($C_6H_5CH_3$) is reduced by a dielectric barrier discharge, where the reactor consists of a coaxial cylindrical electrode system. Packed TiO_2 pellets or coated TiO_2 on the inner electrode surface are used. TiO_2 as catalyst is activated using plasma with coaxial electrodes. The energy efficiency is improved due to synergetic effects between plasma and activated catalyst [67]. The mechanism of toluene destruction involves not only plasma-induced destruction in the gas phase but also the adsorption/desorption of toluene on the TiO_2 as well as catalytic reaction [67].

Abatement of CFC-113 (which is one of the fluorocarbons) was first reported using ferroelectric packed bed discharge [62] and surface discharge [66]. In the surface discharge case [66], CFC-113 with a concentration of 1000 ppm was processed at a destruction rate of 98% for a discharge power of 70 W. Recently, CHF_3 gas was reduced in H_2O/He plasma (13.56 MHz) and disappeared at 700 W. The by-products were CHF_3, CF_4, H_2O, CO_2 and SiF_4 [70]. In CF_4 destruction under identical experimental conditions as in the CH_3 case, the maximum destruction efficiency using H_2-O_2/He as a carrier gas is higher by a factor of approximately 2 than that using O_2/He gas. Hydrogen atoms contribute to the CF_4 destruction. The by-products were CO_2, HF and H_2O [70]. Ar diluted CF_4 as per fluorocarbon was abated using atmospheric pressure microwave plasma (2.45 GHz) with TM_{010} mode. 10 sccm CF_4 with 100 sccm Ar in 2 lpm O_2 and 10 lpm N_2 flow was treated. CO_2, COF_2, H_2O and NO were identified as the by-products [71].

The principal processes of the destruction of toluene are electron and radical dissociation in the discharges, although charge transfer of toluene with ions and recombination of toluene ions may also be responsible. TiO_2 activated by plasma may induce various reactions on the surface of the TiO_2, resulting in an enhanced toluene destruction. TiO_2 plays a role to enhance the destruction efficiency based on the following reactions: (1) photocatalyst process by ultraviolet light emission from plasma [67], (2) direct activation by fast energetic electrons and active species, (3) oxidation by oxygen radicals produced by the destruction of O_3 on TiO_2 catalyst [72] and (4) chemical reactions by OH and HO_2 radicals [72]. Toluene was mostly reduced to CO, CO_2, H_2O by OH radicals, O_3 and O [62, 65, 67, 73]. Ozone generation is dependent on the heat by the gas discharge. In the presence of air or nitrogen, nitrogen atoms are produced in the direct and/ or sensitized cleavage of nitrogen molecules and produce N_2O, NO and NO_2 [68]. N_2O concentration is significant [68]. In air, triplet oxygen molecules are the most reactive oxygen source in the presence or absence of water, and carbon balance can be improved with suppression of by-products due to promoted autoxidation processes [68].

The principal processes of the VOC destruction are electron and radical impact dissociation of molecules. For toluene, the reaction of toluene with OH radicals is effective to make H_2O as a final product [65] and water can be reduced in NTP to give OH radicals and hydrogen atoms. The effect of water was discussed in the destruction of butane. In low voltage application, higher destruction efficiencies were obtained under wet conditions compared with dry conditions. However, at higher voltages, water had almost no or some negative effect on butane destruction efficiency [68]. This is much different from NO destruction. Benzene was reduced using alumina-hybrid and catalyst-hybrid plasma reactors. It was found that Ag-, Cu-, Mo-, Ni-supported Al_2O_3 can suppress the N_2O formation [74].

Carbon tetrachloride (CCl_4) was reduced using catalysis-assisted plasma technology. Catalysts such as Co, Cu, Cr, Ni and V were coated on 1 mm diameter $BaTiO_3$ pellets. For high frequency operation at 18 kHz, the best CCl_4 destruction was achieved with the Ni catalyst although the destruction of CCl_4 is based on the direct electron impact and short-lived reactive species [63, 75]. That is,

$$e + CCl_4 \longrightarrow Cl^- + CCl_3. \qquad (9.8.7.1)$$

CCl_4 is reproduced by three-body reaction through CCl_3,

$$Cl + CCl_3 + M \longrightarrow CCl_4 + M. \qquad (9.8.7.2)$$

On the other hand, O_2 scavenges the CCl_3 through the reaction

$$CCl_3 + O_2 \longrightarrow CCl_3O_2. \qquad (9.8.7.3)$$

Methylene chloride (CH_2Cl_2) was destroyed by a packed bed plasma rector. Because the chlorine in methylene chloride is strongly bonded with carbon, it is much more stable chemically than toluene, and it is expected that higher electron energies are necessary to reduce methylene chloride [62]. Trichloroethylene (C_2HCl_3, or TCE) was reduced in DBD [61] and in a capillary discharge [65]. The majority of the Cl from TCE was converted into HCl, Cl_2, and $COCl_2$ [61] and CO_2, CO, NO_2 are also identified [65]. The destruction efficiency of TCE is smaller in humid mixtures compared to dry mixtures due to interception of reactive intermediates by OH radicals [61]. The reaction to form $COCl_2$ is as follows:

$$C_2HCl_3 + OH \longrightarrow C_2Cl_3 + H_2O \qquad (9.8.7.4)$$

$$C_2HCl_3 + Cl \longrightarrow C_2Cl_3 + HCl \qquad (9.8.7.5)$$

$$C_2Cl_3 + O_2 \longrightarrow COCl_2 + COCl. \qquad (9.8.7.6)$$

TCE reacts with hydroxyl radicals, but the rate coefficient is no larger than that with O atoms. There are intermediates such as CHOCl, CCl_2 and ClO due to O and OH radicals produced by electron impact dissociation of O_2 and H_2O. The ClO radical is attributed with an important role in oxidizing TCE [61, 76]. TCE can be dissociated or ionized by a direct electron impact to form C_2Cl_3, C_2HCl_2, $C_2HCl_3^+$ etc. It was pointed out that negative ions such as Cl^- and C^- might play an important role in the destruction process [65]. These form terminal species such as CO, CO_2, HCl and $COCl_2$ [61]. NO_2 is also produced after the process [65].

9.8.8 Conclusion

Processing of exhaust gases emitted from motor vehicle and different factories and harmful gases emitted from various industries is increasingly necessary to preserve our earth environment, thus improving our living conditions. For practical use of the NTP system, we must make greater effort to increase the process efficiency and reduce unit cost. In order to realize an easy handling unit, not only modification of the conventional process is needed but also development of new systems, in particular new plasma sources, is very important. Combinations of different systems are effective in bringing fruitful processing results.

References

[1] Filmonova E A, Amirov R H, Hong S H, Kim Y H and Song Y H 2002 *Proc. HAKONE VIII, International Symposium on High Pressure Low Temperature Plasma Chemistry*, 337–341

[2] Filmonova E A, Kim Y H, Hong S H and Song Y H 2002 *J. Phys. D: Appl. Phys.* **35** 2795–2807

[3] Kudrjashov S V, Sirotkina E E and Loos D 2000 *Proc. HAKONE VII, International Symposium on High Pressure Low Temperature Plasma Chemistry*, 257–261

[4] Dorai R and Kushner M J 2002 *J. Phys. D: Appl. Phys.* **35** 2954–2968

[5] Chang M B and Yang S C 2001 *Environmental and Energy Engineering* **47** 1226–1233

[6] Dorai R and Kushner M J 2001 *J. Phys. D: Appl. Phys.* **34** 574–583

[7] Hammer Th 2000 Proc. HAKONE VII, International Symposium on High Pressure Low Temperature Plasma Chemistry, 234–241

[8] Aritoshi K, Fujiwara M and Ishida M 2002 *Jpn. J. Appl. Phys.* **41** 7522–7528

[9] Kirkpatrick M, Finney W C and Locke B R 2000 *IEEE Trans. Industry Applications* **36** 500–509

[10] Eichwald O, Yousfi M, Hennad A and Benabdessadok M D 1997 *J. Appl. Phys.* **82** 4781–4794

[11] Cramariuc R, Martin G, Martin D, Cramariuc B, Teodorescu I, Munteanu V and Ghiuta V 2000 *Radiation Phys. Chem.* **57** 501–505

[12] Chang J S, Looy P C, Nagai K, Yoshioka T, Aoki S and Maezawa A 1996 *IEEE Trans. Industry Applications* **32** 131–137

[13] Shimizu K, Hirano T and Oda T 1998 33rd IAS Annual Meeting 1998 IEEE, Industry Applications Conference **3** 1865–1870

[14] Kim H H, Takashima K, Katsura S and Mizuno A 2001 *J. Phys D: Appl. Phys.* **34** 604–613

[15] Hayashi Y, Yanobe T and Itoyama K 1994 IEEE 1994 Annual report, Conference on Electrical Insulation and Dielectric Phenomena, 828–833

[16] Nishida M, Yukimura K, Kambara S and Maruyama T 2001 *J. Appl. Phys.* **90** 2672–2677

[17] Kanazawa S, Chang J S, Round G F, Sheng G, Ohkubo T, Nomoto Y and Adachi T 1998 *Combust. Sci. Tech.* **133** 93–105

[18] Chang J S, Urashima K, Arquilla M and Ito T 1998 *Combust Sci. and Tech.* **133** 31–47

[19] Takaki K, Jani M A and Fujiwara T 1999 *IEEE Trans. Plasma Sci.* **27** 1137–1145

[20] Moon J D, Lee G T and Geum S T 2000 *J. Electrostatics* **50** 1–15

[21] Mok Y S and Ham S W 1998 *Chem. Engineering Sci.* **53** 1667–1678

[22] Puchkarev V 2002 Conference record of 25th International Power Modulator Symposium and 2002 High-Voltage Workshop, 161–164

[23] Tsukamoto S, Namihira T, Wang D, Katsuki S, Akiyama H, Nakashima E, Sato A, Uchida Y and Koike M 2001 *Electrical Engineering in Japan* **134** 28–35

[24] Takaki K, Sasaki T, Kato S, Mukaigawa S and Fujiwara T 2002 Conference record of 25th International Power Modulator Symposium and 2002 High-Voltage Workshop, 575–578

[25] Chae J O, Hwang J W, Jung J Y, Han J H, Hwang H J, Kim S and Demidiouk V I 2001 *Phys. Plasmas* **8** 1403–1410

[26] Gasprik R, Gasprikova M, Yamabe C, Satoh S and Ihara S 1998 *Jpn. J. Appl. Phys.* **37** 4186–4187

[27] Penghui G, Hyashi N, Ihara S, Satoh S and Yamabe C 2002 Proc. HAKONE VIII, International Symposium on High Pressure Low Temperature Plasma Chemistry, 347–350

[28] Yan K, Hui H, Cui M, Miao J, Wu X, Bao C and Li R 1998 *J. Electrostatics* **44** 17–39

[29] Mutaf-Yardimci O, Kennedy L A, Nester S A, Saveliev A V and Fridman A A 1998 Proc. 1998 SAE International Fall Fuels and Lubricants Meeting, Plasma Exhaust Aftertreatment SP-1395 1–6

[30] Moon J-D, Lee G-T and Geum S-T 2000 *J. Electrostatics* **50** 1–15

[31] Park M C, Chang D R, Woo M H, Nam G J and Lee S P 1998 Proc. 1998 SAE International Fall Fuels and Lubricants Meeting, Plasma Exhaust Aftertreatment SP-1395, 93–99

[32] Fujii T and Ree M 2000 *Vacuum* **59** 228–235

[33] Yan K, Higashi D, Kanazawa S, Ohkubo T, Nomoto Y and Chang J S 1998 *Trans. IEE Japan* **118-A** 948–953

[34] Minami K, Akiyama M, Okino A, Watanabe M and Hotta E 2001 Proceedings of 2nd Asia-Pacific International Symposium on the Basis and Application of Plasma Technology, 39–44

[35] Kawasaki T, Kanazawa S, Ohkubo T, Mizeraczyk J and Nomoto Y 2001 *Thin Solid Films* **386** 177–182

[36] Boyle J, Russell A, Yao S C, Zhou Q, Ekmann J, Fu Y and Mathur M 1993 *Fuel* **72** 1419–1427

[37] Ohkubo T, Yan K, Higashi D, Kanazawa S, Nomoto Y and Chang J S 1999 *J. Oxid. Technol.* **4** 1–4

[38] Masuda S, Hosokawa S, Tu X L, Sakakibara K, Kitoh S and Sakai S 1993 *IEEE Trans. Industry Applications* **29** 781–786

[39] Baeva M, Pott A and Uhlenbusch J 2002 *Plasma Source Sci. Technol.* **11** 135–141

[40] Baeva M, Gier H, Pott A, Uhlenbusch J, Höschele J and Steinwandel J 2002 *Plasma Source Sci. Technol.* **11** 1–9

[41] Tsuji M, Tanaka A, Hamagami T, Nakano K and Nishimura Y 2000 *Jpn. J. Appl. Phys.* **39**(2) L933–L935

[42] Scheffler P, Geßner C and Gericke K-H 2000 Proc. HAKONE VII, International Symposium on High Pressure Low Temperature Plasma Chemistry, 407–411

[43] Yamada M, Ehara Y and Ito T 2000 Proc. HAKONE VII, International Symposium on High Pressure Low Temperature Plasma Chemistry, 370–374

[44] Toda K, Takaki K, Kato S and Fujiwara T 2001 *J. Phys. D: Appl. Phys.* **34** 2032–2036

[45] Namihira T, Tsukamoto S, Wang D, Katsuki S, Hackam R, Akiyama H and Uchida Y 2000 *IEEE Trans. Plasma Sci.* **28** 434–442

[46] Kadowaki K, Nishimoto S and Kitani I 2003 *Jpn. J. Appl. Phys.* **42** L688–L690

[47] Dors M and Mizeraczyk J 2000 Proc. HAKONE VII, International Symposium on High Pressure Low Temperature Plasma Chemistry, 375–378

[48] Miessner H, Francke K-P, Rudolph R and Hammer Th 2002 *Catalysis Today* **75** 325–330

[49] Miessner H, Francke K P and Rudolph R 2002 *Appl. Catalysis B: Environmental* **36** 53–62

[50] Ogawa S, Nomura T, Ehara Y, Kishida H and Ito T 2000 Proc. HAKONE VII, International Symposium on High Pressure Low Temperature Plasma Chemistry, 365–369

[51] Higashi M and Fujii K 1997 *Electrical Engineering in Japan* **120** 1–7 [1996 *Trans. IEEJ* **116-A** 868–872]

[52] Lepperhoff G, Hentschel K, Wolters P, Neff W, Pochner K and Trompeter F-J 1998 Proc. 1998 SAE International Fall Fuels and Lubricants Meeting, Plasma Exhaust Aftertreatment SP-1395, 79–86

[53] Mok Y S and Nam I-S 1999 *IEEE Trans Plasma Science* **27** 1188–1196

[54] Chang J S, Urashima K, Tong Y X, Liu W P, Wei H Y, Yang F M and Liu X J 2003 *J. Electrostatics* **57** 313–323

[55] Dorai R, Hassouni K and Kushner M J 2000 *J. Appl. Phys.* **88** 6060–6071

[56] Müller S, Conrads J and Best W 2000 Proc. HAKONE VII, International Symposium on High Pressure Low Temperature Plasma Chemistry, 340–344

[57] Tsukamoto S, Namihira T, Wang D, Katsuki S, Hackam R, Akiyama H, Sato A, Uchida Y and Koike M 2001 *IEEE Trans Plasma Science* **29** 29–36

[58] Khacef A, Nikravech M, Motret O, Lefaucheux P, Viladrosa R, Pouvesle J M and Cormier J M 2000 Proc. HAKONE VII, International Symposium on High Pressure Low Temperature Plasma Chemistry, 360–364

[59] Fujii T, Aoki Y, Yoshioka N and Rea M 2001 *J. Electrostatics* **51/52** 8–14

[60] Nishida M, Yukimura K, Kambara S and Maruyama T 2001 *Jpn. J. Appl. Phys.* **40** 1114–1117

[61] Evans D, Rosocha L A, Anderson G K, Coogan J J and Kushner M J 1993 *J. Appl. Phys.* **74** 5378–5386

[62] Yamamoto T, Ramanathan K, Lawless P A, Ensor D S, Newsome J R, Plaks N and Ramsey G H 1992 *IEEE Trans. Industry Applications* **28** 528–534

[63] Yamamoto T, Mizuno K, Tamori I, Ogata A, Nifuku M, Michalska M and Prieto G 1996 *IEEE Trans. Industry Applications* **32** 100–105

[64] Kozlov K V, Michel P and Wagner H-E 2000 Proc. HAKONE VII, International Symposium on High Pressure Low Temperature Plasma Chemistry, 262–266

[65] Kohno H, Berezin A A, Chang J S, Tamura M, Yamamoto T, Shibuya A and Honda S 1998 *IEEE Trans. Industry Applications* **34** 953–966

[66] Oda T, Takahashi T, Nakano H and Masuda S 1993 *IEEE Trans. Industry Applications* **29** 787–792

[67] Kanazawa S, Li D, Akamine S, Ohkubo T and Nomoto Y 2000 *Trans. Institute of Fluid-Flow Machinery*, No 107, 65–74

[68] Futamura S, Zhang A, Prieto G and Yamamoto T 1998 *IEEE Trans. Industry Applications* **34** 967–974

[69] Proeto G, Prieto O, Gay C R and Yamamoto T 2003 *IEEE Trans. Industry Applications* **39** 72–78

[70] Kogoma M, Abe T and Tanaka K 2002 Proc. HAKONE VIII, International Symposium on High Pressure Low Temperature Plasma Chemistry, 303–307

[71] Hong J, Kim S, Lee K, Lee K, Choi J J and Kim Y-K 2002 Proc. HAKONE VIII, International Symposium on High Pressure Low Temperature Plasma Chemistry, 360–363

[72] Kim H H, Tsunoda K, Katsura S and Mizuno A 1999 *IEEE Trans. Industry Applications* **35** 1306–1310

[73] Ponizovsky A Z, Ponizovsky L Z, Kryutchkov S P, Starobinsky V Ya, Battleson D, Joyce J, Montgomery J, Babko S, Harris G and Shvedchikov A P 2000 Proc. HAKONE VII, International Symposium on High Pressure Low Temperature Plasma Chemistry, 345–349

[74] Ogata A, Yamanouchi K, Mizuno K, Kushiyama S and Yamamoto T 1999 *IEEE Trans. Industry Applications* **35** 1289–1295

[75] Penetrante B M, Bardsley J N and Hsiao M C 1997 *Jpn. J. Appl. Phys.* **36** 5007–5017

[76] Vertriest R, Morent R, Dewulf J, Leys C and Langenhove H V 2002 Proc. HAKONE VIII, International Symposium on High Pressure Low Temperature Plasma Chemistry, 342–346

9.9 Biological Decontamination by Non-equilibrium Atmospheric Pressure Plasmas

In this section, a review of various works on the germicidal effects of atmospheric pressure non-equilibrium plasmas is presented. First, a few of the variety of plasma sources, which have been used by various research groups, will be briefly presented. In-depth discussion of these sources and others can be found in chapter 6. Analysis of the inactivation kinetics for various bacteria seeded in (or on) various media and exposed to the plasma generated by these devices is then outlined. Three basic types of survivor curves have been shown to exist, depending on the type of microorganism, the type of medium, and the type of exposure (direct versus remote) (Laroussi 2002). Lastly, insights into the roles of ultraviolet radiation, active species, heat, and charged particles are presented. The most recent results show that it is the chemically reactive species, such as free radicals, that play the most important role in the inactivation process by atmospheric pressure air plasmas.

It is important to stress to the reader that only experiments carried out at pressures around 1 atm are the subjects of this presentation. For comprehensive studies conducted at low pressures, the reader is referred to Moreau *et al* (2000) and Moisan *et al* (2001). In addition, works that used etching-type gas mixtures, such as O_2/CF_4, or which used plasmas only as a secondary mechanism to assist a chemical-based sterilization method will not be covered. To learn about these, the reader is referred to Lerouge *et al* (2000), Boucher (1980) and Jacobs and Lin (1987).

9.9.1 Non-equilibrium, high pressure plasma generators

Here, a few methods that have been used to generate relatively large volumes of non-equilibrium plasmas, at or near atmospheric pressure (sometimes referred to as 'high' pressure) are briefly presented. This is far from being a comprehensive list of all existing methods. The devices presented here were chosen mainly because they have been used extensively to study the germicidal effects of low-temperature high-pressure plasmas. More detailed analysis of the physics of these devices can be found in chapter 6 of this book.

9.9.1.1 *DBD-based diffuse plasma source*

One of the early developments of diffuse glow discharge plasma at atmospheric pressure was reported by Donohoe (1976). Donohoe used a large gap (cm) pulsed barrier discharge in a mixture of helium and ethylene to polymerize ethylene (Donohoe and Wydeven 1979). Later, Kanazawa *et al* (1988) reported their development of a stable glow discharge at atmospheric pressure by using a dielectric barrier discharge (DBD). The most common configuration of the DBD uses two parallel plate electrodes separated by a variable gap. The experimental set-up of a DBD is shown in chapter 6 (section 6.6, figure 6.4.1). At least one of the two electrodes has to be covered by a dielectric material. After the ignition of the discharge, charged particles are collected on the surface of the dielectric. This charge build-up creates a voltage drop, which counteracts the applied voltage, and greatly decreases the voltage across the gap. The discharge subsequently extinguishes. As the applied voltage increases again (at the second half cycle of the applied voltage) the discharge re-ignites.

Laroussi (1995, 1996) reported the use of the DBD-based glow discharge at atmospheric pressure to destroy cells of *Pseudomonas fluorecens*. He used suspensions of the bacteria in Petri dishes placed on a dielectric-covered lower electrode. The electrodes were placed within a chamber containing helium with an admixture of air. He obtained full destruction of concentrations of 4×10^6/ml in less than 10 min. Subsequently, gram-negative bacteria such as *Escherichia coli*, and gram-positive bacteria such as *Bacillus subtilis* were inactivated successfully by many researchers using various types of high pressure glow discharges (Kelly-Wintenberg *et al* 1998, Herrmann *et al* 1999, Laroussi *et al* 1999, Kuzmichev *et al* 2001).

9.9.1.2 The atmospheric pressure plasma jet

The atmospheric pressure plasma jet (APPJ) (Scutze *et al* 1998) is a capacitively coupled device consisting of two co-axial electrodes between which a gas flows at high rates. Figure 9.9.1 is a schematic of the APPJ. The outer electrode is grounded while the central electrode is excited by rf power at 13.56 MHz. The free electrons are accelerated by the rf field and enter into collisions with the molecules of the background gas. These inelastic collisions produce various reactive species (excited atoms and molecules, free radicals, etc.) which exit the nozzle at high velocity. The reactive species can therefore react with a contaminated surface placed in the proximity (cm) of the nozzle

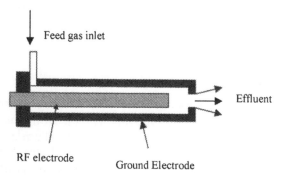

Figure 9.9.1. The atmospheric pressure plasma jet (Scutze *et al* 1998).

(Herrmann *et al* 1999). As in the case of the diffuse DBD, the stability of the APPJ plasma (as well as its non-thermal characteristic) depends on using helium as a carrier gas. Herrmann used the APPJ to inactivate spores of *Bacillus globigii*, a simulant to anthrax (*Bacillus anthracis*) (Herrmann *et al* 1999). They reported the reduction of seven orders of magnitude of the original concentration of *B. globigii* in about 30 s.

9.9.1.3   *The resistive barrier discharge*

The concept of the resistive barrier discharge (RBD) is based on the DBD configuration. However, instead of a dielectric material, a high resistivity sheet is used to cover at least one of the electrodes (see section 6.4, figure 6.4.7). The high resistivity layer plays the role of a distributed ballast which limits the discharge current and therefore prevents arcing. The advantage of the RBD over the DBD is the possibility to use dc power (or low frequency ac, 60 Hz) to drive the discharge. Using helium, large volume diffuse cold plasma at atmospheric pressure can be generated (Laroussi *et al* 2002a).

Using the RBD, up to four orders of magnitude reduction in the original concentration of vegetative *B. subtilis* cells in about 10 min was reported (Richardson *et al* 2000). Endospores of *B. subtilis* were also inactivated, but not as effectively as the vegetative cells. In these experiments, a gas mixture of helium:oxygen 97:3% was used.

9.9.2   Inactivation kinetics

The concept of inactivation or destruction of a population of microorganisms is not an absolute one. This is because it is impossible to determine if and when all microorganisms in a treated sample are destroyed (Block 1992). It is also impossible to provide the ideal conditions, which inactivate all microorganisms: some cells can always survive under otherwise lethal conditions. Therefore, experimental investigation of the kinetics of cell inactivation is paramount in providing a reliable temporal measure of microbial destruction.

9.9.2.1   *Survivor curves and D-value*

Survivor curves are plots of the number of colony forming units (CFUs) per unit volume versus treatment time. They are plotted on a semi-logarithmic scale with the CFUs on the logarithmic vertical scale and time on the linear horizontal scale. Figure 9.9.2 shows an example of a survivor curve obtained by exposing a culture of *E. coli* to an atmospheric pressure glow discharge in a helium/air mixture (Laroussi and Alexeff 1999). A line, such as shown in figure 9.9.2, indicates that the relationship between the

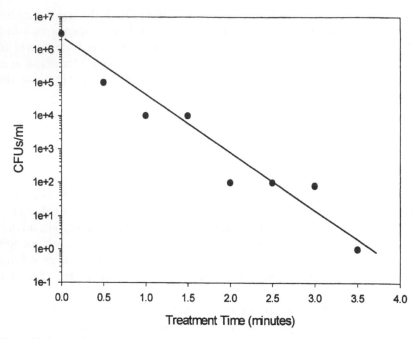

Figure 9.9.2. Survivor curve of *E. coli* exposed to DBD plasma.

concentration of survivors and time is given by

$$\log[N(t)/N_0] = -kt$$

where N_0 is the initial concentration and k is the 'death rate' constant.

One kinetics measurement parameter, which has been used extensively by researchers studying sterilization by plasma, is what is referred to as the 'D' (decimal) value. This parameter was borrowed from studies on heat sterilization. The D-value is the time required to reduce an original concentration of microorganisms by 90%. Since survivor curves are plotted on semi-logarithmic scales, the D-value is determined as the time for a $\log_{10}$ reduction. Sometimes the D-value is referred to as the 'log reduction time' (Block 1992) and expressed as follows:

$$D_v = t/(\log N_0 - \log N_s)$$

where t is the time to destroy 90% of the initial population, N_0 is the initial population, and N_s is the surviving population (Block 1992).

Another parameter, which is of great importance for practical systems, is the inactivation factor (IF). The IF is the percentage kill of a microbial population by a particular treatment (Block 1992). The IF is generally determined for spores (highly resistant microorganisms), by taking the ratio of the initial count to the final extrapolated count (Block 1992). Since the IF depends on the initial count (before treatment, what is referred to as the 'bioburden'), its value reveals the expected number of viable microorganisms after the treatment. Therefore, the IF of a treatment method directly reflects its sterilizing effectiveness, given a certain bioburden.

9.9.2.2 *Survivor curves of plasma-based inactivation processes*

To date, the experimental work on the germicidal effects of cold, atmospheric pressure plasmas has shown that survivor curves take different shapes depending on the type of microorganism, the type of the medium supporting the microorganisms, and the method of exposure (*direct exposure*: samples are placed in direct contact with the plasma; *remote exposure*: samples are placed away from the discharge volume or in a second chamber. The reactive species from the plasma, but not the plasma itself, are allowed to diffuse and come in contact with the samples) (Laroussi 2002).

Herrmann (APPJ, remote exposure), Laroussi (diffuse DBD-type discharge, direct exposure), and Yamamoto (corona discharge with H_2O_2, remote exposure) reported a 'single slope' survivor curve (one-line curve) for *B. globigii* on glass coupons (dry samples), for *E. coli* in suspension, and for *E. coli* on glass, respectively (Herrmann *et al* 1999, Laroussi *et al* 2000, Yamamoto *et al* 2001). The D-values ranged from 4.5 s for the *B. globigii* on glass (APPJ), to 15 s for *E. coli* on glass (Corona with H_2O_2 plasma), to 5 min for *E. coli* in liquid suspensions (DBD-type plasma).

Two-slope survivor curves (two consecutive lines with different slopes) were reported by Kelly-Wintenberg (DBD-type, *direct exposure*) for *S. aureus* and *E. coli* on polypropylene samples, and by Laroussi for *Pseudomonas aeruginosa* in liquid suspension (Kelly-Wintenberg *et al* 1998, Laroussi *et al* 2000). The curves show that the D-value of the second line (D_2) was smaller (shorter time) than the D-value of the first line (D_1). Montie also reported the same type of survivor curve for *E. coli* and *B. subtilis* on glass, agar, and polypropelene (all under direct exposure to a DBD-type discharge) (Montie *et al* 2000). Montie claimed that D_1 was dependent on the species being treated and that D_2 was dependent on the type of surface (or medium) supporting the microorganisms (Montie *et al* 2000). A given explanation of the 'bi-phasic' nature of the survivor curve was the following. During the first phase, the active species in the plasma react with the outer membrane of the cells, inducing damaging alterations. After this process is advanced enough, the reactive species can then quickly cause cell death, resulting in a rapid second phase (Kelly-Wintenberg *et al* 1998).

Multi-slope survivor curves were also reported for *E. coli* and *P. aeruginosa* on nitrocellulose filter (diffuse DBD-type, direct exposure) and for *B. stearothermophilus* on stainless steel strips (pulsed barrier discharge, remote exposure) (Laroussi *et al* 2000, Kuzmichev *et al* 2001). Each line has a different D-value. Similar survivor curves (three phases) were reported in low pressure studies (Moreau *et al* 2000, Moisan *et al* 2001). Moisan explains that the first phase, which exhibits the shortest D-value, is mainly due to the action of ultraviolet radiation on isolated spores or on the first layer of stacked spores. The second phase, which has the slowest kinetics, is attributed to a slow erosion process by active species. Finally the third phase comes into action after spores and debris have been cleared by phase 2, hence allowing ultraviolet to hit the genetic material of the still living spores. The D-value of this phase was observed to be close to the D-value of the first phase. It is important to note that the explanation given above would not apply to the case of atmospheric pressure air plasmas, which generate a negligible ultraviolet power output at the germicidal wavelengths (200–300 nm).

9.9.3   Analysis of the inactivation factors

This section presents a discussion on the contributions of the various agents emanating from non-equilibrium air plasmas to the killing process. These are the heat, ultraviolet radiation, reactive species, and charged particles. Note that in general various gas mixtures can be used to optimize the generation of one inactivation agent or another and ultimately to optimize the killing efficiency. The following results and discussions, however, are limited to the case of atmospheric pressure air (containing some degree of humidity). As a plasma generation device, a DBD is used.

9.9.3.1   *Heat and its potential effect*

High temperatures can have deleterious effects on the cells of microorganisms. A substantial increase in the temperature of a biological sample can lead to the inactivation of bacterial cells. Therefore, heat-based sterilization techniques were developed and commercially used for applications that do not require medium preservation. In heat-based conventional sterilization methods, both moist heat and dry heat are used. In the case of moist heat, such as in an autoclave, a temperature of 121 °C at a pressure of 15 psi is used. Dry heat sterilization requires temperatures close to 170 °C and treatment times of about 1 h.

To assess if heat plays a role in the case of decontamination by an air plasma, a thermocouple probe was used to measure the temperature increase in a biological sample under plasma exposure. In addition, the gas temperature in the discharge can be measured by evaluating the rotational band of the 0–0 transition of the second positive system of nitrogen. Figure 9.9.3 shows that the gas temperature and the sample temperatures in a DBD air plasma undergo only a small increase above room temperature (Laroussi and Leipold 2003). Based on these measurements no substantial thermal effects are expected.

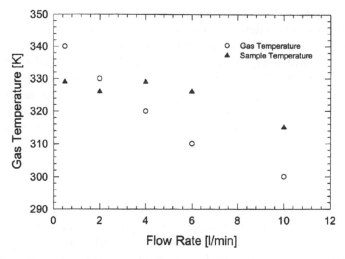

Figure 9.9.3. Gas and sample temperature versus air flow rate at a power of 10 W.

9.9.3.2 Ultraviolet radiation and its potential effect

Among ultraviolet effects on cells of bacteria is the dimerization of thymine bases in their DNA strands. This inhibits the bacteria's ability to replicate properly. Wavelengths in the 220–280 nm range and doses of several mW s/cm^2 are known to have the optimum effect. Figure 9.9.4 shows the emission spectrum between 200 and 300 nm from a DBD air plasma

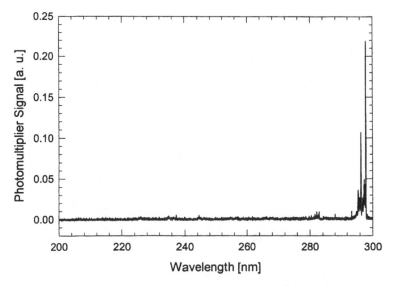

Figure 9.9.4. Emission spectrum of an air plasma in the ultraviolet region.

(Laroussi and Leipold 2003). ultraviolet emission at wavelengths greater than 300 nm was also detected. The spectrum is dominated by N_2 rotational bands 0–0 transition (337 nm) and NO_β transition around 304 nm. Measurements of the ultraviolet power density by a calibrated ultraviolet detector, in the 200–310 nm band, showed that less than 1 mW/cm² was emitted, under various plasma operating conditions. Therefore, according to these measurements, the ultraviolet radiation has no significant direct influence on the decontamination process of low temperature air plasmas. This is consistent with the results of several investigators (Laroussi 1996, Herrmann *et al* 1999, Kuzmichev *et al* 2001).

9.9.3.3   Charged particles and their potential effects

Mendis suggested that charged particles may play a very significant role in the rupture of the outer membrane of bacterial cells. By using a simplified model of a cell, they showed that the electrostatic force caused by charge accumulation on the outer surface of the cell membrane could overcome the tensile strength of the membrane and cause its rupture (Mendis *et al* 2000, Laroussi *et al* 2003). They claim that this scenario is more likely to occur for gram-negative bacteria, the membrane of which possesses an irregular surface. Experimental work by Laroussi and others has indeed shown that cell lysis is one outcome of the exposure of gram-negative bacteria to plasma under direct exposure (Laroussi *et al* 2002b). However, it is not clear if the rupture of the outer membrane is the result of the charging mechanism or a purely chemical effect. Figure 9.9.5 shows SEM micrographs of controls and plasma-treated *E coli* cells (Laroussi *et al* 2002b). The micrograph of the plasma-treated cells shows gross morphological damage.

9.9.3.4   Reactive species and their inactivation role

In high-pressure non-equilibrium discharges, reactive species are generated through electron impact excitation and dissociation. They play an important

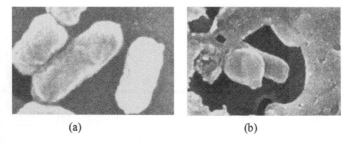

(a)   (b)

Figure 9.9.5. SEM micrographs of controls (a) and plasma-treated bacteria (b) *E. coli* cells. The plasma-treated cells show gross morphological damage.

role in all plasma–surface interactions. Among the radicals generated in air plasmas, oxygen-based and nitrogen-based species such as atomic oxygen, ozone (O_3), NO, NO_2, and the hydroxyl radical (OH) have direct impact on the cells of microorganisms, especially when they come in contact with their outer structures such as the outer membrane. Membranes are made of lipid bilayers, an important component of which is unsaturated fatty acids. The unsaturated fatty acids give the membrane a gel-like nature. This allows the transport of the biochemical by-products across the membrane. Since unsaturated fatty acids are susceptible to attacks by hydroxyl radical (OH) (Montie *et al* 2000), the presence of this radical can therefore compromise the function of the membrane lipids. This will ultimately affect their vital role as a barrier against the transport of ions and polar compounds in and out of the cells (Bettelheim and March 1995). Imbedded in the lipid bilayer are protein molecules, which also control the passage of various compounds. Proteins are basically linear chains of aminoacids. Aminoacids are also susceptible to oxidation when placed in the radical-rich environment of the plasma. Therefore, oxygen-based and nitrogen-based species are expected to play a crucial role in the inactivation process.

The following are measurements of nitrogen dioxide (NO_2), hydroxyl (OH), and ozone (O_3) obtained from a DBD operated in atmospheric pressure air (Laroussi and Leipold 2003). Figure 9.9.6 shows the concentration of NO_2 in the DBD, as measured by a calibrated gas detection system. The presence of OH was measured by means of emission spectroscopy, looking for the rotational spectrum of OH A–X (0–0) transition. This molecular band has a branch at about 306.6 nm (R branch) and another one at 309.2 nm (P branch). Figure 9.9.7 shows the emission spectrum in

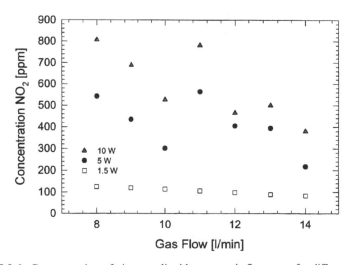

Figure 9.9.6. Concentration of nitrogen dioxide versus air flow rate, for different powers.

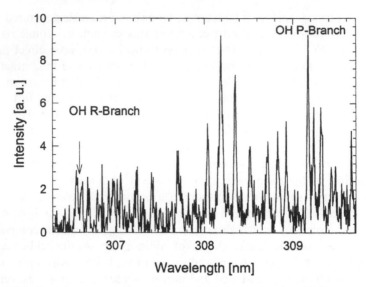

Figure 9.9.7. Emission spectra from a humid air discharge showing OH lines.

the range between 306 and 310 nm and it indicates the OH band heads. Figure 9.9.8 shows the relative concentration of OH in the discharge as a function of power and air flow rate. Ozone concentration produced by the DBD in atmospheric air was measured for varying flow rate and at various

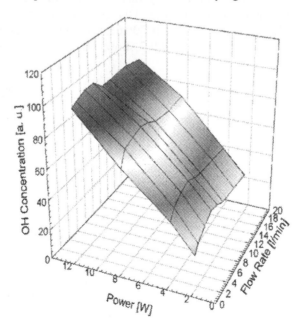

Figure 9.9.8. Relative concentration of OH versus power and air flow rate.

power levels by ultraviolet absorption spectroscopy and by a chemical titration method. Concentrations up to 2000 ppm could be obtained. Ozone germicidal effects are caused by its interference with cellular respiration.

9.9.4 Conclusions

Research on the interaction of both low-pressure and high-pressure non-equilibrium plasmas with biological media has reached a stage of maturity, which indicates that this emerging field promises to yield valuable technological novelty. In the medical field, the use of plasma to sterilize heat-sensitive re-usable tools in a rapid, safe, and effective way is bound to replace the present method which relies on the use of ethylene oxide, a toxic gas. In the food industry, the use of plasmas to sterilize packaging will lead to safer food with a longer shelf life. In space applications, plasma is considered as a potential method to decontaminate spacecraft on planetary missions. The goal in this application is to avoid transporting microorganisms from Earth to the destination planet (or moon). Air plasma is also a potential technology that can be used for the destruction of biological warfare agents.

Extensive research on the use of high-pressure low-temperature plasmas to inactivate microorganisms is a relatively recent event. There are still a lot of basic issues that need more in depth investigations. Among these are the effects of plasma on the biochemical pathways of bacteria. A clear understanding of these will lead to new applications other than sterilization/decontamination. However, for practical devices intended for the destruction of pathogens, all the available results indicate that non-equilibrium plasmas generated in atmospheric pressure air offer a very efficient decontamination method. This is mainly due to the efficient production of oxygen-based and nitrogen-based reactive species, which interact directly with the cells and can cause them irreversible damage.

References

Bettleheim F A and March J. 1995 Introduction to General, Organic, and Biochemistry 4th edition (Saunders College Pub.)

Block S S 1992 'Sterilization' in *Encyclopedia of Microbiology*, vol 4, pp 87–103 (Academic Press)

Boucher (Gut) R M 1980 'Seeded gas plasma sterilization method' US Patent 4,207,286

Donohoe K G 1976 'The development and characterization of an atmospheric pressure non-equilibrium plasma chemical reactor' PhD Thesis, California Institute of Technology, Pasadena, CA

Donohoe K G and Wydeven T 1979 'Plasma polymerization of ethylene in an atmospheric pressure discharge' *J. Appl. Polymer Sci.* **23** 2591–2601

Herrmann H W, Henins I, Park J and Selwyn G S 1999 'Decontamination of chemical and biological warfare (CBW) agents using an atmospheric pressure plasma jet' *Phys. Plasmas.* **6**(5) 2284–2289

Jacobs P T and Lin S M 1987 'Hydrogen peroxide plasma sterilization system' US Patent 4,643,876

Kanazawa S, Kogoma M, Moriwaki T and Okazaki S 1988 'Stable glow plasma at atmospheric pressure' *J. Appl. Phys. D: Appl. Phys.* **21** 838–840

Kelly-Wintenberg K, Montie T C, Brickman C, Roth J R, Carr A K, Sorge K, Wadworth L C and Tsai P P Y 1998 'Room temperature sterilization of surfaces and fabrics with a one atmosphere uniform glow discharge plasma' *J. Industrial Microbiology and Biotechnology* **2** 69–74

Kuzmichev A I, Soloshenko I A, Tsiolko V V, Kryzhanovsky V I, Bazhenov V Yu, Mikhno I L and Khomich V A 2001 'Feature of sterilization by different type of atmospheric pressure discharges' in *Proc. Int. Symp. High Pressure Low Temperature Plasma Chem.* (HAKONE VII), pp. 402–406, Greifswald, Germany

Laroussi M 1995 'Sterilization of tools and infectious waste by plasmas' *Bull. Amer. Phys. Soc. Div. Plasma Phys.* **40**(11) 1685–1686

Laroussi M 1996 'Sterilization of contaminated matter with an atmospheric pressure plasma' *IEEE Trans. Plasma Sci.* **24**(3) 1188–1191

Laroussi M 2002 'Non-thermal decontamination of biological media by atmospheric pressure plasmas: review, analysis and prospects' *IEEE Trans. Plasma Sci.* **30**(4) 1409–1415

Laroussi M and Alexeff I 1999 'Decontamination by non-equilibrium plasmas' in *Proc. Int. Symp. Plasma Chem.*, pp 2697–2702, Prague, Czech Rep., August

Laroussi M and Leipold F 2003 'Mechanisms of inactivation of bacteria by an air plasma' in *Proc. Int. Colloq. Plasma Processing*, Juan les Pins, France, June

Laroussi M, Alexeff I and Kang W 2000 'Biological decontamination by non-thermal plasmas' *IEEE Trans. Plasma Sci.* **28**(1) pp. 184–188

Laroussi M, Alexeff I, Richardson J P and Dyer F F 2002a 'The resistive barrier discharge' *IEEE Trans. Plasma Sci.* **30**(1) 158–159

Laroussi M, Mendis D A and Rosenberg M 2003 'Plasma interaction with microbes' *New Journal of Physics* **5** 41.1–41.10

Laroussi M, Richardson J P and Dobbs F C 2002b 'Effects of non-equilibrium atmospheric pressure plasmas on the heterotrophic pathways of bacteria and on their cell morphology' *Appl. Phys. Lett.* **81**(4) 772–774

Laroussi M, Sayler G S, Galscock B B, McCurdy B, Pearce M, Bright N and Malott C 1999 'Images of biological samples undergoing sterilization by a glow discharge at atmospheric pressure' *IEEE Trans. Plasma Sci.* **27**(1) 34–35

Lerouge S, Werthheimer M R, Marchand R, Tabrizian M and Yahia L'H 2000 'Effects of gas composition on spore mortality and etching during low-pressure plasma sterilization' *J. Biomed. Mater. Res.* **51** 128–135

Mendis D A, Rosenberg M and Azam F 2000 'A note on the possible electrostatic disruption of bacteria' *IEEE Trans. Plasma Sci.* **28**(4) 1304–1306

Moisan M, Barbeau J, Moreau S, Pelletier J, Tabrizian M and Yahia L'H 2001 'Low temperature sterilization using gas plasmas: a review of the experiments, and an analysis of the inactivation mechanisms' *Int. J. Pharmaceutics* **226** 1–21

Montie T C, Kelly-Wintenberg K and Roth J R 2000 'An overview of research using the one atmosphere uniform glow discharge plasma (OAUGDP) for sterilization of surfaces and materials' *IEEE Trans. Plasma Sci.* **28**(1) 41–50

Moreau S, Moisan M, Barbeau J, Pelletier J, Ricard A 2000 'Using the flowing afterglow of a plasma to inactivate *Bacillus subtilis* spores: influence of the operating conditions' *J. Appl. Phys.* **88** 1166–1174

Richardson J P, Dyer F F, Dobbs F C, Alexeff I and Laroussi M 2000 'On the use of the resistive barrier discharge to kill bacteria: recent results' in *Proc. IEEE Int. Conf. Plasma Science*, New Orleans, LA, p 109

Scutze A, Jeong J Y, Babyan S E, Park J, Selwyn G S and Hicks R F 1998 'The atmospheric pressure plasma jet: a review and comparison to other plasma sources' *IEEE Trans. Plasma Sci.* **26**(6) 1685–1694

Yamamoto M, Nishioka M and Sadakata M 2001 'Sterilization using a corona discharge with H_2O_2 droplets and examination of effective species' in *Proc. 15th Int. Symp. Plasma Chem.*, Orleans, France, vol II, pp 743–751

9.10 Medical Applications of Atmospheric Plasmas

This section concludes the chapter devoted to practical aspects of atmospheric plasmas. At this point, the reader is provided with state of the art information on available plasma sources and their applications in inorganic/material technology, gas cleaning, combustion, etc. The remaining issue is the role of plasma in health care.

Several biomedical applications of plasmas have been already identified, including surface functionalization of scaffolds, deposition of bio-compatible coatings, and bacterial decontamination. For *in vivo* treatment, plasma-based devices have been successfully used in wound sealing and non-specific tissue removal. Since the modern plasma sources have become quite friendly and 'bio-compatible', the area of applications is expanding rapidly and many novel medical techniques are under preparation. The most recent development is *in vivo* bacterial sterilization and tissue modification at the cellular level. All these techniques will be described in this section.

9.10.1 A bio-compatible plasma source

A plasma can be considered 'bio-compatible' when it combines therapeutic action with minimum damage to the living tissue. In non-specific tissue removal, the penetration depth and the degree of devitalization must be controllable. In refined/selective tissue modification there are more restrictions on the thermal, electrical and chemical properties of the plasma. In this paragraph the necessary safety requirements will be briefly discussed.

9.10.1.1 Thermal properties of a non-equilibrium plasma

Surface processing of materials usually involves non-thermal plasmas. 'Non-thermal' does not imply that such plasmas cannot inflict thermal damage; it means that they are non-equilibrium systems with electron temperature 100 to 1000 times higher than neutral gas temperature. In table 9.10.1.1 typical

Table 9.10.1.1.

Plasma source	Type	Gas	T (K)	Ref.
Atmospheric pressure plasma jet (APPJ)	RF capacitively coupled	Helium, argon	400	Park *et al* (2002)
Atmospheric glow	AC/DC glow above water	Air	800–1500	Lu and Laroussi (2003)
Cold arc-plasma jet	AC 10–40 kHz	Air, N_2, O_2	520	Toshifuji *et al* (2003)
Microwave torch	2.45 GHz	Argon + O_2	2200	Moon and Choe (2003)
AC plasma	AC	Helium + O_2	800–900	Moon and Choe (2003)
DBD	Dielectric barrier	$N_2 + O_2 + NO$	300	Baeva *et al* (1999)
Pulsed DBD	Dielectric barrier	Argon + H_2O	350–450	Motret *et al* (2000)
Atmospheric glow	DC glow with micro-hollow cathode electrode	Air	2000	Mohamed *et al* (2002)
Plasma needle	RF capacitively coupled, mm size	Helium + N_2	350–700	Stoffels *et al* (2002)
	RF micro-plasma	Helium (+H_2O)	300	Stoffels *et al* (2003)

gas temperatures in several types of non-thermal plasmas are given. Most of these results have been obtained using spectroscopic methods: optical emission and CARS (Baeva *et al* 1999). Moon and Choe (2003) have calibrated optical emission spectroscopy against thermocouples. Stoffels *et al* (2002, 2003) has also used both methods; some details are given in section 9.10.3 where the plasma needle is characterized.

Most of these sources can be used for non-specific treatment, like burning and coagulation (see section 9.10.2). For this purpose the temperature may be quite high as long as there is no carbonization or deep damage. In other applications, like specific treatment *without* tissue devitalization, temperature is an essential issue. The tissue may be warmed up to at most a few degrees above the ambient temperature, and exposure time must be limited to several minutes. Discharges suitable for this kind of treatment are the micro-plasmas (plasma needle) and possibly some kinds of DBDs.

9.10.1.2   The influence of electricity

The influence of electric fields on living cells and tissues has been elaborately studied in relation to electrosurgery and related techniques. High electric fields are surely a matter of concern for the health of the patient, because they may interact with the nervous system, disturb the heartbeat, and cause damage to the individual cells.

Much attention has been given to alternating (high-frequency) currents passing through the body. For detailed data the reader should refer to works like Gabriel *et al* (1996) (dielectric properties and conductivity of tissues), Reilly (1992) (nerve and muscle stimulation) and Polk and Postow (1995) (electroporation and other field-induced effects). These studies have revealed that the sensitivity of nerves and muscles decreases with increasing ac frequency. The threshold current that causes irritation is as high as 0.1 A at 100 kHz. It implies that for medical applications high-frequency sources should be employed. At present, most of the electrosurgical equipment operates at 300 kHz or higher; the plasma needle is sustained by rf excitation. Under these conditions no undesired effects are induced.

9.10.1.3 *Toxicity*

Plasma is a rich source of radicals and other active species. Reactive oxygen species (ROS) (O, OH and HO_2, peroxide anions O_2^- and HO_2^-, ozone and hydrogen peroxide) may cause severe cell and tissue damage, known under a common name of oxidative stress. On the cellular level, several effects leading to cell injury have been identified: lipid peroxidation (damage to the membrane), DNA damage, and protein oxidation (decrease in the enzyme activity). On the other hand, free radicals have various important functions, so they are also produced by the body. For example, macrophages generate ROS to destroy the invading bacteria, and endothelial cells (inner artery wall) produce nitric oxide (NO) to regulate the artery dilation. The natural level of radical concentration lies in the μM range (Coolen 2000).

The density of radical species in the plasma can be determined using a variety of plasma diagnostics. However, for applications in biology/medicine, standard gas-phase plasma characterization is not very relevant. Instead, one has to identify radical species that penetrate the solution and enter the cell. Biochemists have some standard methods for radical detection, e.g. laser-induced fluorescence in combination with confocal microscopy. Special organic probes are used, which become fluorescent after reaction with free radicals. This yields detection limits below 0.01 μM in a solution, and allows three-dimensional profiling with a resolution of about 0.2 μm.

9.10.2 *In vivo* treatment using electric and plasma methods

9.10.2.1 *Electrosurgery*

From very early times it was believed that electricity might have some healing properties. In the 17th century some cases of improving the heart function, waking up from swoon, etc. were reported. About 200 years later the technology of artificial generation of electricity was ready for advanced medical applications. In 1893, d'Arsonval discovered that high-frequency current passing through the body does not cause nerve and muscle stimulation (d'Arsonval 1893). Soon after, high-frequency devices were introduced for cutting of tissues.

At present, electrosurgery has a solid, established name in medicine: the electrical cutting device replaces the scalpel in virtually all kinds of surgery. A detailed list of applications can be found in the database of ERBE (http://www.erbe-med.de), a leading company producing equipment for electric, cryogenic and plasma surgery. The electrosurgical tools manufactured by ERBE are powered by high-frequency generators, either at 330 kHz or at 1 MHz. The reason for using these frequencies has been already discussed in the previous section: they are well above 100 kHz, the lower limit for electric safety. The devices can supply reasonably high powers—up to 200 or 450 W, dependent on the type and application. The power can be (automatically) regulated during the operation, to obtain the desired depth of the incision. Various electrode designs and configurations are used: a monopolar high-frequency powered pin (in this case the current is flowing through the patient's body), a bipolar coaxial head, and a tweezers-like design (see figure 9.10.1). In the latter case the arms of the tweezers have opposite polarities, and the distance between their tips can be varied. The quality of cuts for all these configurations is about the same.

The features that have made electric devices so successful and desired are: good cutting reproducibility, high precision, good control of depth,

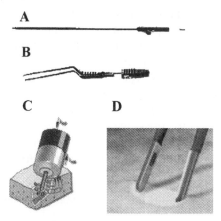

Figure 9.10.1. Electrosurgery devices and techniques developed by ERBE (http://www.erbe-med.de/): (a) a monopolar cutting device, (b) bipolar cutting/coagulation tweezers, (c) tissue cutting using coaxial bipolar device, (d) tissue coagulation using bipolar tweezers.

and the possibility of local coagulation. The latter is especially important in achieving hemostasis and thus preventing blood loss, formation of thrombus, and contamination of tissues during surgery. Electrical coagulation is also used on its own, when no incision is necessary—for this purpose a bipolar tweezers-like device is used (see figures 9.10.1b,d). The current flowing through the tissue induces ohmic heating that allows for fast and superficial coagulation. This method is often used to seal small blood vessels.

9.10.2.2 Argon plasma coagulation

The step from electric to plasma surgery is readily made. The electric methods discussed above are based on local tissue heating. Devitalization by heat is a rather unsophisticated effect, which can be achieved by exposure to any heat source. Atmospheric plasma generated by a high-power electric discharge is one of the options. Needless to say, for these applications it is not required that the gas temperature in the plasma be low. On the contrary, controlled burning of the diseased tissue is an essential part of the therapy. The aim of the treatment is coagulation and stopping the bleeding, and sometimes even total desiccation and devitalization of the tissue.

An adequate discharge has been developed by ERBE, and the corresponding surgical technique is called argon plasma coagulation (APC). The design of the APC source resembles somewhat the APPJ (Park *et al* 2002), because the latter is also a plasma generated in a tube with flowing argon. The APC source has not been characterized, but considering the parameters (frequency of 350 kHz, operating voltage of several kV and power input of 50 W) it seems to be a classical ac atmospheric jet. The gas temperature within the plasma can easily reach several hundreds of degrees Celsius.

A schematic view of an APC device (figure 9.10.2) shows a tube through which argon is supplied. The flow rate is adjustable between 0.1 and 0.9 l/min. The powered electrode is placed coaxially inside the tube (monopolar

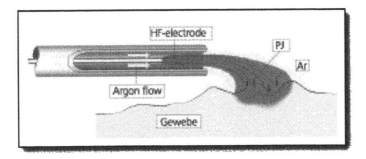

Figure 9.10.2. An argon plasma coagulation device, developed by ERBE. Argon flow is blown through the tube, in which the high frequency electrode is placed. The plasma flame stretches out of the tube.

configuration). Like in monopolar electrosurgery, the patient is placed on a conducting sheet and the high-frequency current flows through the body. The APC electrode generates argon plasma, which stretches about 2–10 mm from the tip. Since the plasma is conductive, the current can flow to the tissue, but the electrode does not touch it. This is one of the most important advantages of APC: the energy is transferred in a non-contact way, so the problems with tissue sticking to the metal device, heavy burning and tearing can be avoided. Another unique feature of APC is its self-limiting character. Since the desiccated tissues have a lower electrical conductivity than the bleeding ones, the plasma beam will turn away from already coagulated spots toward bleeding or still inadequately coagulated tissue in the area receiving treatment. The argon plasma beam acts not only in a straight line (axially) along the axis of the electrode, but also laterally and radially and 'around the corner' as it seeks conductive bleeding surfaces. This automatically results in evenly applied, uniform surface coagulation. The tissues are not subjected to surface carbonization and deep damage, and the penetration depth is at most 3–4 mm. It should be mentioned that the action 'around the corner' is typical for all plasmas, but it cannot be achieved in e.g. laser surgery. Superficial scanning of irregular surfaces, small penetration depths, and low equipment costs, make plasma devices competitive with lasers.

It is not entirely clear what causes the coagulation of the treated tissue. It may be the heat transferred directly from the hot gas as well as the heat generated within the tissue by ohmic heating. It is also plausible that argon ions bombarding the tissue contribute to desiccation.

Although the exact physical mechanism of coagulation is not yet completely understood, the APC device has been successfully applied in many kinds of surgery. The most obvious application is open surgery— promoting hemostasis in wounds and bleeding ulcers. Treatment of various skin diseases has been discussed by Brand *et al* (1998). Devitalization of mucosal lesions in the oral cavity (e.g. leucoplakia) has been also performed. However, the most obvious techniques are not necessarily the most frequently applied ones. Since ERBE has developed a flexible endoscopic probe, the way to minimally invasive internal surgery has been opened. The area of interest is enormous, and most of the APC applications involve endoscopy. In gastroenterology there are many situations where large bleeding areas must be devitalized. APC treatment has been used to destroy gastric and colon carcinoma or to remove their remains after conventional surgery, to reduce tissue ingrowth into supporting metal stents (e.g. stents placed in the esophagus), to treat watermelon stomach and colitis. APC techniques are also frequently used for various operations in the tracheo-bronchial system—removal of tumors, opening of various blockages (stenoses) in the respiratory tract (e.g. scar stenoses), etc. In the nasal cavity, APC can reduce hyperplasia of nasal concha (which causes respiratory problems) and hemorrhaging. More examples and detailed information about the medical procedures can be found on the website of ERBE. In all mentioned cases, the physicians are positive about the immediate body reaction and post-treatment behavior. Of course, during the operation the surgeon has to be careful not to cause membrane/tissue perforation by applying high powers and/or prolonging the treatment too much. When the treatment is performed correctly, the devitalized (necrotic) tissue dissolves and the healing proceeds without complications.

9.10.2.3 Spark erosion and related techniques

Spark erosion is a special and unconventional application of plasma in surgery. It is remarkable for two reasons: first, as an attempt to treat athero-sclerosis, a complex cardiovascular disease that plagues most of the Western world, and second, as an example to show that a quite powerful discharge can be induced in vulnerable places, like blood vessels. In the following passage a brief description of athcrosclerosis, its pathogenesis and current treatment methods will be given, followed by a discussion of the spark erosion technique.

Atherosclerosis is a chronic inflammatory disease, where lipid-rich plaque accumulates in arteries. The consequences are plaque rupture and/ or obstruction of the arteries. The occluded artery cannot supply blood to a tissue. This results in ischemic damage and infarct (necrosis). For example, direct obstruction of a coronary artery causes irreversible damage to a part of the heart muscle, and a myocardial infarct (heart attack). Plaque rupture produces thrombus that can cause vascular embolization and infarct far away from the actual site of plaque. Complications include stroke and gangrene of extremities. At present it is the principal cause of death in the Western world (Ross 1999).

Atherosclerotic obstructions are usually removed surgically (Guyton and Hall 2000), by inflating and stretching the artery (balloon angioplasty). In severe cases an additional blood vessel must be inserted (bypass operation). However, there is no universal cure, because restenosis after balloon angioplasty occurs within six months in 30–40% of treated cases, and the bypasses are less stable than original arteries.

In surgical treatment the plaque must be removed, but in a way that causes least damage to the artery, so as to minimize restenosis. Recently, laser methods have been applied with reasonable success. However, as mentioned earlier, lasers cannot act 'around the corner', which in this case is essential. In 1985 Slager presented a new concept, which lies between electrosurgery and plasma treatment (Slager *et al* 1985). This technique, called spark erosion, is based on plaque vaporization by electric heating. The tool developed by Slager is similar to the monopolar device used in APC, but no feed gas is used. Instead, the electrode is immersed directly in

Figure 9.10.3. A crater in the atherosclerotic plaque, produced by tissue ablation using the spark erosion technique (Slager *et al* 1985).

the blood stream and directed towards the diseased area. Alternating current (250 kHz) is applied to the electrode tip in a pulsed way, with a pulse duration of 10 ms. The voltages are up to 1.2 kV. Under these conditions, the tissue is rapidly heated and vaporized. The produced vapor isolates the electrode from the tissue, so that further treatment is performed in a non-contact way. After vaporization, electric breakdown in the vapor occurs and a small (<1mm) spark is formed. Spark erosion allows removing substantial amounts of plaque—craters produced can have dimensions of up to 1.7 mm. The crater edges are smooth and the coagulation layer does not exceed 0.1–0.2 mm (see figure 9.10.3).

It is not yet clear whether spark erosion will become competitive with lasers and mechanical methods in treatment of atherosclerosis. One possible problem is formation of vapor bubbles, which may lead to vascular embolization. Nevertheless, the spark-producing electrode can be used in open-heart operations, e.g. in surgical treatment of hypertrophic obstructive cardiomyopathy (Maat *et al* 1994). The cutting performance is similar to electrosurgery but, as in plasma techniques, the treatment is essentially non-contact.

Compared to argon plasma coagulation, thermal effects in spark surgery are minor. The spark plasma is much smaller than the argon plasma, so that heating is more local. Since there is no gas flow, no heat is transferred by convection, and pulsed operation suppresses the thermal load. The physical characterization of spark-like discharges was performed by Stalder *et al* (2001) and Woloszko *et al* (2002). The spark generated by these authors was similar to the discharge employed by Slager, but they focused on the plasma interactions with electrolyte solution. The electron density in such plasmas is in the order of $10^{18}\,\mathrm{m}^{-3}$, and the electron temperature is about $4\,\mathrm{eV}$. The gas temperature is about $100°\,\mathrm{C}$ above the ambient.

9.10.3 Plasma needle and its properties

In the medical techniques described above the action of plasma is not refined—it is based on local burning/vaporization of the tissue. Using the analogy to material science, APC and spark erosion can be compared to cutting and welding. However, plasmas are capable of much more sophisticated surface treatment than mere thermal processing. If the analogy to material science holds, it is expected that fine tissue modification can be achieved using advanced plasma techniques.

However, the construction of non-thermal *and* atmospheric plasma sources suitable for fine tissue treatment is not trivial. Moreover, most plasmas must be confined in reactors, so they cannot be applied directly and with high precision to a diseased area. In the following section another approach will be presented: a flexible and non-destructive micro-plasma for direct and specific treatment of living tissues.

9.10.3.1 *Plasma needle*

Small-sized atmospheric plasmas are usually non-thermal. This is simply a consequence of their low volume to surface ratio. Energy transfer from electrons to gas atoms/molecules occurs in the volume, and the resulting heat is lost by conduction through the plasma boundary surface. A simple balance between electron-impact heating and thermal losses can be made for a spherical glow with a radius L:

$$\frac{m_e}{m_a}\nu_{ea}n_e k_B T_e\,\frac{4}{3}\,\pi L^3 = \kappa\,\frac{\Delta T}{L}\,4\pi L^2$$

where $m_{e,a}$ is the electron/atomic mass, ν_{ea} is the electron–atom collision frequency and κ is the thermal conductivity of the gas. This allows estimation of a typical plasma size:

$$L = \sqrt{\frac{m_a}{m_e}\,\frac{3\kappa\Delta T}{\nu_{ea}n_e k_B T_e}}.$$

Dependent on the plasma conditions, the typical length scales of non-thermal plasmas with $\Delta T < 10°\,\mathrm{C}$ are of the order of $1\,\mathrm{mm}$.

A plasma needle (Stoffels *et al* 2002) fulfills the requirements of being small, precise in operation, flexible and absolutely non-thermal. This is a capacitively coupled rf ($13.56\,\mathrm{MHz}$) discharge created at the tip of a sharp needle. The experimental scheme, including a photograph of the flexible hand-held plasma torch, is shown in figure 9.10.4. Like most atmospheric

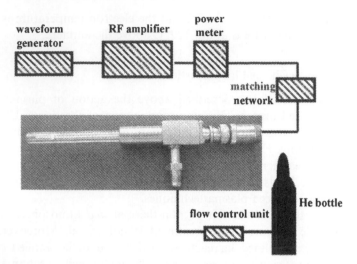

Figure 9.10.4. A schematic view of the plasma needle set-up. In the photograph of the flexible torch: rf voltage (right throughput) is supplied to the electrode (needle), confined in a plastic tube, through which helium is blown (bottom throughput).

discharges, the needle operates most readily in helium: the voltage needed for ignition is only 200 V peak-to-peak. In fact, using helium as a carrier gas has other advantages. The thermal conductivity (144 W/m/K) is very high, and consequently the plasma temperature can be maintained low. Moreover, helium is light and inert, and possible tissue damage due to ion bombardment and toxic chemicals can be thus excluded. The therapeutic working of the plasma depends on the additives. As said in section 9.10.1, small doses of active species may be beneficial, while large doses inflict damage. In case of a plasma needle, the amount of active species is easy to regulate. The right dose can be administered by adjusting the plasma power, distance to the tissue, treatment time and gas composition. So far, helium plasmas with about 1% of air have been used.

The glow can be applied directly to the tissues. In figure 9.10.5 one can see how the plasma interacts with human skin: it spreads over the surface without causing any damage or discomfort.

Prior to tests with living cells and tissues the needle has been characterized in terms of electrical properties, temperature and thermal fluxes. In figure 9.10.6(a) the temperature versus plasma power is shown for a needle with 1 mm diameter: the power lies in the range of several watts and the temperatures rise far above the tolerance limits for biological materials. For a thinner needle (0.3 mm) the power dissipation is only 10–100 mW and the temperature increase is at most a couple of degrees (figure 9.10.6(b). Thus, the needle geometry is important for its operation.

The flux of radicals emanated by the plasma into a liquid sample has been determined using a fluorescent probe (see section 9.10.1). In figure 9.10.7 the

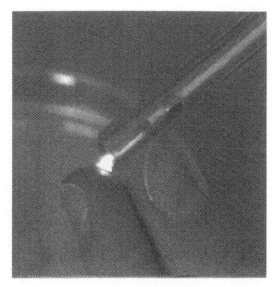

Figure 9.10.5. Plasma generated in the flexible torch stretches out to reach the skin.

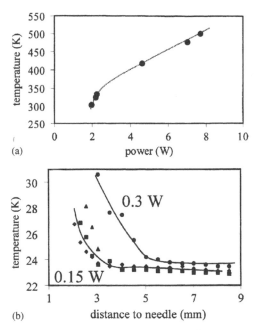

Figure 9.10.6. (a) Temperature of the plasma determined using a spectroscopic method for a 1 mm thick needle. (b) Temperature of the surface (thermocouple) as a function of the distance between the needle and the thermocouple for a 0.3 mm thick needle.

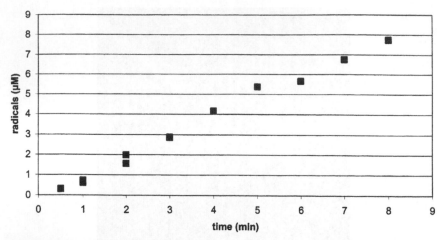

Figure 9.10.7. Active radical concentration in a 400 μl water sample treated with the plasma needle, as a function of exposure time. The plasma power is about 50 mW, the needle-to-surface distance is 1.5 mm.

concentration of ROS as a function of exposure time is shown for a helium plasma with 1% air. The estimated radical density in the gas phase is $10^{19}\,\mathrm{m}^{-3}$. The ROS concentration in the liquid lies in the μM range. This amount can trigger cell reactions, but it is too low to cause tissue damage.

9.10.4   Plasma interactions with living objects

Interactions of non-thermal plasmas with living objects are an entirely new area of research. Of course, the ultimate goal of this research is introducing plasma treatment as a novel medical therapy. However, living organisms are so complicated that one has to begin with a relatively simple and predictable model system, like a culture of cells. In the following section it will be shown that even the simplest biological models can exhibit complex reactions when exposed to an unknown medium.

9.10.4.1   Apoptosis versus necrosis

The essential difference between the non-thermal plasma needle and APC or spark erosion lies in the manner in which the cells are affected. In fine surgery cell damage should be minimal. Cell death should be induced only when necessary, and then it should fit in the natural pathway, in which the body renews and repairs its tissues.

Cell death is the consequence of irreversible cell injury. It can be classified in two types described below.

- *Necrosis*, or accidental cell death. Necrosis is defined as the consequence of a catastrophic injury to the mechanisms that maintain the integrity of the cell. There are many factors that cause necrosis: cell swelling and rupture due to electrolyte imbalance, mechanical stress, heating or freezing, and contact with aggressive chemicals (e.g. acids, formaldehyde, alcohols). In necrotic cells the membrane is damaged, and the cytoplasm leaks to the outside. Since the content of the cell is harmful to the tissue, the organism uses its immune reaction to dispose of the dangerous matter, and an inflammatory reaction is induced. In surgery, mechanical, thermal or laser methods always cause severe injury and necrosis. The necrotic tissue is eventually removed by the organism, but the inflammation slows down the healing and may cause complications, the most common being restenosis and scar formation.

* *Apoptosis*, or programmed cell death. Apoptosis is an internal mechanism of self-destruction, which is activated under various circumstances. This kind of 'cell suicide' is committed by cells which are damaged, dangerous to the tissue, or simply no longer functional. Thus, apoptosis takes place in developmental morphogenesis, in natural renewal of tissues, in DNA-damaged, virus-infected or cancer cells, etc. Presumably, any moderate yet irreversible cell damage can also activate apoptosis. Known factors are ultraviolet exposure, oxidative stress (section 9.10.1) and specific chemicals. The role of radicals and ultraviolet has given rise to the hypothesis, that plasma treatment may also induce apoptosis.

Since the intracellular mechanism of apoptosis is rather complex, no details will be given here. The reader may refer to textbooks on cell biology (Alberts 1994) or more specific articles (Cohen 1997). The morphological changes in the cell during apoptosis are easy to recognize. In early apoptosis, the DNA in the nucleus undergoes condensation and fragmentation and the cell membrane displays blebs. Later, the cell is fragmented in membrane-bound elements (apoptotic bodies). Note that the membrane retains its integrity, so no cytoplasm leakage and no inflammatory reaction occur. The apoptotic bodies are engulfed by macrophages or neighboring cells and the cell vanishes in a neat manner.

It is clear that apoptosis is preferred to necrosis. Selective induction of apoptosis can make a pathological tissue disappear virtually without a trace. Such refined surgery is the least destructive therapeutic intervention. No inflammation, no complications in healing and no scar formation/stenosis is expected. In the next paragraph plasma induction of apoptosis and other cell reactions (without necrosis) will be discussed.

9.10.4.2   Plasma needle and cell reactions

A fundamental study on a model system is necessary to identify and classify the possible ways in which the plasma can affect mammalian cells. Stoffels *et al* (2003) used two model systems: the Chinese hamster ovarian cells (CHO-K1) and the human cells MR65. CHO-K1 cells are fibroblasts, a basal cell type that can differentiate in other cells, like muscle cells, chondrocytes, adipocytes, etc. Fibroblasts are sturdy and easy to culture, which makes them a good model at the beginning of a new study. They are also actively involved in wound repair, so their reactions to plasma treatment may be of interest in plasma-aided wound healing. The MR65 cells are human epithelial cells, originating from non-small cell lung carcinoma (NSCLC). The NSCLC is one of the most chemically resistant tumors. The usage of MR65 has a twofold advantage: (a) information on epithelial cells brings one closer to medical applications, like healing of skin ailments, and (b) induction of apoptosis in tumor cells is anyway one of the major objectives of plasma treatment. Cells were treated using the plasma needle under various conditions and observed using phase contrast microscopy or fluorescent staining in combination with confocal microscopy. Initially, basic viability staining was used: propidium iodide (PI) and cell tracker green (CTG). Propidium iodide stains the DNA of necrotic cells red, while cell tracker green stains the cytoplasm of viable cells green. Apoptosis in tumor cells was assayed using the M30 antibody. Antibody assays are very specific. M30 recognizes a molecule, which is a product of enzymatic reaction that occurs solely in apoptosis—a caspase-cleaved cytoskeletal protein. When M30 binds to this product, a fluorescent complex is formed. The diagnosis is unambiguous. Next to specific antibody assays, cells were observed to detect morphological changes characteristic for apoptosis. Various cell reactions are briefly described below.

Plasma treatment of living cells can have many consequences. Naturally, a high dose leads to accidental cell death (necrosis). Typically, necrosis occurs when the plasma power is higher than 0.2 W and the exposure time is longer than 10 s (per treated spot). In terms of energy dose, this corresponds to $20 \, J/cm^2$, which is very high. However, even upon such harsh treatment the cells are not disintegrated, but they retain their shape and internal structure. A typical necrotic spot in a CHO-K1 sample is shown in figure 9.10.8. Note that the dead cells (red stained) are separated from the living cells (green) by a characteristic void. This void is ascribed to local loss of cell adhesion.

A moderate cell damage can activate the apoptotic pathway. In MR65 apoptosis occurs under the threshold dose for necrosis. Simultaneously, cell adhesion is disturbed. Typical images of plasma-treated cells are shown in figure 9.10.9. The whole cytoplasm of the cell is stained using the M30 antibody, which detects the enzymatic activity that is displayed during apoptosis. The percentage of apoptosis after treatment is up to 10%; the plasma conditions still have to be optimized.

When the power and treatment time is substantially reduced (to 50 mW and 1 s per spot), neither necrosis nor apoptosis occur. Instead, the

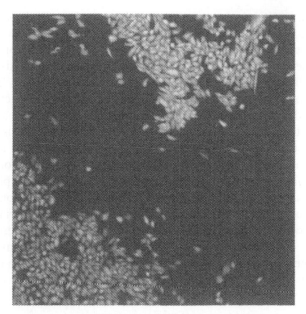

Figure 9.10.8. A sample of CHO-K1 cells after plasma treatment: a necrotic zone (red stained with PI), an empty space and the viable zone (green stained with CTG).

cells round up and (partly) detach from the sample surface: voids like in figure 9.10.8 (but without necrotic zone) are created in the sheet of cells. The cells remain unharmed and after 2–4 h the attachment is restored. It seems that plasma treatment induces a temporary disturbance in the cell

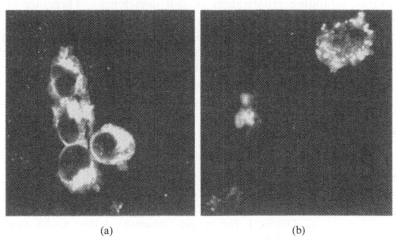

(a)   (b)

Figure 9.10.9. Apoptosis induced in MR65 cells by plasma treatment, assayed by the M30 antibody method: (a) early apoptosis (caspase activity in the cytoplasm, first changes in the cell shape), (b) late apoptosis (formation of apoptotic bodies).

metabolism, which is expressed (among others) by loss of adhesion. Further discussion of possible causes is given elsewhere (Stoffels *et al* 2003).

Cell detachment without severe damage is a refined way of cell manipulation. The loosened cells can be removed (peeled) from a tissue but, as they are still alive, no inflammatory response can be induced. The area of plasma action is always well defined: the influenced cells are strictly localized and the borders between affected and unaffected zones are very sharp. Thus, plasma treatment can be performed locally and with high precision.

The last but very important feature of plasma treatment is related to plasma sterilization. The latter is a well-known effect, demonstrated by many authors (Moisan *et al* 2001, Laroussi 2002) and even implemented in practice. Parallel to plasma-cell interactions, bacterial decontamination using a plasma needle was studied. It appeared that bacteria are much more vulnerable to plasma exposure than eukaryotic cells. Bacterial inactivation to 10^{-4} of the original population can be achieved in 1–2 min at plasma power lower than 10 mW, while under the same conditions the mammalian cells remain uninfluenced. This demonstrates the ability of a non-thermal plasma to *selectively* sterilize infected tissues.

9.10.4.3 *Motivation for the future*

Minimal destructive surgery using non-thermal plasmas is still in its infancy. So far several potentially useful cell reactions have been identified, but the way to clinical implementation will probably be long and painstaking. However, one thing can be stated for sure—non-thermal plasma can be used for controlled, high-precision cell removal *without* necrosis, be it by apoptosis, inhibiting proliferation or cell detachment. There are strong indications that no inflammatory reaction will be induced. After the necessary tests are completed, an enormous area of applications will open. Removal of cancer and other pathological tissues, cosmetic surgery, aiding wound healing, *in vivo* sterilization and preparation of dental cavities without drilling are just a few examples. The plasma needle can be also operated in a catheter (like in APC) and used endoscopically. An enormous effort must be invested in developing all these therapies, but considering the benefit for human health, it is certainly rewarding.

References

Alberts B 1994 *Molecular Biology of the Cell* (New York: Garlands Publishing)

Baeva M, Dogan A, Ehlbeck J, Pott A and Uhlenbusch J 1999 'CARS diagnostic and modeling of a dielectric barrier discharge' *Plasma Chem. Plasma Proc.* **19**(4) 445–466

Brand C U, Blum A, Schlegel A, Farin G and Garbe C 1998 'Application of argon plasma coagulation in skin surgery' *Dermatology* **197** 152–157

Cohen G M 1997 'Caspases: the executioners of apoptosis' *Biochem. J.* **326** 1–16

Coolen S 2000 'Antipirine hydroxylates as indicators for oxidative damage' PhD Thesis, Eindhoven University of Technology

D'Arsonval A 1893 'Action physiologique des courants alternatifs à grand fréquence' *Archives Physiol. Norm. Path.* **5** 401–408

Gabriel S, Lau R W and Gabriel C 1996 'The dielectric properties of biological tissues. 2. Measurements in the frequency range 10 Hz to 20 GHz' *Phys. Med. Biol.* **41**(11) 2251–2269

Guyton A C and Hall J E 2000 *Textbook of Medical Physiology* (W B Saunders Company)

Laroussi M 2002 'Non-thermal decontamination of biological media by atmospheric pressure plasmas: review, analysis, and prospects' *IEEE Trans. Plasma Sci.* **30**(4) 1409–1415

Lu X P and Laroussi M 2003 'Ignition phase and steady-state structures of a non-thermal air plasma' *J. Phys. D: Appl. Phys.* **36**(6) 661–665

Maat L P W M, Slager C J, Van Herwerden L A, Schuurbiers J C H, Van Suylen R J, Kofflard MJM, Ten Cate FJ and Bos E 1994 'Spark erosion myectomy in hypertrophic obstructive cardiomyopathy' *Annals Thoracic Surgery* **58**(2) 536–540

Mohamed A A H, Block R and Schoenbach K H 2002 'Direct current glow discharges in atmospheric air' *IEEE Trans. Plasma Sci.* **30**(1) 182-183

Moisan M, Barbeau J, Moreau S, Pelletier J, Tabrizian M and Yahia L'H 2001 'Low temperature sterilization using gas plasmas: a review of the experiments, and an analysis of the inactivation mechanisms' *Int. J. Pharmaceutics* **226** 1–21

Moon S Y and Choe W 2003 'A comparative study of rotational temperatures using diatomic OH, O_2 and N_2^+ molecular spectra emitted from atmospheric plasmas' *Spectrochimica Acta B: Atomic Spectroscopy* **58**(2/3) 249–257

Motret O, Hibert C, Pellerin S and Pouvesle J M 2000 'Rotational temperature measurements in atmospheric pulsed dielectric barrier discharge—gas temperature and molecular fraction effects' *J. Phys. D: Appl. Phys.* **33**(12) 1493–1498

Park J, Henins I, Herrmann H W, Selwyn G S and Hicks R F 2001 'Discharge phenomena of an atmospheric pressure radio-frequency capacitive plasma source' *J. Appl. Phys.* **89**(1) 20–28

Polk C and Postow E (eds) 1995 *Handbook of Biological Effects of Electromagnetic Fields* (Boca Raton: CRC Press)

Reilly J P 1992 *Electrical Stimulation and Electropathology* (Cambridge: Cambridge University Press)

Ross R 1999 'Atherosclerosis—an inflammatory disease' *New England J. Med.* **340**(2) 115–126

Slager C J, Essed C E, Schuurbiers J C H, Bom N, Serruys P W and Meester G T 1985 'Vaporization of atherosclerotic plaques by spark erosion' *J. American College of Cardiology* **5**(6) 1382–1386

Stalder K R, Woloszko J, Brown I G and Smith C D 2001 'Repetitive plasma discharges in saline solutions' *Appl. Phys. Lett.* **79** 4503–4505

Stoffels E, Flikweert A J, Stoffels W W and Kroesen G M W 2002 'Plasma needle: a non-destructive atmospheric plasma source for fine surface treatment of (bio)materials' *Plasma Sources Sci. Technol.* **11** 383–388

Stoffels E, Kieft I E and Sladek R E J 2003 'Superficial treatment of mammalian cells using plasma needle' *J. Phys. D: Appl. Phys.* **36** 2908–2913

Toshifuji J, Katsumata T, Takikawa H, Sakakibara T and Shimizu I 2003 'Cold arc-plasma jet under atmospheric pressure for surface modification' *Surface and Coatings Technology* **171**(1–3) 302–306

Woloszko J, Stalder K R and Brown I G 2002 'Plasma characteristics of repetitively-pulsed electrical discharges in saline solutions used for surgical procedures' *IEEE Trans. Plasma Sci.* **30** 1376–1383

Appendix

This Appendix contains three sections with results pertaining to section 5.3.3 which were inadvertently omitted from the manuscript. They have been added in the proof stage as an Appendix.

(C)   Vibrational distribution of N_2 ground state
The V–T, V–V and V–V$'$ rates of the foregoing section were implemented in the model and the vibrational distribution of the N_2 ground and excited electronic states was determined by solving a system of kinetic equations at steady state in which the vibrational levels of the N_2 ground and excited electronic states are the unknowns. The total concentration of N_2 was determined with the two-temperature kinetic [12] model and fixed by replacing the vibrational level $v = 0$ of the ground electronic state by the mass conservation equation. The total populations of the other species were fixed and determined with the two-temperature kinetic model, and their internal distribution was calculated according to a Boltzmann distribution at the vibrational temperature $T_v = T_g$ and at the electronic temperature $T_{el} = T_e$. We now present our calculations of the vibrational distribution of the N_2 ground state at $T_g = 2000$ K and for different electron temperatures.

For electron temperatures T_e lower than 6000 K, the vibrational distribution is very close to a Boltzmann distribution at the gas temperature $T_g = 2000$ K. Figures A.1 and A.2 show the calculated vibrational distributions for a gas temperature of 2000 K and an electron temperature of 9000 K and 16 000 K respectively. The Boltzmann distributions at $T_v = T_g$ and $T_v = T_e$ are also shown on these figures.

For $T_e = 9000$ K, the vibrational excitation introduced by VE transfer is mainly redistributed via V–T relaxation of N_2 by collision with N_2, and via N_2–N_2 V–V exchange. The N_2–O_2 and N_2–NO V–V$'$ processes do not significantly affect the populations of N_2 levels. We checked that this conclusion remains valid if we assume a different internal distribution for the O_2 and NO molecules.

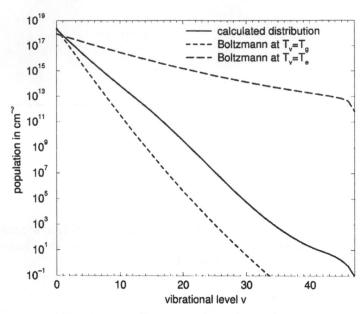

Figure A.1. $N_2(X, v)$ vibrational distribution function at $T_g = 2000\,\text{K}$ and $T_e = 9000\,\text{K}$, $P = 1\,\text{atm}$.

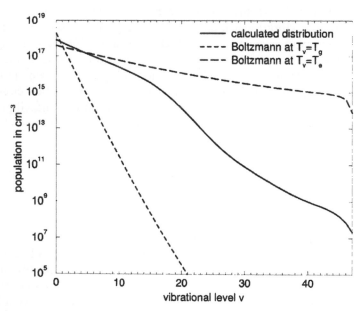

Figure A.2. $N_2(X, v)$ vibrational distribution function at $T_g = 2000\,\text{K}$ and $T_e = 16\,000\,\text{K}$, $P = 1\,\text{atm}$.

Indeed, the rates of N_2–NO exchange are faster than those of N_2–N_2 exchange above $v = 3$ (see figure 5.3.11 in section 5.3.3), but the total concentration of NO is two orders of magnitude lower than the concentration of N_2 and the rates for N_2–O_2 V–V exchange are fast for $v > 20$ but the population of those levels is mainly governed by V–T transfer processes. The vibrational distribution at $T_e = 9000\,K$ lies between the Boltzmann distributions at $T_v = T_g$ and $T_v = T_e$, but remains closer to a distribution at $T_v = T_g$.

For $T_e = 16\,000\,K$, almost 25% of the O_2 molecules are dissociated and the vibrational excitation is mainly redistributed by V–T relaxation of N_2 by collision with O atoms and by N_2–N_2 V–V exchange. At this electron temperature, the vibrational distribution of the first 15 levels is close to the Boltzmann distribution at $T_v = T_e$.

(D)   Inelastic electron energy losses in air plasmas
Electron inelastic energy losses can now be calculated by summing the contributions of all electron impact collisional processes

$$\dot{Q}_{\text{inel}} = \sum_{\text{processes}} \left[\sum_i \sum_f \frac{dn_f}{dt} (E_f - E_i) \right] \tag{A.1}$$

where $E_f - E_i$ represents the internal energy gained by heavy species during the collision (E_f must be greater than E_i) and dn_f/dt is the net volumetric rate of production of heavy species in the final energy level f. In an atmospheric pressure air plasma characterized by a gas temperature between 1000 and 3000 K and electron temperatures up to 17 000 K, the dominant contribution to electron inelastic energy losses is the electron-impact vibrational excitation of N_2 ground state. The electron impact vibrational excitation cross-sections of O_2 and NO ground states are two orders of magnitude lower than those of N_2, and therefore the contribution of these molecules is negligible.

The total rate of energy loss can be expressed as

$$\dot{Q}_{\text{e-V}} = \sum_{v_1} \sum_{v_2 > v_1} \dot{Q}_{v_1 v_2} \tag{A.2}$$

where v_1 and v_2 are the initial and final vibrational levels of the transition, and where the elementary rate $\dot{Q}_{v_1 v_2}$ is written as

$$\dot{Q}_{v_1 v_2} = \left(k_{v_1 v_2} [N_2(X, v_1)] - k_{v_2 v_1} [N_2(X, v_2)] \right) n_e \Delta E_{v_2 v_1}. \tag{A.3}$$

In equation (A.3), n_e is the concentration of electrons, $k_{v_1 v_2}$ and $k_{v_2 v_1}$ are the excitation and de-excitation rate coefficients and $\Delta E_{v_2 v_1}$ stands for the difference of energy between the two vibrational levels v_2 and v_1. $\dot{Q}_{v_1 v_2}$ depends strongly on the N_2 ground state internal distribution. Vibrational population distributions calculated with the method presented in the foregoing section are used in equation (A.3) to determine the electron inelastic energy losses.

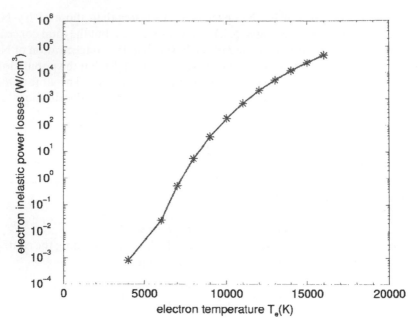

Figure A.3. Predicted inelastic electron power losses in atmospheric pressure air at 2000 K.

The predicted inelastic power losses are shown in figure A.3. At low electron temperatures and densities, the vibrational levels of N_2 ground state are close to a Boltzmann distribution at $T_v = T_g$. The excited vibrational levels have low population.

Therefore, the power lost by e–V excitation is not balanced by the power regained from V–e super-elastic de-excitation. As the electron temperature increases, the electron density also increases and eventually the vibrational population distribution tends toward $T_v = T_e$. The net power losses do not increase as rapidly because of the increased importance of super-elastic collisions. It is sometimes convenient to define an electron 'energy loss factor' as the ratio of total (elastic + inelastic) energy losses to the elastic energy losses

$$\delta_e = \frac{\dot{Q}_{el} + \dot{Q}_{inel}}{\dot{Q}_{el}} \tag{A.4}$$

where $\dot{Q}_{el}$ is the volumetric power lost by free electrons through elastic collisions, and $\dot{Q}_{el}$ is the sum of contributions of collisions between electrons and heavy species h = N_2, O_2 and O:

$$\dot{Q}_{el} = n_e \sum_h 3k(T_e - T_h)\frac{m_e}{m_h}\bar{\nu}_{eh}. \tag{A.5}$$

In equation (A.5), k is the Boltzmann constant, m_e and m_h are the masses of electron and heavy species respectively, T_h is the kinetic temperature of the

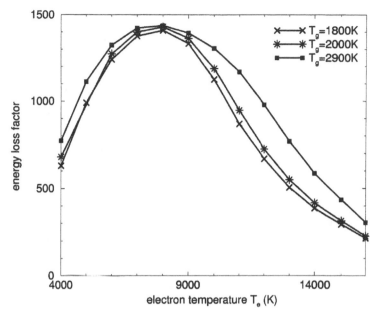

Figure A.4. Energy loss factor δ_e at $T_g = 1800$, 2000 and 2900 K, as a function of T_e.

heavy species (equal to T_g), and $\bar{\nu}_{eh}$ represents the average frequency of collisions between the electrons and heavy particle h. $\bar{\nu}_{eh}$ can be expressed in terms of the number density of neutral species n_h, the electron velocity $g_e = \sqrt{8kT_e/\pi m_e}$ and the average elastic collision cross-section $\bar{Q}_{eh}^{el}$:

$$\bar{\nu}_{eh} \cong n_h g_e \bar{Q}_{eh}^{el}. \tag{A.6}$$

Figure A.4 shows the calculated electron energy loss factor as a function of the electron temperature for two values of the gas temperature, $T_g = 1800$ and 2900 K. As can be seen from this figure, the inelastic loss factor is a relatively weak function of the gas temperature. It increases up to $T_e = 8000$ K as the net rate of production of N_2 molecules in vibrational level $v_2 > v_1$ increases with T_e, and then decreases due to the transition $T_v \cong T_g$ to $T_v \cong T_e$. When T_v becomes close to T_e, the forward and reverse rates are practically balanced and the net rate of energy lost by VE transfer approaches zero.

(E) Predicted DC discharge characteristics in atmospheric pressure air
The results of the previous subsections enable us to convert the 'S-shaped' curve of n_e vs. T_e into electric field vs. current density discharge characteristics. This result is obtained by combining Ohm's law and the electron energy equation. The latter incorporates the results of the collisional–radiative model to account for non-elastic energy losses from the free electrons to the molecular species. The predicted discharge characteristics for atmospheric pressure air at

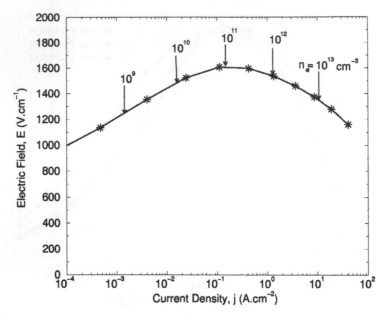

Figure A.5. Predicted discharge characteristics for atmospheric pressure air at 2000 K.

2000 K are shown in figure A.5. These discharge characteristics exhibit variations that reflect both the S-shaped dependence of electron number density versus T_e, and the dependence of the inelastic energy loss factor on the electron temperature and number density. We have used these predicted characteristics as a starting point to design the DC glow discharge experiments presented in section 5.2. If these predictions are correct, the production of 10^{13} electron/cm^3 requires an electric field of $\sim$1.35 kV/cm, and a current density of $\sim$10.4 A/cm^2. Thus the power required to produce 10^{13} electrons/cc in air at 2000 K is approximately 14 kW/cm^3.

Index